DESIGN OF ROTATING ELECTRICAL MACHINES

DESIGN OF ROTATING ELECTRICAL MACHINES

Second Edition

Juha Pyrhönen
Lappeenranta University of Technology, Finland

Tapani Jokinen
Aalto University, School of Electrical Engineering, Finland

Valéria Hrabovcová
Faculty of Electrical Engineering, University of Žilina, Slovakia

WILEY

This edition first published 2014

Registered office
John Wiley & Sons Ltd, The Atrium, Southern Gate, Chichester, West Sussex, PO19 8SQ, United Kingdom

For details of our global editorial offices, for customer services and for information about how to apply for permission to reuse the copyright material in this book please see our website at www.wiley.com.

Library of Congress Cataloging-in-Publication Data

Pyrhönen, Juha.
Design of rotating electrical machines / Juha Pyrhönen, Tapani Jokinen, Valéria Hrabovcová. – Second edition.
pages cm
Includes bibliographical references and index.
ISBN 978-1-118-58157-5 (hardback)
1. Electric machinery–Design and construction. 2. Electric generators–Design and construction. 3. Electric motors–Design and construction. 4. Rotational motion. I. Jokinen, Tapani, 1937– II. Hrabovcova, Valeria. III. Title.
TK2331.P97 2013
621.31′042–dc23

2013021891

A catalogue record for this book is available from the British Library.

ISBN: 978-1-118-58157-5

Typeset in 10/12pt Times by Aptara Inc., New Delhi, India

1 2014

Contents

Preface xi

About the Authors xiii

Abbreviations and Symbols xv

1 Principal Laws and Methods in Electrical Machine Design **1**
1.1 Electromagnetic Principles 1
1.2 Numerical Solution 8
1.3 The Most Common Principles Applied to Analytic Calculation 12
1.3.1 Flux Line Diagrams 16
1.3.2 Flux Diagrams for Current-Carrying Areas 22
1.4 Application of the Principle of Virtual Work in the Determination of Force and Torque 25
1.5 Maxwell's Stress Tensor; Radial and Tangential Stress 32
1.6 Self-Inductance and Mutual Inductance 36
1.7 Per Unit Values 42
1.8 Phasor Diagrams 45
Bibliography 47

2 Windings of Electrical Machines **48**
2.1 Basic Principles 49
2.1.1 Salient-Pole Windings 49
2.1.2 Slot Windings 53
2.1.3 End Windings 54
2.2 Phase Windings 54
2.3 Three-Phase Integral Slot Stator Winding 57
2.4 Voltage Phasor Diagram and Winding Factor 64
2.5 Winding Analysis 72
2.6 Short Pitching 74
2.7 Current Linkage of a Slot Winding 81
2.8 Poly-Phase Fractional Slot Windings 94
2.9 Phase Systems and Zones of Windings 97
2.9.1 Phase Systems 97
2.9.2 Zones of Windings 99

2.10	Symmetry Conditions	101
	2.10.1 Symmetrical Fractional Slot Windings	101
2.11	Base Windings	104
	2.11.1 First-Grade Fractional Slot Base Windings	104
	2.11.2 Second-Grade Fractional Slot Base Windings	105
	2.11.3 Integral Slot Base Windings	106
2.12	Fractional Slot Windings	108
	2.12.1 Single-Layer Fractional Slot Windings	108
	2.12.2 Double-Layer Fractional Slot Windings	117
2.13	Single- and Double-Phase Windings	124
2.14	Windings Permitting a Varying Number of Poles	127
2.15	Commutator Windings	129
	2.15.1 Lap Winding Principles	133
	2.15.2 Wave Winding Principles	136
	2.15.3 Commutator Winding Examples, Balancing Connectors	139
	2.15.4 AC Commutator Windings	143
	2.15.5 Current Linkage of the Commutator Winding and Armature Reaction	144
2.16	Compensating Windings and Commutating Poles	146
2.17	Rotor Windings of Asynchronous Machines	149
2.18	Damper Windings	152
	Bibliography	153
3	**Design of Magnetic Circuits**	**155**
3.1	Air Gap and its Magnetic Voltage	161
	3.1.1 Air Gap and Carter Factor	161
	3.1.2 Air Gaps of a Salient-Pole Machine	166
	3.1.3 Air Gap of Nonsalient-Pole Machine	172
3.2	Equivalent Core Length	173
3.3	Magnetic Voltage of a Tooth and a Salient Pole	176
	3.3.1 Magnetic Voltage of a Tooth	176
	3.3.2 Magnetic Voltage of a Salient Pole	180
3.4	Magnetic Voltage of Stator and Rotor Yokes	180
3.5	No-Load Curve, Equivalent Air Gap and Magnetizing Current of the Machine	183
3.6	Magnetic Materials of a Rotating Machine	186
	3.6.1 Characteristics of Ferromagnetic Materials	189
	3.6.2 Losses in Iron Circuits	194
3.7	Permanent Magnets in Rotating Machines	203
	3.7.1 History and Development of Permanent Magnets	203
	3.7.2 Characteristics of Permanent Magnet Materials	205
	3.7.3 Operating Point of a Permanent Magnet Circuit	210
	3.7.4 Demagnetization of Permanent Magnets	217
	3.7.5 Application of Permanent Magnets in Electrical Machines	219
3.8	Assembly of Iron Stacks	226
	Bibliography	227

4	**Inductances**	**229**
4.1	Magnetizing Inductance	230
4.2	Leakage Inductances	233
	4.2.1 Division of Leakage Flux Components	235
4.3	Calculation of Flux Leakage	238
	4.3.1 Skewing Factor and Skew Leakage Inductance	239
	4.3.2 Air-Gap Leakage Inductance	243
	4.3.3 Slot Leakage Inductance	248
	4.3.4 Tooth Tip Leakage Inductance	259
	4.3.5 End Winding Leakage Inductance	260
	Bibliography	264
5	**Resistances**	**265**
5.1	DC Resistance	265
5.2	Influence of Skin Effect on Resistance	266
	5.2.1 Analytical Calculation of Resistance Factor	266
	5.2.2 Critical Conductor Height in Slot	276
	5.2.3 Methods to Limit the Skin Effect	277
	5.2.4 Inductance Factor	278
	5.2.5 Calculation of Skin Effect in Slots Using Circuit Analysis	279
	5.2.6 Double-Sided Skin Effect	287
	Bibliography	292
6	**Design Process of Rotating Electrical Machines**	**293**
6.1	Eco-Design Principles of Rotating Electrical Machines	293
6.2	Design Process of a Rotating Electrical Machine	294
	6.2.1 Starting Values	294
	6.2.2 Main Dimensions	297
	6.2.3 Air Gap	305
	6.2.4 Winding Selection	309
	6.2.5 Air-Gap Flux Density	310
	6.2.6 The No-Load Flux of an Electrical Machine and the Number of Winding Turns	311
	6.2.7 New Air-Gap Flux Density	316
	6.2.8 Determination of Tooth Width	317
	6.2.9 Determination of Slot Dimensions	318
	6.2.10 Determination of the Magnetic Voltages of the Air Gap, and the Stator and Rotor Teeth	323
	6.2.11 Determination of New Saturation Factor	326
	6.2.12 Determination of Stator and Rotor Yoke Heights and Magnetic Voltages	326
	6.2.13 Magnetizing Winding	327
	6.2.14 Determination of Stator Outer and Rotor Inner Diameter	329
	6.2.15 Calculation of Machine Characteristics	329
	Bibliography	330

7	**Properties of Rotating Electrical Machines**	**331**
7.1	Machine Size, Speed, Different Loadings and Efficiency	331
	7.1.1 Machine Size and Speed	331
	7.1.2 Mechanical Loadability	333
	7.1.3 Electrical Loadability	337
	7.1.4 Magnetic Loadability	338
	7.1.5 Efficiency	340
7.2	Asynchronous Motor	342
	7.2.1 Current Linkage and Torque Production of an Asynchronous Machine	342
	7.2.2 Impedance and Current Linkage of a Cage Winding	349
	7.2.3 Characteristics of an Induction Machine	356
	7.2.4 Equivalent Circuit Taking Asynchronous Torques and Harmonics into Account	361
	7.2.5 Synchronous Torques	367
	7.2.6 Selection of the Slot Number of a Cage Winding	369
	7.2.7 Construction of an Induction Motor	371
	7.2.8 Cooling and Duty Types	373
	7.2.9 Examples of the Parameters of Three-Phase Industrial Induction Motors	378
	7.2.10 Asynchronous Generator	380
	7.2.11 Wound Rotor Induction Machine	382
	7.2.12 Asynchronous Motor Supplied with Single-Phase Current	383
7.3	Synchronous Machines	388
	7.3.1 Inductances of a Synchronous Machine in Synchronous Operation and in Transients	390
	7.3.2 Loaded Synchronous Machine and Load Angle Equation	400
	7.3.3 RMS Value Phasor Diagrams of a Synchronous Machine	407
	7.3.4 No-Load Curve and Short-Circuit Test	417
	7.3.5 Asynchronous Drive	419
	7.3.6 Asymmetric-Load-Caused Damper Currents	423
	7.3.7 Shift of Damper Bar Slotting from the Symmetry Axis of the Pole	424
	7.3.8 V Curve of a Synchronous Machine	426
	7.3.9 Excitation Methods of a Synchronous Machine	426
	7.3.10 Permanent Magnet Synchronous Machines	427
	7.3.11 Synchronous Reluctance Machines	456
7.4	DC Machines	468
	7.4.1 Configuration of DC Machines	468
	7.4.2 Operation and Voltage of a DC Machine	470
	7.4.3 Armature Reaction of a DC machine and Machine Design	474
	7.4.4 Commutation	475
7.5	Doubly Salient Reluctance Machine	479
	7.5.1 Operating Principle of a Doubly Salient Reluctance Machine	479
	7.5.2 Torque of an SR Machine	480
	7.5.3 Operation of an SR Machine	481

	7.5.4	*Basic Terminology, Phase Number and Dimensioning of an SR Machine*	485
	7.5.5	*Control Systems of an SR Motor*	489
	7.5.6	*Future Scenarios for SR Machines*	491
	Bibliography		492
8	**Insulation of Electrical Machines**		**495**
8.1	Insulation of Rotating Electrical Machines		497
8.2	Impregnation Varnishes and Resins		503
8.3	Dimensioning of an Insulation		506
8.4	Electrical Reactions Ageing Insulation		509
8.5	Practical Insulation Constructions		510
	8.5.1	*Slot Insulations of Low-Voltage Machines*	511
	8.5.2	*Coil End Insulations of Low-Voltage Machines*	512
	8.5.3	*Pole Winding Insulations*	512
	8.5.4	*Low-Voltage Machine Impregnation*	513
	8.5.5	*Insulation of High-Voltage Machines*	513
8.6	Condition Monitoring of Insulation		515
8.7	Insulation in Frequency Converter Drives		518
	Bibliography		521
9	**Losses and Heat Transfer**		**523**
9.1	Losses		524
	9.1.1	*Resistive Losses*	524
	9.1.2	*Iron Losses*	526
	9.1.3	*Additional Losses*	526
	9.1.4	*Mechanical Losses*	527
	9.1.5	*Decreasing Losses*	529
	9.1.6	*Economics of Energy Savings*	533
9.2	Heat Removal		534
	9.2.1	*Conduction*	534
	9.2.2	*Radiation*	538
	9.2.3	*Convection*	541
9.3	Thermal Equivalent Circuit		548
	9.3.1	*Analogy between Electrical and Thermal Quantities*	548
	9.3.2	*Average Thermal Conductivity of a Winding*	549
	9.3.3	*Thermal Equivalent Circuit of an Electrical Machine*	550
	9.3.4	*Modeling of Coolant Flow*	560
	9.3.5	*Solution of Equivalent Circuit*	565
	9.3.6	*Cooling Flow Rate*	568
	Bibliography		568
Appendix A			**570**
Appendix B			**572**
Index			**575**

Preface

Electrical machines are almost entirely used in producing electricity, and there are very few electricity-producing processes where rotating machines are not used. In such processes, at least auxiliary motors are usually needed. In distributed energy systems, new machine types play a considerable role: for instance, the era of permanent magnet machines has commenced.

About half of all electricity produced globally is used in electric motors, and the share of accurately controlled motor drives applications is increasing. Electrical drives provide probably the best control properties for a wide variety of processes. The torque of an electric motor may be controlled accurately, and the efficiencies of the power electronic and electromechanical conversion processes are high. What is most important is that a controlled electric motor drive may save considerable amounts of energy. In the future, electric drives will probably play an important role also in the traction of cars and working machines. Because of the large energy flows, electric drives have a significant impact on the environment. If drives are poorly designed or used inefficiently, we burden our environment in vain. Environmental threats give electrical engineers a good reason for designing new and efficient electric drives.

Finland has a strong tradition in electric motors and drives. Lappeenranta University of Technology and Aalto University have found it necessary to maintain and expand the instruction given in electric machines. The objective of this book is to provide students in electrical engineering with an adequate basic knowledge of rotating electric machines, for an understanding of the operating principles of these machines as well as developing elementary skills in machine design. Although, due to the limitations of this material, it is not possible to include all the information required in electric machine design in a single book, this material will serve as a manual for a machine designer in the early stages of his or her career. The bibliographies at the end of chapters are intended as sources of references and recommended background reading. The Finnish tradition of electrical machine design is emphasized in this monograph through the important contributions of Professor Tapani Jokinen, who has spent decades in developing the Finnish machine design profession. Equally important is the view of electrical machine design provided by Professor Valéria Hrabovcová from Slovak Republic, which also has a strong industrial tradition.

In the second edition, some parts of the first edition have been rewritten to make the text proceed more logically and many printing errors have been corrected. Especially, permanent magnet machine and synchronous reluctance machine chapters are now much more comprehensive including new research results. Also the Eco-design principles and economical considerations in machine design are shortly introduced.

The authors are thankful for Dr. Hanna Niemelä for translating the original Finnish material for the first edition.

We express our gratitude to the following persons, who have kindly provided material for this book: Professor Antero Arkkio (Aalto University), Dr Jorma Haataja, Dr Tanja Hedberg (ITT Water and Wastewater AB), Mr Jari Jäppinen (ABB), Dr Hanne Jussila (LUT), Dr Panu Kurronen (The Switch Oy), Dr Janne Nerg (LUT), Dr Markku Niemelä (ABB), Dr Asko Parviainen (AXCO Motors), Dr Sami Ruoho (Teollisuuden Voima), Dr Marko Rilla (Visedo), Dr Pia Salminen (LUT), Dr Ville Sihvo (MAN Turbo), Mr Pavel Ponomarev, Mr Juho Montonen, Ms Julia Alexandrova, Dr. Henry Hämäläinen and numerous other colleagues. Dr Hanna Niemelä's contribution to the first edition and the publication process of the original manuscript is particularly acknowledged.

Juha Pyrhönen
Tapani Jokinen
Valéria Hrabovcová

About the Authors

Juha Pyrhönen is a Professor in the Department of Electrical Engineering at Lappeenranta University of Technology, Finland. He is engaged in the research and development of electric motors and drives. He is especially active in the fields of permanent magnet synchronous machines and drives and solid-rotor high-speed induction machines and drives. He has worked on many research and industrial development projects and has produced numerous publications and patents in the field of electrical engineering.

Tapani Jokinen is a Professor Emeritus in the School of Electrical Engineering at Aalto University, Finland. His principal research interests are in AC machines, creative problem solving and product development processes. He has worked as an electrical machine design engineer with Oy Strömberg Ab Works. He has been a consultant for several companies, a member of the Board of High Speed Tech Ltd and Neorem Magnets Oy, and a member of the Supreme Administrative Court in cases on patents. His research projects include, among others, the development of superconducting and large permanent magnet motors for ship propulsion, the development of high-speed electric motors and active magnetic bearings, and the development of finite element analysis tools for solving electrical machine problems.

Valéria Hrabovcová is a Professor of Electrical Machines in the Department of Power Electrical Systems, Faculty of Electrical Engineering, at the University of Žilina, Slovak Republic. Her professional and research interests cover all kinds of electrical machines, electronically commutated electrical machines included. She has worked on many research and development projects and has written numerous scientific publications in the field of electrical engineering. Her work also includes various pedagogical activities, and she has participated in many international educational projects.

Abbreviations and Symbols

A	linear current density, [A/m]
$\boldsymbol{A}$	magnetic vector potential, [Vs/m]
A	magnetic vector potential scalar value, [Vs/m]
A	temperature class 105 °C
AC	alternating current
AM	asynchronous machine
A1-A2	armature winding of a DC machine
A_{1n}, A_{2n}, A_{3n}	factors for defining permanent magnet flux density
a	number of parallel paths in windings without commutator: per phase, in windings with a commutator: per half armature, diffusivity
$\boldsymbol{B}$	magnetic flux density, vector [Vs/m^2], [T]
B	magnetic flux density scalar value, [Vs/m^2]
B_{r}	remanent flux density, [T]
B_{sat}	saturation flux density, [T]
B	temperature class 130 °C
B1-B2	commutating pole winding of a DC machine
b	width, [m]
$b_{0\mathrm{c}}$	conductor width [m]
b_{c}	conductor width [m]
b_{d}	tooth width, [m]
b_{dr}	rotor tooth width, [m]
b_{ds}	stator tooth width, [m]
b_{r}	rotor slot width, [m]
b_{s}	stator slot width, [m]
b_0	slot opening, [m]
b_{v}	width of ventilation duct, [m]
C	capacitance, [F], machine constant, integration constant, fabrication cost, [€]
C	temperature class >180 °C
C1-C2	compensating winding of a DC machine
C_{f}	friction coefficient
C_{M}	torque coefficient
C_{s}	saving cost per year, [€/a]

c	specific heat capacity, [J/kgK], capacitance per unit of length, factor, divider, constant
C_{diff}	increase of the purchase cost, [€]
c_{e}	energy cost, [€/kWh]
c_p	specific heat capacity of air in constant pressure
C_{pw}	cost per one kilowatt of loss over the life of motor, [€/kW]
c_{th}	heat capacity
CTI	Comparative Tracking Index
c_{v}	specific volumetric heat, [kJ/Km^3]
D	electric flux density [C/m^2], diameter [m]
DC	direct current
DOL	direct- on-line
D_{s}	inner diameter of the stator, [m]
D_{se}	outer diameter of the stator, [m]
D_{r}	outer diameter of the rotor, [m]
D_{ri}	inner diameter of the rotor, [m]
D1-D2	series magnetizing winding of a DC machine
d	thickness, [m]
d_{t}	thickness of the fringe of a pole shoe, [m]
E	electromotive force (emf), [V], RMS, electric field strength, [V/m], scalar, elastic modulus, Young's modulus, [Pa], bearing load
E_{a}	activation energy, [J]
$\boldsymbol{E}$	electric field strength, vector, [V/m]
E	electric field strength scalar value, [V/m]
E	temperature class 120 °C
E	irradiation intensity [W/m^2]
E1-E2	shunt winding of a DC machine
e	electromotive force [V], instantaneous value $e(t)$
e	Napier's constant
emf	electromotive force, [V]
F	force, [N], scalar
$\boldsymbol{F}$	force, [N], vector
F	temperature class 155 °C
FEA	Finite Element Analysis
F_{g}	geometrical factor
F_{m}	magnetomotive force $\oint \boldsymbol{H} \cdot \mathrm{d}\boldsymbol{l}$, [A], (mmf)
F1-F2	separate magnetizing winding of a DC machine or a synchronous machine
f	frequency, [Hz], Moodyfriction factor
f_{Br}	factor for defining permanent magnet radial flux density
$f_{\mathrm{B}\theta}$	factor for defining permanent magnet tangential flux density
g	coefficient, constant,thermal conductance per unit length
G	electrical conductance
G_{th}	thermal conductance
$\boldsymbol{H}$	magnetic field strength, [A/m]
H	magnetic field strength scalar value, [A/m]
H_{c}, $H_{\mathrm{c}B}$	coercivity related to flux density, [A/m]

H_{cJ}	coercivity related to magnetization, [A/m]
H	temperature class 180 °C
H_n	number of partial discharges
h	height, [m]
h_{0c}	conductor height [m]
h_c	conductor height [m]
h_d	tooth height [m]
h_p	height of a subconductor, [m]
h_{p2}	height of pole body, [m]
h_{ys}	height of stator yoke, [m]
h_{yr}	height of rotor yoke, [m]
h_s	stator slot height, [m]
I	electric current, [A], RMS, brush current, second moment of inertia of an area, [m^4]
IM	induction motor
I_{ns}	counter-rotating current (negative sequence component), [A]
I_o	current of the upper bar, [A]
I_u	current of the lower bar, slot current, slot current amount, [A]
I_s	conductor current
IC	classes of electrical machines
IEC	International Electrotechnical Commission
Im	imaginary part
i	current, [A], instantaneous value $i(t)$, per unit value of current, [pu], annual rate of interest
J	moment of inertia, [kgm^2], current density [A/m^2], magnetic polarization
J_{0PM}	current density on the PM surface, [A/m^2]
J_{PM}	eddy current density, [A/m^2]
$\boldsymbol{J}$	Jacobian matrix
J_{ext}	moment of inertia of load, [kgm^2]
J_M	moment of inertia of the motor, [kgm^2]
J_{sat}	saturation of polarization, [Vs/m^2]
$\boldsymbol{J}_s$	surface current, vector, [A/m]
J_s	surface current vector scalar value, [A/m]
j	difference of the numbers of slots per pole and phase in different layers
j	imaginary unit
K	transformation ratio, constant, number of commutator segments
K_{Br}	factor for defining permanent magnet radial flux density
$K_{B\theta}$	factor for defining permanent magnet tangential flux density
K_L	inductance ratio
k	connecting factor (coupling factor), correction coefficient, safety factor, ordinal of layers, roughness coefficient
k_E	machine-related constant
k_C	Carter factor
k_{Cu}, k_{Fe}	space factor for copper, space factor for iron
k_d	distribution factor, correction factor, saliency factor in d- axis
k_q	saliency factor in q- axis

k_{dsat}	saliency factor taking into account saturation in d- axis
k_{qpar}	saliency factor taking into account parallel magnetic lines in q- axis
$k_{Fe,n}$	correction factor
k_k	short circuit ratio
k_L	skin effect factor for the inductance
k_p	pitch factor
k_{pw}	pitch factor due to coil side shift, present worth factor of an equal payment series
k_R	skin effect factor for the resistance
k_{sat}	saturation factor
k_{sq}	skewing factor
k_{th}	coefficient of heat transfer, [W/m^{2}K]
k_v	pitch factor of the coil side shift in a slot
k_w	winding factor
k_σ	safety factor in the yield
L	self inductance [H]
L	characteristic length, characteristic surface length, tube length [m]
LC	inductor-capacitor
L_d	tooth tip leakage inductance, synchronous inductance in d- axis [H]
L_q	synchronous inductance in q- axis [H]
L_d/ L_q	inductance ratio
L_k	short-circuit inductance, [H]
L_m	magnetizing inductance, [H]
L_{md}	magnetizing inductance of an m-phase synchronous machine, in d-axis,[H]
L_{mq}	magnetizing inductance of an m-phase synchronous machine, in q-axis, [H]
L_{mn}	mutual inductance, [H]
L_{mp}	magnetizing inductance of single-phase winding, [H]
L_{pd}	main inductance of a single phase, [H]
L_{sq}	skew leakage inductance, [H]
L_u	slot inductance, [H]
L_w	end winding leakage inductance, [H]
L_δ	air-gap leakage inductance, [H]
$L_{m\delta}$	magnetizing inductance of synchronous machines with non-salient poles, [H]
L'	transient inductance, [H]
L''	subtransient inductance, [H]
L1, L2, L3	network phases
l	length [m], closed line, distance, inductance per unit of length, relative inductance (inductance per unit value), gap spacing between the electrodes
l	unit vector collinear to the integration path
l'	effective core length, [m]
l_{ew}	average conductor length of winding overhang, [m]
l_p	wetted perimeter of tube, [m]
l_{pu}	inductance as a per unit value
l_w	length of coil ends, [m]
l_{sub}	length of one sub- stack, [m]

M	mutual inductance [H], magnetization [A/m]
M_{sat}	saturation magnetization, [A/m]
m	number of phases, mass, [kg]
m_{c}	mutual coupling factor
m_0	constant
mmf	magnetomotive force, [A]
N	number of turns in a winding, number of turns in series
N_{f1}	number of coil turns in series in a single pole
Nu	Nusselt number
N_{u1}	number of bars of a coil side in the slot
N_{p}	number of turns of one pole pair
N_{k}	number of turns of compensating winding
N_{v}	number of conductors in each side
N	Non-drive end
N	set of integers
N_{even}	set of even integers
N_{odd}	set of odd integers
$\boldsymbol{n}$	normal unit vector of the surface
n	rotation speed (rotation frequency), [1/s], ordinal of the harmonic (sub), ordinal of the critical rotation speed, integer, exponent, years of saving (motor life time)
n_{v}	number of ventilation ducts
n_{U}	number of section of flux tube in sequence
n_{Φ}	number of flux tube
P	power, losses [W]
P_{in}	input power, [W]
PAM	Pole-Amplitude-Modulation
PM	permanent magnet
PMSM	permanent magnet synchronous motor (or machine)
PWM	Pulse Width Modulation
P_1, P_{ad}, P_{LL}	additional loss, [W]
P_{ew}	end winding losses, [W]
Pr	Prandtl number
P_{ρ}	friction loss, [W]
P_{diff}	reduction of the purchase cost, [€]
P_{PM}	eddy current loss in permanent magnet, [W]
p	number of pole pairs, ordinal, losses per core length, resistive losses per core length, [W/m], pressure, [Pa]
p_{Al}	aluminium content
p^*	number of pole pairs of a base winding
pd	partial discharge
Q	electric charge, [C], number of slots, reactive power, [VA]
Q_{av}	average number of slots of a coil group
Q_{p}	number of slots per pole
Q_{o}	number of free slots
Q'	number of radii in a voltage phasor graph

Q^*	number of slots of a base winding
Q_{th}	quantity of heat
q	number of slots per pole and phase, instantaneous charge, $q(t)$, [C]
q_k	number of slots in a single zone
q_m	mass flow, [kg/s]
q_{th}	density of the heat flow, [W/m^2]
R	resistance, [Ω], gas constant, 8.314472 [J/K·mol], thermal resistance, reactive parts
R_{bar}	bar resistance, [Ω]
RM	reluctance machine
RMS	root mean square
R_m	reluctance, [A/Vs = 1/H]
R_{th}	thermal resistance, [K/W]
Re	real part
Re	Reynolds number
Re_{crit}	critical Reynolds number
RR	Resin Rich impregnation method
r	radius, [m], thermal resistance per unit length, per unit resistance [pu], coefficient of radiation
$\boldsymbol{r}$	radius unit vector
S1-S8	duty types
S	apparent power, [VA], cross-sectional area
SM	synchronous motor
SR	switched reluctance
SyRM	synchronous reluctance machine
S_c	cross-sectional area of conductor, [m^2]
S_p	pole surface area, [m^2]
S_r	rotor surface area facing the air gap, [m^2]
$\boldsymbol{S}$	Poynting's vector, [W/m^2], unit vector of the surface
s	slip, skewing measured as an arc length
s_b	slip at maximum torque
s_{sp}	skewing expressed as a number of slot pitches
T	torque, [Nm], absolute temperature, [K], period, operating time of the motor per year, [h/a]
Ta	Taylor number
Ta_m	modified Taylor number
T_b	pull-out torque, peak (maximum) torque [Nm]
t_c	commutation period, [s]
TEFC	totally enclosed fan-cooled
T_J	mechanical time constant, [s]
T_{mec}	mechanical torque, [Nm]
T_{pb}	payback time
T_s	temperature of the plane
T_u	pull-up torque, [Nm]
T_v	counter torque, [Nm]
T_l	locked rotor torque, [Nm]

TC	tooth coil
t	time, [s], number of phasors of a single radius, largest common divider, lifetime of insulation
$\boldsymbol{t}$	tangential unit vector
t_{c}	commutation period, [s]
t_{r}	rise time, [s]
$t*$	number of layers in a voltage vector graph for a base winding
U	voltage, [V], RMS
U	depiction of a phase
U_{contact}	contact voltage drop, [V]
U_{m}	magnetic voltage, [A]
U_{r}	resistive voltage, [V]
U_{sj}	peak value of the impulse voltage, [V]
U_{v}	coil voltage, [V]
U1	terminal of the head of the U-phase of the machine
U2	terminal of the end of the U-phase of the machine
u	voltage, instantaneous value $u(t)$, [V], number of coil sides in a layer, per unit value of voltage, [pu]
u_{b1}	blocking voltage of the oxide layer, [V]
u_{c}	commutation voltage, [V]
u_{m}	mean fluid velocity in tube, [m/s]
V	volume, [m^3], electric potential
V	depiction of a phase
V_{m}	scalar magnetic potential, [A]
VPI	Vacuum Pressure Impregnation
V1	terminal of the head of the V-phase of the machine
V2	terminal of the end of the V-phase of the machine
v	speed, velocity, [m/s]
$\boldsymbol{v}$	vector
W	energy, [J], coil span (width), average coil span [m]
W	depiction of a phase
W_{fc}	energy stored in the magnetic field in SR machines
W_{d}	energy returned through the diode to the voltage source in SR drives
W_{mt}	energy converted into mechanical work when the transistor is conducting in SR drives
W_{md}	energy converted to mechanical work while de-energizing the phase in SR drives
W_{R}	energy returning to the voltage source in SR drives
W'	co-energy, [J]
W1	terminal of the head of the W-phase of the machine
W2	terminal of the end of the W-phase of the machine
W_{Φ}	magnetic energy, [J]
w	length, [m], energy per volume unit
w_{PM}	permanent magnet width, [m]
X	reactance, [Ω]
x	coordinate, length, ordinal number, coil span decrease [m]

x_m	relative value of reactance
Y	admittance, [S]
Y	temperature class 90°C
y	coordinate, length, step of winding
y_m	winding step in an AC commutator winding
y_n	coil span in slot pitches
y_Q	coil span of full-pitch winding in slot pitches, pole pitch expressed in number of slots per pole
y_v	coil span decrease in slot pitches
y_1	step of span in slot pitches, back end connector pitch
y_2	step of connection in slot pitches, front end connector pitch
y_C	commutator pitch in number of commutator segments
Z	impedance, [Ω], number of bars, number of positive and negative phasors of the phase
Z_M	characteristic impedance of the motor, [Ω]
Z_s	surface impedance, [Ω]
Z_0	characteristic impedance, [Ω]
z	coordinate, length, integer, total number of conductors in the armature winding
z_a	number of adjacent conductors
z_b	number of brushes
z_c	number of coils
z_{cs}	number of conductors in half slot
z_p	number of parallel-connected conductors
z_Q	number of conductors in a slot
z_t	number of conductors on top each other
α	angle, [rad], [°], coefficient, temperature coefficient, relative pole width of the pole shoe, convection heat transfer coefficient, [W/K], skew angle, [rad], [°]
$1/\alpha$	depth of penetration
α_{DC}	relative pole width coefficient for DC machines
α_i	ratio of the arithmetical average of the flux density to its peak value
α_m	mass transfer coefficient, [(mol/sm^2)/ (mol/m^3) = m/s]
α_{PM}	relative permanent magnet width
α_{SM}	relative pole width coefficient for synchronous machines
α_r	heat transfer coefficient of radiation
α_{str}	angle between the phase winding
α_{th}	heat transfer coefficient [W/(m^{2}K)]
α_{ph}	angle between the phase winding
α_u	slot angle, [rad], [°]
α_z	phasor angle, zone angle, [rad], [°]
α_ρ	angle of single phasor, [rad], [°]
β	angle [rad], [°]
β	absorptivity
Γ	energy ratio, integration route
Γ_c	interface between iron and air

γ	angle, [rad], [°], coefficient
γ_c	commutation angle, [rad], [°]
γ_D	switch conducting angle, [rad], [°]
δ	air gap (length), penetration depth [m], dielectric loss angle, [rad], [°], dissipation angle, [rad], [°], load angle, [rad], [°]
δ_c	the thickness of concentration boundary layer, [m]
$\delta_d(x)$	air gap profile function in d- axis, [m]
$\delta_q(x)$	air gap profile function in q- axis, [m]
δ_e	equivalent air gap (slotting taken into account), [m]
δ_{ef}	effective air gap (influence of iron and slotting taken into account)
δ_{PM}	depth of penetration in PM, [m]
δ_v	velocity boundary layer, [m]
δ_T	temperature (thermal) boundary layer, [m]
δ '	load angle, [rad], [°], corrected air gap, [m]
δ_0	minimum air gap, [m]
δ_{0e}	air gap in the middle of the pole corrected with Carter factor, [m]
δ_{de}	equivalent d- axis air gap, [m]
δ_{qe}	equivalent q- axis air gap, [m]
Δ_2	damping factor
ε	permittivity [F/m], position angle of the brushes, [rad], [°], stroke angle, [rad], [°], amount of short pitching
ε_{sp}	amount of short pitching expressed as slot pitches
ε_{th}	emissivity
ε_{thr}	relative emissivity
ε_0	permittivity of vacuum $8.854 \cdot 10^{-12}$ [F/m]
ζ	phase angle, [rad], [°], harmonic factor, saliency ratio, phase angle of the rotor impedance
ζ_d	harmonic factor in d-axis
ζ_q	harmonic factor in q-axis
η	efficiency, empirical constant, experimental pre-exponential constant
η	reflectivity, thermal conductivity
Θ	current linkage, [A], temperature rise (difference) [K]
Θ_k	compensating current linkage, [A]
Θ_Σ	total current linkage [A]
θ	angle, position, [rad], [°]
ϑ	angle, [rad], [°]
κ	angle, [rad], [°], factor for reduction of slot opening
κ	transmissivity
Λ	permeance, [Vs/A], [H]
Λ'	specific permeance, [Vs/A/m^2]
Λ'_0	average of specific permeance, [Vs/A/m^2]
λ	thermal conductivity [W/m·K], permeance factor, proportionality factor, inductance factor, inductance ratio
μ	permeability [Vs/Am, H/m], number of pole pairs operating simultaneously per phase, friction coefficient
μ_r	relative permeability

μ	dynamic viscosity, [Pa·s, kg/(s·m)]
μ_0	permeability of vacuum, $4{\cdot}\pi{\cdot}10^{-7}$ [Vs/Am, H/m]
ν	ordinal of harmonic, Poisson's ratio, reluctivity, [Am/Vs, m/H], pole pair number of harmonics, kinematic viscosity of the coolant
ν	pulse velocity
ξ	reduced conductor height
ρ	resistivity, [Ωm], electric charge density, [C/m^2], density, [kg/m^3], reflection factor, ordinal number of a single phasor
ρ_A	absolute overlap ratio
ρ_E	effective overlap ratio
ρ_ν	transformation ratio for IM impedance, resistance, inductance
σ	specific conductivity, electric conductivity [S/m], leakage factor, ratio of the leakage flux to the main flux
σ_δ	air gap harmonic leakage factor
σ_F	tension, [Pa]
σ_{Fn}	normal tension, [Pa]
σ_{Ftan}	tangential tension, [Pa]
σ_{mec}	mechanical stress, [Pa]
σ_{SB}	Stefan-Boltzmann constant, 5.670400×10^{-8} W·m^{-2}·K^{-4}
τ	relative time, span of the lamination thickness on one pole pitch
τ_p	pole pitch, [m]
τ_{q2}	pole pitch on the pole surface, [m]
τ_r	rotor slot pitch, [m]
τ_s	stator slot pitch, [m]
τ_u	slot pitch, [m]
τ_v	zone distribution
τ'_d	direct transient short-circuit time constant, [s]
τ'_{d0}	direct transient open-circuit time constant, [s]
τ''_d	direct subtransient short-circuit time constant, [s]
τ''_{d0}	direct subtransient open-circuit time constant, [s]
τ''_q	quadrature subtransient short-circuit time constant, [s]
τ''_{q0}	quadrature subtransient open-circuit time constant, [s]
υ	factor, kinematic viscosity, μ/ρ, [Pa·s/(kg/m^3)]
Φ	magnetic flux, [Vs], [Wb]
Φ_{th}	thermal power flow, heat flow rate [W]
Φ_δ	air gap flux, [Vs], [Wb]
ϕ	magnetic flux, instantaneous value $\phi(t)$, [Vs], electric potential [V]
φ	phase shift angle, [rad], [°]
φ'	function for skin effect calculation
Ψ	magnetic flux linkage, [Vs]
ψ	electric flux, [C]
ψ_e	electric flux, [C]
ψ_m	air gap flux linkage [Vs]
ψ_{mp}	magnetic flux linkage of phase winding [Vs]
ψ	function for skin effect calculation

χ	length/diameter ratio, shift of a single pole pair
Ω	mechanical angular speed [rad/s]
ω	electric angular velocity [rad/s], angular frequency [rad/s]
Δ	difference, drop
ΔT	temperature rise (difference) [K], [°C]
∇T	temperature gradient [K/m], [°C/m]
Δp	pressure drop [Pa]

Subscripts

0	section
1	primary, fundamental component, beginning of a phase, locked rotor torque
2	secondary, end of a phase
Al	aluminum
a	armature, shaft
ad	additional (loss)
av	average
B	brush
b	base value, peak value of torque, blocking, damper bar
bar	bar
bearing	bearing (losses)
C	capacitor
Cu	copper
Cuw	End winding conductor
conv	convection
c	conductor, commutation
cf	centrifugal
cp	commutating poles
contact	brush contact
cr, crit	critical
DC	direct current
D	direct, damper
d	tooth, direct, tooth tip leakage flux
diff	difference
E	emf
e	equivalent
ef	effective
el	electric
em	electromagnetic
ew	end winding
ext	external
F	force
Fe	iron
f	field
Ft	eddy current

Hy	hysteresis
i	internal, insulation, ordinal
in	input
k	compensating, short circuit, ordinal
lam	laminations
LL	additional load losses
M	motor
max	maximum
m	mutual, main, magnetizing
mag	magnetizing, magnetic
mec	mechanical
min	minimum
mut	mutual
mp	single-phase magnetizing
N	rated
n	nominal, normal
ns	negative-sequence component
o	starting, upper, over
opt	optimal
out	output
PM	permanent magnet
p	pole, primary, subconductor, pole leakage flux, operational harmonic
p1	pole shoe
p2	pole body
ph	phasor, phase
ps	positive-sequence component
pu	per unit
q	quadrature, zone
r	rotor, remanence, relative, damper ring short circuit
res	resultant
S	surface
s	stator
sj	impulse wave
sat	saturation
str	phase section
sq	skew
syn	synchronous
tan	tangential
test	test
th	thermal
tot	total
u	slot, lower, under, bottom, slot leakage flux, pull-up torque
v	zone, coil side shift in a slot, coil
x	x-direction
y,	y-direction, yoke
ya	armature yoke

yr	rotor yoke
ys	stator yoke
w	end winding
z	z-direction, phasor of voltage phasor graph
ρ	ordinal number of single phasor
ρ	friction loss
ρw	windage (loss)
δ	air gap
Φ	flux
ν	harmonic
σ	flux leakage
γ	ordinal of a subconductor
μ	harmonic ordinal

Superscripts

^	peak/maximum value, amplitude
'	imaginary, apparent, reduced, virtual, referred to the stator
*	base winding, complex conjugate

Boldface symbols are used for vectors with components parallel to the unit vectors *i*, *j*, and *k*.

$\boldsymbol{A}$	vector potential, $\boldsymbol{A} = \boldsymbol{i}A_{\mathrm{x}} + \boldsymbol{j}A_{\mathrm{k}} + \boldsymbol{k}A_{\mathrm{z}}$
$\boldsymbol{B}$	flux density, $\boldsymbol{B} = \boldsymbol{i}B_{\mathrm{x}} + \boldsymbol{j}B_{\mathrm{k}} + \boldsymbol{k}B_{\mathrm{z}}$
$\underline{I}$	complex phasor of the current
$\bar{I}$	bar above the symbol denotes average value

1

Principal Laws and Methods in Electrical Machine Design

1.1 Electromagnetic Principles

A comprehensive command of electromagnetic phenomena relies fundamentally on Maxwell's equations. The description of electromagnetic phenomena is relatively easy when compared with various other fields of physical sciences and technology, since all the field equations can be written as a single group of equations. The basic quantities involved in the phenomena are the following five vector quantities and one scalar quantity:

Electric field strength	$\boldsymbol{E}$	[V/m]
Magnetic field strength	$\boldsymbol{H}$	[A/m]
Electric flux density	$\boldsymbol{D}$	[C/m^2]
Magnetic flux density	$\boldsymbol{B}$	[Vs/m^2], [T]
Current density	$\boldsymbol{J}$	[A/m^2]
Electric charge density, $\mathrm{d}Q/\mathrm{d}V$	ρ	[C/m^3]

The presence of an electric and magnetic field can be analyzed from the force exerted by the field on a charged object or a current-carrying conductor. This force can be calculated by the Lorentz force (Figure 1.1), a force experienced by an infinitesimal charge $\mathrm{d}Q$ moving at a speed $\boldsymbol{v}$. The force is given by the vector equation

$$\mathrm{d}\boldsymbol{F} = \mathrm{d}Q(\boldsymbol{E} + \boldsymbol{v} \times \boldsymbol{B}) = \mathrm{d}Q\boldsymbol{E} + \frac{\mathrm{d}Q}{\mathrm{d}t}\mathrm{d}\boldsymbol{l} \times \boldsymbol{B} = \mathrm{d}Q\boldsymbol{E} + i\mathrm{d}\boldsymbol{l} \times \boldsymbol{B}. \tag{1.1}$$

In principle, this vector equation is the basic equation in the computation of the torque for various electrical machines. The latter part of the expression in particular, formulated with a current-carrying element of a conductor of the length $\mathrm{d}\boldsymbol{l}$, is fundamental in the torque production of electrical machines.

Design of Rotating Electrical Machines, Second Edition. Juha Pyrhönen, Tapani Jokinen and Valéria Hrabovcová.

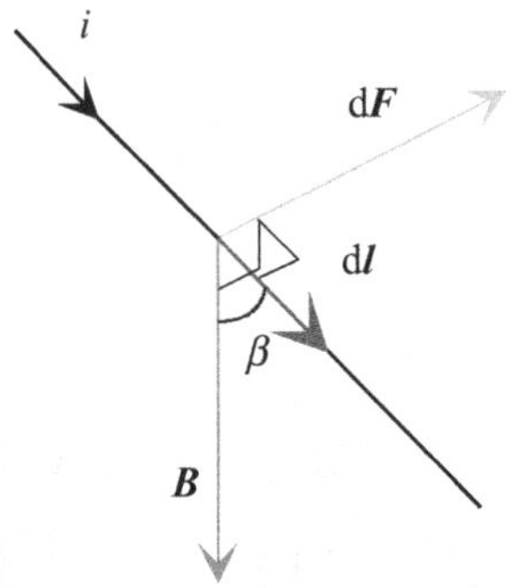

Figure 1.1 Lorentz force d$\boldsymbol{F}$ acting on a differential length d$\boldsymbol{l}$ of a conductor carrying an electric current i in the magnetic field $\boldsymbol{B}$. The angle β is measured between the conductor and the flux density vector $\boldsymbol{B}$. The vector product $i\,\mathrm{d}\boldsymbol{l} \times \boldsymbol{B}$ may now be written in the form $i\,\mathrm{d}\boldsymbol{l} \times \boldsymbol{B} = i\mathrm{d}lB \sin\beta$.

Example 1.1: Calculate the force exerted on a conductor 0.1 m long carrying a current of 10 A at an angle of 80° with respect to a field density of 1 T.

Solution: Using (1.1) we get directly for the magnitude of the force

$$F = |i\boldsymbol{l} \times \boldsymbol{B}| = 10\ \mathrm{A} \cdot 0.1\ \mathrm{m} \cdot \sin 80^\circ \cdot 1\ \mathrm{Vs/m^2} = 0.98\ \mathrm{V\ As/m} = 0.98\ \mathrm{N}.$$

In electrical engineering theory, the other laws, which were initially discovered empirically and then later introduced in writing, can be derived from the following fundamental laws presented in complete form by Maxwell. To be independent of the shape or position of the area under observation, these laws are presented as differential equations.

A current flowing from an observation point reduces the charge of the point. This law of conservation of charge can be given as a divergence equation

$$\nabla \cdot \boldsymbol{J} = -\frac{\partial \rho}{\partial t}, \tag{1.2}$$

which is known as the continuity equation of the electric current.

Maxwell's actual equations are written in differential form as

$$\nabla \times \boldsymbol{E} = -\frac{\partial \boldsymbol{B}}{\partial t}, \tag{1.3}$$

$$\nabla \times \boldsymbol{H} = \boldsymbol{J} + \frac{\partial \boldsymbol{D}}{\partial t}, \tag{1.4}$$

$$\nabla \cdot \boldsymbol{D} = \rho, \tag{1.5}$$

$$\nabla \cdot \boldsymbol{B} = 0. \tag{1.6}$$

The curl relation (1.3) of an electric field is Faraday's induction law, which describes how a changing magnetic flux creates an electric field around it. The curl relation (1.4) for magnetic field strength describes the situation where a changing electric flux and current produce magnetic field strength around them. This is Ampère's law. Ampère's law also yields a law for conservation of charge (1.2) by a divergence Equation (1.4), since the divergence of the curl is identically zero. In some textbooks, the curl operation may also be expressed as $\nabla \times \boldsymbol{E} = \mathrm{curl}\boldsymbol{E} = \mathrm{rot}\boldsymbol{E}$.

An electric flux always flows from a positive charge and passes to a negative charge. This can be expressed mathematically by the divergence Equation (1.5) of an electric flux. This law is also known as Gauss's law for electric fields. Magnetic flux, however, is always a circulating flux with no starting or end point. This characteristic is described by the divergence Equation (1.6) of the magnetic flux density. This is Gauss's law for magnetic fields. The divergence operation in some textbooks may also be expressed as $\nabla \cdot \boldsymbol{D} = \mathrm{div}\,\boldsymbol{D}$.

Maxwell's equations often prove useful in their integral form: Faraday's induction law

$$\oint_l \boldsymbol{E} \cdot \mathrm{d}\boldsymbol{l} = -\frac{\mathrm{d}}{\mathrm{d}t}\int_S \boldsymbol{B} \cdot \mathrm{d}\boldsymbol{S} = -\frac{\mathrm{d}\Phi}{\mathrm{d}t} \tag{1.7}$$

states that the change of a magnetic flux Φ penetrating an open surface S is equal to a negative line integral of the electric field strength along the line l around the surface. Mathematically, an element of the surface S is expressed by a differential operator $\mathrm{d}\boldsymbol{S}$ perpendicular to the surface. The contour line l of the surface is expressed by a differential vector $\mathrm{d}\boldsymbol{l}$ parallel to the line.

Faraday's law together with Ampère's law are extremely important in electrical machine design. At its simplest, the equation can be employed to determine the voltages induced in the windings of an electrical machine. The equation is also necessary, for instance, in the determination of losses caused by eddy currents in a magnetic circuit, and when determining the skin effect in copper. Figure 1.2 illustrates Faraday's law. There is a flux Φ penetrating through a surface S, which is surrounded by the line l.

The arrows in the circles point the direction of the electric field strength $\boldsymbol{E}$ in the case where the flux density $\boldsymbol{B}$ inside the observed area is increasing. If we place a short-circuited metal wire around the flux, we will obtain an integrated voltage $\oint_l \boldsymbol{E} \cdot \mathrm{d}\boldsymbol{l}$ in the wire, and

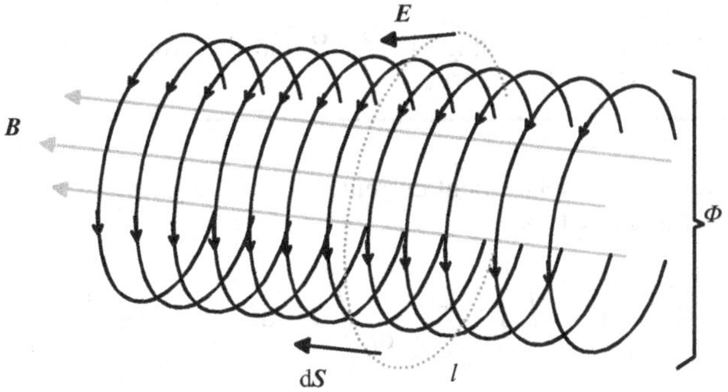

Figure 1.2 Illustration of Faraday's induction law. A typical surface S, defined by a closed line l, is penetrated by a magnetic flux Φ with a density $\boldsymbol{B}$. A change in flux density creates an electric field strength $\boldsymbol{E}$. The circles illustrate the behavior of $\boldsymbol{E}$. $\mathrm{d}\boldsymbol{S}$ is a vector perpendicular to the surface S.

consequently also an electric current. This current creates its own flux that will oppose the flux penetrating through the coil.

If there are several turns N of winding (cf. Figure 1.2), the flux does not link all these turns ideally, but with a ratio of less than unity. Hence we may denote the effective turns of winding by $k_w N$, ($k_w < 1$). Equation (1.7) yields a formulation with an electromotive force e of a multiturn winding. In electrical machines, the factor k_w is known as the winding factor (see Chapter 2). This formulation is essential to electrical machines and is written as

$$e = -k_w N \frac{d}{dt} \int_S \boldsymbol{B} \cdot d\boldsymbol{S} = -k_w N \frac{d\Phi}{dt} = -\frac{d\Psi}{dt}. \tag{1.8}$$

Here, we introduce the flux linkage $\Psi(t) = k_w N\Phi(t) = Li(t)$, one of the core concepts of electrical engineering. It may be noted that the inductance L describes the ability of a coil to produce flux linkage Ψ. Later, when calculating the inductance, the effective turns, the permeance Λ or the reluctance R_m of the magnetic circuit are needed ($L = (k_w N)^2 \Lambda = (k_w N)^2 / R_m$).

Example 1.2: There are 100 turns in a coil having a cross-sectional area of 0.01 m^2. There is an alternating peak flux density of 1 T linking the turns of the coil with a winding factor of $k_w = 0.9$. Calculate the electromotive force induced in the coil when the flux density variation has a frequency of 100 Hz.

Solution: Using Equation (1.8) we, as $\omega = 2\pi f$, get

$$e = -\frac{d\Psi}{dt} = -k_w N \frac{d\Phi}{dt} = -k_w N \frac{d}{dt} \hat{B} S \sin \omega t$$

$$= -0.9 \cdot 100 \cdot \frac{d}{dt}\left(1 \frac{\mathrm{Vs}}{\mathrm{m}^2} \cdot 0.01\ \mathrm{m}^2 \sin \frac{100}{\mathrm{s}} \cdot 2\pi t\right)$$

$$e = -90 \cdot 2\pi \mathrm{V} \cos \frac{200}{\mathrm{s}} \pi t = -565\ \mathrm{V} \cos \frac{200}{\mathrm{s}} \pi t.$$

Hence, the peak value of the voltage is 565 V and the effective value of the voltage induced in the coil is $565\ \mathrm{V}/\sqrt{2} = 400\ \mathrm{V}$.

Ampère's law involves a displacement current that can be given as the time derivative of the electric flux ψ_e. Ampère's law

$$\oint_l \boldsymbol{H} \cdot d\boldsymbol{l} = \int_S \boldsymbol{J} \cdot d\boldsymbol{S} + \frac{d}{dt} \int_S \boldsymbol{D} \cdot d\boldsymbol{S} = i(t) + \frac{d\Psi_e}{dt} \tag{1.9}$$

indicates that a current $i(t)$ penetrating a surface S and including the change of electric flux has to be equal to the line integral of the magnetic flux $\boldsymbol{H}$ along the line l around the surface S. Figure 1.3 depicts an application of Ampère's law.

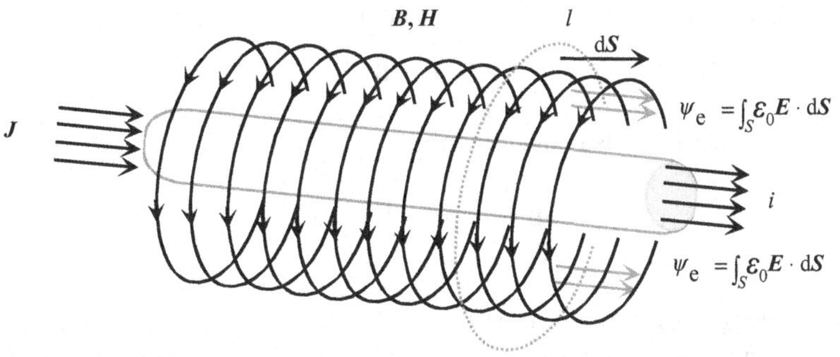

Figure 1.3 Application of Ampère's law in the surroundings of a current-carrying conductor. The line l defines a surface S, the vector $\mathrm{d}\boldsymbol{S}$ being perpendicular to it.

The term

$$\frac{\mathrm{d}}{\mathrm{d}t}\int_S \boldsymbol{D}\cdot \mathrm{d}\boldsymbol{S} = \frac{\mathrm{d}\Psi_e}{\mathrm{d}t}$$

in (1.9) is known as Maxwell's displacement current, which ultimately links the electromagnetic phenomena together. The displacement current is Maxwell's historical contribution to the theory of electromagnetism. The invention of displacement current helped him to explain the propagation of electromagnetic waves in a vacuum in the absence of charged particles or currents. Equation (1.9) is quite often presented in its static or quasi-static form, which yields

$$\oint_l \boldsymbol{H}\cdot \mathrm{d}\boldsymbol{l} = \int_S \boldsymbol{J}\cdot \mathrm{d}\boldsymbol{S} = \sum i(t) = \Theta(t) \qquad (1.10)$$

The term "quasi-static" indicates that the frequency f of the phenomenon in question is low enough to neglect Maxwell's displacement current. The phenomena occurring in electrical machines meet the quasi-static requirement well, since, in practice, considerable displacement currents appear only at radio frequencies or at low frequencies in capacitors that are deliberately produced to take advantage of the displacement currents.

The quasi-static form of Ampère's law is a very important equation in electrical machine design. It is employed in determining the magnetic voltages of an electrical machine and the required current linkage. The instantaneous value of the current sum $\sum i(t)$ in Equation (1.10), that is the instantaneous value of current linkage $\Theta(t)$, can, if desired, be assumed to involve also the apparent current linkage of a permanent magnet $\Theta_{PM} = H_c' h_{PM}$. Thus, the apparent current linkage of a permanent magnet depends on the calculated coercive force H_c' of the material (see Chapter 3) and on the thickness h_{PM} of the permanent magnet.

The corresponding differential form of Ampère's law (1.10) in a quasi-static state ($\mathrm{d}D/\mathrm{d}t$ neglected) is written as

$$\nabla \times \boldsymbol{H} = \boldsymbol{J}. \qquad (1.11)$$

The continuity Equation (1.2) for current density in a quasi-static state is written as

$$\nabla \cdot \boldsymbol{J} = 0. \tag{1.12}$$

Gauss's law for electric fields in integral form

$$\oint_S \boldsymbol{D} \cdot \mathrm{d}\boldsymbol{S} = \int_V \rho_V \mathrm{d}V \tag{1.13}$$

indicates that a charge inside a closed surface S that surrounds a volume V creates an electric flux density $\boldsymbol{D}$ through the surface. Here $\int_V \rho_V \mathrm{d}V = q(t)$ is the instantaneous net charge inside the closed surface S. Thus, we can see that in electric fields, there are both sources and drains. When considering the insulation of electrical machines, Equation (1.13) is required. However, in electrical machines, it is not uncommon that charge densities in a medium prove to be zero. In that case, Gauss's law for electric fields is rewritten as

$$\oint_S \boldsymbol{D} \cdot \mathrm{d}\boldsymbol{S} = 0 \quad \text{or} \quad \nabla \cdot \boldsymbol{D} = 0 \Rightarrow \nabla \cdot \boldsymbol{E} = 0 \tag{1.14}$$

In uncharged areas, there are no sources or drains in the electric field either.

Gauss's law for magnetic fields in integral form

$$\oint_S \boldsymbol{B} \cdot \mathrm{d}\boldsymbol{S} = 0 \tag{1.15}$$

states correspondingly that the sum of a magnetic flux penetrating a closed surface S is zero; in other words, the flux entering an object must also leave the object. This is an alternative way of expressing that there is no source for a magnetic flux. In electrical machines, this means for instance that the main flux encircles the magnetic circuit of the machine without a starting or end point. Similarly, all other flux loops in the machine are closed. Figure 1.4 illustrates the surfaces S employed in integral forms of Maxwell's equations, and Figure 1.5, respectively, presents an application of Gauss's law for a closed surface S.

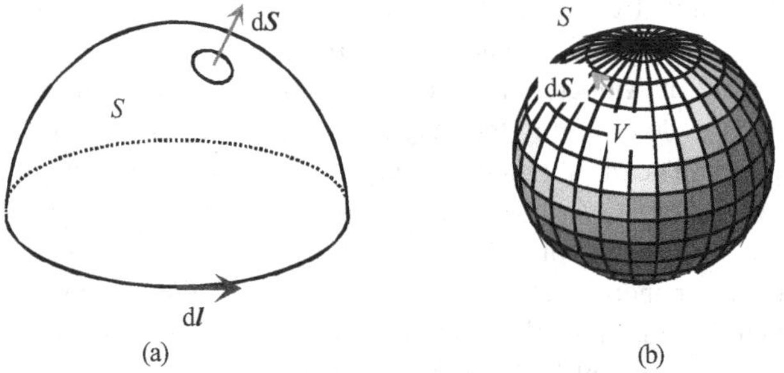

Figure 1.4 Surfaces for the integral forms of the equations for electric and magnetic fields. (a) An open surface S and its contour l, (b) a closed surface S, enclosing a volume V. d$\boldsymbol{S}$ is a differential surface vector that is everywhere normal to the surface.

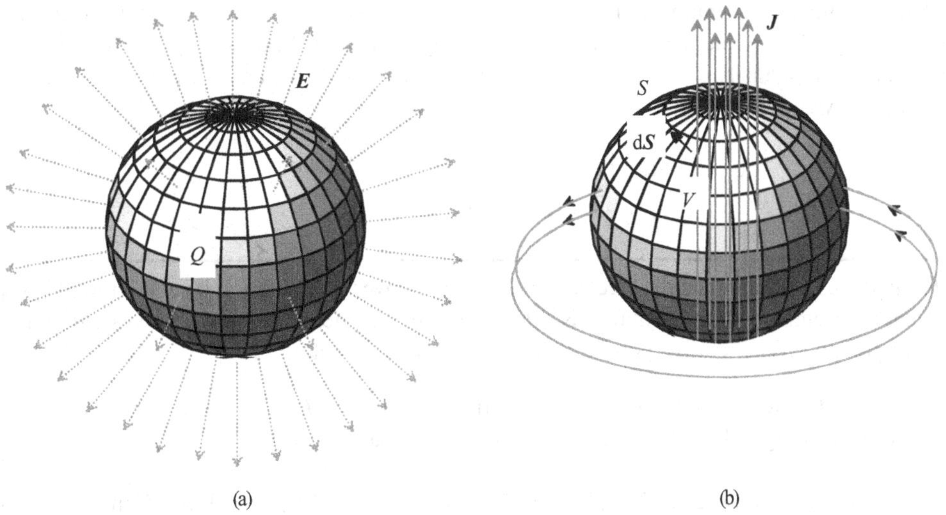

Figure 1.5 Illustration of Gauss's law for (a) an electric field and (b) a magnetic field. The charge Q inside a closed object acts as a source and creates an electric flux with the field strength $\boldsymbol{E}$. Correspondingly, a magnetic flux created by the current density $\boldsymbol{J}$ outside a closed surface S passes through the closed surface (penetrates into the sphere and then comes out). The magnetic field is thereby sourceless (div $\boldsymbol{B} = 0$).

The permittivity, permeability and conductivity ε, μ and σ of the medium determine the dependence of the electric and magnetic flux densities and current density on the field strength. In certain cases, ε, μ and σ can be treated as simple constants; then the corresponding pair of quantities ($\boldsymbol{D}$ and $\boldsymbol{E}$, $\boldsymbol{B}$ and $\boldsymbol{H}$, or $\boldsymbol{J}$ and $\boldsymbol{E}$) are parallel. Media of this kind are called isotropic, which means that ε, μ and σ have the same values in different directions. Otherwise, the media have different values of the quantities ε, μ and σ in different directions, and may therefore be treated as tensors; these media are defined as anisotropic. In practice, the permeability in ferromagnetic materials is always a highly nonlinear function of the field strength H: $\mu = f(H)$.

The general formulations for the equations of a medium can in principle be written as

$$\boldsymbol{D} = f(\boldsymbol{E}), \tag{1.16}$$

$$\boldsymbol{B} = f(\boldsymbol{H}), \tag{1.17}$$

$$\boldsymbol{J} = f(\boldsymbol{E}). \tag{1.18}$$

The specific forms for the equations have to be determined empirically for each medium in question. By applying permittivity ε [F/m], permeability μ [Vs/(Am)] and conductivity σ [S/m], we can describe materials by the following equations:

$$\boldsymbol{D} = \varepsilon \boldsymbol{E}, \tag{1.19}$$

$$\boldsymbol{B} = \mu \boldsymbol{H}, \tag{1.20}$$

$$\boldsymbol{J} = \sigma \boldsymbol{E}. \tag{1.21}$$

The quantities describing the medium are not always simple constants. For instance, the permeability of ferromagnetic materials is strongly nonlinear. In anisotropic materials, the direction of flux density deviates from the field strength, and thus ε and μ can be tensors. In a vacuum the values are

$$\varepsilon_0 = 8.854 \cdot 10^{-12}\,\mathrm{F/m},\ \ \mathrm{As/Vm}\ \text{and}$$

$$\mu_0 = 4\pi \cdot 10^{-7}\,\mathrm{H/m},\ \ \mathrm{Vs/Am}.$$

Example 1.3: Calculate the electric field density D over an insulation layer 0.3 mm thick when the potential of the winding is 400 V and the magnetic circuit of the system is at earth potential. The relative permittivity of the insulation material is $\varepsilon_r = 3$.

Solution: The electric field strength across the insulation is $E = 400$ V/0.3 mm $= 1330$ kV/m. According to Equation (1.19), the electric field density is

$$\boldsymbol{D} = \varepsilon \boldsymbol{E} = \varepsilon_r \varepsilon_0 \boldsymbol{E} = 3 \cdot 8.854 \cdot 10^{-12}\,\mathrm{As/Vm} \cdot 1330\,\mathrm{kV/m} = 35.4\,\mathrm{\mu C/m^2}$$

Example 1.4: Calculate the displacement current over the slot insulation of the previous example at 50 Hz when the insulation surface is 0.001 m^2.

Solution: The electric field over the insulation is $\psi_e = DS = 0.0354$ µAs.
The time-dependent electric field over the slot insulation is

$$\psi_e(t) = \hat{\psi}_e \sin \omega t = 0.0354\,\mathrm{\mu As} \sin(314t)$$

Differentiating with respect to time gives

$$\frac{\mathrm{d}\psi_e(t)}{\mathrm{d}t} = \omega \hat{\psi}_e \cos \omega t = 11\,\mathrm{\mu A} \cos(314t)$$

The amplitude is 11 µA and the effective (RMS) current over the insulation is hence $11/\sqrt{2} = 7.86$ µA.

Here we see that the displacement current is insignificant from the viewpoint of the machine's basic functionality. However, when a motor is supplied by a frequency converter and the transistors create high frequencies, significant displacement currents may run across the insulation and bearing current problems, for instance, may occur.

1.2 Numerical Solution

The basic design of an electrical machine, that is the dimensioning of the magnetic and electric circuits, is usually carried out by applying analytical equations. However, accurate performance of the machine is usually evaluated using different numerical methods. With these numerical methods, the effect of a single parameter on the dynamical performance of the machine can be effectively studied. Furthermore, some tests, which are not even feasible in laboratory circumstances, can be virtually performed. The most widely used numerical method is the finite element method (FEM), which can be used in the analysis of two- or three-dimensional electromagnetic field problems. The solution can be obtained for static, time-harmonic or transient problems. In the latter two cases, the electric circuit describing the power supply of the machine is coupled with the actual field solution. When applying FEM in the electromagnetic analysis of an electrical machine, special attention has to be paid to the relevance of the electromagnetic material data of the structural parts of the machine as well as to the construction of the finite element mesh.

Because most of the magnetic energy is stored in the air gap of the machine and important torque calculation formulations are related to the air-gap field solution, the mesh has to be sufficiently dense in this area. The rule of thumb is that the air-gap mesh should be divided into three layers to achieve accurate results. In the transient analysis, that is in time-stepping solutions, the selection of the size of the time step is also important in order to include the effect of high-order time harmonics in the solution. A general method is to divide one time cycle into 400 steps, but the division could be even denser than this, in particular with high-speed machines.

There are five common methods to calculate the torque from the FEM field solution. The solutions are (1) the Maxwell stress tensor method, (2) Arkkio's method, (3) the method of magnetic coenergy differentiation, (4) Coulomb's virtual work and (5) the magnetizing current method. The mathematical torque formulations related to these methods will shortly be discussed in Sections 1.4 and 1.5.

The magnetic fields of electrical machines can often be treated as a two-dimensional case, and therefore it is quite simple to employ the magnetic vector potential in the numerical solution of the field. In many cases, however, the fields of the machine are clearly three dimensional, and therefore a two-dimensional solution is always an approximation. In the following, first, the full three-dimensional vector equations are applied.

The magnetic vector potential $\boldsymbol{A}$ is given by

$$\boldsymbol{B} = \nabla \times \boldsymbol{A}; \tag{1.22}$$

Coulomb's condition, required to define unambiguously the vector potential, is written as

$$\nabla \cdot \boldsymbol{A} = 0. \tag{1.23}$$

The substitution of the definition for the magnetic vector potential in the induction law (1.3) yields

$$\nabla \times \boldsymbol{E} = -\nabla \times \frac{\partial}{\partial t}\boldsymbol{A}. \tag{1.24}$$

Electric field strength can be expressed by the vector potential $\boldsymbol{A}$ and the scalar electric potential ϕ as

$$\boldsymbol{E} = -\frac{\partial \boldsymbol{A}}{\partial t} - \nabla\phi \tag{1.25}$$

where ϕ is the reduced electric scalar potential. Because $\nabla \times \nabla\phi = 0$, adding a scalar potential causes no problems with the induction law. The equation shows that the electric field strength vector consists of two parts, namely a rotational part induced by the time dependence of the magnetic field, and a nonrotational part created by electric charges and the polarization of dielectric materials.

Current density depends on the electric field strength

$$\boldsymbol{J} = \sigma \boldsymbol{E} = -\sigma \frac{\partial \boldsymbol{A}}{\partial t} - \sigma \nabla \phi. \tag{1.26}$$

Ampère's law and the definition for vector potential yield

$$\nabla \times \left(\frac{1}{\mu} \nabla \times \boldsymbol{A} \right) = \boldsymbol{J}. \tag{1.27}$$

Substituting (1.26) into (1.27) gives

$$\nabla \times \left(\frac{1}{\mu} \nabla \times \boldsymbol{A} \right) + \sigma \frac{\partial \boldsymbol{A}}{\partial t} + \sigma \nabla \phi = 0. \tag{1.28}$$

The latter is valid in areas where eddy currents may be induced, whereas the former is valid in areas with source currents $\boldsymbol{J} = \boldsymbol{J}_{\mathrm{s}}$, such as winding currents, and areas without any current densities $\boldsymbol{J} = 0$.

In electrical machines, a two-dimensional solution is often the obvious one; in these cases, the numerical solution can be based on a single component of the vector potential $\boldsymbol{A}$. The field solution ($\boldsymbol{B}$, $\boldsymbol{H}$) is found in an xy plane, whereas $\boldsymbol{J}$, $\boldsymbol{A}$ and $\boldsymbol{E}$ involve only the z-component. The gradient $\nabla\phi$ only has a z-component, since $\boldsymbol{J}$ and $\boldsymbol{A}$ are parallel to z, and (1.26) is valid. The reduced scalar potential is thus independent of x- and y-components. ϕ could be a linear function of the z-coordinate, since a two-dimensional field solution is independent of z. The assumption of two-dimensionality is not valid if there are potential differences caused by electric charges or by the polarization of insulators. For two-dimensional cases with eddy currents, the reduced scalar potential has to be set as $\phi = 0$.

In a two-dimensional case, the previous equation is rewritten as

$$-\nabla \cdot \left(\frac{1}{\mu} \nabla A_z \right) + \sigma \frac{\partial A_z}{\partial t} = 0. \tag{1.29}$$

Outside eddy current areas, the following is valid:

$$-\nabla \cdot \left(\frac{1}{\mu} \nabla A_z \right) = J_z. \tag{1.30}$$

The definition of vector potential yields the following components for flux density:

$$B_x = \frac{\partial A_z}{\partial y}, \quad B_y = -\frac{\partial A_z}{\partial x}. \tag{1.31}$$

Hence, the vector potential remains constant in the direction of the flux density vector. Consequently, the iso-potential curves of the vector potential are flux lines. In the two-dimensional case, the following formulation can be obtained from the partial differential equation of the vector potential:

$$-\boldsymbol{k}\left[\frac{\partial}{\partial x}\left(\nu\frac{\partial A_z}{\partial x}\right)+\frac{\partial}{\partial y}\left(\nu\frac{\partial A_z}{\partial y}\right)\right]=\boldsymbol{k}J. \tag{1.32}$$

Here ν is the reluctivity of the material. This again is similar to the equation for a static electric field

$$\nabla\cdot(\nu\nabla\boldsymbol{A})=-\boldsymbol{J}. \tag{1.33}$$

Further, there are two types of boundary conditions. Dirichlet's boundary condition indicates that a known potential, here the known vector potential

$$\boldsymbol{A}=\text{constant}, \tag{1.34}$$

can be achieved for a vector potential for instance on the outer surface of an electrical machine. The field is parallel to the contour of the surface. Similar to the outer surface of an electrical machine, also the center line of the machine's pole can form a symmetry plane. Neumann's homogeneous boundary condition determined with the vector potential

$$\nu\frac{\partial\boldsymbol{A}}{\partial\boldsymbol{n}}=0 \tag{1.35}$$

can be achieved when the field meets a contour perpendicularly. Here $\boldsymbol{n}$ is the normal unit vector of a plane. A contour of this kind is for instance part of a field confined to infinite permeability iron or the center line of the pole clearance.

The magnetic flux penetrating a surface is easy to calculate with the vector potential. Stoke's theorem yields for the flux

$$\Phi=\int_S\boldsymbol{B}\cdot\mathrm{d}\boldsymbol{S}=\int_S(\nabla\times\boldsymbol{A})\cdot\mathrm{d}\boldsymbol{S}=\oint_l\boldsymbol{A}\cdot\mathrm{d}\boldsymbol{l} \tag{1.36}$$

This is an integral around the contour l of the surface S. These phenomena are illustrated with Figure 1.6. In the two-dimensional case of the illustration, the end faces' share of the integral is zero, and the vector potential along the axis is constant. Consequently, for a machine of length l we obtain a flux

$$\Phi_{12}=l(A_1-A_2). \tag{1.37}$$

This means that the flux Φ_{12} is the flux between vector equipotential lines A_1 and A_2.

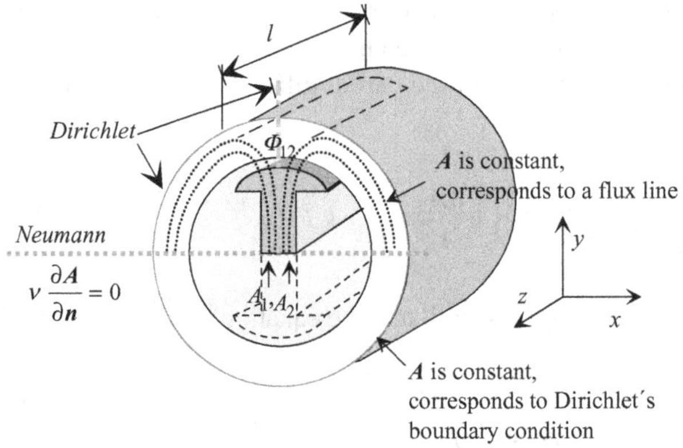

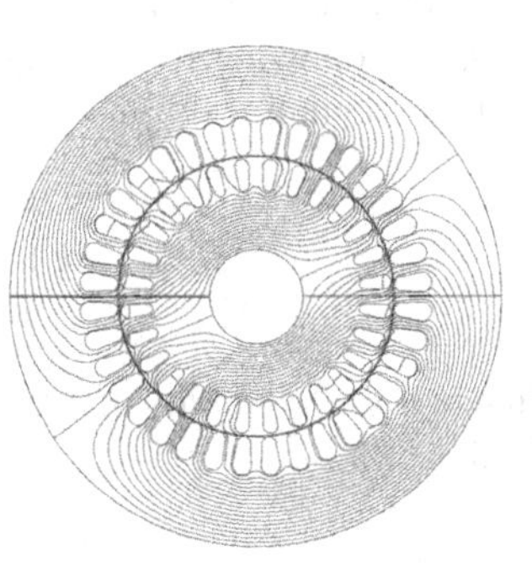

Figure 1.6 Left, a two-dimensional field and its boundary conditions for a salient-pole synchronous machine are illustrated. Here, the constant value of the vector potential $\boldsymbol{A}$ (e.g. the machine's outer contour) is taken as Dirichlet's boundary condition, and the zero value of the derivative of the vector potential with respect to normal is taken as Neumann's boundary condition. In the case of magnetic scalar potential, the boundary conditions with respect to potential would take opposite positions. Because of symmetry, the zero value of the normal derivative of the vector potential corresponds to the constant magnetic potential V_{m}, which in this case would be a known potential and thus Dirichlet's boundary condition. Right, a vector-potential-based field solution of a two-pole asynchronous machine assuming a two-dimensional field is presented.

1.3 The Most Common Principles Applied to Analytic Calculation

The design of an electrical machine involves the quantitative determination of the magnetic flux of the machine. Usually, phenomena in a single pole are analyzed. In the design of a magnetic circuit, the precise dimensions for individual parts are determined, the required current linkage for the magnetic circuit and also the required magnetizing current are calculated, and the magnitude of losses occurring in the magnetic circuit is estimated.

If the machine is excited with permanent magnets, the permanent magnet materials have to be selected and the main dimensions of the parts manufactured from these materials have to be determined. Generally, when calculating the magnetizing current for a rotating machine, the machine is assumed to run at no load: that is, there is a constant current flowing in the magnetizing winding. The effects of load currents are analyzed later.

The design of a magnetic circuit of an electrical machine is based on Ampère's law (1.4) and (1.8). The line integral calculated around the magnetic circuit of an electrical machine, that is the sum of magnetic potential differences $\sum U_{\mathrm{m},i}$, must be equal to the surface integral of the current densities over the surface S of the magnetic circuit. (The surface S here indicates the surface penetrated by the main flux.) In practice, in electrical machines, the current usually flows in the windings. The surface integral of the current density corresponding to the sum of these currents (flowing in the windings) is the current linkage Θ. Now Ampère's law can be rewritten as

$$U_{\mathrm{m,tot}} = \sum U_{\mathrm{m},i} = \oint_l \boldsymbol{H} \cdot \mathrm{d}\boldsymbol{l} = \int_S \boldsymbol{J} \cdot \mathrm{d}\boldsymbol{S} = \Theta = \sum i \tag{1.38}$$

The sum of magnetic potential differences U_m around the complete magnetic circuit is equal to the sum of the magnetizing currents in the circuit, that is the current linkage Θ. In simple applications, the current sum may be given as $\sum i = k_w Ni$, where $k_w N$ is the effective number of turns and i the current flowing in them. In addition to the windings, this current linkage may also involve the effect of the permanent magnets. In practice, when calculating the magnetic voltage, the machine is divided into its components, and the magnetic voltage U_m between points a and b selected suitably in the magnetic circuit is determined as

$$U_{m,ab} = \int_a^b \boldsymbol{H} \cdot d\boldsymbol{l}. \tag{1.39}$$

In electrical machines, the field strength is often in the direction of the component in question, and thus Equation (1.39) can simply be rewritten as

$$U_{m,ab} = \int_a^b H \, dl. \tag{1.40}$$

Further, if the field strength is constant in the area under observation, we get

$$U_{m,ab} = Hl. \tag{1.41}$$

In the determination of the required current linkage Θ of a machine's magnetizing winding, the simplest possible integration path is selected in the calculation of the magnetic voltages. This means selecting a path that encloses the magnetizing winding. This path is defined as the main integration path and it is also called the main flux path of the machine (see Chapter 3). In salient-pole machines, the main integration path crosses the air gap in the middle of the pole shoes.

Example 1.5: Consider a C-core inductor with a 1 mm air gap. In the air gap, the flux density is 1 T. The ferromagnetic circuit length is 0.2 m and the relative permeability of the core material at 1 T is $\mu_r = 3500$. Calculate the field strengths in the air gap and the core. How many turns N of wire carrying a 10 A direct current are needed to magnetize the choke to 1 T? Fringing in the air gap is neglected and the winding factor is assumed to be $k_w = 1$.

Solution: According to (1.20), the magnetic field strength in the air gap is

$$H_\delta = B_\delta / \mu_0 = 1 \text{ Vs/m}^2 \left(4\pi \cdot 10^{-7} \text{ Vs/Am}\right) = 795 \text{ kA/m}$$

The corresponding magnetic field strength in the core is

$$H_{Fe} = B_{Fe} / (\mu_r \mu_0) = 1 \text{ Vs/m}^2 / \left(3500 \cdot 4\pi \cdot 10^{-7} \text{ Vs/Am}\right) = 227 \text{ A/m}$$

The magnetic voltage in the air gap (neglecting fringing) is

$$U_{m,\delta} = H_\delta \delta = 795 \text{ kA/m} \cdot 0.001 \text{ m} = 795 \text{ A}$$

The magnetic voltage in the core is

$$U_{m,Fe} = H_{Fe} l_{Fe} = 227\ \text{A/m} \cdot 0.2\ \text{m} = 45\ \text{A}.$$

The magnetomotive force (mmf) F_m of the magnetic circuit is

$$F_m = \oint_l \boldsymbol{H} \cdot d\boldsymbol{l} = U_{m,tot} = \sum U_{m,i} = U_{m,\delta} + U_{m,Fe} = 795\ \text{A} + 45\ \text{A} = 840\ \text{A}$$

The current linkage Θ of the choke has to be of equal magnitude with the mmf $U_{m,tot}$,

$$\Theta = \sum i = k_w N i = U_{m,tot}.$$

We get

$$N = \frac{U_{m,tot}}{k_w i} = \frac{840\ \text{A}}{1 \cdot 10\ \text{A}} = 84\ \text{turns}.$$

In machine design, not only does the main flux have to be analyzed, but also all the leakage fluxes of the machine have to be taken into account.

In the determination of the no-load curve of an electrical machine, the magnetic voltages of the magnetic circuit have to be calculated with several different flux densities. In practice, for the exact definition of the magnetizing curve, a computation program that solves the different magnetizing states of the machine is required.

According to their magnetic circuits, electrical machines can be divided into two main categories: in salient-pole machines, the field windings are concentrated pole windings, whereas in nonsalient-pole machines, the magnetizing windings are spatially distributed in the machine. The main integration path of a salient-pole machine consists for instance of the following components: a rotor yoke (yr), pole body (p2), pole shoe (p1), air gap (δ), teeth (d) and stator yoke (ys). For this kind of salient-pole machine or DC machine, the total magnetic voltage of the main integration path therefore consists of the following components

$$U_{m,tot} = U_{m,yr} + 2U_{m,p2} + 2U_{m,p1} + 2U_{m,\delta} + U_{m,d} + U_{m,ys} \tag{1.42}$$

In a nonsalient-pole synchronous machine and induction motor, the magnetizing winding is contained in slots. Therefore both stator (s) and rotor (r) have teeth areas (d)

$$U_{m,tot} = U_{m,yr} + 2U_{m,dr} + 2U_{m,\delta} + U_{m,ds} + U_{m,ys} \tag{1.43}$$

With Equations (1.42) and (1.43), we must bear in mind that the main flux has to flow twice across the teeth area (or pole arc and pole shoe) and air gap.

In a switched reluctance (SR) machine, where both the stator and rotor have salient poles (double saliency), the following equation is valid:

$$U_{m,tot} = U_{m,yr} + 2U_{m,rp2} + 2U_{m,rp1}(\alpha) + 2U_{m,\delta}(\alpha) + 2U_{m,sp1}(\alpha) + 2U_{m,sp2} + U_{m,ys} \tag{1.44}$$

This equation proves difficult to employ, because the shape of the air gap in an SR machine varies constantly when the machine rotates. Therefore the magnetic voltage of both the rotor and stator pole shoes depends on the position of the rotor α.

The magnetic potential differences of the most common rotating electrical machines can be presented by equations similar to Equations (1.42)–(1.44).

In electrical machines constructed of ferromagnetic materials, only the air gap can be considered magnetically linear. All ferromagnetic materials are both nonlinear and often anisotropic. In particular, the permeability of oriented electrical steel sheets varies in different directions, being highest in the rolling direction and lowest in the perpendicular direction. This leads to a situation where the permeability of the material is, strictly speaking, a tensor.

The flux is a surface integral of the flux density. Commonly, in electrical machine design, the flux density is assumed to be perpendicular to the surface to be analyzed. Since the area of a perpendicular surface $\boldsymbol{S}$ is S, we can rewrite the equation simply as

$$\Phi = \int B\,\mathrm{d}S. \tag{1.45}$$

Further, if the flux density B is also constant, we obtain

$$\Phi = BS. \tag{1.46}$$

Using the equations above, it is possible to construct a magnetizing curve for each part (part being between points a and b) of the machine

$$\Phi_{ab} = f(U_{\mathrm{m},ab}), \quad B = f(U_{\mathrm{m},ab}). \tag{1.47}$$

In the air gap, the permeability is constant $\mu = \mu_0$. Thus, we can employ magnetic conductivity, that is permeance Λ, which leads us to

$$\Phi_{ab} = \Lambda_{ab} U_{\mathrm{m},ab}. \tag{1.48}$$

If the air gap field is homogeneous, we get

$$\Phi_{ab} = \Lambda_{ab} U_{\mathrm{m},ab} = \frac{\mu_0 S}{\delta} U_{\mathrm{m},ab}. \tag{1.49}$$

Equations (1.38) and (1.42)–(1.44) yield the magnetizing curve for a machine

$$\Phi_\delta = f(\Theta), \quad B_\delta = f(\Theta) \tag{1.50}$$

where the term Φ_δ is the air-gap flux. The absolute value for flux density B_δ is the maximum flux density in the air gap in the middle of the pole shoe, when slotting is neglected. The magnetizing curve of the machine is determined in the order Φ_δ, $B_\delta \rightarrow B \rightarrow H \rightarrow U_\mathrm{m} \rightarrow \Theta$ by always selecting a different value for the air-gap flux Φ_δ, or for its density, and by calculating the magnetic voltages in the machine and the required current linkage Θ. With the current linkage, it is possible to determine the current I flowing in the windings. Correspondingly, with

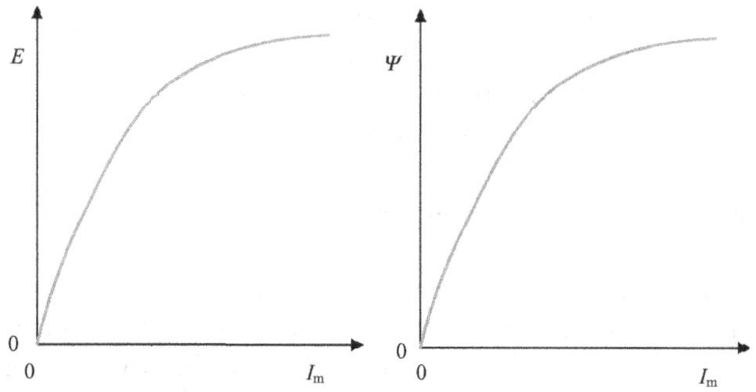

Figure 1.7 Typical no-load curve for an electrical machine expressed by the electromotive force E or the flux linkage Ψ as a function of the magnetizing current I_m. The E curve as a function of I_m has been measured when the machine is running at no load at a constant speed. In principle, the curve resembles a BH curve of the ferromagnetic material used in the machine. The slope of the no-load curve depends on the BH curve of the material, the (geometrical) dimensions and particularly on the length of the air gap.

the air-gap flux and the winding, we can determine the electromotive force (emf) E induced in the windings. Now we can finally plot the actual no-load curve of the machine (Figure 1.7)

$$E = f(I). \tag{1.51}$$

1.3.1 Flux Line Diagrams

Let us consider areas with an absence of currents. A spatial magnetic flux can be assumed to flow in a flux tube. A flux tube can be analyzed as a tube of a quadratic cross-section $\Delta \boldsymbol{S}$. The flux does not flow through the walls of the tube, and hence $\boldsymbol{B} \cdot \mathrm{d}\boldsymbol{S} = 0$ is valid for the walls. As depicted in Figure 1.8, we can see that the corners of the flux tube form the flux lines.

When calculating a surface integral along a closed surface surrounding the surface of a flux tube, Gauss's law (1.15) yields

$$\oint \boldsymbol{B} \cdot \mathrm{d}\boldsymbol{S} = 0 \tag{1.52}$$

Since there is no flux through the side walls of the tube in Figure 1.8, Equation (1.52) can be rewritten as

$$\oint \boldsymbol{B}_1 \cdot \mathrm{d}\Delta\boldsymbol{S}_1 = \oint \boldsymbol{B}_2 \cdot \mathrm{d}\Delta\boldsymbol{S}_2 = \oint \boldsymbol{B}_3 \cdot \mathrm{d}\Delta\boldsymbol{S}_3 \tag{1.53}$$

indicating that the flux of the flux tube is constant

$$\Delta\Phi_1 = \Delta\Phi_2 = \Delta\Phi_3 = \Delta\Phi. \tag{1.54}$$

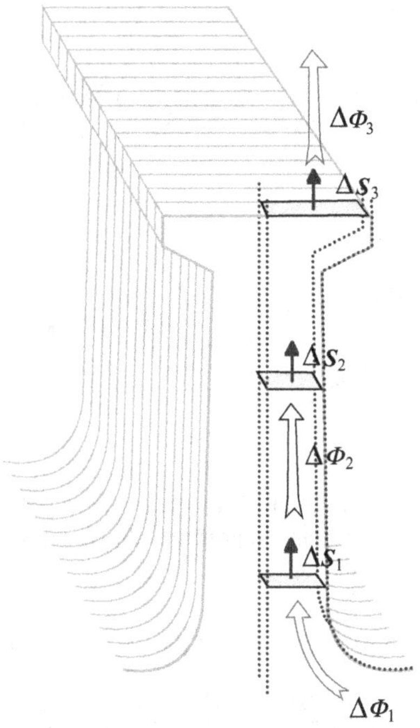

Figure 1.8 Laminated tooth and a coarse flux tube running in a lamination. The cross-sections of the tube are presented with surface vectors $\Delta \boldsymbol{S}_i$. There is a flux $\Delta \Phi$ flowing in the tube. The flux tubes follow the flux lines in the magnetic circuit of the electrical machine. Most of the tubes constitute the main magnetic circuit, but a part of the flux tubes forms leakage flux paths. If a two-dimensional field solution is assumed, two-dimensional flux diagrams as shown in Figure 1.6 may replace the flux tube approach.

A magnetic equipotential surface is a surface with a certain magnetic scalar potential V_m. When traveling along any route between two points a and b on this surface, we must get

$$\int_a^b \boldsymbol{H} \cdot \mathrm{d}\boldsymbol{l} = U_{\mathrm{m},ab} = V_{\mathrm{m}a} - V_{\mathrm{m}b} = 0. \quad (1.55)$$

When observing a differential route, this is valid only when $\boldsymbol{H} \cdot \mathrm{d}\boldsymbol{l} = 0$. For isotropic materials, the same result can be expressed as $\boldsymbol{B} \cdot \mathrm{d}\boldsymbol{l} = 0$. In other words, the equipotential surfaces are perpendicular to the lines of flux.

If we select an adequately small area ΔS of the surface S, we are able to calculate the flux

$$\Delta \Phi = B \Delta S. \quad (1.56)$$

The magnetic potential difference between two equipotential surfaces that are close enough to each other (H is constant along the integration path l) is written as

$$\Delta U_{\mathrm{m}} = Hl. \quad (1.57)$$

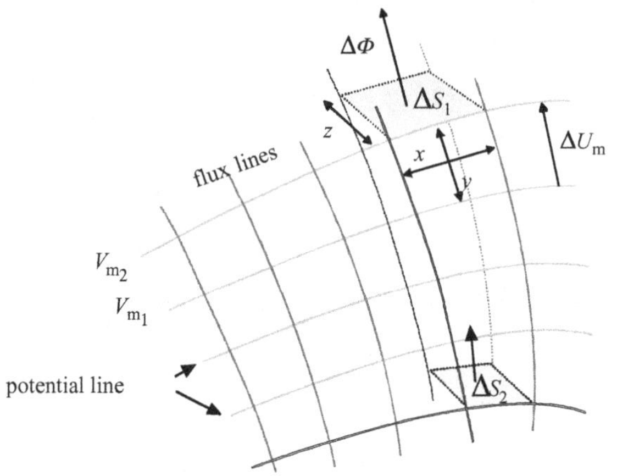

Figure 1.9 Flux lines and potential lines in a three-dimensional area with a flux flowing across an area where the length dimension z is constant. In principle, the diagram is thus two dimensional. Such a diagram is called an orthogonal field diagram.

The above equations give the permeance Λ of the cross-section of the flux tube

$$\Lambda = \frac{\Delta\Phi}{\Delta U_{\mathrm{m}}} = \frac{B \cdot \mathrm{d}S}{Hl} = \mu\frac{\mathrm{d}S}{l}. \tag{1.58}$$

The flux line diagram (Figure 1.9) comprises selected flux and potential lines. The selected flux lines confine flux tubes, which all have an equal flux $\Delta\Phi$. The magnetic voltage between the chosen potential lines is always the same, ΔU_{m}. Thus, the magnetic conductivity of each section of the flux tube is always the same, and the ratio of the distance of flux lines x to the distance of potential lines y is always the same. If we set

$$\frac{x}{y} = 1, \tag{1.59}$$

the field diagram forms, according to Figure 1.9, a grid of quadratic elements.

In a homogeneous field, the field strength H is constant at every point of the field. According to Equations (1.57) and (1.59), the distance of all potential and flux lines is thus always the same. In that case, the flux diagram comprises squares of equal size.

When constructing a two-dimensional orthogonal field diagram, for instance for the air gap of an electrical machine, certain boundary conditions have to be known to be able to draw the diagram. These boundary conditions are often solved based on symmetry, or also because the potential of a certain potential surface of the flux tube in Figure 1.8 is already known. For instance, if the stator and rotor length of the machine is l, the area of the flux tube can, without significant error, be written as $\mathrm{d}S = l\mathrm{d}x$. The interface of the iron and air is now analyzed according to Figure 1.10a. We get

$$\mathrm{d}\Phi_y = B_{y\delta}l\mathrm{d}x - B_{y\mathrm{Fe}}l\mathrm{d}x = 0 \Rightarrow B_{y\delta} = B_{y\mathrm{Fe}}. \tag{1.60}$$

Here, $B_{y\delta}$ and $B_{y\mathrm{Fe}}$ are the flux densities of air in the y-direction and of iron in the y-direction.

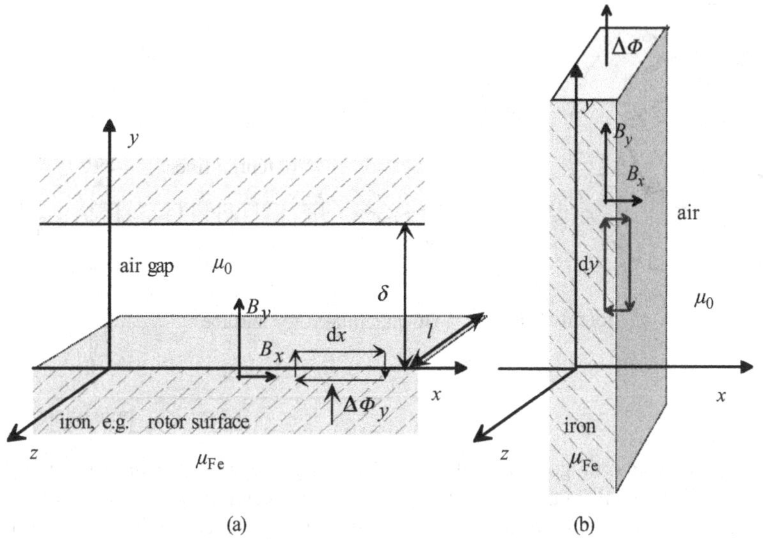

Figure 1.10 (a) Interface of air δ and iron Fe. The x-axis is tangential to the rotor surface. (b) Flux traveling on iron surface.

In Figure 1.10a, the field strength has to be continuous in the x-direction on the iron–air interface. If we consider the interface in the x-direction and, based on Ampère's law, assume a section dx of the surface has no current, we get

$$H_{x\delta}\mathrm{d}x - H_{x\mathrm{Fe}}\mathrm{d}x = 0 \tag{1.61}$$

and thus

$$H_{x\delta} = H_{x\mathrm{Fe}} = \frac{B_{x\mathrm{Fe}}}{\mu_{\mathrm{Fe}}} \tag{1.62}$$

By assuming that the permeability of iron is infinite, $\mu_{\mathrm{Fe}} \to \infty$, we get $H_{x\mathrm{Fe}} = H_{x\delta} = 0$ and thereby also $B_{x\delta} = \mu_0 H_{x\delta} = 0$.

Hence, if we set $\mu_{\mathrm{Fe}} \to \infty$, the flux lines leave the ferromagnetic material perpendicularly into the air. Simultaneously, the interface of iron and air forms an equipotential surface. If the iron is not saturated, its permeability is very high, and the flux lines can be assumed to leave the iron almost perpendicularly in currentless areas. In saturating areas, the interface of the iron and air cannot strictly be considered an equipotential surface. The magnetic flux and the electric flux refract on the interface.

In Figure 1.10b, the flux flows in the iron in the direction of the interface. If the iron is not saturated ($\mu_{\mathrm{Fe}} \to \infty$) we can set $B_x \approx 0$. Now, there is no flux passing from the iron into air. When the iron is about to become saturated ($\mu_{\mathrm{Fe}} \to 1$), a significant magnetic voltage occurs in the iron. Now, the air adjacent to the iron becomes an appealing route for the flux, and part of the flux passes into the air. This is the case for instance when the teeth of electrical machines saturate: a part of the flux flows across the slots of the machine, even though the permeability of the materials in the slot is in practice equal to the permeability of a vacuum.

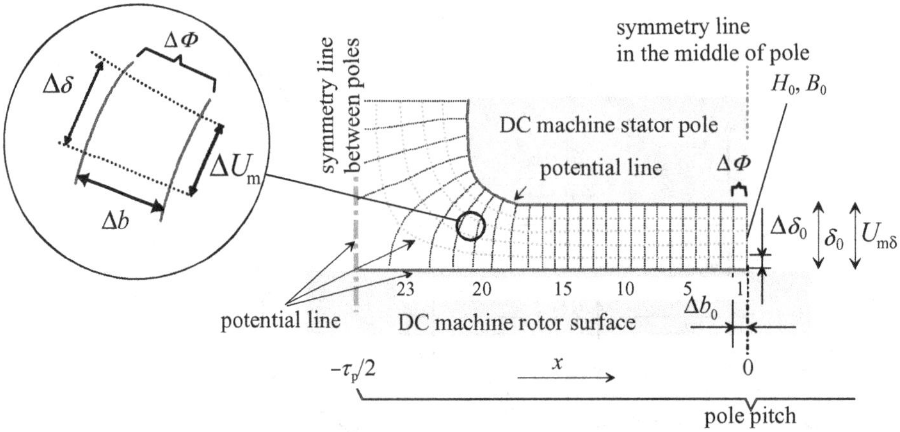

Figure 1.11 Drawing an orthogonal field diagram in an air gap of a DC machine in the edge zone of a pole shoe. Here, a differential equation for the magnetic scalar potential is solved by drawing. Dirichlet's boundary conditions for magnetic scalar potentials created on the surfaces of the pole shoe and the rotor and on the symmetry plane between the pole shoes. The center line of the pole shoe is set at the origin of the coordinate system. At the origin, the element is dimensioned as $\Delta\delta_0$, Δb_0. The $\Delta\delta$ and Δb in different parts of the diagram have different sizes, but the $\Delta\Phi$ remains the same in all flux tubes. The pole pitch is τ_p. There are about 23.5 flux tubes from the pole surface to the rotor surface in the figure.

The lines of symmetry in flux diagrams are either potential or field lines. When drawing a flux diagram, we have to know if the lines of symmetry are flux or potential lines. Figure 1.11 is an example of an orthogonal field diagram, in which the line of symmetry forms a potential line; this could depict for instance the air gap between the contour of a magnetizing pole of a DC machine and the rotor.

The solution of an orthogonal field diagram by drawing is best started at those parts of the geometry where the field is as homogeneous as possible. For instance, in the case of Figure 1.11, we start from the point where the air gap is at its narrowest. Now, the surface of the magnetizing pole and the rotor surface that is assumed to be smooth form the potential lines together with the surface between the poles. First, the potential lines are plotted at equal distances and, next, the flux lines are drawn perpendicularly to potential lines so that the area under observation forms a grid of quadratic elements. The length of the machine being l, each flux tube created this way carries a flux of $\Delta\Phi$.

With the field diagram, it is possible to solve various magnetic parameters of the area under observation. If n_Φ is the number (not necessarily an integer) of contiguous flux tubes carrying a flux $\Delta\Phi$, and ΔU_m is the magnetic voltage between the sections of a flux tube (n_U sections in sequence), the permeance of the entire air gap Λ_δ assuming that $\Delta b = \Delta\delta$ can be written as

$$\Lambda_\delta = \frac{\Phi}{U_{m\delta}} = \frac{n_\Phi \Delta\Phi}{n_U \Delta U_m} = \frac{n_\Phi \Delta b l \mu_0 \dfrac{\Delta U_m}{\Delta\delta}}{n_U \Delta U_m} = \frac{n_\Phi}{n_U}\mu_0 l \tag{1.63}$$

The magnetic field strength in the enlarged element of Figure 1.11 is

$$H = \frac{\Delta U_m}{\Delta\delta}, \tag{1.64}$$

and correspondingly the magnetic flux density

$$B = \mu_0 \frac{\Delta U_m}{\Delta \delta} = \frac{\Delta \Phi}{\Delta b l}. \tag{1.65}$$

With Equation (1.56), it is also possible to determine point by point the distribution of flux density on a potential line; in other words, on the surface of the armature or the magnetizing pole. With the notation in Figure 1.11, we get

$$\Delta \Phi_0 = B_0 \Delta b_0 l = \Delta \Phi(x) = B(x) \Delta b(x) l. \tag{1.66}$$

In the middle of the pole, where the air-gap flux is homogeneous, the flux density is

$$B_0 = \mu_0 H_0 = \mu_0 \frac{\Delta U_m}{\Delta \delta_0} = \mu_0 \frac{U_{m,\delta}}{\delta_0} \tag{1.67}$$

Thus, the magnitude of flux density as a function of the x-coordinate is

$$B(x) = \frac{\Delta b_0}{\Delta b(x)} B_0 = \frac{\Delta b_0}{\Delta b(x)} \mu_0 \frac{U_{m,\delta}}{\delta_0} \tag{1.68}$$

Example 1.6: What is the permeance of the main flux in Figure 1.11 when the air gap $\delta = 0.01$ m and the stator stack length is $l = 0.1$ m? How much flux is created with $\Theta_f = 1000$ A?

Solution: In the center of the pole, the orthogonal flux diagram is uniform and we see that $\Delta\delta_0$ and Δb_0 have the same size; $\Delta\delta_0 = \Delta b_0 = 2$ mm. The permeance of the flux tube in the center of the pole is

$$\Lambda_0 = \mu_0 \frac{\Delta b_0 l}{\Delta \delta_0} = \mu_0 \frac{0.002\ \text{m} \cdot 0.1\ \text{m}}{0.002\ \text{m}} = 4\pi \cdot 10^{-8} \frac{\text{Vs}}{\text{A}}$$

As we can see in Figure 1.11, about 23.5 flux tubes travel from half of the stator pole to the rotor surface. Each of these flux tubes transmits the same amount of flux, and hence the permeance of the whole pole seen by the main flux is

$$\Lambda = 2 \cdot 23.5 \cdot \Lambda_0 = 47 \Lambda_0 = 47 \cdot 4\pi \cdot 10^{-8} \frac{\text{Vs}}{\text{A}} = 5.9 \frac{\mu\text{Vs}}{\text{A}}$$

If we have $\Theta_f = 1000$ A current linkage magnetizing the air gap, we get the flux

$$\Phi = \Lambda \Theta_f = 5.9 \frac{\mu\text{Vs}}{\text{A}} \cdot 1000\ \text{A} = 5.9\ \text{mVs}.$$

1.3.2 Flux Diagrams for Current-Carrying Areas

Let us first consider a situation in which the machine currents are expressed with an equivalent linear current density A [A/m] which covers the area under observation. In principle, the linear current density corresponds to the surface current $\boldsymbol{J}_s = \boldsymbol{n} \times \boldsymbol{H}_0$ induced in a conducting medium by an alternating field strength $\boldsymbol{H}_0$ outside the surface. The surface normal unit vector is denoted by $\boldsymbol{n}$.

In an electrical machine with windings, the "artificial surface current," that is the local value of the linear current density A, may for instance be calculated as a current sum flowing in a slot divided by the slot pitch. Equivalent linear current density can be employed in approximation, because the currents flowing in the windings of electrical machines are usually situated close to the air gap, and the current linkages created by the currents excite mainly the air gaps. Thus, we can set $\mu = \mu_0$ in the observed area of equivalent linear current density. The utilization of equivalent linear current density simplifies the manual calculation of the machine by idealizing the potential surfaces, and does not have a crucial impact on the field diagram in the areas outside the area of linear current density. Figure 1.12 illustrates an equivalent linear current density.

The value for equivalent linear current density A is expressed per unit length in the direction of observation. The linear current density A corresponds to the tangential magnetic field strength $H_{y\delta}$. Assuming the permeability of iron to be infinite, Ampère's law yields for the element dy of Figure 1.12a

$$\oint \boldsymbol{H} \cdot \mathrm{d}\boldsymbol{l} = \mathrm{d}\Theta = H_{y\mathrm{air}}\mathrm{d}y - H_{y\mathrm{Fe}}\mathrm{d}y = A\mathrm{d}y \tag{1.69}$$

Further, this gives us

$$H_{y\mathrm{air}} = A \quad \text{and} \quad B_{y\mathrm{air}} = \mu_0 A. \tag{1.70}$$

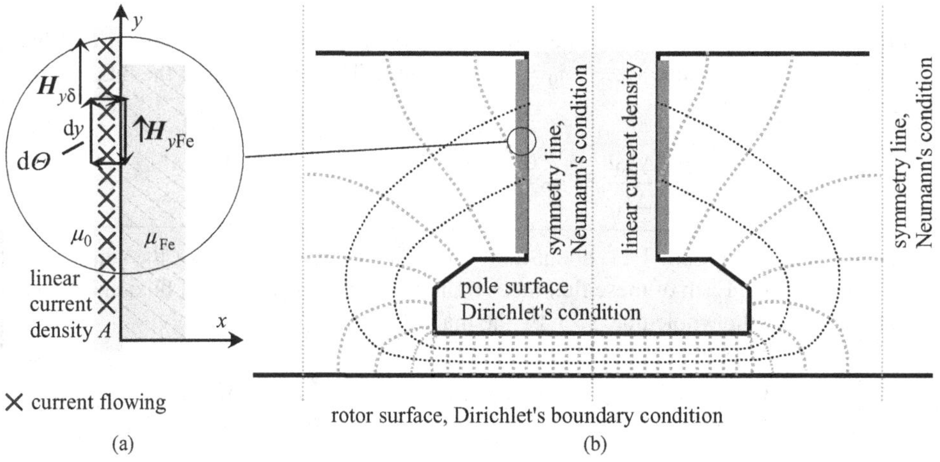

Figure 1.12 (a) General representation of linear current density A [A/m] and (b) its application to the field diagram of a magnetizing pole of a DC machine. It is important to note that in the area of the pole body, the potential lines now pass from air to iron. Dirichlet's boundary conditions indicate here a known equiscalar potential surface.

Equation (1.70) indicates that in the case of Figure 1.12 we have a tangential flux density on the pole body surface. The tangential flux density makes the flux lines travel inclined to the pole body surface and not perpendicular to it as in currentless areas.

If we assume that the phenomenon is observed on the stator inner surface or on the rotor outer surface, the x-components may be regarded as tangential components and the y-components as normal components. In the air gap δ, there is a tangential field strength $\boldsymbol{H}_{x\delta}$ along the x-component, and a corresponding component of flux density $\boldsymbol{B}_{x\delta}$ created by the linear current density A. This is essential when considering the force density, the tangential stress $\sigma_{F\text{tan}}$, that generates torque (Maxwell stresses will be discussed later). On iron surfaces with linear current density, the flux lines no longer pass perpendicularly from the iron to the air gap, as Figure 1.12 depicting the field diagram of a DC machine's magnetizing pole also illustrates. The influence of a magnetizing winding on the pole body is illustrated with the linear current density. Since the magnetizing winding is evenly distributed over the length of the pole body (the linear current density being constant), it can be seen that the potential changes linearly in the area of linear current density in the direction of the height of the pole. As evidence of this we can see that in Figure 1.12 the potential lines starting in the air gap enter the area of linear current density at even distances.

In areas with current densities $\boldsymbol{J}$, the potential lines become gradient lines. This can be seen in Figure 1.13 at points a, b and c. We could assume that the figure illustrates for instance a nonsalient-pole synchronous machine field winding bar carrying a DC density. The magnetic potential difference between V_{m4} and V_{m0} equals the slot current.

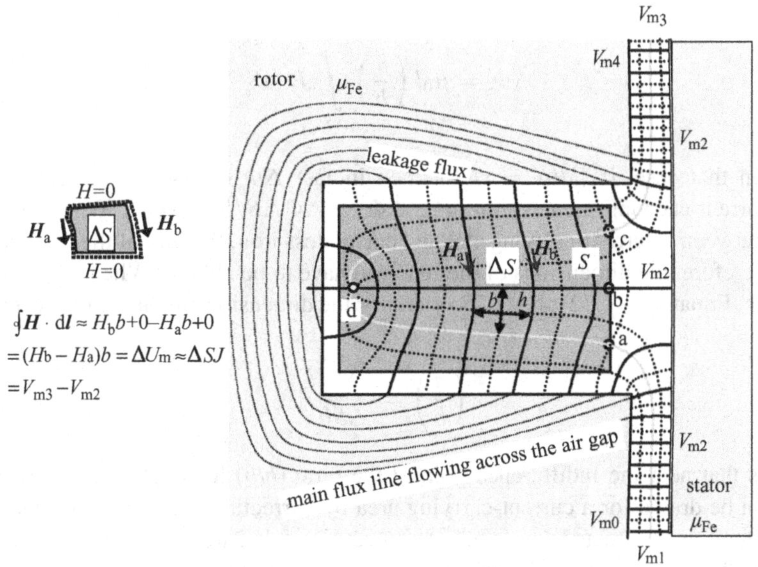

Figure 1.13 Current-carrying conductor in a slot and its field diagram. The illustration on the left demonstrates the closed line integral around the surface ΔS; also some flux lines in the iron are plotted. Note that the flux lines traveling across the slot depict leakage flux.

The gradient lines meet the slot leakage flux lines orthogonally, which means that $\int \boldsymbol{H} \cdot \mathrm{d}\boldsymbol{l} = 0$ along a gradient line. In the figure, we calculate a closed line integral around the area ΔS of the surface S

$$\oint \boldsymbol{H} \cdot \mathrm{d}\boldsymbol{l} = V_{\mathrm{m3}} - V_{\mathrm{m2}} = \int_{\Delta S} \boldsymbol{J} \cdot \mathrm{d}\boldsymbol{S} \tag{1.71}$$

where we can see that when the current density $\boldsymbol{J}$ and the difference of magnetic potentials ΔU_{m} are constant, the area ΔS of the surface S also has to be constant. In other words, the selected gradient lines confine areas of equal size from the surface S with a constant current density. The gradient lines meet at a single point d, which is called an indifference point. If the current-carrying area is confined by an area with infinite permeability, the border line is a potential line and the indifference point is located on this border line. If the permeability of iron is not infinite, then d is located in the current-carrying area, as in Figure 1.13. If inside a current-carrying area the line integral is defined for instance around the area ΔS, we can see that the closer to the point d we get, the smaller become the distances between the gradient lines. In order to maintain the same current sum in the observed areas, the heights of the areas ΔS have to be changed.

Outside the current-carrying area, the following holds:

$$\Delta \Phi = \Lambda \Delta U_{\mathrm{m}} = \mu_0 l \frac{h}{b} \int_S \boldsymbol{J} \cdot \mathrm{d}\boldsymbol{S}. \tag{1.72}$$

Inside the area under observation, when a closed line integral according to Equation (1.71) is written only for the area ΔS ($<S$), the flux of a flux tube in a current-carrying area becomes

$$\Delta \Phi' = \mu_0 l \left(\frac{h}{b}\right)' \int_{\Delta S} \boldsymbol{J} \cdot \mathrm{d}\boldsymbol{S} \tag{1.73}$$

and thus, in that case, if $(h/b)' = (h/b)$, then in fact $\Delta \Phi' < \Delta \Phi$. If the current density $\boldsymbol{J}$ in the current-carrying area is constant, $\Delta \Phi' = \Delta \Phi \Delta S / S$ is valid. When crossing the boundary between a current-carrying slot and currentless iron, the flux of the flux tube cannot change. Therefore, the dimensions of the line grid have to be altered. When $\boldsymbol{J}$ is constant and $\Delta \Phi' = \Delta \Phi$, Equations (1.72) and (1.73) yield for the dimensions in the current-carrying area

$$\left(\frac{h}{b}\right)' = \frac{Sh}{\Delta S b}. \tag{1.74}$$

This means that near the indifference point d the ratio (h/b) increases. An orthogonal field diagram can be drawn for a current-carrying area by correcting the equivalent linear current density by iterating the created diagram. For a current-carrying area, gradient lines are extended from potential lines up to the indifference point. Now, bearing in mind that the gradient lines have to divide the current-carrying area into sections of equal size, next the orthogonal flux lines are plotted by simultaneously paying attention to changing dimensions. The diagram is altered iteratively until Equation (1.74) is valid to the required accuracy.

1.4 Application of the Principle of Virtual Work in the Determination of Force and Torque

When investigating electrical equipment, the magnetic circuit of which changes form during operation, the easiest method is to apply the principle of virtual work in the estimation of force and torque. Examples of this kind of equipment are doubly-salient-pole reluctance machines, various relays and so on.

Faraday's induction law presents the voltage induced in the winding, which creates a current that tends to resist the changes in flux. The voltage equation for the winding is written as

$$u = Ri + \frac{\mathrm{d}\Psi}{\mathrm{d}t} = Ri + \frac{\mathrm{d}}{\mathrm{d}t}Li, \tag{1.75}$$

where u is the voltage connected to the winding terminals, R is the resistance of the winding and Ψ is the winding flux linkage, and L the self-inductance of the winding consisting of its magnetizing inductance and leakage inductance: $L = \Psi/i = N\Phi/i = N^2\Lambda = N^2/R_{\mathrm{m}}$ (see also Section 1.6). If the number of turns in the winding is N and the flux is Φ, Equation (1.75) can be rewritten as

$$u = Ri + N\frac{\mathrm{d}\Phi}{\mathrm{d}t}. \tag{1.76}$$

The required power in the winding is written correspondingly as

$$ui = Ri^2 + Ni\frac{\mathrm{d}\Phi}{\mathrm{d}t}, \tag{1.77}$$

and the energy

$$\mathrm{d}W = P\mathrm{d}t = Ri^2\mathrm{d}t + Ni\mathrm{d}\Phi. \tag{1.78}$$

The latter energy component $Ni\,\mathrm{d}\Phi$ is reversible, whereas $Ri^2\,\mathrm{d}t$ turns into heat. Energy cannot be created or destroyed, but may only be converted to different forms. In isolated systems, the limits of the energy balance can be defined unambiguously, which simplifies the energy analysis. The net energy input is equal to the energy stored in the system. This result, the first law of thermodynamics, is applied to electromechanical systems, where electrical energy is stored mainly in magnetic fields. In these systems, the energy transfer can be represented by the equation

$$\mathrm{d}W_{\mathrm{el}} = \mathrm{d}W_{\mathrm{mec}} + \mathrm{d}W_{\Phi} + \mathrm{d}W_{\mathrm{R}} \tag{1.79}$$

where

$\mathrm{d}W_{\mathrm{el}}$ is the differential electrical energy input,
$\mathrm{d}W_{\mathrm{mec}}$ is the differential mechanical energy output,
$\mathrm{d}W_{\Phi}$ is the differential change of magnetic stored energy,
$\mathrm{d}W_{\mathrm{R}}$ is the differential energy loss.

Here the energy input from the electric supply is written equal to the mechanical energy together with the stored magnetic field energy and heat loss. Electrical and mechanical energy have positive values in motoring action and negative values in generator action. In a magnetic system without losses, the change of electrical energy input is equal to the sum of the change of work done by the system and the change of stored magnetic field energy

$$\mathrm{d}W_{\mathrm{el}} = \mathrm{d}W_{\mathrm{mec}} + \mathrm{d}W_{\Phi}, \tag{1.80}$$

$$\mathrm{d}W_{\mathrm{el}} = ei\mathrm{d}t. \tag{1.81}$$

In the above, e is the instantaneous value of the induced voltage, created by changes in the energy in the magnetic circuit. Because of this electromotive force, the external electric circuit converts power into mechanical power by utilizing the magnetic field. This law of energy conversion combines a reaction and a counter-reaction in an electrical and mechanical system. The combination of Equations (1.80) and (1.81) yields

$$\mathrm{d}W_{\mathrm{el}} = ei\mathrm{d}t = \frac{\mathrm{d}\Psi}{\mathrm{d}t}i\mathrm{d}t = i\mathrm{d}\Psi = \mathrm{d}W_{\mathrm{mec}} + \mathrm{d}W_{\Phi}. \tag{1.82}$$

Equation (1.82) lays a foundation for the energy conversion principle. Next, its utilization in the analysis of electromagnetic energy converters is discussed.

As is known, a magnetic circuit (Figure 1.14) can be described by an inductance L determined from the number of turns of the winding, the geometry of the magnetic circuit and the permeability of the magnetic material. In electromagnetic energy converters, there are air gaps

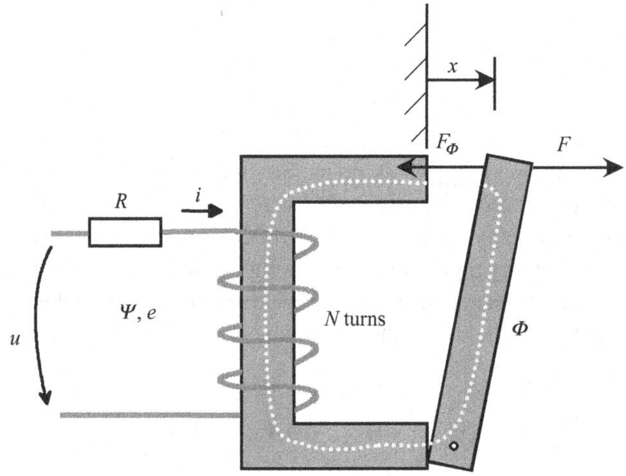

Figure 1.14 Electromagnetic relay connected to an external voltage source u. The mass of a moving yoke is neglected and the resistance of the winding is assumed to be concentrated in the external resistor R. There are N turns in the winding and a flux Φ flowing in the magnetic circuit; a flux linkage $\Psi \approx N\Phi$ is produced in the winding. The negative time derivative of the flux linkage is an emf, e. The force F pulls the yoke open. The force is produced by a mechanical source. A magnetic force F_Φ tries to close the air gap.

that separate the moving magnetic circuit parts from each other. In most cases – because of the high permeability of iron parts – the reluctance R_m of the magnetic circuit consists mainly of the reluctances of the air gaps. Thus, most of the energy is stored in the air gap. The wider the air gap, the more energy can be stored. For instance, in induction motors this can be seen from the fact that the wider is the gap, the higher is the magnetizing current needed.

According to Faraday's induction law, Equation (1.82) yields

$$\mathrm{d}W_\mathrm{el} = i\mathrm{d}\Psi. \tag{1.83}$$

The computation is simplified by neglecting for instance the magnetic nonlinearity and iron losses. The inductance of the device now depends only on the geometry and, in our example, on the distance x creating an air gap in the magnetic circuit. The flux linkage is thus a product of the varying inductance and the current

$$\Psi = L(x)i. \tag{1.84}$$

A magnetic force F_Φ is determined as

$$\mathrm{d}W_\mathrm{mec} = F_\Phi\,\mathrm{d}x. \tag{1.85}$$

From Equations (1.83) and (1.85), we may rewrite Equation (1.80) as

$$\mathrm{d}W_\Phi = i\mathrm{d}\Psi - F_\Phi\,\mathrm{d}x. \tag{1.86}$$

Since it is assumed that there are no losses in the magnetic energy storage, $\mathrm{d}W_\Phi$ is determined from the values of Ψ and x. $\mathrm{d}W_\Phi$ is independent of the integration path A or B, and the energy equation can be written as

$$W_\Phi(\Psi_0, x_0) = \int\limits_{\text{path A}} \mathrm{d}W_\Phi + \int\limits_{\text{path B}} \mathrm{d}W_\Phi. \tag{1.87}$$

With no displacement allowed ($\mathrm{d}x = 0$), Equations (1.86) now (1.87) yield

$$W_\Phi(\Psi, x_0) = \int_0^\Psi i(\Psi, x_0)\mathrm{d}\Psi. \tag{1.88}$$

In a linear system, Ψ is proportional to current i, as in Equations (1.84) and (1.88). We therefore obtain

$$W_\Phi(\Psi, x_0) = \int_0^\Psi i(\Psi, x_0)\mathrm{d}\Psi = \int_0^\Psi \frac{\Psi}{L(x_0)}\mathrm{d}\Psi = \frac{1}{2}\frac{\Psi^2}{L(x_0)} = \frac{1}{2}L(x_0)i^2. \tag{1.89}$$

The magnetic field energy can also be represented by the energy density $w_\Phi = W_\Phi/V = BH/2$ [J/m^3] in a magnetic field integrated over the volume V of the magnetic field. This gives

$$W_\Phi = \int_V \frac{1}{2}(HB)\mathrm{d}V. \tag{1.90}$$

Assuming the permeability of the magnetic medium constant and substituting $B = \mu H$ gives

$$W_\Phi = \int_V \frac{1}{2}\frac{B^2}{\mu}\mathrm{d}V. \tag{1.91}$$

This yields the relation between the stored energy in a magnetic circuit and the electrical and mechanical energy in a system with a lossless magnetic energy storage. The equation for differential magnetic energy is expressed in partial derivatives

$$\mathrm{d}W_\Phi(\Psi, x) = \frac{\partial W_\Phi}{\partial \Psi}\mathrm{d}\Psi + \frac{\partial W_\Phi}{\partial x}\mathrm{d}x. \tag{1.92}$$

Since Ψ and x are independent variables, Equations (1.86) and (1.92) have to be equal at all values of $\mathrm{d}\Psi$ and $\mathrm{d}x$, which yields

$$i = \frac{\partial W_\Phi(\Psi, x)}{\partial \Psi}, \tag{1.93}$$

where the partial derivative is calculated by keeping x constant. The force created by the electromagnet at a certain flux linkage level Ψ can be calculated from the magnetic energy

$$F_\Phi = -\frac{\partial W_\Phi(\Psi, x)}{\partial x}. \tag{1.94a}$$

The minus sign is due to the coordinate system in Figure 1.14. The corresponding equation is valid for torque as a function of angular displacement θ while keeping flux linkage Ψ constant

$$T_\Phi = -\frac{\partial W_\Phi(\Psi, \theta)}{\partial \theta}. \tag{1.94b}$$

Alternatively, we may employ coenergy (see Figure 1.15a), which gives us the force directly as a function of current. The coenergy W'_Φ is determined as a function of i and x as

$$W'_\Phi(i, x) = i\Psi - W_\Phi(\Psi, x). \tag{1.95}$$

In the conversion, it is possible to apply the differential of $i\Psi$

$$\mathrm{d}(i\Psi) = i\mathrm{d}\Psi + \Psi\mathrm{d}i. \tag{1.96}$$

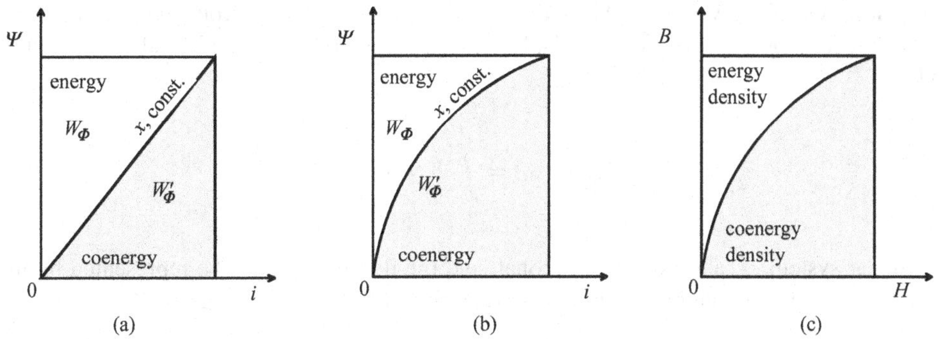

Figure 1.15 Determination of energy and coenergy with current and flux linkage (a) in a linear case (L is constant), (b) and (c) in a nonlinear case (L saturates as a function of current). If the figure is used to illustrate the behavior of the relay in Figure 1.14, the distance x remains constant.

Equation (1.95) now yields

$$\mathrm{d}W'_{\Phi}(i, x) = \mathrm{d}(i\Psi) - \mathrm{d}W_{\Phi}(\Psi, x). \tag{1.97}$$

By substituting Equations (1.86) and (1.96) into Equation (1.97) we obtain

$$\mathrm{d}W'_{\Phi}(i, x) = \Psi \mathrm{d}i + F_{\Phi}\mathrm{d}x. \tag{1.98}$$

The coenergy W'_{Φ} is a function of two independent variables, i and x. This can be represented by partial derivatives

$$\mathrm{d}W'_{\Phi}(i, x) = \frac{\partial W'_{\Phi}}{\partial i}\mathrm{d}i + \frac{\partial W'_{\Phi}}{\partial x}\mathrm{d}x. \tag{1.99}$$

Equations (1.98) and (1.99) have to be equal at all values of di and dx. This gives us

$$\Psi = \frac{\partial W'_{\Phi}(i, x)}{\partial i}, \tag{1.100}$$

$$F_{\Phi} = \frac{\partial W'_{\Phi}(i, x)}{\partial x}. \tag{1.101a}$$

Correspondingly, when the current i is kept constant, the torque is

$$T_{\Phi} = \frac{\partial W'_{\Phi}(i, \theta)}{\partial \theta}. \tag{1.101b}$$

Equation (1.101) gives a mechanical force or a torque directly from the current i and displacement x, or from the angular displacement θ. The coenergy can be calculated with i and x

$$W'_{\Phi}(i_0, x_0) = \int_0^i \Psi(i, x_0)\,\mathrm{d}i. \tag{1.102}$$

In a linear system, Ψ and i are proportional, and the flux linkage can be represented by the inductance depending on the distance, as in Equation (1.84). The coenergy is

$$W'_{\Phi}(i, x) = \int_0^i L(x)\,i\,\mathrm{d}i = \frac{1}{2}L(x)i^2. \tag{1.103}$$

Using Equation (1.91), the magnetic energy can be expressed also in the form

$$W_{\Phi} = \int_V \frac{1}{2}\mu H^2 \mathrm{d}V. \tag{1.104}$$

In linear systems, the energy and coenergy are numerically equal, for instance $0.5Li^2 = 0.5\Psi^2/L$ or $(\mu/2)H^2 = (1/2\mu)B^2$. In nonlinear systems, Ψ and i or B and H are not proportional. In a graphical representation, the energy and coenergy behave in a nonlinear way according to Figure 1.15.

The area between the curve and flux linkage axis can be obtained from the integral $i\,\mathrm{d}\Psi$, and it represents the energy stored in the magnetic circuit W_{Φ}. The area between the curve and the current axis can be obtained from the integral $\Psi\mathrm{d}i$, and it represents the coenergy W'_{Φ}. The sum of these energies is, according to the definition,

$$W_{\Phi} + W'_{\Phi} = i\Psi. \tag{1.105}$$

In the device in Figure 1.14, with certain values of x and i (or Ψ), the field strength has to be independent of the method of calculation; that is, whether it is calculated from energy or coenergy – graphical presentation illustrates the case. The moving yoke is assumed to be in a position x so that the device is operating at the point a, Figure 1.16a. The partial derivative in Equation (1.92) can be interpreted as $\Delta W_{\Phi}/\Delta x$, the flux linkage Ψ being constant and $\Delta x \to 0$. If we allow a change Δx from position a to position b (the air gap becomes smaller), the stored energy change $-\Delta W_{\Phi}$ will be as shown in Figure 1.16a by the shaded area, and the energy thus becomes smaller in this case. Thus, the force F_{Φ} is the shaded area divided by Δx when $\Delta x \to 0$. Since the energy change is negative, the force will also act in the negative x-axis direction. Conversely, the partial derivative can be interpreted as $\Delta W'_{\Phi}/\Delta x$, i being constant and $\Delta x \to 0$.

The shaded areas in Figures 1.16a and b differ from each other by the amount of the small triangle abc, the two sides of which are Δi and $\Delta\Psi$. When calculating the limit, Δx is allowed to approach zero, and thereby the areas of the shaded sections also approach each other.

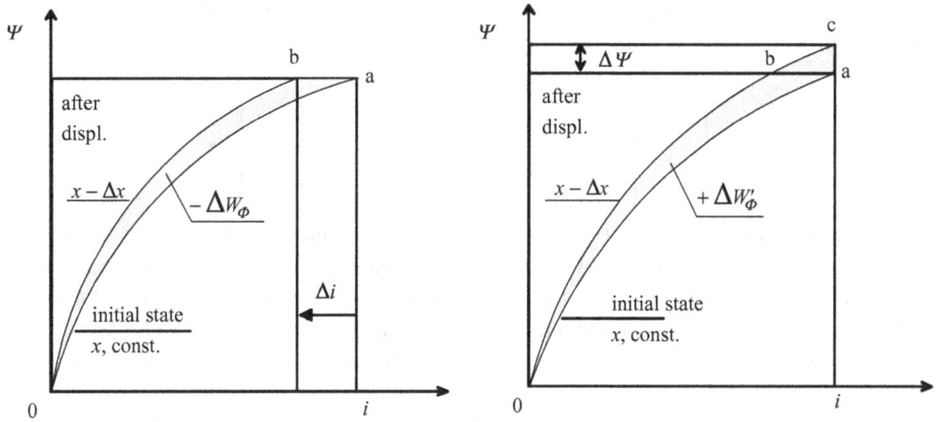

(a) Ψ is constant, i decreases (b) i is constant, flux linkage increases

Figure 1.16 Influence of the change Δx on energy and coenergy: (a) the change of energy, when Ψ is constant; (b) the change of coenergy, when i is constant.

Equations (1.94) and (1.101) give the mechanical force or torque of electric origin as partial derivatives of the energy and coenergy functions $W_\Phi(\Psi, x)$ and $W'_\Phi(x, i)$.

Physically, the force depends on the magnetic field strength $\boldsymbol{H}$ in the air gap; this will be studied in the next section. According to the study above, the effects of the field can be represented by the flux linkage Ψ and the current i. The force or the torque caused by the magnetic field strength tends to act in all cases in the direction where the stored magnetic energy decreases with a constant flux, or the coenergy increases with a constant current. Furthermore, the magnetic force tends to increase the inductance and drive the moving parts so that the reluctance of the magnetic circuit finds its minimum value.

Using finite elements, torque can be calculated by differentiating the magnetic coenergy W' with respect to movement, and by maintaining the current constant:

$$T = l\frac{\mathrm{d}W'}{\mathrm{d}\alpha} = \frac{\mathrm{d}}{\mathrm{d}\alpha}\int_V \int_0^H (\boldsymbol{B} \cdot \mathrm{d}\boldsymbol{H})\,\mathrm{d}V. \tag{1.106}$$

In numerical modeling, this differential is approximated by the difference between two successive calculations:

$$T = \frac{l\left(W'(\alpha + \Delta\alpha) - W'(\alpha)\right)}{\Delta\alpha}. \tag{1.107}$$

Here, l is the machine length and $\Delta\alpha$ represents the displacement between successive field solutions. The adverse effect of this solution is that it needs two successive calculations.

Coulomb's virtual work method in FEM is also based on the principle of virtual work. It gives the following expression for the torque:

$$T = \int_{\Omega} l \left[\left(-\boldsymbol{B}^{\mathrm{t}}\boldsymbol{J}^{-1}\right) \frac{\mathrm{d}\boldsymbol{J}}{\mathrm{d}\varphi} H + \int_{0}^{H} B\mathrm{d}H |\boldsymbol{J}|^{-1} \frac{\mathrm{d}|\boldsymbol{J}|}{\mathrm{d}\varphi} \right] \mathrm{d}\Omega, \tag{1.108}$$

where the integration is carried out over the finite elements situated between fixed and moving parts, having undergone a virtual deformation. In Equation (1.108), l is the length, $\boldsymbol{J}$ denotes the Jacobian matrix, $\mathrm{d}\boldsymbol{J}/\mathrm{d}\varphi$ is its differential representing element deformation during the displacement $\mathrm{d}\varphi$, $|\boldsymbol{J}|$ is the determinant of $\boldsymbol{J}$ and $\mathrm{d}\,|\boldsymbol{J}|/\mathrm{d}\varphi$ is the differential of the determinant, representing the variation of the element volume during displacement $\mathrm{d}\varphi$. Coulomb's virtual work method is regarded as one of the most reliable methods for calculating the torque and it is favored by many important commercial suppliers of FEM programs. Its benefit compared with the previous virtual work method is that only one solution is needed to calculate the torque.

1.5 Maxwell's Stress Tensor; Radial and Tangential Stress

Maxwell's stress tensor is probably the most generic idea of producing magnetic stresses, forces and torque. We discussed previously that the linear current density A on a metal surface creates tangential field strength components on the metal surfaces. Such tangential field strength components are essential in both tangential stress generation and torque generation in rotating-field electrical machines.

In numerical methods, Maxwell's stress tensor is often employed in the calculation of forces and torque. The idea is based on Faraday's statement according to which stress occurs in the flux lines. Figures 1.17 depict the flux solution for an air gap of an asynchronous machine,

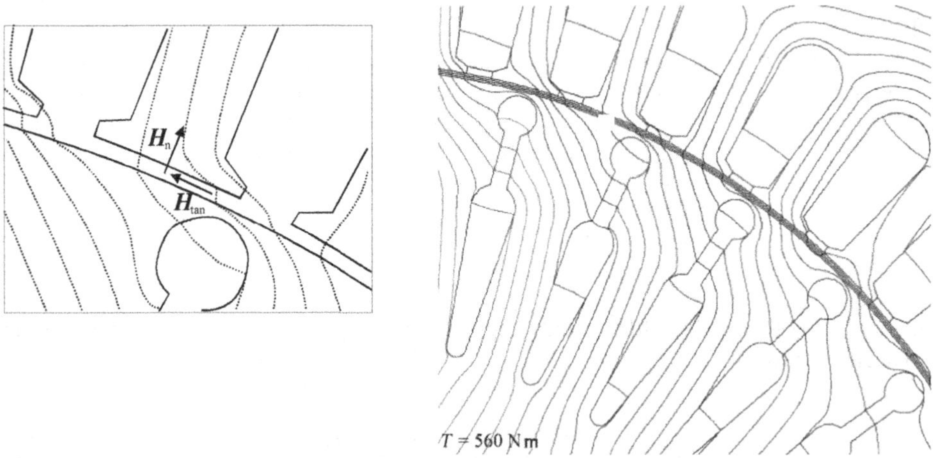

Figure 1.17 Flux solution of a loaded 30 kW, four-pole, 50 Hz induction motor, the machine rotating counterclockwise as a motor. The figure depicts a heavy overload. The tangential field strength in this case is very large and produces a high torque. The enlarged figure shows the tangential and normal components of the field strength in principle. Reproduced by permission of Janne Nerg.

when the machine is operating under a heavy load. Such a heavy load condition is selected in order to illustrate clearly the tangential routes of the flux lines. When we compare Figure 1.17 with Figure 1.6, we can see the remarkable difference in the behavior of the flux lines in the vicinity of the air gap.

In the figures, the flux lines cross the air gap somewhat tangentially so that if we imagine the flux lines to be flexible, they cause a notable torque rotating the rotor counterclockwise. According to Maxwell's stress theory, the magnetic field strength between objects in a vacuum creates a stress σ_F on the object surfaces, given by

$$\sigma_F = \frac{1}{2}\mu_0 H^2. \tag{1.109}$$

The stress occurs in the direction of lines of force and creates an equal pressure perpendicularly to the lines. When the stress term is divided into its normal and tangential components with respect to the object in question, we obtain

$$\sigma_{F\mathrm{n}} = \frac{1}{2}\mu_0 \left(H_\mathrm{n}^2 - H_\mathrm{tan}^2\right), \tag{1.110}$$

$$\sigma_{F\mathrm{tan}} = \mu_0 H_\mathrm{n} H_\mathrm{tan}. \tag{1.111}$$

Considering torque production, the tangential component $\sigma_{F\mathrm{tan}}$ is of the greatest interest. The total torque exerted on the rotor can be obtained by integrating the stress tensor for instance over a cylinder Γ that confines the rotor. The cylinder is dimensioned exactly to enclose the rotor. The torque is obtained by multiplying the result by the radius of the rotor. Note that no steel may be left inside the surface to be integrated. The torque can be calculated by the following relationship:

$$T = \frac{l}{\mu_0}\int_\Gamma \boldsymbol{r}\times\left((\boldsymbol{B}\cdot\boldsymbol{n})\,\boldsymbol{B}\mathrm{d}S - \frac{B^2\boldsymbol{n}}{2}\right)\mathrm{d}\Gamma, \tag{1.112}$$

where l is the length, $\boldsymbol{B}$ is the flux density vector, $\boldsymbol{n}$ the normal unit vector in the elements and $\boldsymbol{r}$ the lever arm, in other words the vector which connects the rotor origin to the midpoint of the segment $\mathrm{d}\Gamma$. The former term contains the tangential force contributing to the torque. Since $\boldsymbol{n}$ and $\boldsymbol{r}$ are parallel the latter term does not contribute to the torque but represents the normal stress.

Maxwell's stress tensor illustrates well the fundamental principle of torque generation. Unfortunately, because of numerical inaccuracies, for instance in the FEM, the obtained torque must be employed with caution. Therefore, in the FEM analysis, the torque is often solved by other methods, for instance Arkkio's method, which is a variant of Maxwell's stress tensor and is based on integrating the torque over the whole volume of the air gap constituted by the layers of radii r_s and r_r. The method has been presented with the following expression for the torque (Arkkio 1987):

$$T = \frac{l}{\mu_0(r_\mathrm{s} - r_\mathrm{r})}\int_S r B_\mathrm{n} B_\mathrm{tan}\mathrm{d}S, \tag{1.113}$$

in which l is the length, B_n and B_{tan} denote the radial and tangential flux densities in the elements of surface S and formed between radii r_r and r_s. dS is the surface of one element.

The magnetizing current method is yet another variant of Maxwell's stress method used in FEM solvers. This method is based on the calculation of the magnetizing current and the flux density over the element edges that constitute the boundary between the iron or permanent magnet and the air. Here the torque can be determined by the following expression:

$$T = \frac{l}{\mu_0} \int_{\Gamma_c} \left\{ \boldsymbol{r} \times \left[\left(B^2_{\text{tan,Fe}} - B^2_{\text{tan,air}} \right) \boldsymbol{n} - \left(B_{\text{tan,Fe}} B_{\text{tan,air}} - B^2_{\text{n,air}} \right) \boldsymbol{t} \right] \right\} \mathrm{d}\Gamma_c, \quad (1.114)$$

where l is the machine length, Γ_c denotes all the interfaces between the iron or permanent magnet and the air, and $\mathrm{d}\Gamma_c$ is the length of the element edge located at the boundary. The vector $\boldsymbol{r}$ is the lever arm, in other words the vector connecting the rotor origin to the midpoint of $\mathrm{d}\Gamma_c$. B_{tan} and B_n denote the tangential and normal flux densities with respect to $\mathrm{d}\Gamma_c$. The subscript Fe refers to the iron or permanent magnet. The normal unit vector is $\boldsymbol{n}$ and tangential unit vector $\boldsymbol{t}$.

Equation (1.70) states that a linear current density A creates a tangential field strength in an electrical machine: $H_{\text{tan},\delta} = A$ and $B_{\text{tan},\delta} = \mu_0 A$. According to Equation (1.111), the tangential stress in an air gap is given by

$$\sigma_{F\text{tan}} = \mu_0 H_n H_{\text{tan}} = \mu_0 H_n A = B_n A. \quad (1.115)$$

This equation gives a local time-dependent value for the tangential stress when local instantaneous values for the normal flux density B_n and the linear current density A are given. Air-gap flux density and linear current density thereby determine the tangential stress occurring in electrical machines. If we want to emphasize the place and time dependence of the stress, we may write the expression in the form

$$\sigma_{F\text{tan}}(x,t) = \mu_0 H_n(x,t) H_{\text{tan}}(x,t) = \mu_0 H_n(x,t) A(x,t) = B_n(x,t) A(x,t). \quad (1.116)$$

This expression is a very important starting point for the dimensioning of an electrical machine. The torque of the machine may be directly determined by this equation when the rotor dimensions are selected.

Example 1.7: Assume a sinusoidal air-gap flux density distribution having a maximum value of 0.9 T and a sinusoidal linear current density with a maximum value of 40 kA/m in the air gap. To simplify the case, also assume that the distributions are overlapping; in other words, there is no phase shift. This condition may occur on the stator surface of a synchronous machine; however, in the case of an induction machine, the condition never takes place in the steady state, since the stator also has to carry the magnetizing current. In our example, both the diameter and the length of the rotor are 200 mm. What is the power output, if the rotation speed is 1450 min^{-1}?

Solution: Because $\sigma_{F\text{tan}}(x) = \hat{B}_n \sin(x) \hat{A} \sin(x)$, the average tangential stress becomes $\overline{\sigma}_{F\text{tan}}(x) = 0.5\hat{B}_n\hat{A} = 18$ kPa. The active surface area of the rotor is $\pi D l = 0.126\ \text{m}^2$.

When we multiply the rotor surface area by the average tangential stress, we obtain 2270 N. This tangential force occurs everywhere at a radial distance of 0.1 m from the center of the axis, the torque being thus 227 N m. The angular velocity is 151 rad/s, which produces a power of approximately 34 kW. These values are quite close to the values of a real, totally enclosed 30 kW induction machine.

In electrical machines, the tangential stresses typically vary between 10 and 60 kPa depending on the machine construction, operating principle and especially on the cooling. For instance, the values for totally enclosed, permanent magnet synchronous machines vary typically between 20 and 30 kPa. For asynchronous machines, the values are somewhat lower. In induction machines with open-circuit cooling, the value of 50 kPa is approached. The most efficient air-cooled, through-ventilated permanent-magnet machines reach 60 kPa. Using direct cooling methods may, however, give notably higher tangential stresses.

No magnetic force is exerted on the air (a nonmagnetic material ($\mu_r = 1$)), although some stress occurs in the air because of the field strength. Only the part of air-gap flux that is caused by the magnetic susceptibility of the iron circuit creates a force. By applying the stress tensor, we may now write for a normal force

$$F_{F\mathrm{n}} = \frac{B_\delta^2 S_\delta}{2\mu_0}\left(1 - \frac{1}{\mu_\mathrm{r}}\right). \tag{1.117}$$

For iron, $1/\mu_r \ll 1$, and thus, in practice, the latter term in Equation (1.117) is of no significance, unless the iron is heavily saturated.

Example 1.8: Calculate the force between two iron bodies when the area of the air gap between the bodies is 10 cm^2, and the flux density is 1.5 T. The relative permeability of iron is assumed to be 700. It is also assumed that the tangential component of the field strength is zero.

Solution:

$$\sigma_{F\mathrm{n}} = \frac{1}{2}\mu_0\left(H_\mathrm{n}^2\right) = \frac{1}{2}\mu_0\left(\frac{B_\mathrm{n}}{\mu_0}\right)^2 = 8.95 \cdot 10^5 \frac{\mathrm{Vs}}{\mathrm{A\,m}}\left(\frac{\mathrm{A}}{\mathrm{m}}\right)^2$$
$$= 8.95 \cdot 10^5 \frac{\mathrm{VAs}}{\mathrm{m}^3} = 8.95 \cdot 10^5 \frac{\mathrm{N}}{\mathrm{m}^2}.$$

This is the stress in the air gap. The force acting on the iron can be approximated by multiplying the stress by the area of the air gap. Strictly speaking, we should investigate the permeability difference of the iron and air, the force acting on the iron therefore being

$$F_{F\mathrm{n}} = S\sigma_{F\mathrm{n}}\left(1 - \frac{1}{\mu_\mathrm{rFe}}\right) = \left(1 - \frac{1}{700}\right)0.001\ \mathrm{m}^2 \cdot 8.95 \cdot 10^5 \frac{\mathrm{N}}{\mathrm{m}^2} = 894\ \mathrm{N}.$$

From this example, we may conclude that the normal stress is usually notably higher than the tangential stress. In these examples, the normal stress was 894 000 Pa, and the tangential stress 18 000 Pa. Some cases have been reported in which attempts have been made to apply normal stress in rotating machines.

1.6 Self-Inductance and Mutual Inductance

Self-inductances and mutual inductances are the core parameters of electrical machines. Permeance is generally determined by

$$\Lambda = \frac{\Phi}{\Theta} = \frac{\Phi}{Ni} \tag{1.118}$$

and inductance by

$$L = N\frac{\Phi}{i} = \frac{\Psi}{i} = N^2\Lambda. \tag{1.119}$$

Inductance describes a coil's ability to produce flux linkage. Therefore, also its unit H (henry) is equal to Vs/A. Correspondingly, the mutual inductance L_{12} is determined from the flux linkage Ψ_{12}, created in winding 1 by the current i_2 that flows in winding 2,

$$L_{12} = \frac{\Psi_{12}}{i_2}. \tag{1.120}$$

In the special case where the flux Φ_{12}, created by the current of winding 2, penetrates all the turns of windings 1 and 2, the mutual permeance between the windings is written as

$$\Lambda_{12} = \frac{\Phi_{12}}{N_2 i_2}, \tag{1.121}$$

and the mutual inductance as

$$L_{12} = N_1 N_2 \Lambda_{12}. \tag{1.122}$$

Here, N_1 is the number of turns of the winding in which the voltage is induced, and N_2 is the number of turns of the winding that produces the flux.

The energy equation for a magnetic circuit can be written with the flux linkage as

$$W_\Phi = \int_0^t iL\frac{di}{dt}dt = \int_0^t i\frac{d\Psi}{dt}dt = \int_0^\Psi i\,d\Psi. \tag{1.123}$$

If an integral has to be calculated, the volume under observation can be divided into flux tubes. A flux flowing in such a flux tube is created by the influence of N turns of the winding. By taking into account the fact that the field strength $\boldsymbol{H}$ is created by the current i according to the equation $\oint \boldsymbol{H} \cdot \mathrm{d}\boldsymbol{l} = k_\mathrm{w}Ni$, the equation for the sum of the energies of all flux tubes in the volume observed, that is the total energy of the magnetic circuit that was previously given by the current and flux linkage, may be written as

$$W_\Phi = \int_0^\Psi i\mathrm{d}\Psi = \int_0^\Phi k_\mathrm{w}Ni\mathrm{d}\Phi = \int_0^B \oint \boldsymbol{H} \cdot \mathrm{d}\boldsymbol{l}S\mathrm{d}B = \int_0^B \int_V H\mathrm{d}V\mathrm{d}B = \int_V \int_0^B H\mathrm{d}B\mathrm{d}V \tag{1.124}$$

The volume integration has to be performed over the volume V in which the flux in question is passing. Energy per volume is thus written in the familiar form

$$\frac{\mathrm{d}W_\Phi}{\mathrm{d}V} = \int_0^B H\mathrm{d}B, \tag{1.125}$$

the energy stored in the complete magnetic circuit being

$$W_\Phi = \int_V \int_0^B H\mathrm{d}B\mathrm{d}V. \tag{1.126}$$

As the flux linkage is proportional to the current i, $\Psi = Li$, the energy can be given also as

$$W_\Phi = L\int_0^i i\mathrm{d}i = {}^1\!/_2 Li^2. \tag{1.127}$$

Equation (1.126) yields

$$\frac{\mathrm{d}W_\Phi}{\mathrm{d}V} = \frac{1}{2}HB, \tag{1.128}$$

$$W_\Phi = \frac{1}{2}\int_V HB\mathrm{d}V = \frac{1}{2}\int_V \mu H^2\mathrm{d}V. \tag{1.129}$$

From Equations (1.119), (1.127) and (1.129), we can calculate an ideal overall magnetic permeance for a magnetic circuit of volume V

$$\Lambda = \frac{1}{N^2 i^2}\int_V HB\mathrm{d}V = \frac{1}{N^2 i^2}\int_V \mu H^2\mathrm{d}V. \tag{1.130}$$

Let us now investigate two electric circuits with a common magnetic energy of

$$W_\Phi = \int_0^{\Psi_1} i_1\mathrm{d}\Psi_1 + \int_0^{\Psi_2} i_2\mathrm{d}\Psi_2. \tag{1.131}$$

Also in this case, the magnetic energy can be calculated from Equations (1.125) and (1.126). We can see that the common flux flowing through the flux tube n is

$$\Phi_{\mathrm{n}} = \frac{\Psi_{\mathrm{n1}}}{N_1} = \frac{\Psi_{\mathrm{n2}}}{N_2} = B\,S_{\mathrm{n}}. \tag{1.132}$$

This flux tube is magnetized by the sum current linkage of two windings N_1 and N_2

$$\oint \boldsymbol{H} \cdot \mathrm{d}\boldsymbol{l} = i_1 N_1 + i_2 N_2 \tag{1.133}$$

In a linear system, the fluxes are directly proportional to the sum magnetizing current linkage $i_1 N_1 + i_2 N_2$, and thus we obtain an energy

$$W_{\Phi} = \frac{1}{2}(i_1 \Psi_1) + \frac{1}{2}(i_2 \Psi_2). \tag{1.134}$$

Because the flux linkages are created with two windings together, they can be divided into parts

$$\Psi_1 = \Psi_{11} + \Psi_{12} \quad \text{and} \quad \Psi_2 = \Psi_{22} + \Psi_{21}. \tag{1.135}$$

Now, the flux linkages and inductances can be linked

$$\Psi_{11} = L_{11} i_1, \quad \Psi_{12} = L_{12} i_2, \quad \Psi_{22} = L_{22} i_2, \quad \Psi_{21} = L_{21} i_1. \tag{1.136}$$

In Equation (1.136), there are the self-inductances L_{11} and L_{22}, and the mutual inductances L_{12} and L_{21}. The magnetic energy can now be rewritten as

$$\begin{aligned} W_{\Phi} &= \frac{1}{2}\left(L_{11} i_1^2 + L_{12} i_1 i_2 + L_{22} i_2^2 + L_{21} i_2 i_1\right) \\ &= W_{11} + W_{12} + W_{22} + W_{21}. \end{aligned} \tag{1.137}$$

The magnetic energy of the magnetic field created by the two current circuits can thus be divided into four parts, two parts representing the energy of the self-inductances and two parts representing the energy of the mutual inductances. Correspondingly, the magnetic energy density in a certain volume can be written according to Equation (1.128), after the substitution

$$H = H_1 + H_2 \quad \text{and} \quad B = B_1 + B_2, \tag{1.138}$$

in the form

$$\frac{\mathrm{d}W_{\Phi}}{\mathrm{d}V} = \frac{1}{2}(H_1 B_1 + H_1 B_2 + H_2 B_2 + H_2 B_1). \tag{1.139}$$

Since in this equation $H_1B_2 = H_2B_1$, the energies and inductances have to behave as $W_{12} = W_{21}$ and $L_{12} = L_{21}$. This gives

$$W_{\Phi} = W_{11} + 2W_{12} + W_{22} = \frac{1}{2}L_{11}i_1^2 + L_{12}i_1i_2 + \frac{1}{2}L_{22}i_2^2. \tag{1.140}$$

Equations (1.137) and (1.139) yield

$$W_{12} = \frac{1}{2}L_{12}i_1i_2 = \frac{1}{2}\int_V H_1B_2\mathrm{d}V. \tag{1.141}$$

Now, we obtain for the permeance between the windings $\Lambda_{12} = \Lambda_{21}$, which corresponds to the mutual inductance, by comparing Equation (1.122),

$$\Lambda_{12} = \frac{1}{N_1i_1N_2i_2}\int_V \mu H_1H_2\mathrm{d}V. \tag{1.142}$$

If the field strengths are created by sinusoidal currents with a phase difference γ

$$i_1 = \hat{i}_1 \sin\omega t \quad \text{and} \quad i_2 = \hat{i}_2 \sin(\omega t + \gamma), \tag{1.143}$$

the mutual average energy of the fields created by these currents is obtained from

$$W_{12\mathrm{av}} = \frac{1}{2}L_{12}\frac{1}{2\pi}\int_0^{2\pi} \hat{i}_1\hat{i}_2\mathrm{d}\omega t = \frac{1}{2}L_{12}\hat{i}_1\hat{i}_2\cos\gamma. \tag{1.144}$$

Correspondingly, the permeance between the windings is

$$\Lambda_{12} = \frac{2}{N_1\hat{i}_1N_2\hat{i}_2}\int_V \mu H_1H_2\cos\gamma\mathrm{d}V. \tag{1.145}$$

In these equations, γ is the time-dependent phase angle either between the currents in two windings or between the partial field strengths created by these currents.

Let us study a single-phase transformer of Figure 1.18a. According to Equations (1.135) and (1.136) we may write for the flux linkages of the primary and secondary windings

$$\psi_1 = L_{11}i_1 + L_{12}i_2 \tag{1.146}$$

$$\psi_2 = L_{22}i_2 + L_{12}i_1. \tag{1.147}$$

By adding and subtracting the term $L_{12}i_1$ in Equation (1.146) and correspondingly $L_{12}i_2$ in Equation (1.147) we obtain

$$\psi_1 = (L_{11} - L_{12})i_1 + L_{12}(i_1 + i_2) \tag{1.148}$$

$$\psi_2 = (L_{22} - L_{12})i_2 + L_{12}(i_1 + i_2) \tag{1.149}$$

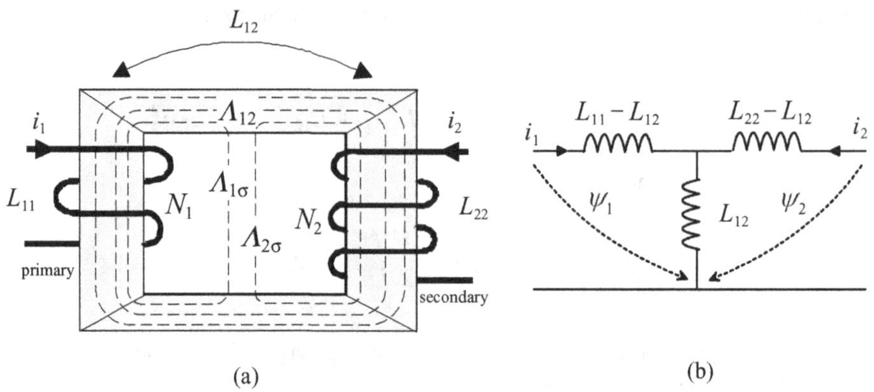

Figure 1.18 (a) Single-phase transformer and (b) its equivalent circuit. The dashed arrows in (b) indicate the route in which the corresponding flux linkages ψ are built. L_{11} and L_{22} are the self inductances of the primary and the secondary and L_{12} is the mutual inductance.

The equivalent circuit of Figure 1.18b follows Equations (1.148) and (1.149). This equivalent circuit is not very useful if primary and secondary voltages differ very much from each other. For more convenient equivalent circuit, we have to reorganize Equations (1.146) and (1.147). The self inductances L_{11} and L_{22} and the mutual inductance L_{12} can be written according to (1.119) and (1.122) as

$$L_{11} = N_1^2(\Lambda_{12} + \Lambda_{1\sigma}) \tag{1.150}$$

$$L_{22} = N_2^2(\Lambda_{12} + \Lambda_{2\sigma}) \tag{1.151}$$

$$L_{12} = N_1 N_2 \Lambda_{12} \tag{1.152}$$

where N_1 and N_2 are the numbers of turns of the primary and secondary windings and Λ_{12}, $\Lambda_{1\sigma}$ and $\Lambda_{2\sigma}$ are the permeances of the main path and the leakage paths. By substituting L_{11}, L_{22} and L_{12} in Equations (1.146) and (1.147) and multiplying by the transformation ratio N_1/N_2 we obtain

$$\begin{aligned}\psi_1 &= N_1^2(\Lambda_{12} + \Lambda_{1\sigma})i_1 + \frac{N_1}{N_2}N_1N_2\Lambda_{12}\frac{N_2}{N_1}i_2 = N_1^2\Lambda_{1\sigma}i_1 + N_1^2\Lambda_{12}i_1 + N_1^2\Lambda_{12}\frac{N_2}{N_1}i_2\\ &= L_{1\sigma}i_1 + L_{\mathrm{m}}(i_1 + i_2')\end{aligned} \tag{1.153}$$

$$\begin{aligned}\frac{N_1}{N_2}\psi_2 &= \left(\frac{N_1}{N_2}\right)^2 N_2^2(\Lambda_{12} + \Lambda_{2\sigma})\frac{N_2}{N_1}i_2 + \frac{N_1}{N_2}N_1N_2\Lambda_{12}i_1 = \left(\frac{N_1}{N_2}\right)^2 N_2^2\Lambda_{2\sigma}\frac{N_2}{N_1}i_2\\ &+ \left(\frac{N_1}{N_2}\right)^2 N_2^2\Lambda_{12}\frac{N_2}{N_1}i_2 + N_1^2\Lambda_{12}i_1 = \Psi_2' = L_{2\sigma}'i_2' + L_{\mathrm{m}}(i_1 + i_2')\end{aligned} \tag{1.154}$$

where

$L_{1\sigma} = N_1^2\Lambda_{1\sigma}$ is the leakage inductance of the primary winding,
$L_{2\sigma}' = \left(\frac{N_1}{N_2}\right)^2 N_2^2\Lambda_{2\sigma} = N_1^2\Lambda_{2\sigma} = \left(\frac{N_1}{N_2}\right)^2 L_{2\sigma}$ is the leakage inductance of the secondary winding referred to the primary,

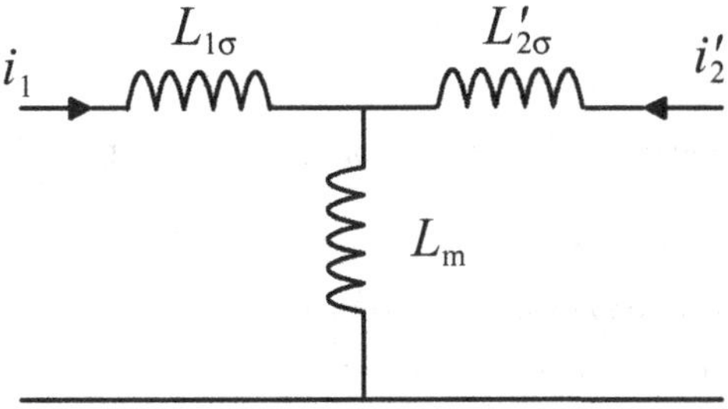

Figure 1.19 Commonly-used equivalent circuit of a single-phase transformer with magnetizing inductance L_m and secondary parameters referred to the primary.

$L_m = N_1^2 \Lambda_{12}$ is the magnetizing inductance,

$i'_2 = \dfrac{N_2}{N_1} i_2$ is the secondary current referred to the primary and

$\psi'_2 = \dfrac{N_1}{N_2} \psi_2$ is the flux linkage of the secondary winding referred to the primary.

We can now draw a new and more common type equivalent circuit for the single-phase transformer. The circuit of Figure 1.19 fulfils Equations (1.153) and (1.154).

Mutual inductances are important in rotating electrical machines. Further discussion of the magnetic circuit properties and inductances, such as magnetizing *in*ductance L_m, will be given in Chapters 3 and 4.

Example 1.9: Calculate the components of the equivalent circuits of Figures 1.18b and 1.19 for the transformer presented in Figure 1.18a. The number of turns are $N_1 = 50$ and $N_2 = 100$. The cross-section area of the iron core is $S = 0.01\ \mathrm{m}^2$. The narrow air gaps in the corner joints of the core and the saturation of the main flux path can be replaced by the effective air gap $\delta_{ef} = 1.05$ mm. The permeance of the leakage paths is 5 % of the main path permeance.

Solution:

$$\Lambda_{12} = \frac{\mu_0 S}{\delta_{ef}} = \frac{4\pi \cdot 10^{-7} \cdot 0.01}{0.00105} \frac{\mathrm{Vs}}{\mathrm{A}} = 0.01197 \frac{\mathrm{mVs}}{\mathrm{A}}$$

$$\Lambda_{1\sigma} = \Lambda_{2\sigma} = 0.05 \cdot \Lambda_{12} = 0.000598 \frac{\mathrm{mVs}}{\mathrm{A}}$$

$$L_{11} = N_1^2 (\Lambda_{12} + \Lambda_{1\sigma}) = 50^2 (0.01197 + 0.000598) \cdot 10^{-3}\ \mathrm{H} = 31.4\ \mathrm{mH}$$

$$L_{22} = N_2^2 (\Lambda_{12} + \Lambda_{2\sigma}) = 100^2 (0.01197 + 0.000598) \cdot 10^{-3}\ \mathrm{H} = 125.7\ \mathrm{mH}$$

$$L_{12} = N_1 N_2 \Lambda_{12} = 50 \cdot 100 \cdot 0.01197 \cdot 10^{-3}\ \mathrm{H} = 59.9\ \mathrm{mH}$$

The components of the equivalent circuit of Figure 1.18b are

$$L_{11} - L_{12} = -28\ \mathrm{mH}\ \text{(the value is negative because } L_{11} < L_{12})$$

$L_{22} - L_{12} = 66$ mH
$L_{12} = 60$ mH

and the components of the equivalent circuit of Figure 1.19 are

$L_{1\sigma} = N_1^2 \Lambda_{1\sigma} = 50^2 \cdot 0.000598 \text{ mH} = 1.50 \text{ mH}$
$L'_{2\sigma} = N_1^2 \Lambda_{2\sigma} = 50^2 \cdot 0.000598 \text{ mH} = 1.50 \text{ mH}$
$L_{\mathrm{m}} = N_1^2 \Lambda_{12} = 50^2 \cdot 0.01197 \text{ mH} = 29.9 \text{ mH}$

Example 1.10: Calculate the secondary voltage of the transformer of Example 1.9 at no-load using the equivalent circuit of Figure 1.18b as the primary voltage is U_1.

Solution: Let us denote the no-load current with I_0. The secondary voltage U_2 is according to Figure 18b:

$$\underline{U}_2 = \underline{I}_0 \mathrm{j}\omega L_{12} = \frac{\underline{U}_1}{\mathrm{j}\omega (L_{11} - L_{12} + L_{12})} \mathrm{j}\omega L_{12} = \frac{L_{12}}{L_{11}} \underline{U}_1$$
$$= \frac{N_1 N_2 \Lambda_{12}}{N_1^2 (\Lambda_{12} + \Lambda_{1\sigma})} \underline{U}_1 = \frac{N_2}{N_1} \frac{1}{\left(1 + \Lambda_{1\sigma}/\Lambda_{12}\right)} \underline{U}_1 .$$

For an ideal transformer $\Lambda_{1\sigma} = 0$ and $U_2 = \dfrac{N_2}{N_1} \underline{U}_1$.

1.7 Per Unit Values

When analysing electrical machines, especially in electric drives, per unit values are often employed. This brings certain advantages to the analysis, since they show directly the relative magnitude of a certain parameter. For instance, if the relative magnetizing inductance of an asynchronous machine is $l_{\mathrm{m}} = 3$, it is quite high. On the other hand, if it is $l_{\mathrm{m}} = 1$, it is rather low. Now it is possible to compare machines, the rated values of which differ from each other.

Relative values can be obtained by dividing each dimension by a base value. When considering electric motors and electric drives, the base values are selected accordingly:

- Peak value for rated stator phase current $\hat{i}_{\mathrm{N}}$ (It is, of course, also possible to select the root mean square (RMS) stator current as a base value, instead. In such a case the voltage also has to be selected accordingly.)
- Peak value for rated stator phase voltage $\hat{u}_{\mathrm{N}}$.
- Rated angular frequency $\omega_{\mathrm{N}} = 2\pi f_{\mathrm{sN}}$.
- Rated flux linkage Ψ_{N}, corresponding also to the rated angular velocity ω_{N}.
- Rated impedance Z_{N}.

- Time in which 1 radian in electrical degrees, $t_N = 1\ \mathrm{rad}/\omega_N$, is traveled at a rated angular frequency. Relative time τ is thus measured as an angle $\tau = \omega_N t$.
- Apparent power S_N corresponding to rated current and voltage.
- Rated torque T_N corresponding to rated power and frequency.

When operating with sinusoidal quantities, the rated current of the machine is I_N and the line-to-line voltage is U_N:

$$\text{The base value for current } I_b = \hat{i}_N = \sqrt{2} I_N. \tag{1.155}$$

$$\text{The base value for voltage } U_b = \hat{u}_N = \sqrt{2}\frac{U_N}{\sqrt{3}}. \tag{1.156}$$

$$\text{Angular frequency } \omega_N = 2\pi f_{sN}. \tag{1.157}$$

$$\text{The base value for flux linkage } \Psi_b = \Psi_N = \frac{\hat{u}_N}{\omega_N}. \tag{1.158}$$

$$\text{The base value for impedance } Z_b = Z_N = \frac{\hat{u}_N}{\hat{i}_N}. \tag{1.159}$$

$$\text{The base value for inductance } L_b = L_N = \frac{\hat{u}_N}{\omega_N \hat{i}_N}. \tag{1.160}$$

$$\text{The base value for capacitance } C_b = C_N = \frac{\hat{i}_N}{\omega_N \hat{u}_N}. \tag{1.161}$$

$$\text{The base value for apparent power } S_b = S_N = \frac{3}{2}\hat{i}_N \hat{u}_N = \sqrt{3} U_N I_N. \tag{1.162}$$

$$\text{The base value for torque } T_b = T_N = \frac{3}{2\omega_N}\hat{i}_N \hat{u}_N \cos\varphi_N = \frac{\sqrt{3} U_N I_N}{\omega_N}\cos\varphi_N. \tag{1.163}$$

The relative values to be used are

$$u_{s,pu} = \frac{u_s}{\hat{u}_N}, \tag{1.164}$$

$$i_{s,pu} = \frac{i_s}{\hat{i}_N}, \tag{1.165}$$

$$r_{s,pu} = \frac{R_s \hat{i}_N}{\hat{u}_N}, \tag{1.166}$$

$$\Psi_{s,pu} = \frac{\omega_N \Psi_s}{\hat{u}_N}, \tag{1.167}$$

$$\omega_{pu} = \frac{\omega}{\omega_N} = \frac{n}{f_N} = n_{pu}, \tag{1.168}$$

where n is the rotational speed per second, and

$$\tau = \omega_N t. \tag{1.169}$$

The relative values of inductances are the same as the relative values of reactances. Thus, we obtain for instance

$$l_{\mathrm{m.pu}} = \frac{L_{\mathrm{m}}}{L_{\mathrm{b}}} = \frac{L_{\mathrm{m}}}{\dfrac{\hat{u}_{\mathrm{N}}}{\omega_{\mathrm{N}}\hat{i}_{\mathrm{N}}}} = \frac{\hat{i}_{\mathrm{N}}}{\hat{u}_{\mathrm{N}}} X_{\mathrm{m}} = x_{\mathrm{m,pu}}, \tag{1.170}$$

where X_{m} is the magnetizing reactance.

We also have a mechanical time constant

$$T_J = \omega_{\mathrm{N}} \left(\frac{\omega_{\mathrm{N}}}{p}\right)^2 \frac{2J}{3\hat{u}_{\mathrm{N}}\hat{i}_{\mathrm{N}} \cos\varphi_{\mathrm{N}}}, \tag{1.171}$$

where J is the moment of inertia. According to (1.171), the mechanical time constant is the ratio of the kinetic energy of a rotor rotating at synchronous speed to the power of the machine.

Example 1.11: A 50 Hz star-connected, four-pole, 400 V induction motor has the following nameplate values: $P_{\mathrm{N}} = 200$ kW, $\eta_{\mathrm{N}} = 0.95$, $\cos\varphi_{\mathrm{N}} = 0.89$, $I_{\mathrm{N}} = 343$ A, $I_{\mathrm{S}}/I_{\mathrm{N}} = 6.9$, $T_{\mathrm{max}}/T_{\mathrm{N}} = 3$, and rated speed 1485 min^{-1}. The no-load current of the motor is 121 A. Give an expression for the per unit inductance parameters of the motor.

Solution:

The base angular frequency $\omega_{\mathrm{N}} = 2\pi f_{\mathrm{sN}} = 314$ 1/s.

The base value for flux linkage $\Psi_{\mathrm{b}} = \hat{\Psi}_{\mathrm{N}} = \dfrac{\hat{u}_{\mathrm{N}}}{\omega_{\mathrm{N}}} = \dfrac{\sqrt{2}\cdot 230\ \mathrm{V}}{314/\mathrm{s}} = 1.036$ V s.

The base value for inductance $L_{\mathrm{b}} = L_{\mathrm{N}} = \dfrac{\hat{\Psi}_{\mathrm{N}}}{\hat{i}_{\mathrm{N}}} = \dfrac{1.036\ \mathrm{V\,s}}{\sqrt{2}\cdot 343\ \mathrm{A}} = 2.14$ mH.

The no-load current of the machine is 121 A, and the stator inductance of the machine is thereby about $L_{\mathrm{s}} = 230\ \mathrm{V}/(121\ \mathrm{A}\cdot 314/\mathrm{s}) = 6.06$ mH. We guess that 97 % of this belongs to the magnetizing inductance. $L_{\mathrm{m}} = 0.97 \cdot 6.06\ \mathrm{mH} = 5.88$ mH.

The per unit magnetizing inductance is now $L_{\mathrm{m}}/L_{\mathrm{b}} = 5.88/2.14 = 2.74 = l_{\mathrm{m,pu}}$ and the stator leakage $l_{\mathrm{s}\sigma} = 0.03 \cdot 6.06\ \mathrm{mH} = 0.18$ mH. $l_{\mathrm{s}\sigma,\mathrm{pu}} = 0.18/2.14 = 0.084$.

We may roughly state that the per unit short-circuit inductance of the motor is, according to the starting current ratio, $l_{\mathrm{k}} \approx 1/(I_{\mathrm{S}}/I_{\mathrm{N}}) = 0.145$. Without better knowledge, we divide the short-circuit inductance 50:50 for the stator and rotor per unit leakages: $l_{\mathrm{s}\sigma,\mathrm{pu}} = l_{\mathrm{r}\sigma,\mathrm{pu}} = 0.0725$. This differs somewhat from the above-calculated $l_{\mathrm{s}\sigma,\mathrm{pu}} = 0.18/2.14 = 0.084$. However, the guess that 97 % of the stator inductance $l_{\mathrm{s,pu}} = l_{\mathrm{s}\sigma,\mathrm{pu}} + l_{\mathrm{m,pu}}$ seems to be correct enough.

The motor per unit slip is $s = (n_{\text{syn}} - n)n_{\text{syn}} = (1500 - 1485)/1500 = 0.00673$. The motor per unit slip at low slip values is directly proportional to the per unit rotor resistance. Thus, we may assume that the rotor per unit resistance is of the same order, $r_{\text{r,pu}} \approx 0.0067$.

The rated efficiency of the motor is 95 %, which gives 5 % per unit losses to the system. If we assume 1 % stator resistance $r_{\text{s,pu}} \approx 0.01$, and 0.5 % excess losses, we have 2.8 % (5 – 1 – 0.5 – 0.67 = 2.8 %) per unit iron losses in the motor. Hence, the losses in the machine are roughly proportional to the per unit values of the stator and rotor resistances. For more detailed information, the reader is referred to Chapter 7.

1.8 Phasor Diagrams

When investigating the operation of electrical machines, sinusoidally alternating currents, voltages and flux linkages are often illustrated with phasor diagrams. These diagrams are based either on generator logic or on motor logic; the principle of the generator logic is that for instance the flux linkage created by the rotor magnetization of a synchronous machine induces an electromotive force in the armature winding of the machine. Here, Faraday's induction law is applied in the form

$$e = -\frac{\mathrm{d}\Psi}{\mathrm{d}t}. \tag{1.172}$$

The flux linkage for a rotating-field machine can be presented as

$$\Psi(t) = \hat{\Psi}\mathrm{e}^{\mathrm{j}\omega t}. \tag{1.173}$$

The flux linkage is derived with respect to time

$$e = -\frac{\mathrm{d}\Psi}{\mathrm{d}t} = -\mathrm{j}\omega\hat{\Psi}\mathrm{e}^{\mathrm{j}\omega t} = \mathrm{e}^{-\mathrm{j}\frac{\pi}{2}}\omega\hat{\Psi}\mathrm{e}^{\mathrm{j}\omega t} = \omega\hat{\Psi}\mathrm{e}^{\mathrm{j}\left(\omega t - \frac{\pi}{2}\right)} \tag{1.174}$$

The emf is thus of magnitude $\omega\hat{\Psi}$ and its phase angle is 90 electrical degrees behind the phasor of the flux linkage. Figure 1.20 illustrates the basic phasor diagrams according to generator and motor logic.

As illustrated in Figure 1.20 for generator logic, the flux linkage Ψ_{m} generated by the rotor of a synchronous machine induces a voltage E_{m} in the armature winding of the machine when the machine is rotating. The stator voltage of the machine is obtained by reducing the proportion of the armature reaction and the resistive voltage loss from the induced voltage. If the machine is running at no load, the induced voltage E_{m} equals the stator voltage U_{s}.

Motor logic represents the opposite case. According to the induction law, the flux linkage can be interpreted as an integral of voltage.

$$\Psi_{\text{s}} \approx \int u_{\text{s}}\mathrm{d}t = \int \hat{u}_{\text{s}}\mathrm{e}^{\mathrm{j}\omega t}\mathrm{d}t = \frac{1}{\mathrm{j}\omega}\hat{u}_{\text{s}}\mathrm{e}^{\mathrm{j}\omega t} = \frac{1}{\omega}\hat{u}_{\text{s}}\mathrm{e}^{\mathrm{j}\left(\omega t - \frac{\pi}{2}\right)} \tag{1.175}$$

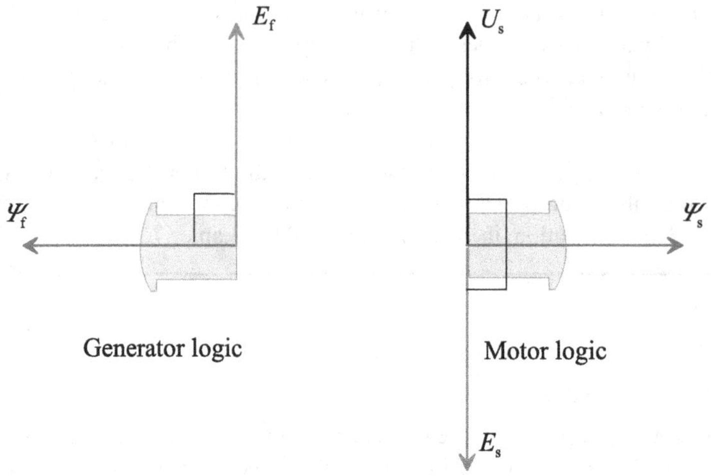

Figure 1.20 Basic phasor diagrams for generator and motor logic.

The phasor of the flux linkage is 90 electrical degrees behind the voltage phasor. Again, differentiating the flux linkage with respect to time produces an emf. In the case of Figure 1.20, a flux linkage is integrated from the voltage, which further leads to the derivation of a back emf now canceling the supply voltage. As is known, this is the case with inductive components. In the case of a coil, a major part of the supply voltage is required to overcome the self-inductance of the coil. Only an insignificant voltage drop takes place in the winding resistances. The resistive losses have therefore been neglected in the above discussion.

Example 1.12: A 50 Hz synchronous generator field winding current linkage creates at no load a stator winding flux linkage of $\hat{\Psi}_s = 15.6$ V s. What is the internal induced phase voltage (and also the stator voltage) of the machine?

Solution: The induced voltage is calculated as

$$e_f = -\frac{d\Psi_s}{dt} = -j\omega\hat{\Psi}_s e^{j\omega t} = -j314/s \cdot 15.6\ V\ s \cdot e^{j\omega t} = 4900\ V \cdot e^{j\omega t}.$$

Thus 4900 V is the peak value of the stator phase voltage, which gives an effective value of the line-to-line voltage:

$$U_{ll} = \frac{4900\ V}{\sqrt{2}}\sqrt{3} = 6000\ V.$$

Hence, we have a 6 kV machine at no load ready to be synchronized to the network.

Example 1.13: A rotating-field motor is supplied by a frequency converter at 25 Hz and 200 V fundamental effective line-to-line voltage. The motor is initially a 400 V star-connected motor. What is the stator flux linkage in the inverter supply?

Solution: The stator flux linkage is found by integrating the phase voltage supplied to the stator

$$\Psi_{\mathrm{s}} \approx \int u_{\mathrm{s}} \mathrm{d}t = \int \hat{u}_{\mathrm{s}} \mathrm{e}^{\mathrm{j}\omega t} \mathrm{d}t = \frac{1}{\mathrm{j}25 \cdot 2\pi} \frac{\sqrt{2} \cdot 200\,\mathrm{V}}{\sqrt{3}} \mathrm{e}^{\mathrm{j}\omega t} = 1.04\,\mathrm{V\,s} \cdot \mathrm{e}^{\mathrm{j}\left(\omega t - \frac{\pi}{2}\right)}$$

Consequently, the flux linkage amplitude is 1.04 V s and it is lagging in a 90° phase shift the voltage that creates the flux linkage.

Bibliography

Arkkio, A. (1987) *Analysis of Induction Motors Based on the Numerical Solution of the Magnetic Field and Circuit Equations*, Dissertation, Electrical Engineering Series No. 59. Acta Polytechnica Scandinavica, Helsinki University of Technology. Available at http://lib.tkk.fi/Diss/198X/isbn951226076X/.

Carpenter, C.J. (1959) *Surface integral methods of calculating forces on magnetised iron parts*, *IEE Monographs*, 342, 19–28.

Johnk, C.T.A. (1975) *Engineering Electromagnetic Fields and Waves*, John Wiley & Sons, Inc., New York.

Sadowski, N., Lefevre, Y., Lajoie-Mazenc, M. and Cros, J. (1992) Finite element torque calculation in electrical machines while considering the movement. *IEEE Transactions on Magnetics*, **28** (2), 1410–13.

Sihvola, A. and Lindell, I. (2004) *Electromagnetic Field Theory 2: Dynamic Fields (Sähkömagneettinen kenttäteoria. 2. dynaamiset kentät)*, Otatieto, Helsinki.

Silvester, P. and Ferrari, R.L. (1988) *Finite Elements for Electrical Engineers*, 2nd edn, Cambridge University Press, Cambridge.

Ulaby, F.T. (2001) *Fundamentals of Applied Electromagnetics*, Prentice Hall, Upper Saddle River, NJ.

Vogt, K. (1996) *Design of Electrical Machines (Berechnung elektrischer Maschinen)*, Wiley-VCH Verlag GmbH, Weinheim.

2

Windings of Electrical Machines

The operating principle of electrical machines is based on the interaction between the magnetic fields and the currents flowing in the windings of the machine. The winding constructions and connections together with the currents and voltages fed into the windings determine the operating modes and the type of the electrical machine. According to their different functions in an electrical machine, the windings are grouped for instance as follows:

- armature windings;
- other rotating-field windings (e.g. stator or rotor windings of induction motors);
- field (magnetizing) windings;
- damper windings;
- commutating windings; and
- compensating windings.

Armature windings are rotating-field windings, into which the rotating-field-induced voltage required in energy conversion is induced. According to IEC 60050-411, the armature winding is a winding in a synchronous, DC or single-phase commutator machine, which, in service, receives active power from or delivers active power to the external electrical system. This definition also applies to a synchronous compensator if the term “active power” is replaced by “reactive power.” The air-gap flux component caused by the armature current linkage is called the armature reaction.

An armature winding determined under these conditions can transmit power between an electrical network and a mechanical system. Magnetizing windings create a magnetic field required in the energy conversion. All machines do not include a separate magnetizing winding; for instance, in asynchronous machines, the stator winding both magnetizes the machine and acts as a winding, where the operating voltage is induced. The stator winding of an asynchronous machine is similar to the armature of a synchronous machine; however, it is not defined as an armature in the IEC standard. In this material, the asynchronous machine stator is therefore referred to as a rotating-field stator winding, not an armature winding. Voltages are also induced in the rotor of an asynchronous machine, and currents that are significant in torque production are created. However, the rotor itself takes only a rotor’s dissipation

Design of Rotating Electrical Machines, Second Edition. Juha Pyrhönen, Tapani Jokinen and Valéria Hrabovcová.

power (I^2R) from the air-gap power of the machine, this power being proportional to the slip; therefore, the machine can be considered stator fed, and, depending on the rotor type, the rotor is called either a squirrel cage rotor or a wound rotor. In DC machines, the function of a rotor armature winding is to perform the actual power transmission, the machine being thus rotor fed. Field windings do not normally participate in energy conversion, double-salient-pole reluctance machines possibly being excluded: in principle, they have nothing but magnetizing windings, but the windings also perform the function of the armature. In DC machines, commutating and compensating windings are windings the purpose of which is to create auxiliary field components to compensate for the armature reaction of the machine and thus improve its performance characteristics. Similar to the previously described windings, these windings do not participate in energy conversion in the machine either. The damper windings of synchronous machines are a special case among different winding types. Their primary function is to damp undesirable phenomena, such as oscillations and fields rotating opposite to the main field. Damper windings are important during the transients of controlled synchronous drives, in which the damper windings keep the air-gap flux linkage instantaneously constant. In the asynchronous drive of a synchronous machine, the damper windings act like the cage windings of asynchronous machines.

The most important windings are categorized according to their geometrical characteristics and internal connections as follows:

- phase windings;
- salient-pole windings; and
- commutator windings.

Windings in which separate coils embedded in slots form a single- or poly-phase winding constitute a large group of AC armature windings. However, a similar winding is also employed in the magnetizing of nonsalient-pole synchronous machines. In commutator windings, individual coils contained in slots form a single or several closed circuits, which are connected together via a commutator. Commutator windings are employed only as armature windings of DC and AC commutator machines. Salient-pole windings are normally concentrated field windings, but may also be used as armature windings in for instance fractional slot permanent magnet machines and in doubly-salient reluctance machines. Concentrated non-overlapping stator windings are used as an armature winding also in small shaded-pole motors.

In the following, the windings applied in electrical machines are classified according to the two main winding types, namely slot windings and salient-pole windings. Both types are applicable to both DC and AC cases, Table 2.1.

2.1 Basic Principles

2.1.1 Salient-Pole Windings

Figure 2.1 illustrates a synchronous machine with a salient-pole rotor. To magnetize the machine, direct current is fed through brushes and slip rings to the windings located on the salient poles. The main flux created by the direct current flows from the pole shoe to the stator and back simultaneously penetrating the poly-phase slot winding of the stator. The dotted lines in the figure depict the paths of the main flux. Such a closed path of a flux forms the magnetic circuit of a machine.

Table 2.1 Different types of windings or permanent magnets used instead of a field winding in the most common machine types

	Stator winding	Rotor winding	Compensating winding	Commutating winding	Damper winding
Salient-pole synchronous machine	Poly-phase distributed rotating-field slot winding	Salient-pole winding	—	—	Short-circuited cage winding
Nonsalient-pole synchronous machine	Poly-phase distributed rotating-field slot winding	Slot winding	—	—	Solid-rotor core or short-circuited cage winding
Synchronous reluctance machine	Poly-phase distributed rotating-field slot winding	—	—	—	Short-circuited cage winding possible
PMSM, $q > 0.5$	Poly-phase distributed rotating-field slot winding	Permanent magnets	—	—	Solid-rotor or short-circuited cage winding, or, for example, aluminum plate on rotor surface
PMSM, $q \leq 0.5$	Poly-phase concentrated pole winding	Permanent magnets	—	—	Damping should be harmful because of excessive losses
Double-salient reluctance machine	Poly-phase concentrated pole winding	—	—	—	—
Induction machine	Poly-phase distributed rotating-field slot winding	Cast or soldered cage winding, squirrel cage winding	—	—	—
Solid-rotor induction machine	Poly-phase distributed rotating-field slot winding	Solid rotor made of steel, may be equipped with squirrel cage or copper coating	—	—	—
Slip-ring asynchronous motor	Poly-phase distributed rotating-field slot winding	Poly-phase distributed rotating-field slot winding	—	—	—
DC machine	Salient-pole winding	Rotating-field commutator slot winding	Slot winding	Salient-pole winding	—

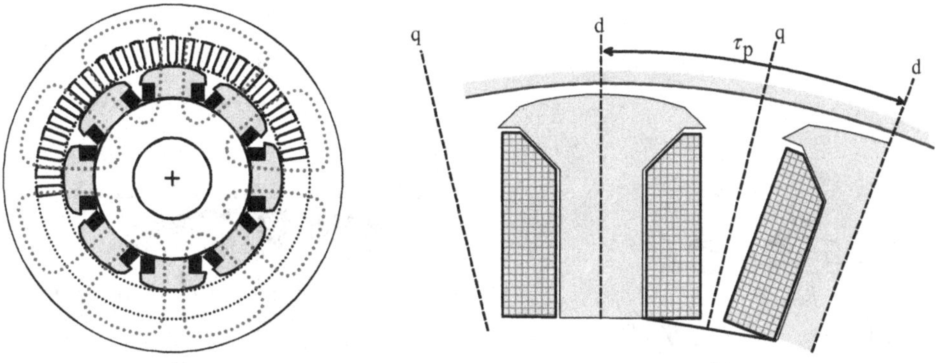

(a) (b)

Figure 2.1 (a) Salient-pole synchronous machine ($p = 4$). The black areas around two pole bodies form a salient-pole winding. (b) Single poles with windings: d, direct axis; q, quadrature axis. In salient-pole machines, these two magnetically different, rotor-geometry-defined axes have a remarkable effect on machine behavior (this issue will be discussed later).

One turn of a coil is a single-turn conductor, through which the main flux traveling in the magnetic circuit passes. A coil is a part of the winding that consists of adjacent series-connected turns between the two terminals of the coil. Figure 2.1a illustrates a synchronous machine with one coil per pole, whereas in Figure 2.1b the locations of the direct (d) and quadrature (q) axes are shown.

A group of coils is a part of the winding that magnetizes the same magnetic circuit. In Figure 2.1a, the coils at the different magnetic poles (N and S alternating) form in pairs a group of coils. The number of field winding turns magnetizing one pole is N_f.

The salient-pole windings located on the rotor or on the stator are mostly used for the DC magnetizing of a machine. The windings are then called magnetizing or sometimes excitation windings. With a direct current, they create a time-constant current linkage Θ. The part of this current linkage consumed in the air gap, that is the magnetic potential difference of the air gap $U_{m\delta}$, may be, for simplicity, regarded as constant between the quadrature axes, and it changes its sign at the quadrature axis q, Figure 2.2.

A significant field of application for salient-pole windings is double-salient reluctance machines. In these machines, a solid salient pole is not utilizable, since the changes of flux are rapid when operating at high speeds. At a simple level, DC pulses are fed to the pole windings with power switches. In the air gap, the direct current creates a flux that tries to turn the rotor in a direction where the magnetic circuit of the machine reaches its minimum reluctance. The torque of the machine tends to be pulsating, and to reach an even torque, the current of a salient-pole winding should be controllable so that the rotor can rotate without jerking.

Salient-pole windings are employed also in the magnetizing windings of DC machines. All series, shunt and compound windings are wound on salient poles. The commutating windings are also of the same type as salient-pole windings.

Example 2.1: Calculate the field winding current that can ensure a maximum magnetic flux density of $B_\delta = 0.82$ T in the air gap of a synchronous machine if there are 95 field winding turns per pole. It is assumed that the air-gap magnetic flux density of the

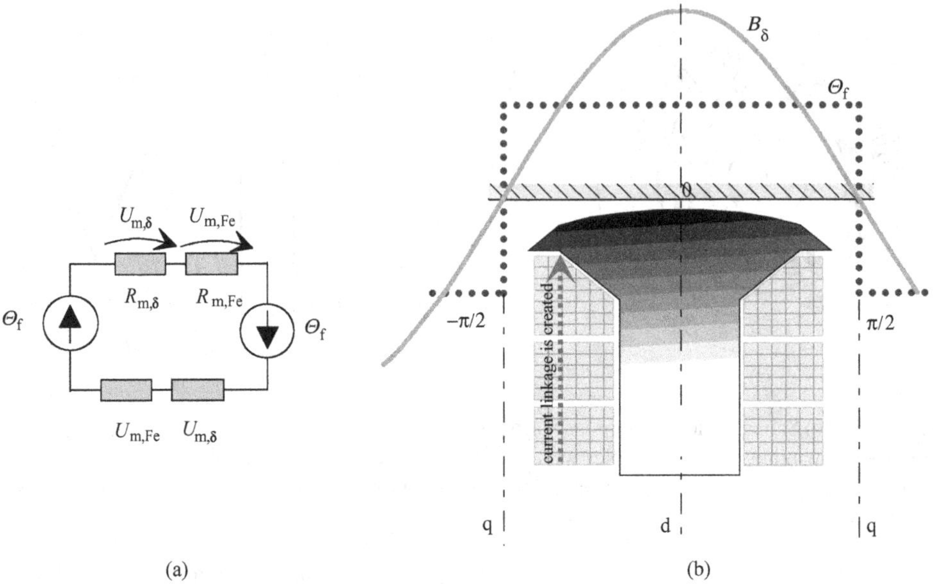

Figure 2.2 (a) Equivalent magnetic circuit. The current linkages Θ_f created by two adjacent salient-pole windings. Part $U_{m,\delta}$ is consumed in the air gap. (b) The behavior of the air-gap flux density B_δ. Due to the appropriate design of the pole shoe, the air-gap flux density varies cosinusoidally even though it is caused by the constant magnetic voltage over the air gap $U_{m,\delta}$. The air-gap magnetic flux density B_δ has its peak value on the d-axis and is zero on the q-axis. The current linkage created by the pole is accumulated by the ampere turns on the pole.

machine is sinusoidal along the pole shoes and the magnetic permeability of iron is infinite ($\mu_{Fe} = \infty$) in comparison with the permeability of air $\mu_0 = 4\pi \times 10^{-7}$ H/m. The minimum length of the air gap is 3.5 mm.

Solution: If $\mu_{Fe} = \infty$, the magnetic reluctance of iron parts and the magnetic voltage in iron is zero. Now, the whole field current linkage $\Theta_f = N_f I_f$ is spent in the air gap to create the required magnetic flux density:

$$\Theta_f = N_f I_f = U_{m,\delta} = H_\delta \delta = \frac{B_\delta}{\mu_0}\delta = \frac{0.82}{4\pi \cdot 10^{-7}} 3.5 \cdot 10^{-3} \text{A}$$

If the number of turns is $N_f = 95$, the field current is

$$I_f = \frac{\Theta_f}{N_f} = \frac{0.82}{4\pi \cdot 10^{-7}} 3.5 \cdot 10^{-3} \frac{1}{95} \text{A} = 24\ \text{A}$$

It should be noted that calculations of this kind are appropriate for an approximate calculation of the current linkage needed. In fact, about 60–90 % of the magnetomotive force (mmf) in electrical machines is spent in the air gap, and the rest in the iron parts. Therefore, in a detailed design of electrical machines, it is necessary to take into account all the iron parts with appropriate material properties. A similar calculation is valid for DC machines, with the exception that in DC machines the air gap is usually constant under the poles.

2.1.2 Slot Windings

Here we concentrate on symmetrical, three-phase AC distributed slot windings, in other words rotating-field windings. However, first, we discuss the magnetizing winding of the rotor of a nonsalient-pole synchronous machine, and finally turn to commutator windings, compensating windings and damper windings. Unlike in the salient-pole machine, since the length of the air gap is now constant, we may create a cosinusoidally distributed flux density in the air gap by producing a cosinusoidal distribution of current linkage with an AC magnetizing winding, Figure 2.3. The cosinusoidal distribution, instead of the sinusoidal one, is used because we want the flux density to reach its maximum on the direct axis, where $\alpha = 0$.

In the case of Figure 2.3, the function of the magnetic flux density approximately follows the curved function of the current linkage distribution $\Theta(\alpha)$. In machine design, an equivalent air gap δ_e is applied, the target being to create a cosinusoidally alternating flux density in the air gap

$$B(\alpha) = \frac{\mu_0}{\delta_e}\Theta(\alpha). \tag{2.1}$$

The concept of equivalent air gap δ_e will be discussed later.

The slot pitch τ_u and the slot angle α_u are the core parameters of the slot winding. The slot pitch is measured in meters, whereas the slot angle is measured in electrical degrees. The number of slots being Q and the diameter of the air gap D, we may write

$$\tau_u = \frac{\pi D}{Q}; \quad \alpha_u = p\frac{2\pi}{Q}. \tag{2.2}$$

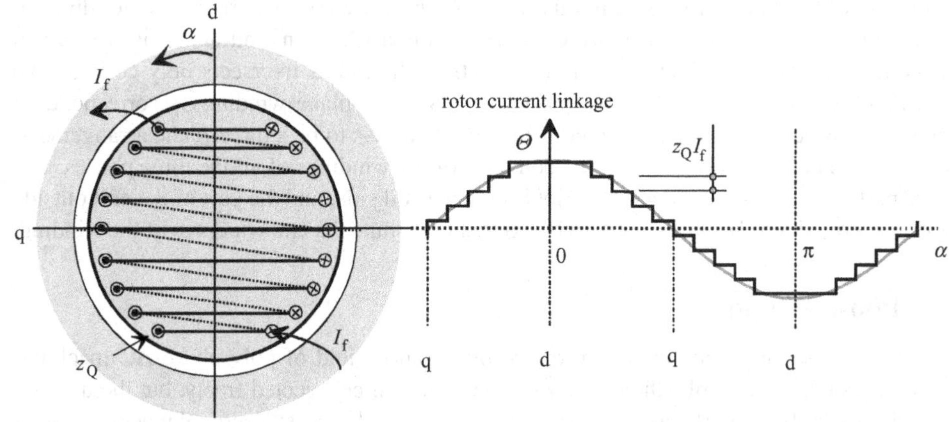

Figure 2.3 Current linkage distribution created by two-pole nonsalient-pole winding and the fundamental of the current linkage. There are z_Q conductors in each slot, and the excitation current in the winding is I_f. The height of a single step of the current linkage is $z_Q I_f$.

As the slot pitch is usually constant in nonsalient-pole windings, the current sum ($z_Q I_f$) in a slot has to be of a different magnitude in different slots (in a sinusoidal or cosinusoidal manner to achieve a sinusoidal or cosinusoidal variation of current linkage along the surface of the air gap). Usually, there is a current of equal magnitude flowing in all turns in the slot, and therefore the number of conductors z_Q in the slots has to be varied. In the slots of the rotor in Figure 2.3, the number of turns is equal in all slots, and a current of equal magnitude is flowing in the slots. We can see that by selecting z_Q slightly differently in different slots, we can improve the stepped waveform of the figure to approach better the cosinusoidal form. The need for this depends on the induced voltage harmonic content in the stator winding. The voltage may be of almost pure sinusoidal waveform despite the fact that the air-gap flux density distribution should not be perfectly sinusoidal. This depends on the stator winding factors for different harmonics. In synchronous machines, the air gap is usually relatively large, and, correspondingly, the flux density on the stator surface changes more smoothly (neglecting the influence of slots) than the stepped current linkage waveform of Figure 2.3. Here, we apply the well-known finding that if two-thirds of the rotor surface are slotted and one-third is left slotless, not only the third harmonic component but any of its multiple harmonics called triplen harmonics are eliminated in the air-gap magnetic flux density, and also the low-order odd harmonics (fifth, seventh) are suppressed.

2.1.3 End Windings

Figure 2.4 illustrates how the arrangement of the coil end influences the physical appearance of the winding. The windings a and b in the figure are of equal value with respect to the main flux, but their leakage inductances diverge from each other because of the slightly different coil ends. When investigating winding of Figure 2.4a, we note that the coil ends form two separate planes at the end faces of the machine. This kind of a winding is therefore called a two-plane winding. The coil ends of this type are depicted in Figure 2.4e. In the winding of Figure 2.4b, the coil ends are overlapping, and therefore this kind of winding is called a diamond winding (lap winding). Figures 2.4c and d illustrate three-phase stator windings that are identical with respect to the main flux, but in Figure 2.4c the coil groups are not divided, and in Figure 2.4d the groups of coils are divided. In Figure 2.4c, an arbitrary radius r is drawn across the coil end. It is shown that at any position, the radius intersects only coils of two phases, and the winding can thus be constructed as a two-plane winding. A corresponding winding constructed with distributed coils (Figure 2.4d) has to be a three-plane arrangement, since now the radius r may intersect the coil ends of the windings of all the three phases.

The part of a coil located in a single slot is called a coil side, and the part of the coil outside the slot is termed a coil end. The coil ends together constitute the end windings of the winding.

2.2 Phase Windings

Next, poly-phase slot windings that produce the rotating field of poly-phase AC machines are investigated. In principle, the number of phases m can be selected freely, but the use of a three-phase supply network has led to a situation in which also most electrical machines are of the three-phase type. Another, extremely common type is two-phase electrical machines that are operated with a capacitor start and run motor in a single-phase network. A symmetrical two-phase winding is in principle the simplest AC winding that produces a rotating field.

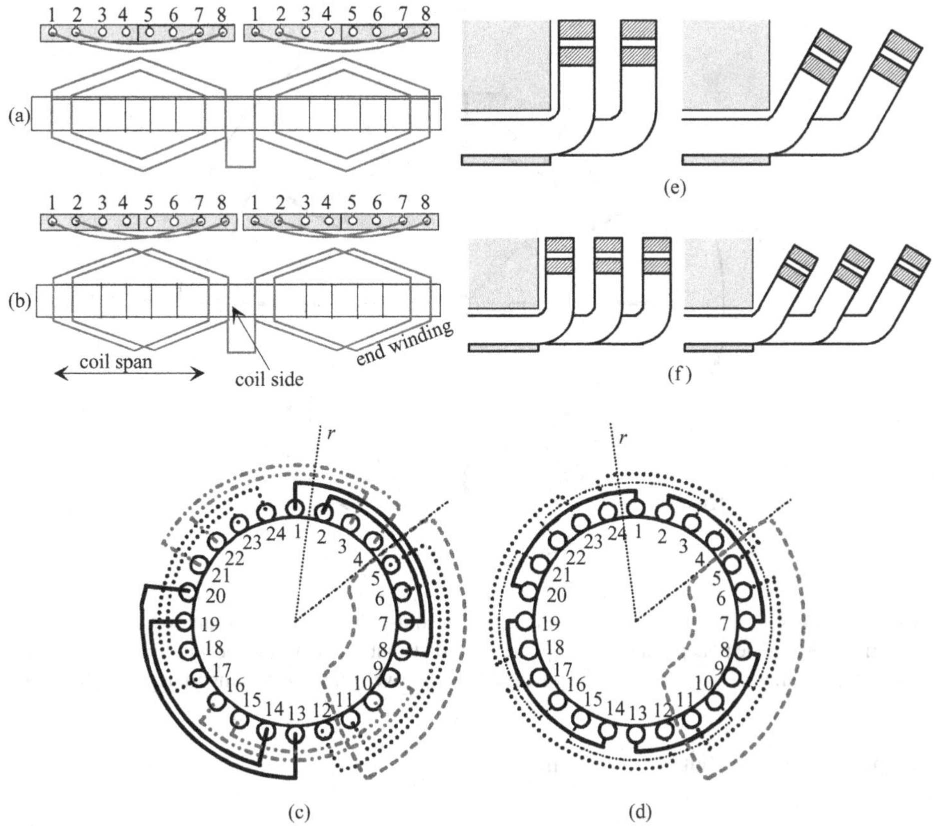

Figure 2.4 (a) Concentric winding and (b) a diamond winding. In a two-plane winding, the coil spans differ from each other. In the diamond winding, all the coils are of equal width. (c) A two-plane, three-phase, four-pole winding with nondivided coil groups. (d) A three-plane, three-phase, four-pole winding with divided coil groups. Figures (c) and (d) also illustrate a single main flux path. (e) Profile of an end winding arrangement of a two-plane winding. (f) Profile of an end winding of a three-plane winding. The radii r in the figures illustrate that in a winding with nondivided groups, an arbitrary radius may intersect only two phases, and in a winding with divided groups, the radius may intersect all three phases. The two- or three-plane windings will result correspondingly.

A configuration of a symmetrical poly-phase winding can be considered as follows: the periphery of the air gap is evenly distributed over the poles so that we can determine a pole arc, which covers 180 electrical degrees and a corresponding pole pitch, τ_p, which is expressed in meters:

$$\tau_p = \frac{\pi D}{2p}. \tag{2.3}$$

Figure 2.5 depicts the division of the periphery of the machine into phase zones of positive and negative values. In the figure, the number of pole pairs $p = 2$ and the number of phases $m = 3$.

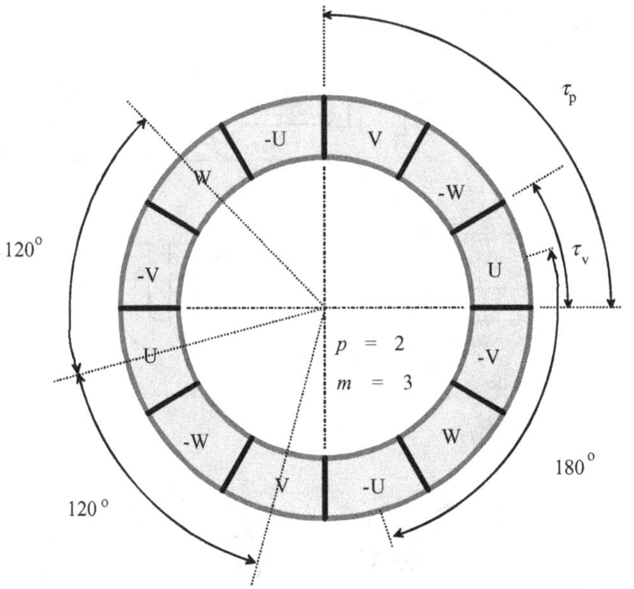

Figure 2.5 Division of the periphery of a three-phase, four-pole machine into phase zones of positive and negative values. The pole pitch is τ_p and phase zone distribution τ_v. When the windings are located in the zones, the instantaneous currents in the positive and negative zones are flowing in opposite directions.

The phase zone distribution is written as

$$\tau_v = \frac{\tau_p}{m}. \tag{2.4}$$

The number of zones will thus be $2pm$. The number of slots for each such zone is expressed by the term q, as a number of slots per pole and phase

$$q = \frac{Q}{2pm}. \tag{2.5}$$

Here Q is the number of slots in the stator or in the rotor. In integral slot windings, q is an integer. However, q can also be a fraction. In that case, the winding is called a fractional slot winding.

The phase zones are distributed symmetrically to different phase windings so that the phase zones of the phases U, V, W are positioned on the periphery of the machine at equal distances in electrical degrees. In a three-phase system, the angle between the phases is 120 electrical degrees. This is illustrated by the periphery of Figure 2.5, where we have 2×360 electrical degrees because of four poles. Now, it is possible to label every phase zone. We start for instance with the positive zone of the phase U. The first positive zone of the phase V will be 120 electrical degrees from the first positive zone of the phase U. Correspondingly, the first positive zone of the phase W will be 120 electrical degrees from the positive zone of the phase V and so on. In Figure 2.5, there are two pole pairs, and hence we need two positive zones for each phase U, V and W. In the slots of each, now labeled phase zones, there are only the coil sides of the labeled phase coil, in all of which the current flows in the same direction. Now, if their direction of current is selected positive in the diagram, the unlabeled zones become negative. Negative zones are labeled by starting from the distance of a pole pitch from the position of the positive zones. Now U and –U, V and –V, W and –W are at distances of 180 electrical degrees from each other.

2.3 Three-Phase Integral Slot Stator Winding

The armature winding of a three-phase electrical machine is usually constructed in the stator, and it is spatially distributed in the stator slots so that the current linkage created by the stator currents is distributed as sinusoidally as possible. The simplest stator winding that produces a noticeable rotating field comprises three coils, the sides of which are divided into six slots, because if $m = 3$, $p = 1$, $q = 1$, then $Q = 2pmq = 6$; see Figures 2.6 and 2.7.

Example 2.2: Create a three-phase, two-pole stator winding with $q = 1$. Distribute the phases in the slots and illustrate the current linkage created based on the instant values of phase sinusoidal currents. Draw a phasor diagram of the slot voltage and sum the voltages of the individual phases. Create a current linkage waveform in the air gap for the time instant t_1 when the phase U voltage is at its positive maximum and for t_2, which is shifted by 30°.

Solution: If $m = 3$, $p = 1$, $q = 1$, then $Q = 2pmq = 6$, which is the simplest case of three-phase windings. The distribution of the phases in the slots will be explained based on Figure 2.6. Starting from slot 1, we insert there the positive conductors of the phase U forming zone U1. The pole pitch expressed by the number of slots per pole, or in other words "the coil span expressed in the number of slot pitches y_Q," is

$$y_Q = \frac{Q}{2p} = \frac{6}{2} = 3.$$

Then, zone U2 will be one pole pitch shifted from U1 and will be located in slot 4, because $1 + y_Q = 1 + 3 = 4$. The beginning of the phase V1 is shifted by 120° from U1, which means slot 3 and its end V2 are in slot 6 ($3 + 3 = 6$). The phase W1 is again shifted from V1 by 120°, which means that it is in slot 5 and its end is in slot 2; see Figure 2.6a. The polarity of instantaneous currents is shown at the instant when the current of the phase U is at its positive maximum value flowing in slot 1, depicted as a cross (the tail of an arrow) in U1 (current flowing away from the observer). Then, U2 is depicted by a dot (the point of the arrow) in slot 4 (current flowing towards the observer). At the same instant in V1 and W1, there are also dots, because the phases V and W are carrying negative current values (see Figure 2.6d), and therefore V2 and W2 are positive, indicated by crosses. In this way, a sequence of slots with inserted phases is as follows: U1, W2, V1, U2, W1, V2, if $q = 1$.

The cross-section of the stator winding in Figure 2.6a shows fictitious coils with current directions resulting in the magnetic field represented by the force lines and arrows.

The phasor diagram in Figure 2.6e includes six phasors. To determine their number, the largest common divider of Q and p denoted t has to be found. In this case, for $Q = 6$ and

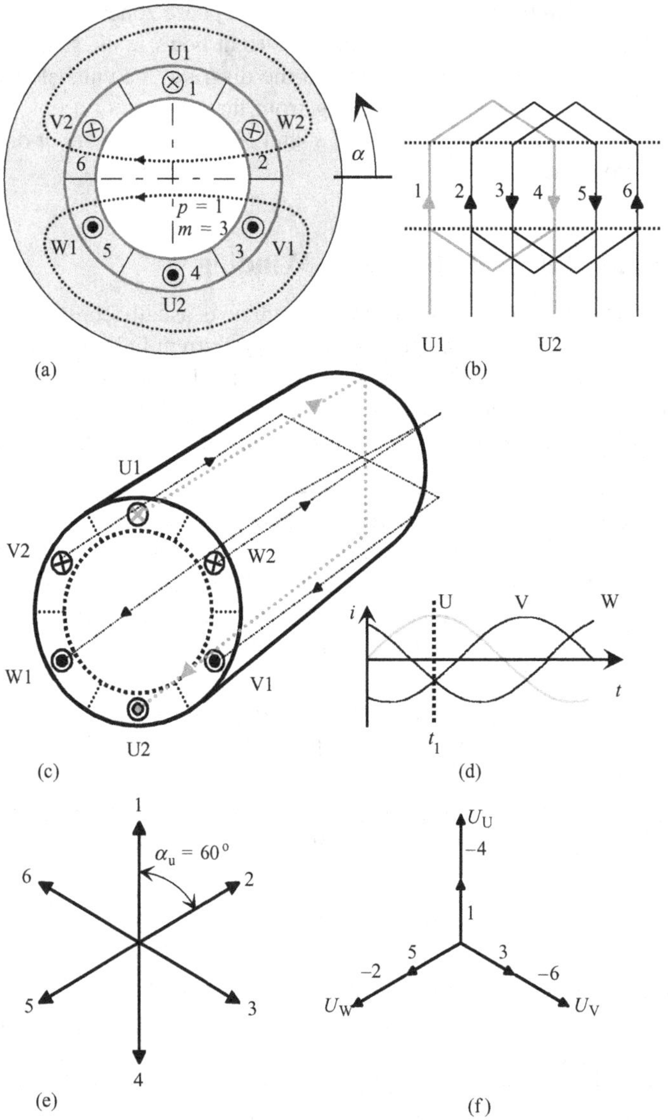

Figure 2.6 The simplest three-phase winding that produces a rotating field. (a) A cross-sectional surface of the machine and a schematic view of the main flux route at the observation instant t_1, (b) a developed view of the winding in a plane and (c) a three-dimensional view of the winding. The figure illustrates how the winding penetrates the machine. The coil end at the rear end of the machine is not illustrated as in reality, but the coil comes directly from a slot to another without traveling along the rear end face of the stator. The ends of the phases U, V and W at the terminals are denoted U1–U2, V1–V2 and W1–W2. (d) The three-phase currents at the observed time instant t_1 when $i_W = i_V = -\frac{1}{2}i_U$, which means ($i_U + i_V + i_W = 0$), (e) a voltage phasor diagram for the given three-phase system, (f) the total phase voltage for individual phases. The voltage of the phase U is created by summing the voltage of slot 1 and the negative voltage of slot 4, and therefore the direction of the voltage phasor in slot 4 is taken opposite and denoted –4. We can see the sum of voltages in both slots and the phase shift by 120° of the V- and W-phase voltages.

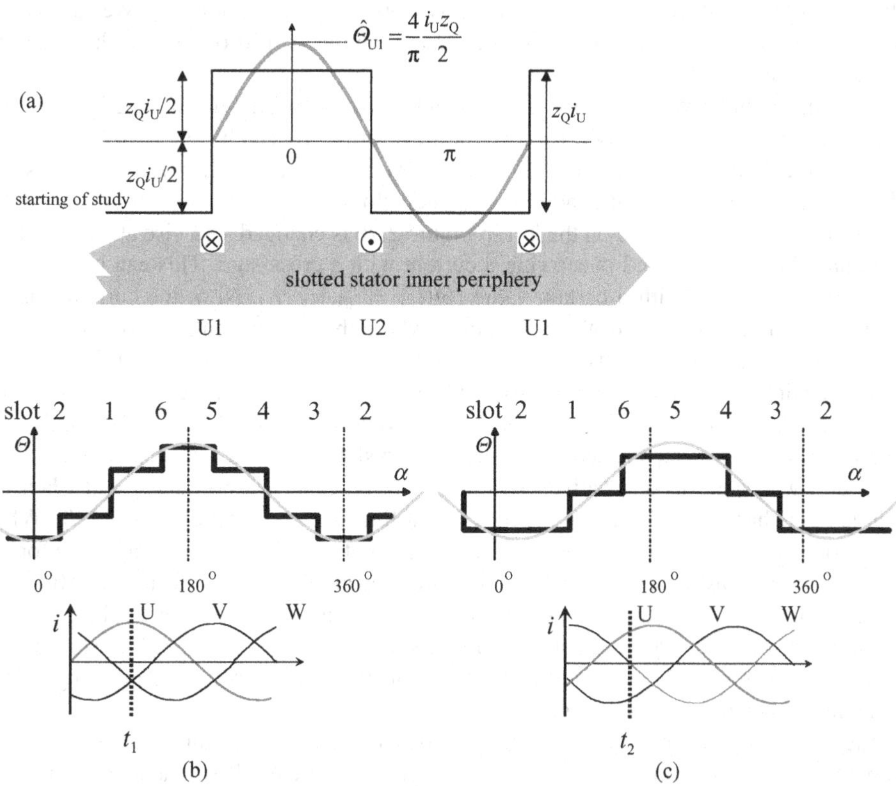

Figure 2.7 Current linkages Θ created by a simple three-phase $q = 1$ winding. (a) Only the phase U is fed by current and observed. A rectangular waveform of current linkage with its fundamental component is shown to explicate the staircase profile of the current linkages below. If all three phases are fed and observed in two different current situations ($i_U + i_V + i_W = 0$) at two time instants t_1 and t_2, see (b) and (c), respectively. The figure also illustrates the fundamental of the staircase current linkage curves. The stepped curves are obtained by applying Ampère's law in the current-carrying teeth zone of the electrical machine. Note that as time elapses from t_1 to t_2, the three-phase currents change and also the position of the fundamental component changes. This indicates clearly the rotating-field nature of the winding. The angle α and the numbers of slots refer to the previous figure, in which we see that the maximum flux density in the air gap lies between slots 6 and 5. This coincides with the maximum current linkage shown in this figure. This is valid if no rotor currents are present.

$p = 1, t = 1$ and therefore the number of phasors is $Q/t = 6$. The angle between the voltage phasors in the adjacent slots is given by the expression

$$\alpha_u = \frac{360° p}{Q} = \frac{360° \cdot 1}{6} = 60°,$$

which results in the numbering of the voltage phasors in slots as shown in Figure 2.6e. Now the total phase voltage for individual phases has to be summed. The voltage of the phase U is created by the positive voltage in slot 1 and the negative voltage in slot 4. The direction of the voltage phasor in slot 4 is taken opposite and denoted –4. We can see the sum of voltages in both slots of the phase U, and the phase shift of 120° of the V and W phase voltages in Figure 2.6f.

The current linkage waveforms for this winding are illustrated in Figure 2.7b and c for the time instants t_1 and t_2, between which the waveforms proceed by 30°. The procedure of drawing the figure can be described as follows. We start observation at $\alpha = 0$. We assume the same constant number of conductors z_Q in all slots.

The current linkage value on the left in Figure 2.7b is changed stepwise at slot 2, where the phase W is located and is carrying a current with a cross sign. This can be drawn as a positive step of Θ with a certain value ($\Theta(t_1) = i_{uW}(t_1)z_Q$). Now, the current linkage curve remains constant until we reach slot 1, where the positive currents of the phase U are located. The instantaneous current in slot 1 is the phase U peak current. The current sum is again indicated with a cross sign. The step height is now twice the height in slot 2, because the peak current is twice the current flowing in slot 2. Then, in slot 6, there is again a positive half step caused by the phase V. In slot 5, there is a current sum indicated by a dot, which means a negative Θ step. This is repeated with all slots, and when the whole circle has been closed, we get the current linkage waveform of Figure 2.7b. When this procedure is repeated for one period of the current, we obtain a traveling wave for the current linkage waveform. Figure 2.7c shows the current linkage waveform after 30°. Here we can see that if the instantaneous value of a slot current is zero, the current linkage does not change, and the current linkage remains constant; see slots 2 and 5. We can also see that the Θ profiles in b and c are not similar, but the form is changed depending on the time instant at which it is investigated.

Figure 2.7 shows that the current linkage produced with such a simple winding deviates considerably from a sinusoidal waveform. Therefore, in electrical machines, more coil sides are usually employed per pole and phase.

Example 2.3: Consider an integral slot winding, where $p = 1$ and $q = 2$, $m = 3$. Distribute the phase winding into the slots, produce an illustration of the windings in the slots, draw a phasor diagram and show the phase voltages of the individual phases. Create a waveform of the current linkage for this winding and compare it with that in Figure 2.7.

Solution: The number of slots needed for this winding is $Q = 2pmq = 2 \times 3 \times 2 = 12$. The cross-sectional area of such a stator with 12 slots and embedded conductors of individual phases is illustrated in Figure 2.8a. The distribution of the slots into the phases is made in the same order as in Example 2.2, but now $q = 2$ slots per pole and phase. Therefore, the sequence of the slots for the phases is as follows: U1, U1, W2, W2, V1, V1, U2, U2, W1, W1, V2, V2. The direction of the current in the slots will be determined in the same way as above in Example 2.2. The coils wound in individual phases are shown in Figure 2.8b. The pole pitch expressed in number of slot pitches is

$$y_Q = \frac{Q}{2p} = \frac{12}{2} = 6.$$

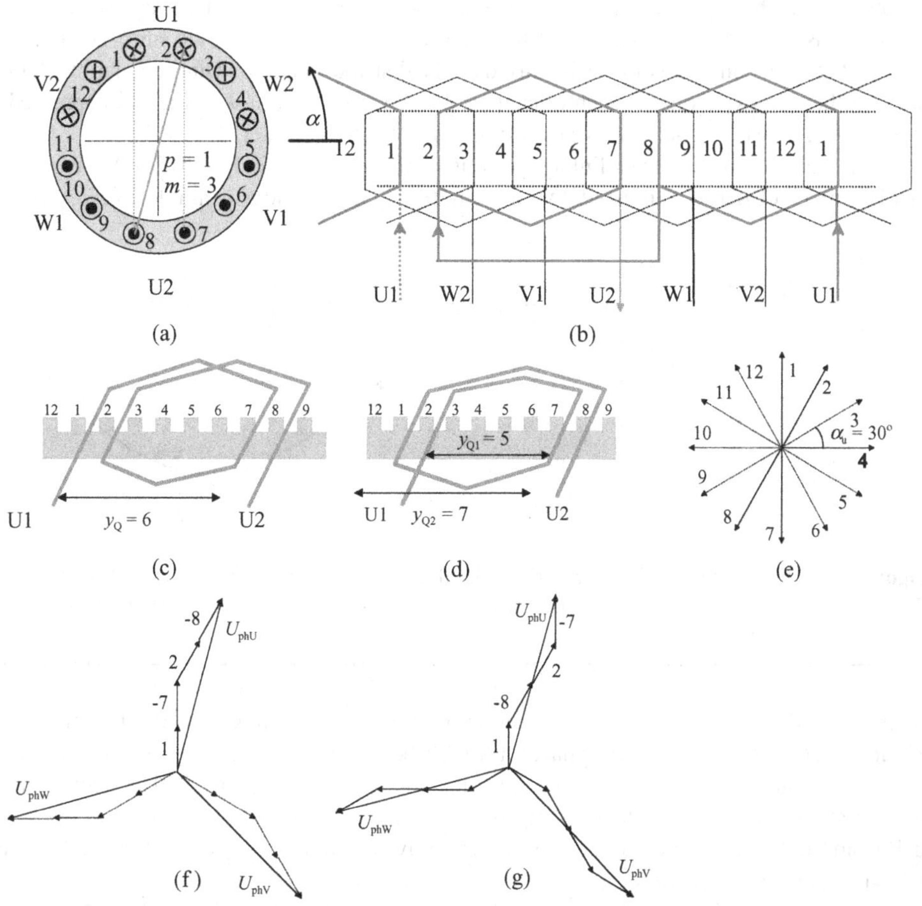

Figure 2.8 Three-phase, two-pole winding with two slots per pole and phase: (a) a stator with 12 slots, the number of slots per pole and phase $q = 2$, (b) divided coil groups, (c) full-pitch coils of the phase U, (d) average full-pitch coils of the phase U, (e) a phasor diagram with 12 phasors, one for each slot, (f) the sum phase voltage of individual phases corresponding to figure (c), (g) the sum phase voltage of individual phases corresponding to figure (d).

Figure 2.8c shows how the phase U is wound to keep the full pitch equal to six slots. In Figure 2.8d, the average pitch is also six, but the individual steps are $y_Q = 5$ and 7, which give the same average result for the value of induced voltage.

The phasor diagram has 12 phasors, because $t = 1$ again. The angle between two phasors of adjacent slots is

$$\alpha_u = \frac{360^\circ p}{Q} = \frac{360^\circ \cdot 1}{12} = 30^\circ.$$

The phasors are numbered gradually around the circle. Based on this diagram, the phase voltage of all phases can be found. Figures 2.8f and g show that the voltages are the same independent of how the separate coil sides are connected in series. In comparison with the previous example, the geometrical sum is now less than the algebraic sum. The phase shifting between coil side voltages is caused by the distribution of the winding in more than one slot, here in two slots for each pole. This reduction of the phase voltage is expressed by means of a distribution winding factor; this will be derived later.

The waveform of the current linkage for this winding is given in Figure 2.9. We can see that it is much closer to a sinusoidal waveform than in the previous example with $q = 1$.

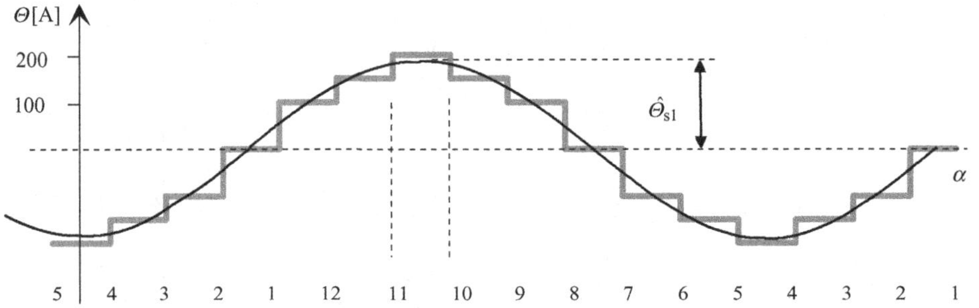

Figure 2.9 Current linkage $\Theta_s = f(\alpha)$ created by the winding on the surface of the stator bore of Figure 2.8 at a time $i_W = i_V = -\frac{1}{2}i_U$. The fundamental Θ_{s1} of Θ_s is given as a sinusoidal curve. The numbering of the slots is also given.

In undamped permanent magnet synchronous motors, such windings can also be employed, the number of slots per pole and phase of which being clearly less than one, for instance $q = 0.4$. In that case, a well-designed machine looks like a rotating-field machine when observed at its terminals, but the current linkage produced by the stator winding deviates so much from the fundamental that, because of excessive harmonic losses in the rotor, no other rotor type comes into question.

When comparing Figure 2.9 ($q = 2$) with Figure 2.7 ($q = 1$), it is obvious that the higher the term q (slots per pole and phase), the more sinusoidal the current linkage of the stator winding.

As we can see in Figure 2.7a, the current linkage amplitude of the fundamental component for one full-pitch coil is

$$\hat{\Theta}_{1U} = \frac{4}{\pi}\frac{z_Q \hat{i}_U}{2}. \tag{2.6}$$

If the coil winding is distributed into more slots, and $q > 1$ and $N = pqz_Q/a$, the winding factor must be taken into account:

$$\hat{\Theta}_{1U} = \frac{4}{\pi}\frac{Nk_{w1}\hat{i}_U}{2}. \tag{2.7}$$

In a $2p$-pole machine ($2p > 2$), the current linkage for one pole is

$$\hat{\Theta}_{1U} = \frac{4}{\pi}\frac{Nk_{w1}\hat{i}_U}{2p}. \tag{2.8}$$

This expression can be rearranged with the number of conductors in a slot. In one phase, there are 2*N* conductors, and they are embedded in the slots belonging to one phase *Q*/*m*. Therefore, the number of conductors in one slot will be

$$z_Q = \frac{2N}{Q/m} = \frac{2mN}{2pqm} = \frac{N}{pq} \tag{2.9}$$

and

$$\frac{N}{p} = qz_Q. \tag{2.10}$$

Then *N*/*p* presented in Equation (2.8) and in the following can be introduced by qz_Q:

$$\hat{\Theta}_{1U} = \frac{4}{\pi}\frac{Nk_{w1}\hat{i}_U}{2p} = \frac{4}{\pi}qz_Q\frac{k_{w1}\hat{i}_U}{2}. \tag{2.11}$$

Equation (2.8) can also be expressed by the effective value of sinusoidal phase current if there is a symmetrical system of phase currents:

$$\hat{\Theta}_{1U} = \frac{4}{\pi}\frac{Nk_{w1}}{2p}\sqrt{2}I. \tag{2.12}$$

For an *m*-phase rotating-field stator or rotor winding, the amplitude of current linkage is *m*/2 times as high

$$\hat{\Theta}_1 = \frac{m}{2}\frac{4}{\pi}\frac{Nk_{w1}}{2p}\sqrt{2}I \tag{2.13}$$

and for a three-phase stator or rotor winding, the current linkage amplitude of the fundamental component for one pole is

$$\hat{\Theta}_1 = \frac{3}{2}\frac{4}{\pi}\frac{Nk_{w1}}{2p}\sqrt{2}I = \frac{3}{\pi}\frac{Nk_{w1}}{p}\sqrt{2}I. \tag{2.14}$$

For a stator current linkage amplitude $\hat{\Theta}_{s\nu}$ of the harmonic ν of the current linkage of a poly-phase ($m > 1$) rotating-field stator winding (or rotor winding), when the effective value of the stator current is I_s, we may write

$$\hat{\Theta}_{s\nu} = \frac{m}{2}\frac{4}{\pi}\frac{k_{w\nu}N_s}{p\nu}\frac{1}{2}\sqrt{2}I_s = \frac{mk_{w\nu}N_s}{\pi p\nu}\hat{i}_s \tag{2.15}$$

Example 2.4: Calculate the amplitude of the fundamental component of stator current linkage, if $N_s = 200$, $k_{w1} = 0.96$, $m = 3$, $p = 1$ and $i_{sU}(t) = \hat{i} = 1$ A, the effective value for a sinusoidal current being $I_s = (1/\sqrt{2})\text{A} = 0.707$ A.

Solution: For the fundamental, we obtain

$$\hat{\Theta}_1 = \frac{3}{2}\frac{4}{\pi}\frac{Nk_{w1}}{2p}\sqrt{2}I_s = \frac{3}{\pi}\frac{Nk_{w1}}{p}\sqrt{2}I_s = \frac{3}{\pi}\cdot\frac{200\cdot 0.96}{1}\sqrt{2}\cdot 0.707\text{A} = 183\text{ A}.$$

2.4 Voltage Phasor Diagram and Winding Factor

Since the winding is spatially distributed in the slots on the stator surface, the flux (which is proportional to the current linkage Θ) penetrating the winding does not intersect all windings simultaneously, but with a certain phase shift. Therefore, the eletromotive force (emf) of the winding is not calculated directly with the number of turns N_s, but the winding factors $k_{w\nu}$ corresponding to the harmonics are required. The emf of the fundamental induced in the turn is calculated with the flux linkage Ψ by applying Faraday's induction law $e = -Nk_{w1}\mathrm{d}\Phi/\mathrm{d}t = -\mathrm{d}\Psi/\mathrm{d}t$ (see Equations 1.3, 1.7 and 1.8). We can see that the winding factor correspondingly indicates the characteristics of the winding to produce harmonics, and it has thus to be taken into account when calculating the current linkage of the winding (Equation 2.15). The common distribution of all the current linkages created by all the windings together produces a flux density distribution in the air gap of the machine, which, when moving with respect to the winding, induces voltages in the conductors of the winding. The phase shift of the induced emf in different coil sides is investigated with a voltage phasor diagram. The voltage phasor diagram is presented in electrical degrees. If the machine is for instance a four-pole one, $p = 2$, the voltage vectors have to be distributed along two full circles in the stator bore. Figure 2.10a illustrates the voltage phasor diagram of a two-pole winding of Figure 2.8.

In Figure 2.10a, phasors 1 and 2 are positive and 7 and 8 are negative for the phase under consideration. Hence, phasors 7 and 8 are turned by 180 ° to form a bunch of phasors. For harmonic ν (excluding slot harmonics that have the same winding factor as the fundamental) the directions of the phasors of the coil sides vary more than in the figure, because the slot angles α_u are replaced with the angles $\nu\alpha_u$.

According to Figure 2.10b, when calculating the geometric sum of the voltage phasors for a phase winding, the symmetry line for the bunch of phasors, where the negative phasors have been turned opposite, must be found. The angles α_ρ of the phasors with respect to this symmetry line may be used in the calculation of the geometric sum. Each phasor contributes to the sum with a component proportional to $\cos\alpha_\rho$.

We can now write a general presentation for the winding factor $k_{w\nu}$ of a harmonic ν, by employing the voltage phasor diagram

$$k_{w\nu} = \frac{\sin\frac{\nu\pi}{2}}{Z}\sum_{\rho=1}^{z}\cos\alpha_\rho. \tag{2.16}$$

Here Z is the total number of positive and negative phasors of the phase in question, ρ is the ordinal number of a single phasor, and ν is the ordinal number of the harmonic under

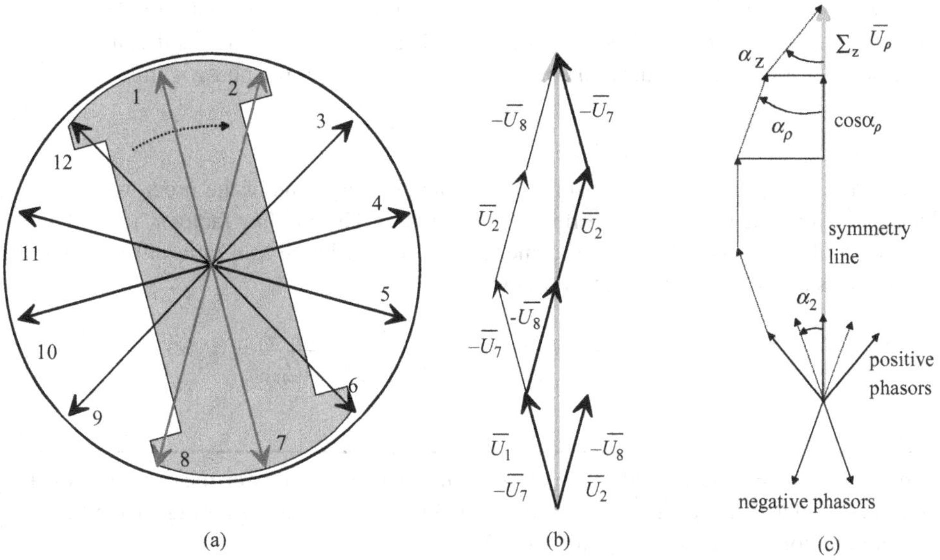

Figure 2.10 (a), (b) Fundamental voltage phasor diagram for the winding of Figure 2.8 $Q_s = 12$, $p = 1$, $q_s = 2$. A maximum voltage is induced in the bars in slots 1 and 7 at the moment depicted in the figure, when the rotor is rotating clockwise. The figure also illustrates the calculation of the voltage in a single coil with the radii of the voltage phasor diagram. (c) General application of the voltage phasor diagram in the determination of the winding factor (fractional slot winding since the number of phasors is uneven). The phasors of negative coil sides are turned 180 °, and then the summing of the resulting bunch of phasors is calculated according to Equation (2.16). A symmetry line is drawn in the middle of the bunch, and each phasor forms an angle α_ρ with the symmetry line. The geometric sum of all the phasors lies on the symmetry line.

observation. The coefficient $\nu\pi/2$ in the equation only influences the sign (of the factor). The angle of a single phasor α_ρ can be found from the voltage phasor diagram drawn for the specific harmonic, and it is the angle between an individual phasor and the symmetry line drawn for a specific harmonic (cf. Figure 2.10b). This voltage phasor diagram solution is universal and may be used in all cases, but the numerical values of Equation (2.16) do not always have to be calculated directly from this equation, or with the voltage phasor diagram at all. In simple cases, we may apply equations introduced later. However, the voltage phasor diagram forms the basis for the calculations, and therefore its utilization is discussed further when analyzing different types of windings.

If in Figure 2.10a we are considering a currentless stator of a synchronous machine, a maximum voltage can be induced in the coil sides 1 and 7 at the middle of the pole shoe, when the rotor is rotating at no load inside the stator bore (which corresponds to the peak value of the flux density, but the zero value of the flux penetrating the coil), where the derivative of the flux penetrating the coil reaches its peak value, the voltage induction being at its highest at that moment. If the rotor rotates clockwise, a maximum voltage is induced in coil sides 2 and 8 in a short while, and so on. The voltage phasor diagram then describes the amplitudes of voltages induced in different slots and their temporal phase shift.

The series-connected coils of the phase U travel, for example, from slot 1 to slot 8 (coil 1) and from slot 2 to slot 7 (coil 2; see Figure 2.8c). Thus a voltage, which is the difference of the phasors $\underline{U}_1$ and $\underline{U}_8$, is induced in coil 1. The total voltage of the phase is thus

$$\underline{U}_{\mathrm{U}} = \underline{U}_1 - \underline{U}_8 + \underline{U}_2 - \underline{U}_7. \tag{2.17}$$

The figure also indicates the possibility of connecting the coils in the order 1–7 and 2–8, which gives the same voltage but a different end winding. The winding factor k_{w1} based on the distribution of the winding for the fundamental is calculated here as a ratio of the geometric sum and the sum of absolute values as follows:

$$k_{\mathrm{w1}} = \frac{\text{geometric sum}}{\text{sum of absolute values}} = \frac{\underline{U}_1 - \underline{U}_8 + \underline{U}_2 - \underline{U}_7}{|\underline{U}_1| + |\underline{U}_8| + |\underline{U}_2| + |\underline{U}_7|} = 0.966 \leq 1. \tag{2.18}$$

Example 2.5: Equation (2.16) indicates that the winding factor for the harmonics may also be calculated using the voltage phasor diagram. Derive the winding factor for the seventh harmonic of the winding in Figure 2.8.

Solution: We now draw a new voltage phasor diagram based on Figure 2.10 for the seventh harmonic, Figure 2.11.

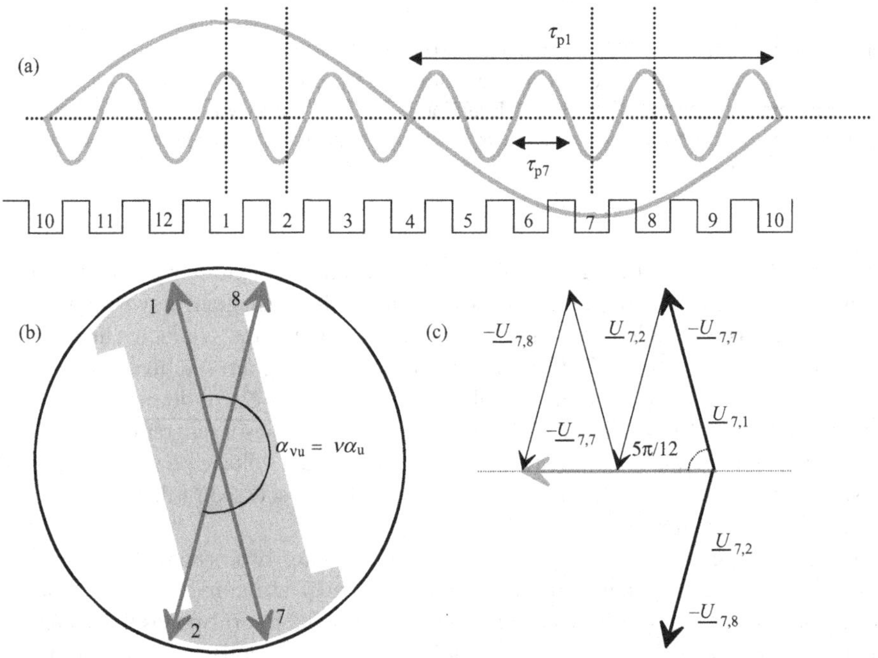

Figure 2.11 Deriving the harmonic winding factor: (a) the fundamental and the seventh harmonic field in the air gap over the slots, (b) voltage phasors for the seventh harmonic of a full-pitch $q = 2$ winding (slot angle $\alpha_{\mathrm{u7}} = 210°$) and (c) the symmetry line and the sum of the voltage phasors. The phasor angles α_ρ with respect to the symmetry line are $\alpha_\rho = 5\pi/12$ or $-5\pi/12$.

Slots 1 and 2 belong to the positive zone of the phase U and slots 7 and 8 to the negative zone measured by the fundamental. In Figure 2.11, we see that the pole pitch of the seventh harmonic is one-seventh of the fundamental pole pitch. Deriving the phasor sum for the seventh harmonic is started for instance with the voltage phasor of slot 1. This phasor remains in its original position. Slot 2 is physically and by the fundamental located 30° clockwise from slot 1, but as we are now studying the seventh harmonic, the slot angle measured in degrees for it is $7 \times 30° = 210°$, which can also be seen in the figure. The phasor for slot 2 is, hence, located 210° clockwise from phasor 1. Slot 7 is located at $7 \times 180° = 1260°$ from slot 1. Since $1260° = 3 \times 360° + 180°$ phasor 7 remains opposite to phasor 1. Phasor 8 is located 210° clockwise from phasor 7 and will find its place 30° clockwise from phasor 1. By turning the negative zone phasors by π and applying Equation (2.17) we obtain

$$k_{w7} = \frac{\sin\frac{7\pi}{2}}{4}\sum_{\rho=1}^{4}\cos\alpha_\rho$$

$$= \frac{\sin\frac{7\pi}{2}}{4}\left(\cos\frac{-5\pi}{12} + \cos\frac{+5\pi}{12} + \cos\frac{-5\pi}{12} + \cos\frac{+5\pi}{12} = -0.2588.\right)$$

It is not necessary to apply the voltage phasor diagram, but also simple equations may be derived to directly calculate the winding factor. In principle, we have three winding factors: a distribution factor, a pitch factor and a skewing factor. The winding factor derived from the shifted voltage phasors in the case of a distributed winding is called the distribution factor, denoted by the subscript "d." This factor is always $k_{d1} \leq 1$. The value $k_{d1} = 1$ can be reached when $q = 1$, in which case the geometric sum equals the sum of absolute values, see Figure 2.6f. If $q \neq 1$, then $k_{d1} < 1$. In fact, this means that the total phase voltage is reduced by this factor (see Example 2.6).

If each coil is wound as a full-pitch winding, the coil pitch is in principle the same as the pole pitch. However, the voltage of the phase with full-pitch coils is reduced because of the winding distribution with the factor k_d. If the coil pitch is shorter than the pole pitch and the winding is not a full-pitch winding, the winding is called a short-pitch winding, or a chorded winding (see Figure 2.15 and 2.16). Note that the winding in Figure 2.8d is not a short-pitch winding, even though the coil may be realized from slot 1 to slot 8 (longer than pole pitch) and not from slot 1 to slot 7 (equivalent to pole pitch). A real short pitching is obviously employed in the two-layer windings. Short pitching is another reason why the voltage of the phase winding may be reduced. The factor of such a reduction is called the pitch factor k_p. The total winding factor is given as

$$k_w = k_d \cdot k_p. \tag{2.19}$$

Equations to calculate the distribution factor k_d will now be derived; see Figure 2.12. The equations are based on the geometric sum of the voltage phasors in a similar way as in Figures 2.10 and 2.11.

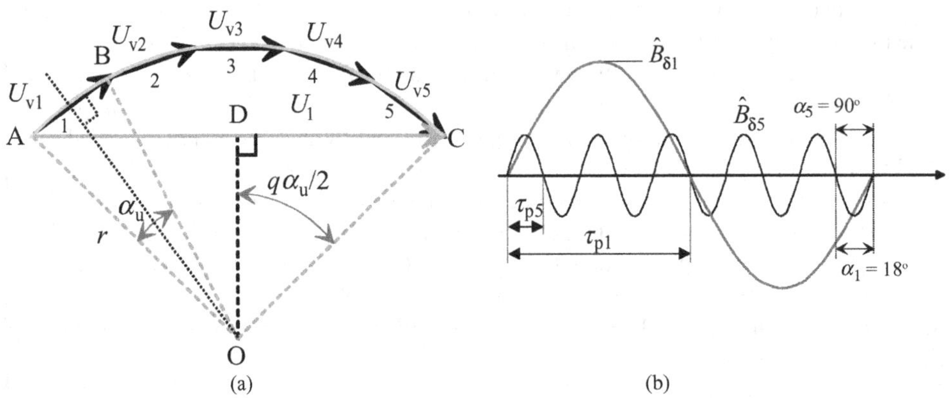

Figure 2.12 (a) Determination of the distribution factor with a polygon with $q = 5$, (b) the pole pitch for the fundamental and the fifth harmonic. The same physical angles for the fifth harmonic and the fundamental are shown as an example. Voltages $U_{v1} - U_{v5}$ represent corresponding coil voltages.

The distribution factor for the fundamental component is given as

$$k_{d1} = \frac{\text{geometric sum}}{\text{sum of absolute values}} = \frac{U_1}{qU_{coil1}}. \tag{2.20}$$

According to Figure 2.12, we may write for the triangle ODC

$$\sin\frac{q\alpha_u}{2} = \frac{\frac{U_1}{2}}{r} \Rightarrow U_1 = 2r\sin\frac{q\alpha_u}{2}, \tag{2.21}$$

and for the triangle OAB

$$\sin\frac{\alpha_u}{2} = \frac{\frac{U_{v1}}{2}}{r} \Rightarrow U_{v1} = 2r\sin\frac{\alpha_u}{2}. \tag{2.22}$$

We may now write for the distribution factor

$$k_{d1} = \frac{U_1}{qU_{v1}} = \frac{2r\sin\frac{q\alpha_u}{2}}{q2r\sin\frac{\alpha_u}{2}} = \frac{\sin\frac{q\alpha_u}{2}}{q\sin\frac{\alpha_u}{2}}. \tag{2.23}$$

This is the basic expression for the distribution factor for the calculation of the fundamental in a closed form. Since the harmonic components of the air-gap magnetic flux density are present, the calculation of the distribution factor for the νth harmonic will be carried out applying the angle $\nu\alpha_u$; see Figure 2.11b and 2.12b:

$$k_{d\nu} = \frac{\sin \nu \dfrac{q\alpha_u}{2}}{q \sin \nu \dfrac{\alpha_u}{2}}. \tag{2.24}$$

Example 2.6: Repeat Example 2.5 using Equation (2.24).

Solution for $\nu = 7$:

$$k_{d\nu} = \frac{\sin \nu \dfrac{q\alpha_u}{2}}{q \sin \nu \dfrac{\alpha_u}{2}} = \frac{\sin 7 \dfrac{2\dfrac{\pi}{6}}{2}}{2 \sin 7 \dfrac{\pi}{6 \cdot 2}} = \frac{\sin \dfrac{7\pi}{6}}{2 \sin \dfrac{7\pi}{12}} = -0.2588$$

The result is the same as above.

The expression for the fundamental may be rearranged as

$$k_{d1} = \frac{\sin\left(\dfrac{Q}{2pm}\dfrac{2\pi p}{2Q}\right)}{q \sin\left(\dfrac{2\pi p}{2 \cdot 2pmq}\right)} = \frac{\sin\left(\dfrac{\pi}{2m}\right)}{q \sin\left(\dfrac{\pi}{2mq}\right)}. \tag{2.25}$$

For three-phase machines, $m = 3$, the expression is as follows:

$$k_{d1} = \frac{\sin(\pi/6)}{q \sin \dfrac{\pi/6}{q}} = \frac{1}{2q \sin \dfrac{\pi/6}{q}}. \tag{2.26}$$

This simple expression of the distribution factor for the fundamental is most often employed for practical calculations.

Example 2.7: Calculate the phase voltage of a three-phase, four-pole synchronous machine with a stator bore diameter of 0.30 m, a length of 0.5 m and a speed of rotation 1500 min^{-1}. The excitation creates the air-gap fundamental flux density $\hat{B}_{\delta 1} = 0.8$ T. There are 36 slots, in which a one-layer winding with three conductors in each slot is embedded.

Solution: According to the Lorentz law, an instantaneous value of the induced electric field strength in a conductor is $\boldsymbol{E} = \boldsymbol{v} \times \boldsymbol{B}_0$. In one conductor embedded in a slot of an AC machine, we may get the induced voltage by integrating, $e_{1c} = B_\delta l' v$, where B_δ is the local air-gap flux density value of the rotating magnetic field, l' is the effective length of the stator iron stack, and v is the speed at which the conductor travels in the magnetic field, see Figure 2.13.

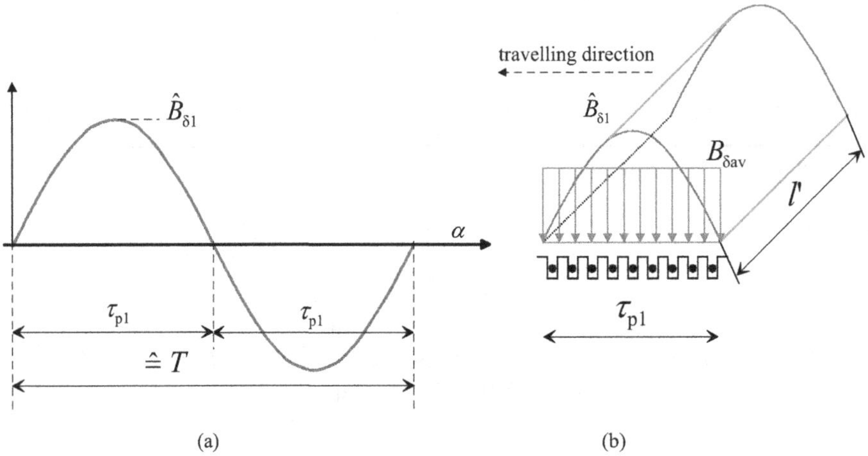

Figure 2.13 (a) Flux density variation in the air gap. One flux density period travels two pole pitches during one time period *T*. (b) The flux distribution over one pole and the conductors in slots.

During one period *T*, the magnetic flux density wave travels two pole pitches, as shown in the figure above. The speed of the magnetic field moving in the air gap is

$$v = \frac{2\tau_p}{T} = 2\tau_p f.$$

The effective value of the induced voltage in one conductor is

$$E_{1c} = \frac{\hat{B}_\delta}{\sqrt{2}} l' 2\tau_p f.$$

Contrary to a transformer, where approximately the same value of magnetic flux density penetrates all the winding turns, Figure 2.13 shows that in AC rotating-field machines, the conductors are subject to the sinusoidal waveform of flux density, and each conductor has a different value of magnetic flux density. Therefore, an average value of the magnetic flux density is calculated to unify the value of the magnetic flux for all conductors. The average value of the flux density equals the maximum value of the flux penetrating a full-pitch winding:

$$\hat{\Phi} = B_{\delta av} l' \tau_p.$$

B_δ is spread over the pole pitch τ_p, and we get for the average value

$$B_{\delta av} = \frac{2}{\pi}\hat{B}_\delta \Rightarrow \hat{B}_\delta = \frac{\pi}{2} B_{\delta av}.$$

Now, the RMS value of the voltage induced in a conductor written by means of average magnetic flux density is

$$E_{1c} = \frac{\pi}{2\sqrt{2}} B_{\delta av} l' 2\tau_p f = \frac{\pi}{\sqrt{2}} \hat{\Phi} f.$$

The frequency f is found from the speed n

$$f = \frac{pn}{60} = \frac{2 \cdot 1500}{60} = 50 \text{ Hz}.$$

Information about three conductors in each slot can be used for calculating the number of turns N in series. In one phase, there are $2N$ conductors, and they are embedded in the slots belonging to one phase Q/m. Therefore, the number of conductors in one slot z_Q will be

$$z_Q = \frac{2N}{Q/m} = \frac{2Nm}{2pqm} = \frac{N}{pq}.$$

In this case, the number of turns in series in one phase is $N = 3pq = 3 \cdot 2 \cdot 3 = 18$, where $q = Q/2pm = 36/(4 \cdot 3) = 3$. The effective value of the induced voltage in one slot is $(N/pq)E_c$. The number of such slots is $2pq$. The linear sum of the voltages of all conductors belonging to the same phase must be reduced by the winding factor to get the phase voltage

$$E_{ph} = \frac{N}{pq} E_{1c} q 2 p k_w.$$

The final expression for the effective value of the induced voltage in the AC rotating machine is

$$E_{ph} = E_{1c} \frac{N}{pq} q 2 p k_{w1} = \frac{\pi}{\sqrt{2}} \hat{\Phi} f \frac{N}{pq} q 2 p k_{w1} = \sqrt{2} \pi f \hat{\Phi} N k_{w1}.$$

In this example, there is a full-pitch one-layer winding, and therefore $k_p = 1$, and only k_d must be calculated (see Equation (2.26)):

$$k_{w1} = k_{d1} = \frac{1}{2q \sin \frac{30^\circ}{q}} = \frac{1}{2 \cdot 3 \sin \frac{30^\circ}{3}} = 0.960$$

The maximum value of the magnetic flux is

$$\hat{\Phi} = \frac{2}{\pi} \hat{B}_\delta \tau_p l' = \frac{2}{\pi} 0.8 \cdot 0.236 \cdot 0.5 = 0.060 \text{ Wb}$$

where the pole pitch τ_p is

$$\tau_p = \frac{\pi D_s}{2p} = \frac{\pi \cdot 0.3}{4} = 0.236 \text{ m}.$$

An effective value of the phase induced voltage is

$$E_{\text{ph}} = \sqrt{2}\pi f \hat{\Phi} N k_{\text{d}} = \sqrt{2}\pi \cdot 50 \cdot 0.060 \cdot 18 \cdot 0.960 = 230 \text{ V}.$$

Example 2.8: The stator of a four-pole, three-phase induction motor has 36 slots, and it is fed by 3 × 400/230 V, 50 Hz. The diameter of the stator bore is $D_{\text{s}} = 15$ cm and the length $l' = 20$ cm. A two-layer winding is embedded in the slots. Besides this, there is a one-layer, full-pitch search coil embedded in two slots. The number of turns in the search coil is four. In the no-load condition, a voltage of 11.3 V has been measured at its terminals. Calculate the air-gap flux density, if the voltage drop on the impedance of the search coil can be neglected.

Solution: To be able to investigate the air-gap flux density, the data of the search coil can be used. This coil is embedded in two electrically opposite slots, and therefore the distribution factor $k_{\text{d}} = 1$; because it is a full-pitch coil, also the pitch factor $k_{\text{p}} = 1$, and therefore $k_{\text{w}} = 1$.

Now, the maximum value of magnetic flux is

$$\hat{\Phi} = \frac{U_{\text{c}}}{\sqrt{2}\pi f N_{\text{c}} k_{\text{wc}}} = \frac{11.3}{\sqrt{2}\pi \cdot 50 \cdot 4 \cdot 1.0} \text{Wb} = 0.0127 \text{ Wb}.$$

The amplitude of the air-gap flux density is

$$\hat{B}_{\delta} = \frac{\pi}{2} B_{\text{av}} = \frac{\pi}{2} \frac{\hat{\Phi}}{\frac{\pi D_{\text{s}}}{2p} l'} = \frac{\pi}{2} \frac{0.0127}{\frac{\pi \cdot 0.15}{4} 0.2} \text{T} = 0.847 \text{ T}.$$

2.5 Winding Analysis

The winding analysis starts with the analysis of a single-layer stator winding, in which the number of coils is $Q_{\text{s}}/2$. In the machine design, the following setup is advisable: the periphery of the air gap of the stator bore (diameter D_{s}) is distributed evenly in all poles, that is $2p$ of equal parts, which yields the pole pitch τ_{p}. Figure 2.14 illustrates the configuration of a two-pole slot winding ($p = 1$).

The pole pitch is evenly distributed in all stator phase windings, that is in m_{s} equal parts. Now we obtain a zone distribution τ_{sv}. In Figure 2.14, the number of stator phases is $m_{\text{s}} = 3$. The number of zones thus becomes $2pm_{\text{s}} = 6$. The number of stator slots in a single zone is q_{s}, which is the number of slots per pole and phase in the stator. By using stator values in the general equation (2.5), we obtain

$$q_{\text{s}} = \frac{Q_{\text{s}}}{2pm_{\text{s}}}. \tag{2.27}$$

If q_{s} is an integer, the winding is termed an integral slot winding, and if q_{s} is a fraction, the winding is called a fractional slot winding.

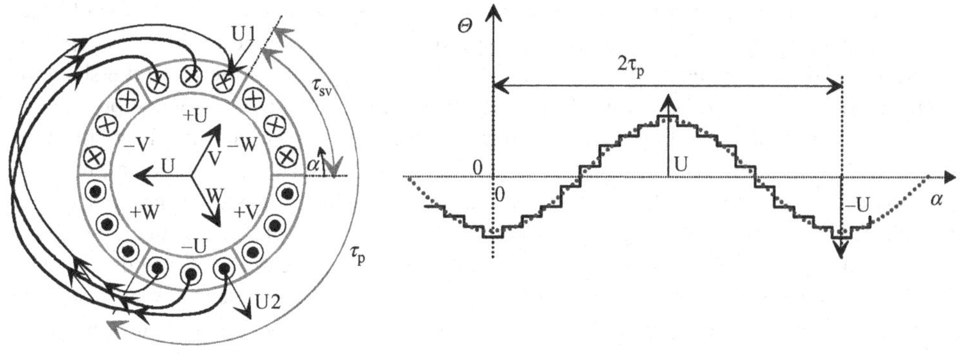

(a) (b)

Figure 2.14 (a) Three-phase stator diamond winding $p = 1$, $q_s = 3$, $Q_s = 18$. Only the coil end on the side of observation is visible in the U-phase winding. The figure also illustrates the positive magnetic axes of the phase windings U, V and W. The current linkage creates a flux in the direction of the magnetic axis when the current is penetrating the winding at its positive terminal, for example U1. The current directions of the figure depict a situation in which the current of the windings V and W is a negative half of the current in the winding U. (b) The created current linkage distribution Θ_s is shown at the moment when its maximum is in the direction of the magnetic axis of the phase U. The small arrows in (a) at the end winding indicate the current directions and the transitions from coil to coil. The same winding will be observed in Figure 2.23.

The phase zones are labeled symmetrically to the phase windings, and the directions of currents are determined so that we obtain a number of m_s magnetic axes at a distance of $360°/m_s$ from each other. The phase zones are labeled as stated in Section 2.2. The positive zone of the phase U, that is a zone where the current of the phase U is flowing away from the observer, is set as an example (Figure 2.14). Now the negative zone of the phase U is at a distance of 180 electrical degrees; in other words, electrically on the opposite side. The conductors of respective zones are connected so that the current flows as desired. This can be carried out for instance as illustrated in Figure 2.14. In the figure, it is assumed that there are three slots in each zone, $q_s = 3$. The figure shows that the magnetic axis of the phase winding U is in the direction of the arrow drawn in the middle of the illustration. Because this is a three-phase machine, the directions of the currents of the phases V and W have to be such that the magnetic axes of the phases V and W are at a distance of 120° (electrical degrees) from the magnetic axis of the phase U. This can be realized by setting the zones of the V and W phases and the current directions according to Figure 2.14.

The way in which the conductors of different zones are connected produces different mechanical winding constructions, but the air gap remains similar irrespective of the mechanical construction. However, the arrangement of connections has a significant influence on the space requirements for the end windings, the amount of copper and the production costs of the winding. The connections also have an effect on certain electrical properties, such as the leakage flux of the end windings.

The poly-phase winding in the stator of a rotating-field machine creates a flux wave when a symmetric poly-phase current flows in the winding. A flux wave is created for instance when the current linkage of Figure 2.14 begins to propagate in the direction of the positive α-axis, and the currents of the poly-phase winding are alternating sinusoidally as a function of time. We have to note, however, that the propagation speeds of the harmonics created by the winding are different from the speed of the fundamental ($n_{s\nu} = \pm\, n_{s1}/\nu$), and therefore the shape of the current linkage curve changes as a function of time. However, the fundamental propagates in the air gap at a speed defined by the fundamental of the current and by the number of pole pairs. Furthermore, the fundamental is usually dominating (when $q \geq 1$), and thus the operation of the machine can be analyzed basically with the fundamental. For instance, in a three-phase winding, time-varying sinusoidal currents with a 120° phase shift in time create a temporally and positionally alternating flux in the windings that are distributed at distances of 120 electrical degrees. The flux distribution propagates as a wave on the stator surface. (See e.g. Figure 7.8 illustrating the fundamental $\nu = 1$ of a six-pole and a two-pole machine.)

2.6 Short Pitching

In a double-layer diamond winding, the slot is divided into an upper and a lower part, and there is one coil side in each half slot. The coil side at the bottom of the slot belongs to the bottom layer of the slot, and the coil side adjacent to the air gap belongs to the upper layer. The number of coils is now the same as the number of slots Q_s of the winding; see Figure 2.15b.

A double-layer diamond winding is constructed like the single-layer winding. As illustrated in Figure 2.15, there are two zone rings, the outer illustrating the bottom layer and the inner the upper layer, Figure 2.15a. The distribution of zones does not have to be identical in the upper and bottom layers. The zone distribution can be shifted by a multiple of the slot pitch. In Figure 2.15a, a single zone shift equals a single slot pitch. Figure 2.15b illustrates one of the coils of the phase U. By comparing the width of the coil with the coil span of the

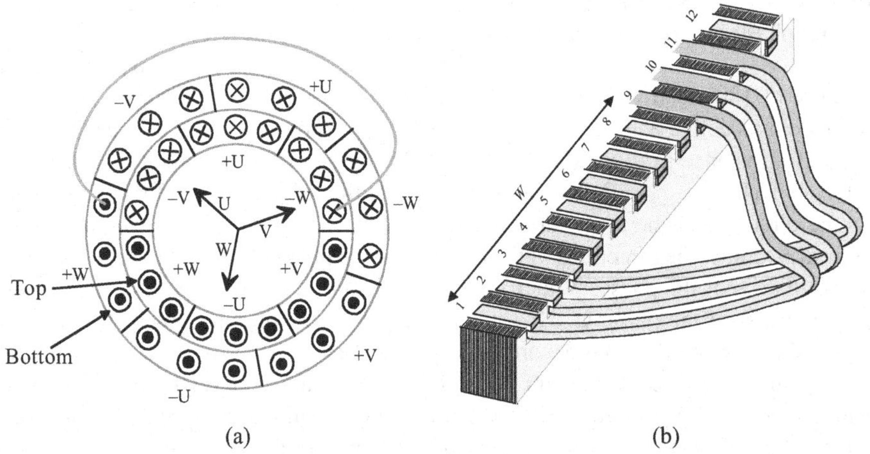

Figure 2.15 (a) Three-phase, double-layer diamond winding $Q_s = 18$, $q_s = 3$, $p = 1$. One end winding is shown to illustrate the coil span. The winding is created from the previous winding by dividing the slots into upper and bottom layers and by shifting the bottom layers clockwise by a single slot pitch. The magnetic axes of the new short-pitched winding are also shown. (b) A coil end of a double-layer winding produced from preformed copper, with a coil span W or expressed as numbers of slot pitches y. The coil ends start from the left at the bottom of the slot and continue to the right to the top of the slot.

winding in Figure 2.14, we can see that the coil is now one slot pitch narrower; the coil is said to be short pitched. Because of short pitching, the coil end has become shorter, and the copper consumption is thus reduced. On the other hand, the flux linking the coil decreases somewhat because of short pitching, and therefore the number of coil turns at the same voltage has to be higher than for a full-pitch winding. The short pitching of the coil end is of more significance than the increased number of coil turns, and as a result the consumption of coil material decreases.

Short pitching also influences the harmonics content of the flux density of the air gap. A correctly short-pitched winding produces a more sinusoidal current linkage distribution than a full-pitch winding. In a salient-pole synchronous generator, where the flux density distribution is basically governed by the shape of pole shoes, a short-pitch winding produces a more sinusoidal pole voltage than a full-pitch winding.

Figure 2.16 illustrates the basic difference of a short-pitch winding and a full-pitch winding.

The short-pitch construction is now investigated in more detail. Short pitching is commonly created by winding step shortening (Figure 2.17b), coil side shift in a slot (Figure 2.17c) and coil side transfer to another zone (Figure 2.17d). In Figure 2.17, the zone graphs illustrate the configurations of a full-pitch winding and of the short-pitch windings constructed by the above-mentioned methods. Of these methods, the step shortening can be considered to be created from a full-pitch winding by shifting the upper layer left for a certain number of slot pitches.

Coil side shift in a slot is generated by changing the coil sides of the upper and bottom layers in certain slots of a short-pitch winding. For instance, if in Figure 2.17b the coil sides

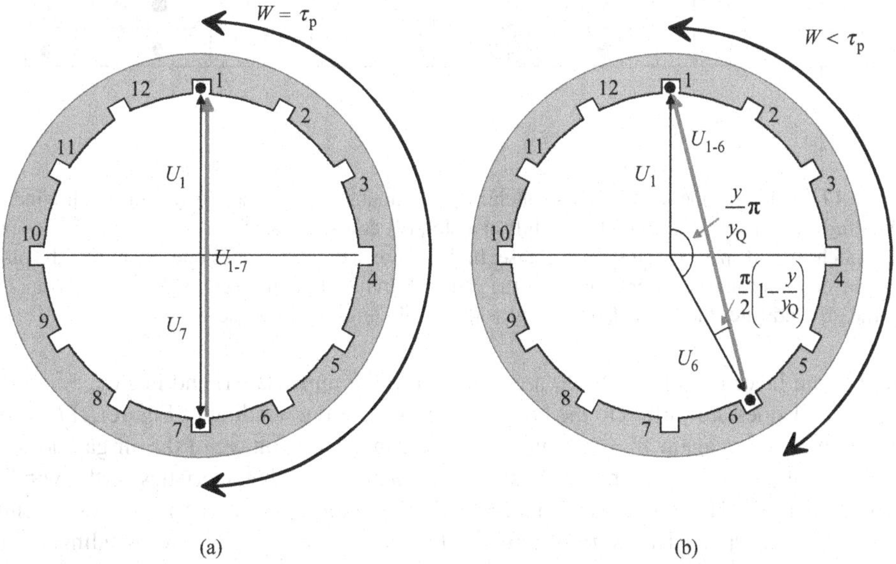

Figure 2.16 Cross-sectional areas of two machines with 12 slots. The basic differences of (a) a two-pole full-pitch winding and (b) a two-pole short-pitch winding. In the short-pitch winding, the width of a single coil W is less than the pole pitch τ_p or, expressed as the number of slots, the short pitch y is less than a full pitch y_Q. The coil voltage $U_{1\text{–}6}$ is lower than $U_{1\text{–}7}$. The short-pitched coil is located on the chord of the periphery, and therefore the winding type is also called a chorded winding. The coil without short pitching is located on the diameter of the machine.

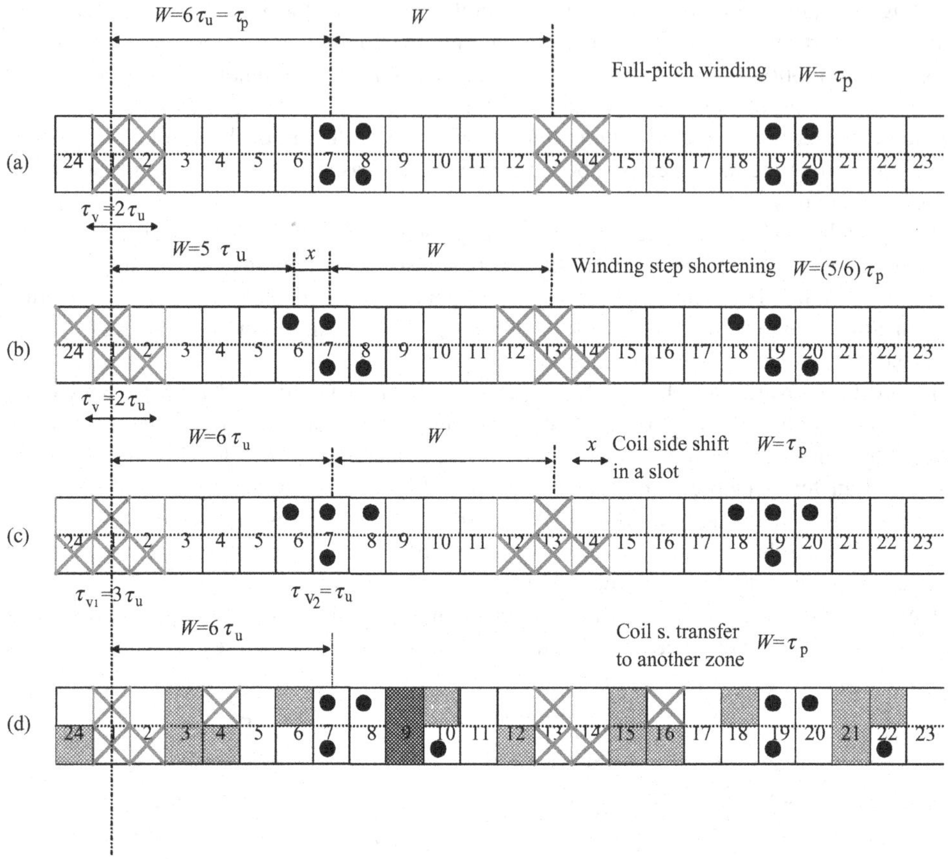

Figure 2.17 Different methods of short pitching for a double-layer winding: (a) full-pitch winding, (b) winding step shortening, (c) coil side shift in a slot, (d) coil side transfer to another zone. τ_v, zone; τ_u, slot pitch; W coil span; x, coil span decrease. In the figure, a cross indicates one coil end of the phase U, and the dot indicates the other coil end of the phase U. In the graph for a coil side transfer to another zone, the grid indicates the parts of slots filled with the windings of the phase W.

of the bottom layer in slots 8 and 20 are removed to the upper layer, and in slots 12 and 24 the upper coil sides are shifted to the bottom layer, we get the winding of Figure 2.17c. Now, the width of a coil is again $W = \tau_p$, but because the magnetic voltage of the air gap does not depend on the position of a single coil side, the magnetizing characteristics of the winding remain unchanged. The windings with a coil side shift in a slot and winding step shortening are equal in this respect. The average number of slots per pole and phase for windings with a coil side shift in a slot is

$$q_s = \frac{Q_s}{2pm} = \frac{q_1 + q_2}{2}; \quad q_1 = q + j, \quad q_2 = q - j \tag{2.28}$$

where j is the difference of q (the numbers of slots per pole and phase) in different layers. In Figure 2.17, $j = 1$.

Coil side transfer to another zone (Figure 2.17d) can be considered to be created from the full-pitch winding of Figure 2.17a by transferring the side of the upper layers of 2 and 14 to a foreign zone W. There is no general rule for the transfer, but practicality and the purpose of use decide the solution to be selected. The method may be adopted in order to cancel a certain harmonic from the current linkage of the winding.

The above-described methods can also be employed simultaneously. For instance, if we shift the upper layer of the coil with a coil side transfer of Figure 2.17d left for the distance of one slot pitch, we receive a combination of a winding step shortening and a coil side transfer. This kind of winding is double short pitched, and it can eliminate two harmonics. This kind of short pitching is often employed in machines where the windings may be rearranged during drive to form another pole number.

When different methods are compared, we have to bear in mind that when considering the connection, a common winding step shortening is always the simplest to realize, and the consumption of copper is often the lowest. A winding step shortening is an advisable method down to $W = 0.8\tau_p$ without an increase in copper consumption. In short two-pole machines, the ends are relatively long, and therefore it is advisable to use even shorter pitches to make the end winding area shorter. Short pitchings down to $W = 0.7\tau_p$ are used in two-pole machines with prefabricated coils. The most crucial issue concerning short pitching is, however, how completely we wish to eliminate the harmonics. This is best investigated with winding factors.

The winding factor $k_{w\nu}$ has already been determined with Figure 2.10 and Equation (2.16). When the winding is short pitched, and the coil ends are not at a distance of 180 electrical degrees from each other, we can easily understand that short pitching reduces the winding factor of the fundamental. This is described by a pitch factor $k_{p\nu}$. Further, if the number of slots per pole and phase is higher than one, we can see that in addition to the pitch factor $k_{p\nu}$, the distribution factor $k_{d\nu}$ is required as was discussed above. Thus, we can consider the winding factor to consist of the pitch factor $k_{p\nu}$ and the distribution factor $k_{d\nu}$ and, in some cases, of a skewing factor (cf. Equation 2.35).

The full pitch can be expressed in radians as π, as a pole pitch τ_p, or as the number of slot pitches y_Q covering the pole pitch. The pitch expressed in the number of slots is y, and now the relative shortening is y/y_Q. Therefore, the angle of the short pitch is $(y/y_Q)\pi$. A complement to π, which is the sum of angles in a triangle, is

$$\left(\pi - \frac{y}{y_Q}\pi\right) = \pi\left(1 - \frac{y}{y_Q}\right)$$

This value will be divided equally between the other two angles as in Figure 2.16b:

$$\frac{\pi}{2}\left(1 - \frac{y}{y_Q}\right)$$

Also the pitch factor is defined as the ratio of the geometric sum of phasors and the sum of the absolute values of the voltage phasors, see Figure 2.16b. The pitch factor is

$$k_p = \frac{U_{total}}{2U_{slot}}. \tag{2.29}$$

When

$$\cos\left(\frac{\pi}{2}\left(1-\frac{y}{y_Q}\right)\right)$$

is expressed in the triangle being analyzed, it will be found that it equals the pitch factor defined by

$$\cos\left(\frac{\pi}{2}\left(1-\frac{y}{y_Q}\right)\right)=\frac{U_{total}/2}{U_{slot}}=\frac{U_{total}}{2U_{slot}}=k_p. \tag{2.30}$$

On rearranging the final expression for the pitch factor will be obtained:

$$k_p=\sin\left(\frac{y}{y_Q}\frac{\pi}{2}\right)=\sin\left(\frac{W}{\tau_p}\frac{\pi}{2}\right). \tag{2.31}$$

In a full-pitch winding, the pitch is equal to the pole pitch, $y=y_Q$, $W=\tau_p$ and the pitch factor is $k_p=1$. If the pitch is less than y_Q, $k_p<1$.

In a general presentation, the distribution factor k_d and the pitch factor k_p have to be valid also for the stator harmonics. We may write for the νth harmonic, the pitch factor $k_{p\nu}$ and the distribution factor $k_{d\nu}$

$$k_{p\nu}=\sin\left(\nu\frac{W}{\tau_p}\frac{\pi}{2}\right)=\sin\left(\nu\frac{y}{y_Q}\frac{\pi}{2}\right), \tag{2.32}$$

$$k_{d\nu}=\frac{\sin\left(\nu q\alpha_u/2\right)}{q\,\sin\left(\nu\alpha_u/2\right)}=\frac{2\sin\left(\nu\dfrac{\pi}{2m}\right)}{\dfrac{Q}{mp}\sin\left(\nu\dfrac{\pi p}{Q}\right)}=\frac{\sin\left(\nu\dfrac{\pi}{2m}\right)}{q\sin\left(\nu\dfrac{\pi}{2mq}\right)}. \tag{2.33}$$

Here α_u is the slot angle, $\alpha_u=p2\pi/Q$.

The skewing factor will be developed in Chapter 4, but it is shown here as

$$k_{sq\nu}=\frac{\sin\left(\nu\dfrac{s}{\tau_p}\dfrac{\pi}{2}\right)}{\nu\dfrac{s}{\tau_p}\dfrac{\pi}{2}}=\frac{\sin\left(\nu\dfrac{\pi}{2}\dfrac{s_{sp}}{mq}\right)}{\nu\dfrac{\pi}{2}\dfrac{s_{sp}}{mq}}. \tag{2.34}$$

Here the skew is measured as s/τ_p (cf. Figure 4.10). s_{sp} is the skewing denoted as the number of slot pitches. The skewing factor is used, for example, in calculation of the induced voltage in a winding, which is skewed as to the direction of the winding creating the magnetic field. We can always choose the stator slots as the reference for which the direction of the rotor slots are considered independent of if the rotor or the stator or both are skewed towards the shaft line of the machine. So, the skewing factor is included in the rotor winding of the squirrel cage motor with skewed rotor slots and it is no longer included in the stator winding factor.

The winding factor is thus the product of pitch, and distribution factors and the rotor winding factor may also include the skewing factor:

$$k_{w\nu} = k_{p\nu}k_{d\nu}\left[k_{sq\nu}\right] = \sin\left(\nu\frac{W}{\tau_p}\frac{\pi}{2}\right)\frac{\sin\left(\nu\dfrac{\pi}{2m}\right)}{q\sin\left(\nu\dfrac{\pi}{2mq}\right)}\left[\frac{\sin\left(\nu\dfrac{s}{\tau_p}\dfrac{\pi}{2}\right)}{\nu\dfrac{s}{\tau_p}\dfrac{\pi}{2}}\right]. \tag{2.35}$$

Example 2.9: Calculate a winding factor for the two-layer winding $Q = 24$, $2p = 4$, $m = 3$, $y = 5$ (see Figure 2.17b).

Solution: The number of slots per phase per pole is

$$q = \frac{Q}{2pm} = \frac{24}{4 \cdot 3} = 2.$$

The distribution factor for the three-phase winding is

$$k_{d1} = \frac{\sin\left(\dfrac{\pi}{2m}\right)}{q\sin\left(\dfrac{\pi}{2mq}\right)} = \frac{\sin\left(\dfrac{\pi}{6}\right)}{2\sin\left(\dfrac{\pi}{12}\right)} = 0.966.$$

The number of slots per pole, or in the other words the pole pitch expressed as the number of slots, is

$$y_Q = \frac{Q}{2p} = \frac{24}{4} = 6.$$

If the pitch is five slots, it means that it is a short pitch, and it is necessary to calculate the pitch factor:

$$k_{p1} = \sin\left(\frac{y}{y_Q}\frac{\pi}{2}\right) = \sin\left(\frac{5}{6}\frac{\pi}{2}\right) = 0.966.$$

The winding factor is

$$k_{w1} = k_{p1}k_{d1} = 0.966 \cdot 0.966 = 0.933.$$

Example 2.10: A two-pole alternator has on the stator a three-phase two-layer winding embedded in 72 slots, two conductors in each slot, with a short pitch of 29/36. The diameter of the stator bore is $D_s = 0.85$ m, the effective length of the stack is $l' = 1.75$ m. Calculate the fundamental component of the induced voltage in one phase of the stator winding, if the amplitude of the fundamental component of the air gap flux density is $\hat{B}_{1\delta} = 0.92$ T and the speed of rotation is 3000 min^{-1}.

Solution: The effective value of the induced phase voltage will be calculated from the expression $E_{1ph} = \sqrt{2}\pi\, f\, \hat{\Phi}_1\, N_s k_{1w}$ at the frequency of $f = pn/60 = 1 \cdot 3000/60 = 50$ Hz. The pole pitch is $\tau_p = \pi\, D_s/2p = \pi \cdot 0.85\ \mathrm{m}/2 = 1.335$ m. The maximum value of the magnetic flux is $\hat{\Phi}_1 = (2/\pi)\hat{B}_{1\delta}\tau_p l' = (2/\pi) \cdot 0.92\ \mathrm{T} \cdot 1.335\ \mathrm{m} \cdot 1.75\ \mathrm{m} = 1.368$ Wb. The number of slots per pole and phase is $q_s = Q_s/2pm = 72/2 \cdot 3 = 12$. The number of turns in series N_s will be determined from the number of conductors in the slot $z_Q = N/pq = 2 \Rightarrow N = 2pq = 2 \cdot 1 \cdot 12 = 24$. There is a two-layer distributed short-pitch winding, and therefore both the distribution and pitch factors must be calculated (cf. Equation 2.26):

$$k_{d1} = \frac{1}{2q_s \sin \dfrac{\pi/6}{q_s}} = \frac{1}{2 \cdot 12 \sin \dfrac{\pi/6}{12}} = 0.955,$$

$$k_{p1} = \sin\left(\frac{y}{y_Q}\frac{\pi}{2}\right) = \sin\left(\frac{29}{36}\frac{\pi}{2}\right) = 0.954,$$

because the full pitch is $y_Q = Q/2p = 72/2 = 36$. The winding factor yields $k_{w1} = k_{d1}k_{p1} = 0.955 \cdot 0.954 = 0.91$ and the induced phase voltage of the stator is $E_{ph} = \sqrt{2}\pi\, f\, \hat{\Phi}_1\, N_s k_w = \sqrt{2}\pi \cdot 50\frac{1}{s} \cdot 1.368\ \mathrm{Vs} \cdot 24 \cdot 0.91 = 6637$ V.

On the other hand, as shown previously, the winding constructed with the coil side shift in a slot in Figure 2.17c proved to have an identical current linkage as the winding step shortening 2.17b, and therefore also its winding factor has to be same. The distribution factor $k_{d\nu}$ is calculated with an average number of slots per pole and phase $q = 2$, and thus the pitch factor $k_{p\nu}$ has to be the same as above, although no actual winding step shortening has been performed. For coil side shift in a slot, Equation (2.32) is not valid as such for the calculation of pitch factor (because $\sin(\nu\pi/2) = 1$).

When comparing magnetically equivalent windings of Figure 2.17 that apply winding step shortening and coil side shift in a slot, it is shown that an equivalent reduction x of the coil span for a winding with coil side shift is

$$x = \tau_p - W = \frac{1}{2}(q_1 - q_2)\,\tau_u. \tag{2.36}$$

The substitution of slot pitch $\tau_u = 2p\tau_p/Q$ in the equation yields

$$\frac{W}{\tau_p} = 1 - \frac{p}{Q}(q_1 - q_2). \tag{2.37}$$

In other words, if the number of slots per pole and phase of the different layers of coil side shift are q_1 and q_2, the winding corresponds to the winding step shortening in the ratio of

W/τ_p. By substituting (2.37) in Equation (2.32) we obtain a pitch factor $k_{pw\nu}$ of the coil side shift in a slot

$$k_{pw\nu} = \sin\left\{\nu\left[1 - \frac{p}{Q}(q_1 - q_2)\right]\frac{\pi}{2}\right\}. \tag{2.38}$$

In the case of the coil side shift of Figure 2.17c, $q_1 = 3$ and $q_2 = 1$. We may assume the winding to be a four-pole construction as a whole ($p = 2$, $Q_s = 24$). In the figure, a basic winding is constructed of the conductors of the first 12 slots (the complete winding may comprise an undefined number of sets of 12-slot windings in series), and thus we obtain for the fundamental winding factor

$$k_{pw1} = \sin\left\{\left[1 - \frac{2}{24}(3 - 1)\right]\frac{\pi}{2}\right\} = \sin\left(\frac{5}{6}\frac{\pi}{2}\right) = 0.966.$$

which is the same result as Equation (2.32) in the case of a winding step shortening of Figure 2.17b.

Because we may often apply both the winding step shortening and coil side transfer in a different zone in the same winding, the winding factor has to be rewritten as

$$k_{w\nu} = k_{d\nu} k_{pw\nu} k_{p\nu}. \tag{2.39}$$

With this kind of doubly short-pitched winding, we may eliminate two harmonics, as stated earlier. The elimination of harmonics implies that we select a double-short-pitched winding for which, for instance, $k_{pw5} = 0$ and $k_{p7} = 0$. Now we can eliminate the undesirable fifth and seventh harmonics. However, the fundamental winding factor will get smaller. The distribution factor $k_{d\nu}$ is now calculated with the average number of slots per pole and phase $q = (q_1 + q_2)/2$.

When analyzing complicated short-pitch arrangements, it is often difficult to find a universal equation for the winding factor. In that case, a voltage phasor diagram can be employed, as mentioned earlier in the discussion of Figure 2.10. Next, the coil side transfer to another zone of Figure 2.17 is discussed.

Figure 2.18a depicts the fundamental voltage phasor diagram of the winding with coil side transfer to another zone in Figure 2.17, assuming that there are $Q = 24$ slots and $2p = 4$ poles. The slot angle is now $\alpha_u = 30$ electrical degrees, and thus the phase shift of the emf is 30°. Figure 2.18b depicts the phasors of the phase U in a polygon according to Figure 2.17d. The resultant voltage U_{phU} and its ratio to the sum of the absolute values of the phasors is the fundamental winding factor. By also drawing the resultants of the windings V and W from point 0, we obtain an illustration of the symmetry of a three-phase machine. The other harmonics (ordinal ν) are treated equally, but the phase shift angle of the phasors is now $\nu\alpha_u$.

2.7 Current Linkage of a Slot Winding

The current linkage of a slot winding refers to a function $\Theta = f(\alpha)$, created by the winding and its currents in the equivalent air gap of the machine. The winding of Figure 2.8 and its

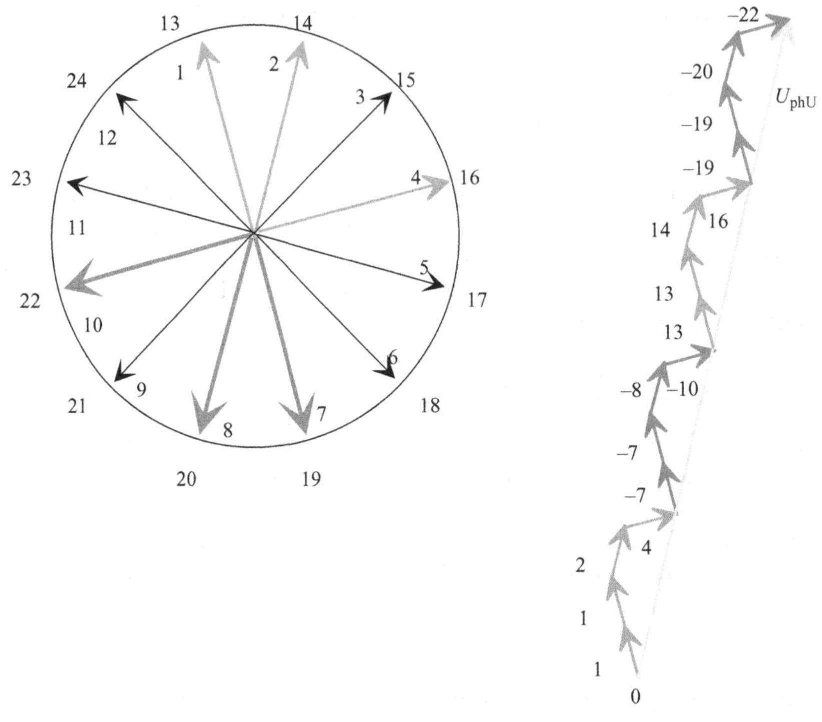

Figure 2.18 Voltage phasor diagram of a four-pole winding with a coil side transfer (Figures 2.17d) and (b) the sum of slot voltage phasors of a single phase. Note that the voltage phasor diagram for a four-pole machine ($p = 2$) is doubled, because it is drawn in electrical degrees. The parts of the figure are in different scales. In the voltage phasor diagram, there are in principle two layers (one for the bottom layer and another for the top layer), but only one of them is illustrated here. Now two consequent phasors, for example 2 and 14, when placed one after another, form a single radius of the voltage phasor diagram. The winding factor is thus defined by the voltage phasor diagram by comparing the geometrical sum with the sum of absolute values.

current linkage in the air gap are investigated at a time when $i_W = i_V = -{}^1/_2 i_U$, Figure 2.9. The curved function in the figure is drawn at time $t = 0$. The phasors of the phase currents rotate at an angular speed ω, and thus after a time $2\pi/(3\omega) = 1/(3f)$ the current of the phase V has reached its positive peak, and the function of current linkage has shifted three slot pitches to the right. After a time $2/(3f)$, the shift is six slot pitches, and so on. The curved function shifts constantly in the direction $+\alpha$. To be exact, the curved waveform is of the form presented in the figure only at times $t = c/(3f)$, when the factor c has values $c = 0, 1, 2, 3, \ldots$. As time elapses, the waveform proceeds constantly. Fourier analysis of the waveform, however, produces harmonics that remain constant.

Figure 2.19 illustrates the current linkage $\Theta(\alpha)$ produced by a single coil. A flux that passes through the air gap at an angle γ returns at an angle $\beta = 2\pi - \gamma$. In the case of a nonfull-pitch

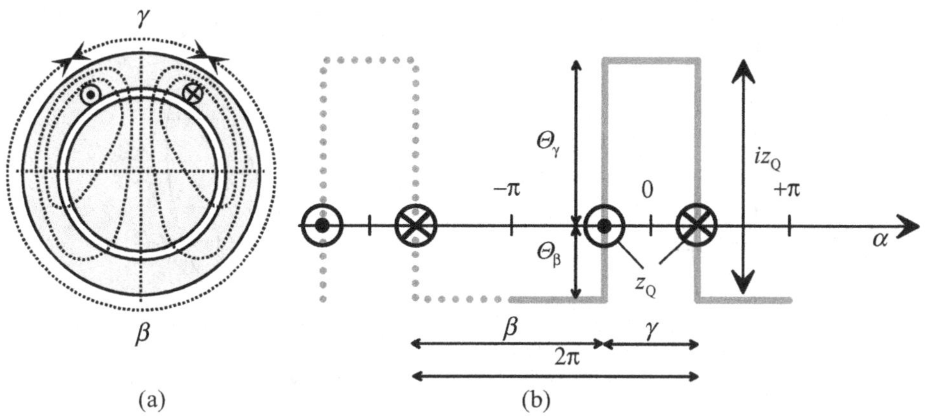

Figure 2.19 (a) Currents and schematic flux lines of a short-pitch coil in a two-pole system. (b) Two wavelengths of the current linkage created by a single, narrow coil.

winding, the current linkage is distributed in the ratio of the permeances of the paths. This gives us a pair of equations

$$z_{Q}i = \Theta_{\gamma} + \Theta_{\beta}; \quad \frac{\Theta_{\gamma}}{\Theta_{\beta}} = \frac{\beta}{\gamma} \tag{2.40}$$

from which we obtain two constant values for the current linkage waveform $\Theta(\alpha)$

$$\Theta_{\gamma} = \frac{\beta}{\pi} Z_{Q}i; \quad \Theta_{\beta} = \frac{\gamma}{\pi} Z_{Q}i. \tag{2.41}$$

The Fourier series of the function $\Theta(\alpha)$ of a single coil is

$$\Theta(\alpha) = \hat{\Theta}_1 \cos\alpha + \hat{\Theta}_3 \cos 3\alpha + \hat{\Theta}_5 \cos 5\alpha + \cdots + \hat{\Theta}_{\nu} \cos \nu\alpha + \cdots. \tag{2.42}$$

The magnitude of the νth term of the series is obtained from the equation by substituting the function of the current linkage waveform for a single coil

$$\Theta_{\nu} = \frac{2}{\pi} \int_{0}^{\pi} \Theta(\alpha) \cos(\nu\alpha)\, d\alpha = \frac{2}{\nu\pi} z_{Q}i \sin\left(\frac{\nu\gamma}{2}\right). \tag{2.43}$$

As $\gamma/W = 2\pi/2\tau_{p}$, and hence $\gamma/2 = (W/\tau_{p}) \cdot (\pi/2)$, the last factor of Equation (2.43)

$$\sin\left(\frac{\nu\gamma}{2}\right) = \sin\left(\nu \frac{W}{\tau_{p}} \frac{\pi}{2}\right) = k_{p\nu} \tag{2.44}$$

is the pitch factor of the harmonic ν for the coil observed. For the fundamental $\nu = 1$, we obtain k_{p1}. We can thus see that the fundamental is just a special case of the general harmonic ν.

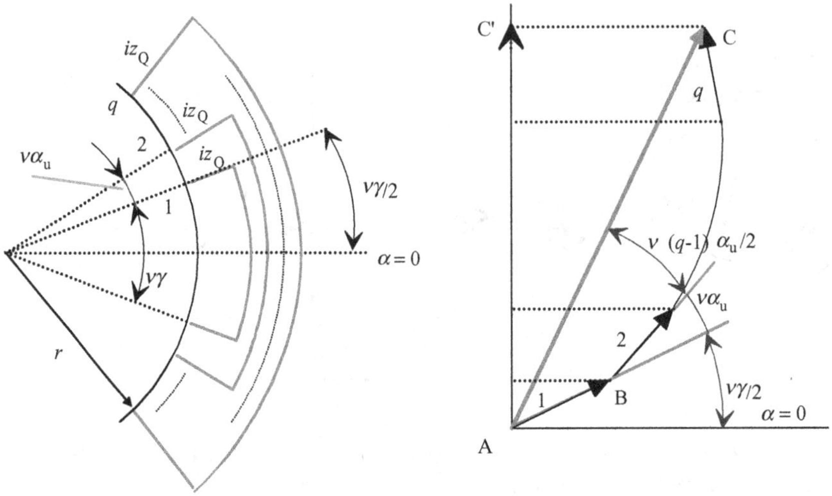

Figure 2.20 Concentric coils of a single pole; the calculation of the winding factor for a harmonic ν.

While the electrical angle of the fundamental is α, the corresponding angle for the harmonic ν is always $\nu\alpha$. If now $\nu\gamma/2$ is a multiple of the angle 2π, the pitch factor becomes $k_{p\nu} = 0$. Thus, the winding does not produce such harmonics, neither are voltages induced in the winding by the influence of possible flux components at this distribution. However, voltages are induced in the coil sides, but over the whole coil these voltages compensate each other. Thus, with a suitable short pitching, it is possible to eliminate harmful harmonics.

In Figure 2.20, there are several coils $1 \ldots q$ in a single pole of a slot winding. The current linkage of each coil is $z_Q i$. The coil angle (in electrical degrees) for the harmonic ν of the narrowest coil is $\nu\gamma$, the next being $\nu(\gamma + 2\alpha_u)$ and the broadest $\nu(\gamma + 2(q-1)\alpha_u)$. For an arbitrary coil g, the current linkage is, according to Equation (2.43),

$$\Theta_{\nu g} = \frac{2}{\nu\pi} z_Q i \sin\left[\nu\left(\frac{\gamma}{2} + (q-1)\,\alpha_u\right)\right]. \tag{2.45}$$

When all the harmonics of the same ordinal generated by all coils of one phase are summed, we obtain per pole

$$\begin{aligned}\Theta_{\nu\text{tot}} &= \sum_{g=1}^{q} \Theta_{\nu g} = \frac{2}{\nu\pi} q z_Q i \sum_{g=1}^{q} k_{w\nu} \\ &= \frac{2}{\nu\pi} z_Q i \left\{\sin\frac{\nu\gamma_1}{2} + \sin\nu\left(\frac{\gamma_1}{2} + \alpha_u\right) + \cdots + \sin\nu\left[\frac{\gamma_1}{2} + (q-1)\,\alpha_u\right]\right\}.\end{aligned} \tag{2.46}$$

The sum $k_{w\nu}$ in braces can be calculated for instance with the geometrical figure of Figures 2.20 and 2.12. The line segment $\overline{AC}$ is written as

$$\overline{AC} = 2r \sin\frac{\nu q \alpha_u}{2}. \tag{2.47}$$

The arithmetical sum of unit segments is

$$q\overline{AB} = q2r \sin \frac{\nu \alpha_{\mathrm{u}}}{2} \tag{2.48}$$

and we can find for the harmonic ν

$$k_{\mathrm{d}\nu} = \frac{\overline{AC}}{q\overline{AB}} = \frac{\sin \dfrac{\nu q \alpha_{\mathrm{u}}}{2}}{q \sin \dfrac{\nu \alpha_{\mathrm{u}}}{2}}. \tag{2.49}$$

This equals the distribution factor of Equation (2.33). Now, we see that the line segment $\overline{AC} = qk_{\mathrm{d}\nu\overline{AB}}$. We use Figure 2.12a again. The angle BAC is obtained from Figure 2.12 as the difference of angles OAB and OAC. It is $\nu(q-1)\alpha_{\mathrm{u}}/2$. The projection $\overline{AC'}$ is thus

$$\overline{AC}' = \overline{AC} \sin \frac{\nu\left(\gamma + (q-1)\,\alpha_{\mathrm{u}}\right)}{2} = \overline{AC} k_{\mathrm{d}\nu}. \tag{2.50}$$

Here we have a pitch factor influencing the harmonic ν, because in Figure 2.20, $\nu[\gamma + (q-1)\alpha_{\mathrm{u}}]$ is the average coil width angle. Equation (2.46) is now reduced to

$$\Theta_{\nu \mathrm{tot}} = \frac{2}{\pi} \frac{k_{\mathrm{w}\nu}}{\nu} q z_{\mathrm{Q}} i \tag{2.51}$$

where

$$k_{\mathrm{w}\nu} = k_{\mathrm{p}\nu} k_{\mathrm{d}\nu}. \tag{2.52}$$

This is the winding factor of the harmonic ν. It is an important observation. The winding factor was originally found for the calculation of the induced voltages. Now we understand that the same distribution and pitch factors also affect the current linkage harmonic production. By substituting the harmonic current linkage Θ_ν in Equation (2.42) with the current linkage $\Theta_{\nu \mathrm{tot}}$, we obtain the harmonic ν generated by the current linkage of q coils

$$\Theta_\nu = \Theta_{\nu \mathrm{tot}} \cos \nu a. \tag{2.53}$$

This is valid for a single-phase coil. The harmonic ν created by a poly-phase winding is calculated by summing all the harmonics created by different phases. By its nature, the distribution factor is zero if $\nu\alpha_{\mathrm{u}} = \pm c2\pi$ (since $\sin(\nu q \alpha_{\mathrm{u}}/2) = \sin(\pm c\pi q) = 0$) (i.e. the coil sides are at the same magnetic potential), when the factor $c = 0, 1, 2, 3, 4, \ldots$. It therefore allows only the harmonics

$$\nu \neq c \frac{2\pi}{\alpha_{\mathrm{u}}}. \tag{2.54}$$

The slot angle and the phase number are interrelated

$$q\alpha_u = \frac{\pi}{m}. \tag{2.55}$$

The distribution factor is zero if $\nu = \pm c2m$. The winding thus produces harmonics

$$\nu = +1 \pm c2m. \tag{2.56}$$

Example 2.11: Calculate which ordinals of the harmonics can be created by a three-phase winding.

Solution: A symmetrical three-phase winding may create harmonics calculated from Equation (2.56), by inserting $m = 3$. These are listed in Table 2.2.

Table 2.2 Ordinals of the harmonics created by a three-phase winding ($m = 3$)

c	0	1	2	3	4	5	6	7...	
ν	+1	+7	+13	+19	+25	+31	+37	+43...	Positive sequence
	—	-5	−11	−17	−23	−29	−35	−41...	Negative sequence

We see that $\nu = -1$, and all even harmonics and harmonics divisible by three are missing. In other words, a symmetrical poly-phase winding does not produce for instance a harmonic propagating in the opposite direction at the fundamental frequency. Instead, a single-phase winding $m = 1$ creates also a harmonic, the ordinal of which is $\nu = -1$. This is a particularly harmful harmonic, and it impedes the operation of single-phase machines. For instance, a single-phase induction motor, because of the field rotating in the negative direction, does not start without assistance because the positive and negative sequence fields are equally strong.

Example 2.12: Calculate the pitch and distribution factors for $\nu = 1, 5, 7$ if a chorded stator of an AC machine has 18 slots per pole and the first coil is embedded in slots 1 and 16. Calculate also the relative harmonic current linkages.

Solution: The full pitch would be $y_Q = 18$ and a full-pitch coil should be embedded in slots 1 and 19. If the coil is located in slots 1 and 16, the coil pitch is shorted to $y = 15$. Therefore, the pitch factor for the fundamental will be

$$k_{p1} = \sin\frac{W}{\tau_p}\frac{\pi}{2} = \sin\frac{y}{y_Q}\frac{\pi}{2} = \sin\frac{15}{18}\frac{\pi}{2} = 0.966$$

and correspondingly for the fifth and seventh harmonics

$$k_{p5} = \sin\left(\nu\frac{y}{y_Q}\frac{\pi}{2}\right) = \sin\left(-5\frac{15}{18}\frac{\pi}{2}\right) = -0.259$$

$$k_{p7} = \sin\left(\nu\frac{y}{y_Q}\frac{\pi}{2}\right) = \sin\left(7\frac{15}{18}\frac{\pi}{2}\right) = 0.259.$$

Notice that the ordinal of the harmonic is substituted in the winding factor equation together with its sign. The same applies to the current linkage equation (2.15) (Jokinen 1972).

The number of slots per pole and phase is $q = 18/3 = 6$ and the slot angle is $\alpha_u = \pi/18$. Now, the distribution factor is

$$k_{d\nu} = \frac{\sin\dfrac{\nu q \alpha_u}{2}}{q\sin\dfrac{\nu\alpha_u}{2}},\quad k_{d1} = \frac{\sin\dfrac{1\cdot 6\pi/18}{2}}{6\sin\dfrac{1\pi/18}{2}} = 0.956,\quad k_{d-5} = \frac{\sin\dfrac{-5\cdot 6\pi/18}{2}}{6\sin\dfrac{-5\pi/18}{2}} = 0.197,$$

$$k_{d7} = \frac{\sin\dfrac{7\cdot 6\pi/18}{2}}{6\sin\dfrac{7\pi/18}{2}} = -0.145.$$

$k_{w1} = k_{d1}k_{p1} = 0.956\cdot 0.966 = 0.923$, $k_{w-5} = k_{d-5}\,k_{p-5} = 0.197\cdot(-0.259) = -0.051$, $k_{w7} = k_{d7}\,k_{p7} = -0.145\cdot 0.259 = -0.038$.

The winding (one phase of the winding) creates current linkages (2.51)

$$\Theta_{\nu\mathrm{tot}} = \frac{2}{\pi}\frac{k_{w\nu}}{\nu} q z_Q i.$$

Calculating $k_{w\nu}/\nu$ for the harmonics 1, –5, 7

$$\frac{k_{w1}}{1} = 0.923,\ \frac{k_{w-5}}{-5} = \frac{-0.051}{-5} = 0.01,\ \frac{k_{w7}}{7} = \frac{-0.038}{7} = -0.0054.$$

Here we can see that because of the chorded winding, the current linkages of the fifth and seventh harmonics will be reduced to 1.1 and 0.54 % of the fundamental, as the fundamental is also reduced to 92.3 % of the full sum of the absolute values of slot voltages.

Example 2.13: A rotating magnetic flux created by a three-phase 50 Hz, 600 min^{-1} alternator has a spatial distribution of magnetic flux density given by the expression

$$B = \hat{B}_1\cos\vartheta + \hat{B}_3\cos 3\vartheta + \hat{B}_5\cos 5\vartheta = 0.9\cos\vartheta + 0.25\cos 3\vartheta + 0.18\cos 5\vartheta\ [\mathrm{T}].$$

The alternator has 180 slots, the winding is wound with two layers, and each coil has three turns with a span of 15 slots. The armature diameter is 135 cm and the effective length of the iron core 0.50 m. Write an expression for the instantaneous value of the induced voltage in one phase of the winding. Calculate the effective value of phase voltage and also the line-to-line voltage of the machine.

Solution: The number of pole pairs is given by the speed and the frequency:

$$f = \frac{pn}{60} \Rightarrow p = \frac{60f}{n} = \frac{60\cdot 50}{600} = 5$$

and the number of poles is 10. The area of one pole is

$$\tau_p l' = \frac{\pi D}{2p} l' = \frac{\pi 1.35}{10} 0.50 = 0.212 \text{ m}^2.$$

From the expression for the instantaneous value of the magnetic flux density, we may derive $\hat{B}_1 = 0.9$ T, $\hat{B}_3 = 0.25$ T and $\hat{B}_5 = 0.18$ T. The fundamental of the magnetic flux on the τ_p is $\hat{\Phi}_1 = (2/\pi)\hat{B}_1\tau_p l' = (2/\pi) \cdot 0.9 \text{ T} \cdot 0.212 \text{ m}^2 = 0.1214$ Vs. To be able to calculate the induced voltage, it is necessary to make a preliminary calculation of some parameters:

The number of slots per pole and phase is $q = Q/2pm = 180/10 \cdot 3 = 6$.

The angle between the voltages of adjacent slots is $\alpha_u = p\,2\pi/Q = \pi/18$.

The distribution and pitch factors for each harmonic is

$$k_{d1} = \frac{\sin\left(q\frac{\alpha_u}{2}\right)}{q\sin\left(\frac{\alpha_u}{2}\right)} = \frac{\sin\left(6 \cdot \frac{\pi}{36}\right)}{6\sin\left(\frac{\pi}{36}\right)} = 0.9561, \; k_{d3} = \frac{\sin\left(3 \cdot 6 \cdot \frac{\pi}{36}\right)}{6\sin\left(3 \cdot \frac{\pi}{36}\right)} = 0.644,$$

$$k_{d5} = \frac{\sin\left(-5 \cdot 6 \cdot \frac{\pi}{36}\right)}{6\sin\left(-5 \cdot \frac{\pi}{36}\right)} = 0.\,197.$$

The number of slots per pole is $Q_p = 180/10 = 18$, which would be a full pitch. The coil span is 15 slots, which means the chorded pitch $y = 15$, and the pitch factors are

$$k_{p1} = \sin\left(\frac{y}{y_Q}\frac{\pi}{2}\right) = \sin\left(\frac{15}{18}\frac{\pi}{2}\right) = 0.9659,$$

$$k_{p3} = \sin\left(3\frac{y}{y_Q}\frac{\pi}{2}\right) = \sin\left(3 \cdot \frac{15}{18}\frac{\pi}{2}\right) = -0.707,$$

$$k_{p5} = \sin\left(-5\frac{y}{y_Q}\frac{\pi}{2}\right) = \sin\left(-5 \cdot \frac{15}{18}\frac{\pi}{2}\right) = -0.259$$

which results in the following winding factors:

$$k_{w1} = k_{d1} \cdot k_{p1} = 0.9561 \cdot 0.9659 = 0.9236, \; k_{w3} = 0.644 \cdot (-0.707) = -0.455,$$

$$k_{w5} = 0.197 \cdot (-0.259) = -0.0510.$$

Now it is possible to calculate the effective values of the induced voltages of the harmonics. The phase number of turns is determined as follows: the total number of coils in a 180-slot machine in a two-layer winding is 180. This means that the number of coils per phase is 180/3 = 60, each coil has three turns, and therefore $N = 60 \times 3 = 180$:

$$E_1 = \sqrt{2}\pi f \hat{\Phi}_1 N k_{w1} = \sqrt{2}\pi \cdot 50\,\frac{1}{\text{s}} \cdot 0.1214 \text{ Vs} \cdot 180 \cdot 0.9234 = 4482 \text{ V}.$$

The induced voltage of harmonics can be written similarly as follows:

$$E_\nu = \sqrt{2}\pi\, f_\nu \hat{\Phi}_\nu N\, k_{w\nu} = \sqrt{2}\pi f_\nu \frac{2}{\pi}\hat{B}_\nu \tau_{p\nu} N\, k_{w\nu} = \sqrt{2}\pi\, |\nu|\, f_1 \frac{2}{\pi}\hat{B}_\nu \frac{\tau_{p1}}{\nu} N k_{w\nu}$$
$$= \sqrt{2}\pi\, f_1 \frac{2}{\pi}\hat{B}_\nu \frac{\tau_{p1}}{\text{sign}(\nu)} N\, k_{w\nu}.$$

Notice that by reducing ν, its sign remains in the induced voltage equation. The ratio of the νth harmonic and fundamental is

$$\frac{E_\nu}{E_1} = \frac{\sqrt{2}\pi\, f_1 \frac{2}{\pi}\hat{B}_\nu \frac{\tau_{p1}}{\text{sign}(\nu)} N k_{w\nu}}{\sqrt{2}\pi\, f_1 \frac{2}{\pi}\hat{B}_1 \tau_{p1} N k_{w1}} = \frac{\hat{B}_\nu \frac{k_{w\nu}}{\text{sign}(\nu)}}{\hat{B}_1 k_{w1}}$$

and the results for E_ν:

$$E_3 = \frac{\hat{B}_3 k_{w3}}{\hat{B}_1 k_{w1}} E_1 = \frac{0.25 \cdot (-0.455)}{0.9 \cdot 0.9236} 4482\ \text{V} = -613\ \text{V},$$
$$E_5 = \frac{\hat{B}_5 k_{w5}}{\hat{B}_1 k_{w1}} E_1 = \frac{0.18 \cdot (-0.0510)/(-1)}{0.9 \cdot 0.9236} 4482\ \text{V} = 49.5\ \text{V}.$$

Finally, the expression for the instantaneous value of the induced voltage is

$$e(t) = \hat{E}_1 \sin \omega t + \hat{E}_3 \sin 3\omega t + \hat{E}_5 \sin 5\omega t$$
$$= \sqrt{2} E_1 \sin \omega t + \sqrt{2} E_3 \sin 3\omega t + \sqrt{2} E_5 \sin 5\omega t$$
$$e\,(t) = \sqrt{2} \cdot 4482 \sin \omega t - \sqrt{2} \cdot 613 \sin 3\omega t + \sqrt{2} \cdot 49.5 \sin 5\omega t$$
$$= 6339 \sin \omega t - 867 \sin 3\omega t + 70 \sin 5\omega t$$

The total value of the effective phase voltage is

$$E_{ph} = \sqrt{E_1^2 + E_3^2 + E_5^2} = \sqrt{4482^2 + 613^2 + 49.5^2}\ \text{V} = 4524\ \text{V}$$

and its line-to-line voltage is

$$E = \sqrt{3}\sqrt{E_1^2 + E_5^2} = \sqrt{3}\sqrt{4482^2 + 49.5^2}\ \text{V} = 7764\ \text{V}.$$

The third harmonic component does not appear in the line-to-line voltage, which will be demonstrated later on.

Example 2.14: Calculate the winding factors and per unit magnitudes of the current linkage for $\nu = 1, 3, -5$ if $Q = 24$, $m = 3$, $q = 2$, $W/\tau_p = y/y_Q = 5/6$.

Solution: The winding factor is used to derive the per unit magnitude of the current linkage. In Figure 2.21, we have a current linkage distribution of the phase U of a short-pitch winding

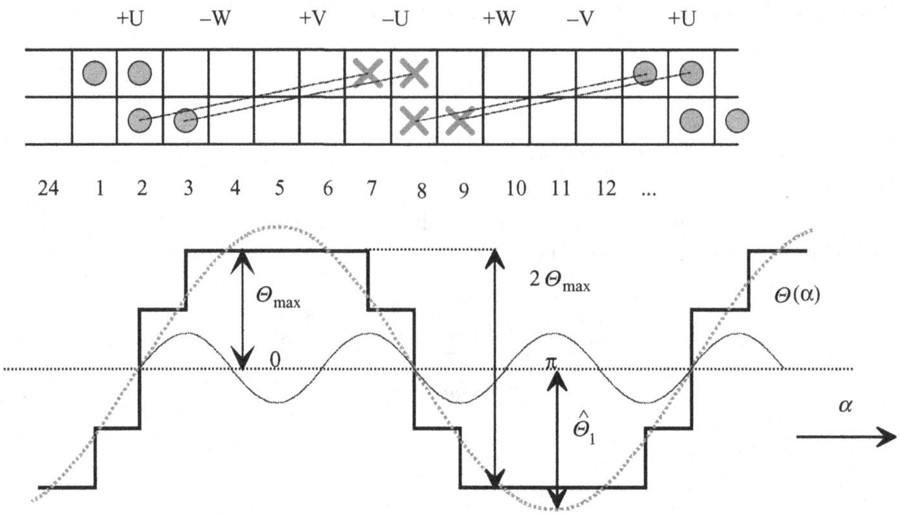

Figure 2.21 Short-pitch winding ($Q_s = 24$, $p = 2$, $m = 3$, $q_s = 2$) and the analysis of its current linkage distribution of the phase U. The distribution includes a notable amount of the third harmonic. In the figure, the fundamental and third harmonic are illustrated by dotted lines.

($Q = 24, m = 3, q = 2, 2p = 4, W/\tau_p = y/y_Q = 5/6$), as well as its fundamental and the third harmonic at time $t = 0$, when $i_U = \hat{i}$. The total maximum height of the current linkage of a pole pair is at that moment $qz_Q\hat{i}$. Half of the magnetic circuit (involving a single air gap) is influenced by half of this current linkage. The winding factors for the fundamental and lowest harmonics and the amplitudes of the current linkages according to Equations (2.51) and (2.52) and Example 2.13 are:

$\nu = 1$	$k_{w1} = k_{p1}k_{d1} = 0.966 \cdot 0.966 = 0.933$	$\hat{\Theta}_1 = 1.188\Theta_{max}$
$\nu = 3$	$k_{w3} = k_{p3}k_{d3} = -0.707 \cdot 0.707 = -0.5$	$\hat{\Theta}_3 = -0.212\Theta_{max}$
$\nu = -5$	$k_{w\text{-}5} = k_{p\text{-}5}k_{d\text{-}5} = -0.259 \cdot 0.259 = -0.067$	$\hat{\Theta}_{-5} = 0.017\Theta_{max}$

The minus sign for the third harmonic amplitude means that, if starting at the same phase, the third harmonics will have a negative peak value as the fundamental is at its positive peak, see Figures 2.21 and 2.22. The fifth harmonic and the fundamental have their positive maxima at the same time.

The amplitudes of the harmonics are calculated with (2.51); see also (2.15). For instance, for the fundamental we obtain:

$$\hat{\Theta}_1 = \frac{2}{\pi}\frac{k_{w\nu}}{\nu}qz_Q i = \frac{2}{\pi}\frac{0.931}{1}2z_Q i = 1.188z_Q i = 1.188\Theta_{max}.$$

Only the fundamental and the third harmonic are illustrated in Figure 2.21.

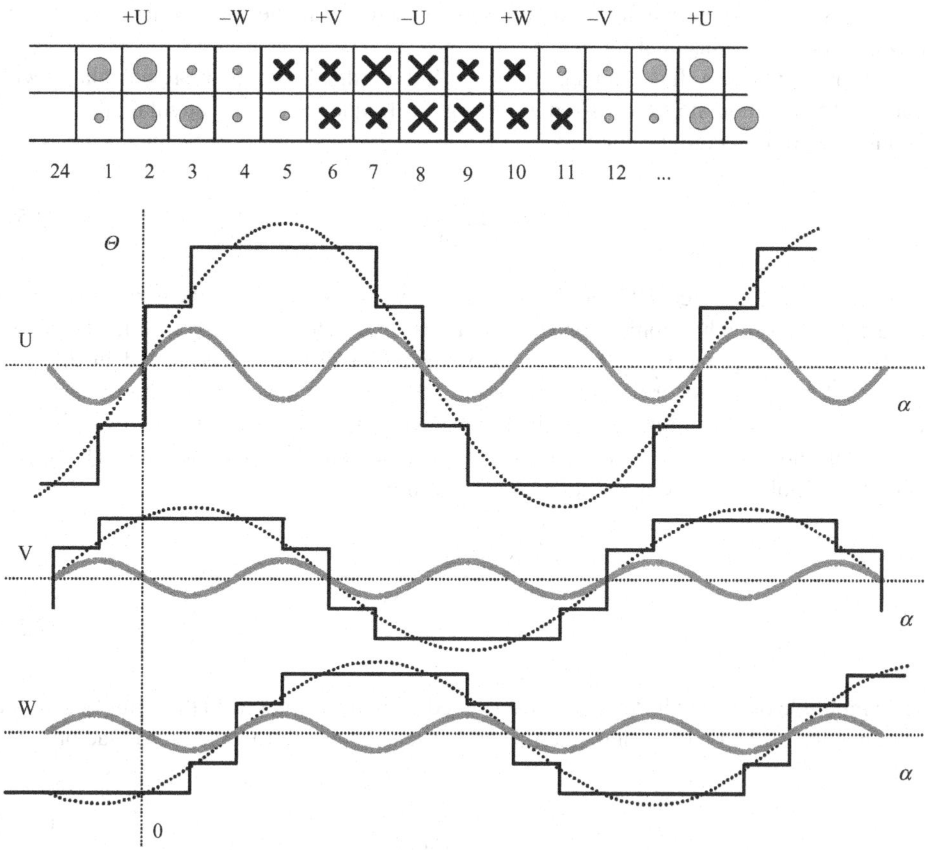

Figure 2.22 Compensation of the third harmonic in a three-phase winding. There are currents $i_U = -2i_V = -2i_W$ flowing in the winding. We see that when we sum the third harmonics of the phases V and W with the harmonic of the phase U, the harmonics compensate each other.

The amplitude of harmonics is often expressed as a percentage of the fundamental. In this case, the amplitude of the third harmonic is 17.8 % (0.212/1.185) of the amplitude of the fundamental. However, this is not necessarily harmful in a three-phase machine, because in the harmonic current linkage created by the windings together, the third harmonic is compensated. The situation is illustrated in Figure 2.22, where the currents $i_U = 1$ and $i_V = i_W = -1/2$ flow in the winding of Figure 2.21. In salient-pole machines, the third harmonic may, however, cause circulating currents in delta connection, and therefore star connection in the armature is preferred.

In single- and two-phase machines, the number of slots is preferably selected higher than in three-phase machines, because in these coils, at certain instants, only a single-phase coil alone creates the whole current linkage of the winding. In such a case, the winding alone should produce as sinusoidal a current linkage as possible. In single- and double-phase windings, it is sometimes necessary to fit a different number of conductors in the slots to make the stepped line $\Theta\ (\alpha)$ approach sinusoidal form.

A poly-phase winding thus produces harmonics, the ordinals of which are calculated with Equation (2.56). When the stator is fed at an angular frequency ω_s, the angular speed of the harmonic ν with respect to the stator is

$$\omega_{\nu s} = \frac{\omega_s}{\nu}. \tag{2.57}$$

The situation is illustrated in Figure 2.23, which shows that the shape of the harmonic current linkage changes as the harmonic propagates in the air gap. The deformation of the harmonic indicates the fact that harmonic amplitudes propagate at different speeds and in different directions. A harmonic according to Equation (2.57) induces the voltage of the fundamental frequency in the stator winding. The ordinal of the harmonic indicates how many wavelengths of a harmonic are fitted in a distance $2\tau_p$ of a single pole pair of the fundamental. This yields the number of pole pairs and the pole pitch of a harmonic

$$p_\nu = \nu p, \tag{2.58}$$

$$\tau_{p\nu} = \frac{\tau_p}{\nu}. \tag{2.59}$$

The amplitude of the νth harmonic is determined with the ordinal from the amplitude of the current linkage of the fundamental, and it is calculated in relation to the winding factors

$$\hat{\Theta}_\nu = \hat{\Theta}_1 \frac{k_{w\nu}}{\nu k_{w1}}. \tag{2.60}$$

The winding factor of the harmonic ν can be determined with Equations (2.32) and (2.33) by multiplying the pitch factor $k_{p\nu}$ and the distribution factor $k_{d\nu}$:

$$k_{w\nu} = k_{p\nu} k_{d\nu} = \frac{\sin\left(\nu \frac{W}{\tau_p} \frac{\pi}{2}\right) \sin\left(\nu \frac{\pi}{2m}\right)}{q \sin\left(\nu \frac{\pi}{2mq}\right)}. \tag{2.61}$$

Compared with the angular velocity ω_{1s} of the fundamental component, a harmonic current linkage wave propagates in the air gap at a fractional angular velocity ω_{1s}/ν. The synchronous speed of the harmonic ν is also at the same very angular speed ω_{1s}/ν. If a motor is running at about synchronous speed, the rotor is traveling much faster than the harmonic wave. If we have an asynchronously running motor with a slip $s = (\omega_s - p\Omega_r)/\omega_s$ (ω_s is the stator angular frequency and Ω_r is the rotor mechanical angular rotating frequency), the slip of the rotor with respect to the νth stator harmonic is given by

$$s_\nu = 1 - \nu(1 - s). \tag{2.62}$$

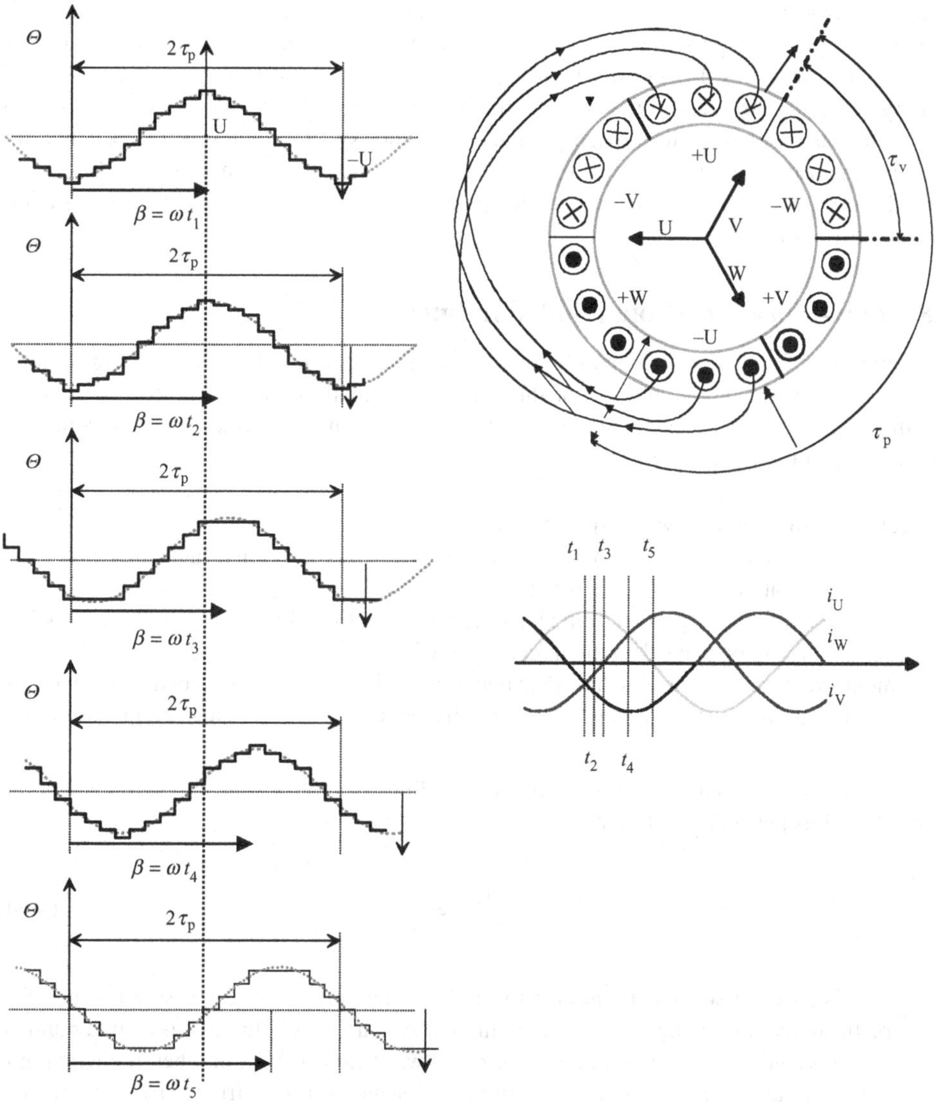

Figure 2.23 Propagation of a harmonic current linkage and the deformations caused by harmonics. If there is a current flowing only in the stator winding, we are able to set the peak of the air-gap flux density at β. The flux propagates but the magnetic axis of the winding U remains stable.

The angular frequency of the νth harmonic in the rotor is thus

$$\omega_{\nu r} = \omega_s(1 - \nu(1 - s)). \tag{2.63}$$

If we have a synchronous machine running with slip $s = 0$, we immediately observe from Equations (2.62) and (2.63) that the angular frequency created by the fundamental component of the flux density of the stator winding is zero in the rotor coordinate. However, harmonic current linkage waves pass the rotor at different speeds. If the shape of the pole shoe is such that the rotor produces flux density harmonics, they propagate at the speed of the rotor and thereby generate pulsating torques with the stator harmonics traveling at different speeds. This is a particular problem in low-speed permanent magnet synchronous motors, in which the rotor magnetization often produces a quadratic flux density and the number of slots in the stator is small, for instance $q = 1$ or even lower, the amplitudes of the stator harmonics being thus notably high.

2.8 Poly-Phase Fractional Slot Windings

If the number of slots per pole and phase q of a winding is a fraction, the winding is called a fractional slot winding. Windings of this type are normally either concentric or diamond windings with one or two layers. Some advantages of fractional slot windings when compared with integer slot windings are:

- great freedom of choice with respect to the number of slots;
- opportunity to reach a suitable magnetic flux density with the given dimensions;
- multiple alternatives for short pitching;
- if the number of slots is predetermined, the fractional slot winding can be applied to a wider range of numbers of poles than the integral slot winding;
- segment structures of large machines are better controlled by using fractional slot windings;
- opportunity to improve the voltage waveform of a generator by removing certain harmonics.

The greatest disadvantage of fractional slot windings is subharmonics, when the denominator of q (slots per pole and phase) is $n \neq 2$

$$q = \frac{Q}{2pm} = \frac{z}{n}. \tag{2.64}$$

Now, q is reduced so that the numerator and the denominator are the smallest possible integers, the numerator being z and the denominator n. If the denominator n is an odd number, the winding is said to be a first-grade winding, and when n is an even number, the winding is of the second grade. The most reliable fractional slot winding is constructed by selecting $n =$ 2. An especially interesting winding of this type can be designed for fractional slot permanent magnet machines by selecting $q = 1/2$. In this case the fractional slot winding becomes a non-overlapping winding with concentrated coils. The winding can be called simply a tooth-coil winding.

In integral slot windings, the base winding is of the length of two pole pitches (the distance of the fundamental wavelength), whereas in the case of fractional slot windings, a distance of several fundamental wavelengths has to be traveled before the phasor of a voltage phasor diagram again meets the exact same point of the waveform. The difference between an integral slot and fractional slot winding is illustrated in Figure 2.24.

In a fractional slot winding, we have to proceed a distance of p' pole pairs before a coil side of the same phase again meets exactly the peak value of the flux density. Then, we need

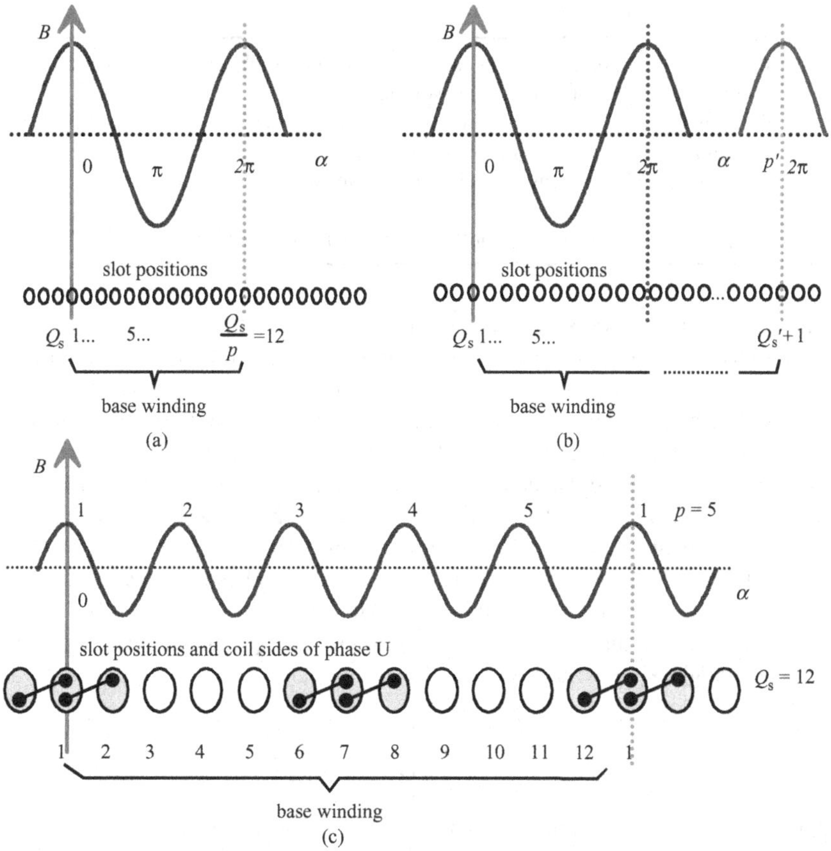

Figure 2.24 Basic differences of (a) an integral slot stator winding and (b) a fractional slot winding. The number of stator slots is Q_s. In an integral slot winding, the length of the base winding is Q_s/p slots ((a): 12 slots, $q_s = 2$), but in a fractional slot winding, the division is not equal ((b): $q_s < 2$). In the observed integer slot winding, the base winding length $Q_s = 12$ and, after that, the magnetic conditions for the slots repeat themselves equally; observe slots 1 and 13. In the fractional slot winding, the base winding is notably longer and contains Q'_s slots. Figure (c) illustrates an example of a fractional slot non-overlapping (tooth-coil) winding with $Q_s = 12$ and $p = 5$. Such a winding may be used in tooth-coil permanent magnet machines, where $q = 0.4$. In a two-layer system, each of the stator phases carries four coils. The coil sides are located in slots 12–1, 1–2, 6–7 and 7–8. The air gap flux density is mainly created by the rotor poles.

a number of Q'_s phasors of the voltage phasor diagram, pointing in different directions. Now, we can write

$$Q'_s = p'\frac{Q_s}{p}, \quad Q'_s < Q_s, \quad p > p'. \tag{2.65}$$

Here the voltage phasors $Q'_s + 1, 2Q'_s + 1, 3Q'_s + 1$ and $(t - 1)Q'_s + 1$ are in the same position in the voltage phasor diagram as the voltage phasor of slot 1. In this position, the cycle of the

Table 2.3 Parameters of voltage phasor diagrams

t	The largest common divider of Q and p, the number of phasors of a single radius, the number of layers of a voltage phasor diagram
$Q' = Q/t$	The number of radii, or the number of phasors of a single turn in a voltage phasor diagram (the number of slots in a base winding)
$p' = p/t$	The number of revolutions around a single layer when numbering a voltage phasor diagram
$(p/t) - 1$	The phasors skipped in the numbering of the voltage phasor diagram

voltage phasor diagram is always started again. Either a new periphery is drawn, or more slot numbers are added to the phasors of the initial diagram. In the numbering of a voltage phasor diagram, each layer of the diagram has to be circled p' times. Thus, t layers are created in the voltage phasor diagram. In other words, in each electrical machine there are t electrically equal slot sequences, the slot number of which is $Q_s{}' = Q_s/t$ and the number of pole pairs $p' = p/t$. To determine t, we have to find the smallest integers Q_s' and p'. Thus, t is the largest common divider of Q_s and p. If $Q_s/(2pm) \in \mathbf{N}$ ($\mathbf{N}$ is the set of integers, $\mathbf{N}_{\text{even}}$ the set of even integers and $\mathbf{N}_{\text{odd}}$ the set of odd integers), we have an integral slot winding, and $t = p$, $Q_s' = Q_s/p$ and $p' = p/p = 1$. Table 2.3 shows some parameters of a voltage phasor diagram. To generalize the representation, the subscript "s" is left out of what follows.

If the number of radii in the voltage phasor diagram is $Q' = Q/t$, the angle of adjacent radii, that is the phasor angle α_z, is written as

$$\alpha_z = \frac{2\pi}{Q}t. \tag{2.66}$$

The slot angle α_u is correspondingly a multiple of the phasor angle α_z

$$\alpha_u = \frac{p}{t}\alpha_z = p'\alpha_z. \tag{2.67}$$

When $p = t$, we obtain $\alpha_u = \alpha_z$, and the numbering of the voltage phasor diagram proceeds continuously. If $p > t$, $\alpha_u > \alpha_z$, a number of $(p/t) - 1$ phasors have to be skipped in the numbering of slots. In that case, a single layer of a voltage phasor diagram has to be circled (p/t) times when numbering the slots. When considering the voltage phasor diagrams of harmonics ν, we see that the slot angle of the νth harmonic is $\nu\alpha_u$. Also the phasor angle is $\nu\alpha_z$. The voltage phasor diagram of the νth harmonic differs from the voltage phasor diagram of the fundamental with respect to the angles, which are ν-fold.

Example 2.15: Create voltage phasor diagrams for two different fractional slot windings: (a) $Q = 27$ and $p = 3$, (b) $Q = 30$, $p = 4$.

Solution: (a) $Q = 27$, $p = 3$, $Q/p = 9 \in \mathbf{N}$, $q_s = 1.5$, $t = p = 3$, $Q' = 9$, $p' = 1$, $\alpha_u = \alpha_z = 40°$.

There are, therefore, nine radii in the voltage phasor diagram, each having three phasors. Because $\alpha_u = \alpha_z$, no phasors are skipped in the numbering, Figure 2.25a.

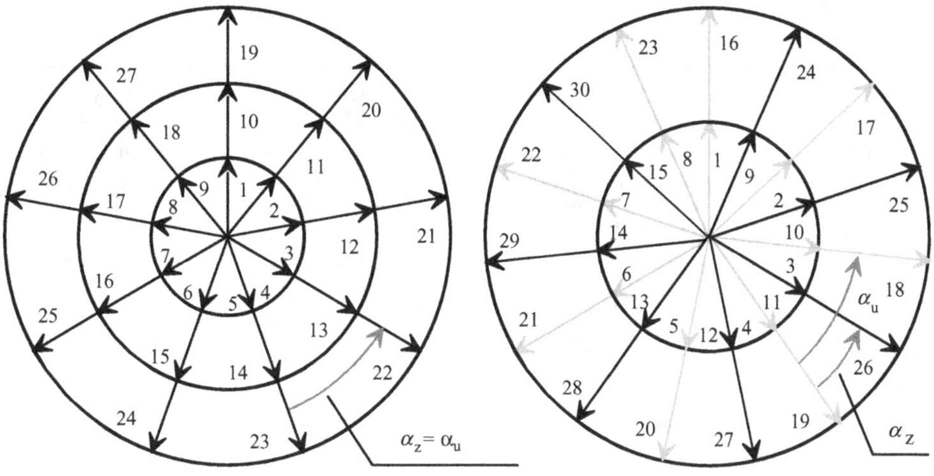

(a) (b)

Figure 2.25 Voltage phasor diagrams for two different fractional slot windings. On the left, the numbering is continuous, whereas on the right, certain phasors are skipped. (a) $Q = 27$, $p = 3$, $t = 3$, $Q' = 9$, $p' = 1$, $\alpha_u = \alpha_z = 40°$; (b) $Q = 30$, $p = 4$, $t = 2$, $Q' = 15$, $p' = 2$, $\alpha_u = 2\alpha_z = 48°$; α_u is the angle between voltages in the slots in electrical degrees and the angle α_z is the angle between two adjacent phasors in electrical degrees.

(b) $Q = 30$, $p = 4$, $Q/p = 7.5 \notin \mathbf{N}$, $q_s = 1.25$, $t = 2 \neq p$, $p' = 2$, $Q' = Q/t = 30/2 = 15$, $\alpha_z = 360°/15 = 24°$, $\alpha_u = 2\alpha_z = 2 \times 24° = 48°$, $(p/t) - 1 = 1$.

In this case, there are 15 radii in the voltage phasor diagram, each having two phasors. Because $\alpha_u = 2\alpha_z$, the number of phasors skipped will be $(p/t) - 1 = 1$. Both of the layers of the voltage phasor diagram have to be circled twice in order to number all the phasors, Figure 2.25b.

2.9 Phase Systems and Zones of Windings

2.9.1 Phase Systems

Generally speaking, windings may involve single or multiple phases, the most common case being a three-phase winding, which has been discussed here also. However, various other winding constructions are possible, as illustrated in Table 2.4.

On a single magnetic axis of an electrical machine, only one axis of a single-phase winding may be located. If another phase winding is placed on the same axis, no genuine poly-phase system is created, because both windings produce parallel fluxes. Therefore, each phase system that involves an even number of phases is reduced to involving only half of the original number of phases m' as illustrated in Table 2.4. If the reduction produces a system with an odd number of phases, we obtain a radially symmetric poly-phase system, also known as a normal system.

Table 2.4 Phase systems of the windings of electrical machines. The fourth column introduces radially symmetric winding alternatives.

Number of phases m	Nonreduced winding systems have separate windings for positive and negative magnetic axes	Reduced system: loaded star point needs a neutral line unless radially symmetric (e.g. $m = 6$)	Normal system: nonloaded star point and no neutral line, unless $m = 1$
1	 $m' = 2$	—	
2	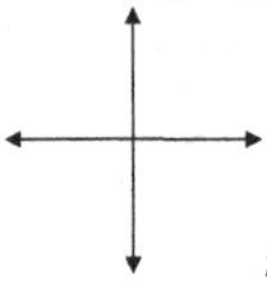 $m' = 4$	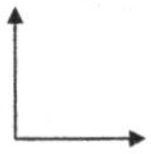	—
3	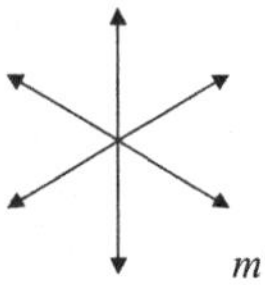 $m' = 6$	—	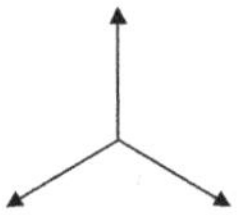
4	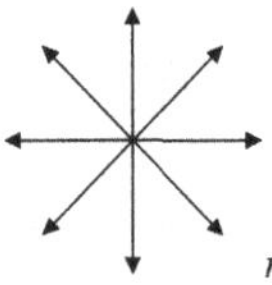 $m' = 8$	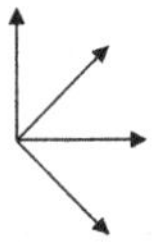	—
5	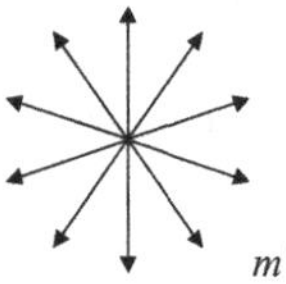 $m' = 10$	—	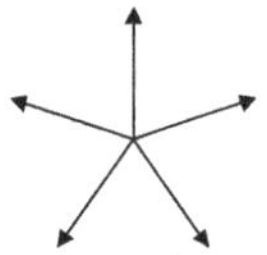
6…	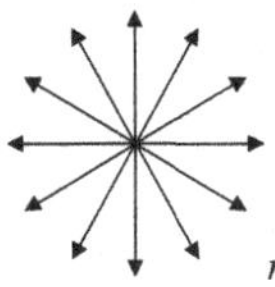 $m' = 12$	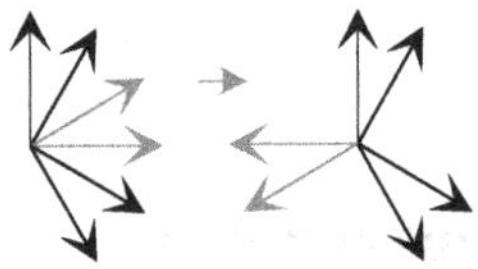	—
12	 $m' = 24$	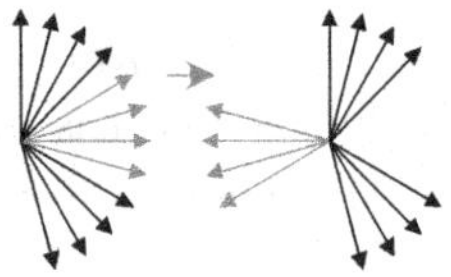	—
Reading instructions for Table 2.4	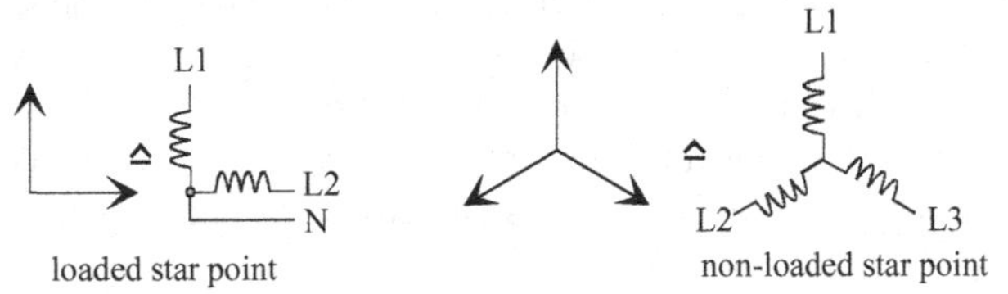		

If the reduction produces a system with an even number of phases, the result is called a reduced system. In this sense, an ordinary four-phase system is reduced to a two-phase system, as illustrated in Table 2.4. For an m-phase normal system, the phase angle is

$$\alpha_{\mathrm{ph}} = 2\pi/m. \quad (2.68)$$

Correspondingly, for a reduced system, the phase angle is

$$\alpha_{\mathrm{ph}} = \pi/m. \quad (2.69)$$

For example, in a three-phase system $\alpha_{\mathrm{ph3}} = 2\pi/3$ and for a two-phase system $\alpha_{\mathrm{ph2}} = \pi/2$.

If there is even a single odd number as a multiplicand of the phase number in the reduced system, a radially symmetric winding can be constructed again by turning the direction of the suitable phasors by 180 electrical degrees in the system, as shown in Table 2.4 for a six-phase system ($6 = 2 \times 3$). With this kind of a system, a nonloaded star point is created exactly as in a normal system. In a reduced system, the star point is normally loaded, and thus for instance the star point of a reduced two-phase machine requires a conductor of its own, which is not required in a normal system. Without a neutral conductor, a reduced two-phase system becomes a single-phase system, because the windings cannot operate independently, but the same current that produces the current linkage is always flowing in them, and together they form only a single magnetic axis. An ordinary three-phase system also becomes a single-phase system if the voltage supply of one phase ceases for some reason.

Of the winding systems in Table 2.4, the three-phase normal system is dominant in industrial applications. Five- and seven-phase windings have been suggested for frequency converter use to increase the system output power at a low voltage. Six-phase motors are used in large synchronous motor drives. In some larger high-speed applications, six-phase windings are also useful. In practice, all phase systems divisible by three are practical in inverter supplies. Each of the three-phase partial systems is supplied by its own three-phase frequency converter having a temporal phase shift $2\pi/m'$, in a 12-phase system, for example $\pi/12$. For example, a 12-phase system is supplied with four three-phase converters having a $\pi/12$ temporal phase shift.

Single-phase windings may be used in single-phase synchronous generators and also in small induction motors. In the case of a single-phase-supplied induction motor, however, the motor needs starting assistance, which is often realized as an auxiliary winding with a phase shift of $\pi/2$. In such a case, the winding arrangement starts to resemble the two-phase reduced winding system, but since the windings are usually not similar, the machine is not purely a two-phase machine.

2.9.2 Zones of Windings

In double-layer windings, both layers have separate zones: an upper-layer zone and a bottom-layer zone, Figure 2.26. This double number of zones also means a double number of coil groups when compared with single-layer windings. In double-layer windings, one coil side is always located in the upper layer and the other in the bottom layer. In short-pitched double-layer windings, the upper layer is shifted with respect to the bottom layer, as shown in Figures 2.15

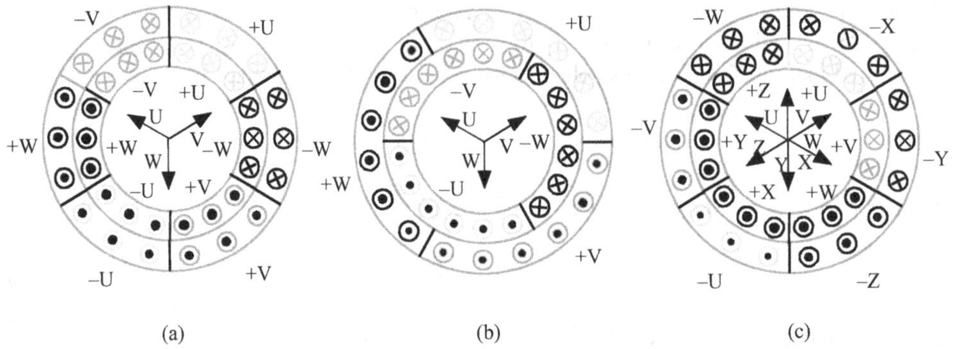

Figure 2.26 Zone formation of double-layer windings, $m = 3$, $p = 1$. (a) a normal-zone span, (b) a double-zone span, (c) the zone distribution of a six-phase radially symmetric winding with a double-zone span. The tails and heads of the arrows correspond to a situation in which there are currents $I_U = -2I_V = -2I_W$ flowing in the windings. The winding in (a) corresponds to a single-layer winding, which is obtained by unifying the winding layers by removing the insulation layer between the layers.

and 2.17. The span of the zones can be varied between the upper and bottom layer, as shown in Figure 2.17 (zone variation). With double-layer windings, we can easily apply systems with a double-zone span, which usually occur only in machines where the windings may be rearranged during the drive to produce another number of poles. In fractional slot windings, zones of varying spans are possible. This kind of zone variation is called natural zone variation.

In a single-layer winding, each coil requires two slots. For each slot, there is now half a coil. In double-layer windings, there are two coil sides in each slot, and thus, in principle, there is one coil per slot. The total number of coils z_c is thus

$$\text{for single-layer windings}: z_c = \frac{Q}{2}, \tag{2.70}$$

$$\text{for double-layer windings}: z_c = Q. \tag{2.71}$$

The single- and double-layer windings with double-width zones form m coil groups per pole pair. Double-layer windings with a normal zone span form $2m$ coil groups per pole pair. The total numbers of coil groups are thus pm and $2pm$ respectively. Table 2.5 lists some of the core parameters of phase windings.

Table 2.5 Phase winding parameters

Winding	Number of coils z_c	Number of coil groups	Average zone span	Average zone angle α_{zav}
Single-layer	$Q/2$	pm	τ_p/m	π/m
Double-layer, normal zone span	Q	$2pm$	τ_p/m	π/m
Double-layer, double-zone span	Q	pm	$2\tau_p/m$	$2\pi/m$

2.10 Symmetry Conditions

A winding is said to be symmetrical if, when fed from a symmetrical supply, it creates a rotating magnetic field. Both of the following symmetry conditions must be fulfilled:

(a) **The first condition of symmetry**: Normally, the number of coils per phase winding has to be an integer:

$$\text{for single-layer windings}: \frac{Q}{2m} = pq \in \mathrm{N}, \tag{2.72}$$

$$\text{for double-layer windings}: \frac{Q}{m} = 2pq \in \mathrm{N}. \tag{2.73}$$

The first condition is met easier by double-layer windings than by single-layer windings, thanks to a wider range of alternative constructions.

(b) **The second condition of symmetry**: In poly-phase machines, the angle α_{ph} between the phase windings has to be an integral multiple of the angle α_{z}. Therefore for normal systems, we can write

$$\frac{\alpha_{\mathrm{ph}}}{\alpha_{\mathrm{z}}} = \frac{2\pi Q}{m 2\pi t} = \frac{Q}{mt} \in \mathbf{N} \tag{2.74}$$

and for reduced systems

$$\frac{\alpha_{\mathrm{ph}}}{\alpha_{\mathrm{z}}} = \frac{\pi Q}{m 2\pi t} = \frac{Q}{2mt} \in \mathbf{N}. \tag{2.75}$$

Let us now consider how the symmetry conditions are met with integral slot windings. The first condition is always met, since p and q are integers. The number of slots in integral slot windings is $Q = 2pqm$. Now the largest common divider t of Q and p is always p. When we substitute $p = t$ into the second symmetry condition, we can see that it is always met, since

$$\frac{Q}{mt} = \frac{Q}{mp} = 2q \in \mathbf{N}. \tag{2.76}$$

Integral slot windings are thus symmetrical. Because $t = p$, also $\alpha_{\mathrm{u}} = \alpha_{\mathrm{z}}$, and hence the numbering of the voltage phasor diagram of the integral slot winding is always consecutive, as can be seen for instance in Figures 2.10 and 2.18.

2.10.1 Symmetrical Fractional Slot Windings

Fractional slot windings are not necessarily symmetrical. A successful fulfillment of symmetry requirements starts with the correct selection of the initial parameters of the winding. First, we have to select q (slots per pole and phase) so that the fraction presenting the number of slots per pole and phase

$$q = \frac{z}{n} \tag{2.77}$$

is indivisible. Here the denominator n is a quantity typical of fractional slot windings.

(a) **The first condition of symmetry**: For single-layer windings (Equation 2.72), it is required that in the equation

$$\frac{Q}{2m} = pq = p\frac{z}{n}, \quad \frac{p}{n} \in \mathbf{N}. \tag{2.78}$$

Here z and n constitute an indivisible fraction and thus p and n have to be evenly divisible. We see that when designing a winding, with the pole pair number p usually as an initial condition, we can select only certain integer values for n. Correspondingly, for double-layer windings (Equation 2.73), the first condition of symmetry requires that in the equation

$$\frac{Q}{m} = 2pq = 2p\frac{z}{n}, \quad \frac{2p}{n} \in \mathbf{N}. \tag{2.79}$$

On comparing Equation (2.78) with Equation (2.79), we can see that we achieve a wider range of alternative solutions for fractional slot windings by applying a double-layer winding than a single-layer winding. For instance, for a two-pole machine $p = 1$, a single-layer winding can be constructed only when $n = 1$, which leads to an integral slot winding. On the other hand, a fractional slot winding, for which $n = 2$ and $p = 1$, can be constructed as a double-layer winding.

(b) **The second condition of symmetry**: To meet the second condition of symmetry (Equation 2.74), the largest common divider t of Q and p has to be defined. This divider can be determined from the following equation:

$$Q = 2pqm = 2mz\frac{p}{n} \text{ and } p = n\frac{p}{n}.$$

According to Equation (2.78), $p/n \in \mathbf{N}$, and thus this ratio is a divider of both Q and p. Because z is indivisible by n, the other dividers of Q and p can be included only in the figures $2m$ and n. These dividers are denoted generally by c and thus

$$t = c\frac{p}{n}. \tag{2.80}$$

Now the second condition of symmetry can be rewritten for normal poly-phase windings in a form that is in harmony with Equation (2.74):

$$\frac{Q}{mt} = \frac{2mz\frac{p}{n}}{mc\frac{p}{n}} = \frac{2z}{c} \in \mathbf{N}. \tag{2.81}$$

The divider c of n cannot be a divider of z. The only possible values for c are $c = 1$ or $c = 2$.

For normal poly-phase systems, m is an odd integer. For reduced poly-phase systems, according to Equation (2.75), it is written as

$$\frac{Q}{2mt} = \frac{2mz\frac{p}{n}}{2mc\frac{p}{n}} = \frac{z}{c} \in \mathbf{N}. \tag{2.82}$$

For c, this allows only the value $c = 1$.

As shown in Table 2.4, for normal poly-phase windings, the phase number m has to be an odd integer. The divider $c = 2$ of $2m$ and n cannot be a divider of m. For reduced poly-phase systems, m is an even integer, and thus the only possibility is $c = 1$. The second condition of symmetry can now be written simply in the form: n and m cannot have a common divider $n/m \notin \mathbf{N}$. If $m = 3$, n cannot be divisible by three, and the second condition of symmetry reads

$$\frac{n}{3} \notin \mathbf{N}. \tag{2.83}$$

Conditions (2.78) and (2.83) automatically determine that if p includes only the figure 3 as its factor ($p = 3, 9, 27, \ldots$), a single-layer fractional slot winding cannot be constructed at all.

Table 2.6 lists the symmetry conditions of fractional slot windings.

As shown, it is not always possible to construct a symmetrical fractional slot winding for certain numbers of pole pairs. However, if some of the slots are left without a winding, a fractional slot winding can be carried out. In practice, only three-phase windings are realized with empty slots.

Free slots Q_o have to be distributed on the periphery of the machine so that the phase windings become symmetrical. The number of free slots has thus to be divisible by three, and the angle between the corresponding free slots has to be 120°. The first condition of symmetry is now written as

$$\frac{Q - Q_o}{6} \in \mathbf{N}. \tag{2.84}$$

Table 2.6 Conditions of symmetry for fractional slot windings

Number of slots per pole and phase $q = z/n$, where z and n cannot be mutually divisible	
Type of winding, number of phases	Condition of symmetry
Single-layer windings	$p/n \in \mathbf{N}$
Double-layer windings	$2p/n \in \mathbf{N}$
Two-phase $m = 2$	$n/m \notin \mathbf{N} \rightarrow n/2 \notin \mathbf{N}$
Three-phase $m = 3$	$n/m \notin \mathbf{N} \rightarrow n/3 \notin \mathbf{N}$

The second condition of symmetry is

$$\frac{Q}{3t} \in \mathbf{N}. \tag{2.85}$$

Furthermore, it is also required that

$$\frac{Q_o}{3} \in \mathbf{N}_{odd}. \tag{2.86}$$

Usually, the number of free slots is selected to be three, because this enables the construction of a winding, but does not leave a considerable amount of the volume of the machine without utilization. For normal zone width windings with free slots, the average number of slots of a coil group is obtained from the equation

$$Q_{av} = \frac{Q - Q_o}{2pm} = \frac{Q}{2pm} - \frac{Q_o}{2pm} = q - \frac{Q_o}{2pm}. \tag{2.87}$$

2.11 Base Windings

It was shown previously that in fractional slot windings, a certain coil side of a phase winding occurs at the same position with the air gap flux always after $p' = p/t$ pole pairs, if the largest common divider t of Q and p is greater than one. In that case, there are t electrically equal slot sequences containing Q' slots in the armature, each of which includes a single layer of the voltage phasor diagram. Now it is worth considering whether it is possible to connect a system of t equal sequences of slots containing a winding as t equal independent winding sections. This is possible when all the slots Q' of the slot sequence of all t electrically equal slot groups meet the first condition of symmetry. The second condition of symmetry does not have to be met.

If Q'/m is an even number, both a single-layer and a double-layer winding can be constructed in Q' slots. If Q'/m is an odd number, only a double-layer winding is possible in Q' slots. When constructing a single-layer winding, q has to be an integer. Thus, two slot pitches of t with $2Q'$ slots are all that is required for the smallest, independent, symmetrical single-layer winding. The smallest independent symmetrical section of a winding is called a base winding. When a winding consists of several base windings, the current and voltage of which are due to geometrical reasons always of the same phase and magnitude, it is possible to connect these basic windings in series and in parallel to form a complete winding. Depending on the number of Q'/m, that is whether it is an even or odd number, the windings are defined either as first- or second-grade windings.

Table 2.7 lists some of the parameters of base windings.

2.11.1 *First-Grade Fractional Slot Base Windings*

In first-grade base windings,

$$\frac{Q'}{m} = \frac{Q}{mt} \in \mathbf{N}_{even}. \tag{2.88}$$

Table 2.7 Some parameters of fractional slot base windings

	Base winding of first grade	Base winding of second grade	
Parameter q	$q = z/n$		
Denominator n	$n \in \mathbf{N}_{\text{odd}}$	$n \in \mathbf{N}_{\text{even}}$	
Parameter Q'/m	$\dfrac{Q'}{m} \in \mathbf{N}_{\text{even}}$	$\dfrac{Q'}{m} \in \mathbf{N}_{\text{odd}}$	
Parameter Q/tm	$\dfrac{Q}{tm} \in \mathbf{N}_{\text{even}}$	$\dfrac{Q}{tm} \in \mathbf{N}_{\text{odd}}$	
Divider t, the largest common divider of Q and p	$t = \dfrac{p}{n}$	$t = \dfrac{2p}{n}$	
Slot angle α_u expressed with voltage phasor angle α_z	$\alpha_u = n\alpha_z = n\dfrac{2\pi}{Q}t$	$\alpha_u = \dfrac{n}{2}\alpha_z = n\dfrac{\pi}{Q}t$	
Type of winding	Single-layer windings double-layer windings	Single-layer windings	Double-layer windings
Number of slots Q^* of a base winding	$Q^* = \dfrac{Q}{t}$	$Q^* = 2\dfrac{Q}{t}$	$Q^* = \dfrac{Q}{t}$
Number of pole pairs p^* of a base winding	$p^* = \dfrac{p}{t} = n$	$p^* = 2\dfrac{p}{t} = n$	$p^* = \dfrac{p}{t} = \dfrac{n}{2}$
Number of layers t^* in a voltage phasor diagram for a base winding	$t^* = 1$	$t^* = 2$	$t^* = 1$

The asterix * indicates the values of the base winding.

There are Q^* slots in a first-grade base winding, and the following is valid for the parameters of the winding:

$$Q^* = \frac{Q}{t}, \ p^* = \frac{p}{t} = n, \ t^* = 1. \tag{2.89}$$

Both conditions of symmetry (2.72)–(2.75) are met under these conditions.

2.11.2 Second-Grade Fractional Slot Base Windings

A precondition of the second-grade base windings is that

$$\frac{Q'}{m} = \frac{Q}{mt} \in \mathbf{N}_{\text{odd}}. \tag{2.90}$$

According to Equations (2.81) and (2.82), Equation (2.90) is valid for normal poly-phase windings when $c = 2$ only for even values of n. Thus we obtain $t = 2p/n$ and $\alpha_u = n\alpha_z/2$. The first condition of symmetry is met with the base windings of the second grade only when $Q^* = 2Q'$. Now we obtain

$$\frac{Q^*}{2m} = \frac{Q'}{m} = \frac{Q}{mt} \in \mathbf{N}. \tag{2.91}$$

The second-grade single-layer base winding thus comprises two consequent tth parts of a total winding. Their parameters are written as

$$Q^* = 2\frac{Q}{t},\ p^* = 2\frac{p}{t} = n,\ t^* = 2. \tag{2.92}$$

With these values, also the second condition of symmetry is met, since

$$\frac{Q^*}{mt^*} = \frac{2Q'}{2m} = \frac{Q}{mt} \in \mathbf{N}. \tag{2.93}$$

The second-grade double-layer base winding meets the first condition of symmetry immediately when the number of slots is $Q^* = Q'$. Hence

$$\frac{Q^*}{m} = \frac{Q'}{m} = \frac{Q}{mt} \in \mathbf{N}. \tag{2.94}$$

The parameters are now

$$Q^* = \frac{Q}{t},\ p^* = \frac{p}{t} = \frac{n}{2},\ t^* = 1. \tag{2.95}$$

The second condition of symmetry is now also met.

2.11.3 Integral Slot Base Windings

For integral slot windings, $t = p$. Hence, we obtain for normal poly-phase systems

$$\frac{Q}{mt} = \frac{Q}{mp} = 2q \in \mathbf{N}_{\text{even}}. \tag{2.96}$$

For a base winding of the first grade, we may write

$$Q^* = \frac{Q}{p},\ p^* = \frac{p}{p} = 1,\ t^* = 1. \tag{2.97}$$

Since also the integral slot windings of reduced poly-phase systems form the base windings of the first grade, we can see that all integral slot windings are of the first grade, and that integral slot base windings comprise only a single pole pair. The design of integral slot windings is therefore fairly easy. As we can see in Figure 2.17, the winding construction is repeated without change always after one pole pair. Thus, to create a complete integral slot winding, we connect a sufficient number of base windings to a single pole pair either in series or in parallel.

Example 2.16: Create a voltage phasor diagram of a single-layer integral slot winding, for which $Q = 36, p = 2, m = 3$.

Solution: The number of slots per pole and phase is

$$q = \frac{Q}{2pm} = 3.$$

A zone distribution, Figure 2.27, and a voltage phasor diagram, Figure 2.28, are constructed for the winding.

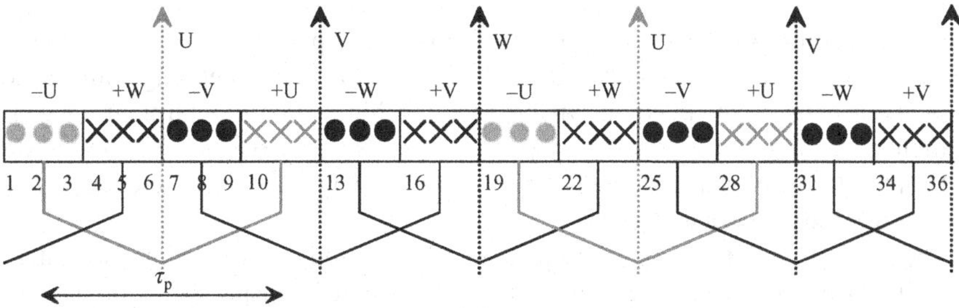

Figure 2.27 Zone distribution for a single-layer winding. $Q = 36, p = 2, m = 3, q = 3$.

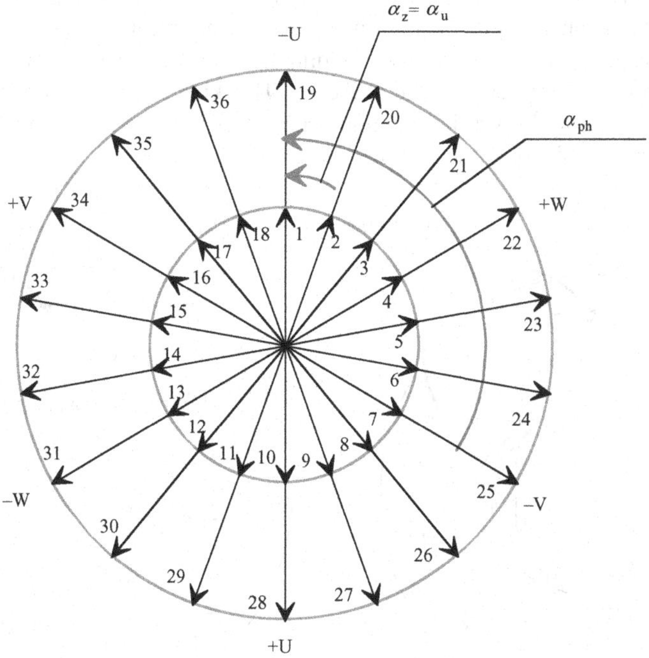

Figure 2.28 Complete voltage phasor diagram for a single-layer winding. $p = 2, m = 3, Q = 36, q = 3, t = 2, Q' = 18, \alpha_z = \alpha_u = 20°$. The second layer of the voltage phasor diagram repeats the first layer and it may, therefore, be omitted. The base winding length is 18 slots.

A double-layer integral slot winding is now easily constructed by selecting different phasors of the voltage phasor diagram of Figure 2.28, for instance for the upper layer. This way, we can immediately calculate the influence of different short pitchings. The voltage phasor diagram of Figure 2.28 is applicable to the definition of the winding factors for the short-pitched coils of Figure 2.15. Only the zones labeled in the figure will change place. Figure 2.28 is directly applicable to the full-pitch winding of Figure 2.14.

2.12 Fractional Slot Windings

2.12.1 Single-Layer Fractional Slot Windings

Fractional slot windings with extremely small fractions are popular in brushless DC machines and permanent magnet synchronous machines (PMSMs). Machines operating with sinusoidal voltages and currents are regarded as synchronous machines even though their air-gap flux density might be rectangular. Figure 2.29 depicts the differences between concentrated single-layer and double-layer fractional-slot non-overlapping windings in a case where the permanent magnet rotor has four poles and $q = 1/2$.

When the number of slots per pole and phase q of a fractional slot machine is greater than one, the coil groups of the winding have to be of the desired slot number q on average. In principle, the zone distribution of the single-layer fractional slot windings is carried out based on either the voltage phasor diagram or the zone diagram. The use of a voltage phasor diagram has often proved to lead to an uneconomical distribution of coil groups, and therefore it is usually advisable to apply a zone diagram in the zone distribution. The average slot number per pole and per phase of a fractional slot winding is hence q, which is a fraction that gives the average number of slots per pole and phase q_{av}. This kind of average number of slots can naturally be realized only by varying the number of slots in different zones. The number of slots in a single zone is denoted by q_k. Now

$$q_k \neq q_{av} = q \notin \mathbf{N}. \tag{2.98}$$

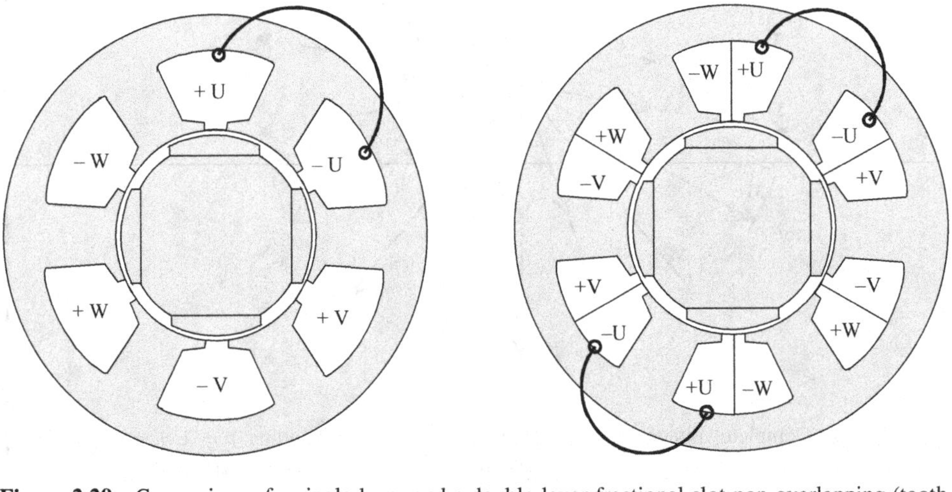

Figure 2.29 Comparison of a single-layer and a double-layer fractional slot non-overlapping (tooth-coil) windings with concentrated coils. $Q_s = 6$, $m = 3$, $p = 2$, $q = 1/2$.

q_{av} has thus to be an average of the different values of q_k. Then we write

$$q = g + \frac{z'}{n}, \tag{2.99}$$

where g is an integer, and the quotient is indivisible so that $z' < n$. Now we have an average number of slots per pole and phase $q = q_{\text{av}}$, when the width of z zones in n coil groups is set to $g + 1$ and the width of $n - z$ zones is g:

$$q_{\text{av}} = \frac{1}{n}\sum_{k=1}^{n} q_{\text{k}} = \frac{z'(g+1) + (n - z')g}{n} = g + \frac{z'}{n} = q. \tag{2.100}$$

The divergences from a totally symmetrical winding are smallest when the same number of slots per pole and phase occurs in consequent coil groups as seldom as possible. The best fractional slot winding is found with $n = 2$, when the number of slots per pole and phase varies constantly when traveling from one zone to another. To meet Equation (2.100), at least n groups of coil are required. Now we obtain the required number of coils

$$q_{\text{av}} nm = qp^* m = \frac{Q^*}{2}. \tag{2.101}$$

This number corresponds to the size of a single-layer base winding. Thus we have shown again that a base winding is the smallest independent winding for single-layer windings. When the second condition of symmetry for fractional slot windings is considered, it makes no difference how the windings are distributed in the slots (n zones, q_k coil sides in each), if only the desired average q_{av} is reached (e.g. $1 + 2 + 1 + 1 + 2$ gives an average of $1\frac{2}{3}$). The nm coil groups of a base winding have to be distributed in m phase windings so that each phase gets n single values of q_k (a local number of slots), in the same order in each phase. The coil numbers of consequent coil groups run through the single values of q_k n times in m equal cycles. This way, a cycle of coil groups is generated.

The first column of Table 2.8 shows m consequently numbered cycles of coil groups. The second column consists of nm coil groups in running order. The third column lists n single values of q_k m times in running order. Because the consequent coil groups belong to consequent phases, we get a corresponding running phase cycle in the fourth column. During a single cycle, the adjacent fifth column goes through all the phases U, V, W, ..., m of the machine. The sixth column repeats the numbers of coil groups.

Example 2.17: Compare two single-layer windings, an integral slot winding and a fractional slot winding having the same number of poles. The parameters for the integral slot winding are $Q = 36$, $p = 2$, $m = 3$, $q = 3$ and for the single-layer fractional slot winding $Q = 30$, $p = 2$, $m = 3$, $q = 2\frac{1}{2}$.

Table 2.8 Order of coil groups for symmetrical single-layer fractional slot windings

Cycle of coil groups 1 ... m	Number of coil group. This column runs m times from 1 to n (the divider of the fraction $q = z/n$)	Local number of slots per pole and phase (equals local number of slots per pole and phase q_k)	Phase cycle. All the phases are introduced once	Phases from 1 to m (for a three-phase system we have U, V, W)	Number of coil group
1	1	Q_1	1	U	1
1	2	Q_2	1	V	2
1	3	Q_3	1	W	3
1	|	|	1	|	|
1	K	q_k	1	m	m
1	|	|	2	U	$m+1$
1	|	|	2	V	$m+2$
1	|	|	2	W	$m+3$
1	N	q_n	2	|	|
2	$n+1$	Q_1	2	|	|
2	$n+2$	|	2	|	|
2	|	|		|	|
2	$n+k$	q_k		|	|
	dn	q_n		|	|
$d+1$	$dn+1$	Q_1	c	|	|
	|	|	c	|	|
	|	|	C	m	Cm
	$dn+k$	q_k	$c+1$	U	$cm+1$
	|	|		V	|
	|	|		W	|
m	$(m-1)n+k$	q_k		|	|
m	|	|		|	|
m	|	|		|	|
m	|	|		|	|
m	|	|		|	|
m	mn	q_n	N	m	Nm

Solution: For the fractional slot winding,

$$q_{av} = \frac{1}{n}\sum_{k=1}^{n} q_k = \frac{z'(g+1)+(n-z')g}{n} = g + \frac{z'}{n} = q.$$

We see that in this case $g = 2$, $z' = 1$, $n = 2$. Now, a group of coils with $z' = 1$ is obtained, in which there are $q_1 = g + 1 = 3$ coils, and another $n - z' = 1$ group of coils with $q_2 = g = 2$ coils. As $p = p^* = n = 2$, we have here a base winding with three ($m = 3$) cycles of coil groups of both the coil numbers q_1 and q_2. They occur in $n = 2$ cycles of three phases.

Table 2.9 Example of Table 2.8 applied to Figure 2.30. For the fractional slot winding $q = z/n = 5/2$

Cycle of coil groups	Number of coil group	Number of coils q_k	Phase cycle	Phase	Number of coil group
1	1	$3 = q_1$	1	U	1
1	$2\ (= n)$	$2 = q_2$	1	V	2
2	$2 + 1 = 3$	3	1	W	3
2	$2 + 2 = 4$	2	2	U	4
$3\ (= m)$	$(2 + 2) + 1 = 5$	3	2	V	5
$3\ (= m)$	$(2 + 2) + 2 = 6 = nm = 2{\cdot}3$	2	2	W	$6 = nm$

In Table 2.9, the example of Table 2.8 is applied to Figure 2.30. The above information presented in Table 2.9 can be presented simply as:

q_k	3	2	3	2	3	2
Phase	U	V	W	U	V	W

Each phase comprises a single coil group with two coils, and one coil group with three coils. Figure 2.30 compares the above integral slot winding and a fractional slot winding.

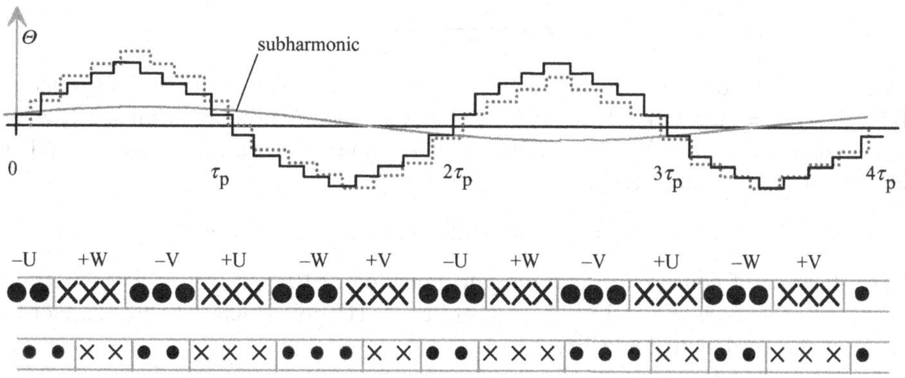

Figure 2.30 Zone diagrams and current linkage distributions of two different windings ($q = 3$, $q = 2^1/_2$). The integral slot winding is fully symmetrical, but the current linkage distribution of the fractional slot winding (dotted line) differs somewhat from the distribution of the integral slot winding (solid line). The current linkage of the fractional slot winding clearly contains a subharmonic, which has a double pole pitch compared with the fundamental.

Fractional slot windings create more harmonics than integral slot windings. By dividing the ordinal number ν of the harmonics of a fractional slot winding by the number of pole pairs p^*, we obtain

$$\nu' = \frac{\nu}{p^*}. \tag{2.102}$$

In integral slot windings, such relative ordinal numbers of the harmonics are the following odd integers: $\nu' = 1, 3, 5, 7, 9, \ldots$. For fractional slot windings, when $\nu = 1, 2, 3, 4, 5, \ldots$, the relative ordinal number gets the values $\nu' = 1/p*$, $\nu' = 2/p^*$, $\nu' = 3/p^*, \ldots$; in other words, values for which $\nu' < 1$, $\nu' \notin \mathbf{N}$ or $\nu' \in \mathbf{N}_{\text{even}}$. The lowest harmonic created by an integral slot winding is the fundamental ($\nu' = 1$), but a fractional slot winding can also produce subharmonics ($\nu' < 1$). Other harmonics also occur, the ordinal number of which is a fraction or an even integer. These harmonics cause additional forces, unintended torques and losses. These additional harmonics are the stronger, the greater is the zone variation; in other words, the divergence of the current linkage distribution from the respective distribution of an integral slot winding. In poly-phase windings, not all the integer harmonics are present. For instance, in the spectrum of three-phase windings, those harmonics are absent, the ordinal number of which is divisible by three, because $\alpha_{\text{ph},\nu} = \alpha_{\text{ph},1} = \nu\ 360°/m = \nu\ 120°$, and thus, because of the displacement angle of the phase windings $\alpha_{\text{ph}} = 120°$, they do not create a voltage between different phases.

Example 2.18: Design a single-layer fractional slot winding of the first grade, for which $Q = 168$, $p = 20$, $m = 3$. What is the winding factor of the fundamental?

Solution: The number of slots per pole and phase is

$$q = \frac{168}{2.20.3} = 1\frac{2}{5}.$$

We have a fractional slot winding with $n = 5$ as a divider. The conditions of symmetry (Table 2.6) $p/n = 20/5 = 4 \in \mathbf{N}$ and $n/3 = 5/3 \notin \mathbf{N}$ are met. According to Table 2.7, when n is an odd number, $n = 5 \in \mathbf{N}_{\text{odd}}$, a first-grade fractional slot winding is created. When $t = p/n = 4$, its parameters are

$$Q^* = Q/t = 168/4 = 42, \quad p^* = n = 5, \quad t^* = 1.$$

The diagram of coil groups, according to Equation (2.101), $q_{\text{av}}nm = qp^*m = Q^*/2$, consists of $p^*m = nm = 5 \times 3 = 15$ groups of coils. The coil groups and the phase order are selected according to Table 2.8:

q_k	2	1	2	1	1	2	1	2	1	1	2	1	2	1	1
Phase	U	V	W	U	V	W	U	V	W	U	V	W	U	V	W

$m = 3$ cycles of coil groups with $n = 5$ consequent numbers of coils q_k yield a symmetrical distribution of coil groups for single-phase coils.

q_k	q_1	q_2	q_3	q_4	q_5	q_1	q_2	q_3	q_4	q_5	q_1	q_2	q_3	q_4	q_5
Phase U	2			1			1			1			2		
Phase V		1			1			2			2			1	
Phase W			2			2			1			1			1

In each phase, there is one group of coils q_n. The average number of slots per pole and phase q_{av} of the coil group is written according to Equation (2.100) using the local q_k value order of phase U

$$\frac{1}{5}\sum_{k=1}^{5} q_k = \frac{1}{5}(2+1+1+1+2) = 1\frac{2}{5} = q.$$

We now obtain a coil group diagram according to Figure 2.31 and a winding phasor diagram according to Figure 2.32.

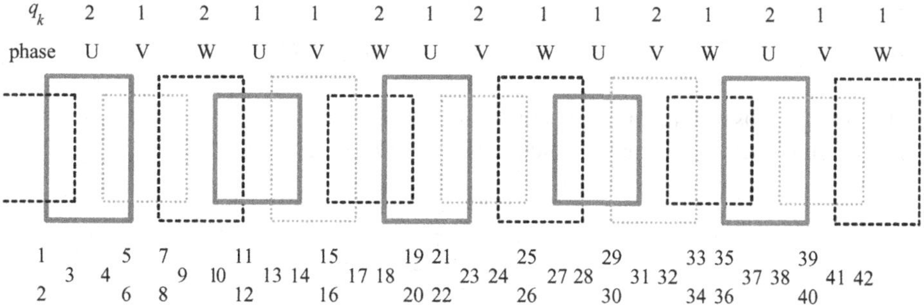

Figure 2.31 Coil group diagram of a single-layer fractional slot winding. $Q^* = Q/t = 168/4 = 42$, $p^* = n = 5$, $t^* = 1$.

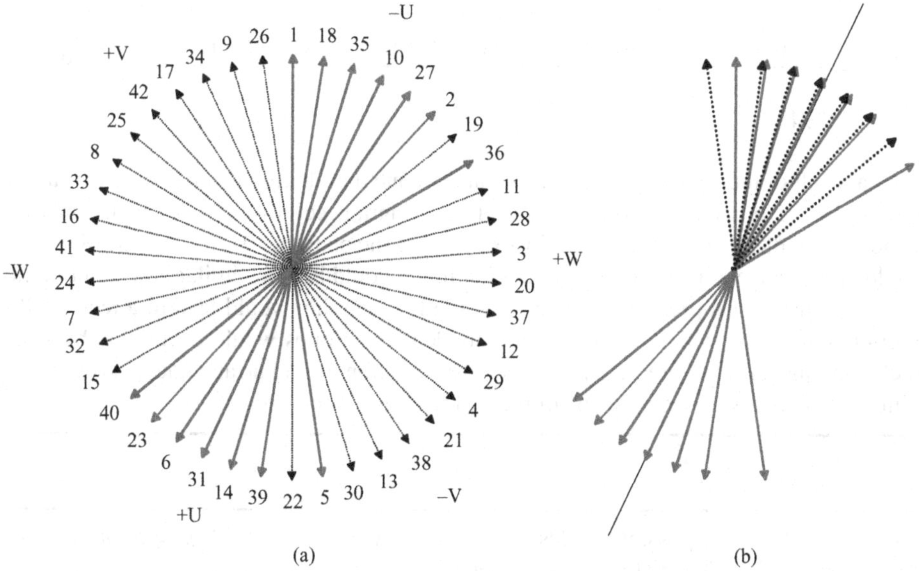

Figure 2.32 (a) Voltage phasor diagram of a first-grade, single-layer base winding. $p^* = 5$, $m = 3$, $Q' = Q^* = 42$, $q = 1^2/_5$, $t^* = 1$, $\alpha_u = 5\alpha_z = 42^6/_7°$, $\alpha_z = 8^4/_7°$. The phasors of the phase U are illustrated with a solid line, (b) the phasors of phase U are turned to form a bunch of phasors for the winding factor calculation and the symmetry line.

When calculating the winding factor for this winding, the following parameters are obtained for the voltage phasor diagram:

Number of layers in the voltage phasor diagram	$t^* = 1$
Number of radii	$Q' = Q^*/t^* = 42$
Slot angle	$\alpha_u = 360° p^*/Q^* = 360° \times 5/42 = 42\tfrac{6}{7}°.$
Phasor angle	$\alpha_z = 360° t^*/Q^* = 360° \times 1/42 = 8\tfrac{4}{7}°.$
Number of phasors skipped in the numbering	$(p^*/t^*) - 1 = 5 - 1 = 4$
Number of phasors for one phase	$Z = Q'/m = 42/3 = 14$

The voltage phasor diagram is illustrated in Figure 2.32.

The winding factor may now be calculated using Equation (2.16):

$$k_{wv} = \frac{\sin \frac{v\pi}{2}}{Z} \sum_{\rho=1}^{Z} \cos \alpha_\rho$$

The number of phasors $Z = 14$ for one phase and the angle between the phasors in the bunch is $\alpha_z = 8\tfrac{4}{7}°$. The fundamental winding factor is found after having determined the angles α_ρ between the phasors and the symmetry line, hence

$$k_{w1} = \frac{(\cos(4 \cdot 8\tfrac{4}{7}°) + \cos(3 \cdot 8\tfrac{4}{7}°) + 2\cos(2 \cdot 8\tfrac{4}{7}°) + 2\cos(8\tfrac{4}{7}°) + 1) \cdot 2}{14} = 0.945$$

$$= 0.945$$

As a result of the winding design based on the zone distribution given above, we have a winding in which, according to the voltage phasor diagram, certain coil sides are transferred to the zone of the neighboring phase. By exchanging the phasors 19–36, 5–22 and 8–33 we would also receive a functioning winding but there would be less similar coils than in the winding construction presented above. This kind of winding would lead to a technically inferior solution, in which undivided and divided coil groups would occur side by side. Such winding solutions are favorable when the variation of coil arrangements is kept to a minimum. This way, the best shape of the end winding is achieved.

Example 2.19: Is it possible to design a winding with (a) $Q = 72$, $p = 5$, $m = 3$, (b) $Q = 36$, $p = 7$, $m = 3$, (c) $Q = 42$, $p = 3$, $m = 3$?

Solution: (a) Using Table 2.6, we check the conditions of symmetry for fractional slot windings. The number of slots per pole and phase is $q = z/n = 72/(2 \times 5 \times 3) = 2\tfrac{2}{5}$, $z = 12$ and $n = 5$, which are not mutually divisible. As $p/n = 5/5 = 1 \in \mathbf{N}$ a single-layer winding should be made and as $n/m = 5/3 \notin \mathbf{N}$ the symmetry conditions are all right. And as $n \in \mathbf{N}_{\text{odd}}$ we will consider a first-grade base winding as follows:

$$Q^* = 72, \quad p^* = 5, \quad m = 3, \quad q = 2^2/_5.$$

That is, $q_{\text{av}} = \frac{1}{5}(3 + 2 + 2 + 2 + 3) = 2\frac{2}{5} = q$. This is a feasible winding.

q_k	3	2	2	2	3	3	2	2	2	3	3	2	2	2	3
Phase	U	V	W	U	V	W	U	V	W	U	V	W	U	V	W

(b) The number of slots per pole and phase is $q = z/n = 36/(2 \cdot 7 \cdot 3) = 6/7$, $z = 6$ and $n = 7$, which are not mutually divisible. As $p/n = 7/7 = 1 \in \mathbf{N}$ a single-layer winding can be made, and as $n/m = 7/3 \notin \mathbf{N}$ the symmetry conditions are all right. And as $n \in \mathbf{N}_{\text{odd}}$ we will consider a first-grade base winding as follows:

$$Q^* = 36, \quad p^* = 7, \quad m = 3, \quad q = {}^6/_7.$$

q_k	1	1	1	0	1	1	1	1	1	1	0	1	1	1	1	1	1	0	1	1	1
Phase	U	V	W	U	V	W	U	V	W	U	V	W	U	V	W	U	V	W	U	V	W

$$q_{\text{av}} = \frac{1}{7}(1 + 1 + 1 + 0 + 1 + 1 + 1) = \frac{6}{7}.$$

The number of slots per pole and phase can thus also be less than one, $q < 1$. In such a case, coil groups with no coils occur. These nonexistent coil groups are naturally evenly distributed in all phases.

(c) The number of slots per pole and phase is $q = z/n = 42/(2 \cdot 3 \cdot 3) = 2\frac{1}{3}$, $z = 7$ and $n = 3$, which are not mutually divisible, the condition $n/3 \notin \mathbf{N}$ is not met, and the winding is not symmetric. If, despite the nonsymmetrical nature, we considered a first-grade base winding, we should get a result as follows:

$$Q^* = 42, \quad p^* = 3, \quad m = 3, \quad q = 2^1/_3.$$

q_k	2	2	3	2	2	3	2	2	3
Phase	U	V	W	U	V	W	U	V	W

We can see that all coil groups with three coils now belong to the phase W. Such a winding is not functional.

Example 2.20: Create a winding with $Q = 60$, $p = 8$, $m = 3$.

Solution: The number of slots per pole and phase is $q = 60/(2 \cdot 8 \cdot 3) = 1\frac{1}{4}$. $z = 5$, $n = 4$. The largest common divider of Q and p is $t = 2p/n = 16/4 = 4$. As t also indicates the number of layers in the phasor diagram we get $Q' = Q/t = 60/4 = 15$ which is the number of radii in the phasor diagram in one layer. $Q'/m = 15/3 = 5 \in \mathbf{N}_{\text{odd}}$. The conditions of symmetry $p/n = 8/4 = 2 \in \mathbf{N}$ and $n/3 = 4/3 \notin \mathbf{N}$ are met. Because $n = 4 \in \mathbf{N}_{\text{even}}$, we have according to the parameters in Table 2.7 a second-grade, single-layer fractional slot winding. We get the base winding parameters

$$Q^* = 2Q/t = 2 \cdot 60/4 = 30, \quad p^* = n = 4, \quad t^* = 2.$$

The second-grade, single-layer fractional slot windings are designed like the first-grade windings. However, the voltage phasor diagram is now doubled. The coil group diagram of the base winding comprises $p^* \times m = n \times m = 4 \times 3 = 12$ coil groups. The coil group phase diagram is selected as follows:

q_k	2	1	1	1	2	1	1	1	2	1	1	1
Phase	U	V	W	U	V	W	U	V	W	U	V	W

A coil group diagram for the base winding corresponding to this case is illustrated in Figure 2.33.

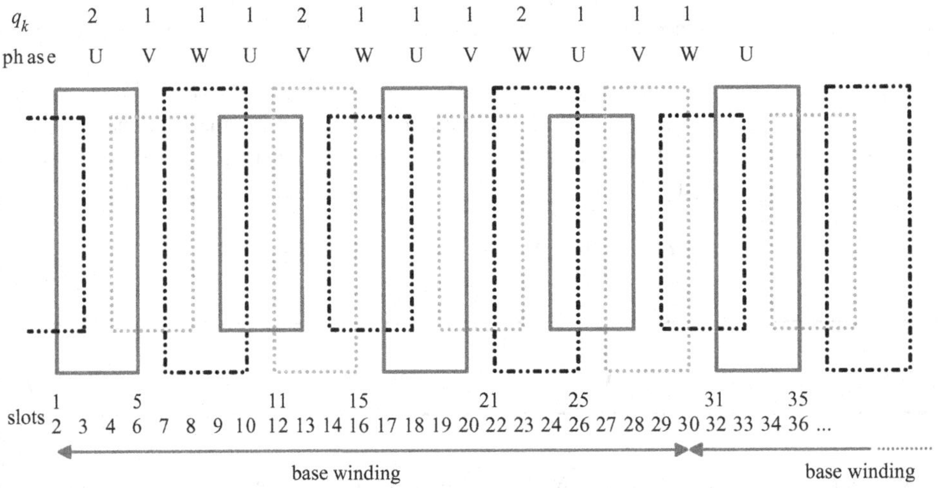

Figure 2.33 Coil group diagram of a base winding for a single-layer fractional slot winding $p = 8$, $m = 3$, $Q = 60$, $q = 1^1/_4$.

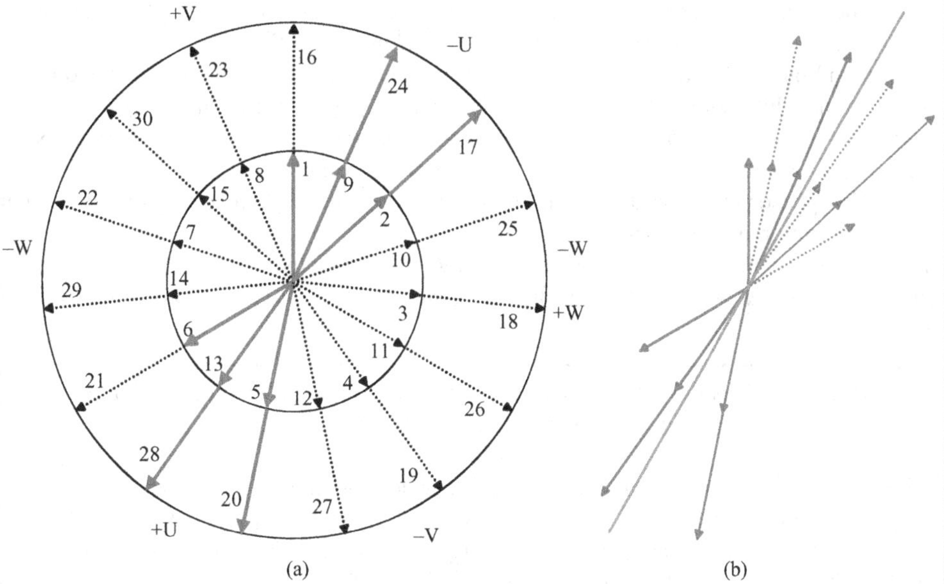

Figure 2.34 (a) Voltage phasor diagram of the base winding $p^* = 4$, $Q^* = 30$, $t^* = 2$, $Q' = 15$, $\alpha_u = 2\alpha_z = 48°$ of a single-layer fractional slot winding $p = 8$, $m = 3$, $Q = 60$, $q = 1\ ^1/_4$. The phasors belonging to the phase U are illustrated with a solid line. (b) The phasors of the phase U are turned for calculating the winding factor and for illustrating a symmetrical bunch of phasors.

A voltage phasor diagram for the base winding is illustrated in Figure 2.34:

Number of layers in the voltage phasor diagram	$t^* = 2$ (second-grade winding)
Number of radii	$Q' = Q^*/t^* = 30/2 = 15$
Slot angle	$\alpha_u = 360°p^*/Q^* = 360° \times 4/30 = 48°$
Phasor angle	$\alpha_z = 360°t^*/Q^* = 360° \times 2/30 = 24°$
Number of phasors skipped in the numbering	$(p^*/t^*) - 1 = 4/2 - 1 = 1$

The number of phasors $Z = 30/3 = 10$ for one phase and the angle between the phasors in the bunch is $\alpha_z = 24°$. After having found the angles α_ρ with respect to the symmetry line, the fundamental winding factor may be calculated using Equation (2.16)

$$k_{w1} = \frac{(2\cos(6°) + 2\cos(6° + 12) + \cos(6° + 24°)) \cdot 2}{10} = 0.951.$$

2.12.2 *Double-Layer Fractional Slot Windings*

In double-layer windings, one of the coil sides of each coil is in the upper layer of the slot, and the other coil side is in the bottom layer. The coils are all of equal span. Consequently, when the positions of the left coil sides are defined, the right sides also will be defined. Here double-layer fractional slot windings differ from single-layer windings. Let us now assume that the left coil sides are positioned in the upper layer. For these coil sides of the upper layer, a voltage phasor diagram of a double-layer winding is valid. Contrary to the voltage phasor diagram illustrated in Figure 2.34, there is only one layer in the voltage phasor diagram of the double-layer fractional slot winding. Therefore, the design of a symmetrical double-layer fractional slot winding is fairly straightforward with a voltage phasor diagram of a base winding. Now, symmetrically distributed closed bunches of phasors are composed of the phasors of single phases. This phasor order produces minimum divergence when compared with the current linkage distribution of the integral slot winding.

First, we investigate first-grade, double-layer fractional slot windings. It is possible to divide the phasors of such a winding into bunches of equal size; in other words, into zones of equal width.

Example 2.21: Design the winding previously constructed as a single-layer winding $Q = 168$, $p = 20$, $m = 3$, $q = 1^2/_5$ now as a double-layer winding.

Solution: We have a fractional slot winding for which the divider $n = 5$. The conditions of symmetry (Table 2.6) $p/n = 20/5 = 4 \in \mathbf{N}$ and $n/3 = 5/3 \notin \mathbf{N}$ are met. According to Table 2.7, if n is an odd number $n = 5 \in \mathbf{N}_{\mathrm{odd}}$, a first-grade fractional slot winding is created. The parameters of the voltage phasor diagram of such a winding are:

Number of layers in the voltage phasor diagram	$t^* = 1$
Number of pole pairs in the base winding	$p^* = 5$
Number of radii	$Q' = Q^*/t^* = 42$
Slot angle	$\alpha_u = 360° p^*/Q^* = 360° \times 5/42 = 42\frac{6}{7}°$
Phasor angle	$\alpha_z = 360° t^*/Q^* = 360° \times 1/42 = 8\frac{4}{7}°$
Number of phasors skipped in the numbering	$(p^*/t^*) - 1 = 5 - 1 = 4$

Since $t^* = 1$, the number of radii Q' is the same as the number of phasors Q^*, and we obtain $Q^*/m = 42/3 = 14$ phasors for each phase, which are then divided into negative Z^- and positive Z^+ phasors. The number of phasors per phase in the first-grade base winding is $Q^*/m = Q/mt \in \mathbf{N}_{\mathrm{even}}$. In normal cases, there is no zone variation, and the phasors are evenly divided into positive and negative phasors. In the example case, the number of phasors of both types is seven, $Z^- = Z^+ = 7$. By employing a normal zone order –U, +W, –V, +U, –W, +V we are able to divide the voltage phasor diagram into zones with seven phasors in each, Figure 2.35.

When the voltage phasor diagram is ready, the upper layer of the winding is set. The positions of the coil sides in the bottom layer are defined when an appropriate coil span is selected. For fractional slot windings, it is not possible to construct a full-pitch winding, because $q \notin \mathbf{N}$. For the winding in question, the full-pitch coil span y_Q of a full-pitch winding would be y in slot pitches

$$y = y_Q = mq = 3 \cdot 1\frac{2}{5} = 4\frac{1}{5} \notin \mathbf{N},$$

which is not possible in practice because the step has, of course, to be an integer number of slot pitches.

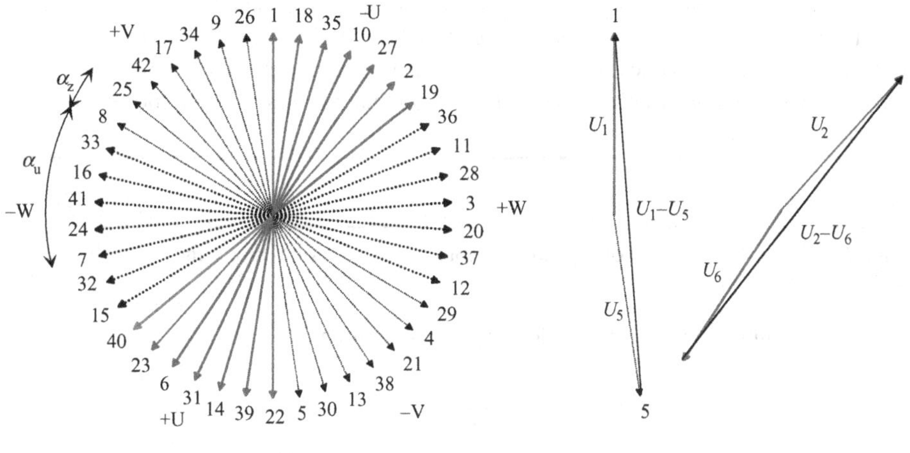

Figure 2.35 (a) Voltage phasor diagram of a first-grade, double-layer base winding $p^* = 5$, $m = 3$, $Q' = Q^* = 42$, $q^* = 1^2/_5$, $t^* = 1$, $\alpha_u = 5\alpha_z = 42^6/_7{}^\circ$, $\alpha_z = 8^4/_7{}^\circ$. (b) A couple of examples of coil voltages in the phase U.

Now the coil span may be decreased by $y_v = 1/5$. The coil span thus becomes an integer, which enables the construction of the winding:

$$y = mq - y_v = 3 \cdot 1\frac{2}{5} - \frac{1}{5} = 4 \in \mathbf{N}.$$

Double-layer fractional slot windings are thus short-pitched windings. When constructing a two-layer fractional slot winding, there are two coil sides in each slot. Hence, we have as many coils as slots in the winding. In this example, we first locate the U-phase bottom coil side in slot 1. The other coil side is placed according to the coil span of $y = 4$ at a distance of four slots in the upper part of slot 5. Similarly, coils run from 2 to 6. The coils to be formed are 1–5, 18–22, 35–39, 10–14, 27–31, 2–6 and 19–23. Starting from the +U zone, we have coils 22–26, 39–1, 14–18, 31–35, 6–10, 23–27 and 40–2. Now, six coil groups with one coil in each and four coil groups with two coils in each are created in each phase. The average is

$$q = \frac{1}{10}(6 \cdot 1 + 4 \cdot 2) = \frac{14}{10} = 1\frac{2}{5}.$$

A section of the base winding of the constructed winding is illustrated in Figure 2.36.

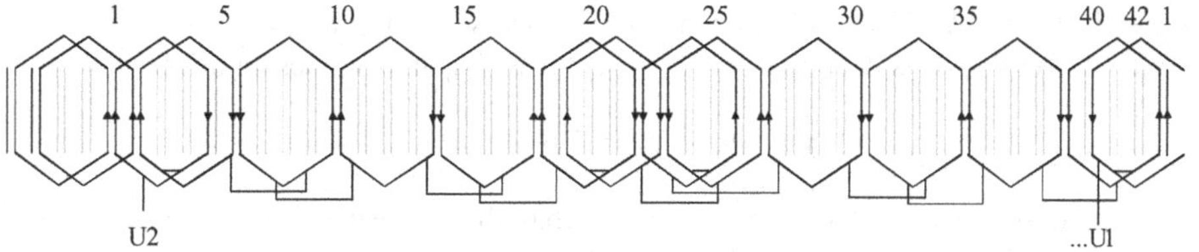

Figure 2.36 Base winding of a fractional slot winding. $p = 20$, $m = 3$, $Q = 168$, $q = 1^2/_5$. The U1 end of the base winding is placed in slot 40.

Next, the configuration of a second-grade, double-layer fractional slot winding is investigated. Because now $Q'/m = Q^*/m = Q/mt \in \mathbf{N}_{\text{odd}}$, a division $Z^- = Z^+$ is not possible. In other words, all the zones of the voltage phasor diagram are not equal. The voltage phasor diagram can nevertheless be constructed so that phasors of neighboring zones are not located inside each other's zones.

Example 2.22: Create a second-grade, double-layer fractional slot winding with $Q = 30$, $p = 2$, $m = 3$.

Solution: The number of slots per pole per phase is written as

$$q = \frac{30}{2 \cdot 2 \cdot 3} = 2^1\!/\!_2.$$

As $n = 2 \in \mathbf{N}_{\text{even}}$, we have a second-grade, double-layer fractional slot winding. Because $t = 2p/n = 2$, its parameters are

$$Q^* = Q/t = 30/2 = 15,$$
$$p^* = n/2 = 2/2 = 1,$$
$$t^* = 1.$$

This winding shows that the base winding of a second-grade, double-layer fractional slot winding can only be the length of one pole pair. The parameters of the voltage phasor diagram are:

Number of layers in the voltage phasor diagram	$t^* = 1 = p^*$
Number of radii	$Q' = Q^*/t^* = 15$
Slot angle	$\alpha_{\text{u}} = 360°p^*/Q^* = 360°/15 = 24°$
Phasor angle	$\alpha_{\text{z}} = 360°t^*/Q^* = 360°/15 = 24°$
Number of phasors skipped in the numbering	$(p^*/t^*) - 1 = 0$

For each phase, we obtain $Q'/m = Q^*/m = 15/3 = 5$ phasors. This does not allow an equal number of negative and positive phasors. If a natural zone variation is employed, we have to set either $Z^+ = Z^- + 1$ or $Z^+ = Z^- - 1$. In the latter case, we obtain $Z^- = 3$ and $Z^+ = 2$. With the known zone variation, electrical zones are created in the voltage phasor diagram, for which the number of phasors varies: $Z^- = 3$ phasors in zone –U, $Z^+ = 2$ phasors in zone +W, $Z^- = 3$ phasors in zone –V, and so on, Figure 2.37.

When the coil span is decreased by $y_{\text{v}} = 1/2$, the coil span becomes an integer

$$y = mq - y_v = 7^1\!/\!_2 - 1/2 = 7.$$

The winding diagram of Figure 2.38 shows that all the positive coil groups consist of three coils, and all the negative coil groups consist of two coils, which yields an average of $q = 2\,1/2$. Since all negative and all positive coil groups comprise an equal number of coils, the winding can be constructed as a wave winding. A wave is created that passes through the winding three times in one direction and two times in the opposite direction. The waves are connected in series to create a complete phase winding.

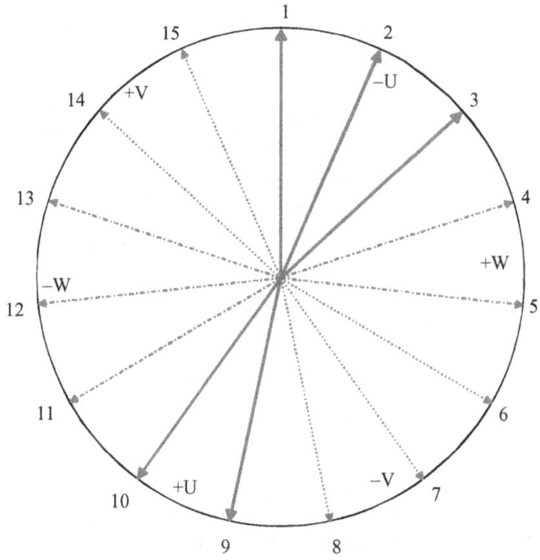

Figure 2.37 Voltage phasor diagram of a second-grade, double-layer fractional slot winding. $p^* = 1$, $m = 3$, $q = 2\frac{1}{2}$, $Q' = Q^*/t^* = 15$, $\alpha_u = 360°p^*/Q^* = 360°/15 = 24°$, $\alpha_z = 360°t^*/Q^* = 360°/15 = 24°$, $(p^*/t^*) - 1 = 0$.

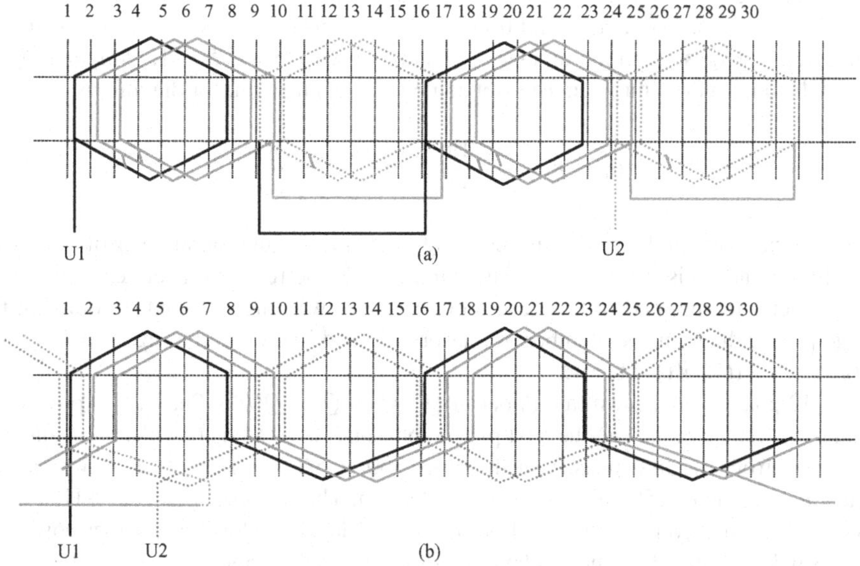

Figure 2.38 Winding diagram of a double-layer fractional slot winding. $p = 2$, $m = 3$, $Q = 30$, $q = 2\frac{1}{2}$ (a) Lap winding, (b) wave winding. To simplify the illustration, only a single phase is shown.

Example 2.23: Create a fractional slot non-overlapping (tooth-coil) winding for a three-phase machine, where the number of stator slots is 12, and the number of rotor poles is 10, Figure 2.39.

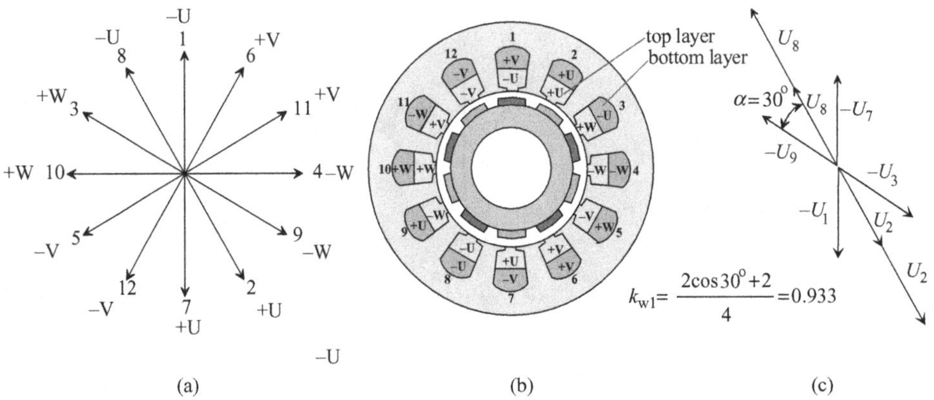

Figure 2.39 (a) Phasors of a 12-slot, 10-pole machine, (b) the double-layer winding of a 12-slot, 10-pole machine, (c) the phasors of the phase U for the calculation of the winding factor.

Solution: The number of slots per pole and phase is $q = 12/(3 \times 10) = 2/5 = z/n = 0.4$. Hence, $n = 5$. We should thereby find a base winding of the first grade. According to Table 2.6, $p/n \in \mathbf{N}$. In this case $5/5 \in \mathbf{N}$. In a three-phase machine $n/m \notin \mathbf{N} \to n/3 \notin \mathbf{N}$. Now $5/3 \notin \mathbf{N}$ and the symmetry conditions are met. Let us next consider the parameters in Table 2.7. The largest common divider of Q and p is $t = p/n = 5/5 = 1$, $Q/tm = 12/(1 \times 3) = 4$ which is an even number. The slot angle in the voltage phasor diagram is

$$\alpha_u = n\alpha_z = n\frac{2\pi}{Q}t = 5\frac{2\pi}{12}1 = \frac{5\pi}{6}.$$

The number slots in the base windings is $Q^* = Q/t = 12$ and the number of pole pairs in the base winding is $p/t = n = 5$. The winding may be realized as either a single- or double-layer winding, and in this case a double-layer winding is found. In drawing the voltage phasor diagram, the number of phasors skipped in the numbering is $(p^*/t^*) - 1 = ((p/t)/t^*) - 1 = ((5/1)/1) - 1 = 4$.

First, 12 phasors are drawn (a number of Q', when $Q' = Q^*/t^*$). Phasor 1 is positioned to point straight upwards, and the next phasor, phasor 2, is located at an electrical angle of $360 \times p/Q$ from the first phasor, in this case $360 \times 5/12 = 150°$. Phasor 3 is, again, located at an angle of 150° from phasor 2 and so on. The first coil 1–2 (–U, +U) will be located on the top layer of slot 1 and on the bottom layer of slot 2. The other coil (+U, –U) 2–3 will be located on the top layer of slot 2 and on the bottom layer of slot 3. The phase coils are set in the order U, –V, W, –U, V, –W. In the example, a single-phase zone comprises four slots, and thus a single winding zone includes two positive and two negative slots.

Based on the voltage phasor diagram of Figure 2.39a and the winding construction of Figure 2.39b, the fundamental winding factor of the machine can be solved, Figure 2.39c. First, the polarity of the coils of phase U in Figure 2.39b is checked and the respective phasors are drawn. In slots 1, 2 and 3, there are four coil sides of the phase U in total, and the number of phasors will thus be four. Now the angles between the phasors and their cosines are calculated. This yields a winding factor of 0.933.

Example 2.24: Create a tooth-coil winding for a three-phase machine, in which the number of slots is 21, and the number of rotor poles is 22, Figure 2.40.

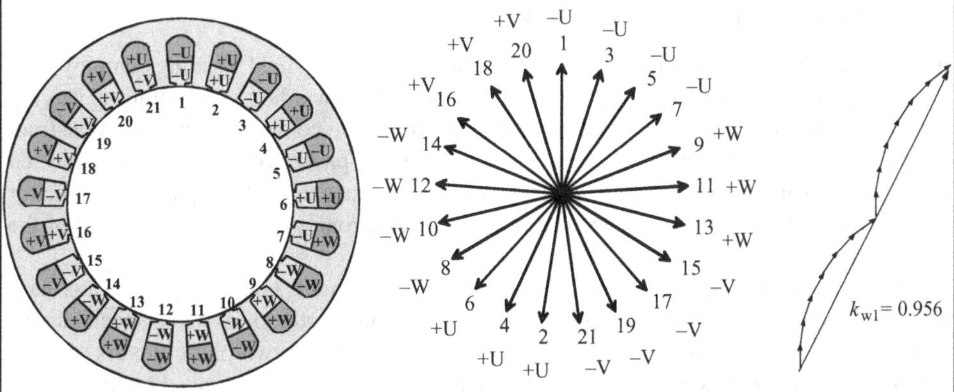

(a) (b) (c)

Figure 2.40 (a) Winding of a 21-slot, 22-pole machine, (b) the phasors of a 21-slot, 22-pole machine and (c) the phasors of the phase U for the calculation of the winding factor.

Solution: The number of slots per pole and phase is thus only $q = 21/66 = z/n = 7/22 = 0.318$. As $n \in \mathbf{N}_{\text{even}}$, we have a fractional slot winding of the second grade. Although a winding of this kind meets the symmetry conditions, it is not an ideal construction, because in the winding all the coils of a single phase are located on the same side of the machine. Such a coil system may produce harmful unbalanced magnetic forces in the machine.

In Figure 2.40b, 21 phasors are drawn (a number of Q', when $Q' = Q^*/t^*$). Phasor 1 is placed at the top and the next phasor at a distance of $360 \times p/Q$ from it. In this example, the distance is thus $360 \times 11/21 = 188.6°$. Phasor 2 is thus set at an angle of 188.6° from phasor 1. The procedure is repeated with phasors 3, 4, ... The phase coils are set in the order –W, U, –V, W, –U, V. Here a single phase consists of seven slots, and therefore we cannot place an equal number of positive and negative coils in one phase. In one phase, there are three positive and four negative slots. Note that we are now generating just the top winding layer, and when the bottom winding is also inserted, we have an equal number of positive and negative coils.

In Figure 2.40a, the coils are inserted in the bottom layer of the slots according to the phasors of Figure 2.40b. Phasor 1 of Figure 2.40b is –U, and it is located in the top layer of slot 1. Correspondingly, phasor 2, +U, is mounted in the top layer of slot 2. The bottom winding of the machine repeats the order of the top winding. When the top coil sides are transferred by a distance of one slot forward and the ± sign of each one is changed, a suitable bottom layer is obtained. The first coil of the phase U will be located in the bottom of slot 21 and on the surface of slot 1, and so on.

Table 2.10 contains some parameters of double-layer fractional slot windings, when the number of slots $q \leq 0.5$ (Salminen 2004).

2.13 Single- and Double-Phase Windings

The above three-phase windings are the most common rotating-field windings employed in poly-phase machines. Double- and single-phase windings, windings permitting a varying number of poles, and naturally also commutator windings are common in machine construction. Of commutator AC machines, nowadays only single-phase-supplied, series-connected commutator machines occur, for instance as motors of electric tools. Poly-phase commutator AC machines will eventually disappear as the power electronics enables easy control of the rotation speed of different motor types.

Since there is no two-phase supply network, two-phase windings occur mainly as auxiliary and main windings of machines supplied from a single-phase network. In some special cases, for instance small auxiliary automotive drives such as fan drives, two-phase motors are also used in power electronic supply. A two-phase winding can also be constructed on the rotor of low-power slip-ring asynchronous motors. As is known, a two-phase system is the simplest possible winding that produces a rotating field, and it is therefore most applicable to rotating-field machines. In a two-phase supply, however, there exist time instants when the current of either of the windings is zero. This means that each of the windings should alone be capable of creating as sinusoidal a supply as possible to achieve low harmonic content in the air gap and low losses in the rotor. This makes the design of high-efficiency two-phase winding machines more demanding than three-phase machines.

The design of a two-phase winding is based on the principles already discussed in the design of three-phase windings. However, we must always bear in mind that in the case of a reduced poly-phase system, when constructing the zone distribution, the signs of the zones do not vary in the way that they do in a three-phase system, but the zone distribution will be –U, –V, +U, +V. In a single-phase asynchronous machine, the number of coils of the main winding is usually higher than the number of coils of the auxiliary winding.

Example 2.25: Create a 5/6 short-pitched, double-layer, two-phase winding of a small electrical machine, $Q = 12, p = 1, m = 2, q = 3$.

Solution: The required winding is illustrated in Figure 2.41, where the rules mentioned above are applied.

Table 2.10 Winding factors k_{w1} of the fundamental and numbers of slots per pole and phase q for double-layer, three-phase fractional slot concentrated windings ($q \leq 0.5$). The boldface figures are the highest values in each column. Reproduced by permission of Pia Salminen

		Number of poles	2p									
Q_s		4	6	8	10	12	14	16	20	22	24	26
6	k_{w1}	**0.866**	—[a]	**0.866**	0.5	—[a]	0.5	0.866	0.866	0.5	—[a]	0.5
	q	0.5		0.25	0.2		0.143	0.125	0.1	0.091		0.077
9	k_{w1}		**0.866**	—[b]	—[b]	**0.866**	0.617	0.328	0.328	0.617	**0.866**	0.945
	q		0.5	0.375	0.3	0.25	0.214	0.188	0.15	0.136	0.125	0.115
12	k_{w1}			**0.866**	**0.933**	—[a]	**0.933**	0.866	0.5	0.25	—[a]	0.25
	q			0.5	0.4		0.286	0.25	0.2	0.182		0.154
15	k_{w1}				0.866	—[a]	—[b]	—[b]	0.866	0.711	—[a]	0.39
	q				0.5		0.357	0.313	0.25	0.227		0.192
18	k_{w1}					**0.866**	0.902	**0.945**	**0.945**	0.902	**0.866**	0.74
	q					0.5	0.429	0.375	0.3	0.273	0.25	0.231
21	k_{w1}						0.866	0.89	—[b]	—[b]	—[a]	0.89
	q						0.5	0.438	0.35	0.318		0.269
24	k_{w1}							0.866	0.933	**0.949**	—[a]	**0.949**
	q							0.5	0.4	0.364		0.308

[a]Not recommended, because the denominator n ($q = z/n$) is an integral multiple of the number of phases m.
[b]Not recommended as single base winding because of unbalanced magnetic pull.

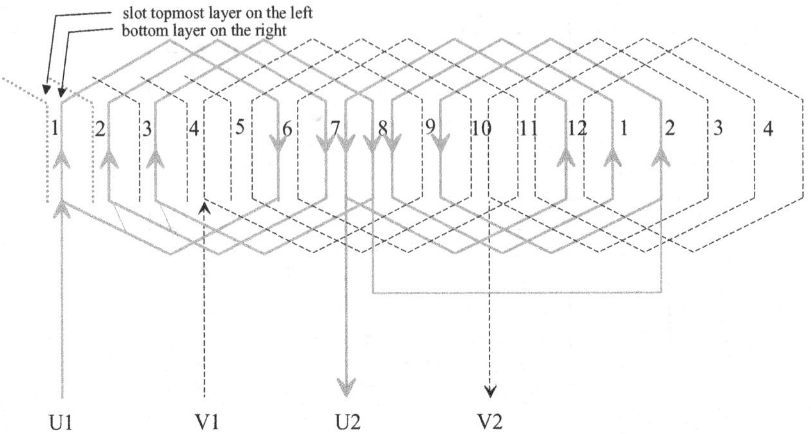

Figure 2.41 Symmetrical 5/6 short-pitched double-layer, two-phase winding, $Q = 12, p = 1, m = 2, q = 3$.

When considering a single-phase winding, we must bear in mind that it does not, as a stationary winding, produce a rotating field, but a pulsating field. A pulsating field can be presented as a sum of two fields rotating in opposite directions. The armature reaction of a single-phase machine thus has a field component rotating against the rotor. In synchronous machines, this component can be damped with the damper windings of the rotor. However, the damper winding copper losses are significant. In single-phase squirrel cage induction motors IMs, the rotor also creates extra losses when damping the negative-sequence field.

Also the magnetizing windings of the rotors of nonsalient-pole machines belong to the group of single-phase windings, as exemplified at the beginning of the chapter. If a single-phase winding is installed on the rotating part of the machine it, of course, creates a rotating field in the air gap of the machine contrary to the pulsating field of a single-phase stator winding.

Large single-phase machines are rare, but for instance single-phase synchronous machines are used to feed the old supply network of $16\frac{2}{3}$ Hz electric locomotives. Since there is only one phase in such a machine, there are only two zones per pole pair, and the construction of an integral slot winding is usually relatively simple. In these machines, damper windings have to cancel the negative-sequence field. This, however, obviously is problematic because lots of losses are generated in the damper.

The core principle also in designing single-phase windings is to aim at as sinusoidal a distribution of the current linkage as possible. This is even more important in single-phase windings than in three-phase windings, the current linkage distribution of which is by nature closer to ideal. The current linkage distribution of a single-phase winding can be made to resemble the current linkage distribution of a three-phase winding instantaneously in a position where a current of one phase of a three-phase winding is zero. At that instant, a third of the slots of the machine are in principle currentless. The current linkage distribution of a single-phase machine can best be made to approach a sinusoidal distribution when a third of the slots are left without conductors, and a different number of turns of coil are inserted in each slot. The magnetizing winding of the nonsalient-pole machine of Figure 2.3 is illustrated as an example of such a winding.

Example 2.26: Create various kinds of zone distributions to approach a sinusoidal current linkage distribution for a single-phase winding with $m = 1, p = 1, Q = 24, q = 12$.

Solution: Figure 2.42 depicts various methods to produce a current linkage waveform with a single-phase winding.

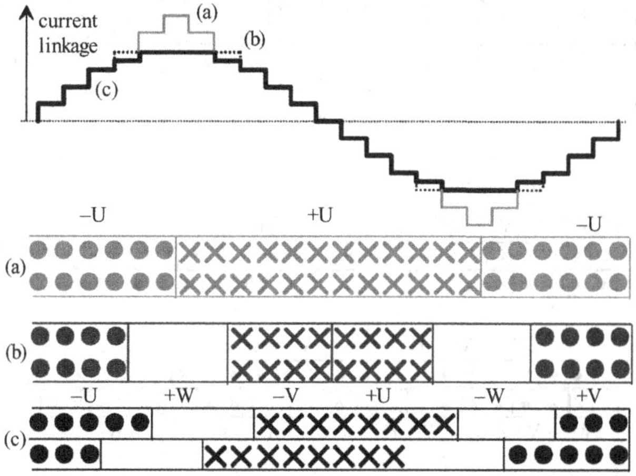

Figure 2.42 Zone diagram of a single-phase winding $p = 1, Q = 24, q = 12$ and current linkage distributions produced by different zone distributions. (a) A single-phase winding covering all slots. (b) A two-thirds winding, with the zones of a corresponding three-phase winding. The three phase zones +W and –W are left without conductors. The distribution of the current linkage is better than in the case (a). (c) A two-thirds short-pitched winding, producing a current linkage distribution closer to an ideal (dark stepped line).

2.14 Windings Permitting a Varying Number of Poles

Here, windings permitting a variable pole number refer to such single- or poly-phase windings that can be connected via terminals to a varying number of poles. Windings of this type occur typically in asynchronous machines, when the rotation speed of the machine has to be varied in a certain ratio. The most common example of such machines is a two-speed motor. The most common connection for this kind of arrangement is a Lindström–Dahlander connection that enables alteration of the pole number of a three-phase machine in the proportion of 1 to 2. Figure 2.43 illustrates the winding diagram of a single phase of a 24-slot machine. It can be arranged as a double-layer diamond winding, which is a typical Dahlander winding. Now the smaller number of pole pairs is denoted by p' and the higher by p'', where $p'' = 2p'$. The winding is divided into two sections U1–U2 and U3–U4, both of which consist of two coil groups with the higher number of poles.

The Dahlander winding is normally realized for the higher pole pair number as a double-layer, double-zone-width winding. The number of coil groups per phase is equal to p'', which

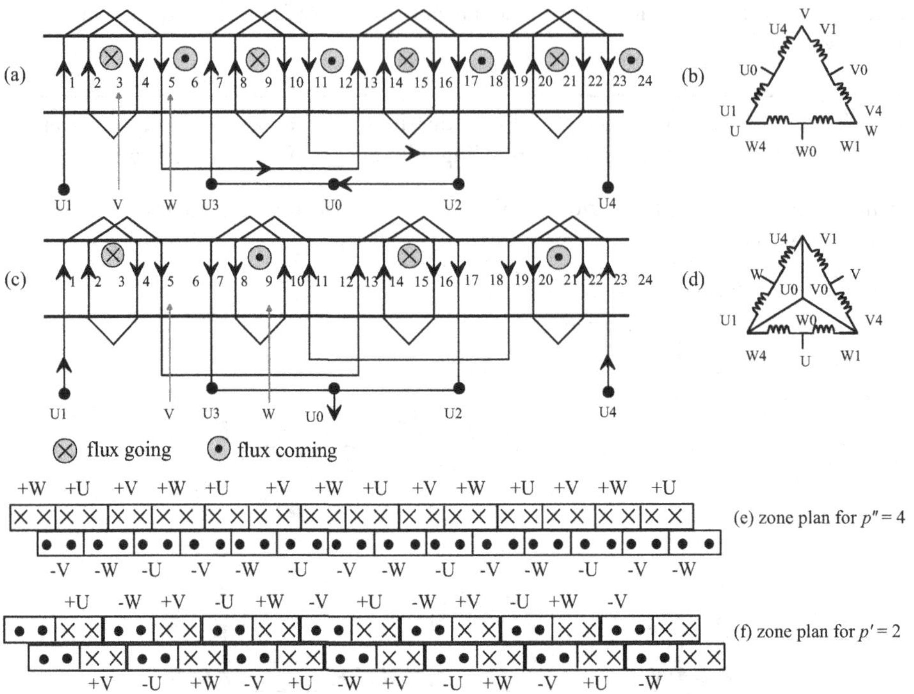

Figure 2.43 Principle of a Dahlander winding. The upper connection (a) produces eight poles and the lower (c) four poles. The number of poles is shown by the flux arrowheads and tails (b) and (d) equivalent connections. When the number of coil turns of the phase varies inversely proportional to the speed, the winding can be supplied with the same voltage at both speeds. Network connections are U, V and W. The figure illustrates only a winding of one phase. In Figure 2.43d between U1 and U4 there is the connection W, between W4 and W1 there is U and between V1 and V4 there is V to keep the same direction of rotation; (e) and (f) zone plans for $p'' = 4$ and $p' = 2$.

is always an even number. Deriving a double-layer, integral slot, full-pitch Dahlander winding (short-pitch or fractional slot Dahlander windings are not possible at all) starts by creating mp'' negative zones with a double width. All the negative zones –U, –V, –W are located in the top layer and all the positive zones in the bottom layers of slots.

The phase U is followed by the winding V at a distance of 120°. For a number of pole pairs $p'' = 4$, the winding V has to be placed at a distance

$$\frac{Q}{3p''}\tau_u = \frac{24}{3 \cdot 4}\tau_u = 2\tau_u$$

from the winding U. The winding V thus starts from slot 3 in the same way as the winding U starts from slot 1. The winding W starts then from slot 5. When considering the pole pair number $p' = 2$, we can see that the winding V is placed at a distance

$$\frac{Q}{3p'}\tau_u = \frac{24}{3 \cdot 2}\tau_u = 4\tau_u$$

from the winding U and thus starts from slot 5, and the winding W from slot 9. External connections have to be arranged to meet these requirements. At its simplest, the shift of the above-mentioned pole pair from one winding to another is carried out according to the right-hand circuit diagrams. To keep the machine rotating in the same direction, the phases U, V and W have to be connected according to the illustration.

There is also another method to create windings with two different pole numbers: pole amplitude modulation (PAM) is a method with which ratios other than 1 : 2 may be found. PAM is based on the following trigonometric equation:

$$\sin p_{\mathrm{b}}\alpha \sin p_{\mathrm{m}}\alpha = \frac{1}{2}\left[\cos(p_{\mathrm{b}} - p_{\mathrm{m}})\alpha - \cos(p_{\mathrm{b}} + p_{\mathrm{m}})\alpha\right]. \tag{2.103}$$

The current linkage is produced as a function of the angle α running over the perimeter of the air gap. A phase winding might be realized with a base pole pair number p_{b} and a modulating pole pair number p_{m}. In practice, this means that if for instance $p_{\mathrm{b}} = 4$ and $p_{\mathrm{m}} = 1$, the PAM method produces pole pairs $4 - 1$ or $4 + 1$. The winding must be created so that one of the harmonics is damped and the other dominates.

2.15 Commutator Windings

A characteristic of poly-phase windings is that the phase windings are, in principle, galvanically separated. The phase windings are connected via terminals to each other, in a star or in a polygon. The armature winding of commutator machines does not start or end at terminals. The winding comprises turns of conductor soldered as a continuum and wound in the slots of the rotor so that the sum of induced voltages is always zero in the continuum. This is possible if the sum of slot voltages is zero. All the coil sides of such a winding can be connected in series to form a continuum without causing a current to flow in the closed ring as a result of the voltages in the coil sides. An external electric circuit is created by coupling the connection points of the coils to the commutator segments. A current is fed to the winding via brushes dragging along the commutator. The commutator switches the coils in turns to the brushes thus acting as a mechanical inverter or rectifier depending on the operating mode of the machine. This is called commutating. In the design of a winding, the construction of a reliable commutating arrangement is a demanding task.

Commutator windings are always double-layer windings. One coil side of each coil is always in the upper layer and the other in the bottom layer approximately at the distance of a pole pair from each other. Because of problems in commutating, the voltage difference between the commutator segments must not be too high, and thus the number of segments and coils has always to be high enough. On the other hand, the number of slots is restricted by the minimum width of the teeth. Therefore, usually more than two coil sides are placed in each slot. In the slot of the upper diagram of Figure 2.44, there are two coil sides, and in the lower diagram the number of coil sides is four. The coil sides are often numbered so that the sides of the bottom layer are even numbers, and the slots of the upper layer are odd numbers. If the number of coils is z_{c}, $2z_{\mathrm{c}}$ coil sides have to be mounted in Q slots, and thus there are $2u = 2z_{\mathrm{c}}/Q$ sides in a slot. The symbol u gives the number of coil sides in one layer. In each side, there are N_{v} conductors. The total number of conductors z in the armature is

$$z = Qz_{\mathrm{Q}} = 2uN_{\mathrm{v}}Q = 2z_{\mathrm{c}}N_{\mathrm{v}}. \tag{2.104}$$

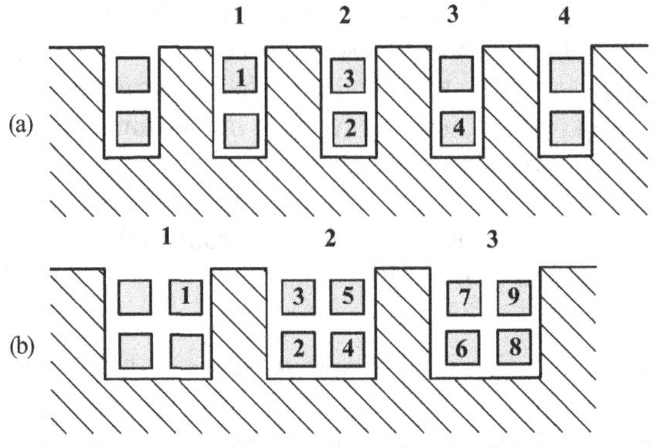

Figure 2.44 Two examples of commutator winding coil sides mounted in the slots. (a) Two coil sides in a slot, one side in a layer, $u = 1$. (b) Four coil sides in a slot, two coil sides in a layer, $u = 2$. Even-numbered coil sides are located at the bottom of the slots. There has to be a large enough number of coils and commutator segments to keep the voltage between commutator segments small enough.

Here

Q is the number of slots,
z_Q is the number of conductors in a slot,
u is the number of coil sides in a layer,
z_c is the number of coils,
N_v is the number of conductors in a coil side, $2uN_v = z_Q$, because
$z_Q = z/Q = 2uN_v\ Q/Q = 2uN_v$, see (2.104).

Commutator windings may be used both in AC and DC machines. Multi-phase commutator AC machines are, however, becoming rare. DC machines, instead, are built and used also in the present-day industry even though DC drives are gradually being replaced by power electronic AC drives. Nevertheless, it is advisable to look briefly also at the DC windings.

The AC and DC commutator windings are in principle equal. For simplicity, the configuration of the winding is investigated with the voltage phasor diagram of a DC machine. Here, it suffices to investigate a two-pole machine, since the winding of machines with multiple poles is repeated unchanged with each pole pair. The rotor of Figure 2.45, with $Q = 16$, $u = 1$, is assumed to rotate clockwise at an angular speed Ω in a constant magnetic field between the poles N and S.

The magnetic field rotates in the positive direction with respect to the conductors in the slots, that is counterclockwise. Now, a coil voltage phasor diagram is constructed for a winding, in which we have already calculated the difference of the coil side voltages given by the coil voltage phasor diagram. By applying the numbering system of Figure 2.44, we have in slot 1 the coil sides 1 and 32, and in slot 9 the coil sides 16 and 17. With this system, the coil voltage phasor diagram can be illustrated as in Figure 2.45b.

Figure 2.45 shows that if the induced emf decides the direction of the armature current, the produced torque is opposite to the direction of rotation (counterclockwise in Figure 2.45), and mechanical power has to be supplied to the machine, which is acting as a generator. Now,

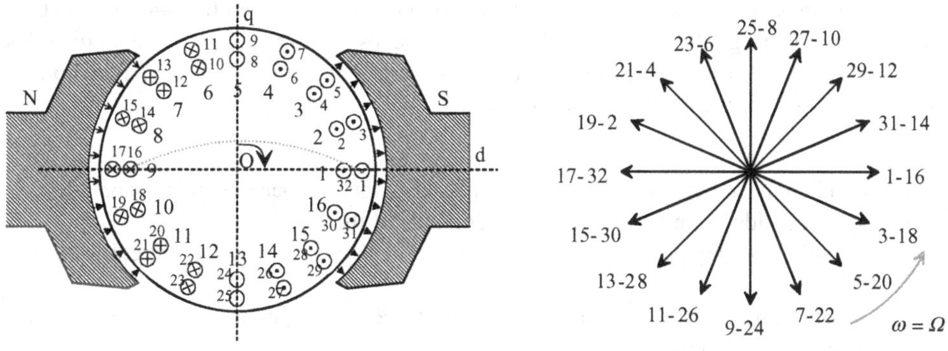

(a) (b)

Figure 2.45 (a) Principle of a two-pole, double-layer commutator armature. The armature rotates at an angular speed Ω clockwise generating an emf in the conductors in the slots. The emf tends to create the current directions illustrated in the figure. (b) A coil voltage phasor diagram of the armature. It is a full-pitch winding, which does not normally occur as a commutator winding. Nevertheless, a full-pitch winding is given here as a clarifying example. $Q = 16$, $u = 1$ (one coil side per layer).

if the armature current is forced to flow against the emf with the assistance of an external voltage or current source, the torque is in the direction of rotation, and the machine acts as a motor.

There are $z_c = Qu = 16 \times 1 = 16$ coils in the winding, the ends of which should next be connected to the commutator. Depending on the way they are connected, different kinds of windings are produced. Each connection point of the coil ends is connected to the commutator. There are two main types of commutator windings: lap windings and wave windings. A lap winding has coils, creating looplike patterns. The ends of the coils are connected to adjacent commutator segments. A wave winding has a wavelike drawing pattern when presented in a plane.

The number of commutator segments is given by

$$K = uQ, \tag{2.105}$$

because each coil side begins and ends at the commutator segment. The number of commutator segments, therefore, depends on the conductor arrangement in the slot, and eventually on the number of coil sides in one layer. Further important parameters of commutator windings are:

y_Q coil span expressed as the number of slots per pole (see Equation (2.111)).

y_1 back-end connector pitch, which is a coil span expressed as the number of coil sides. For the winding, the coil sides of which are numbered with odd figures in the top layer and with even figures in the bottom layer, this is

$$y_1 = 2uy_Q \mp 1, \tag{2.106}$$

where the minus sign stands for the coil side numbering as seen in Figure 2.44, and the plus sign for the numbering where in slot 1 there are coil sides 1, 2, in slot 2 there are coil sides 3, 4, and so on, if $u = 1$; or in the top layer of slot 1 there are coil sides 1, 3 and in the bottom layer there are coil sides 2, 4, and so on, if $u = 2$.

y_2 front-end connector pitch; it is a pitch expressed as the number of coil sides between the right coil side of one coil and the left coil side of the next coil.
y total winding pitch expressed as the number of coil sides between two left coil sides of two adjacent coils.
y_c commutator pitch between the beginning and end of one coil expressed as the number of commutator segments.

The equation for commutator pitch is a basic equation for winding design because this pitch must be an integer

$$y_c = \frac{nK \pm a}{p}, \tag{2.107}$$

where a is the number of parallel paths per half armature in a commutator winding, which means $2a$ parallel paths for the whole armature.

The windings that are most often employed are characterized on the basis of n:

1. If $n = 0$, it results in a lap winding. The commutator pitch will be $y_c = \pm a/p$, which means that a is an integer multiple of p to give an integer for the commutating pitch. For a lap winding $2a = 2p$, this means $a = p$, $y_c = \pm 1$. Such a winding is called a parallel one. The plus sign is for a progressive winding moving from left to right, and the minus sign for a retrogressive winding moving from right to left. If a is a k-multiple of the pole pair number, $a = kp$, then it is a k-multiplex parallel winding. For example, for $a = 2p$, the commutator pitch is $y_c = \pm 2$, and this winding is called a duplex parallel winding.
2. If $n = 1$, it results in a wave winding and a commutator pitch, that is

$$y_c = \frac{K \pm a}{p} = \frac{uQ \pm a}{p} \tag{2.108}$$

must be an integer. The plus sign is for progressive and the minus sign for retrogressive winding. In the wave winding the number of parallel paths is always 2; there is only one pair of parallel paths, irrespective of the number of poles: $2a = 2$, $a = 1$.

Not all the combinations of K, a, p result in an integer. It is a designer's task to choose a proper number of slots, coil sides, number of poles and type of winding to ensure an integer commutator pitch.

If the number of coils equals the number of commutator segments, then, if the coil sides are numbered with odd figures in the top layer and even figures in the bottom layer, we can write

$$y = y_1 + y_2 = 2y_c. \tag{2.109}$$

Therefore, if the commutator pitch is determined, the total pitch expressed as a number of coil sides is given by

$$y = 2y_c \tag{2.110}$$

and after y_1 is determined from the numbers of slots per pole y_Q and number of coil sides in a layer u,

$$y_Q \approx \frac{Q}{2p}, \tag{2.111}$$

$$y_1 = 2u\, y_Q \mp 1. \tag{2.112}$$

The front-end connector pitch can be determined as

$$y_2 = y - y_1. \tag{2.113}$$

2.15.1 Lap Winding Principles

The principles of the lap winding can best be explained by an example.

Example 2.27: Produce a layout of a lap winding for a two-pole DC machine with 16 slots and one coil side in a layer.

Solution: Given $Q = 16$, $2p = 2$, $u = 1$, the number of commutator segments is $K = uQ = 1 \cdot 16 = 16$, and for a lap winding $2a = 2p = 2$. The commutator pitch is $y_c = \pm a/p = \pm 1$. We choose a progressive winding, which means that $y_c = +1$ (the winding proceeds from left to right), and the total pitch is $y = 2y_c = 2$. The coil span y_Q in number of slots is given by the number of slots per pole:

$$y_Q = \frac{Q}{2p} = \frac{16}{2} = 8.$$

The same pitch expressed in the number of coil sides is $y_1 = 2u\, y_Q - 1 = 2 \cdot 1 \cdot 8 - 1 = 15$. The front-end connector pitch is $y_2 = y - y_1 = 2 - 15 = -13$, which is illustrated in Figure 2.46. The coils can be connected in series in the same order that they are inserted in the slots of the rotor. The neighboring coils are connected together, coil 1–16 to coil 3–18, and again to 5–20, and so on. This yields the winding diagram of Figure 2.46.

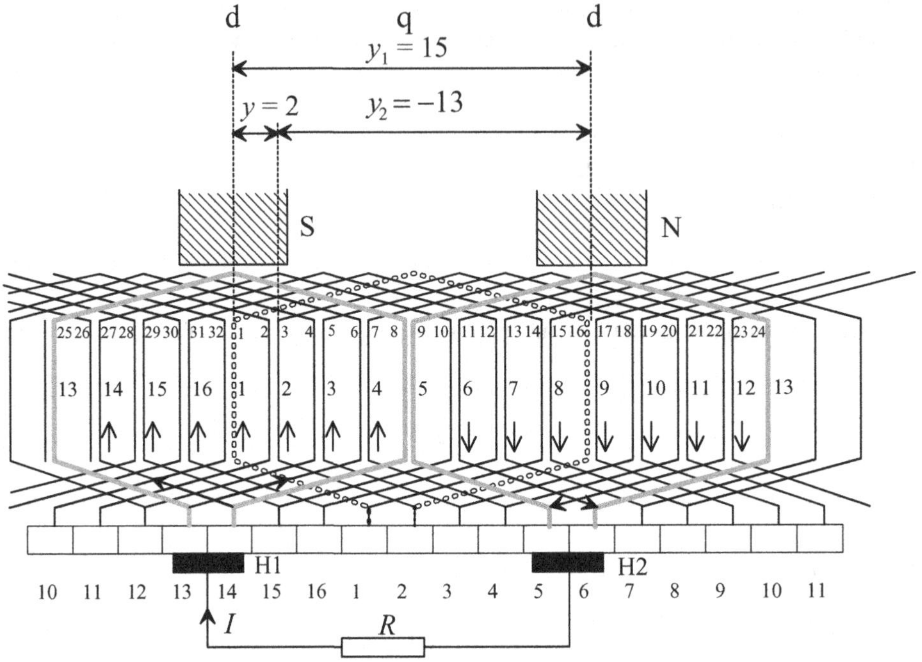

Figure 2.46 Diagram of a full-pitch lap winding. The winding is connected via brushes to an external resistance R. The pole shoes are also illustrated above the winding. In reality, they are placed above the coil sides. The laps illustrated with a thick line have been short-circuited via the brushes during commutation. The direction of current changes during the commutation. The numbers of slots (1–16) are given. The numbers of coil sides (1–32) in the slots are also given, the coil sides 1 and 32 are located in slot 1 and, for example, the coil sides 8 and 9 are located in slot 5. The commutator segments are numbered (1–16) according to the slots. It is said that the brushes are on the quadrature axis; this is nevertheless valid only magnetically. In this figure, the brushes are physically placed close to the direct axes.

When we follow the winding by starting from the coil side 1, we can see that it proceeds by one step of span $y_1 = 15$ coil sides. In slots 1 and 9, a coil with a large enough number of winding turns is inserted. Finally, after the last coil turn the winding returns left by a distance of one step of connection $y_2 = 2 - 15 = -13$ coil sides, to the upper coil side 3. We continue in this way until the complete winding has been gone through, and the coil side 14 is connected to the upper coil side 1 through the commutator segment 1. The winding has now been closed as a continuum. The winding proceeds in laps from left to right; hence the name lap winding.

The pitches of a commutator winding are thus calculated by the number of coil sides, not by the number of slots, because there can be more than two coil sides in one layer, for instance four coil sides in a slot layer ($u = 1, 2, 3, 4, \ldots$).

In the lap winding of Figure 2.46, all the coil voltages are connected in series. The connection in series can be illustrated by constructing a polygon of the coil voltages, Figure 2.47. This

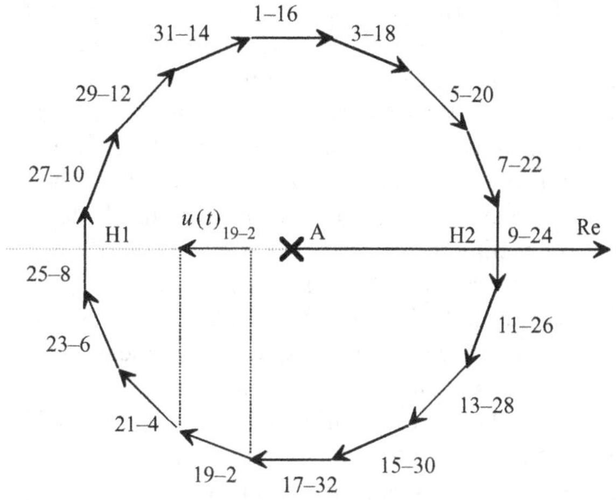

Figure 2.47 Polygon of coil voltages of the winding in Figure 2.46. The sum of all the voltages is zero and hence the coils may be connected in series. The instantaneous value of a coil voltage $u(t)$ will be the projection of the phasor on the real axis, for example $u(t)_{19-2}$.

figure illustrates the phasors at time $t = 0$ in the coil voltage phasor diagram of Figure 2.45. When the rotor rotates at an angular speed Ω, the coil voltage phasor diagram also rotates in a two-pole machine at an angular speed $\omega = \Omega$. Also the polygon rotates around its center at the same angular speed. The real instantaneous value of each coil voltage may be found as a projection of the phasor on the real axis (see figure). According to the figure, the sum of all coil voltages is zero. Therefore, no circulating currents occur in the continuum.

The highest value of the sum of the instantaneous values of coil voltages is equal to the diameter H1–H2 parallel to the real axis. This value remains almost constant as the polygon rotates, and thus the phasor H1–H2 represents a DC voltage without significant ripple. The voltage approaches a constant value when the number of coils approaches infinity. A DC voltage can be connected to an external electric circuit via the brushes that are in contact with the commutator segments. At the moment $t = 0$, as illustrated, the brushes have to be in contact with the commutator segment pairs 5–6 and 13–14 that are connected to coils 9–24 and 25–8. According to Figure 2.46, the magnetic south pole (S) is at slot 1 and magnetic north pole (N) at slot 9. Further, the direction of the magnetic flux is towards the observer at the south pole, and away from the observer at the north pole. As the winding moves left, a positive emf is induced in the conductors under the south pole, and a negative emf under the north pole, in the direction indicated by the arrows.

By following the laps from segment 14 to coil 27–10, we end up at the commutator segment 15, then gradually at segments 16, 1, 2, 3, 4 and at last coil 7–22 is brought to segment 5, touched by the brush H2. We have just described one parallel path created by coils connected in series via commutator segments. An induced emf creates a current in the external part of the electric circuit from the brush H2 to the brush H1, and thus in a generator drive, H2 is a positive brush with the given direction of rotation. Half of the current I in the external part of the circuit flows in the above-described path, and the other half via the coils 23–6 . . . 11–26, via commutator segments 12, 11, . . . 7, to the brush H2 and further to the external part of the circuit. In other words, there are two parallel paths in the winding. In the windings of large machines, there can be several pairs of paths in order to prevent the cross-sectional area of the conductors from increasing impractically. Because the ends of different pairs of paths touch the neighboring commutator segments and have no other galvanic contact, the brushes have to be made wider to keep each pair of paths always in contact with the external circuit.

If for instance in the coil voltage phasor diagram of Figure 2.45 every other coil 1–16, 5–20, 9–24 ... 29–12 is connected in series with the first pair of paths, the lap is closed after the last turn of coil side 12 by connecting the coil to the first coil 1–16 ($12 \rightarrow 1$, from 12 to 1). The coils that remain free are connected in the order 3–18, 7–22 ... 31–14 and the lap is closed at the position $14 \rightarrow 3$. This way, a doubly-closed winding with two paths $2a = 2$ is produced. In the voltage polygon, there are two revolutions, and its diameter, that is the brush voltage, is reduced to half the original polygon of one revolution illustrated in Figure 2.47. The output power of the system remains the same, because the current can be doubled when the voltage is cut in half. In general, the number of pairs of paths a always requires that $a - 1$ phasors are left between the phasors of series-connected coils in a coil voltage phasor diagram. The phasors may be similar. Because u is the number of coil sides per layer, each phasor of the coil voltage phasor diagram represents u coil voltages. This makes it possible to skip completely similar voltage phasors. This takes place for instance when $u = 2$.

The winding of Figure 2.46 is wound clockwise, because the voltages of the coil voltage phasor diagram are connected in series clockwise starting from phasor 1–16. Were coil 1–16 connected via the commutator segment 16 to coil 31–14, the winding would have been wound counterclockwise.

The number of brushes in a lap winding is always the same as the number of poles. Brushes of the same sign are connected together. According to Figure 2.46, the brushes always short-circuit those coils, the coil sides of which are located at the quadrature axis (in the middle, between two stator poles) of the stator, where the magnetic flux density created by the pole magnetization is zero. This situation is also described by stating that the brushes are located at the quadrature axis of the stator independent of the real physical position of the brushes.

2.15.2 Wave Winding Principles

The winding of Figure 2.46 can be turned into a wave winding by bending the coil ends of the commutator side according to the illustration in Figure 2.48, as a solution of Example 2.28 (see also Figure 2.49).

Example 2.28: Produce a layout of a wave winding for a two-pole DC machine with 16 slots and one coil side in a layer.

Solution: Given that $Q = 16$, $2p = 2$, $u = 1$. The number of commutator segments is $K = uQ = 1 \cdot 16 = 16$, and for a wave winding $2a = 2$. The commutator pitch is

$$y_c = \frac{K \pm a}{p} = \frac{16 \pm 1}{1} = 17 \text{ or } 15.$$

We choose $y_c = +15$ (winding proceeds from right to left), and the total winding pitch is $y = 2y_c = 30$. The coil span y_Q in number of slots is given by the number of slots per pole: $y_Q = Q/2p = 16/2 = 8$. The same pitch expressed as number of coil sides is $y_1 = 2u\, y_Q -$

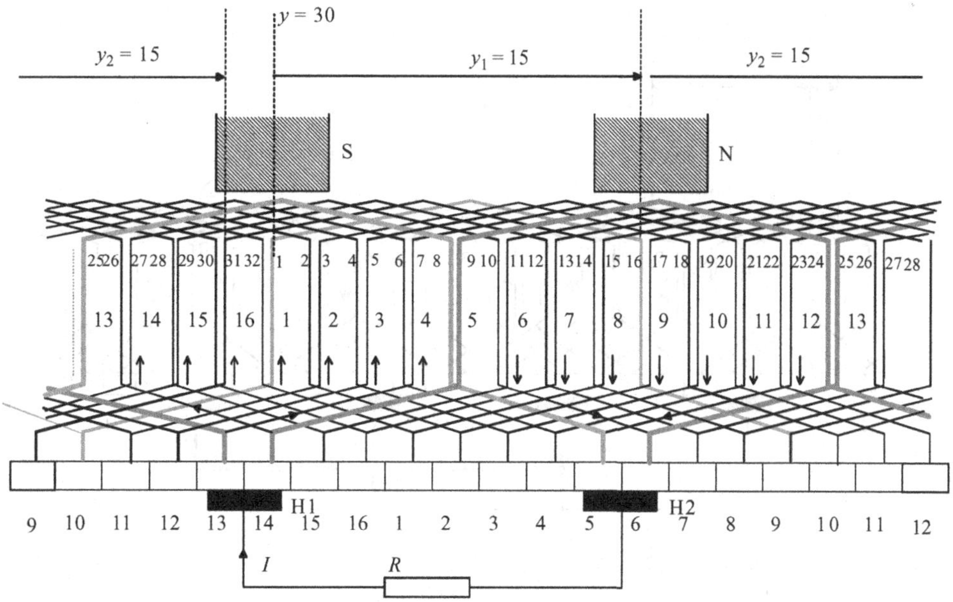

Figure 2.48 Full-pitch lap winding of Figure 2.46 turned into a wave winding. The currents of waveforms indicated with a thick line are commutating at the moment illustrated in the figure. The commutator pitch is $y_c = 15$, which means almost two pole pitches. For instance, the wave coil that starts at the commutator segment 14 ends at segment 13, because $14 + y_c = 14 + 15 = 29$, but there are only 16 commutator segments, and therefore $29 - 16 = 13$.

$1 = 2 \cdot 1 \cdot 8 - 1 = 15$. The front-end connector pitch is $y_2 = y - y_1 = 30 - 15 = 15$, which is shown in Figure 2.48.

In the above wave winding, the upper coil side 1 is connected to the commutator segment 10, and not to segment 1 as in the lap winding. From segment 10, the winding proceeds to the bottom side 18. The winding thus receives a waveform. In the figure, the winding proceeds from right to left, and counterclockwise in the coil voltage phasor diagram. The winding is thus rotated to the left. If the winding were turned to the right, the commutator pitch would be $y_c = 17$, $y = 34$, $y_1 = 15$, $y_2 = 19$. The coil from the bottom side 16 would have to be bent to the right to segment 10, and further to the upper side 3, because $16 + y_2 = 16 + 19 = 35$. But there are only 32 coil sides, and therefore the coil will proceed to $35 - 32 = 3$ or the third coil side. The commutator ends would in that case be even longer, which would be of no use.

The pitch of the winding for a wave winding follows the illustration

$$y = y_1 + y_2. \tag{2.114}$$

The position of brushes in a wave winding is solved in the same way as in a lap winding. When comparing the lap and wave windings, we can see that the brushes short-circuit the same coils in both cases. The differences between the windings are merely structural, and the winding type is selected basically on these structural grounds. As written above, the pitch of

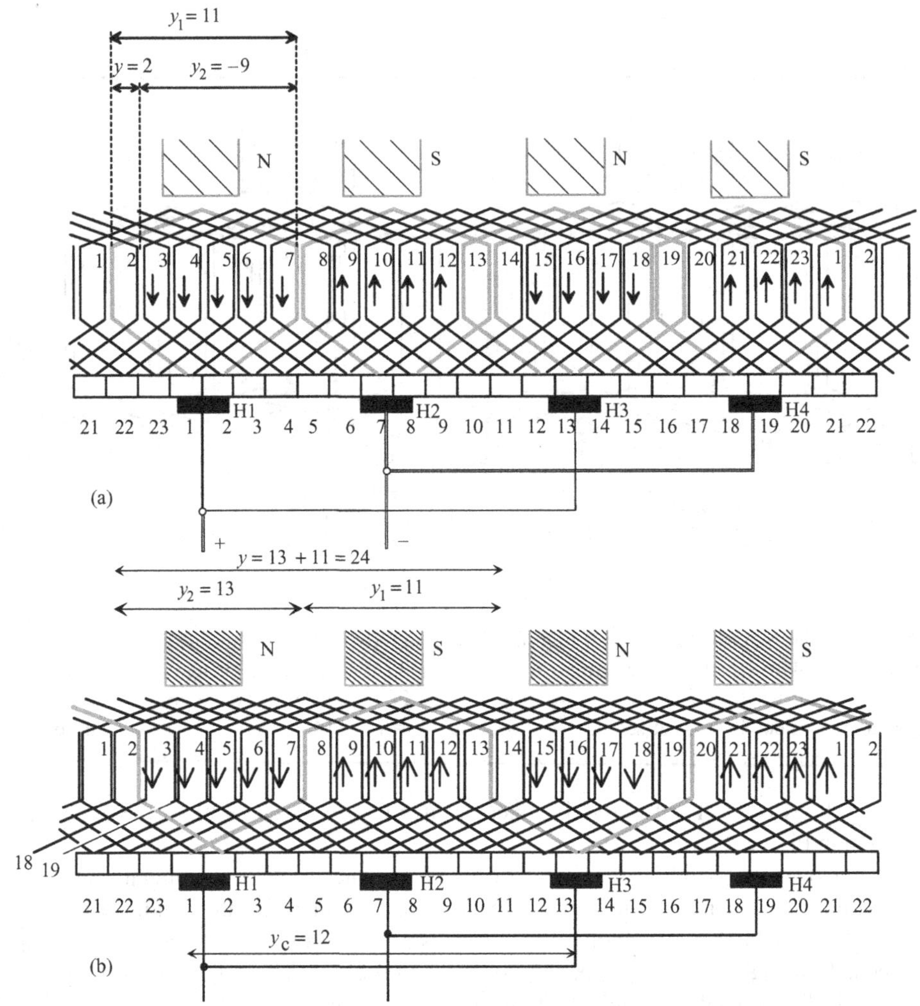

Figure 2.49 (a) Four-pole, double-layer lap winding presented in a plane. The winding moves from left to right and acts as a generator. The coils belonging to the commutator circuit are illustrated by a thick line. This winding is not a full-pitch winding, unlike the previous ones. The illustrated winding commutates better than a full-pitch winding. (b) The same winding developed into a wave winding. The wave under commutation is drawn with a thicker line than the others. Coil side numbering is left away for clarity. The principle, however, is the same as in Fig 2.46.

the winding for regular commutator windings, which equals the commutator pitch, is obtained from

$$y_c = \frac{nK \pm a}{p} = \frac{nuQ \pm a}{p}, \tag{2.115}$$

where the plus sign in the equation is used for progressive winding (from left to right, i.e. clockwise) and the minus is used for retrogressive winding (from right to left, i.e.

counterclockwise); n is zero or a positive integer. If n is zero, it results in a lap winding; if $n = 1$, it results in a wave winding. The commutator pitch y_c must be an integer, otherwise the winding cannot be constructed. Not all combinations of K, p and a result in y_c as an integer, and therefore a designer must solve this problem in its complexity.

2.15.3 Commutator Winding Examples, Balancing Connectors

Nowadays, the conductors are typically inserted in slots that are on the surfaces of the armature. These windings are called drum armature windings. Drum-wound armature windings are in practice always double-layer windings, in which there are two coil sides in the slot on top of each other. Drum armature windings are constructed either as lap or wave windings. As discussed previously, the term "lap winding" describes a winding that is wound in laps along the periphery of the armature, the ends of one coil being connected to adjacent segments, Figure 2.49a.

Example 2.29: Produce a layout of the lap winding for a four-pole DC machine with 23 slots and one coil side in a layer.

Solution: Given that $Q = 23$, $2p = 4$, $u = 1$, the number of commutator segment is $K = uQ = 1 \cdot 23 = 23$ and for a lap winding $2a = 2p = 4$. The commutator pitch is $y_c = \pm a/p = 2/2 = \pm 1$. We choose a progressive winding, which means that $y_c = +1$ (winding proceeds from left to right), and the total winding pitch is $y = 2y_c = 2$. The coil span y_Q as the number of slots is given by the number of slots per pole: $y_Q = Q/2p = 23/4 = 5.75 \Rightarrow$ 6 slots. The same pitch expressed as the number of coil sides is $y_1 = 2uy_Q - 1 = 2 \cdot 1 \cdot 6 - 1 = 11$. The minus sign is used because of the coil side arrangement in the slots according to Figure 2.44a.

The front-end connector pitch is $y_2 = y - y_1 = 2 - 11 = -9$, which is shown in Figure 2.49a.

In the winding of Figure 2.49a, there are 23 armature coils (46 coil sides, two in each slot), with one turn in each, four brushes and a commutator with 23 segments. There are four current paths in the winding ($2a = 4$), which thereby requires four brushes. By following the winding starting from the first brush, we have to travel a fourth of the total winding to reach the next brush of opposite sign.

From segment 1, the left coil side is put to the upper layer 1 in slot 1 (see Figure 2.44a). Then, the right side is put to the bottom layer $1 + y_1 = 1 + 11 = 12$ in slot 7, from where it is led to the segment number $1 + y_c = 1 + 1 = 2$, and then from segment 2 to the upper layer 3 in slot 2, because $12 - 9 = 3$. It proceeds to the bottom layer 14 in slot 8, because $3 + 11 = 14$, and then to segment 3, and continues to the coil side 5 in slot 3 and via 16 in slot 9 to segment 4, and so on.

In the figure, the brushes are broader than the segments of the commutator, the laps illustrated with thick lines being short-circuited via the brushes. In DC machines, a proper commutation requires that the brushes cover several segments. The coil sides of short-circuited coils are approximately in the middle between the poles, where the flux density is small. In these coils, the induced voltage is low, and the created short-circuit current is thus insignificant.

Example 2.30: Produce a layout of a wave winding for a four-pole DC machine with 23 slots and one coil side in a layer.

Solution: Given that $Q = 23$, $2p = 4$, $u = 1$. The number of commutator segments is $K = uQ = 1 \cdot 23 = 23$, and for a wave winding $2a = 2$. The commutator pitch is $y_c = \frac{K \pm a}{p} = \frac{23 \pm 1}{2} = 12$, or 11.

We choose $y_c = +12$ and the total winding pitch is $y = 2y_c = 24$. The coil span y_Q as the number of slots is given by the number of slots per pole: $y_Q = Q/2p = 23/4 = 5.75 \Rightarrow 6$. The same pitch expressed as the number of coil sides is $y_1 = 2uy_Q - 1 = 2 \cdot 1 \cdot 6 - 1 = 11$. The minus sign is used because of the coil side arrangement in the slots according to Figure 2.44a.

The front-end connector pitch is $y_2 = y - y_1 = 24 - 11 = 13$, which is shown in Figure 2.49b. Figure 2.49b illustrates the same winding as in Figure 2.49a but developed for a wave winding. In wave windings, there are only two current paths: $2a = 2$ regardless of the number of poles. A wave winding and a lap winding can also be combined as a frog-leg winding.

We can see in Figure 2.49b that from the commutator segment 1 the coil left side is put to the upper layer 13 in slot 7. Then the right side in the lower layer is put to $13 + y_1 = 13 + 11 = 24$ in slot 13, from where it is led to segment $1 + y_c = 1 + 12 = 13$, and then from segment 13 to the upper layer 37 (slot 19), because $24 + y_2 = 24 + 13 = 37$. We then proceed to the lower layer 2 in slot 2, because $37 + y_1 = 37 + 11 = 48$, which is over the number coil sides of 46 in 23 slots; therefore, it is necessary to make a correction $48 - (2 \times 23) = 48 - 46 = 2$. From here we continue to segment 2, because $13 + y_c = 13 + 12 = 25$, after the correction $25 - 23 = 2$, and so on.

When passing through a wave winding from one brush to another brush of the opposite sign, half of the winding and half of the segments of the commutator are gone through. The current thus has only two paths irrespective of the number of poles. As a matter of fact, in a wave winding, only one pair of brushes is required, which is actually enough for small machines. Nevertheless, usually as many brushes are required as there are poles in the machine. This number is selected in order to reach a maximum brush area with the shortest commutator possible. One coil of a wave winding is always connected to the commutator at about a distance of two pole pitches.

A wave winding is a more common solution than a lap winding for small (<50 kW) machines, since it is usually more cost effective than a lap winding. In a machine designed for a certain speed, a number of pole pairs and a flux, the wave winding requires less turns than a lap winding, excluding a two-pole machine. Correspondingly, the cross-sectional area of conductors in a wave winding has to be larger than the area of a lap winding. Therefore, in a machine of a certain output, the copper consumption is the same irrespective of the type of winding.

The previous windings are simple examples of various alternative constructions for commutator windings. In particular, when numerous parallel paths are employed, we must ensure that the voltages in the paths are equal, or else compensating currents will occur flowing through the brushes. These currents create sparks and wear out the commutator and the brushes. The commutator windings have to be symmetrical to avoid extra losses.

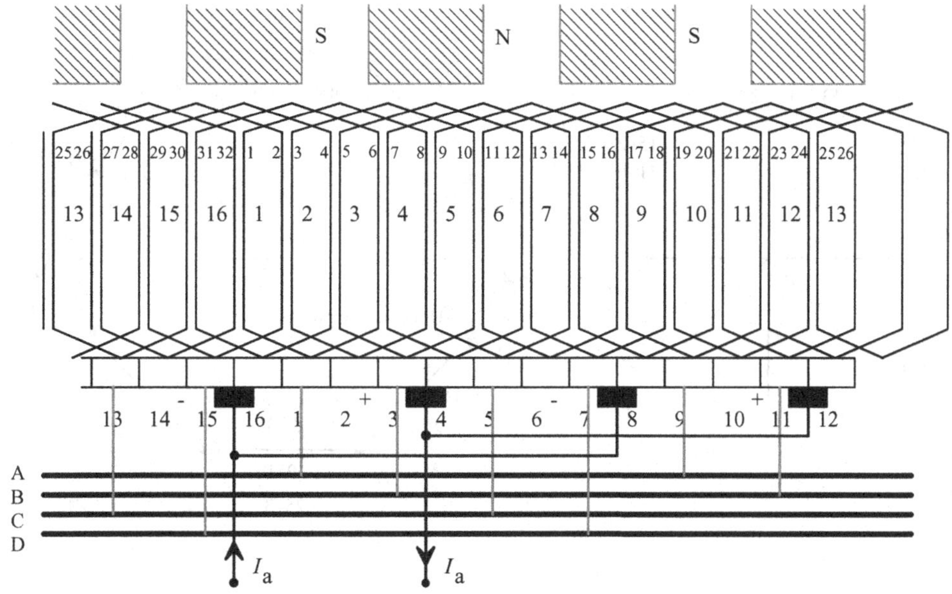

Figure 2.50 Balancing connectors or equalizer bars (bars A, B, C and D) of a lap winding. For instance, the coil sides 29 and 13 are located in similar magnetic positions if the machine is symmetric. Hence, the commutator segments 15 and 7 may be connected together with a balancing connector.

If the number of parallel pairs of paths is a, there are also a revolutions in the voltage polygon. If the revolutions completely overlap the voltage polygon, the winding is symmetrical. In addition to this condition, the diameter H1–H2 has to split the polygon into two equal halves at all times. These conditions are usually met when both the number of slots Q and the number of poles $2p$ are evenly divisible by the number of parallel paths $2a$. Figure 2.50 illustrates the winding diagram of a four-pole machine. The number of slots is $Q = 16$, and the number of parallel paths is $2a = 4$. Hence, the winding meets the above conditions of symmetry. The coil voltage phasor diagram and the voltage polygon are depicted in Figure 2.51. Since $a = 2$, there has to be one phasor $a - 1 = 1$ of the coil voltage phasor diagram between the consequent phasors of the polygon. When starting with phasor 1–8, the next phasor in the voltage polygon is 3–10. In between, there is phasor 17–24, which is of the same phase as the previous one, and so on. The first circle around the voltage polygon ends up at the point of phasor 15–22, in the winding diagram, at the commutator segment 9. However, the winding is not yet closed at this point, but continues for a second, similar revolution formed by phasors 17–24 . . . 31–6. The winding is fully symmetrical, and the coils short-circuited by the brushes placed on the quadrature axes.

The potential at different positions of the winding is now investigated with respect to an arbitrary position, for instance a commutator segment 1. In the voltage polygon, this zero potential is indicated by point A of the polygon. At $t = 0$, the instant depicted by the voltage polygon, the potential of segment 2 amounts to the amplitude of phasor 1–8, otherwise it is a projection on the straight line H1–H2. Respectively, the potential at all other points in the polygon with respect to segment 1 is at every instant the phasor drawn from point A to this

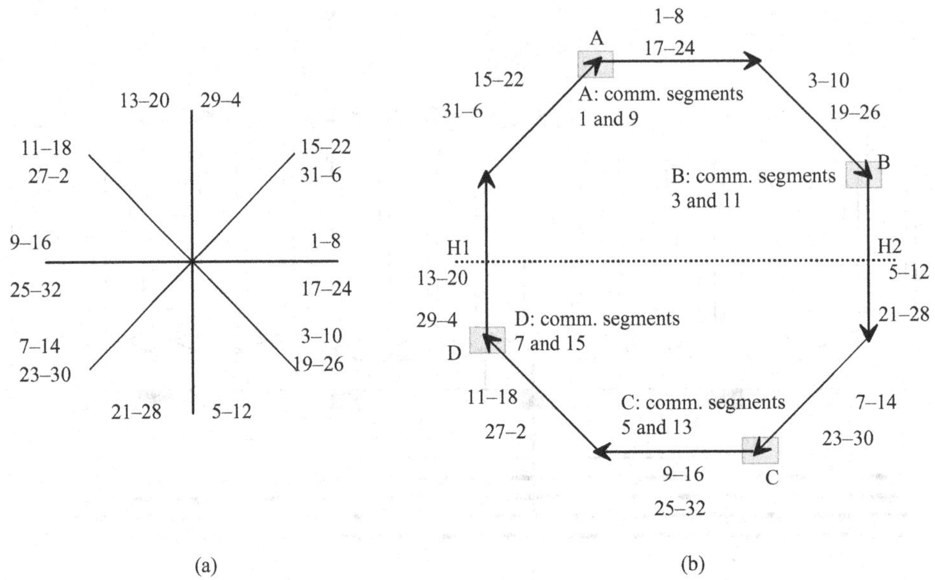

Figure 2.51 (a) Coil voltage phasor diagram of the winding of Figure 2.50. (b) The coil voltage polygon of the winding of Figure 2.50 and the connection points of the balancing connectors A, B, C and D. There are two overlapping polygons in the illustrated voltage polygon. Phasor 1–8 is the first phasor and 3–10 the next phasor of the polygon. Phasor 17–24 is equal to phasor 1–8 because both have their positions in the middle of poles, but it is skipped when constructing the first polygon. The first polygon is closed at the tip of phasor 15–22. The winding continues to form another similar polygon using phasors 17–24 . . . 31–6. The winding is completely symmetrical, and its brush-short-circuited coils are on the quadrature axes. Phasors 3–10 and 19–26 have a common tip point B in the polygons created by the commutator segments 3 and 11, as shown in Figure 2.50 and in figure (b), and the points can thus be connected by balancing connectors. The three other balancing connector points are A, C and D.

point, projected on the straight line H1–H2. Since for instance phasors 3–10 and 19–26 have a common point in the voltage polygon, the potential of the respective segments 3 and 11 of the commutator is always the same, and the potential difference between them is zero at every instant. Thus, these commutator segments can be connected with conductors. All those points that correspond to the common points of the voltage polygon can be interconnected. Figure 2.50 also depicts three other balancing connectors. The purpose of these compensating combinations is to conduct currents that are created by the structural asymmetries of the machine, such as the eccentricity of the rotor. Without balancing connectors, the compensating currents, created for various reasons, would flow through the brushes, thus impeding the commutation. There is an alternating current flowing in the compensating combinations, the resulting flux of which tends to compensate the asymmetry of the magnetic flux caused by the eccentricity of the rotor. From this we may conclude that compensating combinations are not required in machines with two brushes.

The maximum number of compensating combinations is obtained from the number of equipotential points. In the winding of Figures 2.50 and 2.51, we could thus assemble eight combinations; however, usually only a part of the possible combinations is needed to improve the operation of the machine. According to the illustrations, there are four possible combinations: A, B, C and D. In machines that do not commutate easily, it may prove necessary to employ all the possible compensating combinations. In small and medium machines, the compensating combinations are placed behind the commutator. In large machines, ring rails are placed at one end of the rotor, while the commutator is at the other end.

2.15.4 AC Commutator Windings

The equipotential points A, B, C and D of the winding in Figure 2.51 are connected with rails at A, B, C and D in Figure 2.50. The voltages between the rails at time $t = 0$ are illustrated by the respective voltages in the voltage polygon. When the machine is running, the voltage polygon is rotating, and therefore the voltages between the rails form a symmetrical four-phase system. The frequency of the voltages depends on the rotation speed of the machine. The phase windings of this system, with two parallel paths in each, are connected in a square. From the same principle, with tappings, we may create other poly-phase systems connected in a polygon, Figure 2.52.

If z_c coils are connected as a closed commutator winding with a parallel path pairs, the system is transformed with tappings into an m-phase AC system connected in a polygon by coupling the tappings at the distance of a step

$$y_m = \frac{z_c}{ma} \tag{2.116}$$

from each other. In a symmetrical poly-phase system, both y_m and z_c/a are integers. Windings of this type have been employed for instance in rotary converters, the windings of which have been connected both to the commutator and to the slip rings. They convert direct current into alternating current and vice versa. Closed commutator windings cannot be turned into star-connected windings, and only polygons are allowed.

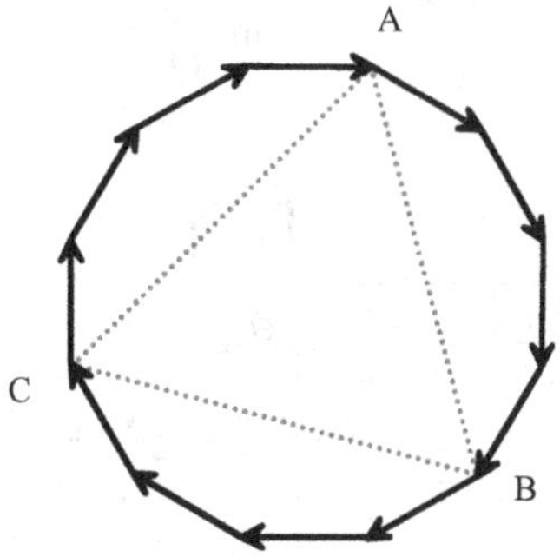

Figure 2.52 Equipotential points A, B and C of the voltage polygon, which represents 12 coils of the commutator winding, are connected as a triangle to form a poly-phase system. The respective voltages are connected via slip rings and brushes to the terminals of the machine.

2.15.5 *Current Linkage of the Commutator Winding and Armature Reaction*

The curved function of the current linkage created by the commutator winding is computed in the way illustrated in Figure 2.9 for a three-phase winding. When defining the slot current I_u, the current of the short-circuited coils can be set to zero. In short-pitched coils, there may be currents flowing in opposite directions in the different coil sides of a single slot. If z_b is the number of brushes, the armature current I_a is divided into brush currents $I = I_a/(z_b/2)$. Each brush current in turn is divided into two paths as conductor currents $I_s = I/2 = I_a/2a$, where a is the total number of path pairs of the winding. In a slot, there are z_Q conductors, and thus the sum current of a slot is

$$I_u = z_Q I_s = \frac{z}{Q}\frac{I_a}{2a}, \tag{2.117}$$

where z is the number of conductors in the complete winding. All the pole pairs of the armature are alike, and therefore it suffices to investigate only one of them, namely a two-pole winding, Figure 2.46. The curved function of the magnetic voltage therefore follows the illustration in Figure 2.53.

The number of brushes in a commutator machine normally equals the number of poles. The number of slots between the brushes is

$$q = \frac{Q}{2p}. \tag{2.118}$$

This corresponds to the number of slots per pole and phase of AC windings. The effective number of slots per pole and phase is always somewhat lower, because a part of the coils is always short-circuited. The distribution factor for an armature winding k_{da} is obtained from Equation (2.33). For a fundamental, and $m = 1$, it is rewritten in the form

$$k_{da1} = \frac{2p}{Q\sin\dfrac{p\pi}{Q}}. \tag{2.119}$$

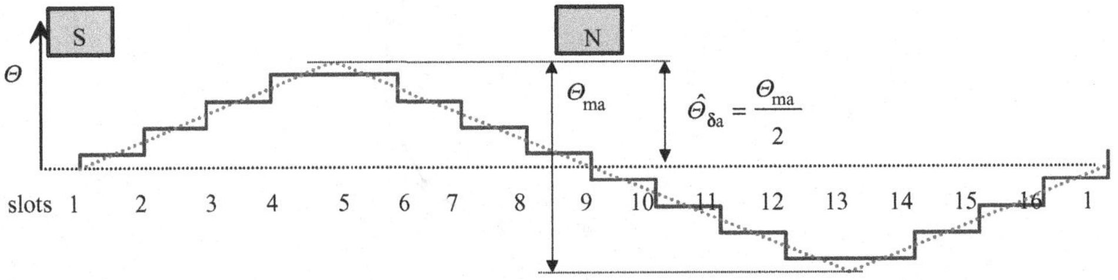

Figure 2.53 Current linkage curve of the winding of Figure 2.46, when the commutation takes place in the coils in slots 5 and 13.

Armature coils are often short pitched, and the pitch factor k_{pa} is thus obtained from Equation (2.32). The fundamental winding factor of a commutator winding is thus

$$k_{wa1} = k_{da1}k_{pa1} \approx \frac{2}{\pi}. \tag{2.120}$$

When the number of slots per pole increases, k_{da1} approaches the limit $2/\pi$. This is the ratio of the voltage circle (polygon) diameter to half of the circle perimeter. In ordinary machines, the ratio of short pitching is $W/\tau_p > 0.8$, and therefore $k_{pa1} > 0.95$. As a result, the approximate value $k_{wa1} = 2/\pi$ is an adequate starting point in the initial manual computation. More thorough investigations have to be based on an analysis of the curved function of the current linkage. In that case, the winding has to be observed in different positions of the brushes. Figure 2.45 shows that at the right side of the quadrature axis q, the direction of each slot current is towards the observer, and on the left, away from the observer. In other words, the rotor becomes an electromagnet with its north pole at the bottom and its south pole at the top. The pole pair current linkage of the rotor is

$$\Theta_{ma} = qI_u = \frac{Q}{2p}\frac{z}{Q}\frac{I_a}{2a} = \frac{z}{4ap}I_a = N_a I_a. \tag{2.121}$$

The term

$$N_a = \frac{z}{4ap} \tag{2.122}$$

in the equation is the number of coil turns per pole pair in a commutator armature in one parallel path, that is turns connected in series, because $z/2$ is the number of all armature turns; $z/2(2a)$ is the number of turns in one parallel path, in other words connected in series, and finally $z/2(2a)p$ is the number of turns per pole pair. The current linkage calculated according to Equation (2.121) is slightly higher than in reality, because the number of slots per pole and phase includes also the slots with short-circuited coil sides. In calculation, we may employ the linear current density

$$A_a = \frac{QI_u}{\pi D} = \frac{2p}{\pi D}N_a I_a = \frac{N_a I_a}{\tau_p}. \tag{2.123}$$

The current linkage of the linear current density is divided into magnetic voltages of the air gaps, the peak value of which is

$$\hat{\Theta}_{\delta a} = \int_0^{\frac{\tau_p}{2}} A_a \mathrm{d}x = \frac{1}{2}A_a\tau_p = \frac{1}{2}\frac{z}{2p}\frac{I_a}{2a} = \frac{N_a I_a}{2} = \frac{\Theta_{ma}}{2}, \tag{2.124}$$

In the diagram, the peak value $\hat{\Theta}_{\delta a}$ is located at the brushes (in the middle of the poles), the value varying linearly between the brushes, as illustrated with the dashed line in Figure 2.53. $\hat{\Theta}_{\delta a}$ is the armature reaction acting along the quadrature axis under one tip of a pole shoe,

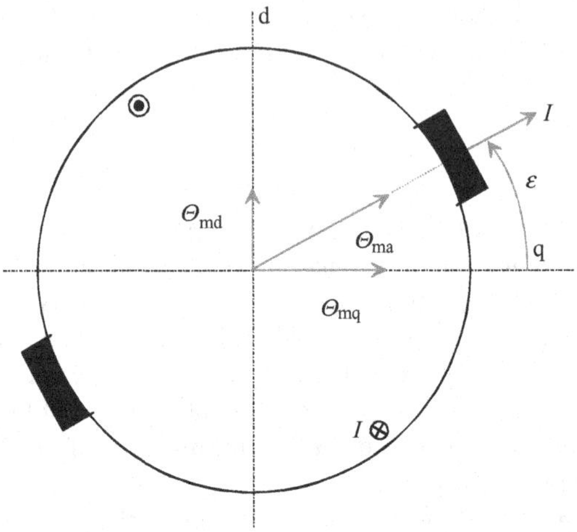

Figure 2.54 Current linkage of a commutator armature and its components. To ensure better commutation, the brushes are not placed on the q-axis.

and it is the current linkage to be compensated. The armature current linkage also creates commutation problems, which means that the brushes have to be shifted from the q-axis by an angle ε to a new position as shown in Figure 2.54.

This figure also gives the positive directions of the current I and the respective current linkage. The current linkage can be divided into two components:

$$\Theta_{md} = \frac{\Theta_{ma}}{2} \sin \varepsilon = \Theta_{\delta a} \sin \varepsilon, \tag{2.125}$$

$$\Theta_{mq} = \frac{\Theta_{ma}}{2} \cos \varepsilon = \Theta_{\delta a} \cos \varepsilon. \tag{2.126}$$

The former is called a direct component and the latter a quadrature component. The direct component magnetizes the machine either parallel or in the opposite direction to the actual field winding of the main poles of the machine. There is a demagnetizing effect if the brushes are shifted in the direction of rotation in generator mode, or in the opposite direction of rotation in motoring operation; the magnetizing effect, on the contrary, is in generating operation opposite the direction of rotation, and in motoring mode in the direction of rotation. The quadrature component distorts the magnetic field of the main poles, but neither magnetizes nor demagnetizes it. This is not a phenomenon restricted to commutator machines – the reaction is in fact present in all rotating machines.

2.16 Compensating Windings and Commutating Poles

As stated above, the armature current linkage (also called armature reaction) has some negative influence on DC machine operation. The armature reaction may create commutation problems and must, therefore, be compensated. There are different methods to mitigate such armature reaction problems: (1) shift the brushes from their geometrical neutral axis to the new magnetically neutral axis; (2) increase the field current to compensate for the main flux decrease caused by the armature reaction; (3) build commutating poles; and (4) build compensating winding.

The purpose of compensating windings in DC machines is to compensate for harmful flux components created by armature windings. Flux components are harmful, because they create an unfavorable air-gap flux distribution in DC machines. The dimensioning of compensating windings is based on the current linkage that has to be compensated by the compensating winding. The conductors of a compensating winding have therefore to be placed close to the surface of the armature, and the current flowing in them has to be opposite to the armature current. In DC machines, the compensating winding is inserted in the slots of the pole shoes. The compensating effect has to be created in the section $\alpha_i \tau_p$ of the pole pitch, as illustrated in Figure 2.55. If z is the total number of conductors in the armature winding, and the current flowing in them is I_s, we obtain an armature linear current density

$$A_a = \frac{zI_s}{D\pi}. \tag{2.127}$$

The total current linkage Θ_Σ of the armature reaction and the compensating winding has to be zero in the integration path. It is possible to calculate the required compensating current linkage Θ_k by evaluating the corresponding current linkage of the armature to be compensated. The current linkage of the armature Θ_a occurring under the compensating winding at the distance $\alpha_i \tau_p/2$, as shown in Figure 2.55, is

$$\Theta_a = zI_s \frac{\alpha_i \tau_p}{2D\pi} = \frac{\alpha_i \tau_p A_a}{2}. \tag{2.128}$$

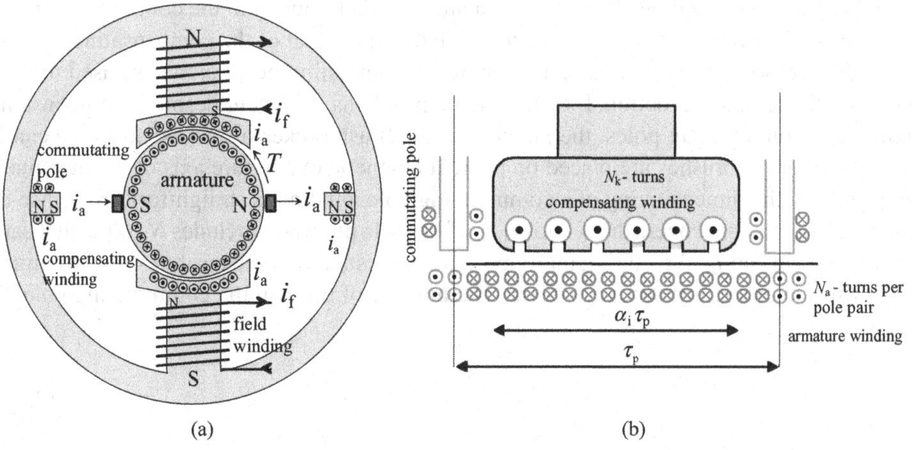

Figure 2.55 (a) Location of the compensating windings and the commutating poles. (b) Definition of the current linkage of a compensating winding.

Since there is an armature current I_a flowing in the compensating winding, we obtain the current linkage of the compensating winding accordingly

$$\Theta_k = -N_k I_a, \tag{2.129}$$

where N_k is the number of turns of the compensating winding. Since the current linkage of the armature winding has to be compensated in the integration path, the common current linkage is written as

$$\Theta_\Sigma = \Theta_k + \Theta_a = -N_k I_a + \frac{a_i \tau_p A_a}{2} = 0, \tag{2.130}$$

where

$$\Theta_a = \frac{1}{2}\alpha_i A_a \tau_p = \frac{1}{2}\alpha_i \frac{z}{2p}\frac{I_a}{2a} = \frac{1}{2}N_a I_a. \tag{2.131}$$

Now, we obtain the number of turns of the compensating winding to be inserted in the pole shoes producing demagnetizing magnetic flux in the q-axis compensating the armature reaction flux:

$$N_k = \frac{\alpha_i \tau_p A_a}{2I_a}. \tag{2.132}$$

Since N_k has to be an integer, Equation (2.132) is only approximately feasible. To avoid large pulsating flux components and noise, the slot pitch of the compensating winding is set to diverge by 10–15 % from the slot pitch of the armature.

Since a compensating winding cannot completely cover the surface of the armature, commutating poles are also utilized to compensate for the armature reaction although their function is just to improve commutation. These commutating poles are located between the main magnetizing poles of the machine. There is an armature current flowing in the commutating poles. The number of turns on the poles is selected such that the effect of the compensating winding is strengthened appropriately. In small machines, commutating poles alone are used to compensate for the armature reaction. If commutation problems still occur despite a compensating winding and commutating poles, the position of the brush rocker of the DC machine can be adjusted so that the brushes are placed on the real magnetic quadrature axis of the machine.

In principle, the dimensioning of a commutating pole winding is straightforward. Since the compensating winding covers the section $a_i \tau_p$ of the pole pitch and includes N_k turns that carry the armature current I_a, the commutating pole winding should compensate for the remaining current linkage of the armature $(1 - a_i)\tau_p$. The number of turns in the commutating pole N_{cp} should be

$$N_{cp} = \frac{1 - \alpha_i}{\alpha_i} N_k. \tag{2.133}$$

When the same armature current I_a flows both in the compensating winding and in the commutating pole, the armature reaction will be fully compensated.

If there is no compensation winding, the commutating pole winding must be dimensioned and the brushes positioned so that the flux in a commutating armature coil is at its maximum, and no voltage is induced in the coil.

2.17 Rotor Windings of Asynchronous Machines

The simplest rotor of an induction machine is a solid iron body, turned and milled to the correct shape. In general, a solid rotor is applicable to high-speed machines and in certain cases also to normal-speed drive. However, the computation of the electromagnetic characteristics of a steel rotor is a demanding task, and it is not discussed here. A solid rotor is characterized by a high resistance and a high leakage inductance of the rotor. The phase angle of the apparent power created by a waveform penetrating a linear material is 45°, but the saturation of the steel rotor reduces the phase angle. A typical value for the phase angle of a solid rotor varies between 30° and 45°, depending on the saturation. The characteristics of a solid-rotor machine are discussed for instance in Peesel (1958), Pyrhönen (1991), Huppunen (2004) and Aho (2007). The performance characteristics of a solid rotor can be improved by slotting the surface of the rotor, Figure 2.56 or by coating the rotor with copper (Lähteenmäki 2002). Axial slots are used to control the flow of eddy currents in a direction that is favorable to torque production. Radial slots increase the length of the paths of the eddy currents created by certain high-frequency phenomena. This way, eddy currents are damped and the efficiency of the machine is improved. The structure of the rotor is of great significance in torque production, Figure 2.57. An advantage of common cage winding rotors is that they produce the highest torque with small values of slip, whereas solid rotors yield a good starting torque.

In small machines, a Ferraris rotor can be employed. It is constructed from a laminated steel core covered with a thin layer of copper. The copper covering provides a suitable path for

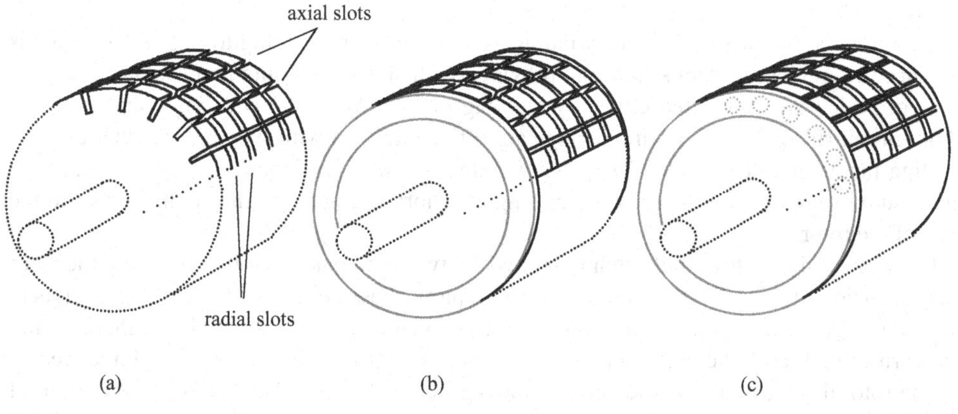

Figure 2.56 Different solid rotors. (a) A solid rotor with axial and radial slots (in this model, short-circuit rings are required). They can be constructed either by leaving the part of the rotor that extends from the stator without slots, or by equipping it with aluminum or copper rings. (b) A rotor equipped with short-circuit rings in addition to slots. (c) A slotted and cage-wound rotor. A completely smooth rotor can also be employed.

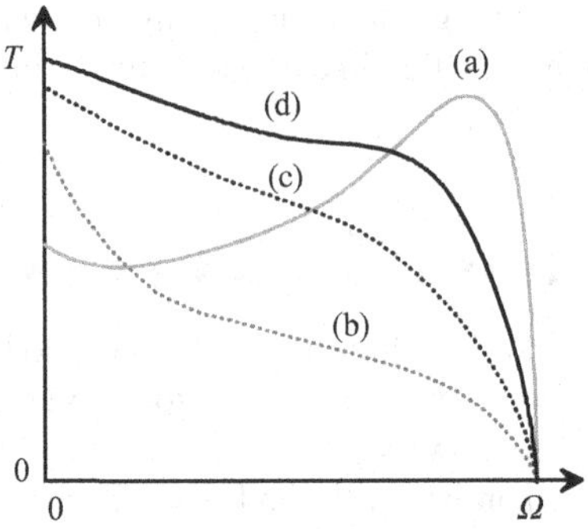

Figure 2.57 Torque curves of different induction rotors as a function of mechanical angular speed Ω: (a) a normal double-cage winding rotor, (b) a smooth solid-rotor without short-circuit rings, (c) a smooth solid rotor equipped with copper short-circuit rings, (d) an axially and radially slitted solid rotor equipped with copper short-circuit rings.

eddy currents induced in it. The copper covering takes up a certain amount of space in the air gap, the electric value of which increases notably because of the covering, since the relative permeability of copper is $\mu_r = 0.999\,992\,6$. As a diamagnetic material, copper is thus even a somewhat weaker path for the magnetic flux than air.

The rotor of an induction machine can be produced as a normal slot winding by following the principles discussed in the previous sections. A wound rotor has to be equipped with the same number of pole pairs as the stator, and therefore it is not in practice suitable for machines permitting a varying number of poles. The phase number of the rotor may differ from the phase number of the stator. For instance, a two-phase rotor can be employed in slip-ring machines with a three-phase stator. The rotor winding is connected to an external circuit via slip rings.

The most common short-circuit winding is the cage winding, Figure 2.58. The rotor is produced from electric steel sheets and provided with slots containing noninsulated bars, the ends of which are connected either by welding or brazing to the end rings, that is to the short-circuit rings. The short-circuit rings are often equipped with fins that together act as a cooling fan as the rotor rotates. The cage winding of small machines is produced from pure aluminum by simultaneously pressure casting the short-circuit rings, the cooling ribs and the bars of the rotor.

Figure 2.59 illustrates a full-pitch winding of a two-pole machine observed from the rotor end. Each coil of the rotor also constitutes a complete phase coil, since the number of slots in the rotor is $Q_r = 6$. The star point 0 forms, based on symmetry, a neutral point. If there is only one turn in each coil, the coils can be connected at this point. The magnetic voltage created by the rotor depends only on the current flowing in the slot, and therefore the connection of the windings at the star point is of no influence. However, the connection of the star point at one end of the rotor turns the winding into a six-phase star connection with one bar, that is half a turn, in each phase. The six-phase winding is then short-circuited also at the other end. Since the shaft of the machine also takes up some room, the star point has to be created with a short-circuit ring as illustrated in Figure 2.58. We can now see in Figure 2.59 a star-connected,

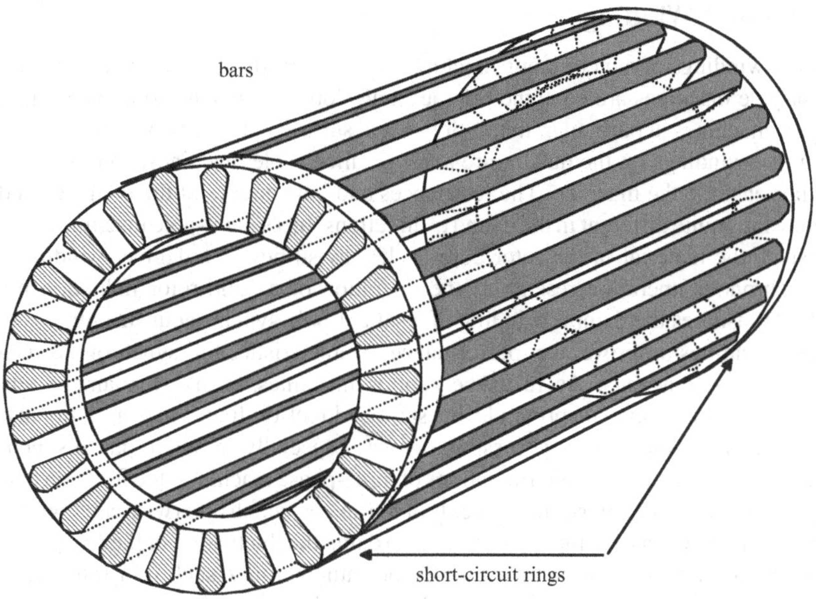

Figure 2.58 Simple cage winding. Cooling fans are not illustrated. $Q_r = 24$.

short-circuited poly-phase winding, for which the number of phase coils is in a two-pole case equal to the number of bars in the rotor: $m_r = Q_r$.

In machine design, it is often assumed that analysis of the fundamental $\nu = 1$ alone gives an adequate description of the characteristics of the machine. However, this is valid for cage windings only if we consider also the conditions related to the number of bars. A cage winding acts differently with respect to different harmonics ν. Therefore, a cage winding has to be analyzed with respect to the general harmonic ν. This is discussed in more detail in Chapter 7, in which different types of machines are investigated separately.

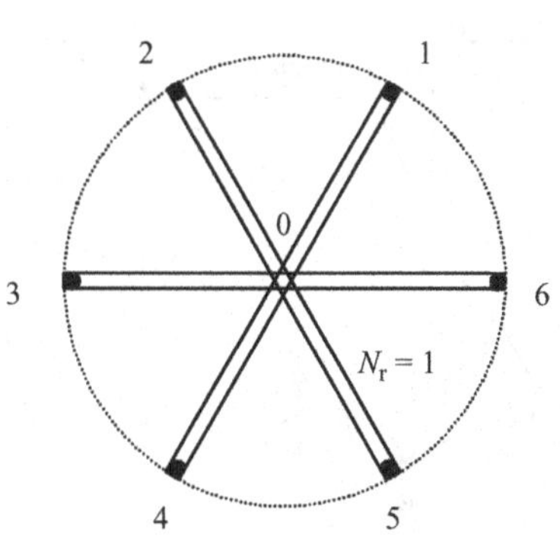

Figure 2.59 Three-phase winding of a two-pole rotor. The number of turns in the phase coil is $N_r = 1$. If the winding is connected in star at point 0 and short-circuited at the other end, a six-phase, short-circuited winding is created, for which the number of turns is $N_r = 1/2$, $k_{wr} = 1$.

2.18 Damper Windings

The damper windings of synchronous machines are usually short-circuit windings, which in nonsalient-pole machines are contained in the same slots with magnetizing windings, and in salient-pole machines in particular, in the slots at the surfaces of pole shoes. There are no bars in the damper windings on the quadrature axes of salient-pole machines, and only the short-circuit rings encircle the machine. The resistances and inductances of the damper winding of the rotor are thus quite different in the d- and q-directions. In a salient-pole machine constructed of solid steel, the material of the rotor core itself may suffice as a damper winding. In that case, asynchronous operation resembles the operation of a solid-rotor induction machine. Figure 2.60 illustrates the damper winding of a salient-pole synchronous machine.

Damper windings improve the performance characteristics of synchronous machines especially during transients. As in asynchronous machines, thanks to damper windings, synchronous machines can in principle be started direct-on-line. Also a stationary asynchronous drive is in some cases a possible choice. Especially in single-phase synchronous machines and in the unbalanced load situations of three-phase machines, the function of damper windings is to damp the counter-rotating fields of the air gap which otherwise cause great losses. In particular, the function of damper windings is to damp the fluctuation of the rotation speed of a synchronous machine when rotating loads with pulsating torques, such as piston compressors.

The effective mechanisms of damper windings are relatively complicated and diverse, and therefore their mathematically accurate design is difficult. That is why damper windings are usually constructed by drawing upon empirical knowledge. However, the inductances and resistances of the selected winding can usually be evaluated with normal methods to define the time constants of the winding.

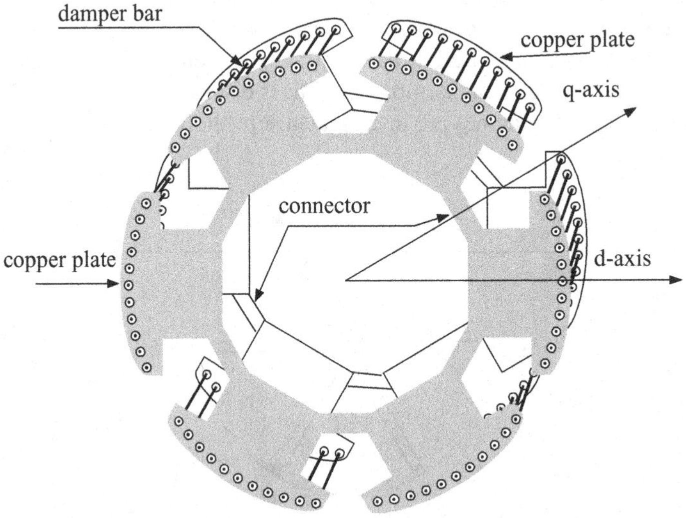

Figure 2.60 Structure of the damper winding of a six-pole salient-pole synchronous machine. The copper end plates are connected with a suitable copper connector to form a ring for the damper currents. Sometimes real rings also connect the damper bars.

When the damper windings of salient-pole machines are placed in the slots, the slot pitch has to be selected to diverge by 10–15 % from the slot pitch of the stator to avoid pulsation of the flux and noise. If the slots are skewed (usually by the amount of a single stator slot pitch), the same slot pitch can be selected both for the stator and the rotor. Damper winding comes into effect only when the bars of the winding are connected with short-circuit rings. If the pole shoes are solid, they may, similar to the solid rotor of a nonsalient-pole machine, act as a damper winding as long as the ends of the pole shoes are connected with durable short-circuit rings. In nonsalient-pole machines, an individual damper winding is seldom used; however, conductors may be mounted under slot wedges, or the slot wedges themselves are used as the bars of the damper winding.

In synchronous generators, the function of damper windings is for instance to damp counter-rotating fields. To minimize losses, the resistance is kept to a minimum in damper windings. The cross-sectional area of the damper bars is selected to be 20–30 % of the cross-sectional bar area of the armature winding. The windings are made of copper. In single-phase generators, damper bar cross-sectional areas larger than 30 % of the stator copper area are employed. The frequency of the voltages induced by counter-rotating fields to the damper bars is doubled when compared with the network frequency. Therefore, it has to be considered whether special actions are required with respect to the skin effect of the damper windings (e.g. utilization of Roebel bars (braided conductors) to avoid the skin effect). The cross-sectional area of the short-circuit rings is selected to be approximately 30–50 % of the cross-sectional area of the damper bars per pole.

The damper bars have to damp the fluctuations of the rotation speed caused by the pulsating torque loads. They also have to guarantee a good starting torque when the machine is starting as an asynchronous machine. Thus, brass bars or small-diameter copper damper bars are employed to increase the rotor resistance. The cross-sectional area of copper bars is typically only 10 % of the cross-sectional area of the copper of the armature winding.

In PMSMs, in axial flux machines in particular, the damper winding may be easily constructed by placing a suitable copper or aluminum plate on the surface of the rotor, on top of the magnets. However, achieving a total conducting surface in the range of 20–30 % of the stator copper surface may be somewhat difficult because the plate thickness easily increases to become too large and limits the air-gap flux density created by the magnets.

Bibliography

Aho, T. (2007) *Electromagnetic Design of a Solid Steel Rotor Motor for Demanding Operational Environments*, Dissertation. Acta Universitatis Lappeenrantaensis 292, Lappeenranta University of Technology. (https://oa.doria.fi/)

Heikkilä, T. (2002) *Permanent Magnet Synchronous Motor for Industrial Inverter Applications – Analysis and Design*, Dissertation. Acta Universitatis Lappeenrantaensis 134, Lappeenranta University of Technology. (https://oa.doria.fi/)

Hindmarsh, J. (1988) *Electrical Machines and Drives. Worked Examples*, 2nd edn, Pergamon Press, Oxford.

Huppunen, J. (2004) *High-Speed Solid-Rotor Induction Machine – Electromagnetic Calculation and Design*, Dissertation. Acta Universitatis Lappeenrantaensis 197, Lappeenranta University of Technology. (https://oa.doria.fi/)

Lähteenmäki, J. (2002) *Design and Voltage Supply of High-Speed Induction Machines,* Acta Polytechnica Scandinavica, Electrical Engineering Series 108, Dissertation Helsinki University of Technology. Available at http://lib.tkk.fi/Diss/2002/isbn951226224X.

IEC 60050-411 (1996) *International Electrotechnical Vocabulary (IEC). Rotating Machines*. International Electrotechnical Commission, Geneva.

Jokinen, T. (1972) *Utilization of harmonics for self-excitation of a synchronous generator by placing an auxiliary winding in the rotor*, Acta Polytechnica Scandinavica, Electrical Engineering Series 32, Dissertation Helsinki University of Technology. Available at http://lib.tkk.fi/Diss/197X/isbn9512260778/.

Peesel, H. (1958) *Behaviour of asynchronous motor with different solid steel rotors (Über das Verhalten eines Asynchronmotors bei verschiedenen Läufern aus massive Stahl)*. Dissertation Technische Universität Braunschweig (Braunschweig Institute of Technology).

Pyrhönen, J. (1991) *The High-Speed Induction Motor: Calculating the Effects of Solid-Rotor Material on Machine Characteristics*, Dissertation, Acta Polytechnica Scandinavica, Electrical Engineering Series 68, (http://urn.fi/URN:ISBN:978-952-214-538-3)

Richter, R. (1954) *Electrical Machines: Induction Machines (Elektrische Maschinen: Die Induktionsmaschinen)*, Vol. **IV**, 2nd edn, Birkhäuser Verlag, Basle and Stuttgart.

Richter, R. (1963) *Electrical Machines: Synchronous Machines and Rotary Converters (Elektrische Maschinen: Synchronmaschinen und Einankerumformer)*, Vol. **II**, 3rd edn, Birkhäuser Verlag, Basle and Stuttgart.

Richter, R. (1967) *Electrical Machines: General Calculation Elements. DC Machines (Elektrische Maschinen: Allgemeine Berechnungselemente. Die Gleichstrommaschinen)*, Vol. **I**, 3rd edn, Birkhäuser Verlag, Basle and Stuttgart.

Salminen, P. (2004) *Fractional Slot Permanent Magnet Synchronous Motors for Low Speed Applications*, Dissertation. Acta Universitatis Lappeenrantaensis 198, Lappeenranta University of Technology (http://urn.fi/URN:ISBN:951-764-983-5).

Vogt, K. (1996) *Design of Electrical Machines (Berechnung elektrischer Maschinen)*, Wiley-VCH Verlag GmbH, Weinheim.

3

Design of Magnetic Circuits

The magnetic circuit of an electrical machine generally consists of ferromagnetic materials and air gaps. In an electrical machine, all the windings and possible permanent magnets participate in the magnetizing of the machine. It must also be noted that in a multiple-pole system, the magnetic circuit has several magnetic paths. Normally, an electrical machine has as many magnetic paths as it has poles. In a two-pole system, the magnetic circuit is symmetrically divided into two paths. A possible magnetic anisotropy occurring in the geometry of the magnetic circuit also influences the magnetic state of the machine. In the literature, a magnetic circuit belonging to one pole is usually analyzed in the design of a complete magnetic circuit. This method is employed here also. In other words, according to Equation (2.15), the amplitude $\hat{\Theta}_{s1}$ of the fundamental component of the current linkage is acting on half of the magnetic path. A complete magnetic path requires two amplitudes; see Figure 2.9.

The design of a magnetic circuit is based on the analysis of the magnetic flux density $\boldsymbol{B}$ and the magnetic field strength $\boldsymbol{H}$ in different parts of the machine. The design of a magnetic circuit is governed by Ampère's law. First, we select a suitable air-gap flux density B_δ to the machine. Next, we calculate the corresponding field strength values H in different parts of the machine. The mmf F_m of the circuit equals the current linkage Θ of the circuit

$$F_m = \oint \boldsymbol{H} \cdot d\boldsymbol{l} = \sum i = \Theta. \tag{3.1}$$

In a running electrical machine, the sum current linkage is produced by all the currents and possible permanent magnet materials. In the basic design of a magnetic circuit, only the winding, the main task of which is to magnetize the machine, is considered the source of magnetizing current linkage; that is, the machine is observed when running at no load. In DC machines and synchronous machines, the machine is magnetized by magnetizing windings (field windings) or permanent magnets, and the armature winding is kept currentless. In a synchronous machine, the armature winding may nevertheless take part in the determination of the magnetic state of the machine also when the machine is running at no load, if the air-gap flux created by the rotor magnetization does not induce a stator emf exactly equal to the stator voltage. The influence of an armature winding, that is the armature reaction, is investigated later in design, when the performance characteristics of the machine are being calculated. In an induction machine, the magnetizing winding and the armature winding are not separated, and therefore the magnetizing of the machine is carried out by the stator winding. However, according to the IEC, the induction machine stator is not regarded as an armature even though it is similar to a synchronous machine armature.

Our objective now is to solve the magnetic potential differences

$$U_{\mathrm{m},i} = \int \boldsymbol{H}_i \cdot \mathrm{d}\boldsymbol{l}_i \tag{3.2}$$

in different parts of the magnetic circuit and the required current linkage Θ corresponding to their sum $\sum U_{\mathrm{m},i}$. In traditional machines, this is a fairly straightforward task, since the main parameters of the magnetic circuit have in principle been set quite early in the machine design and the need for magnetizing current does not necessarily alter the dimensions of the magnetic circuit.

In magnetic circuits with permanent magnets instead, the situation is somewhat more complicated especially in machines where magnets are embedded inside the iron core. In that case, the leakage flux of permanent magnets is high and the leakage flux saturates the iron bridges, which enclose the magnets inside the iron core. The division of the flux created by permanent magnets into the main flux and the leakage flux is difficult to carry out analytically. In practice, the magnetic field has to be solved numerically using a finite element program for determining the magnetic fields. To increase the air-gap flux density of a permanent magnet machine, we can embed the magnets in V-form and have magnets wider than the pole pitch of the machine. This way, it is possible to increase the air-gap flux density of a permanent magnet machine even higher than the remanent flux density of the permanent magnet material.

In electrical machines, the term "magnetic circuit" refers to those sections of the machine through which the main flux of the machine is flowing. In stator and rotor yokes, the main flux splits up into two paths. Now, an electrical machine actually includes as many magnetic paths as there are magnetic poles, that is $2p$ pieces. Figure 3.1 illustrates, in addition to the cross-section of a magnetic circuit of a six-pole induction machine and a four-pole synchronous reluctance machine, a single magnetic path of the machine defined by the curves 1–2–3–4–1. The same figure also depicts the direct (d) and quadrature (q) axes of the magnetic circuit. Since we now have a rotating-field machine, the flux density wave created by the poly-phase winding of the stator rotates along the inner surface of the stator, and the depicted d- and q-axes rotate fixed to the peak value of the magnetic flux.

Rotor saliency makes the machine's magnetic circuit anisotropic in the machine cross-sectional plane (normal to shaft). Differences in reluctance, and correspondingly in stator magnetizing inductance, may be found depending on the rotor position with respect to the stator. In traditional machines, the d-axis is usually placed so that when the magnetic flux joins the rotor d-axis, the minimum reluctance for the whole magnetic system is found. Vice versa, the q-axis normally represents the maximum reluctance of the magnetic circuit. In such a machine, the stator d-axis inductance is larger than the stator q-axis inductance $L_\mathrm{d} > L_\mathrm{q}$. In synchronous machines, the field windings are wound around the rotor d-axes. In permanent magnet machines, however, the permanent magnet material itself belongs to

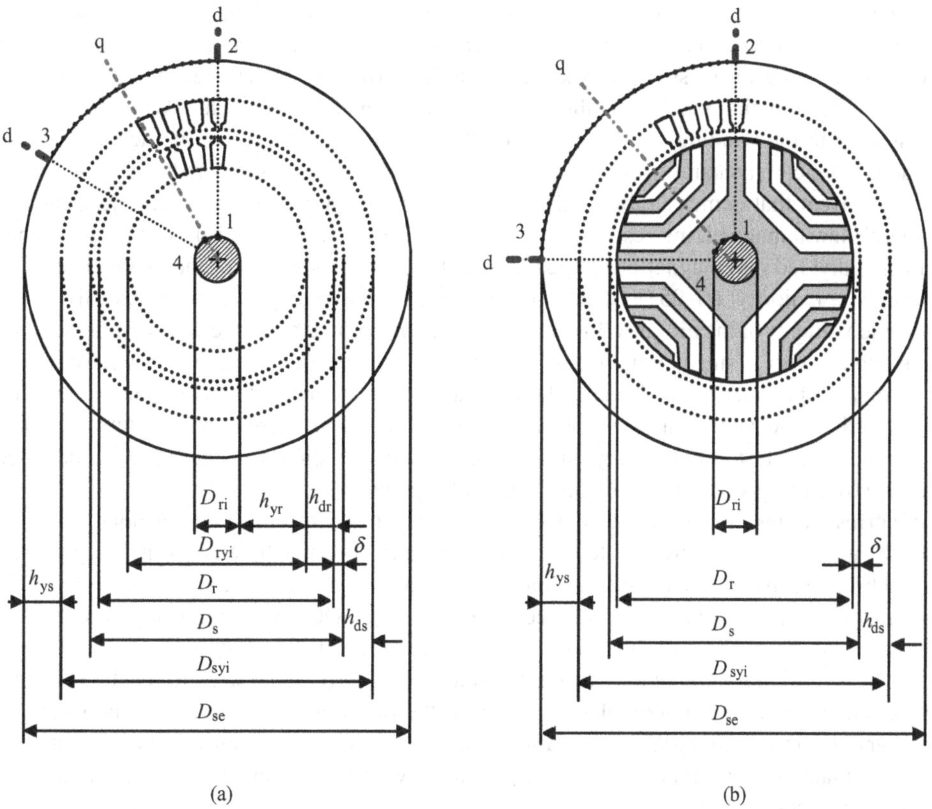

Figure 3.1 Main dimensions of the cross-section of an electrical machine, and the instantaneous positions of the magnetic axes and the magnetic circuit of (a) a six-pole poly-phase induction machine and (b) a four-pole synchronous reluctance machine. The magnetic circuits can be considered to rotate with the stator flux of the machine. The d-axis of the rotor of a synchronous machine remains stationary at every instant, but in an induction machine, the d-axis is only virtual and turns with respect to the rotor at the speed of slip.

the d-axis magnetic circuit, which makes the d-axis reluctance high. Depending on the rotor construction, in permanent magnet machines, we may have "inverse saliency," where the d-axis reluctance is higher than the q-axis reluctance. Correspondingly, $L_d < L_q$. In permanent magnet machines it is rare to have normal saliency with $L_d > L_q$.

In some cases, for instance for control purposes, the d- and q-axes may also be defined for magnetically isotropic induction machines. From the machine design point of view, the division to d- and q-axes in magnetically isotropic machines is not as important as in anisotropic machines.

In nonsalient synchronous machines, the division to d- and q-axes is natural, since despite the fact that the reluctances and hence also the inductances are about equal, the rotor is anisotropic because of the field winding wound around the rotor d-axis.

If the main flux of the machine is assumed to flow along the d-axis, the magnetic circuit of the figure comprises half of the main flux, created by the current penetrating the magnetic circuit in the area in question. When the machine is running under load, the magnetizing current of an induction machine is the sum of the stator and rotor currents. When computing this sum, both the windings of the stator and the rotor have to be taken into account so that in each conductor penetrating the area $\boldsymbol{S}$ (here the slots of the stator and the rotor), there flows a current that is measurable at the stator and at the rotor. We now obtain according to Ampère's law a magnetizing resultant current linkage Θ, which, as a result of the occurrence of flux, is divided into magnetic voltages $U_{\mathrm{m},i}$ in the different sections of the magnetic circuit. Normally, a majority, 60–95 % of the sum of the magnetic voltages, consists usually of the magnetic voltages of the air gaps. In the design of a magnetic circuit, we start by magnetizing the machine with a single winding, for instance a magnetizing winding. Later, when the performance characteristics of the machine are analyzed, other windings and their effects are considered. In the rotor of a synchronous reluctance machine, there is no winding, and thus the torque production is based only on the saliency effects. Saliency is created in the depicted machine by cutting suitable sections from the rotor plate.

A corresponding illustration of the magnetic circuit and the main dimensions of a six-pole DC machine and an eight-pole salient-pole synchronous machine is given in Figure 3.2. A synchronous machine can operate magnetized either with the direct current in the field winding or with permanent magnets, or completely without DC magnetizing as a synchronous reluctance machine. However, in the rotor circuit of the reluctance machine, special attention has to be paid to maximizing the ratio of the reluctances of q- and d-axes to reach maximum torque. Here, the torque is completely based on the difference of inductances between the direct and quadrature axes. When the ratio of inductances $L_\mathrm{d}/L_\mathrm{q}$ is about 7–10, approximately the same machine constants can be applied with a synchronous reluctance machine as with induction motors.

Doubly-salient reluctance machines (switched reluctance, SR machines) differ both structurally and by their performance characteristics from traditional electrical machines. However, there are also some similarities. Figure 3.3 illustrates two doubly-salient reluctance machines that differ from each other by their ratios of poles (8/6 and 6/4); in these machines, both the stator and the rotor have salient poles. A crucial difference when compared with traditional machines is that the stator and rotor have different numbers of magnetizing poles. Such a machine cannot operate without power electronics or other switches, and therefore it has to be designed in accordance with the accompanying electronics.

The definition of the current linkage is essential for the design of the magnetic circuit to create the desired flux density and the respective magnetizing current. The current linkage needed per pole pair is solved by applying Equation (1.10) and by calculating the line integral of the field strength $\boldsymbol{H}$ along a suitable integration path (mmf), for instance along the route 1–2–3–4–1 of Figure 3.2:

$$F_\mathrm{m} = \hat{U}_{\mathrm{m,tot}} = \oint_l \boldsymbol{H} \cdot \mathrm{d}\boldsymbol{l} = \int_\mathrm{S} \boldsymbol{J} \cdot \mathrm{d}\boldsymbol{S} = \sum_\mathrm{S} I = \hat{\Theta}_\mathrm{tot}. \tag{3.3}$$

In rotating-field machines, the peak value of the magnetic voltage $\hat{U}_{\mathrm{m,tot}}$ is usually calculated by following the flux line at the peak value of the air-gap flux density around the magnetic

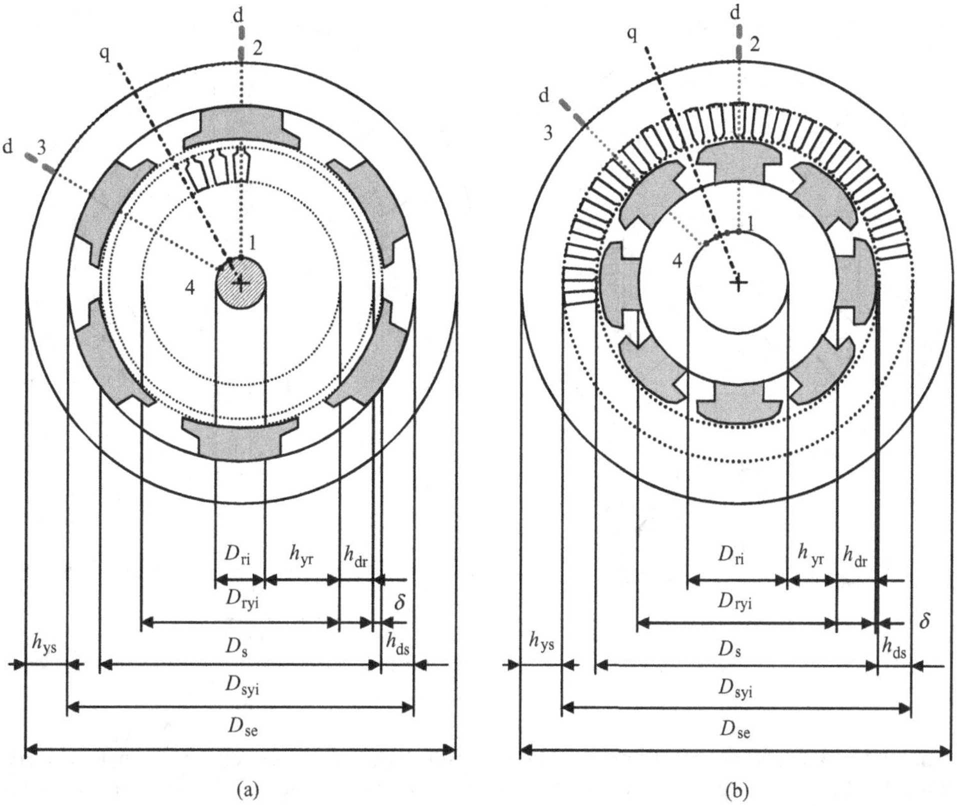

Figure 3.2 Main dimensions of the cross-sections of a six-pole DC machine (or an outer pole synchronous machine) and an eight-pole salient-pole synchronous machine, and the instantaneous positions of one of the magnetic axes and the magnetic circuit. In a DC machine or a synchronous machine, the position of the magnetic axes can be considered more easily than with an asynchronous machine, since the magnetic axes d and q are defined by the position of the rotor. DC machines are usually constructed as external (inward-projecting) pole machines, and the poles of the stator chiefly define the position of the magnetic axes of the machine. The field windings of both machine types are placed around the poles lying on the d-axis.

circuit. $\boldsymbol{l}$ is the unit vector parallel to the integration path, $\boldsymbol{S}$ is the unit vector of the surface of the cross-section of the electrical machine (in practice, e.g. either the teeth area of an induction machine or the teeth area of the stator of a synchronous machine, and the pole bodies of the rotor are observed), and finally, $\boldsymbol{J}$ is the density of the current penetrating the magnetic circuit. The task is simplified by calculating the sum of all the currents I penetrating the magnetic circuit. The sum of all currents is called current linkage and denoted $\hat{\Theta}_{tot}$. Equation (3.3) describes how the current linkage of the machine has to equal the mmf $\oint_l \boldsymbol{H} \cdot \mathrm{d}\boldsymbol{l}$ of the machine.

When computing the peak magnetic voltage $\hat{U}_{m,tot}$ over a single magnetic circuit of the machine, the task can be divided so that each section of the magnetic circuit is analyzed for

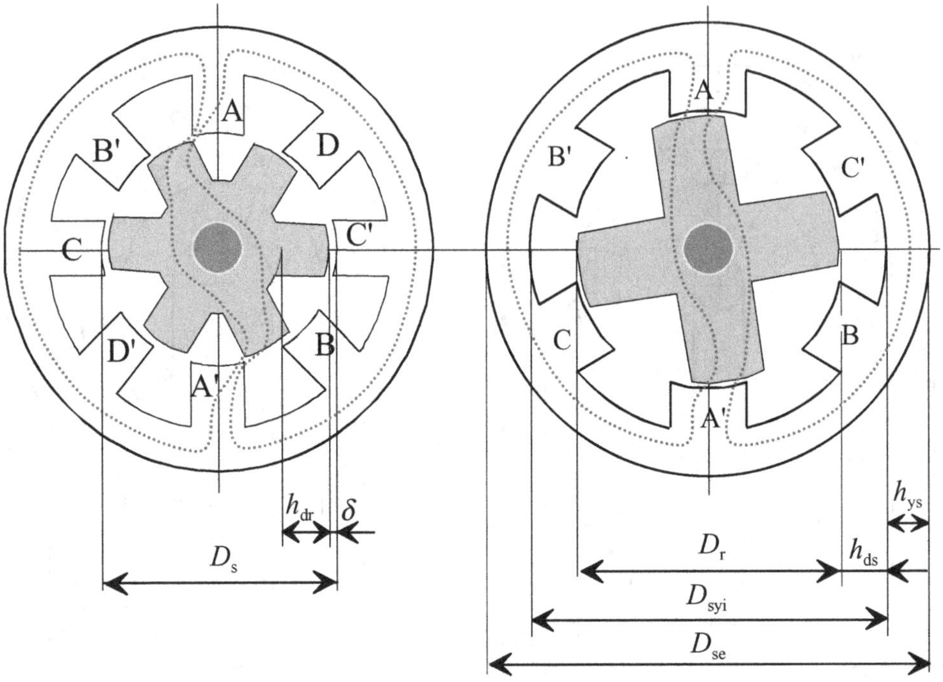

Figure 3.3 Basic types of an SR machine. (a) In an 8/6 machine, there are eight stator poles and six rotor poles. Correspondingly, there are six stator poles and four rotor poles in a 6/4 motor (b). The pole numbers of the stator and the rotor always differ from each other. The rotor of both machines turns clockwise when the poles A and A′ are magnetized. The illustration of an eight-pole machine shows the path of the main flux when the poles A and A′ are being magnetized.

instance at the peak value of the flux density. Now, the total magnetic voltage over a complete pole pair is written as

$$\hat{U}_{\mathrm{m,tot}} = \sum_i \int \boldsymbol{H}_i \cdot \mathrm{d}\boldsymbol{l}_i = \sum_i \hat{U}_{\mathrm{m},i}. \tag{3.4}$$

To be able to calculate the current linkage required by the iron parts of the magnetic circuit, the magnetizing curve of the material in question has to be known. The magnetizing curve illustrates the flux density reached in the material as a function of the magnetic field strength $B = f(H)$. First, the magnetic flux density B_i in each section of the magnetic circuit of the machine is calculated with the selected air-gap flux density. Next, the field strength H_i is checked from the BH curve of the material in question. Finally, in simple cases, the result is multiplied by the length of the section parallel to the magnetic path l_i, which yields the magnetic voltage of the section in question $\hat{U}_{\mathrm{m},i} = l_i H_i$. In the appendices, BH curves are given for some typical electric sheets measured with DC magnetization.

Since Equation (2.15) gives the height of the amplitude of the current linkage wave of a rotating-field winding, and such an amplitude magnetizes half of a single magnetic circuit, we usually treat only half of a single magnetic circuit in the computation of magnetic voltages. For instance, the stator of an asynchronous machine has to produce the amplitude $\hat{\Theta}_{s1}$ of the fundamental current linkage. Correspondingly, on the rotor pole of a nonsalient-pole synchronous machine, there has to be a current producing a similar current linkage

$$\hat{\Theta}_{s1} = \hat{U}_{m,\delta e} + \hat{U}_{m,sd} + \hat{U}_{m,rd} + {}^{1}\!/_{2}\hat{U}_{m,sy} + {}^{1}\!/_{2}\hat{U}_{m,ry}. \tag{3.5a}$$

$\hat{U}_{m,\delta e}$ denotes the magnetic voltage of a single air gap, $\hat{U}_{m,d}$ the magnetic voltages of the teeth, and $\hat{U}_{m,ys} + \hat{U}_{m,yr}$ the magnetic voltages of the stator and rotor yokes. The subscripts s and r denote the stator and the rotor. Equation (3.5a) represents only one pole of the magnetic circuit. Two current linkage amplitudes $\hat{U}_{m,tot} = 2\hat{\Theta}_{s1}$ are thus acting on the total magnetic path.

In an internal salient-pole machine (e.g. an ordinary synchronous machine), the correlation of the magnetic voltage and a single pole is

$$\hat{\Theta}_{rp} = \hat{U}_{m,\delta e} + \hat{U}_{m,sd} + \hat{U}_{m,rp} + {}^{1}\!/_{2}\hat{U}_{m,sy} + {}^{1}\!/_{2}\hat{U}_{m,ry}. \tag{3.5b}$$

$\hat{U}_{m,rp}$ is the magnetic voltage of the salient pole of the rotor. In an external salient-pole machine (e.g. an ordinary DC machine or an external pole synchronous machine) the current linkage of a single pole can be written correspondingly as

$$\hat{\Theta}_{sp} = \hat{U}_{m,\delta e} + \hat{U}_{m,sp} + \hat{U}_{m,rd} + {}^{1}\!/_{2}\hat{U}_{m,sy} + {}^{1}\!/_{2}\hat{U}_{m,ry}. \tag{3.5c}$$

The magnetizing of doubly salient reluctance machines (SR machines) depends on the constantly changing shape of the magnetic circuit in question. The torque calculation of SR machines is carried out for instance based on the principle of virtual work. The machine is always seeking the maximum inductance of the magnetic circuit, which is reached when the poles of the rotor have turned to the position of the magnetic poles of the stator.

3.1 Air Gap and its Magnetic Voltage

The air gap of an electrical machine has a significant influence on the mmf of the magnetic circuit. Nonsalient-pole machines and salient-pole machines have different types of air gaps that greatly influence machine performance.

3.1.1 Air Gap and Carter Factor

To be able to calculate the magnetic voltage manually over an air gap, the geometry of the air gap has to be simplified. Often in an electrical machine, the surfaces of both the stator and the rotor are split with slots. The flux density always decreases at the slot opening (Figure 3.4), and therefore it is not easy to define the average flux density of the slot pitch between the stator and the rotor.

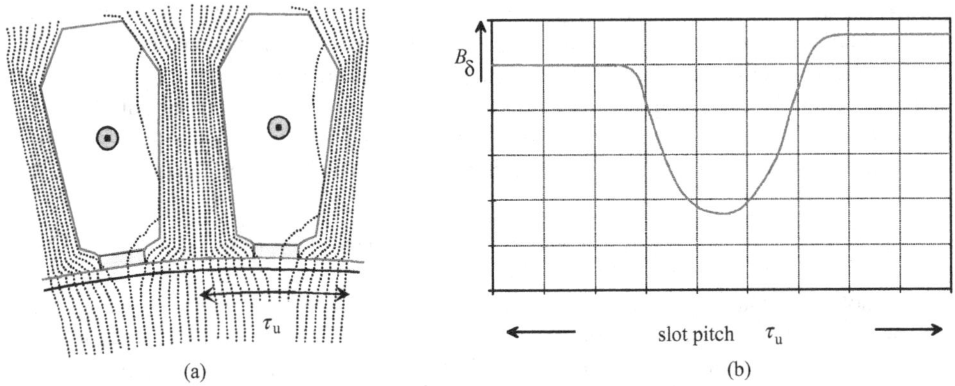

Figure 3.4 (a) Flux diagram under a stator slot along one slot pitch, and (b) the behavior of the air-gap flux density B_δ along a slot pitch. At the slot opening, there is a local minimum of flux density. The flux density on the right side of the slot is slightly higher than on the left side, since a small current is flowing in the slot towards the observer.

However, in 1901 F.W. Carter provided a solution to the problem of manual calculation (Carter 1901). On average, according to Carter's principle, the air gap seems to be longer than its physical measure. The length of the physical air gap δ increases with the Carter factor k_C. The first correction is carried out by assuming the rotor to be smooth. We obtain

$$\delta_{es} = k_{Cs}\delta. \tag{3.6}$$

The Carter factor k_{Cs} is based on the dimensions in Figure 3.5. When determining the Carter factor, the real flux density curve is replaced with a rectangular function so that the flux remains constant under the teeth and is zero at the slot opening; in other words, the shaded areas $S_1 + S_1$ in Figure 3.5 are equal to S_2. The equivalent slot opening b_e, in which the flux density is zero, is

$$b_e = \kappa b_1, \tag{3.7a}$$

where

$$\kappa = \frac{2}{\pi}\left(\text{arc tan}\frac{b_1}{2\delta} - \frac{2\delta}{b_1}\ln\sqrt{1 + \left(\frac{b_1}{2\delta}\right)^2}\right) \approx \frac{\dfrac{b_1}{\delta}}{5 + \dfrac{b_1}{\delta}}. \tag{3.7b}$$

The Carter factor is

$$k_C = \frac{\tau_u}{\tau_u - b_e} = \frac{\tau_u}{\tau_u - \kappa b_1}. \tag{3.8}$$

The Carter factor is also the ratio of the maximum flux density B_{max} to the average flux density B_{av}

$$k_C = \frac{B_{max}}{B_{av}}. \tag{3.9}$$

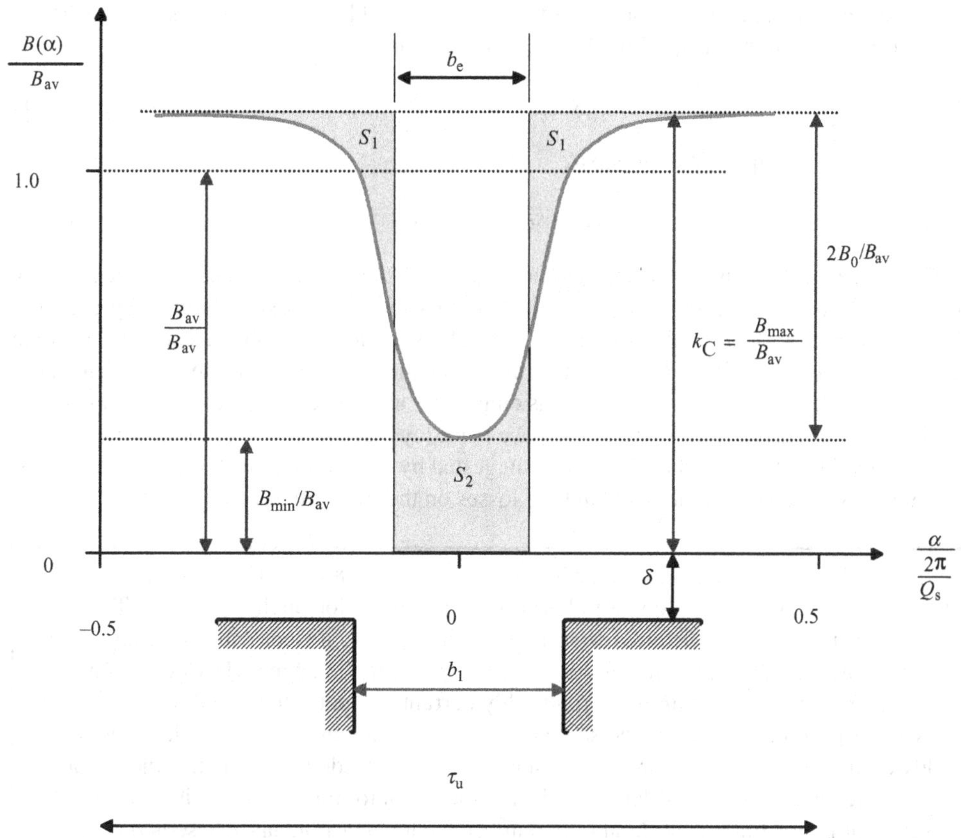

Figure 3.5 Distribution of air-gap flux density $B_\delta(\alpha)$ in a distance of one slot pitch τ_u. α is the angle revolving around the periphery of the machine. b_e is the equivalent slot opening.

The variation of the flux density assuming no eddy currents damping the flux variation is

$$\beta = \frac{B_0}{B_{max}} = \frac{(B_{max} - B_{min})}{2B_{max}} = \frac{1 + u^2 - 2u}{2(1 + u^2)}, \tag{3.10a}$$

$$\frac{B_{min}}{B_{max}} = \frac{2u}{1 + u^2}, \tag{3.10b}$$

$$u = \frac{b_1}{2\delta} + \sqrt{1 + \left(\frac{b_1}{2\delta}\right)^2}. \tag{3.10c}$$

When both the stator and rotor surfaces are provided with slots, we calculate k_{Cs} first by assuming the rotor surface to be smooth. The calculations are repeated by applying the calculated air gap δ_{es} and the slot pitch of the rotor τ_r, and by assuming the stator surface to be smooth. We then obtain k_{Cr}. Finally, the total factor is

$$k_{C,tot} \approx k_{Cs} \cdot k_{Cr}, \tag{3.11}$$

which gives us the equivalent air gap δ_e

$$\delta_e \approx k_{C,tot}\delta \approx k_{Cr}\delta_{es}. \tag{3.12}$$

The influence of slots in the average permeance of the air gap is taken into account by replacing the real air gap by a longer equivalent air gap δ_e. The result obtained by applying the above equations is not quite accurate, yet usually sufficient in practice. The most accurate result is obtained by solving the field diagram of the air gap with the finite element method. In this method, a dense element network is employed, and an accurate field solution is found as illustrated in Figure 3.4. If the rotor surface lets eddy currents run, the slot-opening-caused flux density dips are damped by the values suggested by the Carter factor. In such a case, eddy currents may create remarkable amounts of losses on the rotor surfaces.

Example 3.1: An induction motor has an air gap $\delta = 0.8$ mm. The stator slot opening is $b_1 = 3$ mm, the rotor slots are closed and the stator slot pitch is 10 mm. The rotor magnetic circuit is manufactured from high-quality electrical steel with low eddy current losses. Calculate the Carter-factor-corrected air gap of the machine. How deep is the flux density dip at a slot opening if the rotor eddy currents do not affect the dip (note that the possible squirrel cage is designed so that the slot harmonic may not create large opposing eddy currents)? How much three-phase stator current is needed to magnetize the air gap to 0.9 T fundamental peak flux density? The number of stator turns in series is $N_s = 100$, the number of pole pairs is $p = 2$, and the number of slots per pole and phase is $q = 3$. The winding is a full-pitch one.

Solution:

$$\kappa \approx \frac{b_1/\delta}{5 + b_1/\delta} = \frac{3/0.8}{5 + 3/0.8} = 0.429,$$

$$b_e = \kappa b_1 = 0.429 \cdot 3 = 1.29,$$

$$k_{Cs} = \frac{\tau_u}{\tau_u - b_e} = \frac{10}{10 - 1.29} = 1.148,$$

$$u = \frac{b_1}{2\delta} + \sqrt{1 + \left(\frac{b_1}{2\delta}\right)^2} = \frac{3}{2 \cdot 0.8} + \sqrt{1 + \left(\frac{3}{2 \cdot 0.8}\right)^2} = 3.984,$$

$$\beta = \frac{1 + u^2 - 2u}{2(1 + u^2)} = 0.179.$$

The depth of the flux density dip on the rotor surface is $2B_0$:

$$2B_0 = 2\beta B_{max} = 2\beta k_C B_{av} = 2 \cdot 0.179 \cdot 1.148 B_{av} = 0.41 B_{av}.$$

The equivalent air gap is $\delta_e \approx k_{Cs}\delta = 1.148 \cdot 0.8\text{mm} = 0.918\text{mm}$.

At 0.9 T the peak value of the air gap field strength is

$$\hat{H}_\delta = \frac{0.9\ \text{T}}{4\pi 10^{-7}\ \text{Vs/Am}} = 716\ \text{kA/m}.$$

The magnetic voltage of the air gap is

$$\hat{U}_{\text{m},\delta\text{e}} = \hat{H}_\delta \delta_\text{e} = 716\ \text{kA/m} \cdot 0.000918\ \text{m} = 657\text{A}.$$

According to Equation (2.15), the amplitude of the stator current linkage is

$$\hat{\Theta}_{\text{s1}} = \frac{m}{2}\frac{4}{\pi}\frac{k_{\text{ws1}}N_\text{s}}{2p}\sqrt{2}I_{\text{sm}} = \frac{mk_{\text{ws1}}N_\text{s}}{\pi p}\sqrt{2}I_{\text{sm}}.$$

To calculate the amplitude, we need the fundamental winding factor for a machine with $q = 3$. For a full-pitch winding with $q = 3$, we have the electrical slot angle

$$a_\text{u} = \frac{360^\circ}{2mq} = \frac{360^\circ}{2 \cdot 3 \cdot 3} = 20^\circ.$$

We have six voltage phasors per pole pair: three positive and three negative phasors. When the phasors of negative coil sides are turned 180°, we have two phasors at an angle of −20°, two phasors at an angle of 0° and two phasors at +20°. The fundamental winding factor will hence be

$$k_{\text{wsl}} = \frac{2\cos(-20^\circ) + 2\cos(0^\circ) + 2\cos(+20^\circ)}{6} = 0.960.$$

Since this is a full-pitch winding the same result is found by calculating the distribution factor according to Equation (2.23)

$$k_{\text{ds1}} = k_{\text{ws1}} = \frac{\sin q_\text{s}\dfrac{a_{\text{us}}}{2}}{q_\text{s}\sin\dfrac{a_{\text{us}}}{2}} = \frac{\sin 3\dfrac{20}{2}}{3\sin\dfrac{20}{2}} = 0.960.$$

We may now calculate the stator current needed to magnetize the air gap

$$I_{\text{sm}} = \frac{\hat{\Theta}_{\text{s1}}\pi p}{mk_{\text{ws1}}N_\text{s}\sqrt{2}} = \frac{657 \cdot \pi \cdot 2}{3 \cdot 0.960 \cdot 100\sqrt{2}}\ \text{A} = 10.1\ \text{A}.$$

If an analytic equation is required to describe the flux distribution in case of an air gap slotted on only one side, an equivalent approximation introduced by Heller and Hamata (1977) can be employed. This equation yields a flux density distribution in the case of a smooth rotor in

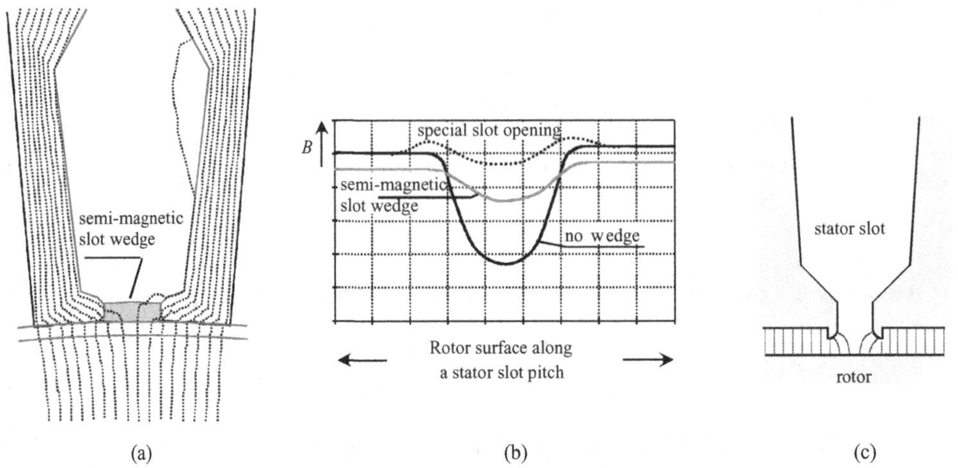

Figure 3.6 (a) Slot opening of a semi-closed stator slot of Figure 3.4 has been filled with a semi-magnetic filling $\mu_r = 5$). The flux drop on the rotor surface, caused by the slot, is remarkably reduced when the machine is running with a small current at no load. (b) Simultaneously, the losses on the rotor surface are reduced and the efficiency is improved. The edges of the stator slot can also be shaped according to (c), which yields the best flux density distribution. (b) The curve at the top.

a distance of one slot. If the origin is set at the center of the stator slot, the Heller and Hamata equations are written for a stator (see Figure 3.5)

$$B(\alpha) = \left(1 - \beta - \beta \cos \frac{\pi}{0.8\alpha_0}\alpha\right) B_{max}, \text{ when } 0 < \alpha < 0.8\alpha_0$$

and

$$B(\alpha) = B_{max} \text{ elsewhere, when } 0.8\alpha_0 < \alpha < \alpha_d. \tag{3.13}$$

Here $\alpha_0 = 2b_1/D$ and $\alpha_d = 2\pi/Q_s = 2\tau_u/D$.

Drops in the flux density caused by stator slots create losses on the rotor surface. Correspondingly, the rotor slots have the same effect on the stator surface. These losses can be reduced by partially or completely closing the slots, by reshaping the slot edges so that the drop in the flux density is eliminated, or by using a semi-magnetic slot wedge, Figure 3.6.

3.1.2 Air Gaps of a Salient-Pole Machine

Next, three different air gaps of a salient-pole machine are investigated. The importance of these air gaps lies in the following: the first air gap is employed in the calculation when the machine is magnetized with a rotor field winding; the second is required for the calculation of the direct-axis armature reaction, and thereby the armature direct-axis inductance; and finally, the third air gap gives the armature quadrature-axis reaction and the quadrature-axis inductance. The air gaps met by the rotor field winding magnetizing are shaped with pole

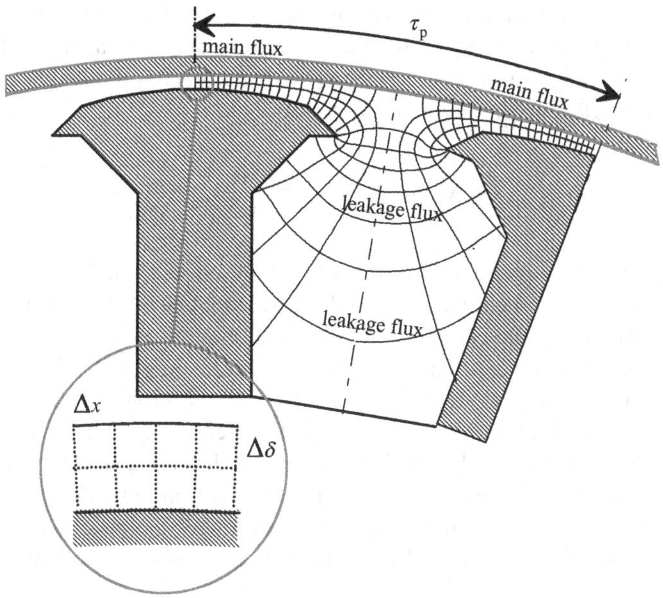

Figure 3.7 Field diagram of the rotor pole of an internal salient-pole synchronous machine with DC magnetizing in an area of half a pole pitch $\tau_p/2$. The figure also indicates that the amount of leakage flux in this case is about 15 %, which is a typical number for pole leakage. Typically, a designer should be prepared for about 20 % leakage flux when designing the field winding.

shoes such that as sinusoidal a flux density distribution as possible is obtained in the air gap of the machine.

The air-gap flux density created by the rotor field winding magnetizing of a salient-pole machine can be investigated with an orthogonal field diagram. The field diagram has to be constructed in an air gap of accurate shape. Figure 3.7 depicts the air gap of a salient-pole machine, into which the field winding wound around the salient-pole body creates a magnetic flux. The path of the flux can be solved with a magnetic scalar potential. The magnetic field refracts on the iron surface. However, in the manual calculation of the air gaps of a synchronous machine, the permeability of iron is assumed to be so high that the flux lines leave the equipotential iron surface perpendicularly.

If the proportion Θ_δ of the current linkage Θ_f of the rotor acts upon the air gap, each duct in the air gap takes, with the notation in Figure 3.7, a flux

$$\Delta\Phi_\delta = \mu_0\,\Theta_\delta\,l\frac{\Delta x}{n\Delta\delta}, \tag{3.14}$$

where l is the axial length of the pole shoe and n the number of square elements in the radial direction. If the origin of the reference frame is fixed to the middle of the pole shoe, we may write in the cosine form

$$\frac{\Delta\Phi_\delta}{l\Delta x} = \frac{\mu_0\Theta_\delta}{n\Delta\delta} = \hat{B}_\delta\cos\theta. \tag{3.15}$$

In the field diagram consisting of small squares, the side of a square equals the average width Δx. The magnetic flux density of the stator surface can thus be calculated by the average width of squares touching the surface. On the other hand, $n\Delta\delta$ is the length of the flux line from the pole shoe surface to the stator surface:

$$n\Delta\delta = \frac{\mu_0 \Theta_\delta}{\hat{B}_\delta \cos\theta} = \frac{\delta_{0e}}{\cos\theta}, \tag{3.16}$$

where δ_{0e} is the air gap in the middle of the pole, corrected with the Carter factor. Now the pole shoe has to be shaped such that the length of the flux density line of the field diagram is inversely proportional to the cosine of the electrical angle θ. The air gap length behaves as $\delta(\theta) = \delta_{0e}/\cos\theta$. Of course, infinite air gap at $\theta = \pi/2$ is impossible but the leakage flux from pole to pole makes the flux density at $\theta = \pi/2$ zero with rotor excitation. A pole shoe shaped in this way creates a cosinusoidal magnetic flux density in the air gap, the peak value of which is $\hat{B}_\delta$. The maximum value of flux penetrating through a full-pitch winding coil is called the peak value of flux, although it is not a question of amplitude here. The peak value of the flux is obtained by calculating a surface integral over the pole pitch and the length of the machine. In practice, the flux of a single pole is calculated. Now, the peak value is denoted $\hat{\Phi}_m$

$$\hat{\Phi}_m = \int_0^{l'} \int_{-\frac{\tau_p}{2}}^{\frac{\tau_p}{2}} \hat{B}_\delta \cos\left(\frac{x}{\tau_p}\pi\right) dx dl', \tag{3.17}$$

where l' is the equivalent core length $l' \approx l + 2\delta$ in a machine without ventilating ducts (see Section 3.2), τ_p is the pole pitch, $\tau_p = \pi D/(2p)$, x is the coordinate, the origin of which is in the middle of the pole, and $\theta = x\pi/\tau_p$.

When the flux density is cosinusoidally distributed, we obtain by integration for the air-gap flux

$$\hat{\Phi}_m = \frac{2}{\pi}\hat{B}_\delta \tau_p l'. \tag{3.18}$$

By reformulating the previous equation, we obtain

$$\hat{\Phi}_m = \frac{Dl'}{p}\hat{B}_\delta = \mu_0 \frac{Dl'}{p\delta_{0e}}\hat{\Theta}_\delta. \tag{3.19}$$

Next, the air gap experienced by the stator winding current linkage is investigated. The stator winding is constructed such that its current linkage is distributed fairly cosinusoidally on the stator surface. The stator current linkage creates an armature reaction in the magnetizing inductances. As a result of the armature reaction, this cosinusoidally distributed current linkage creates a flux of its own in the air gap. Because the air gap is shaped so that the flux density created by the rotor pole is cosinusoidal, it is obvious that the flux density created by the stator is not cosinusoidal, see Figure 3.8. When the peak value of the fundamental current linkage

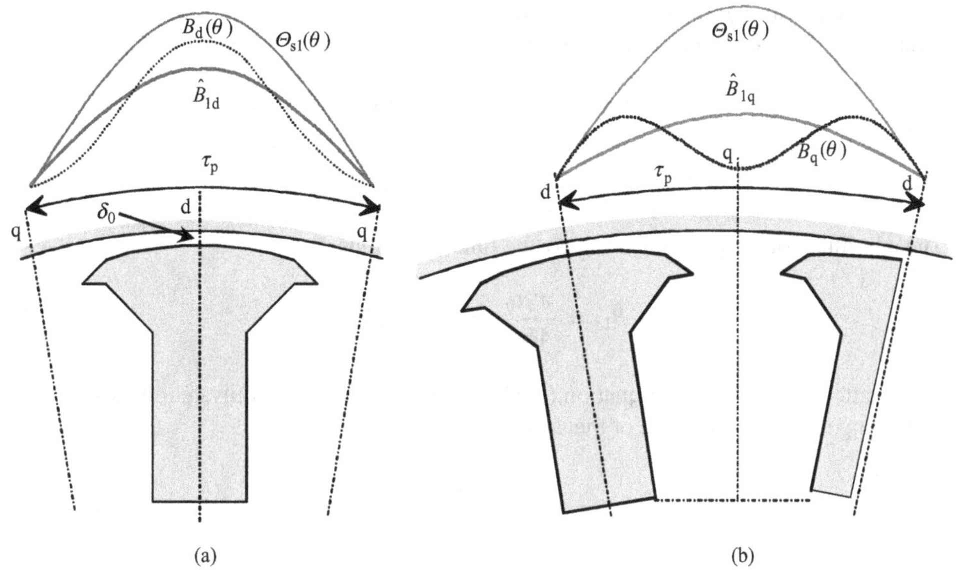

Figure 3.8 (a) Cosine-squared flux density $B_d(\theta)$, created in the shaped air gap by a cosinusoidal stator current linkage Θ_{s1} occurring on a direct axis of the stator, where the peak value of the fundamental component of $B_d(\theta)$ is $\hat{B}_{1d}$. (b) The cosinusoidal current linkage distribution on the quadrature axis creates a flux density curve $B_q(\theta)$. The peak value of the fundamental component of $B_q(\theta)$ is $\hat{B}_{1q}$.

of the stator is on the d-axis, we may write

$$\Theta_{s1}(\theta) = \hat{\Theta}'_d \cos\theta. \tag{3.20}$$

The amplitude of the current linkage is calculated by Equation (2.15). The permeance $d\Lambda$ of the duct at the position θ is

$$d\Lambda = \mu_0 \frac{dS}{n\Delta\delta} = \mu_0 \frac{Dl d\theta}{2p} \frac{\cos\theta}{\delta_{0e}}. \tag{3.21}$$

The magnetic flux density at the position θ is

$$B_d(\theta) = \frac{d\Phi}{dS} = \frac{\mu_0}{\delta_{0e}} \hat{\Theta}'_d \cos^2\theta. \tag{3.22}$$

The distribution of the air-gap flux density created by the stator current is proportional to the square of the cosine when the current linkage of the stator is on the d-axis. To be able to calculate the inductance of the fundamental, this density function has to be replaced by a cosine function with an equal flux. Thus, we calculate the factor of the fundamental of the Fourier series. The condition for keeping the flux unchanged is

$$\frac{\mu_0}{\delta_{0e}}\hat{\Theta}'_{\mathrm{d}}\int_{-\pi/2}^{+\pi/2}\cos^2\theta\mathrm{d}\theta = B_{1\mathrm{d}}\int_{-\pi/2}^{+\pi/2}\cos\theta\mathrm{d}\theta. \tag{3.23}$$

The amplitude of the corresponding cosine function is thus

$$\hat{B}_{1\mathrm{d}} = \frac{\pi\mu_0}{4\delta_{0e}}\hat{\Theta}'_{\mathrm{d}} = \frac{\mu_0}{\delta_{\mathrm{de}}}\hat{\Theta}'_{\mathrm{d}} \tag{3.24}$$

In the latter presentation of Equation (3.24), the air gap δ_{de} is an equivalent d-axis air gap experiencing the current linkage of the stator. Its theoretical value is

$$\delta_{\mathrm{de}} = \frac{4\delta_{0e}}{\pi}. \tag{3.25}$$

Figure 3.8a depicts this situation. In reality, the distance from the stator to the rotor on the edge of the pole cannot be extended infinitely, and therefore the theoretical value of Equation (3.25) is not realized as such (it is only an approximation). However, the error is only marginal, because when the peak value of the cosinusoidal current linkage distribution is at the direct axis, the current linkage close to the quadrature axis is very small. Equation (3.25) gives an interesting result: the current linkage of the stator has to be higher than the current linkage of the rotor, if the same peak value of the fundamental component of the flux density is desired with either the stator or rotor magnetization.

Figure 3.8b illustrates the definition of the quadrature air gap. The peak value of the stator current linkage distribution is assumed to be on the quadrature axis of the machine. Next, the flux density curve is plotted on the quadrature axis, and the flux Φ_{q} is calculated similarly as in Equation (3.19). The flux density amplitude of the fundamental component corresponding to this flux is written as

$$\hat{B}_{1\mathrm{q}} = \frac{p\Phi_{\mathrm{q}}}{Dl} = \frac{\mu_0}{\delta_{\mathrm{qe}}}\hat{\Theta}'_{\mathrm{q}}, \tag{3.26}$$

where δ_{qe} is the equivalent quadrature air gap. The current linkages are set equal: $\hat{\Theta}_{\mathrm{f}} = \hat{\Theta}'_{\mathrm{d}} = \hat{\Theta}'_{\mathrm{q}}$, the equivalent air gaps behaving inversely proportional to the flux density amplitudes

$$\hat{B}_{\delta} : \hat{B}_{1\mathrm{d}} : \hat{B}_{1\mathrm{q}} = \frac{1}{\delta_{0e}} : \frac{1}{\delta_{\mathrm{de}}} : \frac{1}{\delta_{\mathrm{qe}}}. \tag{3.27}$$

Direct and quadrature equivalent air gaps are calculated from this (inverse) proportion. A direct-axis air gap is thus approximately $4\delta_{0e}/\pi$. A quadrature-axis air gap is more problematic to solve without numerical methods, but it varies typically between (1.5–2–3) $\times\ \delta_{\mathrm{de}}$. According to Schuisky (1950), in salient-pole synchronous machines, a quadrature air gap is typically 2.4-fold when compared with a direct air gap in salient-pole machines.

If the machine is designed to produce the theoretical air gap producing sinusoidal air gap flux density at no load the theoretical value for the q-axis air gap is according to Heikkilä (2002) $\delta_{qe} \approx 3\pi\delta_e / \left(4\sin^2\left(\frac{\alpha\pi}{2}\right)\right)$. Here α indicates as a per unit value the pole pitch area where we may assume the $\delta_0/\cos\theta$ form is valid. The equation assumes an infinite air gap at the q-axis but in practice it is impossible. A practical value of $\alpha = 0.9$ results in $\delta_{qe} = 1.77\delta_{de}$.

The physical air gap on the center line of the magnetic pole is set to δ_0. The slots in the stator create an apparent lengthening of the air gap when compared with a completely smooth stator. This lengthening is evaluated with the Carter factor. On the d-axis of the rotor pole, the length of an equivalent air gap is now δ_{0e} in respect of the pole magnetization. In this single air gap, the pole magnetization has to create a flux density $\hat{B}_\delta$. The required current linkage of a single pole is

$$\hat{\Theta}_f = \frac{\delta_{0e}\,\hat{B}_\delta}{\mu_0}. \tag{3.28}$$

The value for current linkage on a single rotor pole is $\Theta_f = N_f I_f$, when the DC field winding current on the pole is I_f and the number of turns in the coil is N_f. The flux linkage and the inductance of the rotor can now be easily calculated. When the pole shoes are shaped according to the above principles, the flux of the phase windings varies at no load as a sinusoidal function of time, $\Phi_m(t) = \hat{\Phi}_m \sin\omega t$, when the rotor rotates at an electric angular frequency ω. By applying Faraday's induction law as presented in Equation (1.8), we can calculate the induced voltage. The applied form of the induction law, which takes the geometry of the machine into account, is written with the flux linkage Ψ as

$$e_1(t) = -\frac{d\Psi(t)}{dt} = -k_{w1}N\frac{d\Phi_m(t)}{dt}, \tag{3.29}$$

and the fundamental component of the voltage induced in a single pole pair of the stator is written as

$$e_{1p}(t) = -N_p k_{w1}\omega\hat{\Phi}_m \cos\omega t. \tag{3.30}$$

Here N_p is the number of turns of one pole pair of the phase winding. The winding factor k_{w1} of the fundamental component takes the spatial distribution of the winding into account. The winding factor indicates that the peak value of the main flux $\hat{\Phi}_m$ does not penetrate all the coils simultaneously, and thus the main flux linkage of a pole pair is $\hat{\Psi}_m = N_p k_{w1}\hat{\Phi}_m$. By applying Equation (3.18) we obtain for the voltage of a pole pair

$$e_{1p}(t) = -N_p k_{w1}\omega\frac{2}{\pi}\hat{B}_\delta\tau_p l'\cos\omega t, \tag{3.31}$$

the effective value of which is

$$E_{1p} = \frac{1}{\sqrt{2}}\,\omega\,N_p k_{w1}\frac{2}{\pi}\,\hat{B}_\delta\,\tau_p\,l' = \frac{1}{\sqrt{2}}\omega\hat{\Psi}_{mp}. \tag{3.32}$$

The maximum value of the air-gap flux linkage $\hat{\Psi}_{mp}$ of a pole pair is found at instants when the main flux best links the phase winding observed. In other words, the magnetic axis of the winding is parallel to the main flux in the air gap.

The voltage of the stator winding is found by connecting an appropriate amount of pole pair voltages in series and in parallel according to the winding construction.

Previously, the air gaps δ_{de} and δ_{qe} were determined for the calculation of the direct and quadrature stator inductances. For the calculation of the inductance, we have also to define the current linkage required by the iron. The influence of the iron can easily be taken into account by correspondingly increasing the length of the air gap, $\delta_{def} = (\hat{U}_{m,\delta de}/(\hat{U}_{m,\delta de} + \hat{U}_{m,Fe}))\delta_{de}$. We now obtain the effective air gaps δ_{def} and δ_{qef}. With these air gaps, the main inductances of the stator can be calculated in the direct and quadrature directions

$$L_{pd} = \frac{2}{\pi}\mu_0\frac{D_\delta l'}{p\delta_{def}}(k_{w1}N_p)^2, \; L_{pq} = \frac{2}{\pi}\mu_0\frac{D_\delta l'}{p\delta_{qef}}(k_{w1}N_p)^2. \tag{3.33}$$

N_p is the number of turns (N_s/p) of a pole pair. The main inductance is the inductance of a single stator phase. To obtain the single-phase equivalent circuit magnetizing inductance, for instance for a three-phase machine, the main inductance has to be multiplied by 3/2 to take the effects of all three windings into account. When deriving the equations, Equation (2.15) for the current linkage of a stator is required, and also the equation for a flux linkage of a single pole pair of the stator

$$\Psi_{mp} = -k_{w1}N_p\frac{2}{\pi}\;\hat{B}_\delta\tau_p\;l'$$

which is included in the previous voltage equations. The peak value for the air-gap flux density is calculated with an equivalent air gap and a stator current linkage, which leads to Equation (3.33). The inductances will be discussed in detail later in Section 4.1 and also in Chapter 7.

3.1.3 Air Gap of Nonsalient-Pole Machine

In nonsalient-pole machines, unlike in salient-pole machines, the shape of the air-gap flux density cannot be adjusted by shaping the air gaps. The rotor of a salient-pole machine is a steel cylinder provided with slots. The magnetizing winding is inserted in these slots. The air gap of such a machine is in principle equal at all positions, and thus, to create a sinusoidal flux density distribution, the magnitude of the current linkage acting upon different positions of the air gap has to be varied. To get a correct result, the conductors of the magnetizing winding have to be divided accordingly among the slots of the rotor, Figure 3.9.

The air gap of the nonsalient-pole machine is equal in all directions. The current linkage of the stator meets the same air gap as the current linkage of the rotor, and thus, with the notation in Equation (3.27), we obtain

$$\delta_{0e} \approx \delta_{de} \approx \delta_{qe}. \tag{3.34}$$

Because both the stator and rotor are slotted, the Carter factor is applied twice. The main inductance of a nonsalient-pole machine can be calculated from Equation (3.33) by employing

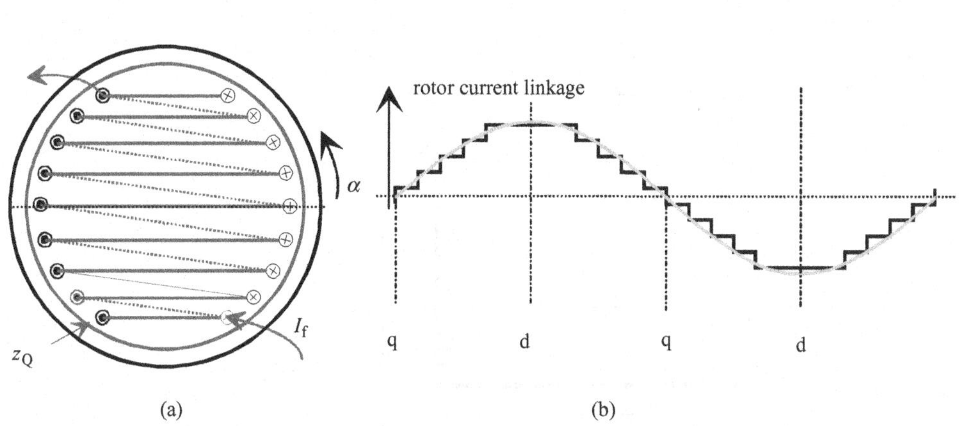

Figure 3.9 (a) Magnetizing winding of a nonsalient-pole rotor and (b) the distribution of the current linkage on the surface of the rotor cylinder, when the number of conductors is equal in all slots. The form of the current linkage can be improved by selecting the number of turns in the slots or the slot positions better than in the above example.

an effective air gap δ_{ef}, which takes also the effect of the iron into account and increases δ_{0e} with the proportion of the iron. A nonsalient-pole machine has only a single main inductance, because all the air gaps are equal. In practice, the slots in the rotor make the inductance of the quadrature axis slightly smaller than the inductance of the direct axis in these machines.

Asynchronous machines are usually isotropic, and therefore only a single equivalent air-gap length is defined for them, similarly as for nonsalient-pole machines. The main inductance is calculated from Equation (3.33). Usually, DC machines are external pole, salient-pole machines, and therefore their air gaps have to be determined by a method similar to the solution presented previously for a salient-pole machine. The above determination of the air gaps for a salient-pole machine is valid also for a synchronous salient-pole reluctance machine. With respect to an SR machine, the concept of an air gap has to be redefined, since the whole operation of the machine is based on the deformations of the magnetic circuit. The air gap is changing constantly when the machine rotates. The difference in the direct and quadrature inductances defines the average torque produced by the machine.

The magnetic voltage over the air gap is usually calculated with the smallest air gap and at the peak value of the flux density. If the equivalent air gap in the middle of the pole is δ_e, we obtain

$$\hat{U}_{m,\delta e} = \frac{\hat{B}_\delta}{\mu_0}\delta_e. \tag{3.35}$$

3.2 Equivalent Core Length

At an end of the machine, and at possible radial ventilating ducts, the effects of field fringings that may occur have to be taken into account. This is done by investigating the field diagrams of the ventilating ducts and of the ends of the machine. Figure 3.10 illustrates the influence of the edge field at the machine end on the equivalent length l' of the machine.

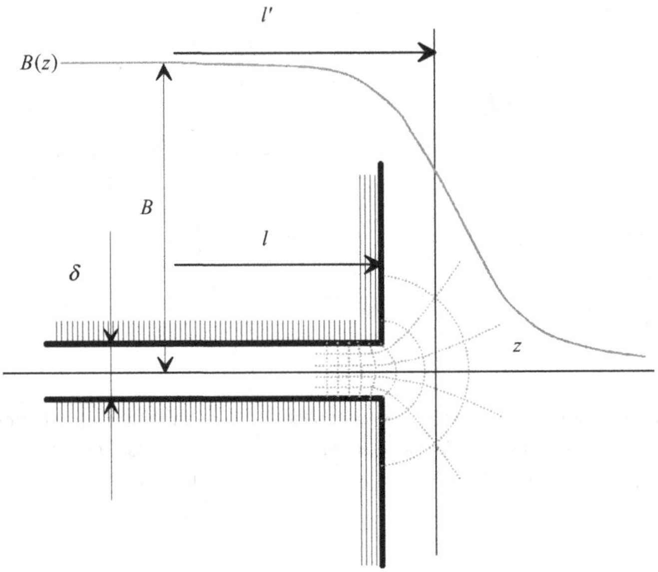

Figure 3.10 Orthogonal field diagram for the determination of the edge field at the end of the machine.

The flux density of the machine changes in the direction of the shaft as a function of the z-coordinate $B = B(z)$. The flux density remains approximately constant over a distance of the sheet core and decreases gradually to zero along the shaft of the machine as an effect of the edge field. This edge field is included in the main flux of the machine, and thus it participates also in torque production. In manual calculations, the lengthening of the machine caused by the edge field can be approximated by the equation

$$l' \approx l + 2\delta. \tag{3.36}$$

This correction is of no great significance, and therefore, when desired, the real length l is accurate enough in the calculations. In large machines, however, there are ventilating ducts that reduce the equivalent length of the machine, see Figure 3.11.

Here, we can estimate the length of the machine by applying the Carter factor again. By applying Equations (3.7a) and (3.7b) and by substituting the width of the slot opening b with the width of the ventilating duct b_v we obtain

$$b_{ve} = \kappa b_v. \tag{3.37}$$

The number of ventilating ducts being n_v in the machine (in Figure 3.11, $n_v = 3$), the equivalent length of the machine is approximated by

$$l' \approx l - n_v b_{ve} + 2\delta. \tag{3.38}$$

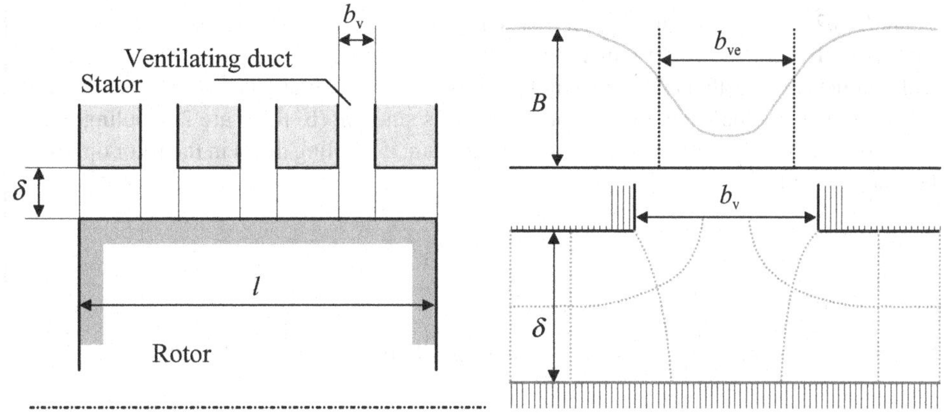

Figure 3.11 Influence of radial ventilating ducts on the equivalent length of the machine and the behavior of the flux density in the vicinity of the ventilating duct.

If there are radial ventilating ducts both in the rotor and in the stator, as depicted in Figure 3.12, the above method of calculation can in principle be employed. In that case, the flux density curve has to be squared, as was done previously, and the equivalent width of the duct b_{ve}, Equation (3.7a), is substituted in Equation (3.38). In the case of Figure 3.12a, the number of rotor ventilating ducts is equal to the number of ventilating ducts of the stator and the sum of the width of the stator and rotor ventilation ducts has to be substituted for b in Equation (3.7b). In the case of Figure 3.12b, the number of ducts n_v has to be the total number of ducts.

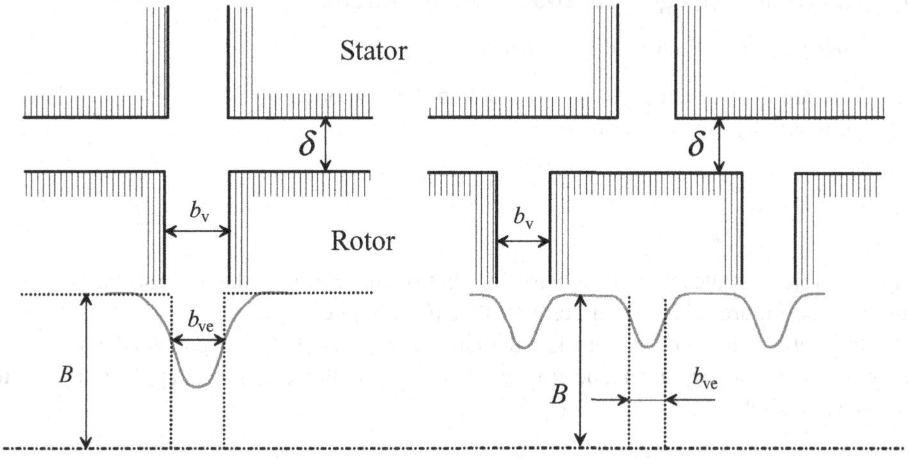

Figure 3.12 (a) In the stator and the rotor, the ventilating ducts are axially at the same positions. (b) The ventilating ducts are at different positions in the stator and in the rotor.

Example 3.2: A synchronous machine has a stator core 990 mm long. There are 25 substacks of laminations 30 mm long and a 10 mm cooling duct after each stack (24 cooling ducts all together). The air-gap length is 3 mm. Calculate the effective stator core length of the machine when (a) the rotor surface is smooth, (b) there are 24 cooling ducts in the rotor opposite the stator ducts and (c) there are 25 cooling ducts in the rotor opposite the stator substacks.

Solution:

(a)

$$\kappa = \frac{\dfrac{b_v}{\delta}}{5+\dfrac{b_v}{\delta}} = \frac{\dfrac{10}{3}}{5+\dfrac{10}{3}} = 0.40,\ b_{ve} = \kappa b_v = 0.40 \cdot 10\ \text{mm} = 4.0\ \text{mm},$$

$$l' \approx l - n_v b_{ve} + 2\delta = 990 - 24 \cdot 4.0 + 2 \cdot 3\ \text{mm} = 900\ \text{mm}.$$

(b)

$$\kappa = \frac{\dfrac{b_{vs}+b_{vr}}{\delta}}{5+\dfrac{b_{vs}+b_{vr}}{\delta}} = \frac{\dfrac{10+10}{3}}{5+\dfrac{10+10}{3}} = 0.571,\ b_{ve} = \kappa b_v = 0.571 \cdot 10\,\text{mm} = 5.71\,\text{mm},$$

$$l' \approx l - n_v b_{ve} + 2\delta = 990 - 24 \cdot 5.71 + 2 \cdot 3\ \text{mm} = 859\ \text{mm}.$$

(c) As in (a), $\kappa = 0.40$, $b_{ve} = 4.0$ mm for the stator and rotor ducts,

$$l' \approx l - n_{vs} b_{ves} - n_{vr} b_{ver} + 2\delta = 990 - 24 \cdot 4.0 - 25 \cdot 4.0 + 2 \cdot 3\ \text{mm} = 800\ \text{mm}.$$

3.3 Magnetic Voltage of a Tooth and a Salient Pole

3.3.1 Magnetic Voltage of a Tooth

When there are Q_s slots in the stator, we obtain the slot pitch of the stator by dividing the air-gap periphery by the number of slots

$$\tau_u = \frac{\pi D}{Q_s}. \tag{3.39}$$

Figure 3.13a illustrates the flux density distribution in an air gap, the other surface of which is smooth, and Figure 3.13b illustrates a tooth and a slot pitch.

The magnetic voltage of a tooth is calculated at a peak of the air-gap fundamental flux density. When a tooth occurs at a peak value of the air-gap flux density, an apparent tooth flux passes the slot pitch

$$\hat{\Phi}'_d = l' \tau_u \hat{B}_\delta. \tag{3.40}$$

If the teeth of the machine are not saturated, almost the complete flux of the slot pitch flows along the teeth, and there is no flux in the slots and the slot insulations. Neglecting the slot

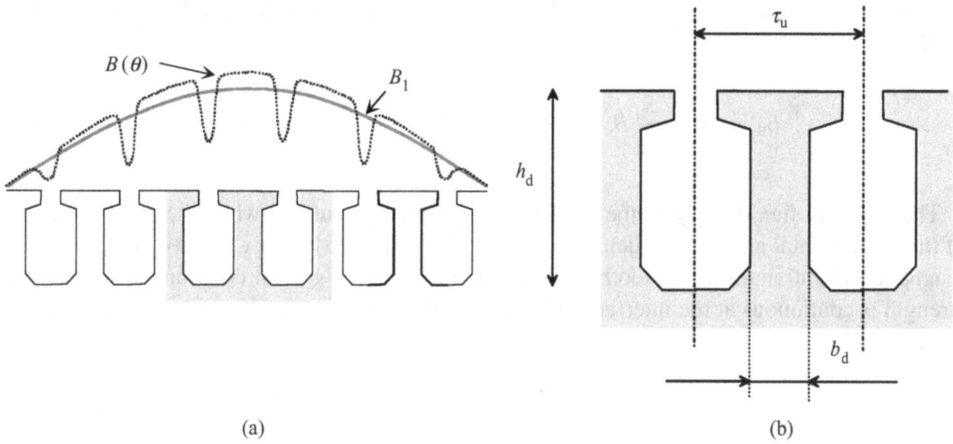

Figure 3.13 (a) Semi-closed slots and the flux density in the air gap. (b) The dimensions of a tooth and a slot: the height h_d of the tooth and the slot, slot pitch τ_u, and the width of the tooth b_d.

opening and taking into account the space factor k_{Fe} of iron, we obtain for a tooth with a uniform diameter and cross-sectional area S_d is

$$S_d = k_{Fe}(l - n_v b_v) b_d. \tag{3.41}$$

Here n_v and b_v are the number of ventilating ducts and their width (see Figure 3.12), and l is the total length of the machine stack. Punching influences the crystal structure of iron, and therefore the permeability on the cutting edges of the tooth is low. Thus, in the calculation of the flux density in a tooth, 0.1 mm has to be subtracted from the tooth width, that is $b_d = b_{real}$ – 0.1 mm in Equation (3.41) and the following equations. During the running of the motor, a relaxation phenomenon occurs, and the magnetic properties recover year by year. The space factor of iron k_{Fe} depends on the relative thickness of the insulation of the electric sheet and on the press fit of the stack. The insulators are relatively thin, their typical thickness being about 0.002 mm, and consequently the space factor of iron can in practice be as high as 98 %. A space factor varies typically between 0.9 and 0.97. Assuming that the complete flux is flowing in the tooth, we obtain its apparent flux density

$$\hat{B}'_d = \frac{\hat{\Phi}'_d}{S_d} = \frac{l'\tau_u}{k_{Fe}(l - n_v b_v) b_d} \hat{B}_\delta. \tag{3.42}$$

In practice, a part of the flux is always flowing through the slot along an area S_u. Denoting this flux by $\hat{\Phi}_u$, we may write for a flux in the tooth iron

$$\hat{\Phi}_d = \hat{\Phi}'_d - \hat{\Phi}_u = \hat{\Phi}'_d - S_u \hat{B}_u. \tag{3.43}$$

By dividing the result by the area of the tooth iron S_d we obtain the real flux density of the tooth iron

$$\hat{B}_d = \hat{B}'_d - \frac{S_u}{S_d}\hat{B}_u, \quad \text{where} \quad \frac{S_u}{S_d} = \frac{l'\tau_u}{k_{Fe}(l - n_v b_v)b_d} - 1. \tag{3.44}$$

The apparent flux density of the tooth iron $\hat{B}'_d$ can be calculated when the peak value $\hat{B}_\delta$ of the fundamental air-gap flux density is known. To calculate the flux density in the slot, the magnetic field strength in the tooth is required. Since the tangential component of the field strength is continuous at the interface of the iron and the air, that is $H_d = H_u$, the flux density of the slot is

$$\hat{B}_u = \mu_0 \hat{H}_d. \tag{3.45}$$

The real flux density in the tooth is thus

$$\hat{B}_d = \hat{B}'_d - \frac{S_u}{S_d}\mu_0 \hat{H}_d. \tag{3.46}$$

Now, we have to find a point that satisfies Equation (3.46) on the *BH* curve of the electric sheet in question. The easiest way is to solve the problem graphically as illustrated in Figure 3.14. The magnetic voltage $\hat{U}_{m,d}$ in the tooth is then approximately $\hat{H}_d h_d$.

When a slot and a tooth are not of equal width, the flux density is not constant, and therefore the magnetic voltage of the tooth has to be integrated or calculated in sections:

$$\hat{U}_{m,d} = \int_0^{h_d} \boldsymbol{H}_d \cdot d\boldsymbol{l}.$$

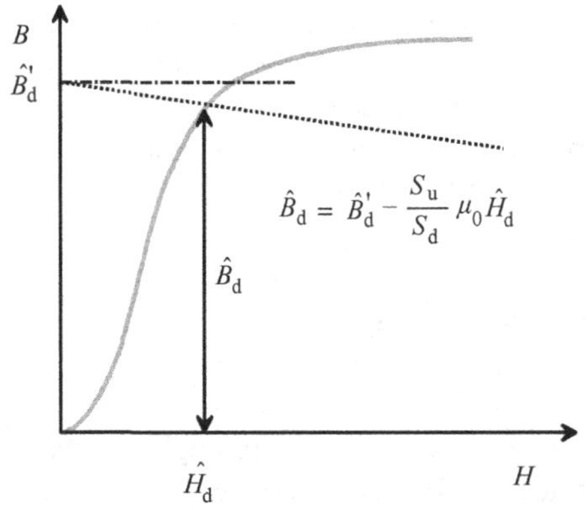

Figure 3.14 Definition of the flux density $\hat{B}_d$ of the tooth with the *BH* curve of the electrical sheet and the dimensions of the tooth.

Example 3.3: The stator teeth of a synchronous machine are 70 mm high and 14 mm wide. Further, the slot pitch $\tau_u = 30$ mm, the stator core length $l = 1$ m, there are no ventilation ducts, the space factor of the core $k_{Fe} = 0.98$, the core material is Surahammars Bruk electrical sheet M400-65A (Figure 3.15), the air gap $\delta = 2$ mm and the fundamental flux density of the air gap $\hat{B}_d = 0.85$ T. Calculate the magnetic voltage over the stator tooth.

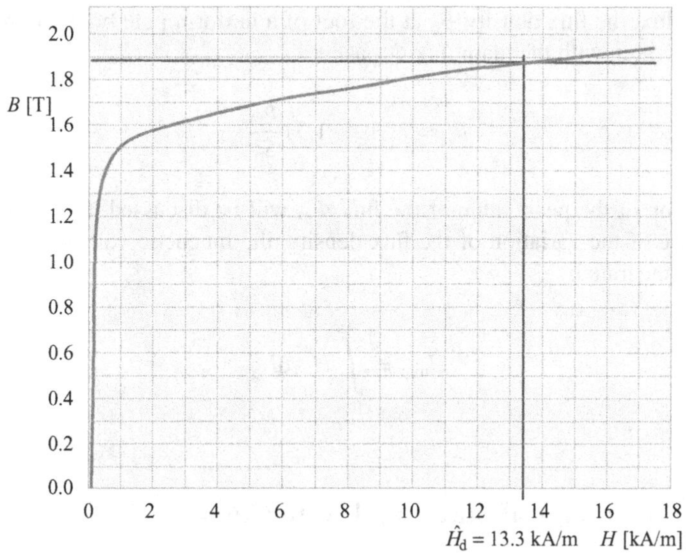

Figure 3.15 *BH* curve of the material used in Example 3.3 and its solution.

Solution: The effective core length $l' = 1000 + 2 \cdot 2$ mm $= 1004$ mm. The apparent flux density of the tooth, Equation (3.42), is

$$\hat{B}'_d = \frac{1004 \cdot 30}{0.98 \cdot 1000 \cdot 14} 0.85\ \text{T} = 1.88\ \text{T}.$$

The intersection of the *BH* curve of the electrical sheet M400-65A and the line (3.46) gives

$$\hat{B}_d = 1.88 - \left(\frac{1004 \cdot 30}{0.98 \cdot 1000 \cdot (14 - 0.1)} - 1 \right) 4 \cdot \pi \cdot 10^{-7} \hat{H}_d$$

The figure above gives the field strength of the teeth $\hat{H}_d = 13.3$ kA/m and the magnetic voltage

$$\hat{U}_{m,d} = \int_0^{h_d} \boldsymbol{H}_d \cdot d\boldsymbol{l} = 13300 \cdot 0.07\ \text{A} = 931\ \text{A}.$$

3.3.2 Magnetic Voltage of a Salient Pole

The calculation of the magnetic voltage of a salient pole is in principle quite similar to the procedure introduced for a tooth. However, there are certain particularities: for instance, the magnetic flux density of the pole shoe is often so small that the magnetic voltage required by it can be neglected, in which case only the magnetic voltage required by the pole body has to be defined. When determining the magnetic voltage of the pole body, special attention has to be paid to the leakage flux of the pole body. Figure 3.7 shows that a considerable amount of the flux of the pole body is leaking. The leakage flux comprises 10–30 % of the main flux. Because of the leakage flux, the flux density $\hat{B}_p$ at the foot of a uniform pole body of cross-sectional area S_p is thus written with the main flux $\hat{\Phi}_m$ as

$$\hat{B}_p = (1.1 \ldots 1.3)\frac{\hat{\Phi}_m}{S_p}. \tag{3.47}$$

The calculation of the peak value of the flux $\hat{\Phi}_m$ will be discussed later at the end of the chapter. Because of the variation of the flux density, the magnetic voltage $\hat{U}_{m,p}$ of the pole body has to be integrated:

$$\hat{U}_{m,p} = \int_0^{h_{dr}} \boldsymbol{H} \cdot d\boldsymbol{l}.$$

3.4 Magnetic Voltage of Stator and Rotor Yokes

Figure 3.16a depicts the flux distribution of a two-pole asynchronous machine when the machine is running at no load. The flux penetrating the air gap and the teeth section is divided into two equal parts at the stator and rotor yokes. At the peak of the air-gap flux density on the d-axis, the flux densities in the yokes are zero. The maximum flux densities for the yokes occur on the q-axis, on which the air-gap flux density is zero. The maximum value for the flux density of the stator yoke can be calculated on the q-axis without difficulty, since half of the main flux is flowing there:

$$\hat{B}_{ys} = \frac{\hat{\Phi}_m}{2S_{ys}} = \frac{\hat{\Phi}_m}{2k_{Fe}(l - n_v b_v)h_{ys}}. \tag{3.48}$$

Here S_{ys} is the cross-sectional area of the stator yoke, k_{Fe} is the space factor of iron and h_{ys} is the height of the yoke, see Figure 3.16b. Respectively, the maximum flux density at the rotor yoke is

$$\hat{B}_{yr} = \frac{\hat{\Phi}_m}{2S_{yr}} = \frac{\hat{\Phi}_m}{2k_{Fe}(l - n_v b_v)h_{yr}}. \tag{3.49}$$

The calculation of the magnetic voltage of the yoke is complicated, since the flux density at the yoke changes constantly over the pole pitch, and the behavior of the field strength is highly nonlinear, see Figure 3.16c.

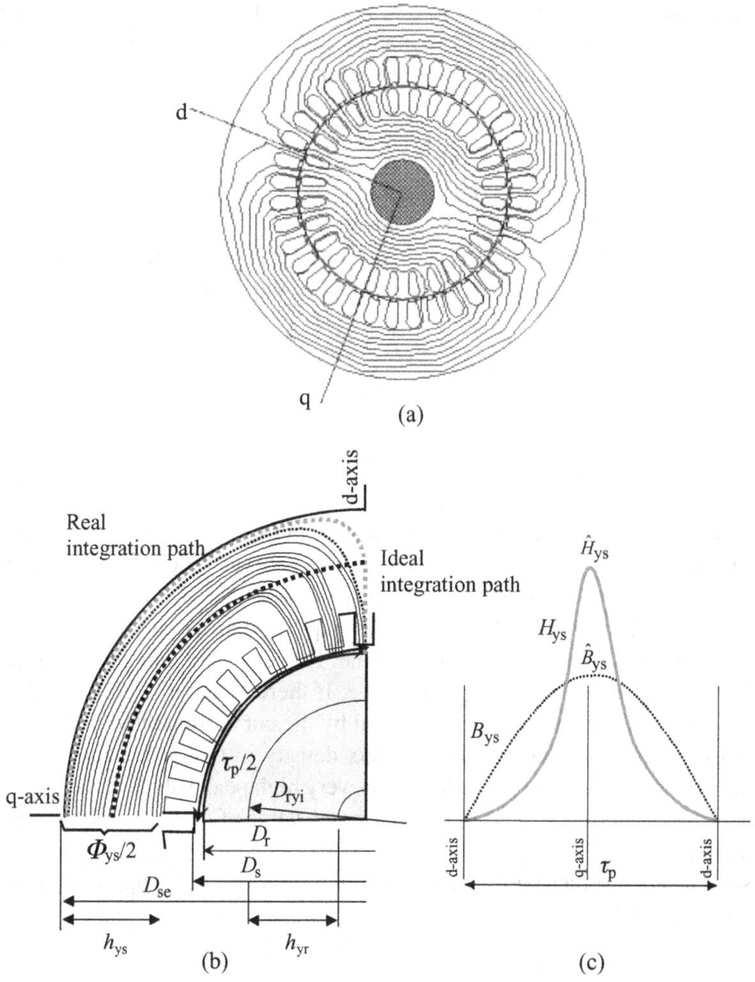

Figure 3.16 (a) Flux diagram for a two-pole induction motor running at no load. The shaft of the machine is far more reluctive than the rotor sheet, and therefore, at the plotted lines, no flux seems to penetrate the rotor shaft. Furthermore, the shaft is often jagged at the rotor bundle, and thus in practice there are air bridges between the rotor iron and the shaft. (b) Flux of the stator yoke in an electrical machine, and the integration path of the magnetic voltage. (c) The behavior of the flux density of the stator yoke, and the strongly nonlinear behavior of the field strength H_{ys}, which explain the difficulties in the definition of the magnetic voltage of the yoke. The ideal integration path is indicated by the thick black dotted line and the real integration path by any flux line, for example the thick gray dotted line.

Magnetic potential difference $\hat{U}_{m,ys}$ over the whole yoke has to be determined by calculating the line integral of the magnetic field strength between the two poles of the yoke integration path

$$\hat{U}_{m,ys} = \int_{d}^{q} \boldsymbol{H} \cdot d\boldsymbol{l}. \tag{3.50}$$

The field diagram has to be known in order to be able to define the integral. Precise calculation is possible only with numerical methods. In manual calculations, the magnetic voltages of the stator and rotor yoke can be calculated from the equations

$$\hat{U}_{\mathrm{m,ys}} = c\hat{H}_{\mathrm{ys}}\tau_{\mathrm{ys}}, \tag{3.51}$$

$$\hat{U}_{\mathrm{m,yr}} = c\hat{H}_{\mathrm{yr}}\tau_{\mathrm{yr}}. \tag{3.52}$$

Here $\hat{H}_{\mathrm{ys}}$ and $\hat{H}_{\mathrm{yr}}$ are the field strengths corresponding to the highest flux density, and τ_{ys} and τ_{yr} are the lengths of the pole pitch in the middle of the yoke (Figures 3.1 and 3.16):

$$\tau_{\mathrm{ys}} = \frac{\pi(D_{\mathrm{se}} - h_{\mathrm{ys}})}{2p}, \tag{3.53a}$$

$$\tau_{\mathrm{yr}} = \frac{\pi(D_{\mathrm{ryi}} - h_{\mathrm{yr}})}{2p}. \tag{3.52b}$$

The coefficient c takes into account the fact that the field strength is strongly nonlinear in the yoke, and that the nonlinearity is the stronger, the more saturated the yoke is at the q-axis. For most places in the yokes, the field strength is notably lower than $\hat{H}_{\mathrm{ys}}$ or $\hat{H}_{\mathrm{yr}}$, see Figure 3.16.

The coefficient is defined by the shape of the air-gap flux density curve, by the saturation of the machine and by the dimensions of the machine. However, the most decisive factor is the maximum flux density in the yoke of the machine. If there is a slot winding in the machine, the magnetic voltage of the yoke can be estimated by the curve illustrated in Figure 3.17.

Figure 3.16 shows clearly that as the peak flux density in the q-axis approaches the iron saturation flux density, the field strength H reaches very high peak values in the yoke. Since the peak H value is possible only on the q-axis, the average value of H decreases and consequently so does the coefficient c.

In the case of a salient-pole stator or rotor, a value of $c = 1$ can be applied for the yoke on the salient-pole side of the machine.

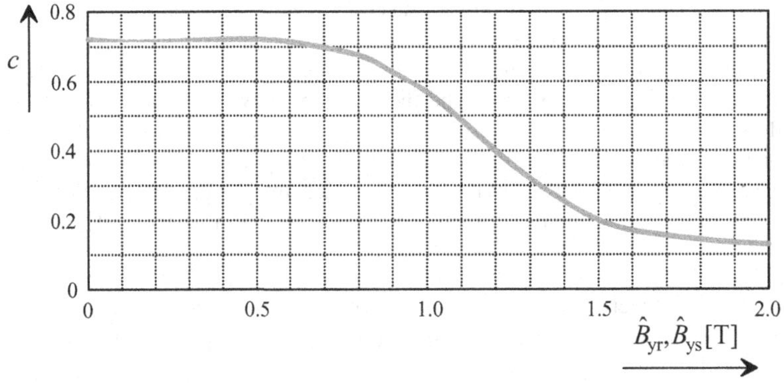

Figure 3.17 Influence of the maximum flux density of the stator or rotor yoke in the definition of the coefficient c, applied in the determination of magnetic voltage.

Example 3.4: In a four-pole machine, the outer diameter $D_{se} = 0.5$ m, the stator air gap diameter is $D_s = 0.3$ m, the machine stator core length is 0.3 m and the air gap is 1 mm. The peak value of the fundamental air-gap flux density is 0.9 T. The stator yoke height $h_{ys} = 0.05$ m. Calculate the stator yoke magnetic voltage when the material is M400-50A.

Solution: Assuming a sinusoidal flux density distribution in the air gap according to Equation (3.18), the peak value of the air gap flux is

$$\hat{\Phi}_m = \frac{2}{\pi}\hat{B}_\delta \tau_p l' = \frac{2}{\pi} \cdot 0.9\,\text{T} \cdot \frac{\pi 0.3\,\text{m}}{2 \cdot 2} \cdot (0.3\,\text{m} + 0.002\,\text{m}) = 0.0408\,\text{Vs}$$

This flux is divided into two halves in the stator yoke. The yoke flux density is therefore

$$\hat{B}_{ys} = \frac{\hat{\Phi}_m}{2S_{ys}} = \frac{\hat{\Phi}_m}{2k_{Fe}(l - n_v b_v)h_{ys}} = \frac{0.0408\,\text{Vs}}{2 \cdot 0.98 \cdot 0.3\,\text{m} \cdot 0.05\,\text{m}} = 1.39\,\text{T}.$$

At this flux density, the maximum field strength in the yoke is about 400 A/m, see Appendix A. The stator yoke length is

$$\tau_{ys} = \frac{\pi(D_{se} - h_{ys})}{2p} = \frac{\pi(0.5\,\text{m} - 0.05\,\text{m})}{2 \cdot 2} = 0.353\,\text{m}$$

According to Figure 3.17, the coefficient $c = 0.26$, and hence we get for the stator yoke magnetic voltage

$$\hat{U}_{m,ys} = c\hat{H}_{ys}\tau_{ys} = 0.26 \cdot 400\,\text{A/m} \cdot 0.353\,\text{m} = 37\,\text{A}.$$

3.5 No-Load Curve, Equivalent Air Gap and Magnetizing Current of the Machine

In the determination of the current linkage required by the magnetic circuit, magnetic voltages corresponding to a certain peak value $\hat{B}_\delta$ of the magnetic flux density of the air gap of the machine have been calculated for all parts of the machine in turn. By selecting several values for $\hat{B}_\delta$ and by repeating the above calculations, we may plot the curves for the magnetic voltages required by the different parts of the machine, and, as their sum, we obtain the current linkage required by the magnetic circuit of the complete machine:

$$\Theta_m = U_{m\delta} + U_{m,sd} + U_{m,rd} + \frac{U_{m,sy}}{2} + \frac{U_{m,ry}}{2}. \quad (3.54)$$

As we can see, the equation sums the magnetic voltages of one air gap, one stator tooth, one rotor tooth and halves of the stator and rotor yokes. This is illustrated in Figure 3.18. The curve corresponds to the no-load curve defined in the no-load test of the machine, in which the flux density axis is replaced by the voltage, and the current linkage axis is replaced by the magnetizing current.

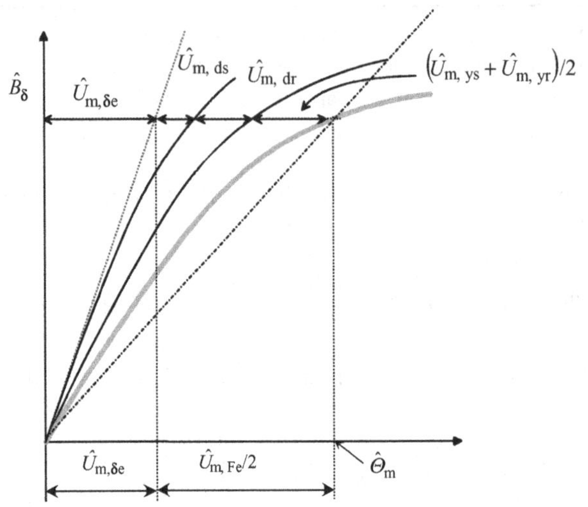

Figure 3.18 No-load curve of the machine (thick line) and its composition of the influences of the stator teeth ds, the air gap δ_e, the rotor teeth dr and the halves of stator and rotor yokes 1/2(ys, yr). In some cases, the whole magnetic circuit is replaced with equivalent air gaps; consequently, the magnetizing curve can be replaced at an operation point by a straight line (dotted line). Note that the proportion of the magnetic voltage of the iron is strongly exaggerated. In well-designed machines, the proportion of the current linkage required by the iron is only a fraction of the current linkage required by the air gaps. The total current linkage required by half a magnetic circuit is denoted $\hat{\Theta}_m$. The current linkage needed in the whole magnetic circuit is $2\hat{\Theta}_m$.

According to Figure 3.18, the magnetic voltage required by half of the iron in the complete magnetic circuit is

$$\frac{\hat{U}_{m,Fe}}{2} = \hat{\Theta}_m - \hat{U}_{m,\delta e}. \tag{3.55}$$

If the circuit has to be linearized (the dotted line in Figure 3.18), the air gap lengthened with the Carter factor is replaced by the effective air gap δ_{ef}, which also takes the reluctance of the iron into account. This air gap can be defined by a proportion

$$\frac{\hat{\Theta}_m}{\hat{U}_{m,\delta e}} = \frac{\delta_{ef}}{\delta_e}. \tag{3.56}$$

When the amplitude $\hat{\Theta}_m$ of the current linkage corresponding to the operating point of the machine has been defined, we are able to calculate the magnetizing current required by the machine. So far, no distinction has been made here between a machine that is magnetized with the alternating current of the rotating-field winding (cf. asynchronous machines) and a machine magnetized with the direct current of the field winding (cf. synchronous machines). In the case of a salient-pole machine (a synchronous machine or a DC machine), the magnetizing field winding direct current I_{fDC} required for one pole is

$$I_{fDC} = \frac{\hat{\Theta}_m}{N_f}. \tag{3.57}$$

Here N_f is the total number of turns on the magnetizing pole of the machine. Pole windings can in turn be suitably connected in series and in parallel to reach a desired voltage level and current.

On the rotor of a nonsalient-pole synchronous machine, there has to be an equal number of coil turns per pole as in a salient-pole machine, that is a number of N_f. These turns are divided to the poles in the same way as in a salient-pole machine. The winding is now inserted in slots as illustrated in Figure 3.9.

In all rotating-field machines (induction machines, various synchronous machines) the stator winding currents have a significant effect on the magnetic state of the machine. However, only induction machines and synchronous reluctance machines are magnetized with the magnetizing current component of the stator current alone. In the cases of separately magnetized machines and permanent magnet machines, the magnitude of the armature reaction is estimated by investigating the stator magnetization. For rotating-field machines, the required effective value of the alternating current I_{sm} is calculated with Equation (2.15). For the sake of convenience, Equation (2.15) is repeated here for the fundamental:

$$\hat{\Theta}_{s1} = \frac{m}{2}\frac{4}{\pi}\frac{k_{ws1}N_s}{2p}\sqrt{2}I_{sm} = \frac{mk_{ws1}N_s}{\pi p}\sqrt{2}I_{sm}. \tag{3.58}$$

$\hat{\Theta}_{s1}$ is the amplitude of the fundamental of the current linkage of the stator winding. k_{ws1} denotes the winding factor of the fundamental of the machine. $N_s/2p$ is the number of turns per pole, when N_s is the number of series-connected turns of the stator (parallel branches being neglected). m is the phase number. A single such amplitude magnetizes half of one magnetic circuit. If the magnetic voltage is defined for half a magnetic circuit as shown in Figure 3.18, and thus includes only one air gap and half of the iron circuit, the magnetizing current of the whole pole pair is calculated with this equation.

Pole pairs in rotating-field machines can also be connected both in series and in parallel, depending on the winding of the machine. In integral slot windings, the base winding is of the length of a pole pair. Thus, in a four-pole machine for instance, the pole pairs of the stator can be connected either in series or in parallel to create a functional construction. In the case of fractional slot windings, a base winding of the length of several pole pairs may be required. These base windings can in turn be connected in series and in parallel as required. The magnetizing current measured at the poles of the machine thus depends on the connection of the pole pairs. The effective value of the magnetizing current for a single pole pair $I_{sm,p}$ is

$$I_{sm,p} = \frac{\hat{\Theta}_{mp}}{\sqrt{2}\frac{m}{2}\frac{4}{\pi}\frac{k_{w1}N_s}{2p}}, \tag{3.59}$$

$$I_{sm,p} = \frac{\hat{\Theta}_{mp}\pi p}{\sqrt{2}mk_{w1}N_s}. \tag{3.60}$$

In a complete magnetic circuit, there are two amplitude peaks, but the same current produces both a positive and a negative amplitude of the current linkage and, hence, $2\hat{\Theta}_{\rm s1}$ altogether magnetizes the whole magnetic circuit. The current linkage produced by the magnetizing current $I_{\rm sm,p}$ for the complete pole pair is

$$2\hat{\Theta}_{\rm s1} = 2\frac{mk_{\rm ws1}N_{\rm s}}{\pi p}\sqrt{2}I_{\rm sm,p} \Leftrightarrow I_{\rm sm,p} = \frac{2\hat{\Theta}_{\rm s1}}{2\frac{mk_{\rm ws1}N_{\rm s}}{\pi p}\sqrt{2}} = \frac{\hat{\Theta}_{\rm s1}\pi p}{\sqrt{2}mk_{\rm ws1}N_{\rm s}}.$$

As we can see, the result is, in practice, the same when we know that $2\hat{\Theta}_{\rm s1}$ must equal $2\hat{\Theta}_{\rm mp}$.

Example 3.5: The sum of magnetic voltages in half of the magnetic circuit of a four-pole induction motor is 1500 A. There are 100 turns in series per stator winding, and the fundamental winding factor is $k_{\rm w1} = 0.925$. Calculate the no-load stator current.

Solution: The current linkage amplitude produced by a stator winding is, according to Equation (2.15),

$$\hat{\Theta}_{\rm s1} = \frac{m}{2}\frac{4}{\pi}\frac{k_{\rm ws1}N_{\rm s}}{2p}\sqrt{2}I_{\rm sm} = \frac{mk_{\rm ws1}N_{\rm s}}{\pi p}\sqrt{2}I_{\rm sm}.$$

Rearranging, we get for the RMS value of the stator magnetizing current

$$I_{\rm sm} = \frac{\hat{\Theta}_{\rm s1}\pi p}{mk_{\rm ws1}N_{\rm s}\sqrt{2}} = \frac{1500\text{A}\cdot\pi\cdot 2}{3\cdot 0.925\cdot 100\sqrt{2}} = 24\ \text{A}$$

3.6 Magnetic Materials of a Rotating Machine

Ferromagnetic materials and permanent magnets are the most significant magnetic materials used in machine construction. In these materials, there are elementary magnets, known as Weiss domains. These Weiss domains are separated from each other by Bloch walls, which are transition regions at the boundaries between magnetic domains. The width of a Bloch wall varies between a few hundred and a thousand atomic spacings, Figure 3.19.

The increase in the magnetic momentum of a body under the influence of an external field strength is a result of two independent processes. First, in a weak external field, those Weiss domains that are already positioned in the direction of the field increase at the expense of the domains positioned in the opposite direction, see Figure 3.20. Second, in a strong magnetic field, the Weiss domains that are in the normal direction turn into the direction of the field.

Turning elementary magnets requires a relatively high field strength. In magnetically soft materials, the Bloch wall displacement processes are almost completed before the Weiss domains begin to turn in the direction of the external field.

Without an external field strength, the Bloch walls are at rest. In practice, the walls are usually positioned at impurities and crystal defects in materials. If the body experiences only a weak external field strength, the Bloch walls are only slightly displaced from rest. If the

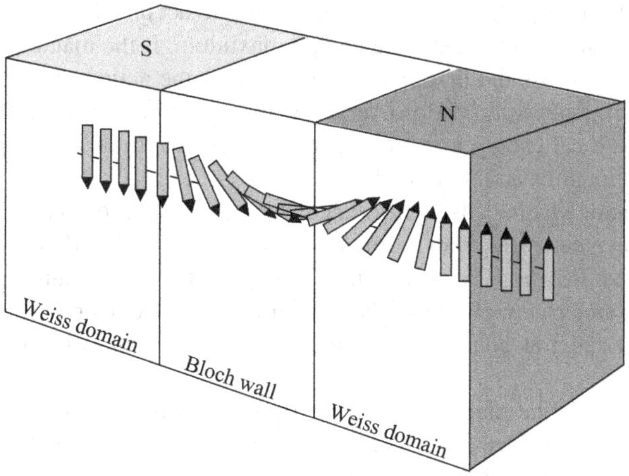

Figure 3.19 Bloch wall separating Weiss domains. For instance, the width of the transition region (Bloch wall) is 300 grid constants (about 0.1 mm).

field strength is removed, the Bloch walls return to their original positions. This process can be observed even with a high-precision microscope.

If we let the field strength increase abruptly, the Bloch walls leave their position at rest and do not return to their original positions even if the field strength is removed. These wall displacements are called Barkhausen jumps, and they result from ferromagnetic hysteresis and Barkhausen noise. It is possible that a Weiss domain takes over one of its neighboring domains in a single Barkhausen jump, particularly if the material includes a few, large crystal defects.

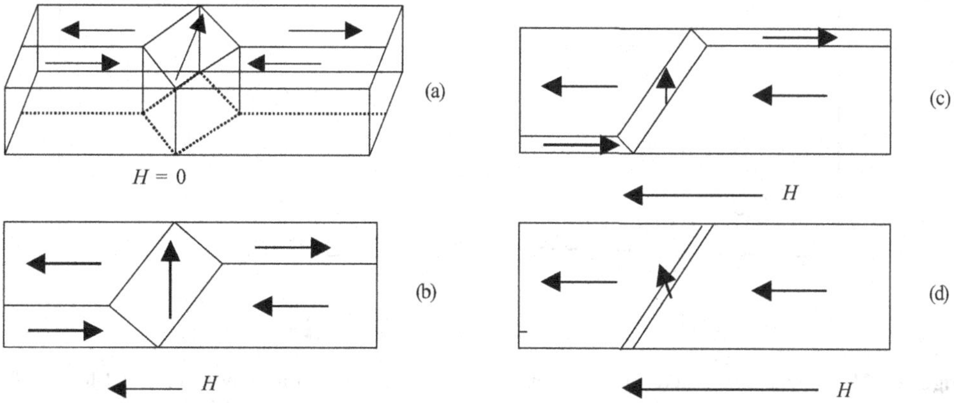

Figure 3.20 In a weak external field, the walls of the Weiss domains are in reversible motion so that the magnetic momentum of the whole body increases. (a) The external field strength is zero. For (b) and (c), as the field strength increases, those Weiss domains which were not originally in the direction of the external field decrease in size. (d) The domain originally in the normal direction has started to turn in the direction of the external field.

If a magnetic field strength that is high enough begins to act on a Bloch wall, it moves from its original position towards the next local energy maximum. If the magnetic field strength is low, the wall does not cross the first energy peak, and as the action of the force ceases, the wall returns to its original position. However, if the field strength is greater than the above, the wall passes the first local energy maximum and cannot return to its original position, unless an opposite field strength is acting upon it.

The displacement of Bloch walls in different materials takes place over a wide range of field strengths. In ferromagnetic materials, some of the walls are displaced at low field strengths, while some walls require a high field strength. The largest Barkhausen jumps occur at medium field strengths. In some cases, all the walls jump at the same field strength so that saturation magnetization is reached at once. Usually, the magnetization takes place in three separate phases.

Figure 3.21 illustrates the magnetizing curve of a ferromagnetic material, with three distinctive phases. In the first phase, the changes are reversible; in the second phase, Barkhausen jumps occur; and in the third phase, orientation of the Weiss domains takes place and all the domains turn in the direction determined by the external field strength. Next, the saturation magnetization and the respective saturation polarization J_s are reached.

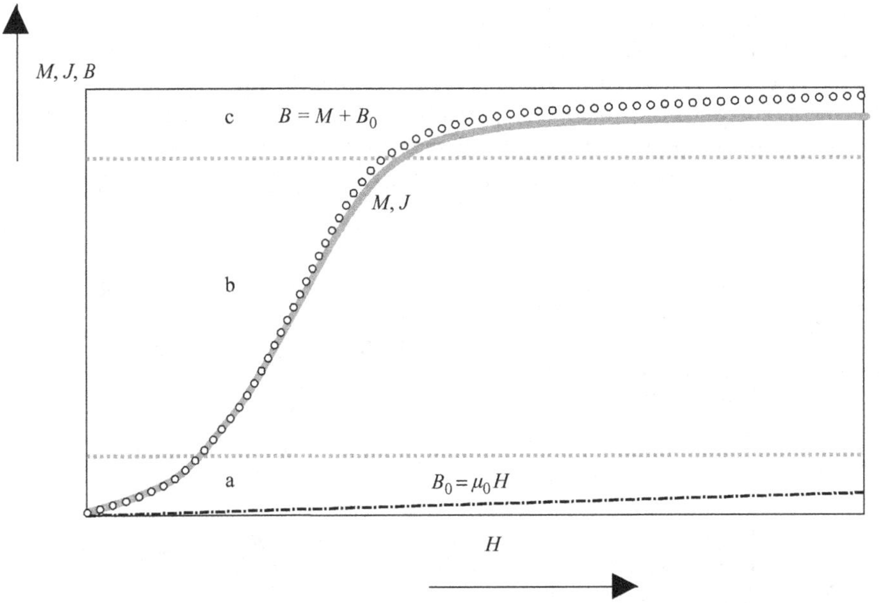

Figure 3.21 Magnetizing curve of a ferromagnetic material; in area a, only reversible Bloch wall displacements take place; in area b, irreversible Barkhausen jumps occur; and in area c, the material saturates when all Weiss domains are settled in parallel positions. Magnetization *M* saturates completely in area c. The polarization curve (*JH*) of the material is equal to the magnetizing curve. A *BH* curve differs from these curves for the amount of the addition caused by the permeability of a vacuum. As is known, the *BH* curve does not actually saturate at a horizontal plane in area c, but continues upwards with a slope defined by μ_0 as the field strength increases.

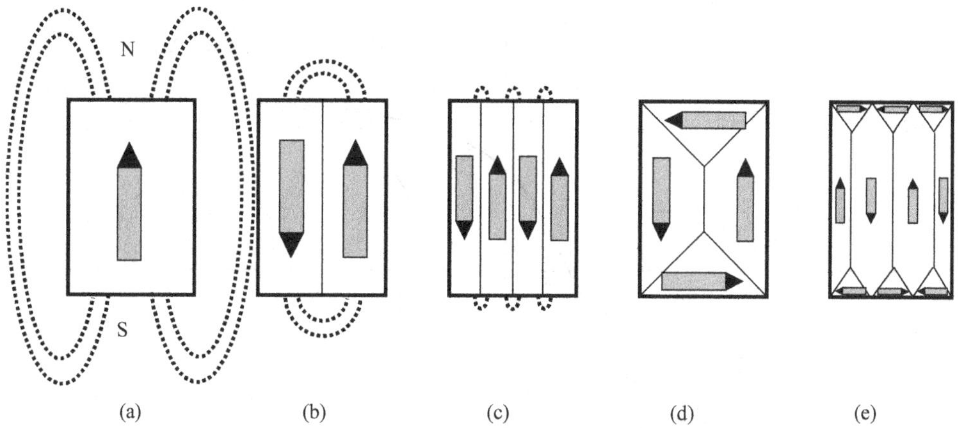

Figure 3.22 Division of a ferromagnetic crystal into Weiss domains in such a way that the energy minimum is realized.

Figure 3.22 helps to comprehend the formation of Weiss domains. The sections of the figure illustrate a ferromagnetic crystal divided into different Weiss domains.

In Figure 3.22a, the crystal comprises only a single Weiss domain that looks like a permanent magnet with N and S poles. In this kind of system, the magnetic energy, $\frac{1}{2}\int BH\mathrm{d}V$, is high. The energy density corresponding to case (a) in Figure 3.22 is for iron of magnitude $\mu_0 M_s^2 = 23\ \mathrm{kJ/m^3}$.

In Figure 3.22b, the magnetic energy has decreased by half when the crystal has been divided into two Weiss domains. In Figure 3.22c, it is assumed that the number of domains is N, and consequently the magnetic energy is reduced to $1/N$th part of case (a).

If the domains are settled as in cases (d) and (e), there is no magnetic field outside the body, and the magnetic energy of the crystal structure is zero. Here the triangular areas are at an angle of 45° with the square areas. No external magnetic field is involved as in Figure 3.22a, b and c. The magnetic flux is closed inside the crystal. In reality, Weiss domains are far more complicated than the cases exemplified here. However, the domains are created inside the body so that the magnetic energy of the body seeks the minimum value.

In the design of electromechanical applications, some of the most valuable information about the magnetizing of a material is obtained from the *BH* curve of the material in question. Figure 3.23 depicts a technical magnetizing curve of a ferromagnetic material. In the illustration, the flux density B is given as a function of field strength H.

3.6.1 Characteristics of Ferromagnetic Materials

The resistivity of pure ferromagnetic metals is usually of a few micro-ohm centimetres ($10^{-8}\ \Omega\mathrm{m}$), as shown in Table 3.1.

Since laminated structures are applied mainly to prevent the harmful effects of eddy currents, it is advisable to select a sheet with maximum resistivity. The different elements alloyed with iron have different effects on the electromagnetic properties of iron. In alloys, the resistivity ρ tends to increase when compared with pure elements. This is an interesting characteristic

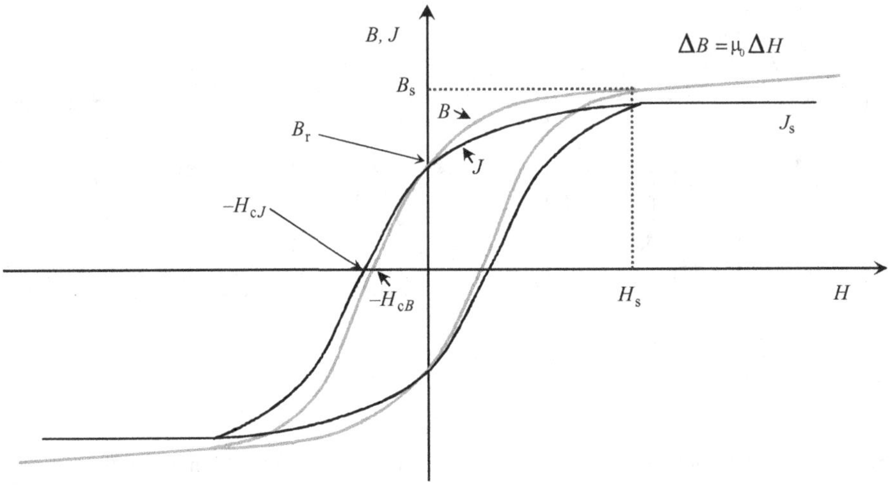

Figure 3.23 Technical magnetizing curve and a corresponding polarization curve of a ferromagnetic material, that is a hysteresis loop. The coercivity related to the flux density $-H_{cB}$ is the field strength required to restore the magnetic field density B from remanent flux density B_r to zero. The remanent flux density B_r is reached when the external magnetic field strength is restored to zero from a very high value. The saturation flux density B_s corresponds the saturation polarization J_s ($B_s = J_s + \mu_0 H_s$).

that has to be taken into account if we wish to reduce the amount of eddy currents in magnetic materials. Figure 3.24 illustrates the increase in resistivity for iron, when a small amount of an other element is alloyed with it. Copper, cobalt and nickel increase the resistivity of iron only marginally, whereas aluminum and silicon give a considerable increase in the resistivity.

Consequently, materials suitable for electric sheets are silicon–iron and aluminum–iron. A silicon-rich alloy makes the material brittle, and thus in practice the amount of silicon is

Table 3.1 Physical characteristics of certain ferromagnetic materials (pure ferromagnetic materials at room temperature are iron, nickel and cobalt)

Material	Composition	Density [kg/m³]	Resistivity [μΩ cm]	Melting point [°C]
Iron	100 % Fe	—	9.6	—
	99.0 % Fe	7874	9.71	1539
	99.8 % Fe	7880	9.9	1539
Ferrosilicon	4 % Si	7650	60	1450
Aluminum–iron	16 % Al, iron for the rest	6500	145	—
Aluminum–ferrosilicon	9.5 % Si, 5.5 % Al, iron for the rest	8800	81	—
Nickel	99.6 % Ni	8890	8.7	—
Cobalt	99 % Co	8840	9	1495
	99.95 % Co	8850	6.3	—

Source: Adapted from Heck (1974).

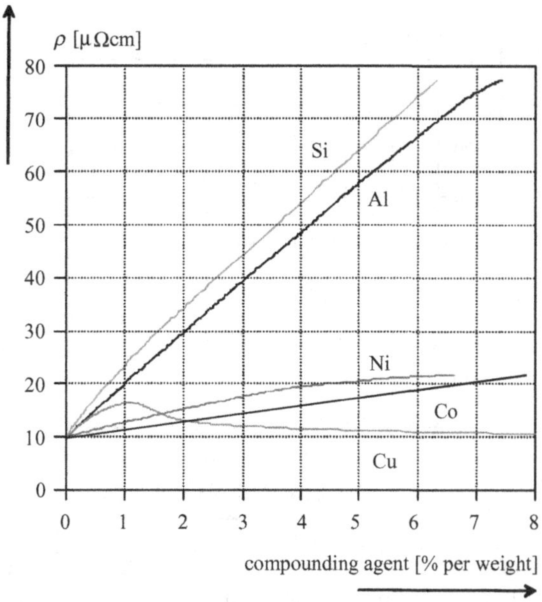

Figure 3.24 Effect of silicon, aluminum, nickel, cobalt and copper alloying on the resistivity of iron. Adapted from Heck (1974).

reduced to a few per cent in the alloy. Electric sheets have been developed with a silicon content of 6 %. An aluminum-rich alloy makes the material very hard (the Vickers hardness HV is about 250 for a material with an aluminum content of 16 % by weight), which may have an influence on the usability of the material. However, the resistivity of the material is so high that it proves a very interesting alternative for certain special applications. In the literature (Heck 1974), an equation is introduced for the resistivity ρ of AlFe alloy as a function of aluminum content p_{Al} (percentage by weight):

$$\rho = (9.9 + 11 p_{Al})\mu\Omega\text{cm}. \tag{3.61}$$

This is valid at a temperature of +20 °C, where the aluminum content is ≤4 % by weight. The temperature coefficient of resistivity decreases sharply when the aluminum content increases, and it is 350×10^{-6}/K when the aluminum content of the alloy is 10 % by weight. When the alloying is in the range where the occurrence of Fe_3Al is possible, for instance if the aluminum content is 12–14 %, the resistivity of the material depends on the cooling method. As shown in Figure 3.25, the resistivity of a material cooled rapidly from 700 °C is notably higher than for a material that is cooled more slowly (30 K/h).

The resistivity reaches a value of 167 μΩcm when the proportion of aluminum is 17 % by weight, but above this content the alloy becomes paramagnetic. Aluminum–iron alloys have been investigated from the beginning of the twentieth century, when it was discovered that the addition of aluminum to iron had very similar effects as the addition of silicon. With small aluminum contents, the coercivity of the AlFe alloys, hysteresis losses and saturation

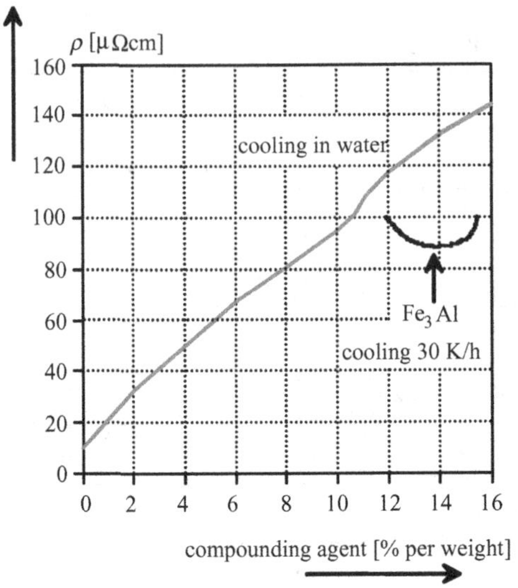

Figure 3.25 Resistivity of aluminum–iron alloy as a function of the proportion of admixing material. The resistivity is also partly dependent on the cooling rate of the material. Fe_3Al compound is created when the cooling takes place slowly, about 30 K/h. Adapted from Heck (1974).

flux density do not differ significantly from the respective properties of SiFe alloys. As the aluminum content increases, the resistivity increases, whereas coercitivity, hysteresis losses and saturation flux density decrease.

Heck (1974) gives an equation for the saturation flux density of AlFe alloys as a function of aluminum content p_{Al} (% by weight):

$$B_{\mathrm{s}} = (2.164 - 0.057 p_{\mathrm{Al}})\,\mathrm{Vs/m^2}. \tag{3.62}$$

According to Heck (1974), the density ρ of the alloy may be written as

$$\rho = (7.865 - 0.117 p_{\mathrm{Al}})\,\mathrm{kg/m^3}. \tag{3.63}$$

Figure 3.26 illustrates half of the hysteresis loop of an AlFe alloy, the proportion of aluminum being 16 atomic per cent. The figure shows that the saturation flux density $B_{\mathrm{s}} = 1.685$ T given by Equation (3.62) is sufficiently accurate.

Iron–aluminum alloys can be employed for instance as laminating materials to reduce the harmful effects of eddy currents in solid parts. Due to its hardness, the alloy has been used in tape recorder magnetic heads, for instance. Aluminum can also be employed together with silicon as an alloying material of iron, but the commercial electric sheets are usually silicon alloys. The alloying of both aluminum and silicon reduces the saturation flux density of iron, but the decrease is not very rapid when compared for instance with carbon, which, already with a content of 0.5 %, makes iron unfit for a magnetic circuit.

The magnetic properties of a material depend on the orientation of the crystals of the material. The crystals may be in random directions, and therefore anisotropy is not discernible in the

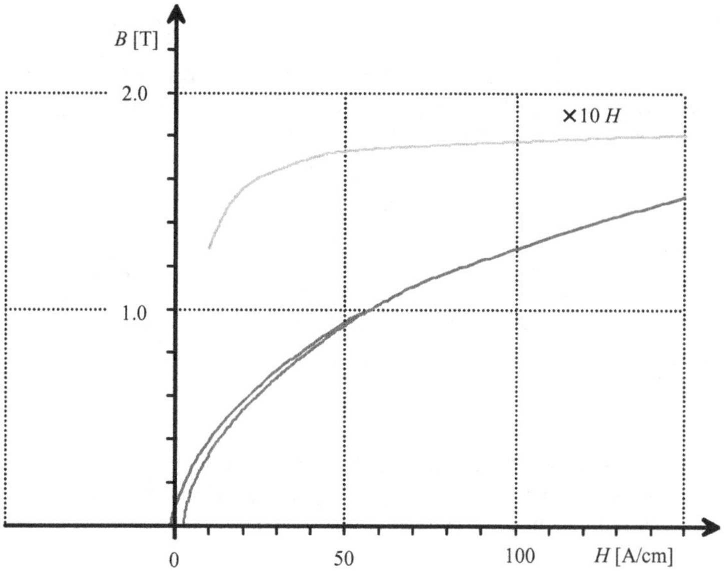

Figure 3.26 Half of a hysteresis loop of an AlFe alloy, the proportion of aluminum being 16 atomic per cent (8.4 % by weight). The resistivity of the material is about 84 μΩ cm at a temperature of +20 °C, and the temperature coefficient of the resistivity is very low, of magnitude 350 ppm (parts per million). On the above curve, the field strength values have to be multiplied by 10.

macroscopic magnetizing curve of the material. However, the crystals may also be positioned so that the anisotropy is discernible at a macroscopic scale. In that case, the magnetizing curve is different depending on the direction of magnetization. The material is then anisotropic, and it is said to have a magnetic texture.

The selection of the most favored crystal orientations is based on several factors. For instance, internal stresses, crystal defects and impurities may ease the orientation when the material is rolled or heat treated in a magnetic field. Finally, all the most favored directions of the crystals are more or less parallel. In that case, the material is said to have a crystal structure or a crystal texture.

This kind of crystal texture is technically important, because the whole body can be treated as a single crystal. There are two significant cases that enable mass production of crystal orientation: one is the Goss texture, common mainly in silicon–irons; and the other is a cubic texture, common in 50 % NiFe alloys. It is also possible to produce silicon-containing irons with a cubic crystal texture. Figure 3.27 depicts the orientation of crystals in these textures with respect to rolling direction.

In a crystal lattice with a Goss texture, only one corner of the cube is parallel to the rolling direction, which is also the main magnetizing direction in technical applications. In a crystal lattice with a cubic texture, the whole side of the cube is parallel to the rolling plane, and thus a favored crystal orientation occurs also in the normal direction. For both textures, a rectangular magnetizing curve is typical, since the saturation flux density is reached without rotation of the Weiss domains. Also, a relatively high coercive force is typical of both textures, since

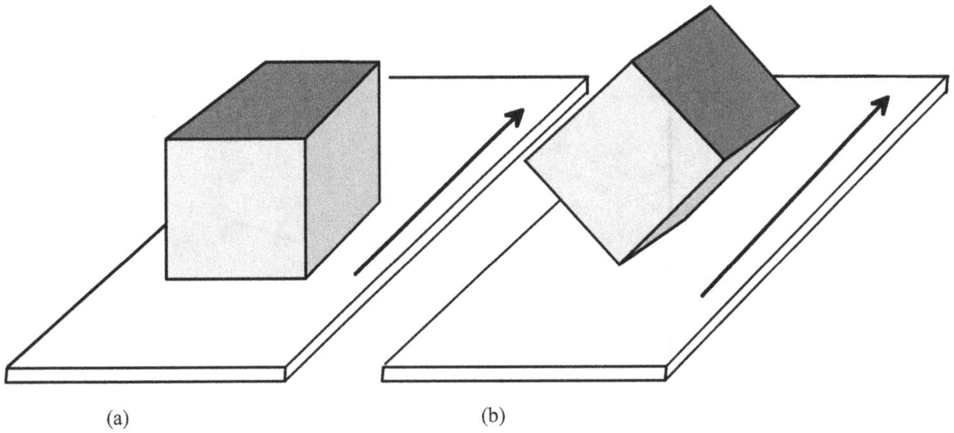

Figure 3.27 (a) Cubic texture and (b) Goss texture. The arrow points in the rolling direction of the sheet. The cubic texture produces nonoriented materials and the Goss texture produces oriented materials. The oriented materials have different magnetic properties in different directions.

the spontaneous magnetizing of the crystals remains easily in the direction of the external magnetic field.

In electrical machine construction, both oriented and nonoriented silicon–iron electric sheets are important materials. Oriented electric sheets are very anisotropic, and their permeability perpendicular to the rolling direction is notably lower than in the longitudinal direction. Oriented sheets can be employed mainly in transformers, in which the direction of the flux has always to be same. In large electrical machines also, an oriented sheet is used, since, due to the large dimensions of the machine, the sheet can be produced from elements in which the direction of flux remains unchanged regardless of the rotation of the flux.

Oriented sheet material can be employed also in small machines, as long as it is ensured during machine construction that the sheets are assembled at random so that the permeance of the machine does not vary in different directions. However, the magnetic properties of an oriented sheet in the direction deviating by 45–90° from the rolling direction are so poor that using an oriented sheet is not necessarily advantageous in rotating electrical machines. Figure 3.28 illustrates the influence of the rolling direction in the iron losses and the permeability of the material.

The main principle in machine construction is that those elements of rotating machines, which experience a rotating field, are produced from nonoriented electric sheets, the properties of which are constant irrespective of the rolling direction. Figure 3.29 depicts the DC magnetizing curves of two nonoriented electric sheets, produced by Surahammars Bruk AB.

3.6.2 Losses in Iron Circuits

In rotating machines, the machine parts are influenced by an alternating flux in different ways. For instance, in an asynchronous machine, all the parts of the machine experience an

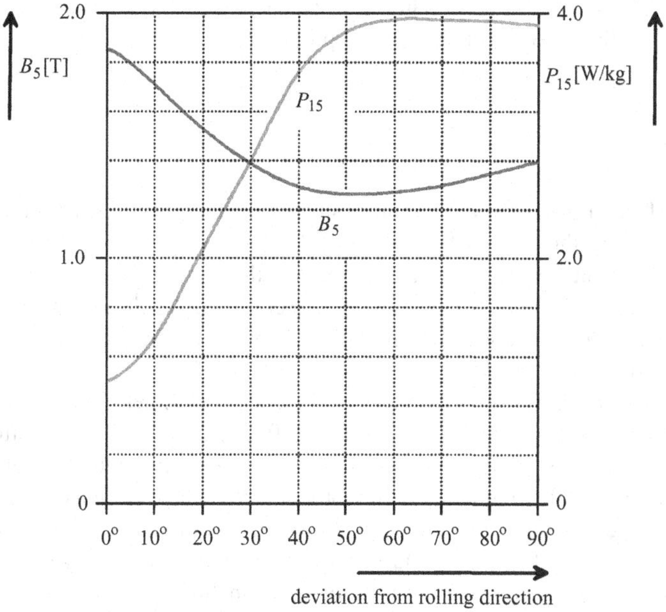

Figure 3.28 Iron losses P_{15} of a transformer sheet M6 at a flux density of 1.5 T, a frequency of 50 Hz and a flux density B_5 with an alternating current, when the effective value of the field strength is 5 A/m. Reproduced by permission of Surahammars Bruk AB.

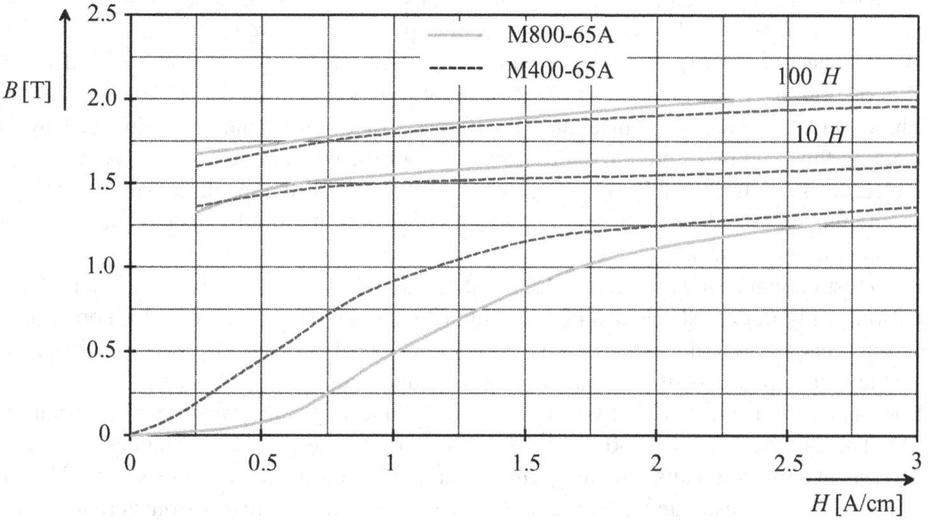

Figure 3.29 DC magnetizing curves of nonoriented electrical sheets, produced by Surahammars Bruk. The silicon content of M400-65A is 2.7 % and of M800-65A about 1 %. The resistivity of M400-65A is 46 μΩ cm, and of M700-65A 25 μΩ cm. As defined in the European Standard EN 10106, the standard grade of the material expresses the iron losses of the sheet at a peak flux density of 1.5 T and a frequency of 50 Hz, and also the thickness of the material. Thus, the grade M800-65A means that the dissipation power is 8 W/kg and the thickness of the sheet is 0.65 mm. Reproduced by permission of Surahammars Bruk AB.

alternating flux. The frequency of the stator is the frequency f_s fed by the network or an inverter. The frequency f_r of the rotor depends on the slip s and is

$$f_r = s \cdot f_s. \tag{3.64}$$

The outer surface of the rotor and the inner surface of the stator experience high harmonic frequencies caused by the slots. Also, the discrete distribution of the windings in the slots creates flux components of different frequencies both in the stator and in the rotor. The variation of the flux in the rotor surface of an asynchronous machine can be restricted with certain measures to a slow variation of the main flux. In that case, the rotor in some constructions can be produced from solid steel.

In synchronous machines (synchronous reluctance machines, separately magnetized machines, permanent magnet machines), the base frequency f_s of the armature core (usually stator) is the frequency of the network (50 Hz in Europe) or the frequency of the supplying frequency converter, and the frequency of the rotor is zero in the stationary state. However, the rotor surface experiences high-frequency alternating flux components because of a changing permeance caused by the stator slots. During transients, the rotor of a synchronous motor is also influenced by an alternating flux. The rotor of a synchronous machine can be produced from solid steel since, in normal use, harmonic frequencies occur only on the surface of the rotor. The amplitudes of these frequencies are quite low because of the large air gaps that are common especially in nonsalient-pole machines.

In DC machines, the frequency of the stator is zero, and the flux varies only during transients, if we neglect the high-frequency flux variation caused by the permeance harmonics on the surface of the pole shoe. DC-magnetized machine parts can be produced from cast steel or thick steel sheet (1–2 mm). However, the armature core experiences a frequency that depends on the rotation speed and the number of pole pairs. In a modification of a DC series-connected machine, an AC commutator machine, all the parts of the machine are influenced by an alternating flux, and therefore the entire iron circuit of the machine has to be produced from thin electric sheet. In a doubly salient reluctance machine, all the machine parts experience pulsating flux components of varying frequencies, and thus in this case also the parts have to be made of thin electric sheet.

The most common thicknesses of electric sheet are 0.2, 0.35, 0.5, 0.65 and 1 mm. There are also notably thinner sheets available for high-frequency purposes. Common nonoriented electric sheets are available at least with a thickness of 0.1 mm. Amorphous iron strips are available with a thickness of 0.05 mm in various widths.

Losses in an iron circuit are of two different types, namely hysteresis losses and eddy current losses. The curves in Figure 3.30 illustrate half of a hysteresis loop for a magnetic material. Hysteresis in a material causes losses in an alternating field. First, a power loss caused by the hysteresis will be investigated in iron, see Figure 3.30. When H increases from zero at point 1 to H_{max} at point 2, an energy per volume w absorbed in a unit volume is

$$w_1 = \int_{-B_r}^{B_{max}} H \, dB. \tag{3.65}$$

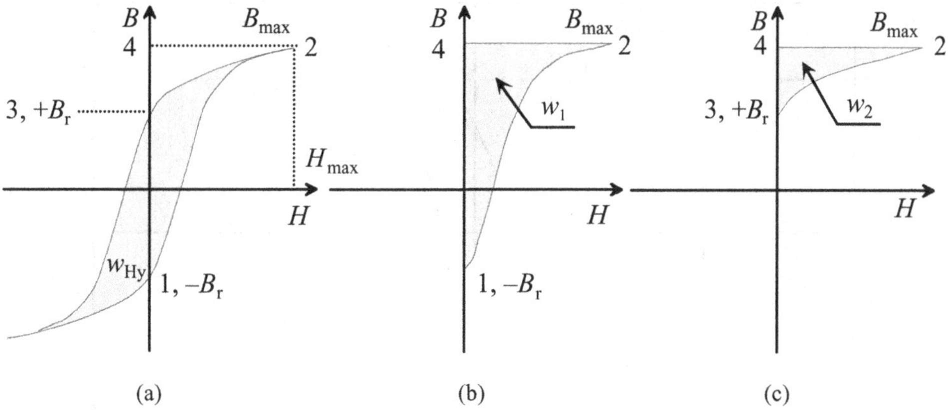

Figure 3.30 Determination of hysteresis loss: (a) entire hysteresis curve, (b) w_1, magnetic energy per volume stored (Area 1 - 2- 4 -1) when moving from 1 to 2, (c) w_2, magnetic energy per volume returned (Area 2 - 3 - 4- 2) when moving from 2 to 3.

Correspondingly, when $H \rightarrow 0$, the dissipated energy is

$$w_2 = \int_{B_{max}}^{B_r} H \, dB. \tag{3.66}$$

The total hysteresis energy is calculated as a line integral, when the volume of the object is V

$$W_{Hy} = V \oint H \, dB. \tag{3.67}$$

The hysteresis energy of Equation (3.67) is obtained by traveling around the hysteresis loop. With an alternating current, the loop is circulated constantly, and therefore the hysteresis dissipation power P_{Hy} depends on the frequency f. When the area of the curve describes the hysteresis energy per volume w_{hy}, we obtain for the hysteresis power loss in volume V

$$P_{Hy} = f V \, w_{Hy}. \tag{3.68}$$

Empirical equations yield an approximation for the hysteresis loss

$$P_{Hy} = \eta V f B_{max}^k, \tag{3.69}$$

where the exponent k varies typically over [1.5, 2.5], η being an empirical constant.

In the case of an alternating flux in the iron core, the alternation of the flux induces voltages in the conductive core material. As a result, eddy currents occur in the core. These currents

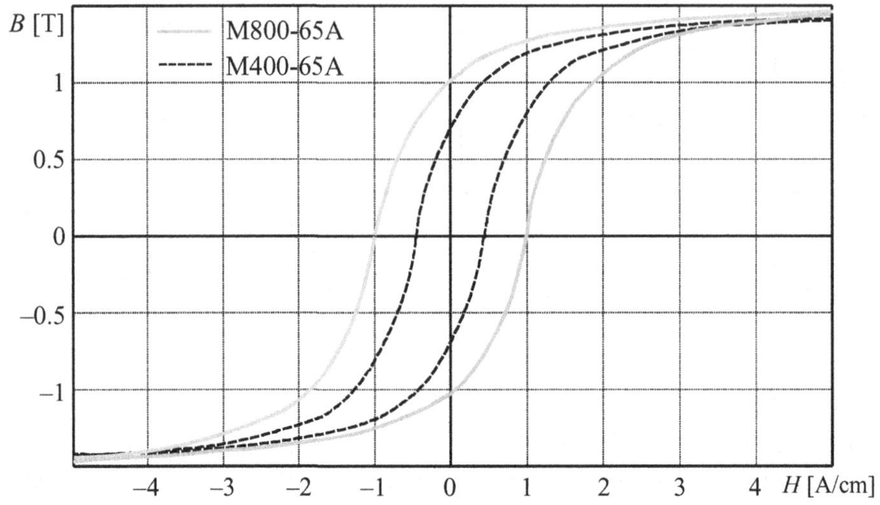

Figure 3.31 Approximate hysteresis curves of electrical sheets produced by Surahammars Bruk AB. M400-65A contains more silicon than M800-65A, which is a common material in small motors. Reproduced by permission of Surahammars Bruk AB.

tend to resist changes in the flux. In solid objects, the eddy currents become massive and effectively restrict the flux from penetrating the material. The effect of eddy currents is limited by using laminations or high-resistivity compounds instead of solid ferromagnetic metal cores. Figure 3.31 depicts the hysteresis curves of two different electric sheets used in laminations, produced by Surahammars Bruk AB.

Although magnetic cores are made of sheet, a thin sheet also enables eddy currents to occur when the flux alternates. The case of Figure 3.32, in which an alternating flux penetrates the core laminate, will now be investigated.

If a maximum flux density $\hat{B}_m$ passes through the region 12341, the peak value for the flux of a parallelogram (broken line) is obtained with the notation in Figure 3.32

$$\hat{\Phi} = 2hx\,\hat{B}_m. \tag{3.70}$$

Since $d \ll h$, the effective value of the voltage induced in this path is, according to the induction law,

$$E = \frac{\omega \hat{B}_m}{\sqrt{2}} 2hx. \tag{3.71}$$

The resistance of the path in question depends on the specific resistivity ρ, the length of the path l and the area S. The lamination is thin compared with its other dimensions. Hence, we may simply write for the resistance of the path l

$$R = \frac{\rho l}{S} \approx \frac{2h\rho}{w\,dx}. \tag{3.72}$$

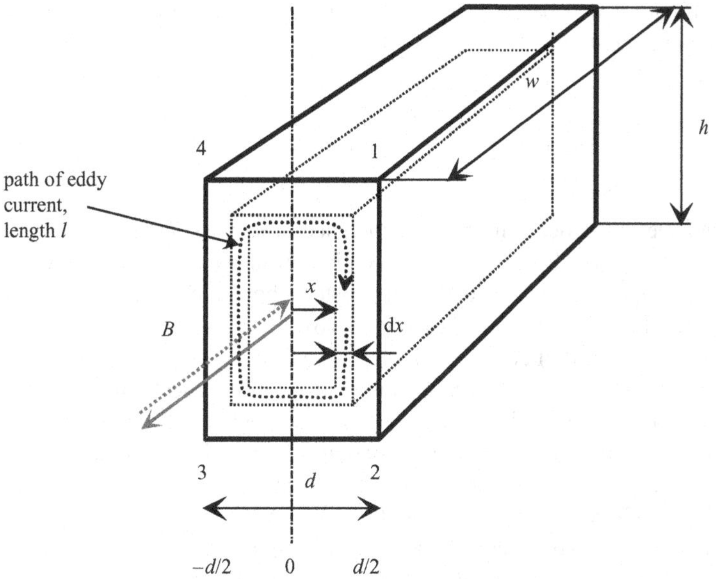

Figure 3.32 Eddy currents in a sheet material. The magnetic flux density B is varying in the directions given by the arrows and the corresponding eddy currents circulate around the magnetic flux. The eddy currents try, according to Lenz's law, to prohibit the flux from penetrating the laminations. The broken line is for electrical sheet M400-65A and the solid line for electrical sheet M800-65A produced by Surahammars Bruk AB.

The flux density in the lamination creates a flux $\Phi = xhB$. The alternating flux creates a voltage $-\mathrm{d}\Phi/\mathrm{d}t$ in the area observed. The induced voltage creates a current

$$\mathrm{d}I = \frac{E}{R} = \frac{\dfrac{2\pi f \cdot \hat{B}_\mathrm{m}}{\sqrt{2}} 2xh}{\dfrac{2h\rho}{w\mathrm{d}x}} = \frac{2\pi f \cdot \hat{B}_\mathrm{m} w x \mathrm{d}x}{\sqrt{2} \cdot \rho}, \tag{3.73}$$

the differential power loss being respectively

$$\mathrm{d}P_\mathrm{Fe,Ft} = E\mathrm{d}I = \frac{(2\pi f \cdot \hat{B}_\mathrm{m})^2 w h x^2 \mathrm{d}x}{\rho}. \tag{3.74}$$

The eddy current loss in the whole sheet is thus

$$P_\mathrm{Fe,Ft} = \int_0^{d/2} \mathrm{d}P_\mathrm{Fe,Ft} = \frac{\left(2\pi f \hat{B}_\mathrm{m}\right)^2 wh}{\rho} \cdot \int_0^{d/2} x^2 \mathrm{d}x. \tag{3.75}$$

Notice that the integration boundaries are set to zero and $d/2$ because the eddy current path resistance R in Eq. (3.72) includes the whole eddy current path ($2h$), and therefore, the other half of the lamination will be taken into account based on symmetry. Since $whd = V$, the volume of the laminate, the eddy current loss is

$$P_{\mathrm{Fe,Ft}} = \frac{wh\pi^2 f^2 d^3 \hat{B}_{\mathrm{m}}^2}{6\rho} = \frac{V \cdot \pi^2 f^2 d^2 \hat{B}_{\mathrm{m}}^2}{6\,\rho}. \tag{3.76}$$

Here we can see the radical influence of the sheet thickness d ($P_{\mathrm{Fe}} \approx d^3$), the peak value of the flux density $\hat{B}_{\mathrm{m}}$ and the frequency f on eddy current losses. Also, the resistivity ρ is of great significance. The measurements for silicon steel show that the eddy current loss is about 50 % higher than the result given by Equation (3.76).

The reason for this difference lies in the large crystal size of silicon steel. In general, we may state that as the crystal size increases, the eddy current losses in the material increase as well. Equation (3.76) can nevertheless be used as a guide when estimating eddy current losses for instance in the surroundings of a given operating point. Manufacturers usually give the losses of their materials per mass unit at a certain peak value of flux density and frequency, for instance $P_{15} = 4$ W/kg, 1.5 T, 50 Hz or $P_{10} = 1.75$ W/kg, 1.0 T, 50 Hz.

Extrapolating the loss to different frequencies is complicated as hysteresis and eddy current losses are not usually given separately. At 50 Hz the hysteresis loss is typically about 75 % of the iron losses. Hysteresis loss is directly proportional to the frequency and eddy current loss proportional to the square of the frequency. If the operating frequency, however, differs significantly from the standard 50 or 60 Hz it is advisable to contact the manufacturer and ask for the losses of the lamination sheets at different frequencies.

Figure 3.33 illustrates the iron losses in two electric sheets of equal thickness and different resistivity. The sheets are produced from the same materials as in the previous examples. The

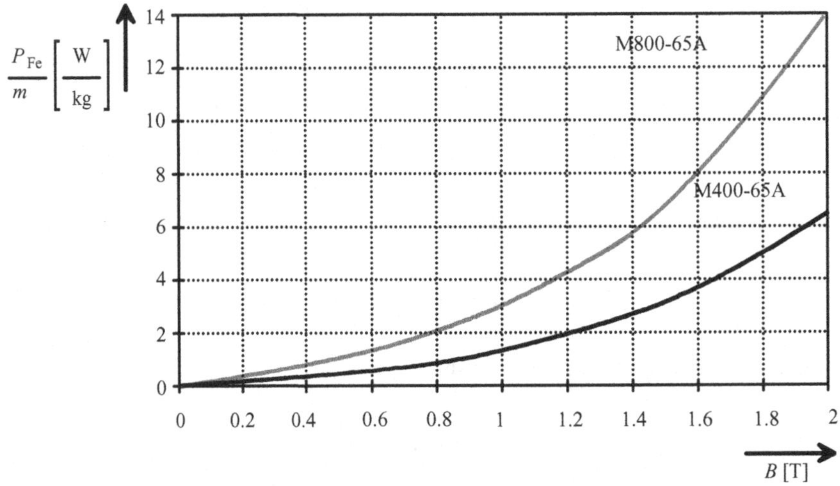

Figure 3.33 Iron losses of two different electrical sheets at an alternating flux of 50 Hz as a function of the maximum value of the flux density. The curves include both the hysteresis loss and the eddy current loss.

thickness of the sheets is 0.65 mm. The manufacturers usually give combined iron losses; in other words, eddy current losses and hysteresis losses are not separated.

In manual calculations, the iron losses are found by dividing the magnetic circuit of the machine into n sections, in which the flux density is approximately constant. Once the masses $m_{\mathrm{Fe},n}$ of the different areas n have been calculated, the losses $P_{\mathrm{Fe},n}$ of the different parts of the machine can be approximated as follows:

$$P_{\mathrm{Fe},n} = P_{10}\left(\frac{\hat{B}_n}{1\,\mathrm{T}}\right)^2 m_{\mathrm{Fe},n} \quad \text{or} \quad P_{\mathrm{Fe},n} = P_{15}\left(\frac{\hat{B}_n}{1.5\,\mathrm{T}}\right)^2 m_{\mathrm{Fe},n}. \tag{3.77}$$

Total losses can be calculated by summing the losses of different sections n. A problem occurring in the calculation of losses in rotating machines is that the loss values P_{15} and P_{10} are valid only for a sinusoidally varying flux density. In rotating machines, however, pure sinusoidal flux variation never occurs alone in any parts of the machine, but there are always rotating fields that have somewhat different losses compared with varying field losses. Also, field harmonics are present, and thus the losses, in practice, are higher than the results calculated above indicate. Furthermore, the stresses created in the punching of the sheet and also the burrs increase the loss index. In manual calculations, these phenomena are empirically taken into account and the iron losses can be solved by taking into account the empirical correction coefficients $k_{\mathrm{Fe},n}$ defined for different sections n, Table 3.2:

$$P_{\mathrm{Fe}} = \sum_n k_{\mathrm{Fe},n} P_{10}\left(\frac{\hat{B}_n}{1\,\mathrm{T}}\right)^2 m_{\mathrm{Fe},n} \quad \text{or} \quad P_{\mathrm{Fe}} = \sum_n k_{\mathrm{Fe},n} P_{15}\left(\frac{\hat{B}_n}{1.5\,\mathrm{T}}\right)^2 m_{\mathrm{Fe},n}. \tag{3.78}$$

The iron losses discussed above are calculated only for a time-varying flux density required by the fundamental of the main flux. In addition to these losses, there are other iron losses of different origin in rotating machines. The most significant of these losses are as follows:

- End losses, which occur when the leakage flux of the machine end penetrates the solid structures of the machine, such as the end shields, creating eddy currents. Calculation of these losses is rather difficult, and in manual calculations it suffices to apply empirical correction coefficients in Equation (3.78) to take the influence of the losses into account.
- Additional losses in the teeth are caused by permeance harmonics that occur when the stator and rotor teeth pass each other rapidly. To be able to calculate the losses, it is necessary to

Table 3.2 Correction coefficients $k_{\mathrm{Fe},n}$ for the definition of iron losses in different sections of different machine types taking the above-mentioned anomalies into account. (Coefficients are valid for AC machines with semiclosed slots and with a sinusoidal supply and for DC machines.)

Machine type	Teeth	Yoke
Synchronous machine	2.0	1.5–1.7
Asynchronous machine	1.8	1.5–1.7
DC machine	2.5	1.6–2.0

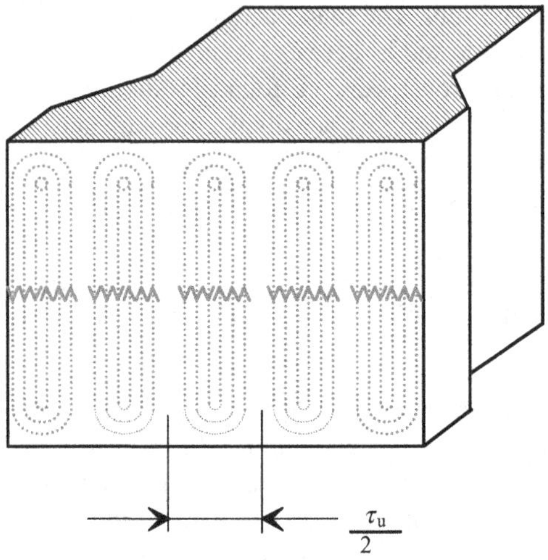

Figure 3.34 Paths of eddy currents in a solid pole shoe. The pattern is repeated at intervals of half a slot pitch.

solve the frequency and the amplitude of the harmonic experienced by a tooth. The losses are calculated similarly as before. These losses are also included in the correction coefficient $k_{\mathrm{Fe},n}$.

- In machines with solid parts, for instance on the surface of pole shoes, harmonics created by slots (cf. Figures 3.5 and 3.6) generate eddy currents that cause surface losses, Figure 3.34. Richter (1967) introduced empirical equations for solving these losses. An accurate analysis of the phenomenon is extremely difficult and requires the solution of the field equations in solid material.

Usually, loss calculations are carried out assuming fundamental flux variation only. Table 3.2 roughly takes into account the harmonic behavior in different parts of the machine by simply multiplying the values calculated by the fundamental. If a more detailed analysis is possible, the eddy current loss in the stator yoke, for instance, may be determined by calculating the tangential and radial flux density components at different frequencies. The eddy current loss in a lamination of thickness d, mass m_{Fe}, conductivity σ_{Fe} and density ρ_{Fe} is calculated from

$$P_{\mathrm{Fe,Ft}} = \frac{\pi^2}{6}\frac{\sigma_{\mathrm{Fe}}}{\rho_{\mathrm{Fe}}} f^2 d^2 m_{\mathrm{Fe}} \sum_{n=1}^{\infty} n^2 (B_{\tan,n}^2 + B_{\mathrm{norm},n}^2) = \frac{\pi^2}{6}\frac{\sigma_{\mathrm{Fe}}}{\sigma_{\mathrm{Fe}}} f^2 d^2 m_{\mathrm{Fe}} (B_{\tan,1}^2 + B_{\mathrm{norm},1}^2) k_{\mathrm{d}}. \tag{3.79}$$

$$k_{\mathrm{d}} = 1 + \sum_{n=2}^{\infty} \frac{\left(B_{\tan,n}^2 + B_{\mathrm{norm},n}^2\right)}{\left(B_{\tan,1}^2 + B_{\mathrm{norm},1}^2\right)}. \tag{3.80}$$

These equations require an accurate flux density solution that may be divided into the components of different harmonics.

For the hysteresis losses, there is a similar approach:

$$P_{\mathrm{Fe,Hy}} = c_{\mathrm{Hy}}\frac{f}{100}m_{\mathrm{Fe}}\sum_{n=1}^{\infty} n^2\left(B_{\tan,n}^2 + B_{\mathrm{norm},n}^2\right) = c_{\mathrm{Hy}}\frac{f}{100}m_{\mathrm{Fe}}\left(B_{\tan,1}^2 + B_{\mathrm{norm},1}^2\right)k_{\mathrm{d}}. \quad (3.81)$$

The hysteresis coefficient $c_{\mathrm{Hy}} = 1.2$–2 [A m^4/V s kg] for anisotropic laminations with 4 % of silicon and 4.4–4.8 [A m^4/V s kg] for isotropic laminations with 2 % of silicon.

Even Equations (3.79–3.81), however, give too low values for the iron losses, and additional core losses have to be taken into account by suitable loss coefficients. For instance, the values of Table 3.2 may be used.

Motors also face the problem of a pulse-width-modulated (PWM) supply. In a PWM supply, the losses of the motor increase in many ways. Also the iron losses increase, especially on the rotor surface. Depending on the PWM switching frequency, the overall efficiency of the motor is typically 1–2 % lower in a PWM supply than in a sinusoidal supply.

3.7 Permanent Magnets in Rotating Machines

Next, permanent magnet materials and their properties are discussed. Magnetically soft materials are employed to facilitate the magnetizing processes, such as the displacement of Bloch walls and the orientation of Weiss domains; see Figure 3.19. With permanent magnets, the requirements are the opposite. The displacement of a permanent magnet from the initial state is difficult. When the Weiss domains have been aligned in parallel orientations with a high external field strength, the material becomes permanently magnetized.

The objective can be met by the following means. The displacement of Bloch walls is prevented by inhomogeneities and the extremely fine structure of the material. The best way to completely prevent the displacement of Bloch walls is to select so fine a structure that each particle of the material comprises only a single Weiss domain in order to create an energy minimum.

In addition, the orientation of Weiss domains has to be impeded. Now, anisotropy is utilized. Both a high crystal anisotropy of the material and a high anisotropy of the shape of the crystals (rod-shaped crystals are selected) significantly impede the orientation of Weiss domains. This leads to an increase of magnetic hardness and coercivity H_{cJ}. With rare-earth permanent magnets, a high crystal anisotropy is reached mainly by employing rare-earth metals as base materials. With hard ferrites, the crystal anisotropy is reached chiefly by orientating the particles under pressure in a magnetic field.

3.7.1 History and Development of Permanent Magnets

Excluding the natural magnet, magnetite (Fe_3O_4), the development and manufacture of permanent magnet materials began in the early twentieth century with the production of carbon, cobalt and wolfram steels. These permanent magnet materials, the magnetic properties of which were rather poor, remained the only permanent magnet materials for decades.

A remarkable improvement in the field was due to the discovery of AlNi and especially AlNiCo materials in the 1930s. Their utilization was at its height in the 1960s. Nowadays they are employed at temperatures of 300 °C and above. AlNiCo magnets are produced either by casting or by sintering. Their typical composition is 50 % Fe, 25 % Co, 14 % Ni, 8 % Al, 3 % Cu/Nd/Si. Impurities of C, Cr, Mn and P have to be avoided. The cast qualities yield a 25 % better magnetic performance than the sintered qualities. AlNiCo magnets are resistant to high temperatures, and do not corrode easily. The energy products are typically 10–80 kJ/m^3.

Ferrites were first introduced in the 1950s, and because of their low price, they are still dominant on the market. There are two commercial alternatives of ferrites available, based on either strontium (18 %) or barium (21 %) carbonate processed into hexa-ferrites. Both isotropic and anisotropic qualities are produced by powder metallurgy methods. The high resistivity of ferrite magnets is a clear advantage.

The next significant step forward was taken in the 1960s, when the compounds of rare-earth metals and cobalt were invented. The most important materials were $SmCo_5$ and Sm_2Co_{17}. Later, better and more complicated variations of these two were discovered, such as $Sm_2(Co, Cu, Fe, Zr)_{17}$. The energy product of SmCo (1:5) magnets (invented in 1969) is 175 kJ/m^3 at maximum, and SmCo (2:17) magnets (invented in the 1980s) typically reach 200 kJ/m^3 (255 kJ/m^3 maximum). After the discovery of these materials, the next significant invention was the neodymium–iron–boron permanent magnets in 1983, which nowadays yield the highest energy product. They are produced by a powder metallurgy process developed by Sumimoto, or by a "melt-spinning" process developed by General Motors. An advantage of NdFeB magnets is that the rare samarium and cobalt have been replaced by the far more common neodymium and iron. The basic type of these materials is $Nd_{15}Fe_{77}B_8$.

At the moment, the most rapidly growing field is polymer-bonded permanent magnets. All the basic types can be produced as polymer-bonded types. Both isotropic and anisotropic qualities are commercially available. In these products, the best properties of rare-earth and ferrite permanent magnet materials and suitable resins are combined. The resin acts as a bonding material for permanent magnet powder. Because of the resin film between the magnet particles, the conductivity of plastic-bonded magnets is extremely low, and thus eddy current losses occur only inside the particles. Losses are in most cases insignificant. The problem with the plastic-bonded magnets is that the magnetic properties are remarkably weaker than in sintered products. The remanent flux density of plastic-bonded materials is about half of that of sintered ones. Polymer-bonded magnets are well machinable, and they can be used in complicated structures. Their heat resistance depends on the bonding agent and varies between 100 °C and 150 °C. They are common in small stepper and DC motors, tachometers, motors of diskette drives, toys, quartz watches and phones. However, ferrite magnets are nowadays used also in some industrial motor applications.

Figure 3.35 illustrates the development of the energy product of permanent magnets from the beginning of the twentieth century (Vacuumschmelze 2003).

Previously, a serious problem with permanent magnet materials was the easy demagnetization of the materials. The best permanent magnet materials are nowadays quite insensitive to external field strengths and the influence of an air gap. Only short-circuit currents in hot machines may constitute a risk of demagnetization in certain structures using rare earth magnets. In Ferrite magnets the polarization has a reversible lowering as a function of temperature similarly as in rare earth magnets. However, the polarization resistance against demagnetizing

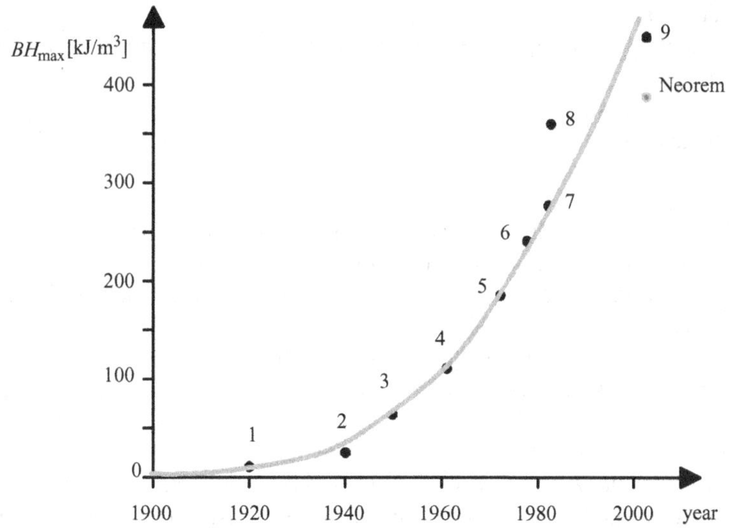

Figure 3.35 Development of the energy product of permanent magnet materials in the twentieth century: 1, cobalt steel; 2, FeCoV; 3, AlNiCo; 4, AlNiCo; 5, $SmCo_5$; 6, $Sm_2(Co, Cu,Fe,Zr)_{17}$; 7, NdFeB; 8, NdFeB; 9, NdFeB. Neorem refers to a commercial material NEOREM 503 i (produced by Neorem Magnets), the energy product of which is about 370 kJ/m^3. Adapted from Vacuumschmelze (2003). http://www.vacuumschmelze.de/dynamic/docroot/medialib/documents/broschueren/dmbrosch/PD002e.pdf.

fields gets better as the temperature rises. Therefore, cold Ferrite magnets are easier to demagnetize than hot Ferrite magnets.

The most significant permanent magnetic materials in commercial production are the following:

- AlNiCo magnets are metallic compounds of iron and several other metals. The most important alloying metals are aluminum, nickel and cobalt.
- Ferrite magnets are made of sintered oxides, barium and strontium hexa-ferrite.
- RECo magnets (rare-earth cobalt magnets) are produced by a powder metallurgy technique, and comprise rare-earth metals (mainly samarium) and cobalt in the ratios of 1 : 5 and 2 : 17. The latter also includes iron, zirconium and copper.
- Neodymium magnets are neodymium–iron–boron magnets, produced by a powder metallurgy technique.

3.7.2 Characteristics of Permanent Magnet Materials

The characteristics of permanent magnets are chiefly described by the following quantities:

- remanence B_r;
- coercivity H_{cJ} (or H_{cB});
- the second quarter of the hysteresis loop;

- energy product $(BH)_{PMmax}$;
- temperature coefficients of B_r and H_{cJ}, reversible and irreversible portions separated;
- resistivity ρ;
- mechanical characteristics;
- chemical characteristics.

In general, we may state that the decisive requirements for good permanent magnet materials are a high saturation polarization, a high Curie temperature, and either a high crystal anisotropy as a material property, or the possibility to shape the anisotropy significantly. As with magnetically soft materials, the hysteresis loop is now an important characteristic curve. Usually, only the section in the second quadrant of the hysteresis loop is given.

In the construction of magnetic circuits with permanent magnets, the geometry of the magnetic circuit can be selected in such a way that a typical maximum energy product is reached for each permanent magnet material, Figure 3.36.

The dependence of the B and J curves is written as $J = B - \mu_0 H$, and therefore either one of the curves may be applied alone to describe the characteristics of a permanent magnet. At the remanence B_r, the curves are united, since $H = 0$. The intersection points on the H-axis are different for the flux density curve and the polarization curve, and the corresponding coercivities are denoted H_{cB} and H_{cJ} respectively. J_s describes the saturation polarization of the material. The quantities related to magnetizing are illustrated in Figure 3.37, which describes a typical hysteresis curve of a neodymium magnet for both the flux density and the polarization.

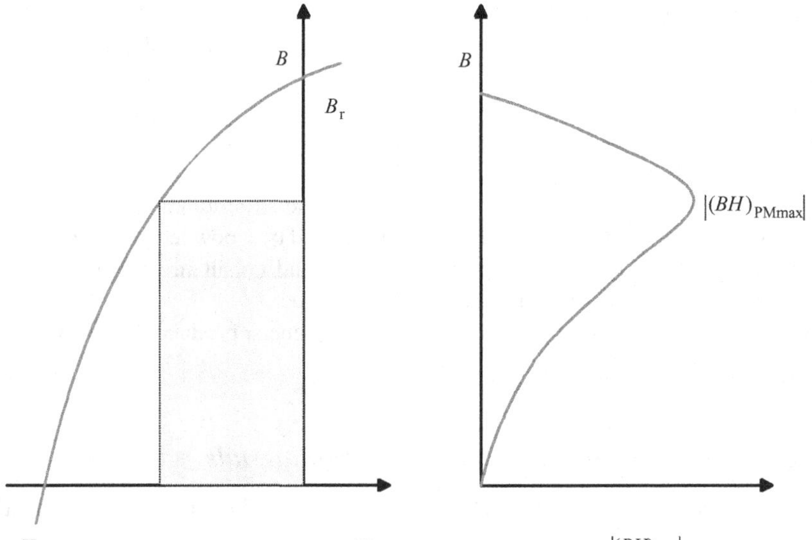

Figure 3.36 Second quarter of the hysteresis curve of a general permanent magnet and the corresponding behavior of $(BH)_{PM}$ product.

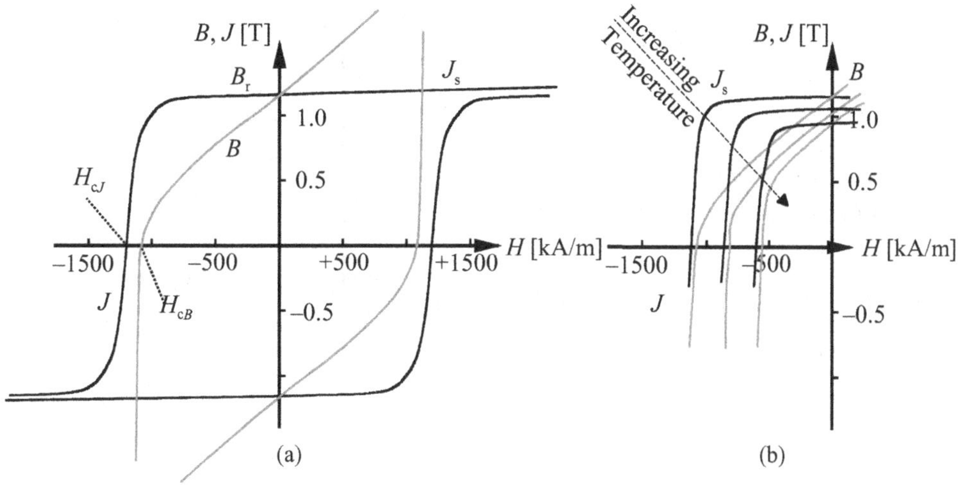

Figure 3.37 (a) Typical hysteresis curve of a neodymium magnet illustrated with the flux density B and the polarization J. (b) The behavior of the polarization and demagnetization curves under increasing temperature T. Typically, the polarization of a neodymium magnet decreases by about 10 % when the temperature increases by 100 K.

3.7.2.1 Samarium-Cobalt Magnets

The remanent flux density of commonly used $SmCo_5$ is 1.05 T at maximum, and its energy product is 210 kJ/m^3. The maximum values for neodymium–iron–boron are 1.5 T and 450 kJ/m^3. In practice, motor-grade materials still remain below an energy product of 400 kJ/m^3.

The heat resistance of SmCo magnet is excellent when compared with neodymium magnets, and they can be used at temperatures up to 250 °C. Furthermore, the corrosion resistance of SmCo magnets is better than that of neodymium magnets, but they are more brittle than neodymium magnets. SmCo magnets are common in applications in which the lightness and the heat resistance are decisive factors, while the price is of little significance. Typical applications are for small stepper motors, cathode-ray tube (CRT) positioning systems, electromechanical actuators, earphones, loudspeakers, etc.

As a single-phase alloy, $SmCo_5$ is rather easily saturated. As shown in Figure 3.38, about 200 kA/m suffices for the saturation of $SmCo_5$, since the Bloch walls are easily displaced in the crystals. Also the coercivity in this case is rather small ($H_{cJ} \approx 150$ kA/m). The coercive force changes only when the external field strength H_{mag} increases so high that all the grains have become magnetized against the inner leakage fields of the material. At complete saturation, there are no longer any Bloch walls in the crystals. Figure 3.38 illustrates the change in the coercive force of $SmCo_5$, when the field strength H_{mag} used for the magnetization of the material is increased.

A second basic material Sm_2Co_{17} behaves completely differently in magnetizing; see Figure 3.39. Here, at low field strengths ($<H_{cJ}$), the material is not magnetized even close to saturation, but a double or triple field strength is required when compared with the coercive force to reach a high remanence. This is because the Bloch walls are oriented along the crystal domain boundaries, and their orientation and displacement are particularly difficult.

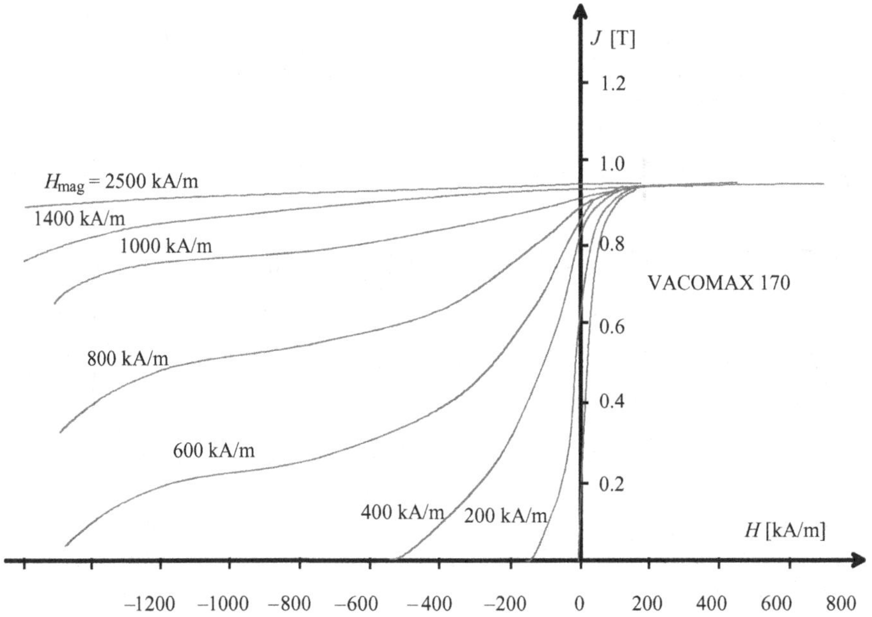

Figure 3.38 Increase in the coercive force of the $SmCo_5$-type VACOMAX 170; the increase is due to the increase in the magnetizing field strength, H_{mag}, as a parameter. The behavior of neodymium magnets in magnetizing is similar to the behavior of the $SmCo_5$. Adapted from Vacuumschmelze (2003). http://www.vacuumschmelze.de/dynamic/docroot/medialib/documents/broschueren/dmbrosch/PD002e.pdf.

3.7.2.2 Neodymium-Iron-Boron Magnets

Sintered NdFeB magnet contains some 30–32 % (weight %) of rare-earth metals, 1 % boron, 0–3 % cobalt and iron as the balance. Different properties are hammered out by different alloys (neodymium and dysprosium contents) and by different pressing methods (orientation). If the dysprosium content is increased, the remanence will drop, but the intrinsic coercivity will rise considerably (Figure 3.40, Ruoho 2011). Cobalt increases corrosion resistance and Curie temperature of the magnet. Dysprosium is more expensive than neodymium hence magnet grades with higher intrinsic coercivity are more expensive than the magnet grades with high remanence. Cobalt is more expensive than iron and hence magnets having high Curie temperature and being corrosion resistive are more expensive.

The resistivity of neodymium magnets depends on the temperature and measurement direction (Figure 3.41, Ruoho 2011). The resistivity in the orientation direction (marked with axial in Figure 3.41) is greater than in transversal direction. IEC standard 60404-8-1 gives the resistivities of magnet materials in axial direction. The eddy currents that air-gap harmonic fields create in magnets circulate in transversal direction, so in eddy-current calculations the resistivity in transversal direction has to be used. A good value for neodymium magnets is 1.35 μΩm (Ruoho 2011).

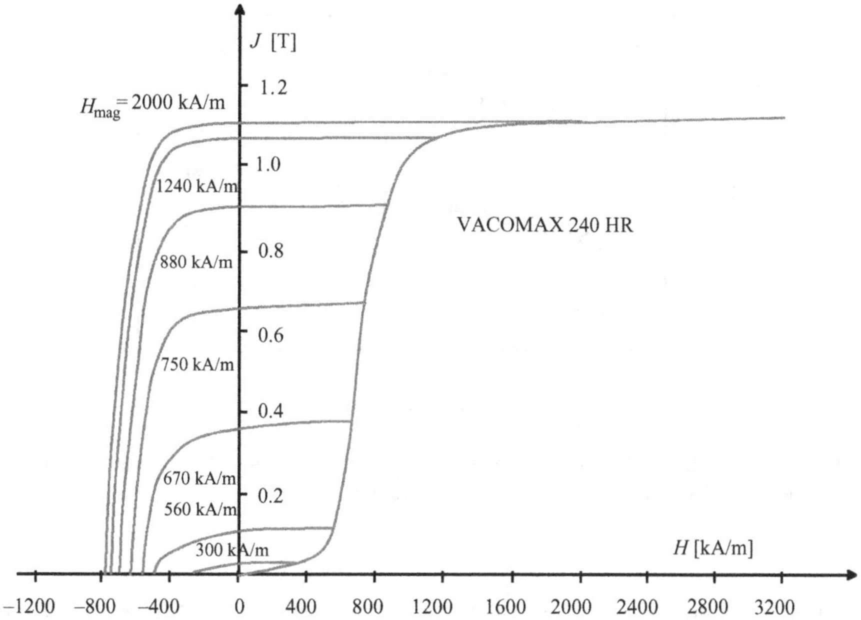

Figure 3.39 Increase in the coercive force of the Sm_2Co_{17}-type VACOMAX 240 HR caused by an increase in the magnetizing field strength H_{mag}. Different qualities may require notably higher magnetizing field strengths, for example up to 4000 kA/m. Adapted from Vacuumschmelze (2003). http://www.vacuumschmelze.de/dynamic/docroot/medialib/documents/broschueren/dmbrosch/PD002e.pdf.

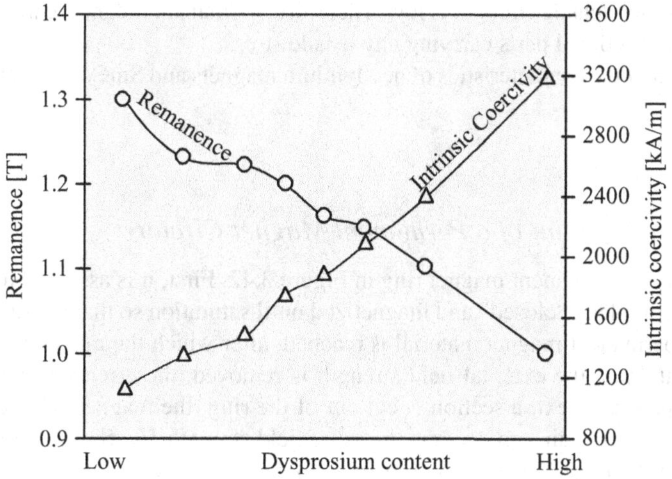

Figure 3.40 The influence of the dysprosium (Dy) content on the remanence and intrinsic coercivity of neodymium magnet. Reproduced by permission of Sami Ruoho.

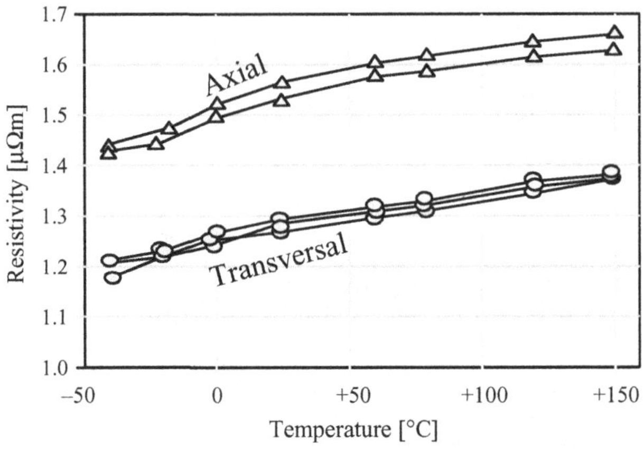

Figure 3.41 The resistivity of neodymium magnets as a function of temperature and the direction of the measurement. Reproduced by permission of Sami Ruoho.

Neodymium magnets are sensitive to changes in temperature. The intrinsic coercive force drops notably when the temperature rises. However, by employing other rare-earth metals as alloying elements for neodymium, the operating temperature can be raised to 180 °C.

The chemical properties of mere neodymium magnets are weak. With different coatings, it is possible to make neodymium magnets relatively moisture resistant. Oxygen and moisture corrode an unprotected neodymium magnet, and the material is transformed to light-colored powder.

The mechanical properties of magnetic materials are weak. Sintered magnets are resistant to pressure, but their tensile strength is low. Therefore, permanent magnets cannot be regarded as machine constructional parts carrying any tensile stress.

A comparison of the characteristics of neodymium magnets and SmCo magnets is presented in Table 3.3.

3.7.3 Operating Point of a Permanent Magnet Circuit

Let us consider the permanent magnet ring in Figure 3.42. First, it is assumed that the ring of Figure 3.42 is complete, "closed" and magnetized until saturation so that the saturation polarization of the permanent magnet material is reached, after which the magnetizing equipment is removed. Although the external field strength is removed, there remains a remanent flux density B_r in the ring. Next, a section is cut out of the ring; the magnet is "opened" with an air gap δ. In the created air gap, we may measure field strength H_δ. Since there is no current flowing through the ring ($\Sigma\, i = 0$), we obtain from Ampère's law

$$\oint \boldsymbol{H} \cdot \mathrm{d}\boldsymbol{l} = H_{\mathrm{PM}} h_{\mathrm{PM}} + H_\delta \delta = 0. \tag{3.82}$$

Table 3.3 Comparison of the characteristics of neodymium magnets and SmCo magnets. Adapted from TDK (2005) www.tdk.co.jp/tefe02/e331.pdf

	Neodymium magnets	SmCo magnets
Composition	Nd, Fe, B, etc.	Sm, Co, Fe, Cu, etc.
Production	Sintering	Sintering
Energy product	199–310 kJ/m^3	255 kJ/m^3
Remanence	1.03–1.3 T	0.82–1.16 T
Intrinsic coercive force, H_{cJ}	875 kA/m to 1.99 MA/m	493 kA/m to 1.59 MA/m
Relative permeability	1.05	1.05
Reversible temperature coefficient of remanence	–0.11 to –0.13 %/K	–0.03 to –0.04 %/K
Reversible temperature coefficient of coercive H_{cJ}	–0.55 to –0.65 %/K	–0.15 to –0.30 %/K
Curie temperature	320 °C	800 °C
Density	7300–7500 kg/m^3	8200–8400 kg/m^3
Coefficient of thermal expansion in magnetizing direction	5.2×10^{-6}/K	5.2×10^{-6}/K
Coefficient of thermal expansion normal to magnetizing direction	-0.8×10^{-6}/K	11×10^{-6}/K
Bending strength	250 N/mm^2	150 N/mm^2
Compression strength	1100 N/mm^2	800 N/mm^2
Tensile strength	75 N/mm^2	35 N/mm^2
Vickers hardness	550–650	500–550
Resistivity	$110–170 \times 10^{-8}$ Ω m	86×10^{-8} Ω m
Conductivity	590 000–900 000 S/m	1160 000 S/m

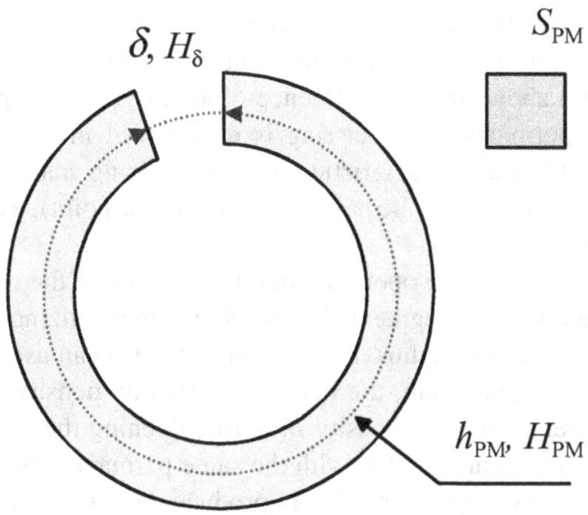

Figure 3.42 Definition of the field strength of an opened permanent magnet ring.

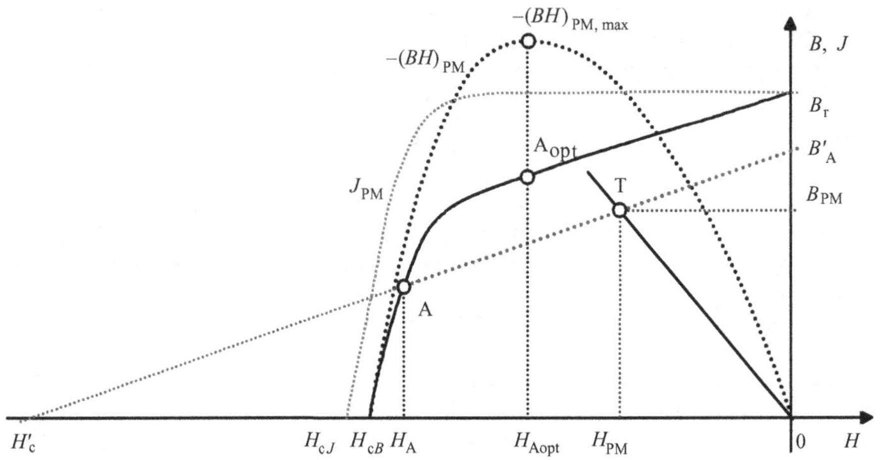

Figure 3.43 Characteristic demagnetization curve (H_{cB}–A–B_r), an operation line segment (A–B'_A), an operating point (T) and the optimum operating point (A_{opt}) of a permanent magnet system. The energy product of the material is indicated by $-(BH)_{PM}$. The polarization curve J_{PM} shows that at higher demagnetizing field strengths, the permanent magnet material loses its polarization. This magnet is in too high a demagnetizing field when the point A is reached. Part of the polarization J is lost. The new remanence is B'_A. For the best utilization of the magnet material, the demagnetizing field strength should not go much beyond H_{Aopt}.

Thus, we obtain for the field strength of the magnet

$$H_{PM} = -\frac{\delta}{h_{PM}} H_\delta \tag{3.83}$$

where h_{PM} is the "height" of the permanent magnet.

We have now defined the field strength H_{PM} caused by the atomic magnets of the ring material. Equation (3.83) shows that the influence of an air gap corresponds to the situation in which an unopened permanent magnet ring is magnetized in a negative direction with the field strength H_{PM}. On the characteristic curve of the ring material (Figure 3.43), the operating point has moved from B_r to the point A (called a base point), where the field strength $H_A = H_{PM}$ Equation (3.83).

If we now shorten the air gap, the operating point moves along the reversible magnetizing line AB'_A, called an operation line segment. The reversible magnetizing curve is not exactly a straight line but very close to it, and thus in our calculations we can use a straight line instead of a curve. If we close the ring again, the new remanent flux density will be B'_A, which is lower than the original remanent flux density B_r before opening the ring. The closing of the magnet may take place for instance either with the same permanent magnet material or with ideal iron, in which no field strength is needed to produce the flux density B'_A.

Permanent magnets are usually hard to machine and expensive, and therefore soft iron is used with them as part of the magnetic circuit, as shown in Figure 3.44. Let us assume that the characteristics of the permanent magnet material in Figure 3.44 are as shown in Figure 3.43

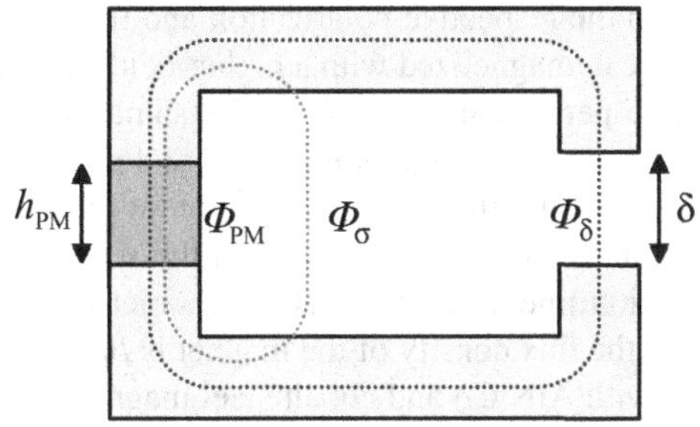

Figure 3.44 Permanent magnet as a part of a magnetic circuit consisting of iron and an air gap. The magnet creates a flux Φ_{PM}, which is divided into an air-gap flux Φ_δ and leakage flux Φ_σ.

and the operation line segment is AB'_A. If the height of the permanent magnet is h_{PM} and the cross-sectional area S_{PM}, the flux of the magnet is

$$\Phi_{PM} = S_{PM}B_{PM} = (1+\sigma)\,S_\delta B_\delta = (1+\sigma)\,S_\delta \mu_0 H_\delta \tag{3.84}$$

where σ is the leakage factor, that is the ratio of the leakage flux to the main flux $\sigma = \Phi_\sigma/\Phi_\delta$, and S_δ the area of the air gap. From Equation (3.84), we obtain taking into account (3.83)

$$B_{PM} = -(1+\sigma)\mu_0 \frac{S_\delta}{\delta}\frac{h_{PM}}{S_{PM}} H_{PM} = -(1+\sigma)\Lambda_\delta \frac{h_{PM}}{S_{PM}} H_{PM} \tag{3.85}$$

where Λ_δ is the permeance of the air gap. Equation (3.85) describes a line passing through the origin and it is called working line. This line meets the curve $H'_c AB'_A$ at the point T (Figure 3.43), which is the operating point of the magnet system.

From the product of Equations (3.83) and (3.84), we may solve

$$H_\delta = \frac{1}{\delta}\sqrt{\frac{|-(HB)_{PM}| \cdot V_{PM}}{(1+\sigma)\cdot \Lambda_\delta}} \tag{3.86}$$

where $V_{PM} = S_{PM}h_{PM}$ is the volume of the permanent magnet. A permanent magnet of a certain size thus yields a higher field in the air gap, the higher the energy product $B_{PM}H_{PM}$ of the magnet material is. Figures 3.36 and 3.43 depict diagrams of this product, which show that it has a maximum value. The magnet material is optimally utilized when the operating point is as close as possible to the point A_{opt}. The maximum of the energy product is reached at this point. If we wish to have a higher flux density than occurs at the optimum point A_{opt}, we have to use more permanent magnet material and the operating point moves toward the remanent flux density B'_A. However, the remanent flux density may not be exceeded by the magnet itself but external current linkage should be used to achieve such a situation which, finally, is not practicable at all.

A permanent magnet keeps its operation line segment once selected, if its circuit is not opened more than the base point A allows. If the magnet is opened more than at the point A, the magnet will be further demagnetized and the absolute value of the calculatory coercive force decreases. If we wish to keep the respective polarization and the remanent flux density B'_A, a permanent magnet may not be demagnetized with a higher field strength than H_A. In practice, because of their low relative permeability ($\mu_r = 1.05$), samarium–cobalt and neodymium magnets form an air gap themselves (cf. Equation (3.83), and physically opening the magnet circuit does not have any significant effect on the operation of the magnet. The characteristic curves of samarium–cobalt and neodymium magnets in the second quadrant are practically straight lines, and the operation line segment joins the characteristic curve. In this case, the magnet is utilized best when the flux density of the magnet is $B_r/2$.

The situation is different with AlNiCo and cobalt steel magnets. They have a high relative permeability, and therefore opening such a magnet has a remarkable influence on the operating point of the magnet. Neodymium and samarium–cobalt magnets tolerate well possible demagnetizing fields, and the operation line of the magnet may remain unchanged unless the polarization of the magnet is lost under a very high demagnetizing field strength. Opening the material with an air gap is not sufficient to demagnetize the magnets, but a negative armature reaction at a high temperature might cause loss of polarization in samarium–cobalt and neodymium magnets, for instance in the case of a stator short circuit creating a large demagnetizing current linkage.

The capability of tolerating negative field strength is a very important property of permanent magnet materials used in electrical machines. Old high-permeability magnet materials allow only a small opening of the magnetic circuit, and a demagnetizing armature reaction is forbidden. Rare-earth magnets, however, can tolerate the opening of the magnet (air gap) and also a demagnetizing armature reaction. In a permanent magnet machine, this means that we may have a negative stator d-axis current in the rotor coordinate system. This allows the possibility of field weakening in permanent magnet machines.

Let us examine how a negative armature reaction affects the working line of a magnet. In Figure 3.45, the demagnetizing coil and its current linkage NI represent a negative armature reaction.

Equation (3.84) is valid also for the circuit in Figure 3.45. Instead of Equation (3.82), Ampère's law has now the form

$$\oint \boldsymbol{H} \mathrm{d}\boldsymbol{l} = H_{PM} h_{PM} + H_\delta \, \delta = -NI \tag{3.87}$$

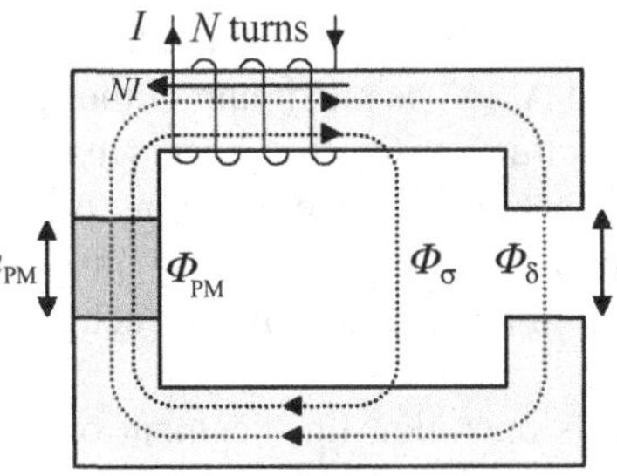

Figure 3.45 Permanent magnet circuit with a demagnetizing current linkage.

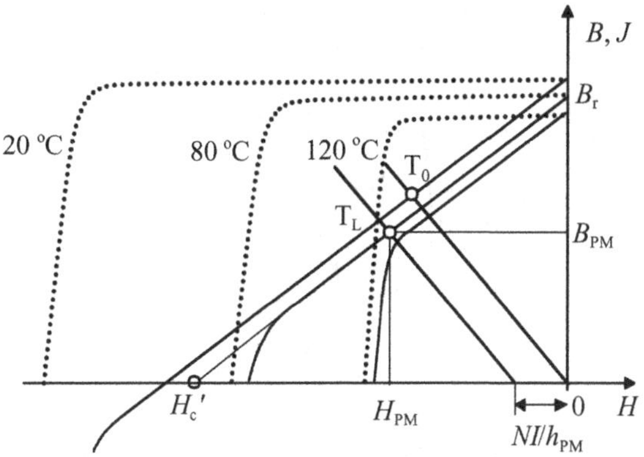

Figure 3.46 The demagnetization curves of a rare-earth magnet. The operating point at no-load (demagnetizing current zero) and at the temperature 20 °C is T_0. At load (demagnetizing current is I) the temperature is increased to 80 °C, the working line is shifted with NI/h_{PM} and the operating point is T_L.

from which we can solve the field strength in the air gap:

$$H_\delta = -\frac{h_{PM}}{\delta} H_{PM} - \frac{NI}{\delta}. \tag{3.88}$$

Inserting H_δ in Equation (3.84), we obtain

$$B_{PM} = -(1+\sigma)\mu_0 \frac{h_{PM} S_\delta}{\delta S_{PM}} \left(H_{PM} + \frac{NI}{h_{PM}} \right). \tag{3.89}$$

Equation (3.89) shows that the working line of the magnet, which was passing through the origin without demagnetizing current linkage, is now shifted with the amount of NI/h_{PM}. This is illustrated in Figure 3.46, in which typical demagnetization curves of rare-earth magnets in three different temperatures and the shift of the working line are illustrated.

In a reluctance network, a permanent magnet can be represented by a current linkage and reluctance (Figure 3.47). The current linkage Θ_{PM} and the magnetic resistance R_{PM} of the magnet are

$$\Theta_{PM} = -H_c' h_{PM} = \frac{B_r}{\mu_{PM}} h_{PM} = \frac{B_r}{\mu_{rPM}\mu_0} h_{PM}, \tag{3.90}$$

$$R_{PM} = \frac{h_{PM}}{\mu_{PM} S_{PM}} \tag{3.91}$$

where H_c' is the calculatory coercive force (the intersection of the continuation of the operation line segment with the field strength axis, Figure 3.46 or 3.43), B_r the remanent flux density, μ_{PM} the permanent magnet material permeability, μ_{rPM} the permanent magnet material relative permeability, h_{PM} the permanent magnet height, and S_{PM} the cross-sectional area of the magnet.

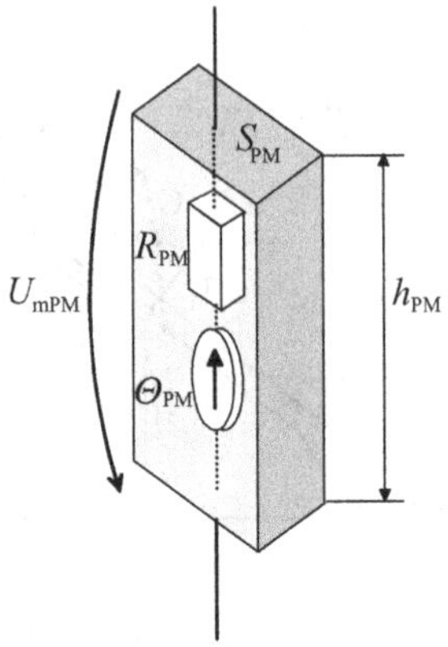

Figure 3.47 Reluctance network of a permanent magnet.

The equation for the magnetic flux of a permanent magnet circuit can be written as

$$\Phi = \Lambda_{tot} H'_c h_{PM} = \Lambda_{tot} \frac{B'_A}{\mu_0 \mu_r} h_{PM}, \tag{3.92}$$

where the permeance of the whole circuit, including the permanent magnet itself, is

$$\Lambda_{tot} = \frac{\Lambda_{PM} \Lambda_{ext}}{\Lambda_{PM} + \Lambda_{ext}}. \tag{3.93}$$

Here Λ_{ext} is the permeance of the external part of the permanent magnet circuit (iron, air gap and leakage) and Λ_{PM} is the permeance of the permanent magnet.

From the study above we may note that, in practice, a permanent magnet material normally operates at a flux density lower than the remanent flux density of the material $B_{PM} < B_r$. If we use some rare-earth magnet material, the operation line joins the demagnetizing curve of the material. Such a permanent magnet can create a flux density equal to its remanent flux density B_r in an infinitesimal air gap. In practical cases, the air gap is finite and also the iron circuit creates a magnetic voltage. Hence, $B_{PM} < B_r$ if the armature reaction is not affecting in the same direction as the permanent magnet, which is not usually the case.

If a high air-gap flux density is sought and the permanent magnet material operates above its optimal point (A_{opt} in Figure 3.43), the utilization of the magnet is inefficient, and an excessive amount of the magnet is required. If, however, there is a possibility to arrange a wide permanent magnet, wider than the air gap, the air-gap flux density may be higher than the remanence of the magnet. This is illustrated in Figure 3.48.

The resistivity of a permanent magnet is a significant factor in the design of an electrical machine. Permanent magnets are often exposed to the permeance and time harmonics of electrical machines, and if the resistivity of the magnet is low, eddy currents and losses are also created. The resistivity of sintered neodymium magnets is about 110–170 × 10^{-8} Ωm. This is only 5–10-fold when compared with steel, and these magnets are thus clearly conductive and produce losses under alternating fields.

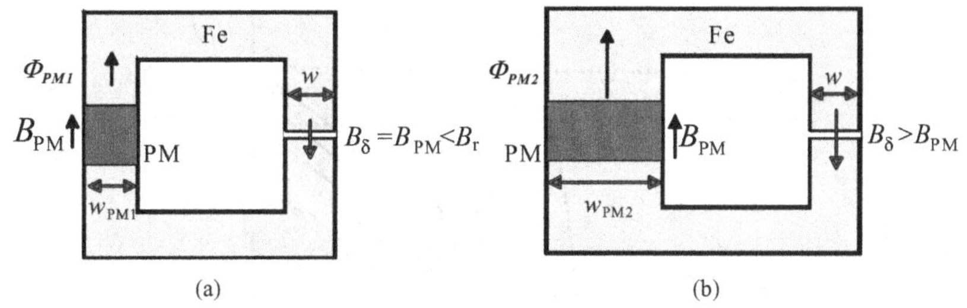

Figure 3.48 Air-gap flux density B_δ compared with the permanent magnet material remanence B_r when using (a) narrow or (b) wide magnets. w_{PM}, the width of the permanent magnet; w, the width of the air gap. $\Phi_{PM2} > \Phi_{PM1}$. This makes it possible that the air-gap flux density B_δ may be even larger than the remanence B_r of the magnet material.

As the resistivity of sintered NdFeB magnets is only a few times as large that of steel, eddy current losses in such magnets may not be omitted in machine design. Especially, the slot harmonics and frequency converter switching harmonics should be taken into account in machines with rotor surface magnets. The slot harmonics may be negligible only in very low-speed machines. Exact calculation of the losses in permanent magnets is, however, difficult and will be studied in more details in Chapter 7 with permanent magnet synchronous machines.

There, however is one clear principle: The losses are almost solely eddy current losses even though there are papers published claiming that there should be a significant amount of hysteresis loss in the magnets, too. Such losses are, however, possible only in special conditions normally not seen in rotating electrical machines.

To produce hysteresis loss the polarization J of a soft magnetic material has to alternate as a result of altering conditions, e.g. by slotting caused harmonics. The main idea of a permanent magnet material is, however, that the polarization J stays constant in the safe operating area. Therefore, in an ideal permanent magnet material there cannot be any hysteresis. In practice, some soft phase can be found in practical permanent magnets. As an indication of this the relative permeability of NdFeB magnets for example is $\mu_r = 1.04 - 1.05$. The value is 4–5 % larger than 1 showing that there will be some minor polarization changes in the material when external field strength changes. Such changes in J are, in principle prone to hysteresis, too. In practice the latest measurements have shown that, in practice, the hysteresis loss in permanent magnets in rotating machinery can be neglected and only eddy currents produce losses in the magnets.

3.7.4 Demagnetization of Permanent Magnets

A permanent magnet demagnetizes partly if the operating point of the magnet falls to the nonlinear part of the magnetization curve (as the point A in Figure 3.43). This can occur if the magnet temperature or the armature reaction is too high. Risky situations e.g. are:

- over-temperature of rare-earth magnets because of e.g. overloading the machine or loss of cooling or dirty cooling channels;

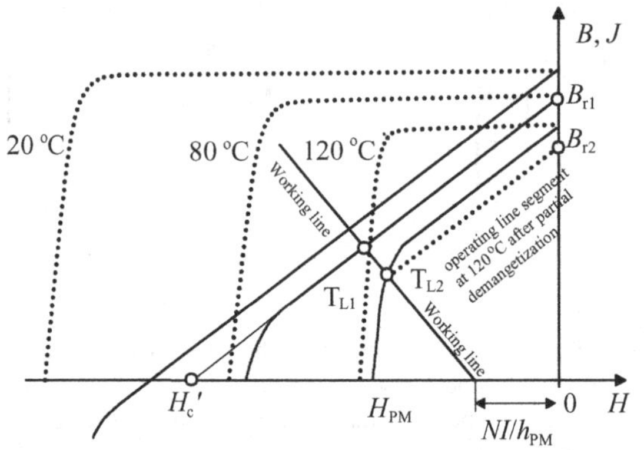

Figure 3.49 Irreversible demagnetization caused by increased temperature, e.g. because of cooling failure. The rated operating point is T_{L1} and the temperature 80 °C. The cooling failure causes the temperature rise to 120 °C and the new operating point is T_{L2}. The loading of the motor is constant and therefore the working line is not changed. After the demagnetization, the operating line segment is T_{L2}–B_{r2}.

- short-circuit in the grid supplying the permanent magnet motor; and
- direct-on-line starting.

Demagnetization by increased temperature is presented in Figure 3.49. It is assumed that a cooling failure increases the temperature of magnets from 80 °C to 120 °C and the load of the motor is constant. The operating point moves below the knee of $B(H)$-characteristics from T_{L1} to T_{L2} and the magnets are demagnetized. The remanence at 120 °C drops to the value B_{r2} and the new operating line segment (called also recoil line) is T_{L2}–B_{r2}. All $B(H)$ and $J(H)$ characteristics in different temperatures go down in proportion of B_{r2} to B_{r1}.

If the demagnetization is less than 10 %, the slope of the new operating line segment is approximately linear. With higher demagnetizations, the operating line segment is slightly bent upwards as shown in Figure 3.50 (Ruoho 2011). The nonlinearity is so small that the operating line segment can be treated as a straight line.

Figure 3.51 presents two cases, where too high current (too high armature reaction) causes the demagnetization. In the first case, the motor is overloaded, the temperature of the magnets increases from 80 °C to 120 °C, and the operating point moves from T_{L1} to T_{L2}. The magnets are irreversible demagnetized and the new operating line segment is T_{L2}–B_{r2}. In the second case, a short-circuit occurs in the frequency converter supplying the motor. The phenomenon is short as fuses react quickly. The temperature of magnets do not change but the short and high armature reaction moves the operating point from T_{L1} to T_{L3} and the new operating line segment is T_{L3}–B_{r3}.

In rotating electrical machines, the magnets are not demagnetized uniformly. For example, with surface mounted magnets if a too high armature reaction is the reason for demagnetization, the leaving edge of magnets of a generator and the front edge of a motor will demagnetize first. With FEM, we can check that no element of the magnet goes down the knee of $B(H)$-characteristics in the worst operational case.

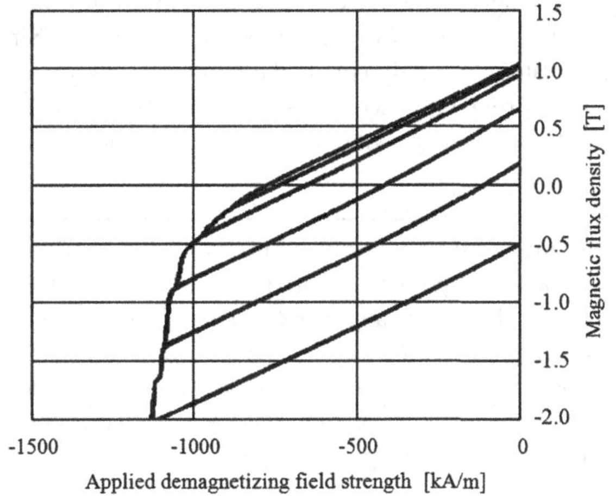

Figure 3.50 Recoil behavior of neodymium magnet sample. The recoil curve bents slightly near the vertical B-axis. Reproduced by permission of Sami Ruoho.

3.7.5 Application of Permanent Magnets in Electrical Machines

Permanent magnets are applied in a wide range of small electrical machines. They are utilized for instance in the excitation of DC machines and synchronous machines, and in hybrid stepper motors. Permanent magnets are increasingly occurring in large machines, too. There are already permanent magnet synchronous machines of several megawatts in for instance direct-driven wind-power generators. Permanent magnets can in some cases be employed as magnet bearings. In that case, the repulsive reaction between two permanent magnets is

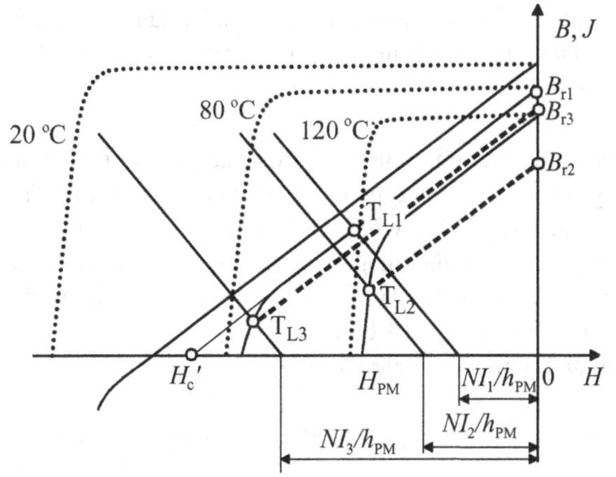

Figure 3.51 Two demagnetization cases are presented in the figure. In the first case, the operating point moves from T_{L1} to T_{L2} because the motor is overloaded. After the demagnetization, the operating line segment is T_{L2}–B_{r2}. In the second case, the operating point moves from T_{L1} to T_{L3} because of a short-circuit in the motor supplying converter. The new operating line segment is T_{L3}–B_{r3}.

utilized. In the design of a magnetic circuit with a permanent magnet material, more iteration is required than in the design of magnetic circuits of ordinary machines. Since the permeability of a permanent magnet material roughly equals the permeance of a vacuum, the permanent magnet material has a very strong influence on the reluctance of the magnetic circuit, and thus also on the inductances of the armature winding of the rotating-field machine.

Example 3.6: Simple magnetic circuits with a small air gap are investigated (Figure 3.52).

One of the circuits is magnetized with a winding N_2. The device is either a reactor with two coils, or a transformer, but the arrangement can also be easily employed to illustrate a rotating machine magnetized with a single magnetizing winding. The other winding corresponds to the armature winding. One of the circuits is magnetized with a permanent magnet material, the coercive force of which is 800 kA/m, and the relative permeability is 1.05. Calculate the properties of the magnetic systems.

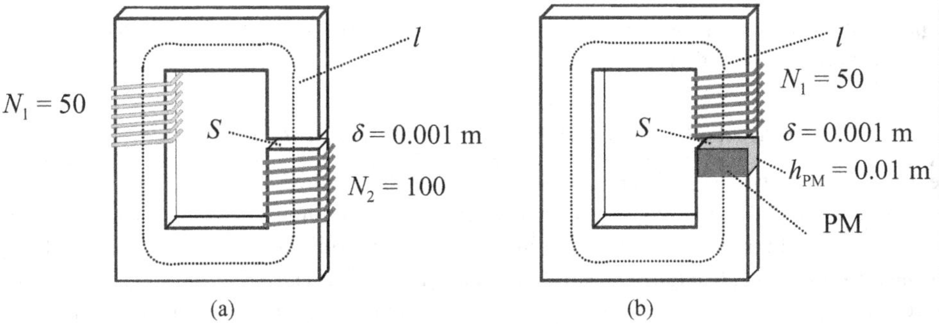

Figure 3.52 Simple magnetic circuits with single main flux magnetic paths. The length of the circuit is $l = 0.35$ m. There is an air gap δ of 1 mm in both circuits. The area of the circuit is $S = 0.01$ m^2. The left-hand circuit (a) is magnetized with the winding N_2. The right-hand circuit (b) is magnetized with an NdFeB permanent magnet. Both the cores have a winding N_1, the inductance of which has to be determined.

Solution: It is first simply assumed that the iron parts of the magnetic circuits take 5 % of the current linkage required for magnetizing the physical air gap. Further, it is assumed that no flux leakage occurs, and the whole magnetic flux flows in the circuit l. Now, we aim for a flux density of 1 T in the air gap. First, a circuit magnetized with winding N_2 is investigated. We may assume that an effective air gap, including the influence of the magnetic voltage of the iron, is $\delta_{ef} = 1.05$ mm.

For a flux density of 1 T in air, a field strength of

$$H_\delta = B_\delta / \mu_0 = 1/(4\pi \cdot 10^{-7})\,\mathrm{A/m} = 796\ \mathrm{kA/m}$$

is required. Because the entire magnetic circuit is replaced by a single air gap $\delta_{ef} = 1.05$ mm, the required current linkage

$$\Theta = H_\delta\, \delta_{ef} = 796 \cdot 1.05\ \mathrm{A} = 836\ \mathrm{A}$$

is obtained, the proportion of the iron being 0.05 × 836 A = 42 A. There are 100 turns in the winding, and hence the current flowing in the circuit has to be $I_2 = 8.36$ A.

The flux of the magnetic circuit is of magnitude

$$\Phi = SB_\delta = 0.01 \cdot 1 \text{ Vs} = 0.01 \text{ Vs}.$$

Respectively, the flux linkage of the winding N_2 is

$$\Psi_2 = \Phi \mathrm{N}_2 = 0.01 \cdot 100 \text{ Vs} = 1 \text{ Vs}.$$

Since there is a current $I_2 = 8.36$ A flowing in the coil, and it creates a flux linkage of 1 Vs, the self-inductance of the coil is approximately

$$L_{22} = \Psi_2 / I_2 = 1/8.36\,\mathrm{H} = 120 \text{ mH}.$$

The flux linkage of the coil N_1 is respectively

$$\Psi_{12} = \Phi \mathrm{N}_1 = 0.01 \cdot 50 \text{ Vs} = 0.5 \text{ Vs}.$$

There is no current flowing in the winding, because the flux Φ is created by the coil N_2 alone. The self-inductance of the coil N_1 could be calculated accordingly. Instead, we apply the reluctance of the magnetic circuit in the calculation, and thus

$$R_{\mathrm{m\,Fe}+\delta} = \delta_{\mathrm{ef}}/\mu S = 0.00105/(4\pi \cdot 10^{-7} \cdot 0.01)\mathrm{A/Vs} = 83.6 \text{ kA/Vs}.$$

The self-inductance is thus

$$L_{11} = N_1^2 / R_{\mathrm{m\,Fe}+\delta} = 50^2/83.6 \text{ mH} = 30 \text{ mH}.$$

Correspondingly, the mutual inductance between the windings can be written for an ideal flux connection in the form

$$L_{12} = L_{21} = N_1 N_2 / R_{\mathrm{mFe}+\delta} = 50 \cdot 100/83.6 \text{ mH} = 60 \text{ mH}.$$

Next, the characteristics of a magnetic circuit with a permanent magnet are investigated. A permanent magnet produces a strong current linkage:

$$\Theta_{\mathrm{PM}} = h_{\mathrm{PM}} H_{\mathrm{c}} = 8000 \text{ A}.$$

This current linkage is very high when compared with the current linkage of the winding N_2 of the previous magnetic circuit; however, most of the current linkage of the permanent magnet is consumed in the reluctance created by the magnet itself. The reluctance of the permanent magnet is

$$R_{\mathrm{PM}} = h_{\mathrm{PM}}/\mu_0 S = 0.01/(1.05 \cdot 4\pi \cdot 10^{-7} \cdot 0.01)\mathrm{A/Vs} = 757.9 \text{ kA/Vs}$$

The reluctances of the physical air gap and the iron are the same as above, $R_{\mathrm{mFe}+\delta} =$ 83.6 kA/(Vs). The reluctance of the entire magnetic circuit is approximately $R_{\mathrm{m,tot}} = R_{\mathrm{PM}} + R_{\mathrm{mFe}+\delta} = 841.5$ kA/Vs. For such a reluctance, a flux of

$$\Phi_{\mathrm{PM}} = \Theta_{\mathrm{PM}}/R_{\mathrm{m,tot}} = 8000/\left(841.5 \cdot 10^3\right)\,\mathrm{Vs} = 0.0095\,\mathrm{Vs}$$

is created by the permanent magnet. The flux density of the permanent magnet is

$$B_{\mathrm{PM}} = \Phi_{\mathrm{PM}}/S = 0.0095/0.01\ \mathrm{Vs/m^2} = 0.95\ \mathrm{T}.$$

Thus, despite the large permanent magnet and its strong current linkage, the flux density remains lower than in the previous case. We can see that with present permanent magnet materials, it is difficult to create flux densities of 1 T in the air gap without a large amount of magnetic material. Increasing the thickness of the magnet alone does not suffice, but the magnets have to be connected in parallel in the magnetic circuit. This can be done for instance by embedding two magnets per pole in a V-shape in the rotor structure.

The inductance of the winding is now

$$L_{11} = N_1^2/R_{\mathrm{m,tot}} = 50^2/841.5\ \mathrm{mH} = 2.97\ \mathrm{mH}.$$

A permanent magnet in the iron circuit considerably increases the reluctance of the magnetic circuit and reduces the inductances of the circuit. These phenomena are common also in rotating machines with permanent magnets. The current linkage required by the iron is yet to be investigated. It is now assumed that the iron is from M400-65A sheet, which requires about 1.1 A/cm for 1 T. The length of the magnetic circuit is 35 cm, and thus a current of 1.1×35 A = 38.5 A is required, which is sufficiently equal to the initial assumption.

Now, the magnetic circuits of Figure 3.52 are investigated further. The left-hand magnetic circuit constitutes a transformer with an air gap, the transformation ratio of which is $K = N_1/N_2 = 1{:}2$. A magnetic circuit of this kind has the equivalent circuit illustrated in Figure 3.53. The leakage inductances are included in the primary and secondary winding,

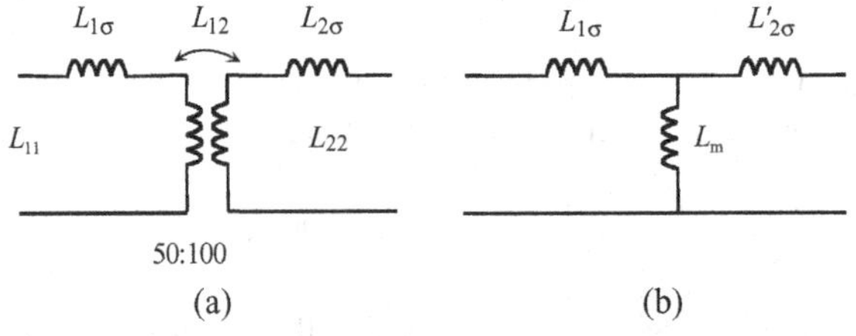

Figure 3.53 (a) Equivalent circuit of a transformer with two windings, illustrated in Figure 3.52, and (b) an equivalent circuit referred to the voltage level of the primary winding. $L_{1\sigma}$ and $L_{2\sigma}$ are the leakage inductances of the primary and secondary windings. In the reduction, the square of the transformation ratio K, $L'_{2\sigma} = K^2 L_{2\sigma}$, is required. The magnetizing inductance L_{m} is proportional to the mutual inductance of the primary and secondary windings, $L_{\mathrm{m}} = KL_{12}$.

although they have not yet been taken into account in the previous discussion. It is now assumed that 95 % of the flux in both windings penetrates all the turns of the other winding, the connecting factor being $k = 0.95$. Thus, a part of the self-inductance is leakage inductance

$$L_{1\sigma} = (1 - k)\,L_{11}, \quad L_{2\sigma} = (1 - k)\,L_{22}.$$

The mutual inductance is now reduced to some extent, since it is determined from the flux linkage created by the magnetizing winding to the other winding. The mutual inductance L_{12} is therefore determined by the flux linkage Ψ_{12} created in the winding N_1 by the current I_2 in the winding N_2:

$$L_{12} = \Psi_{12}/I_2.$$

If now Ψ_{12} is only 95 % of its theoretical maximum value, we can see that also the mutual inductance is reduced correspondingly. The current $I_2 = 8.36$ A of the previous example creates a flux linkage of 100×0.01 Vs $= 1.0$ Vs in the winding N_2, and a flux linkage

$$\Psi_{12} = 0.95\,\Phi N_1 = 0.95 \cdot 0.01 \cdot 50 \text{ Vs} = 0.475 \text{ Vs}$$

in the winding N_1. Next, the mutual inductance

$$L_{12} = L_{21} = \Psi_{12}/I_2 = 0.475 \text{ Vs}/8.36 \text{ A} = 56.8 \text{ mH}$$

can be calculated.

If the connection between the windings is ideal, we obtain

$$L_{12} = \sqrt{L_{11}L_{22}}.$$

Since now only 95 % of the flux of winding 1 penetrates winding 2, we obtain a connecting factor

$$k = L_{12}/\sqrt{L_{11}L_{22}} = 0.95,$$

which in practice is always less than one. The magnetizing inductance for Figure 3.53 becomes

$$L_{\mathrm{m}} = L_{12}K = 56.8 \cdot 0.5 \text{ mH} = 28.4 \text{ mH}.$$

The leakage inductance of the primary winding is thus $L_{1\sigma} = L_{11} - L_{\mathrm{m}} = 30 - 28.4$ mH $=$ 1.6 mH. Correspondingly, the leakage inductance of the secondary winding referred to the primary winding is

$$L'_{2\sigma} = L_{22}K^2 - L_{\mathrm{m}} = (120 \cdot 0.5^2 - 28.4) \text{ mH} = 1.6 \text{ mH}.$$

Now, the equivalent circuits are illustrated referred to the primary winding. They are, in this example, magnetized by a direct current, either the direct current of the winding N_2 or the virtual direct current of the permanent magnet

$$I'_{PM} = \Psi_{PM}/L_m = 0.475\ \text{Vs}/2.97\ \text{mH} = 160\ \text{A}.$$

Here Ψ_{PM} is the permanent magnet flux linkage linking the turns of the primary winding: $\Psi_{PM} = \Phi_{PM}N_1 = 0.0095 \times 50\ \text{Vs} = 0.475\ \text{Vs}$. Since there is no current flowing in the primary winding to create Ψ_{PM}, no leakage of this flux takes place and L_m is applied to calculate the virtual permanent magnet current. Although it is not conventional to discuss a transformer with DC magnetization, in this case we may assume such "transformers" to be applied for instance as DC chokes that emulate the behavior of a synchronous machine in the rotor reference frame. This simplification is a good basis for the analysis of the magnetic circuits of rotating machines, since the equivalent circuits of Figure 3.54 are, in principle, equivalent to the equivalent circuits of electrical machines. The device and the frame of reference in question decide whether the equivalent circuits function with a direct or an alternating current. Often the equivalent circuits of rotating electrical machines are also constructed in such a reference frame where a direct current is flowing in the equivalent circuits in a stationary state. The magnetic circuits of electrical machines differ from the examples presented above only with respect to their geometric complexity. Naturally, in rotating-field machines, at least a two-phase winding is required to produce a rotating magnetic flux that was not present in the simple connections of Figures 3.52–3.54. The equivalent circuit of Figure 3.53 works, with certain preconditions, with transformers and induction machines with an alternating current. Because of the direct currents in the secondary winding of Figure 3.54, the equivalent circuits are suitable with certain supplementary additions for the analysis of synchronous machines.

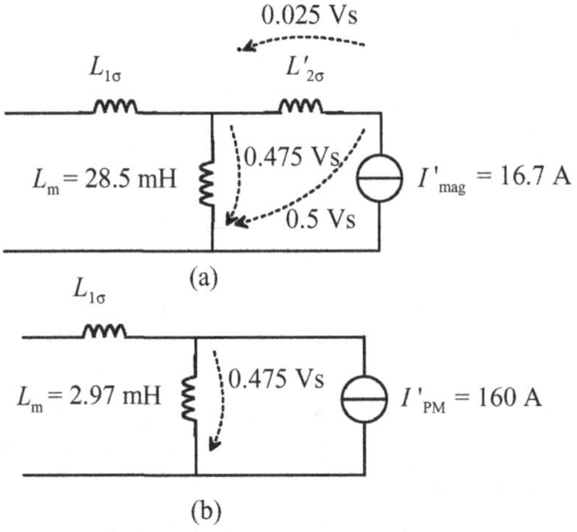

Figure 3.54 Direct current of the winding magnetizes the magnetic circuits of Figure 3.52. (a) The equivalent circuit for the two winding choke, (b) the equivalent circuit for the choke with permanent magnet excitation.

The currents are referred to the primary winding. The virtual direct current of the permanent magnet is calculated by dividing the flux linkage of the magnetizing winding (0.475 Vs) by the magnetizing inductance $L_m = 2.97$ mH (see the equation above where $I'_{PM} = 160$ A). The virtual magnetizing current of the permanent magnet is high, since the reluctance of this magnetic circuit is also relatively high when compared with the magnetic circuit magnetized by the winding N_2. The direct current of the winding N_2 creates a total flux linkage of 0.5 Vs, which is divided into an air-gap flux linkage and a leakage flux linkage. In the case of a permanent magnet, the leakage flux of the permanent magnet is not discussed here. In reality, only a part of the current linkage of the permanent magnet magnetizes the air gap. A small amount of the current linkage directly becomes a leakage flux of the permanent magnet. Its flux leakage cannot be illustrated with a winding, since no "permanent magnet winding" exists (see Figure 3.54):

$$I'_{mag} = I_2/K = 8.36 \text{ A} \cdot 2/1 = 16.7 \text{ A}.$$

Since the permeability of the best permanent magnet materials is approx. 1, such materials have a significant influence on the effective air gap δ_{ef} of the magnetic circuit to which they belong. The relative permeability of neodymium magnets is $\mu_{rPM} = 1.05$, and hence the permanent magnet material itself produces an apparent air gap of $\delta_{PM} = h_{PM}/1.05$, which almost equals the thickness of the magnet itself. Hence, the magnetic air gap of a machine carrying permanent magnets is extremely large, especially if the magnets are fixed on the rotor surface. The magnet itself forms the largest magnetic reluctance of the magnetic circuit and therefore needs a large magnetic voltage itself.

When considering the armature reaction, we can see that the synchronous inductance of a rotor surface permanent magnet machine is low. In some cases, this impedes the application of a voltage source inverter as a supply to these machines, since the current change speeds are high because of the low inductances. Power electronic switches are nevertheless so fast that permanent magnet rotor machines of low inductance can be fed with a voltage source inverter.

Permanent magnets are particularly useful in low-speed machines, since in the rotating-field machines the magnetizing inductance is inversely proportional to the square of the number of pole pairs p ($L_m \sim p^{-2}$, see Section 4.1). A multiple-pole rotating-field machine will have a low magnetizing inductance, and therefore it is generally not very advantageous to construct a low-speed motor with a high number of pole pairs as an asynchronous motor. Such a machine will have a relatively low power factor.

Instead, in the case of a permanent magnet machine, the effect of the low magnetizing inductance is not harmful, since most of the current linkage of the magnetic circuit originates from the permanent magnet. In a permanent magnet machine, the equivalent air gap δ_e is strongly dependent on the positioning of the permanent magnet material and on the thickness of the material. Permanent magnets assembled on the surface of the rotor always produce the lowest inductance, since in practice the thickness of the magnet directly increases the equivalent air gap. A low magnetizing inductance increases the ability of a synchronous machine to produce torque as the maximum torque in synchronous machines is inversely proportional to the synchronous inductance.

When using rare-earth permanent magnets in electrical machines, the relatively large conductivity of neodymium and SmCo magnets is problematic. When the speed of the motor increases, the air-gap harmonics create remarkable losses in the magnets. As neodymium magnets are particularly sensitive to temperature, a very effective cooling for the magnets has to be assured in the machine design. In practice, the highest remanent flux densities in Ferrite magnets are in the range of 0.4 T and they, therefore, have a poor energy product. Ferrite magnets are cheap and have a very low conductivity (resistivity typically $\rho > 10^9\ \Omega\mathrm{m}$), and hence the eddy current losses in ferrites remain low compared with neodymium and SmCo magnets. Because of these properties Ferrite magnets can be used also in some industrial applications. The thermal behavior of ferrites is, however, different compared to NdFeB magnets. They are vulnerable for large demagnetizing field strengths when cold. This differs them clearly from NdFeB magnets that suffer the possible loss of polarization at high temperatures when heavily demagnetized.

3.8 Assembly of Iron Stacks

The assembly of the parts of a magnetic circuit of an electrical machine has a significant influence on the final quality of the machine. Punching of the sheet material is carried out in a single punching action (in case of small machines), or by first producing a suitable disc of the plate material, which is then rotated in an indexing head, and the slots are punched in the disc one by one (in case of larger machines). A high precision is required to produce stacks of even quality. The ready sheets of both the stator and the rotor have to be rotated with respect to each other to avoid virtual saliency caused by the anisotropy of the sheet. Also, the physical thickness of a sheet varies so much that it is advisable to rotate the sheets in the assembly. This way it is possible to construct a stack of even length.

In the assembly of the sheet stack, the stack is straightened with a suitable tool in order to get the slots in the desired position. Here, guides can be employed either at the slot opening or at the bottom of the slot. Next, the stack is pressed into its final length and clamped with a suitable method. The best way to avoid losses is to press the stack straight to the stator frame and to use Seger rings, for instance, as clamps.

In serial production, welding the outer surface of the stator stack has become more and more popular. In this method, small cuts are made in the outer surface of each sheet. When the stack is assembled, these cuts comprise a continuous slot in the stack. The stack is welded at these slots, and a welding bead covers the complete length of the stack. The insulation between the sheets is damaged in the welding process. However, the damage is restricted to a rather small region. Consequently, the iron losses of a welded stack are a few per cent higher than the losses of a key-fitted stack.

Key fitting is a common method of assembly. When a stator is punched, dovetail slots are made on the outer surface of the stator. As the stack is pressed into shape, a key is placed into the dovetail slots. The key bar is somewhat longer than the final length of the stack. The ends of the key are bent 90°, and the stack is thereby clamped. The method is advantageous for the minimization of losses. On the other hand, the key bars impede heat transfer from the stator stack to the frame of the machine, which is a drawback of the key-fitting method.

Sometimes holes are cut in the outer edge of the stack for riveting. In principle, the method is simple and mechanically suitable, but causes extra iron losses in the machine. There remains a narrow bridge of stator yoke between the bar and the stator frame. This bridge provides a path for the flux, which induces a voltage in the rivet. Since the rivet is galvanically connected to the frame, a current flows in the loop formed by the rivet and the frame. This current creates losses in the machine. In principle, the rivets should be insulated from the stack, which is not an easy task.

The stator stack is usually equipped with straight slots to ease automatic winding. Sometimes, skewing is also required in the stator. Therefore, in automatic production, the stator is first insulated and wound, and then it is skewed before clamping the stack and the resin impregnation of the winding.

Rotor stacks are produced in large numbers for induction machines. If there is a cast aluminum cage winding in the machine, the winding is employed to clamp the whole stack. The desired skew is created, and the stack is brought to a die-casting machine. The machine casts aluminum into the slots at high pressure, and simultaneously produces short-circuit rings at the ends of the rotor. The aluminum sets in a few seconds and ties the complete stack.

If the rotor is wound with a round wire or a shaped bar, the stack has to be tied before winding. Now, rivets penetrating through the stack can be easily employed in the rotor yoke.

The stacks wound with a round wire or preformed copper are set in their final positions after the resin impregnation of the winding. The vacuum pressure impregnation (VPI) in particular glues all the sheets firmly together.

If the stack has to be divided into sections for cooling, and a suitable key material is required between the stack sections, the material has to be selected carefully to avoid unnecessary losses. Nonmagnetic stainless steels are appropriate for this purpose. Suitable insulation materials can also be employed, if their mechanical strength is adequate.

Bibliography

Carter, F.W. (1901) Air-gap induction. *Electrical World and Engineering*, **XXXVIII** (22), 884–8.

European Standard EN 10106 (1996) *Cold Rolled Non-Oriented Electrical Steel Sheet and Strip Delivered in the Fully Processed State*, CENELEC, Brussels.

Gieras, J.F., Wang, R.J. and Kamper, M.J. (2008) *Axial Flux Permanent Magnet Brushless Machines*, 2nd edn, Kluwer Academic, Dordrecht.

Heck, C. (1974) *Magnetic Materials and their Applications*, Butterworth, London.

Heikkilä, T. (2002) *Permanent Magnet Synchronous Motor for Industrial Inverter Applications – Analysis and Design*, Dissertation LUT. Available at: http://urn.fi/URN:ISBN:952-214-271-9.

Heller, B. and Hamata, V. (1977) *Harmonic Field Effects in Induction Machines*, Elsevier Scientific, Amsterdam.

IEC 60404-8-1 (2004) *Magnetic Materials – Part 8-1: Specifications for Individual Materials – Magnetically Hard Materials*. International Electrotechnical Commission, Geneva.

Jokinen, T. (1979) *Design of a rotating electrical machine (Pyörivän sähkökoneen suunnitteleminen)*, Lecture notes. Helsinki University of Technology, Laboratory of Electromechanics.

Miller, T.J.E. (1993) *Switched Reluctance Motors and Their Controls*, Magna Physics Publishing and Clarendon Press, Hillsboro, OH and Oxford.

Nerg, J., Pyrhönen, J., Partanen, J. and Ritchie, A.E. (2004) Induction motor magnetizing inductance modeling as a function of torque, in *Proceedings ICEM 2004, XVI International Conference on Electrical Machines, 5–8 September 2004*, Cracow, Poland. Paper 200.

Richter, R. (1967) *Electrical Machines: General Calculation Elements. DC Machines (Elektrische Maschinen: Allgemeine Berechnungselemente. Die Gleichstrommaschinen)*, Vol. **I**, 3rd edn, Birkhäuser Verlag, Basle and Stuttgart.

Ruoho, S. (2011) *Modeling Demagnetization of Sintered NdFeB Magnet Material in Time-Discretized Finite Element Analysis*. Aalto University publication series doctoral dissertations 1/2011. Available at: http://lib.tkk.fi/Diss/2011/isbn9789526040011.

TDK (2005) *Neodymium-Iron-Boron Magnets NEOREC Series*. [online]. Available from http://www.tdk.co.jp/tefe02/e331.pdf (accessed 31 August 2007).

Vacuumschmelze (2003) *Rare-Earth Permanent Magnets VACODYM · VACOMAX PD 002*. 2003 Edition. [online]. Available from http://www.vacuumschmelze.de/dynamic/docroot/medialib/documents/broschueren/dmbrosch/PD002e.pdf (accessed 1 December 2006).

Vogt, K. (1996) *Design of Electrical Machines (Berechnung elektrischer Maschinen)*, Wiley-VCH Verlag GmbH, Weinheim.

4

Inductances

In rotating electrical machines, the total flux consists of the main flux Φ_m and the leakage flux Φ_σ components. The main flux (air-gap flux) of the machine enables electromechanical energy conversion but the proportion of the total flux called the leakage flux does not participate in energy conversion. From the physical point of view a machine has only, for example, the total stator flux, or corresponding (total) stator flux linkage which is the integral of the voltage supply ($\boldsymbol{\psi}_s = \int (\boldsymbol{u}_s - R_s\boldsymbol{i}_s)\,dt$). Also in case of finite element analysis the total flux will result from the calculations and extra work is needed to segregate the flux components. There is, however, a long tradition in electrical machine design according to which the flux components are calculated separately.

The main flux has to cross the air gap of rotating machines; an important function of the main flux is therefore to electromagnetically connect both the stator and the rotor. In this sense, an air-gap flux Φ_m creates an air-gap flux linkage Ψ_m in the investigated winding, and consequently connects different parts of the machine. The leakage fluxes of the stator and rotor do not generally cross the air gap. They contribute to the generation of the total flux linkage of the winding by producing a leakage flux linkage Ψ_σ component to it. Flux leakage occurs in both the stator and the rotor winding. The corresponding leakage flux linkages are the flux leakage $\Psi_{s\sigma}$ of the stator and the flux leakage $\Psi_{r\sigma}$ of the rotor. A leakage flux also occurs in the permanent magnet materials of the machine. Because of the flux leakage, more magnetic material, or correspondingly more magnetizing current, is required than in the fully theoretical case without flux leakage.

The air-gap flux linkage ψ_m corresponds to the magnetizing inductance L_m and the leakage flux linkage Ψ_σ, correspondingly, corresponds to a leakage inductance L_σ. In case of an induction machine the stator inductance is $L_s = L_m + L_{s\sigma}$ – the sum of the magnetizing inductance and the stator leakage inductance. In case of synchronous machines this inductance is called the synchronous inductance. We shall observe the machine inductances in the following, starting with the most important inductance – the magnetizing inductance L_m.

Design of Rotating Electrical Machines, Second Edition. Juha Pyrhönen, Tapani Jokinen and Valéria Hrabovcová.

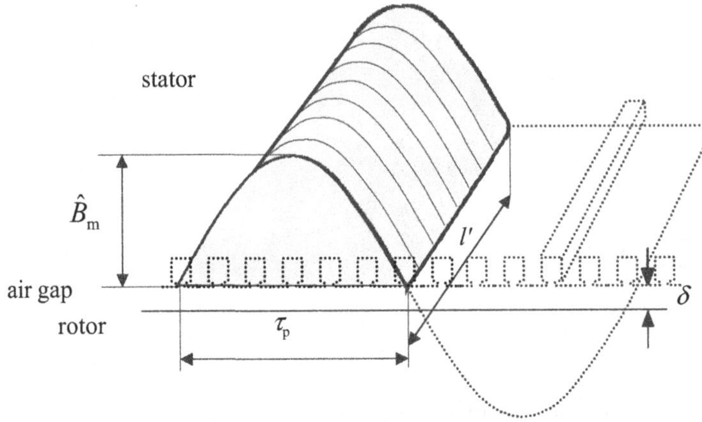

Figure 4.1 Distribution of fundamental flux density over a pole pitch and the length of the machine. The flux of the machine is calculated from this distribution. In practice, the air-gap flux is distorted because of the slotting and the current linkage harmonics. However, the fundamental value of the waveform has to be found in any case.

4.1 Magnetizing Inductance

The winding, the dimensions of the magnetic circuit and the selected magnetic circuit materials define the most important inductance of the machine, namely the magnetizing inductance. Next, a magnetizing inductance of a poly-phase winding is calculated. On the rotor surface, there is a flux density distribution (Figure 4.1) over the pole pitch τ_p and the length of the machine l', from which we may integrate the flux of the machine.

The peak value of the air-gap flux of the machine is the surface integral of the flux density $\boldsymbol{B}$ over one pole surface $\boldsymbol{S}$

$$\hat{\Phi}_m = \int_S \boldsymbol{B} \cdot d\boldsymbol{S} = \alpha_i \tau_p l' \hat{B}_m \tag{4.1}$$

where α_i is the ratio of the arithmetical average flux density B_{av} to peak value of flux density $\hat{B}_m$

$$\alpha_i = \frac{B_{av}}{\hat{B}_m}. \tag{4.2}$$

For a sinusoidal flux density distribution α_i takes the value $2/\pi$.

The expression "peak value of the air-gap flux" describes the maximal flux $\hat{\Phi}_m$ that penetrates a diagonal coil and hence produces the maximum flux linkage of a phase winding $\hat{\Psi}_{mp}$ where the subscript p after m is added for single phase winding.

The expression of the peak value of the air-gap flux may be somewhat misleading, but it is used here since it allows us to write, for example, the important time-dependent function $\psi_m(t) = \hat{\psi}_m \sin \omega t$ of a multiple phase winding. The flux linkage of a single phase – the main inductance – is obtained from the air-gap flux maximum value by multiplying $\hat{\Phi}_{m}$ by the effective turns of winding

$$\hat{\Psi}_{mp} = k_{ws1} N_s \hat{\Phi}_m = k_{ws1} N_s \alpha_i \tau_p l' \hat{B}_m. \tag{4.3}$$

On the other hand, the magnetic flux density of an air gap can be determined by applying the current linkage of the phase; in other words, the current linkage $\hat{\Theta}_s$ of the stator creates a flux in a direct-axis air gap

$$\hat{B}_m = \frac{\mu_0 \cdot \hat{\Theta}_s}{\delta_{ef}}, \tag{4.4}$$

where δ_{ef} is the effective air gap met by the current linkage of the stator winding. Here, the influence of the air gap is included (cf. Equation (3.25)) together with the influence of the iron in the direct-axis direction. The equation is now substituted in Equation (4.3) for $\hat{\Psi}_{mp}$ and we obtain for a flux linkage of a single phase

$$\hat{\Psi}_{mp} = k_{ws1} N_s \alpha_i \frac{\mu_0 \hat{\Theta}_s}{\delta_{ef}} \tau_p l'. \tag{4.5}$$

The current linkage of a single phase winding is

$$\hat{\Theta}_s = \frac{4}{\pi} \frac{k_{ws1} N_s}{2p} \sqrt{2} I_s. \tag{4.6}$$

Substitution yields for the single phase flux linkage

$$\hat{\Psi}_{mp} = k_{ws1} N_s \alpha_i \frac{\mu_0}{\delta_{ef}} \frac{4}{\pi} \frac{k_{w1} N_s}{2p} \tau_p l' \sqrt{2} I_s, \tag{4.7}$$

$$\hat{\Psi}_{mp} = \alpha_i \mu_0 \frac{1}{2p} \frac{4}{\pi} \frac{\tau_p}{\delta_{ef}} l' \left(k_{ws1} N_s\right)^2 \sqrt{2} I_s. \tag{4.8}$$

By dividing the result by the peak value of the current, which in this case is the magnetizing current, we obtain the magnetizing inductance L_{mp} of a single-phase winding (the main inductance)

$$L_{mp} = \alpha_i \mu_0 \frac{1}{2p} \frac{4}{\pi} \frac{\tau_p}{\delta_{ef}} l' \left(k_{ws1} N_s\right)^2 = \alpha_i \frac{2 \mu_0 \tau_p}{\pi p \delta_{ef}} l' \left(k_{ws1} N_s\right)^2. \tag{4.9}$$

This main inductance is the magnetizing inductance of a single-phase winding. In a multiphase winding, the flux is contributed by all the phases of the winding. The magnetizing

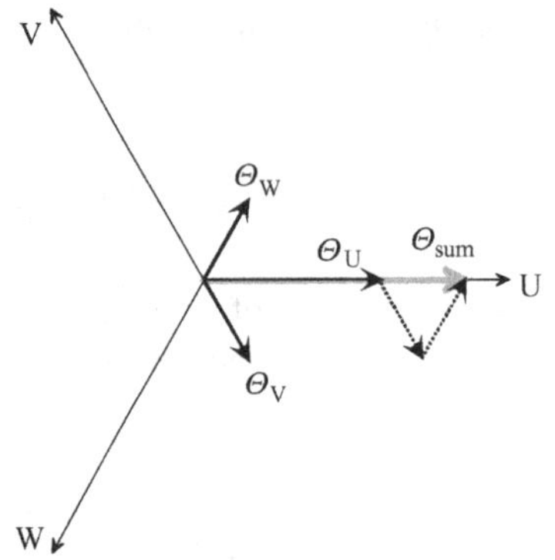

Figure 4.2 Sum of the current linkages of three-phase windings at the time instant when $i_U = +1, i_V = i_W = -1/2$.

inductance of an m-phase machine can be calculated by multiplying the main inductance by $m/2$

$$L_m = \frac{m}{2}\alpha_i\mu_0\frac{1}{2p}\frac{4}{\pi}\frac{\tau_p}{\delta_{ef}}l'(k_{ws1}N_s)^2 = \alpha_i\frac{m\tau_p}{\pi p\delta_{ef}}\mu_0 l'(k_{ws1}N_s)^2 = \alpha_i\frac{mD_\delta}{2p^2\delta_{ef}}\mu_0 l'(k_{ws1}N_s)^2. \tag{4.10}$$

The multiple-phase winding magnetizing (air-gap) flux linkage is now $\Psi_m = L_m I_m$ where I_m is the magnetizing current of the equivalent circuit. According to (4.10) the magnetizing inductance L_m depends on the saturation, that is, on α_i and the effective air gap δ_{ef}, the phase number m, the effective turns of winding $k_{w1} N_s$, the length of the machine l' and the number of pole pairs p. The factor $m/2$ may, in the case of a three-phase machine ($m = 3$), be explained in a simple manner, as follows. Let us, again, consider a time instant when the currents are $i_U = +1$, $i_V = i_W = -\frac{1}{2}$. According to Figure 4.2, the current linkage sum of three phases at this instant is $3/2 = m/2$.

An effective air gap δ_{ef} includes an air gap lengthened with the Carter factor and the effect of iron, which also increases the apparent air gap. The influence of iron varies typically from a few per cent even up to multiples of 10 %. In such a case, the iron circuit is strongly saturated. Figure 3.18 illustrates a case in which the magnetic voltage over the iron circuit is higher than the voltage over the air gap. Such a situation is possible mainly in tightly dimensioned induction machines. On the other hand, in permanent magnet synchronous machines, in which the equivalent air gap includes in the d-direction the length of the permanent magnets, the proportion of iron remains very low.

Magnetizing inductance is not constant but changes as a function of voltage and torque. The voltage dependence can be easily explained. Increasing the voltage increases the flux density, which may saturate the iron parts. The torque dependence can be explained by Faraday's principles. When the torque increases, the tension of the flux lines increases. The lines travel along more saturated paths, and the machine requires more magnetizing current than before. Figure 4.3 illustrates the saturation of the magnetizing inductance of a typical 30 kW, totally enclosed, four-pole induction machine as a function of torque.

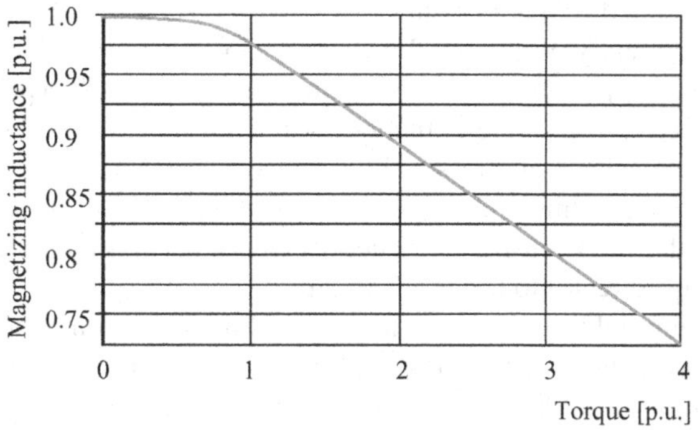

Figure 4.3 Saturation of the magnetizing inductance of an induction machine as a function of torque in a 30 kW, four-pole, 400 V induction machine. The inductance is already reduced by a few per cent by the rated torque of the machine itself. When the machine is overloaded, the decrease in inductance is notable.

Figure 4.4 depicts the paths of the flux with a small load and with an overload. The figure indicates clearly how the flux selects different routes in each case. It is also easy to understand why the line integral $\oint \mathbf{H} \cdot \mathrm{d}\boldsymbol{l}$ gets larger values in the latter case, thus indicating a higher magnetizing current and a lower inductance.

4.2 Leakage Inductances

Flux leakage is often considered a negative phenomenon. However, flux leakage in some cases also has a positive role. For instance, the transient inductance L'_{s} of an asynchronous machine $L'_{\mathrm{s}} = L_{\mathrm{s}\sigma} + L_{\mathrm{r}\sigma}L_{\mathrm{m}}/\,(L_{\mathrm{r}\sigma} + L_{\mathrm{m}})$ consists mainly of the sum of the stator and rotor flux leakage inductances $L'_{\mathrm{s}} \approx L_{\mathrm{s}\sigma} + L_{\mathrm{r}\sigma}$ For instance, if the target is to filter the motor current

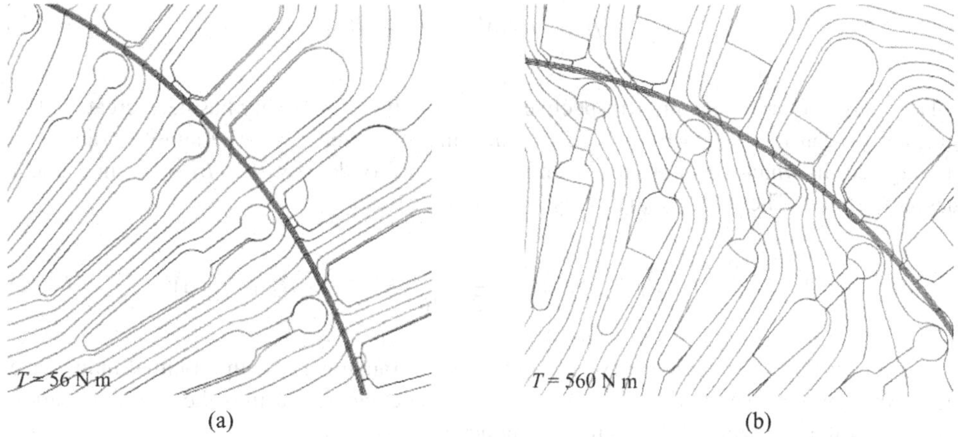

Figure 4.4 Flux diagrams of an induction machine as a function of torque ($0.3T_{\mathrm{n}}$, $2.9T_{\mathrm{n}}$). Reproduced by permission of Janne Nerg.

of a pulse-width-modulated (PWM) AC inverter drive, the stator flux leakage of the machine can be increased intentionally. Without a flux leakage, a PWM supply could not be employed to feed induction machines, for example. Now, a flux leakage is clearly a positive factor. Often, flux leakage is assumed to contribute to the losses in a machine. However, no direct line can be drawn between flux leakage and losses. Flux leakage may cause some extra losses in for instance the frame of the machine. Furthermore, the slot leakage flux increases the skin effect of the conductors in the slots causing more copper losses in the stator. Nevertheless, always linking the flux leakage directly to losses is incorrect.

In traditional rotating-field machines, at least one of the main components of the machine has a distributed winding that has been designed to participate in energy conversion only with the fundamental component of the air-gap flux. The harmonic components of the air-gap flux are considered harmful. In this sense, the harmonic components of the air-gap flux belong to the leakage flux, although they do cross the air gap. The fundamental is called the main flux that creates the main flux linkage in a winding under observation. Depending on the winding assembly, part of the winding may not be connected to the main flux, and flux components resembling a leakage flux are created. These components cross the air gap.

Leakage fluxes thus comprise:

- all the flux components that do not cross the air gap; and
- those components crossing the air gap that do not participate in the formation of the main flux linkage and therefore not participating the electro-mechanical power transformation.

An inductance of a winding or a winding section is obtained by calculating the flux linkage $\Psi = k_w N\Phi$ caused by the current I, bearing in mind that $\Psi = LI$. We also have to remember that $L = N^2\Lambda$. Furthermore, we may utilize the analogy according to which a voltage drop is created over a reactance $\Delta U = XI$. In some cases, the energy of a magnetic field can be employed to determine inductances or reactances. According to Equations (1.90) and (1.91), the energy stored in a magnetic circuit is

$$W_\Phi = {}^1\!/_2 \int_V HB\,\mathrm{d}V = \frac{1}{2\mu}\int_V B^2\mathrm{d}V. \tag{4.11}$$

The energy equation is valid irrespective of the way the flux density has been created. The flux density can be created by one or several winding currents. First, a simple example is investigated, in which two currents I_1 and I_2 create a flux density $B = B_1 + B_2$ in a linear magnetic circuit. The energy equation is written in the form

$$W_\Phi = \frac{1}{2\mu}\int_V (B_1 + B_2)^2\,\mathrm{d}V = \frac{1}{2\mu}\int_V \left(B_1^2 + 2B_1B_2 + B_2^2\right)\mathrm{d}V. \tag{4.12}$$

In electromechanical devices having two winding arrangements with a mutual inductance M, the energy of a magnetic field is stored in the inductances, and thus the energy equation can be rewritten applying self- and mutual inductances L_1, L_2 and M

$$W_\Phi = {}^1\!/_2 L_1 I_1^2 + {}^1\!/_2 M I_1 I_2 + {}^1\!/_2 L_2 I_2^2. \tag{4.13}$$

When Equations (4.12) and (4.13) are set equal, the corresponding terms of both sides are equal. The equation can be divided into three separate terms, the outermost terms now being of equal form. The self-inductance of the winding becomes

$$L = \frac{1}{\mu I^2} \int_V B^2 \mathrm{d}V. \tag{4.14}$$

Since the middle terms in Equations (4.12) and (4.13) are equal, we obtain the mutual inductance

$$M = \frac{2}{\mu I_1 I_2} \int_V B_1 B_2 \, \mathrm{d}V. \tag{4.15}$$

When employing these equations, the volume can be divided into sections and the components of inductance can be calculated. For instance, when calculating self-inductance, it is advisable to divide the volume V into sections of the main and the leakage flux, and to calculate the respective inductances. Next, we analyze the calculation of leakage fluxes and leakage flux linkages.

4.2.1 Division of Leakage Flux Components

The calculation of leakage inductances from the structural dimensions of the machine is a rather demanding task. Some of the leakage inductances are real ones created by a real leakage flux. Imaginary leakage inductances are employed to correct inaccuracies resulting from simplifications made in the calculation. There is often also a physical explanation to imaginary leakage inductances. Next, we discuss the division of leakage fluxes into components crossing and not crossing the air gap.

4.2.1.1 Leakage Fluxes Not Crossing an Air Gap

The flux components that do not cross the air gap self-evidently belong to the leakage flux. Among these components we have:

- slot leakage flux (u);
- tooth tip leakage flux (d);
- end winding leakage flux (w);
- pole leakage flux (p).

Figures 4.5–4.7 illustrate the principles of the main paths of these leakage fluxes.

To determine the leakage flux, we have to define the magnetic field strength on a leakage flux path, or the current linkage and the permeance of the path. Based on the current linkage Θ_σ creating a leakage flux, we may write

$$\Phi_\sigma = \Lambda_\sigma V_\sigma. \tag{4.16}$$

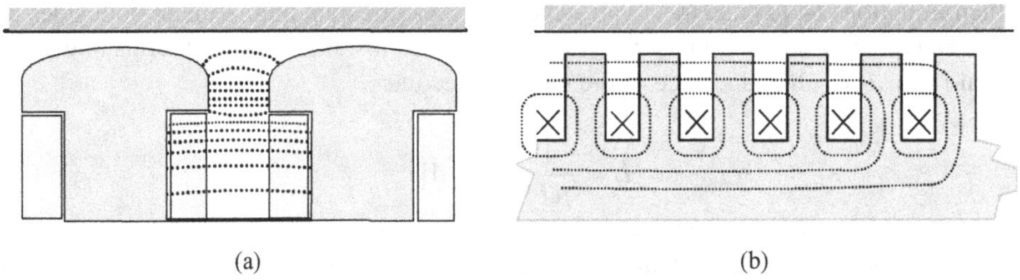

Figure 4.5 (a) Paths of pole leakage fluxes of a salient-pole magnetizing and (b) paths of pole and slot leakage of a nonsalient-pole magnetizing; see also Figure 4.20.

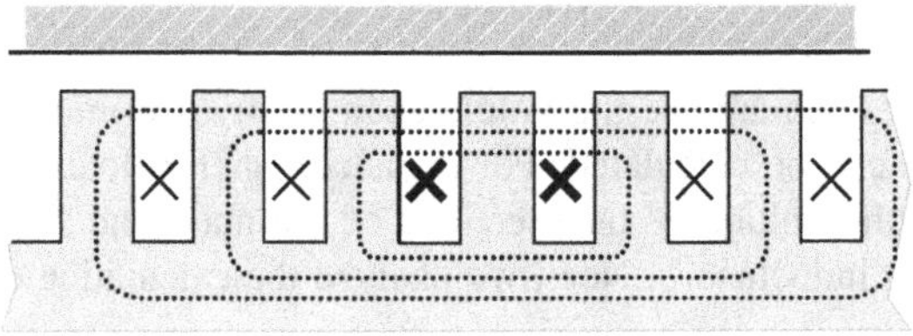

Figure 4.6 Path of the slot leakage flux of a slot winding.

Analytic calculation of leakage fluxes may be difficult, since the geometries are often quite complicated. A three-dimensional geometry of the end windings is particularly difficult to handle in the calculation of leakage fluxes. Therefore, some assisting empirical methods are utilized in the calculation.

4.2.1.2 Leakage Fluxes Crossing an Air Gap

Leakage fluxes that cross the air gap are included in the air-gap flux. An air-gap flux does not completely link the windings of the machine to each other. The reasons for incomplete

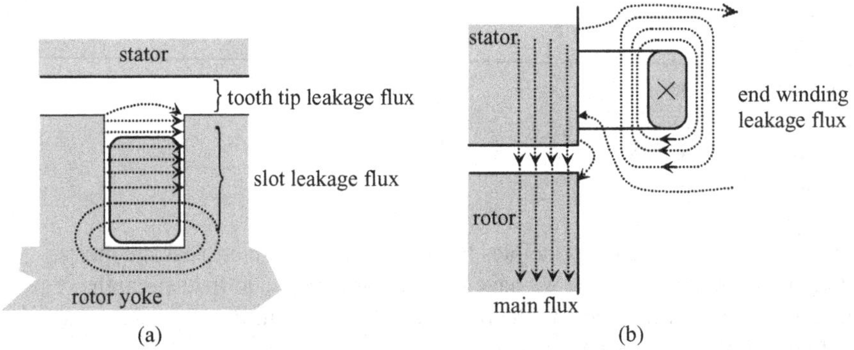

Figure 4.7 Leakage flux components of a coil winding contained in slots and end winding leakage flux, (a) an axial cut-away view of a slot, (b) a side cut-away view of the end winding area.

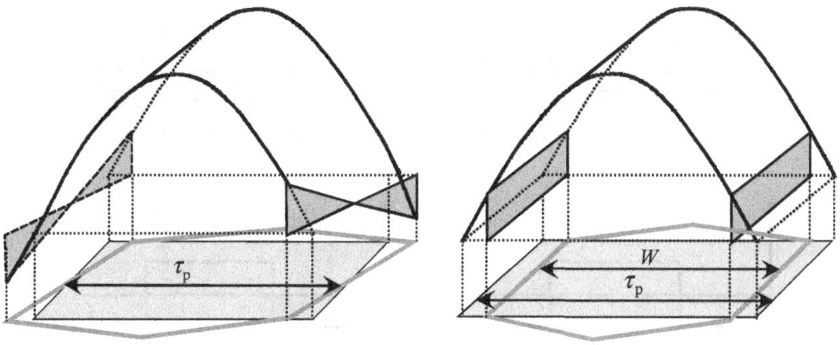

Figure 4.8 As a result of skewing in (a), the coil span W is the same as the pole pitch τ_p but the slots are skewed. As a result of short pitching in (b), the coil span W is smaller than the pole pitch τ_p, and in both cases a part of the air-gap flux remains unlinked to some of the windings of a pole pitch. Short pitching is, however, not regarded as a leakage, but is taken into account in the winding factor. It also has an effect on the harmonics, and consequently on the air-gap inductance.

linking are: short pitching, skewing and the spatial distribution of the windings causing air-gap harmonics that do not participate in the electro-mechanical energy conversion. A weakening of the linking between the stator and rotor windings, caused by short pitching or skewing, Figure 4.8, is not usually regarded as leakage, but it is taken into account with the pitch winding factor k_p and the skewing factor k_{sq} (Chapter 4.3.1). The spatial distribution of the windings causes an air-gap (or harmonic) flux leakage. These leakage flux components crossing the air gap are included in the air-gap flux Φ_m, and they do not separately influence the magnetic potential difference (magnetic voltage U_m) of the path of the main flux, calculated with the air-gap flux.

To analyze the harmonic flux leakage, the simple example in Figure 4.9 is investigated. First, it is assumed that the windings are full-pitch windings (no short pitching) and there is no skewing. Further, it is assumed that no flux leakage occurs in the slots, teeth tips or end windings.

In Figure 4.9a, the current linkage of the stator winding happens to be fully compensated, because a corresponding current flows at an aligned position in the rotor winding. In the latter section (Figure 4.9b) instead, the rotor has moved – as a result of rotation – to a new position and the original rotor current has been divided into two bars. Now, no full compensation takes place, but a small leakage flux penetrates the air gap. When the rotor is rotated further, a position with full compensation is again reached. In practice, there is always some air-gap flux leakage present. The importance of the air-gap leakage is highest in machines with a relatively small air gap, particularly in induction machines and in machines having heavily distorted air-gap flux density distributions. Especially, tooth-coil permanent magnet machines with $q \leq 0.5$ have, in general, the highest air-gap leakages. The air-gap leakage inductance can even be the dominating component in the synchronous inductance of tooth-coil machines. As this is a special feature of permanent magnet machines the topic will be further studied in Chapter 7 together with permanent magnet synchronous machines.

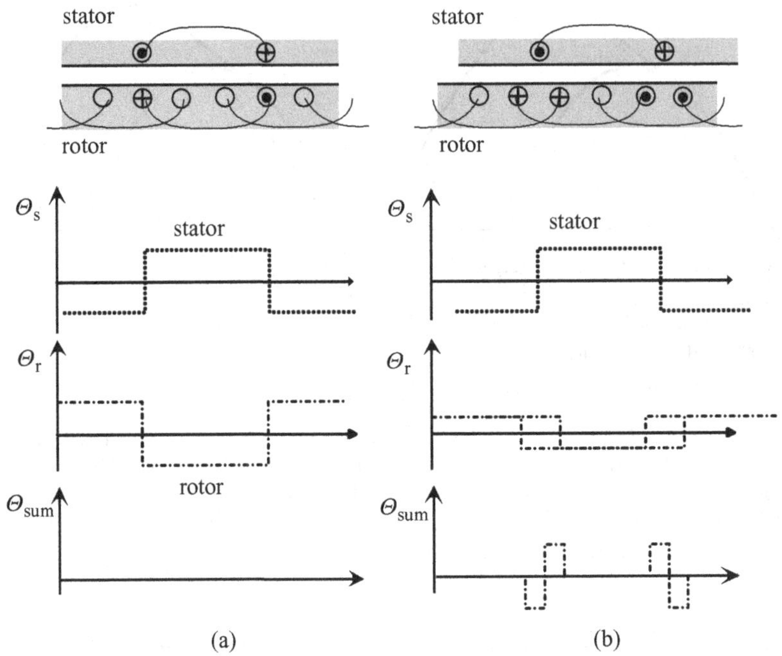

Figure 4.9 Occurrence of an air-gap flux leakage as a result of the spatial distribution of stator and rotor windings in different positions (a) and (b). When the current-carrying parts of the windings are in an aligned position, the resulting sum current linkage is zero, otherwise it deviates from zero.

In basic machine design, only the fundamental component of the main flux linkage is usually employed. In that case, all the harmonic components are included in the harmonic flux linkage. Since the harmonic fields result from the spatial distribution of the winding, a harmonic flux leakage component occurs instead of the average air-gap leakage. The term "air-gap flux leakage" is often employed, and therefore the air-gap inductance L_δ is introduced. It is important to note that the harmonic fields induce voltages of the fundamental frequency in the windings that produce these harmonic fields.

4.3 Calculation of Flux Leakage

Based on the discussion above, the leakage inductance L_σ of a machine can be calculated as the sum of different leakage inductances. Nowadays, when an electrical machine is more and more seldom fed with sinusoidal voltage, to be precise, the term "leakage inductance" should be applied instead of "leakage reactance." According to the electrical motor design tradition, leakage inductance L_σ can be divided into the following partial leakage inductances:

- skew leakage inductance L_{sq}.
- air-gap leakage inductance L_δ,
- slot leakage inductance L_u,

- tooth tip leakage inductance L_d,
- end winding leakage inductance L_w.

The leakage inductance of the machine is the sum of these leakage inductances

$$L_\sigma = L_{sq} + L_\delta + L_u + L_d + L_w. \tag{4.17}$$

4.3.1 Skewing Factor and Skew Leakage Inductance

In rotating-field machines, especially in squirrel cage induction motors, the stator and rotor slots are often assembled in a skewed position with respect to each other in order to reduce the influences of slot harmonics caused by slots. Usually the slots of a stator stack are straight and the slots of the rotor are skewed, as it is shown in Figure 4.10. However, it is of course possible to skew the stator slots instead. We can always choose the stator slots as the reference for which the direction of the rotor slots are considered independent, if the rotor or the stator or both are skewed toward the shaft line of the machine. In Figure 4.10, the orientation of the rotor slots is defined with a peripheral angle α (the angle indicating the peripheral position difference of the ends of the rotor bar). This angle is usually of the same magnitude as one stator slot angle. For the same reason, in synchronous machines, the axial sides of the pole shoes can be similarly skewed. In that case, the damping bars fitted to the rotor surface will also be skewed similarly as the cage winding. In low-speed permanent magnet machines with a low number of slots per pole and phase, either the stator or the permanent magnets are often skewed.

The objective of skewing is to reduce the ability of certain air-gap harmonics, created by the stator, to induce respective voltages in the rotor bars. Correspondingly, the induction effect of a rotor bar on the stator field is weaker than in the case of a straight bar. Both of these effects can be taken into account as follows: the emf calculated for straight slots and induced in the rotor is multiplied by the skewing factor of the rotor, and the magnetic main flux created by the rotor current in the air gap is reduced by the same factor $k_{sq\nu}$. The skewing factor thus

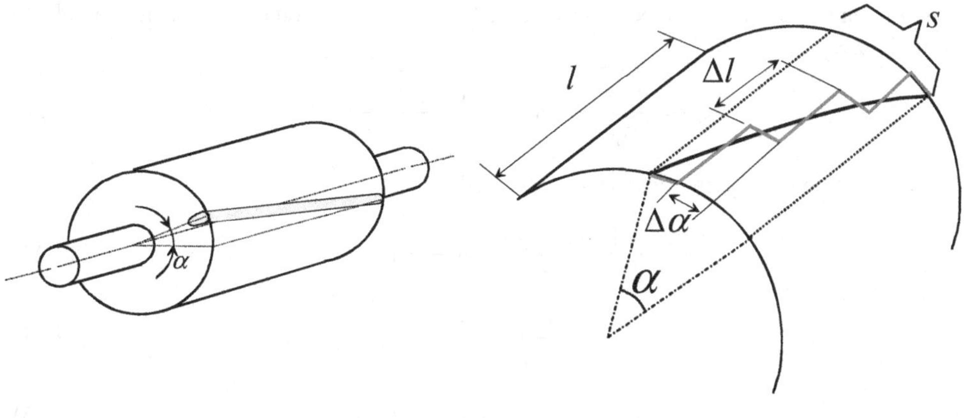

Figure 4.10 Determination of the skew of a rotor slot.

belongs to the winding factors, and can be taken into account as such in the calculations. The value of the skewing factor is derived with Equation (2.24) of the distribution factor $k_{\mathrm{d}\nu}$

$$k_{\mathrm{d}\nu} = \frac{\sin(\nu q \alpha_{\mathrm{u}}/2)}{q \sin(\nu \alpha_{\mathrm{u}}/2)}. \tag{4.18}$$

A skewed bar is considered to be composed of a large number of short, straight bars. The number of these bars is

$$z_1 = \frac{l}{\Delta l} = \frac{\alpha}{\Delta \alpha}. \tag{4.19}$$

The slot angle between these bars is $\Delta\alpha$. This is also the phase shift angle between the emfs of the bars. The emf of a skewed bar can be calculated in the same way as the emf of a coil group divided into q slots. The value $\alpha/\Delta\alpha$ is substituted in Equation (4.18) as a value of q, and a limit value is calculated when $\Delta\alpha$ approaches zero. We now obtain a skewing factor for the νth harmonic

$$k_{\mathrm{sq}\nu} = \lim_{\Delta\alpha \to 0} \frac{\sin\left(\dfrac{\alpha}{\Delta\alpha}\nu\dfrac{\Delta\alpha}{2}\right)}{\dfrac{\alpha}{\Delta\alpha}\sin\left(\nu\dfrac{\Delta\alpha}{2}\right)} = \frac{\sin(\nu\alpha/2)}{\dfrac{\alpha}{\Delta\alpha}\nu\dfrac{\Delta\alpha}{2}} = \frac{\sin(\nu\alpha/2)}{\nu\alpha/2}, \tag{4.20}$$

$$k_{\mathrm{sq}\nu} = \frac{\sin\left(\nu\dfrac{s}{\tau_{\mathrm{p}}}\dfrac{\pi}{2}\right)}{\nu\dfrac{s}{\tau_{\mathrm{p}}}\dfrac{\pi}{2}} = \frac{\sin\left(\nu\dfrac{\pi}{2}\dfrac{s_{\mathrm{sp}}}{mq}\right)}{\nu\dfrac{s}{\tau_{\mathrm{p}}}\dfrac{s_{\mathrm{sp}}}{mq}}. \tag{4.21}$$

Here s is the skewing (Figure 4.10) measured as an arc length, given by $\alpha = s\pi/\tau_{\mathrm{p}}$ and as the number of slot pitches (s_{sp}), given by $\alpha = s_{\mathrm{sp}}\pi/(mq)$ (the pole pitch τ_{p} as the number of slot pitches $= mq$). The skewing factor is applied similarly as a winding factor. For instance, when referring the rotor quantities to the stator, the number of coil turns in a rotor has to be multiplied by the winding factor of the rotor and with the skewing factor.

To eliminate the slot harmonics, the factor $k_{\mathrm{sq}\nu}$ should be zero. The orders of the slot harmonics are $(\pm 2mqc + 1)$, c being an integer $1, 2, 3, \ldots$. The numerator of Equation (4.21) should be zero:

$$\sin\frac{\nu s\pi}{2\tau_{\mathrm{p}}} = \sin\frac{(\pm 2mqc + 1)s\pi}{2\tau_{\mathrm{p}}} = 0. \tag{4.22}$$

This will be true if the argument of the sine function is $k\pi$:

$$\frac{(\pm 2mqc + 1)s\pi}{2\tau_{\mathrm{p}}} = k\pi, \tag{4.23}$$

$$\frac{(\pm 2mqc + 1)s}{2\tau_{\mathrm{p}}} = k, \tag{4.24}$$

$$\frac{s}{\tau_{\mathrm{p}}} = k\frac{2}{\pm 2mqc + 1} \approx \pm k\frac{1}{mqc}. \tag{4.25}$$

This means that for $c = k = 1$

$$\frac{s}{\tau_{\mathrm{p}}} \cong \frac{1}{mq} = \frac{1}{Q/2p} \Rightarrow s = \frac{\tau_{\mathrm{p}}}{Q/2p} = \tau_{\mathrm{u}}. \tag{4.26}$$

Hence, the skewing factor will be effective in canceling the slot harmonics if the skewing is made by one slot pitch.

In the case $s = \tau_{\mathrm{u}}$ the skewing factor is

$$k_{\mathrm{sq}\nu} = \frac{\sin\left(\nu \dfrac{s}{\tau_{\mathrm{p}}} \dfrac{\pi}{2}\right)}{\nu \dfrac{s}{\tau_{\mathrm{p}}} \dfrac{\pi}{2}} = \frac{\sin\left(\nu \dfrac{\pi}{2} \dfrac{1}{mq}\right)}{\nu \dfrac{\pi}{2} \dfrac{1}{mq}}. \tag{4.27}$$

Example 4.1: Calculate the skewing factors for the first stator slot harmonics in a four-pole squirrel cage motor, where the stator has 36 slots and the rotor bars are skewed by one stator slot pitch.

Solution: According to Equation (2.56), a m-phase winding creates harmonics $\nu = 1 \pm 2cm$, where $c = 0, 1, 2, 3, \ldots$ The first slot harmonics are $(1 \pm 2mqc) = 1 \pm 2 \times 3 \times 3 \times c = -17, 19, -35, 37$, when $c = 1, 2$. The skewing factor was defined by (4.21) as

$$k_{\mathrm{sq}\nu} = \frac{\sin\left(\nu \dfrac{\pi}{2} \dfrac{s_{\mathrm{sp}}}{mq}\right)}{\nu \dfrac{\pi}{2} \dfrac{s_{\mathrm{sp}}}{mq}}.$$

Substituting skewing by one stator slot pitch $s_{\mathrm{sp}} = 1$ and $mq = Q_{\mathrm{s}}/(2p) = 36/4 = 9$ we obtain

ν	1	−5	7	−11	13	−17	19	−23	25	−29	31	−35	37
$k_{\mathrm{sq}\nu}$	0.995	0.878	0.769	0.490	0.338	0.06	−0.05	−0.19	−0.22	−0.19	−0.14	−0.03	0.03

As we can see, especially the lowest order slot harmonics −17, 19, −35 and 37 have small skewing factors, and thereby their effects are eliminated to a great degree. The fundamental is reduced only by 0.5 %.

When a winding is assembled and skewed with respect to another winding, the magnetic connection between the windings is weakened. Now, a part of the flux created by the stator winding does not penetrate a skewed rotor winding, although it passes through the air gap. Thus, this part belongs to the leakage flux and can be described by the skew inductance L_{sq}.

To understand how skewing influences the machine characteristics, let us write the voltage equations of an induction machine without and with skewing. As the induction machine properties will be observed later in Chapter 7 it is advisable to study the equivalent circuit of the induction machine first. An induction machine can be described by a simple equivalent circuit containing stator and rotor leakage inductances ($L_{s\sigma}$, $L_{r\sigma}$) and the magnetizing inductance L_m. The leakage inductances are excited by the stator or rotor currents alone but in the magnetizing inductance there flow the sum of the stator and rotor currents. Without skewing the stator and rotor voltage equations are

$$\begin{aligned} \underline{U}_s &= (R_s + j\omega L_{s\sigma})\underline{I}_s + j\omega L_m\left(\underline{I}_s + \underline{I}'_r\right), \\ \underline{U}'_r &= \left(\frac{R'_r}{s} + j\omega L_{r\sigma}\right)\underline{I}'_r + j\omega L_m\left(\underline{I}_s + \underline{I}'_r\right), \end{aligned} \tag{4.28}$$

where the rotor variables are referred to the stator. In squirrel cage motors, the rotor voltage $\underline{U}'_r$ is zero. Skewing influences only the mutual coupling between the stator and the rotor, and therefore the mutual coupling terms $j\omega L_m\underline{I}'_r$ in the stator voltage equation and $j\omega L_m\underline{I}_s$in the rotor voltage equation have to be multiplied by the skewing factor k_{sq}; Now, the voltage equations with skewing are

$$\begin{aligned} \underline{U}_s &= (R_s + j\omega L_{s\sigma})\underline{I}_s + j\omega L_m\underline{I}_s + jk_{sq}\omega L_m\underline{I}'_r, \\ \underline{U}'_r &= \left(\frac{R'_r}{s} + j\omega L'_{r\sigma}\right)\underline{I}'_r + jk_{sq}\omega L_m\underline{I}_s + j\omega L_m\underline{I}'_r. \end{aligned} \tag{4.29}$$

Equations (4.29) can be rearranged to

$$\begin{aligned} \underline{U}_s &= (R_s + j\omega L_{s\sigma})\underline{I}_s + j\omega L_m\underline{I}_s + j\omega L_m k_{sq}\underline{I}'_r, \\ \frac{\underline{U}'_r}{k_{sq}} &= \frac{1}{k_{sq}^2}\left(\frac{R'_r}{s} + j\omega L'_{r\sigma}\right)k_{sq}\underline{I}'_r + j\omega L_m\underline{I}_s + \frac{j}{k_{sq}^2}\omega L_m k_{sq}\underline{I}'_r \end{aligned} \tag{4.30}$$

which, further, can be rearranged to

$$\begin{aligned} \underline{U}_s &= (R_s + j\omega L_{s\sigma})\underline{I}_s + j\omega L_m\underline{I}_s + j\omega L_m k_{sq}\underline{I}'_r, \\ \frac{\underline{U}'_r}{k_{sq}} &= \frac{1}{k_{sq}^2}\left(\frac{R'_r}{s} + j\omega L'_{r\sigma}\right)k_{sq}\underline{I}'_r + j\left(\frac{1}{k_{sq}^2} - 1\right)\omega L_m k_{sq}\underline{I}'_r + j\omega L_m\left(\underline{I}_s + k_{sq}\underline{I}'_r\right) \end{aligned} \tag{4.31}$$

From the rotor voltage equation, we see that skewing results in an additional rotor leakage inductance referred to the stator

$$L_{sq} = \frac{\left(1 - k_{sq}^2\right)}{k_{sq}^2}L_m = \sigma_{sq}L_m. \tag{4.32}$$

where

$$\sigma_{sk} = \frac{1 - k_{sq}^2}{k_{sq}^2} \tag{4.33}$$

is the leakage factor of skewing. Further we see that the skewed rotor voltage referred to the stator is

$$\underline{U}'_{\mathrm{r,sq}} = \frac{\underline{U}'_{\mathrm{r}}}{k_{\mathrm{sq}}} = \frac{1}{k_{\mathrm{sq}}} \frac{m_{\mathrm{s}} k_{\mathrm{ws}} N_{\mathrm{s}}}{m_{\mathrm{r}} k_{\mathrm{wr}} N_{\mathrm{r}}} \underline{U}_{\mathrm{r}} = \frac{m_{\mathrm{s}} k_{\mathrm{ws}} N_{\mathrm{s}}}{m_{\mathrm{r}} k_{\mathrm{sq}} k_{\mathrm{wr}} N_{\mathrm{r}}} \underline{U}_{\mathrm{r}}, \tag{4.34}$$

and the skewed rotor current referred to the stator is

$$\underline{I}'_{\mathrm{r,sq}} = k_{\mathrm{sq}} \underline{I}'_{\mathrm{r}} = k_{\mathrm{sq}} \frac{m_{\mathrm{r}} k_{\mathrm{wr}} N_{\mathrm{r}}}{m_{\mathrm{s}} k_{\mathrm{ws}} N_{\mathrm{s}}} I_{\mathrm{r}} = \frac{m_{\mathrm{r}} k_{\mathrm{sq}} k_{\mathrm{wr}} N_{\mathrm{r}}}{m_{\mathrm{s}} k_{\mathrm{ws}} N_{\mathrm{s}}} I_{\mathrm{r}}. \tag{4.35}$$

The skewed rotor resistance and leakage inductance referred to the stator are

$$R'_{\mathrm{r,sq}} = \frac{1}{k_{\mathrm{sq}}^2} \frac{m_{\mathrm{s}}}{m_{\mathrm{r}}} \left(\frac{k_{\mathrm{ws}} N_{\mathrm{s}}}{k_{\mathrm{wr}} N_{\mathrm{r}}} \right)^2 R_{\mathrm{r}} = \frac{m_{\mathrm{s}}}{m_{\mathrm{r}}} \left(\frac{k_{\mathrm{ws}} N_{\mathrm{s}}}{k_{\mathrm{sq}} k_{\mathrm{wr}} N_{\mathrm{r}}} \right)^2 R_{\mathrm{r}}, \tag{4.36}$$

$$L'_{\sigma\mathrm{r,sq}} = \frac{1}{k_{\mathrm{sq}}^2} \frac{m_{\mathrm{s}}}{m_{\mathrm{r}}} \left(\frac{k_{\mathrm{ws}} N_{\mathrm{s}}}{k_{\mathrm{wr}} N_{\mathrm{r}}} \right)^2 L_{\sigma\mathrm{r}} = \frac{m_{\mathrm{s}}}{m_{\mathrm{r}}} \left(\frac{k_{\mathrm{ws}} N_{\mathrm{s}}}{k_{\mathrm{sq}} k_{\mathrm{wr}} N_{\mathrm{r}}} \right)^2 L_{\sigma\mathrm{r}}. \tag{4.37}$$

Based on this it is possible to understand that the skewing factor can be considered as part of the rotor winding factor.

In synchronous and permanent magnet machines the rotor poles or permanent magnets can be installed in a skewed position while the stator is kept nonskewed or vice versa. This influences the emf $\underline{E}_{\mathrm{f}}$ induced by the flux linkage created by the field winding. The stator voltage equation for a salient pole synchronous machine with skewing is

$$\underline{U}_{\mathrm{s}} = k_{\mathrm{sq}} \underline{E}_{\mathrm{f}} - (R_{\mathrm{s}} + \mathrm{j}\omega L_{\mathrm{s}\sigma}) \underline{I}_{\mathrm{s}} - \mathrm{j}\omega L_{\mathrm{m}} \underline{I}_{\mathrm{s}}. \tag{4.38}$$

4.3.2 Air-Gap Leakage Inductance

According to Faraday's induction law, the air-gap back emf of an electrical machine is induced in the magnetizing inductance L_{m} of the machine as a result of a propagating fundamental component of air-gap flux density. Because of a spatial slotting, the permeance harmonics induce emf components in the fundamental frequency of the winding. The air-gap leakage inductance (i.e. the harmonic leakage inductance components) takes this into account. In integer slot machines the air-gap leakage remains usually low. However, in fractional slot machines, and especially in machines with tooth coils, its influence can be even dominating. This will be studied in more detail with tooth-coil permanent magnet machine in Chapter 7.

The emf E_ν induced by the harmonic ν of the flux density per phase is obtained by applying Faraday's induction law and by taking the geometry of the machine into account

$$E_\nu = \frac{1}{\sqrt{2}} \omega N k_{\mathrm{w}\nu} \hat{\Phi}_\nu. \tag{4.39}$$

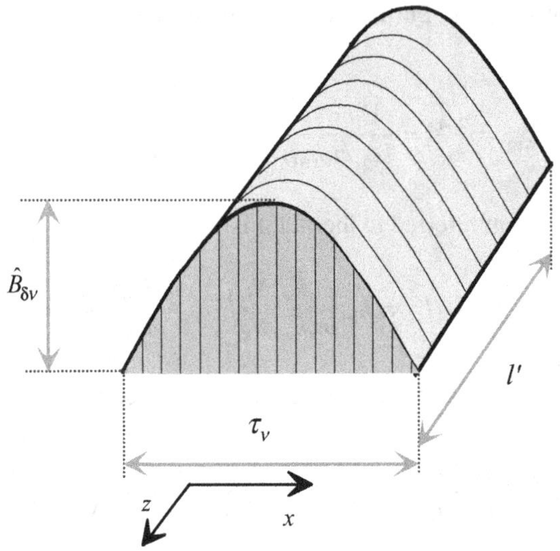

Figure 4.11 Flux distribution created by a harmonic ν over a pole pitch τ_ν corresponding to the harmonic. The maximum flux value is calculated as an integral over the pole.

Next, a sinusoidal harmonic ν is investigated. For this kind of flux, a maximum flux value penetrating the coil is obtained according to Figure 4.11

$$\hat{\Phi}_\nu = \frac{2}{\pi}\hat{B}_{\delta\nu}\tau_\nu l' \tag{4.40}$$

The pole pitch of the harmonic ν is

$$\tau_\nu = \frac{\pi D}{2p\nu} \tag{4.41}$$

Substituting Equations (4.40) and (4.41) into Equation (4.39) we obtain

$$E_\nu = \frac{\omega}{\sqrt{2}}Dl'\frac{N}{p}\frac{k_{\mathrm{w}\nu}}{\nu}\hat{B}_{\delta\nu} \tag{4.42}$$

In a machine with m phases, the magnetizing current I_{m} flowing in the windings creates a peak flux density in the air gap

$$\hat{B}_{\delta\nu} = \frac{\mu_0}{\pi}\frac{m}{\delta}\frac{k_{\mathrm{w}\nu}}{\nu}\frac{N}{p}\sqrt{2}I_{\mathrm{m}} \tag{4.43}$$

By substituting this into Equation (4.42) we obtain

$$E_\nu = \frac{\mu_0}{\pi}\omega\frac{m}{\delta}Dl'\left(\frac{N}{p}\right)^2 I_m\left(\frac{k_{w\nu}}{\nu}\right)^2 \tag{4.44}$$

The sum of the emfs induced by all the harmonics, the fundamental included, is

$$E = \sum_{\nu=-\infty}^{\nu=+\infty} E_\nu. \tag{4.45}$$

The inductance $E/\omega I_m$ is the sum of the magnetizing inductance and the air-gap inductance

$$\frac{E}{\omega I_m} = L_m + L_\delta = \frac{\mu_0}{\pi}\frac{m}{\delta}Dl'\left(\frac{N}{p}\right)^2 \sum_{\nu=-\infty}^{\nu=+\infty}\left(\frac{k_{w\nu}}{\nu}\right)^2. \tag{4.46}$$

The term $\nu = 1$ in the sum equation represents the fundamental component, and thus the magnetizing inductance L_m of the machine, which is calculated from Equation (4.10) substituting for $\alpha_i = 2/\pi$. The remainder of the equation represents the air-gap inductance

$$L_\delta = \frac{\mu_0}{\pi}\frac{m}{\delta}Dl'\left(\frac{N}{p}\right)^2 \sum_{\substack{\nu=-\infty \\ \nu\neq 1}}^{\nu=+\infty}\left(\frac{k_{w\nu}}{\nu}\right)^2. \tag{4.47}$$

The air-gap or the harmonic inductance can also be written in the form

$$L_\delta = \sigma_\delta L_m. \tag{4.48}$$

Here we have the leakage factor σ_δ of the air-gap inductance, defined as

$$\sigma_\delta = \sum_{\substack{\nu=-\infty \\ \nu\neq 1}}^{\nu=+\infty}\left(\frac{k_{w\nu}}{\nu k_{w1}}\right)^2. \tag{4.49}$$

Naturally, in the sum equation, only the harmonics created by the winding occur as addends.

For a three-phase, short-pitch winding, the leakage factor σ_δ can be presented in a closed form (Baffrey 1926)

$$\sigma_\delta = \frac{2\pi^2}{9k_{w1}^2}\frac{5q^2 + 1 + \varepsilon_{sp}^3/(4q) - 3\varepsilon_{sp}^2/2 - \varepsilon_{sp}/(4q)}{12q^2} - 1 \tag{4.50}$$

where ε_{sp} is the difference of the pole-pitch (mq) and the coil width (y) expressed as slot pitches

$$\varepsilon_{sp} = mq - y. \tag{4.51}$$

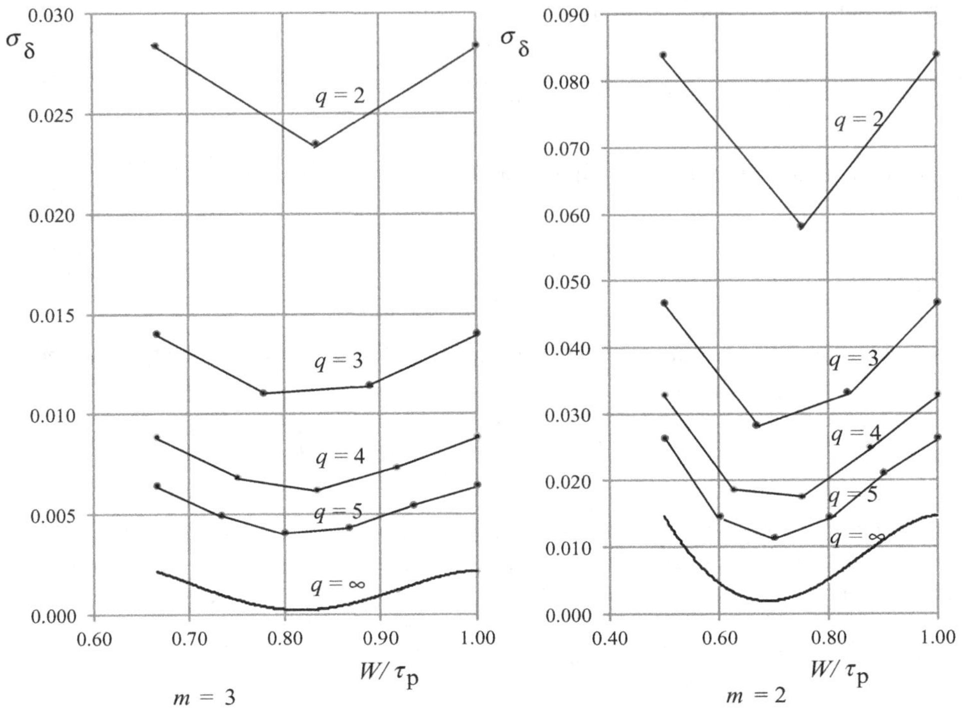

Figure 4.12 Leakage factors σ_δ of three-phase ($m = 3$) and two-phase ($m = 2$) windings with a parameter q (slots per pole and phase), as a function of coil span. It is shown that an increase in q considerably improves the characteristics of the winding. A three-phase winding is notably better with respect to harmonics than a two-phase winding. For $m = 3$ and $q = 1$, $\sigma_\delta = 0.0966$ and for $m = 2$ and $q = 1$, $\sigma_\delta = 0.234$. They are left out of the figure because of their large value.

E.g. if a winding is short-pitched by two slot pitches, $\varepsilon_{sp} = 2$. Equation (4.50) is valid for $\varepsilon_{sp} \geq q$. For a two-phase, short-pitch winding, the leakage factor is respectively

$$\sigma_\delta = \frac{2\pi^2}{4k_{w1}^2}\frac{4q^2 + 2 + \varepsilon_{sp}^3/q - 3\varepsilon_{sp}^2 - \varepsilon_{sp}/q}{24q^2} - 1. \tag{4.52}$$

The leakage factors according to Equations (4.50) and (4.52) are presented in Figure 4.12 as a function of short pitching, with the number of slots per pole and phase q as a parameter. With three-phase windings, the leakage factor is at its minimum when the short pitching is $W/\tau_p \approx 5/6$, which is generally an advantageous short pitching when designing windings of three-phase machines.

Equation (4.48) is valid for a stator winding. When calculating the magnetizing inductance on the rotor side, we obtain it by referring it to the stator $L'_{mr} = L_m$. Thus, the air-gap inductance of a rotor circuit referred to the stator may be written applying a rotor leakage factor

$$L'_{\delta r} = \sigma_{\delta r} L_m. \tag{4.53}$$

For the air-gap leakage factor of a squirrel cage rotor, Richter (1954) gives an equation which takes into account also the possible skewing

$$\sigma_{\delta \mathrm{r}} = \frac{1}{k_{\mathrm{sq}}^2}\left(\frac{p\pi/Q_{\mathrm{r}}}{\sin(p\pi/Q_{\mathrm{r}})}\right)^2 - 1 \tag{4.54}$$

where k_{sq} is the skewing factor, Q_{r} the rotor slot number and p the pole pair number.

Usually, an air-gap leakage inductance is of significance only in machines with a small air gap. This is particularly the case with asynchronous machines. In asynchronous machines with a cage winding, however, the cage damps harmonics, and consequently the air-gap inductance becomes less significant. Richter (1954) gives an equation for the damping factor (Δ_2) with which the stator leakage factor $\sigma_{\mathrm{s}\delta}$ Equations (4.49) and (4.50), has to be multiplied by

$$\Delta_2 \approx 1 - \frac{1}{\sigma_{\mathrm{s}\delta}} \sum_{\nu \neq 1} \left(\frac{k_{\mathrm{w}\nu}}{\nu k_{\mathrm{w}1}} k_{\mathrm{sq}\nu} \frac{\sin\left(\dfrac{\nu\pi p}{Q_{\mathrm{r}}}\right)}{\dfrac{\nu\pi p}{Q_{\mathrm{r}}}} \right)^2 \tag{4.55}$$

where $k_{\mathrm{sq}\nu}$ is the skewing factor for the νth harmonic, Equation (4.21). The terms of the sum expression are mitigated quickly with increasing ordinal ν and therefore it is enough to calculate the sum up to the first slot harmonics ($1 < \nu \leq 2mq_{\mathrm{s}} + 1$). The damping factor is usually of magnitude of about 0.8.

Example 4.2: Calculate the per unit air-gap leakage inductance of a three-phase winding with $p = 2$ and $Q_{\mathrm{s}} = 36$ for a machine with $L_{\mathrm{m,pu}} = 3$. Short pitching by one slot is applied. Take also the damping caused by the squirrel cage rotor into account as $Q_{\mathrm{r}} = 34$ and there is no skewing.

Solution: The number of slots per pole and phase is $q = Q_{\mathrm{s}}/(2pm) = 3$. The pole-pitch $mq = 9$ slot pitches, the coil width $y = 8$ slot pitches and $\varepsilon_{\mathrm{sp}} = 1$. The winding factor of the fundamental is according to Equation (2.35)

$$k_{\mathrm{w}1} = k_{\mathrm{p}1}k_{\mathrm{d}1} = \sin\left(\nu\frac{W}{\tau_{\mathrm{p}}}\frac{\pi}{2}\right)\frac{\sin\left(\nu\dfrac{\pi}{2m}\right)}{q\sin\left(\nu\dfrac{\pi}{2mq}\right)} = \sin\left(\frac{8}{9}\frac{\pi}{2}\right)\frac{\sin\left(\dfrac{\pi}{2\cdot 3}\right)}{3\sin\left(\dfrac{\pi}{2\cdot 3\cdot 3}\right)} = 0.9452.$$

The leakage factor $\sigma_{\mathrm{s}\delta}$ of the stator winding without damping is according to Equation (4.50)

$$\sigma_{\mathrm{s}\delta} = \frac{2\pi^2}{9k_{\mathrm{w}1}^2}\frac{5q^2 + 1 + \varepsilon_{\mathrm{sp}}^3/(4q) - 3\varepsilon_{\mathrm{sp}}^2/2 - \varepsilon_{\mathrm{sp}}/(4q)}{12q^2} - 1$$

$$= \frac{2\pi^2}{9\cdot 0.9452^2}\frac{5\cdot 3^2 + 1 + 1^3/(4\cdot 3) - 3\cdot 1^2/2 - 1/(4\cdot 3)}{12\cdot 3^2} - 1 = 0.0115$$

The damping factor is according to Equation (4.55)

$$\Delta_2 \approx 1 - \frac{1}{\sigma_{s\delta}} \sum_{\nu \neq 1} \left(\frac{k_{w\nu}}{\nu k_{w1}} k_{sq\nu} \frac{\sin\left(\frac{\nu \pi p}{Q_r}\right)}{\frac{\nu \pi p}{Q_r}} \right)^2 .$$

Without skewing, $k_{sq\nu} = 1$ for all ν. For $1 < \nu \leq 2mq_s + 1$ we obtain

ν	1	−5	7	−11	13	−17	19
$k_{w\nu}$	0.9452	−0.13985	0.0606617	0.0606617	−0.13985	0.9452136	0.9452136
$\frac{k_{w\nu}}{\nu k_{w1}} k_{sq\nu} \frac{\sin(\nu\pi p/Q_r)}{\nu\pi p/Q_r}$		0.0255566	0.0068169	−0.002569	−0.003192	−2.299E-18	−0.005415
$\left(\frac{k_{w\nu}}{\nu k_{w1}} k_{sq\nu} \frac{\sin(\nu\pi p/Q_r)}{\nu\pi p/Q_r}\right)^2$		6.53E-04	4.647E-05	6.601E-06	1.019E-05	5.262E-36	2.932E-05

and the damping factor is

$$\Delta_2 \approx 1 - \frac{1}{\sigma_{s\delta}} \sum_{\nu \neq 1} \left(\frac{k_{w\nu}}{\nu k_{w1}} k_{sq\nu} \frac{\sin\left(\frac{\nu \pi p}{Q_r}\right)}{\frac{\nu \pi p}{Q_r}} \right)^2$$

$$= 1 - \frac{1}{0.0115}(6.53 + 0.467 + 0.0066 + 0.102 + 0.0 + 0.293) \cdot 10^{-4} = 0.935$$

As the per unit magnetizing inductance of an induction motor is $L_{m,pu} = 3$, we get for the stator leakage inductance per unit value

$$L_{s\sigma,pu} = \Delta_2 \sigma_{s\delta} L_{m,pu} = 0.935 \cdot 0.0115 \cdot 3 = 0.032.$$

In such a machine, the per unit air-gap leakage inductance is thus 3.2 %.

4.3.3 Slot Leakage Inductance

Slot leakage inductance is an inductance created by a real leakage flux. The total current in a slot is determined by the number of conductors z_Q in the slot and by the current flowing in them. The total current of a slot is thus $z_Q I$. Next, the leakage flux of a slot is investigated with Figure 4.13.

A flux carried by a duct of a height dh is analyzed. First, it is assumed that the current flowing in the slot is evenly distributed in the hatched area. When traveling from the bottom of the conductor to a height h, a current linkage $\Theta = I z_Q h/h_4$ is accumulated. At height h, there is thus a leakage flux density flowing across an area $\mathrm{d}S = \mathrm{d}hl'$ of the duct

$$B(h) = \mu_0 H(h) = \mu_0 \frac{z_Q I \frac{h}{h_4}}{b_4}. \tag{4.56}$$

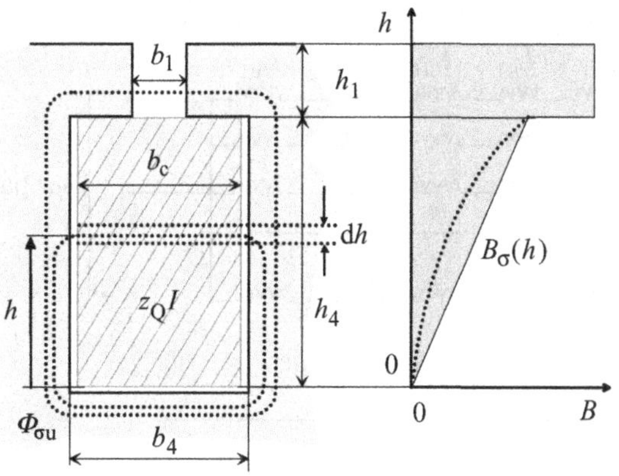

Figure 4.13 Leakage flux of a current-carrying slot of an electrical machine and the density of a leakage flux. Left, conductors in the slot. The current-carrying area is depicted by the current sum $z_Q I$ as a hatched area. Right, the slot leakage flux density B_σ distribution as a function of position h; the solid line depicts the situation with no skin effect in the winding, and the dotted line indicates the instant when the skin effect occurs.

In Figure 4.13, this function is illustrated with a solid line. By substituting $B(h)$ and the volume element $dV = l'b_4 dh$ in Equation (4.14), we obtain, using the energy stored in the field, the inductance

$$L_{u1} = \frac{l'b_4}{\mu_0 I^2} \int_0^{h_4} B^2(h)dh = \mu_0 l' z_Q^2 \frac{h_4}{3b_4} = z_Q^2 \Lambda. \tag{4.57}$$

Here

$$\Lambda = \mu_0 l' \frac{h_4}{3b_4} \tag{4.58}$$

is the magnetic permeance of the slot for the slot leakage. Next, a permeance factor λ_4 defined as magnetic permeance Λ divided by the slot length l' and permeability μ_0 is introduced. Now

$$\lambda_4 = \frac{h_4}{3b_4}. \tag{4.59}$$

$B(h)$ is constant in the area h_1, since the current sum influencing the area does not increase any further, because the current-carrying area has already been passed: $B = \mu_0 z_Q I/b_1$. Therefore, we obtain for this region

$$\lambda_1 = \frac{h_1}{b_1}. \tag{4.60}$$

The total magnetic energy stored in the slot is

$$W_{\Phi u} = 1/2 L_{u1} I^2 = W_{\Phi 1} + W_{\Phi 4} = 1/2 \mu_0 l' z_Q^2 I^2 (\lambda_1 + \lambda_4). \tag{4.61}$$

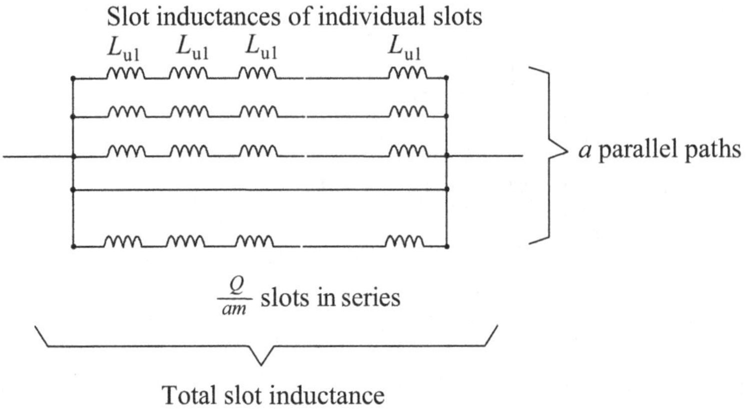

Figure 4.14 Determination of the total slot inductance with the slot inductances of single slots.

The sum $(\lambda_1 + \lambda_4) = \lambda_{\rm u}$ thus defines the leakage inductance $L_{\rm u1}$ of one slot. The slot inductance of a phase winding is obtained as follows.

If the number of parallel symmetrical paths in the winding is a (Figure 4.14), the number of slots in series in a phase winding is $Q/(am)$ and the total slot inductance of a phase winding is

$$L_{\rm u} = \frac{Q}{am}\frac{1}{a}L_{\rm u1} = \mu_0 l' \frac{Q}{m}\left(\frac{z_{\rm Q}}{a}\right)^2 \lambda_{\rm u}. \qquad (4.62)$$

Since the number of conductors in a slot is $z_{\rm Q}$, the number of conductors in series in a phase winding is $z_{\rm Q}Q/(am)$. We need two conductors to form a turn; so the number of series-connected turns N in a phase winding is

$$N = \frac{Q}{2am} z_{\rm Q}. \qquad (4.63)$$

Substituting $z_{\rm Q}$ from Equation (4.63) into Equation (4.62), the slot inductance takes the form

$$L_{\rm u} = \frac{4m}{Q}\mu_0 l' N^2 \lambda_{\rm u}. \qquad (4.64)$$

Example 4.3: Calculate the slot leakage inductance of a three-phase winding with $p = 3$, $Q = 36$ and $z_Q = 20$. The slot shape is according to Figure 4.13: $b_1 = 0.003$ m, $h_1 = 0.002$ m, $b_4 = 0.008$ m and $h_4 = 0.02$ m, $l' = 0.25$ m. There is no short pitching and there are no parallel paths.

Solution: The permeance factor of the wound part of the slot is

$$\lambda_4 = \frac{h_4}{3b_4} = \frac{0.02}{3 \cdot 0.008} = 0.833$$

and the permeance factor of the slot opening is

$$\lambda_1 = \frac{h_1}{b_1} = \frac{0.002}{0.003} = 0.667.$$

The permeance factor of the whole slot is $\lambda_u = \lambda_1 + \lambda_4 = 0.667 + 0.833 = 1.5$. The slot inductance is

$$L_u = \mu_0 l' \frac{Q}{m}\left(\frac{z_Q}{a}\right)^2 \lambda_u = 4\pi \cdot 10^{-7} \cdot 0.25 \cdot \frac{36}{3} \cdot 20^2 \cdot 1.5\,\text{H} = 2.26\,\text{mH}.$$

The slot permeance factor depends on the geometry of the slot. With the method presented above, we may derive slot permeance factors for slots of varying cross-section. Slot permeance factors can be found in the literature (see e.g. Richter 1967; Vogt 1996) calculated for the most common slot types. Figure 4.15 illustrates the different slot shapes and dimensions of single-layer windings.

The following values are obtained for the permeance factors of the single-layer windings of Figure 4.15. For slot types a, b, c, d and e, the equation

$$\lambda_u = \frac{h_4}{3b_4} + \frac{h_3}{b_4} + \frac{h_1}{b_1} + \frac{h_2}{b_4 - b_1}\ln\frac{b_4}{b_1} \tag{4.65}$$

is valid, in which

$$\frac{h_2}{b_4 - b_1}\ln\frac{b_4}{b_1} = \lambda_{u3}.$$

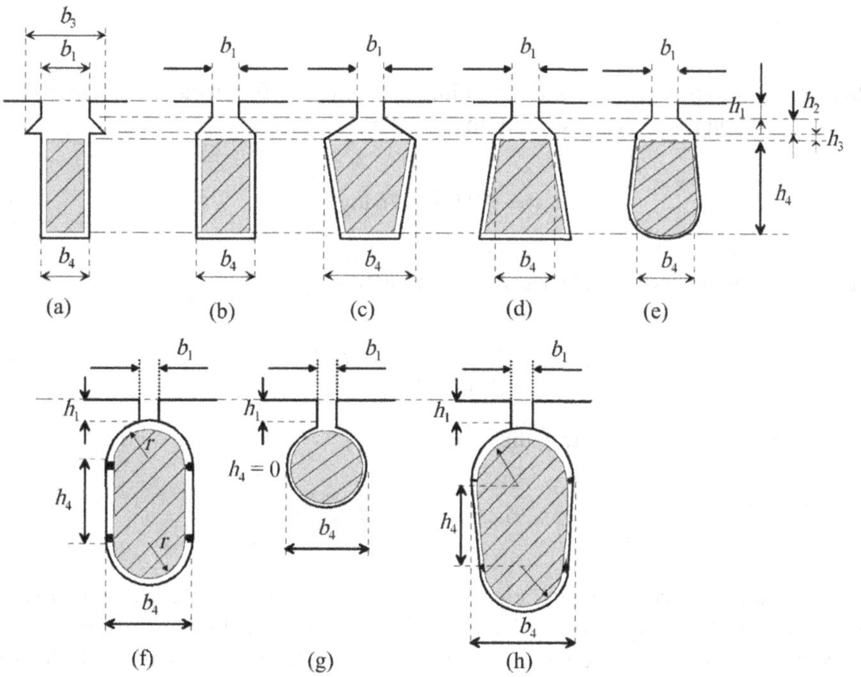

Figure 4.15 Dimensioning of different slot types for the definition of the slot permeance factor.

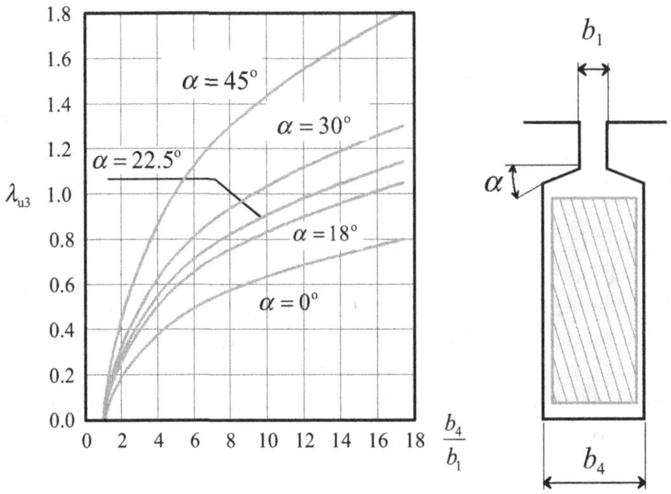

Figure 4.16 Slot permeance factor λ_{u3} as a function of the ratio of the slot width and the slot opening width. Adapted from Richter (1967).

The factor λ_{u3} can also be taken from Figure 4.16.

In the case of Figure 4.15a, b_3 has to be substituted for b_4 in the last term of Equation (4.65). Correspondingly, for slot types f, g and h, we obtain

$$\lambda_u = \frac{h_4}{3b_4} + \frac{h_1}{b_1} + 0.66. \tag{4.66}$$

The last term (0.66) is again λ_{u3}, and, according to Richter (1967), it can be calculated for the upper round section of the slot also from the equation

$$\lambda_{u3} = 0.41 + 0.76 \log \frac{b_4}{b_1}. \tag{4.67}$$

For a slot in Figure 4.15f, Richter gives an alternative equation

$$\lambda_u = \frac{h_4}{3b_4} + \frac{h_1}{b_1} + 0.685, \tag{4.68}$$

and for a round slot in Figure 4.15g, the equation

$$\lambda_u = 0.47 + 0.066\frac{b_4}{b_1} + \frac{h_1}{b_1}. \tag{4.69}$$

In the rotors of induction motors, closed rotor slots are often employed. In that case, the inductance of the slot is highly dependent on the saturation of the bridge closing the slot, and therefore no explicit permeance factor can be solved for the slot. Richter (1954) determines

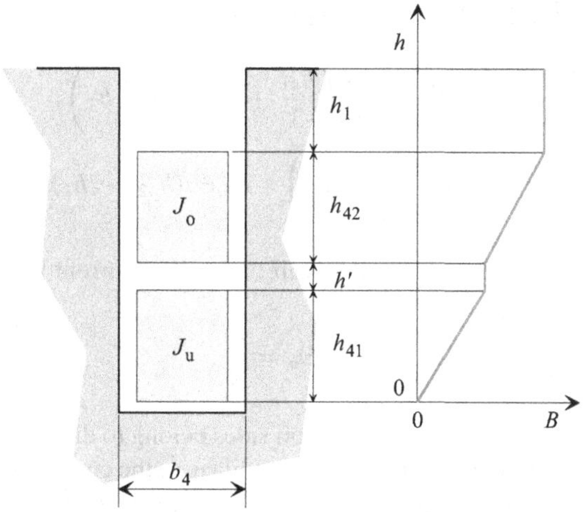

Figure 4.17 Dimensions of a slot of a double-layer winding and the principal behavior of the flux density if no skin effect occurs. In the figure, the current densities J_o and J_u are equal in both current-carrying areas. However, in double-layer windings, there are also slots with coils belonging to two different phases when the current densities are different.

a slot permeance factor for closed rotor slots, but since the selected materials have a strong influence on the factor, the inductance of closed slots has to be estimated numerically as a function of current in the slot.

The slot leakage of a double-layer winding is investigated with Figure 4.17.

The permeance factor for a double-layer winding is determined by calculating the energy of the leakage magnetic field in the slot, Equation (4.12). In the calculation, we have to bear in mind that there are coil sides of different phases in some of the slots.

The current linkages of the bottom (subscript "u" (under)) and upper (subscript "o" (over)) layers increase in proportion to the height (h):

$$\Theta_\mathrm{u}(h) = \frac{h}{h_{41}}\Theta_\mathrm{u}, \tag{4.70}$$

$$\Theta_\mathrm{o}(h) = \Theta_\mathrm{u} + \frac{h}{h_{42}}\Theta_\mathrm{o}, \tag{4.71}$$

where Θ_u and Θ_o are the total current linkages of the bottom and upper layers, respectively. The height h is measured separately from the bottom of the each coil.

The energy of the leakage magnetic field in a slot is, according to Equations (4.12) and (4.56),

$$W_\Phi = \frac{1}{2}\mu_0 l' \frac{1}{b_4}\left[\Theta_\mathrm{u}^2 \int_0^{h41} \left(\frac{h}{h_{41}}\right)^2 \mathrm{d}h + \Theta_\mathrm{u}^2 \int_0^{h'} \mathrm{d}h \right. \\ \left. + \int_0^{h42} \left(\Theta_\mathrm{u} + \Theta_\mathrm{o}\frac{h}{h_{42}}\right)^2 \mathrm{d}h + (\Theta_\mathrm{u} + \Theta_\mathrm{o})^2 \int_0^{h_1} \mathrm{d}h\right]. \tag{4.72}$$

After the integration Equation (4.72) reduces to

$$W_\Phi = \frac{1}{2}\mu_0 l' \frac{1}{b_4}\left[\Theta_u^2\left(\frac{h_{41}}{3} + h' + h_{42} + h_1\right) + \Theta_o^2\left(\frac{h_{42}}{3} + h_1\right) + \Theta_u\Theta_o(h_{42} + 2h_1)\right]. \tag{4.73}$$

The current linkages Θ_u and Θ_o are equal to half of the total current linkage of the slot Θ

$$\Theta_u = \Theta_o = \frac{\Theta}{2}. \tag{4.74}$$

A winding can also have slots in which the coil sides belong to different phases. If the phase shift between the currents of the bottom and upper coils is γ, the current linkage product $\Theta_u\Theta_o$ has to be multiplied by $\cos\gamma$. The product term $\Theta_u\Theta_o$ represents the mutual influence of the bottom and upper currents. Because the phase shift can vary from slot to slot, the average value g over $2q$ coil sides has to be built

$$g = \frac{1}{2q}\sum_{n=1}^{2q}\cos\gamma_n, \tag{4.75}$$

and the current linkage product takes the form

$$\Theta_u\Theta_o = g\left(\frac{\Theta}{2}\right)^2 = \frac{g}{4}\Theta^2. \tag{4.76}$$

Substituting (4.74) and (4.76) into Equation (4.73) gives

$$W_\Phi = \frac{1}{2}\mu_0 l' \frac{1}{b_4}\frac{\Theta^2}{4}\left[\left(\frac{h_{41}}{3} + h' + h_{42} + h_1\right) + \left(\frac{h_{42}}{3} + h_1\right) + g(h_{42} + 2h_1)\right]. \tag{4.77}$$

In two-layer windings, the height of the coil sides is equal:

$$h_{41} = h_{42} = \frac{h_4 - h'}{2}, \tag{4.78}$$

with which Equation (4.77) is reduced to

$$W_\Phi = \frac{1}{2}\mu_0 l' \Theta^2\left[\frac{5+3g}{8}\frac{h_4 - h'}{3b_4} + \frac{1+g}{2}\frac{h_1}{b_4} + \frac{h'}{4b_4}\right]. \tag{4.79}$$

The term in square brackets is the permeance factor λ_u for a double-layer winding

$$\lambda_u = k_1\frac{h_4 - h'}{3b_4} + k_2\frac{h_1}{b_4} + \frac{h'}{4b_4}, \tag{4.80}$$

where

$$k_1 = \frac{5+3g}{8} \tag{4.81}$$

and

$$k_2 = \frac{1+g}{2}. \tag{4.82}$$

Correspondingly, the permeance factor for slot types from a to e in Figure 4.15 is

$$\lambda_u = k_1 \frac{h_4 - h'}{3b_4} + k_2 \left(\frac{h_3}{b_4} + \frac{h_1}{b_1} + \frac{h_2}{b_4 - b_1} \ln \frac{b_4}{b_1} \right) + \frac{h'}{4b_4} \tag{4.83}$$

and for slots types from f to g in Figure 4.15

$$\lambda_u = k_1 \frac{h_4 - h'}{3b_4} + k_2 \left(\frac{h_1}{b_1} + 0.66 \right) + \frac{h'}{4b_4}. \tag{4.84}$$

Equations (4.83) and (4.84) are valid also for a single-layer winding. In that case, $h' = 0$ and $k_1 = k_2 = 1$.

The factors k_1 and k_2 can also be calculated with the aid of the amount of short pitching ε given by

$$\varepsilon = 1 - \frac{W}{\tau_p}. \tag{4.85}$$

Within a zone, there are $q = \tau_p/m$ slots (Figure 4.18), of which $\varepsilon\tau_p$ slots in the upper and bottom coil sides belong to different phases and in the remaining $(\tau_p/m - \varepsilon\tau_p)$ slots, the coil sides belong to the same phase. In a three-phase winding ($m = 3$), the phase shift of the upper and bottom coil currents is 180° – 120° = 60° in the slots, where the coil sides belong to

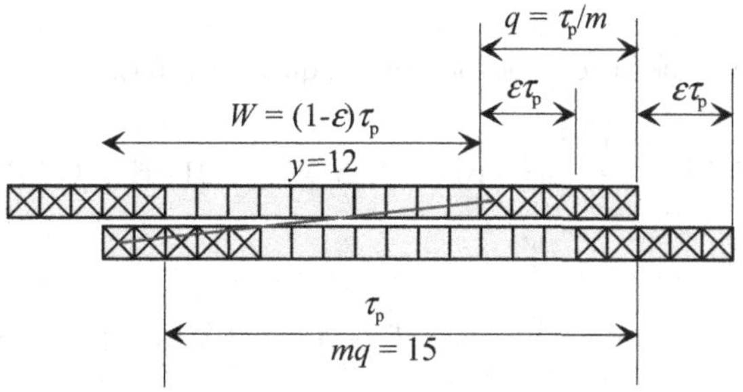

Figure 4.18 Short-pitched winding with $m = 3$, $y = 12$ slots, $mq = 15$ slots and $q = mq/m = 15/3 = 5$.

different phases; so $\cos\gamma = 0.5$ in these slots. In the slots where the coil sides belong to the same phase, $\cos\gamma = 1$. For three-phase windings, Equation (4.75) takes the form

$$g = \frac{1}{2q}\sum_{n=1}^{2q}\cos\gamma_n = \frac{1}{2\frac{\tau_p}{3}}\left[2\varepsilon\tau_p\cdot 0.5 + 2\left(\frac{\tau_p}{3} - \varepsilon\tau_p\right)\cdot 1\right] = 1 - \frac{3}{2}\varepsilon. \tag{4.86}$$

Substituting (4.86) into Equations (4.81) and (4.82), the factors k_1 and k_2 take the form for a three-phase winding

$$k_1 = 1 - \frac{9}{16}\varepsilon \quad \text{and} \quad k_2 = 1 - \frac{3}{4}\varepsilon. \tag{4.87}$$

For a two-phase winding, the factors k_1 and k_2 are

$$k_1 = 1 - \frac{3}{4}\varepsilon \quad \text{and} \quad k_2 = 1 - \varepsilon. \tag{4.88}$$

Example 4.4: Calculate the slot leakage inductance of a double-layer winding with $p = 2$, $Q = 24$, $W/\tau_p = 5/6$ and $N = 40$ (cf. Figure 2.17b). The pole pairs are connected in parallel, $a = 2$. The slot shape is according to Figures 4.15b and 4.17: $b_1 = 0.003$ m, $h_1 = 0.002$ m, $h_2 = 0.001$ m, $h_3 = 0.001$ m, $h' = 0.001$ m, $b_4 = 0.008$ m and $h_{41} = h_{42} = 0.009$ m ($h_4 = 0.019$ m), $l' = 0.25$ m. Compare the result with the slot leakage of a corresponding double-layer full-pitch winding.

Solution: The short pitching $\varepsilon = 1/6$ and $k_1 = 1 - 9/(16\times 6) = 0.906$, $k_2 = 1 - 3/(4\times 6) = 0.875$. The permeance factor is, according to Equation (4.83),

$$\begin{aligned}\lambda_u &= k_1\frac{h_4 - h'}{3b_4} + k_2\left(\frac{h_3}{b_4} + \frac{h_1}{b_1} + \frac{h_2}{b_4 - b_1}\ln\frac{b_4}{b_1}\right) + \frac{h'}{4b_4}\\ &= 0.906\cdot\frac{0.018}{3\cdot 0.008} + 0.875\left(\frac{0.001}{0.008} + \frac{0.002}{0.008} + \frac{0.001}{0.008 - 0.003}\ln\frac{0.008}{0.003}\right) + \frac{0.001}{4\cdot 0.008}\\ &= 1.211\end{aligned}$$

and the slot leakage inductance is, according to Equation (4.64),

$$\begin{aligned}L_u &= \frac{4m}{Q}\mu_0 l' N^2\lambda_u = \frac{4\cdot 3}{24}4\pi\cdot 10^{-7}\cdot 0.25\cdot 40^2\cdot 1.211\ \mathrm{H} = 0.2513\cdot 1.211\ \mathrm{mH}\\ &= 0.304\ \mathrm{mH}\end{aligned}$$

For a full-pitch double-layer winding $k_1 = k_2 = 1$ and Equation (4.83) yields

$$\lambda_u = \frac{0.018}{3\cdot 0.008} + \left(\frac{0.001}{0.008} + \frac{0.002}{0.008} + \frac{0.001}{0.008 - 0.003}\ln\frac{0.008}{0.003}\right) + \frac{0.001}{4\cdot 0.008} = 1.352.$$

The slot leakage inductance is now $L_u = 0.2513 \times 1.352\ \text{mH} = 0.340\ \text{mH}$.

As we can see, the phase shift of the different phase coil sides in the double-layer winding causes a smaller slot leakage inductance for the short-pitched winding compared with the full-pitch winding. The slot leakage inductance in this case is about 10 % smaller for the short-pitched winding.

If the height of a conductor is considerable or a single homogeneous bar forms the winding of a slot (as is the case in the rotor of an induction motor), a skin effect occurs in the conductors with alternating current. The skin effect may be high even at moderate frequencies. It has a strong influence on the rotor resistance during the start-up of a squirrel cage motor. In that case, at the bottom of slot of Figure 4.13, the conductor elements dh are surrounded by a larger flux than the conductor elements in the upper layer of the slot. The inductance of the conductor elements of the upper layer is lower than the inductance of the conductor elements at the bottom, and therefore a time-varying current is distributed so that the current density in the upper section of the slot is higher than the current density at the bottom. The density function of the leakage flux corresponds to the dotted line of Figure 4.13. Thus, the skin effect increases the resistance of the bar and reduces the slot inductance. This will be discussed in more detail in Chapter 5.

Now, the permeance factor (4.59) of the area with height h_4 (or h_{41} and h_{42}) is rewritten in the form

$$\lambda_{4,\mathrm{Ft}} = k_\mathrm{L}\frac{h_4}{3b_4}, \tag{4.89}$$

where k_L is the skin effect factor for the permeance as well as for the inductance (4.57), representing the decrease in the slot inductance caused by the skin effect. $\lambda_{4,\mathrm{Ft}}$ has to be applied in Equations (4.65), (4.66), (4.83) and (4.84) instead of the first term. To define the permeance factor, the reduced conductor height ξ is determined as

$$\xi = h_4\sqrt{\omega\mu_0\sigma\frac{b_\mathrm{c}}{2b_4}}, \tag{4.90}$$

where b_c is the width of a conductor in a slot (see Figure 4.13), ω is the angular frequency of the current investigated, and σ is the specific conductivity of the conductor. For instance, in induction motors the slip s defines the rotor angular frequency $\omega = s\omega_\mathrm{s}$.

The skin effect factor k_L of the permeance can be obtained from the equation

$$k_\mathrm{L} = \frac{1}{z_\mathrm{t}^2}\varphi'(\xi) + \frac{z_\mathrm{t}^2 - 1}{z_\mathrm{t}^2}\psi'(\xi), \tag{4.91}$$

where

$$\varphi'(\xi) = \frac{3}{2\xi}\left(\frac{\sinh 2\xi - \sin 2\xi}{\cosh 2\xi - \cos 2\xi}\right), \tag{4.92}$$

$$\psi'(\xi) = \frac{1}{\xi}\left(\frac{\sinh \xi + \sin \xi}{\cosh \xi + \cos \xi}\right) \tag{4.93}$$

and z_t is the number of conductors on top of each other. For a squirrel cage winding $z_t = 1$, as depicted in Figure 4.13, and the skin effect factor is

$$k_L = \phi'(\xi). \tag{4.94}$$

Usually, in squirrel cage rotors, $h_4 > 2$ cm and for copper bars Equation (4.90) gives $\xi > 2$, in which case $\sinh 2\xi \gg \sin 2\xi$, $\cosh 2\xi \gg \cos 2\xi$ and $\sinh 2\xi \approx \cosh 2\xi$, and hence

$$k_L \approx \frac{3}{2\xi}. \tag{4.95}$$

Example 4.5: Repeat Example 4.3 for an aluminum squirrel cage bar at cold start in a 50 Hz supply.

Solution: The slot shape is, according to Figure 4.13, $b_1 = 0.003$ m, $h_1 = 0.002$ m, $b_4 = 0.008$ m and $h_4 = 0.02$ m, $l' = 0.25$ m and the slot at height h_4 is fully filled with aluminum bar. The conductivity of aluminum at 20 °C is $\sigma_{Al} = 37$ MS/m.

The permeance factor of the wound part of the slot without skin effect is

$$\lambda_4 = \frac{h_4}{3b_4} = \frac{0.02}{3 \cdot 0.008} = 0.833.$$

The permeance factor of the slot opening is

$$\lambda_1 = \frac{h_1}{b_1} = \frac{0.002}{0.003} = 0.667.$$

The reduced height ξ of the conductor is

$$\xi = h_4\sqrt{\omega\mu_0\sigma\frac{b_c}{2b_4}} = 0.02\sqrt{2\pi \cdot 50 \cdot 4\pi \cdot 10^{-7} \cdot 37 \cdot 10^6 \frac{0.008}{2 \cdot 0.008}} = 1.71$$

which is a dimensionless number

$$k_L = \frac{1}{z_t^2}\phi'(\xi) + \frac{z_t^2 - 1}{z_t^2}\psi'(\xi) = \phi'(\xi) + \frac{1-1}{1}\psi'(\xi) = \phi'(\xi)$$

$$= \frac{3}{2\xi} \cdot \left(\frac{\sinh 2\xi - \sin 2\xi}{\cosh 2\xi - \cos 2\xi}\right) = \frac{3}{3.42} \cdot \left(\frac{\sinh 3.42 - \sin 3.42}{\cosh 3.42 - \cos 3.42}\right) = 0.838.$$

The permeance factor of the slot under skin effect is

$$\lambda_u = \lambda_1 + k_L\lambda_4 = 0.667 + 0.838 \cdot 0.833 = 1.37.$$

This is somewhat less than in Example 4.3. The slot leakage inductance of a squirrel cage bar is

$$L_{\mathrm{u,bar}} = \mu_0 l' z_{\mathrm{Q}}^2 \lambda_{\mathrm{u}} = 4\pi \cdot 10^{-7} \cdot 0.25 \cdot 1^2 \cdot 1.37\mathrm{H} = 0.43\ \mu\mathrm{H}$$

4.3.4 Tooth Tip Leakage Inductance

The tooth tip leakage inductance is determined by the magnitude of leakage flux flowing in the air gap outside the slot opening. This flux leakage is illustrated in Figure 4.19. The current linkage in the slot causes a potential difference between the teeth on opposite sides of the slot opening, and as a result part of the current linkage will be used to produce the leakage flux of the tooth tip.

The tooth tip leakage inductance can be determined by applying a permeance factor

$$\lambda_{\mathrm{d}} = k_2 \frac{5\left(\dfrac{\delta}{b_1}\right)}{5 + 4\left(\dfrac{\delta}{b_1}\right)}, \tag{4.96}$$

where $k_2 = (1 + g)/2$ is calculated from Equation (4.82).

The tooth tip leakage inductance of the whole phase winding is obtained by substituting λ_{d} in Equation (4.64):

$$L_{\mathrm{d}} = \frac{4m}{Q} \mu_0 l' \lambda_{\mathrm{d}} N^2. \tag{4.97}$$

In salient-pole machines, we have to substitute the air gap in Equation (4.96) by the air gap at the middle of the pole, where the air gap is at its smallest. If we select the air gap to be infinite, we obtain a limit value of $\lambda_{\mathrm{d}} = 1.25$, which is the highest value for λ_{d}. If δ is small, as

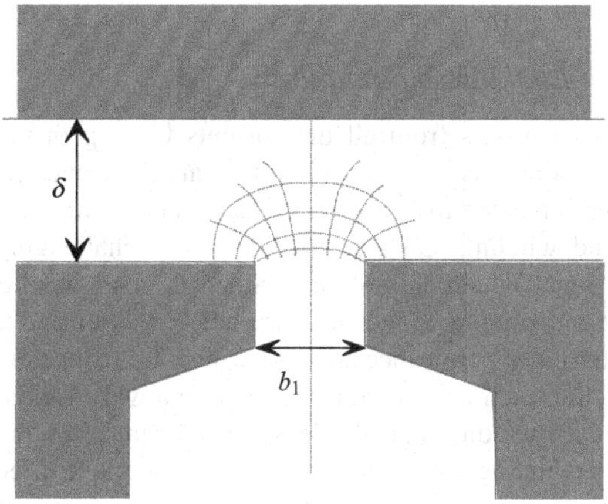

Figure 4.19 Flux leakage creating a tooth tip leakage inductance around a slot opening.

is the case in asynchronous machines in particular, the influence of tooth leakage inductance is insignificant.

Equations (4.96)–(4.97) are no longer valid for the main poles of DC machines. The calculations in the case of DC machines are analyzed for instance by Richter (1967).

Example 4.6: Calculate the toot tip leakage of the machine in Example 4.4. The machine is now equipped with rotor surface permanent magnets. The magnets are neodymium iron boron magnets 8 mm thick, and there is a 2 mm physical air gap, $p = 2$, $Q = 24$, $W/\tau_p = 5/6$ and $N = 40$ (cf. Figure 2.17b). The pole pairs are connected in parallel, $a = 2$. Compare the result with the slot leakage of Example 4.4.

Solution: As the permanent magnets represent, in practice, air ($\mu_{rPM} = 1.05$), we may assume that the air gap in the calculation of the tooth tip leakage is $2 + 8/1.05$ mm $= 9.62$ mm. The factor $k_2 = 1 - 3\varepsilon/4 = 1 - 3/(4 \cdot 6) = 0.875$. We now obtain for the permeance coefficient

$$\lambda_d = k_2 \frac{5\left(\frac{\delta}{b_1}\right)}{5 + 4\left(\frac{\delta}{b_1}\right)} = 0.875 \frac{5\left(\frac{0.00962}{0.003}\right)}{5 + 4\left(\frac{0.00962}{0.003}\right)} = 0.787.$$

The tooth tip inductance is, according to Equation (4.97),

$$L_d = \frac{4m}{Q}\mu_0 l' \lambda_d N^2 = \frac{4 \cdot 3}{24} 4\pi \cdot 10^{-7} \cdot 0.25 \cdot 0.787 \cdot 40^2\,\mathrm{H} = 0.198\,\mathrm{mH}$$

In Example 4.4, the slot leakage of the same machine was 0.340 mH. As the air gap in a rotor surface magnet machine is long, the tooth tip leakage has a significant value, about 70 % of the slot leakage.

4.3.5 End Winding Leakage Inductance

End winding leakage flux results from all the currents flowing in the end windings. The geometry of the end windings is usually difficult to analyze, and, further, all the phases of poly-phase machines influence the occurrence of a leakage flux. Therefore, the accurate determination of an end winding leakage inductance is a challenging task, which would require a three-dimensional numerical solution. Since the end windings are relatively far from the iron parts, the end winding inductances are not very high, and therefore it suffices to employ empirically determined permeance factors λ_{lew} and λ_w. In machines with an alternating current flowing in both the stator and the rotor, measuring always yields a sum of the leakage inductance of the primary winding and the leakage inductance of the secondary winding referred to the primary winding. In a stationary state, there is a direct current flowing in the rotor of a synchronous machine, and thus the end winding inductance of a synchronous machine is determined only by the stator side.

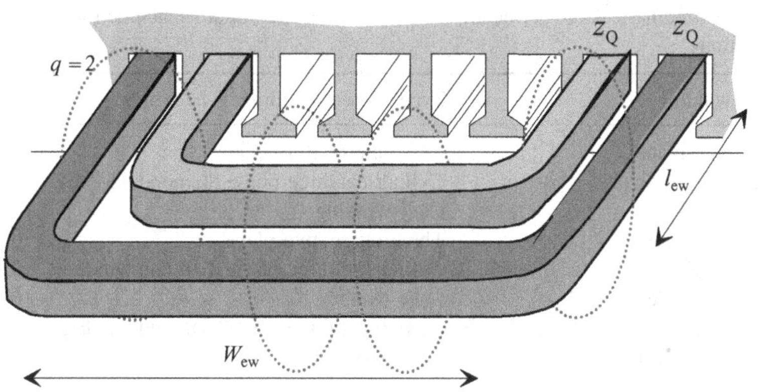

Figure 4.20 Leakage flux and dimensions of an end winding.

When calculating the slot leakage inductance, the proportion of each slot can be calculated separately. The number of conductors in a slot is z_Q (in a two-layer winding $z_Q = 2z_{cs}$, z_{cs} being the amount of conductors in a coil side of a two-layer winding) and the conductors are surrounded by iron. The flux in the end windings is a result of the influence of all the coil turns belonging to a coil group. According to Figure 4.20, the number of coil turns is qz_Q, which has to replace z_Q in the inductance equation (4.57). The average length l_w of the end winding has to be substituted for the length l' of the stack in Equation (4.57). The number of coil groups in series in a phase winding is Q/amq, and the number of parallel current paths is a. The equation for end winding leakage inductance can now be written as

$$\begin{aligned} L_w &= \frac{Q}{amq}\frac{1}{a}\left(qz_Q\right)^2\mu_0 l_w\lambda_w = \frac{Q}{m}q\left(\frac{z_Q}{a}\right)^2\mu_0 l_w\lambda_w \\ &= \frac{4m}{Q}qN^2\mu_0 l_w\lambda_w = \frac{2}{p}N^2\mu_0 l_w\lambda_w. \end{aligned} \tag{4.98}$$

The average length l_w of the end winding and the product $l_w\lambda_w$ can be written in the form (see Figure 4.20)

$$l_w = 2l_{ew} + W_{ew}, \tag{4.99}$$

$$l_w\lambda_w = 2l_{ew}\lambda_{lew} + W_{ew}\lambda_{Wew} \tag{4.100}$$

where l_{ew} is the axial length of the end winding measured from the end of the stack, and W_{ew} is the coil span according to Figure 4.20. λ_{lew} and λ_{Wew} are the corresponding permeance factors, see Tables 4.1 and 4.2.

A permeance factor depends on the structure of the winding (e.g. a single-phase, or a two-plane, three-phase, or a three-plane, three-phase, or a diamond winding constructed of coils of equal shapes), the organization of the planes of the end winding, the ratio of the length of an average coil end to the pole pitch l_w/τ_p, and the rotor type (a nonsalient-pole machine, a salient-pole machine, an armature winding of a DC machine, a cage winding, a three-phase

Table 4.1 End winding leakage permeance factors of an asynchronous machine for various stator and rotor combinations

Type of stator winding	Type of rotor winding	λ_{lew}	λ_{Wew}
Three-phase, three-plane	Three-phase, three-plane	0.40	0.30
Three-phase, three-plane	Cylindrical three-phase diamond winding	0.34	0.34
Three-phase, three-plane	Cage winding	0.34	0.24
Three-phase, two-plane	Three-phase, two-plane	0.55	0.35
Three-phase two-plane	Cylindrical three-phase diamond winding	0.55	0.25
Three-phase, two-plane	Cage winding	0.50	0.20
Cylindrical three-phase diamond winding	Cylindrical three-phase diamond winding	0.26	0.36
Cylindrical three-phase diamond winding	Cage winding	0.50	0.20
Single-phase	Cage winding	0.23	0.13

Table 4.2 Permeance factors of the end windings of a synchronous machine

Cross-section of end winding	Nonsalient pole machine		Salient pole machine	
	λ_{lew}	λ_{Wew}	λ_{lew}	λ_{Wew}
	0.342	0.413	0.297	0.232
	0.380	0.130	0.324	0.215
	0.371	0.166	0.324	0.243
	0.493	0.074	0.440	0.170
	0.571	0.073	0.477	0.187
	0.605	0.028	0.518	0.138

winding). Richter (1954: 161; 1963: 91; 1967: 279) presents in detail some calculated values for permeance factors that are valid for different machine types.

Based on the literature, the Table 4.1 and Table 4.2 can be compiled for the definition of the end winding leakage permeance factors for asynchronous and synchronous machines. With these permeance factors, Equation (4.98) gives the sum of the stator end winding leakage inductance and the rotor to the stator referred end winding leakage inductance. The major part of the sum belongs to the stator (60–80 %).

Example 4.7: The air-gap diameter of the machine in Example 4.4 is 130 mm and the total height of the slots is 22 mm. Calculate the end winding leakage inductance for a three-phase surface-mounted, permanent magnet synchronous machine with $Q = 24$, $q = 2$, $N = 40$, $p = 2$ and $l_w = 0.24$ m. The third end winding arrangement in Table 4.2 is valid for the machine under consideration.

Solution: Let us assume that the average diameter of the end winding is 130 + 22 mm = 152 mm. The perimeter of this diameter is about 480 mm. The pole pitch at this diameter is $\tau'_p = 480$ mm/4 = 120 mm. From this we may assume that the width of the end winding is about the pole pitch subtracted by one slot pitch, $W_{ew} = \tau'_p - \tau'_u = (0.12 - 0.48/24)$m = 0.14 m and $l_{ew} = 0.5(l_w - W_{ew}) = 0.05$ m.

The surface magnet machine may be regarded as a salient-pole machine, Hence, in calculating end winding leakage the permeance factors are $\lambda_{lew} = 0.324$ and $\lambda_{Wew} = 0.243$:

$$l_w\lambda_w = 2l_{ew}\lambda_{lew} + W_{ew}\lambda_{Wew} = 2 \cdot 0.05\,\text{m} \cdot 0.324 + 0.14\,\text{m} \cdot 0.243 = 0.0664\,\text{m}$$

$$L_w = \frac{4m}{Q}qN^2\mu_0 l_w\lambda_w = \frac{4 \cdot 3}{24}2 \cdot 40^2 \cdot 4\pi \cdot 10^{-7} \cdot 0.0664\,\text{H} = 0.134\,\text{mH}.$$

The slot leakage of the 5/6 short-pitched winding in Example 4.4 was $L_u = 0.304$ mH, the tooth tip leakage of the PMSM was $L_d = 0.198$ mH (in Example 4.6), and the end winding leakage was $L_w = 0.134$ mH. The slot leakage is usually the most important leakage flux component as these results may indicate. The air-gap leakage is dependent on the magnetizing inductance and cannot now be compared with these other leakages.

For the leakage inductance of a short-circuit ring of a cage winding in an induction machine or in a damper winding, Liwschitz-Garik and Wipple (1961) introduced the equation

$$L_{rw\sigma} = \mu_0 \frac{Q_r}{2p^2 m_s}\left[\frac{2}{3}\left(l_{bar} - l'_r\right) + \nu\frac{\pi D'_r}{2p}\right] \tag{4.101}$$

where l_{bar} is the length of the rotor bar, l'_r is the equivalent length of the rotor and D'_r is the average diameter of the short-circuit ring. If there is no space between the end rings and the rotor iron stack $l_{bar} = l'_r$. The factor $\nu = 0.36$ when $p = 1$, and $\nu = 0.18$ when $p > 1$. Equation (4.101) gives the leakage inductance per one rotor bar and it includes the end ring segment of one rotor slot pitch and the bar outside the stack. It includes both ends of the rotor. $L_{rw\sigma}$ is referred to the stator by multiplying Equation (4.101) by a referring factor developed later in Chapter 7. Calculation of the total impedance of a cage winding is discussed in Chapter 7.

Example 4.8: Calculate the rotor short-circuit ring leakage for a four-pole, three-phase induction motor with $Q_r = 34$, $l_{bar} = 0.14$ m, $l_r = 0.120$ m, $D'_r = 0.11$ m, $\delta = 0.0004$ m.

Solution: Inserting $l' = l + 2\delta = 0.1208$ m in (4.101) gives

$$L_{rw\sigma} = \mu_0 \frac{Q_r}{2p^2 m_s}\left[\frac{2}{3}(l_{bar} - l'_r) + \nu\frac{\pi D'_r}{2p}\right]$$
$$= 4\pi \cdot 10^{-7}\frac{34}{2^2 \cdot 3}\left[\frac{2}{3}(0.140 - 0.1208) + 0.18\frac{\pi \cdot 0.11}{2 \cdot 2}\right]\text{H} = 0.05\,\mu\text{H}.$$

This value must be referred to the stator to get the single-phase equivalent-circuit leakage inductance. This is studied in Chapter 7.

Bibliography

Baffrey, R. (1926) Influence of short-pitching on AC motors overloading (Über den Enfluss der Schrittverkürzung auf die Überlastungsfähigkeit von Drehstrommotoren). *Archiv für Elektrotechnik*, **16**(2), 97–113.

Liwschitz-Garik, M. and Whipple, C. C. (1961) *Alternating Current Machines*, Van Nostrand, Princeton, NJ.

Richter, R. (1954) *Electrical Machines: Induction Machines (Elektrische Maschinen: Die Induktionsmaschinen)*, Vol. **IV**, 2nd edn, Birkhäuser Verlag, Basle and Stuttgart.

Richter, R. (1963) *Electrical Machines: Synchronous Machines and Rotary Converters (Elektrische Maschinen: Synchronmaschinen und Einankerumformer)*, Vol. **II**, 3rd edn, Birkhäuser Verlag, Basle and Stuttgart.

Richter, R. (1967) *Electrical Machines: General Calculation Elements. DC Machines (Elektrische Maschinen: Allgemeine Berechnungselemente. Die Gleichstrommaschinen)*, Vol. **I**, 3rd edn, Birkhäuser Verlag, Basle and Stuttgart.

Vogt, K. (1996) *Design of Electrical Machines (Berechnung elektrischer Maschinen)*, Wiley-VCH Verlag GmbH, Weinheim.

5

Resistances

Together with inductances, resistances define the characteristics of an electrical machine. From an efficiency point of view, resistances are important electrical machine parameters. In many cases, the resistive losses form a dominant loss component in a machine. In electrical machines, the conductors are usually surrounded by a ferromagnetic material, which encourages flux components to travel through the windings. This, again, may cause large skin effect problems if the windings are not correctly designed.

5.1 DC Resistance

The resistance of a winding can first be defined as a DC resistance. According to Ohm's law, the resistance R_{DC} depends on the total length of a conductor in a coil l_c, the number of parallel paths a in windings without a commutator, per phase, or $2a$ in windings with a commutator, the cross-sectional area of the conductor S_c and the conductivity of the conductor material σ_c

$$R_{DC} = \frac{l_c}{\sigma_c a S_c}. \tag{5.1}$$

Resistance is highly dependent on the running temperature of the machine, and therefore a designer should be well aware of the warming-up characteristics of the machine before defining the resistances. Usually, we may first investigate the resistances of the machine at the design temperature or at the highest allowable temperature for the selected winding type.

Windings are usually made of copper. The specific conductivity of pure copper at room temperature (+20 °C) is $\sigma_{Cu} = 58 \times 10^6$ S/m and the conductivity of commercial copper wire $\sigma_{Cu} = 57 \times 10^6$ S/m. The temperature coefficient of resistivity for copper is $\alpha_{Cu} = 3.81 \times 10^{-3}$/K. The respective parameters for aluminum are: conductivity $\sigma_{Al} = 37 \times 10^6$ S/m and the temperature coefficient of resistivity $\alpha_{Al} = 3.7 \times 10^{-3}$/K.

An accurate definition of the winding length in an electrical machine is a fairly difficult task. Salient-pole machines are a relatively simple case: the conductor length can be defined fairly easily when the shape of the pole body and the number of coil turns are known. Instead, the winding length of slot windings is difficult, especially if coils of different length are employed in the machine. Preliminary calculations can be made by applying the following empirical equations.

The average length l_{av} of a coil turn of a slot winding of low-voltage machines with round enameled wires is given approximately as

$$l_{av} \approx 2l + 2.4W + 0.1 \text{ m}. \tag{5.2}$$

Here l is the length of the stator stack of the machine and W is the average coil span, both expressed in meters. For large machines with prefabricated windings, the following approximation is valid:

$$l_{av} \approx 2l + 2.8W + 0.4 \tag{5.3}$$

and

$$l_{av} \approx 2l + 2.9W + 0.3 \text{ m}, \quad \text{when } U = 6\ldots 11 \text{ kV}. \tag{5.4}$$

Using the length of the winding, the DC resistance may be calculated according to Equation (5.1) by taking all the turns and parallel paths into account.

5.2 Influence of Skin Effect on Resistance

Alternating current in a conductor and currents in the neighboring conductors create an alternating flux in the conductor material which causes skin and proximity effects. In a conductor internal circulating currents are seen and in case of parallel conductors also circulating currents between different strands become possible. Circulating currents between parallel conductors should be avoided by correct geometrical arrangement of the windings. In this presentation the skin and proximity effects will be dealt together and called the skin effect.

In electrical machines, the skin effect occurs chiefly in the area of the slots, but it is also present to a lesser extent in end windings. The calculation of the skin effect must be performed separately in the slot areas and in the end winding areas because the magnetic properties of the slot area and the end winding area are totally different.

5.2.1 Analytical Calculation of Resistance Factor

Let us consider a solid conductor in a slot surrounded on three sides by ferromagnetic material, the permeability of the material being infinity, Figure 5.1. The current i flowing in the conductor creates a magnetic field strength $\boldsymbol{H}$ and leakage flux across the slot and conductor. The leakage flux surrounding the bottom part of the conductor is greater than the flux surrounding the upper part of the conductor, whereupon the impedance of the conductor decreases from the bottom up. Correspondingly, the current density $\boldsymbol{J}$ increases from the bottom up as shown in Figure 5.1.

Since the permeability of iron is assumed infinite, the leakage flux lines cross the slot rectilinearly. The vectors of the current density $\boldsymbol{J}$ and electric field strength $\boldsymbol{E}$ in the conductor

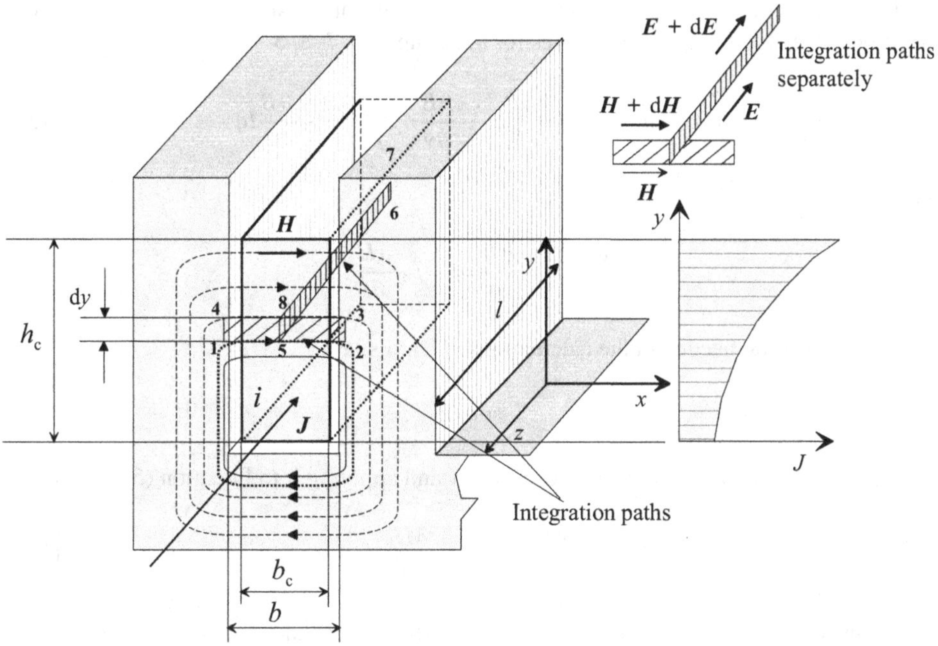

Figure 5.1 Definition of the skin effect in a conductor placed in a slot. The current i creates a leakage flux across the conductor resulting in an uneven distribution of current density J in the conductor.

have only a z-component. The magnetic field strength $\boldsymbol{H}$ and the flux density $\boldsymbol{B}$ have only an x-component across the conductor. Applying Ampère's law (Equation 1.4) to the route 1–2–3–4–1 having the area b dy we obtain

$$\oint \boldsymbol{H} \cdot \mathrm{d}\boldsymbol{l} = Hb - \left(H + \frac{\partial H}{\partial y}\mathrm{d}y\right) b = J b_{\mathrm{c}} \mathrm{d}y \tag{5.5}$$

from which

$$-\frac{\partial H}{\partial y} = \frac{b_{\mathrm{c}}}{b} J. \tag{5.6}$$

Applying Faraday's induction law (Equation 1.7) to the route 5–6–7–8–5 in Figure 5.1, we understand that the slot leakage flux penetrating through the conductor induces voltages and eddy currents trying to oppose the flux penetrating through the conductor. As a result there will be internal "circulating" currents in the conductor summing to the conductor current I. As a result the total current density distribution becomes uneven as Figure 5.1 shows. We now derive the differential equation to solve for the current density distribution in the solid conductor in the slot. Faraday's law gives for the route 5–6–7–8–5

$$\oint \boldsymbol{E} \cdot \mathrm{d}\boldsymbol{l} = -El + \left(E + \frac{\partial E}{\partial y}\mathrm{d}y\right) l = -\frac{\partial B}{\partial t} l\mathrm{d}y \tag{5.7}$$

from which

$$\frac{\partial E}{\partial y} = -\frac{\partial B}{\partial t} = -\mu_0 \frac{\partial H}{\partial t}. \tag{5.8}$$

The third equation needed in the calculations is Ohm's law

$$J = \sigma_\mathrm{c} E. \tag{5.9}$$

Differentiating Equation (5.9) with respect to y and applying it to Equation (5.8) yields

$$\frac{\partial J}{\partial y} = -\mu_0 \sigma_\mathrm{c} \frac{\partial H}{\partial t}. \tag{5.10}$$

With sinusoidally varying quantities, Equations (5.6), (5.8) and (5.10) change to complex form:

$$-\frac{\partial \underline{H}}{\partial y} = \frac{b_\mathrm{c}}{b} \underline{J}, \tag{5.11}$$

$$\frac{\partial \underline{E}}{\partial y} = -\mathrm{j}\omega\mu_0 \underline{H}, \tag{5.12}$$

$$\frac{\partial \underline{J}}{\partial y} = -\mathrm{j}\omega\mu_0\sigma_\mathrm{c} \underline{H}. \tag{5.13}$$

Differentiating Equation (5.13) with respect to y and using Equation (5.11) we obtain

$$\frac{\partial^2 \underline{J}}{\partial y^2} = -\mathrm{j}\omega\mu_0\sigma_\mathrm{c} \frac{\partial \underline{H}}{\partial y} = \mathrm{j}\omega\mu_0\sigma_\mathrm{c} \frac{b_\mathrm{c}}{b} \underline{J}, \tag{5.14}$$

$$\frac{\partial^2 \underline{J}}{\partial y^2} - \mathrm{j}\omega\mu_0\sigma_\mathrm{c} \frac{b_\mathrm{c}}{b} \underline{J} = 0. \tag{5.15}$$

Equation (5.15) has a solution in the form

$$\underline{J} = C_1 \mathrm{e}^{(1+\mathrm{j})\alpha y} + C_2 \mathrm{e}^{-(1+\mathrm{j})\alpha y}, \tag{5.16}$$

where

$$\alpha = \sqrt{\frac{1}{2}\omega\mu_0\sigma_\mathrm{c}\frac{b_\mathrm{c}}{b}}. \tag{5.17}$$

The inverse of α is called the depth of penetration. Usually, α is used in defining a dimensionless number ξ

$$\xi = \alpha h_c = h_c \sqrt{\frac{1}{2}\omega\mu_0\sigma_c \frac{b_c}{b}}, \tag{5.18}$$

which is called reduced conductor height (the units of α are 1/m and so ξ is a dimensionless number, while h_c is the real conductor height in meters).

The integration constants C_1 and C_2 are determined by the following boundary conditions:

on the boundary $y = 0$ the magnetic field strength $\underline{H} = \underline{H}_0 = 0$,
on the boundary $y = h_c$ the magnetic field strength $\underline{H} = \underline{H}_c = \frac{-i}{b} = \sqrt{2}\frac{\underline{I}}{b}$,

where $\underline{I}$ is the effective value of the total current flowing in the conductor.

Differentiating the solution in Equation (5.16), substituting it into Equation (5.13) and using the boundary conditions, we obtain for the integration constants

$$C_1 = C_2 = \frac{\mathrm{j}\omega\mu_0\sigma_c}{(1+\mathrm{j})b\alpha(\mathrm{e}^{(1+\mathrm{j})\alpha h_c} - \mathrm{e}^{-(1+\mathrm{j})\alpha h_c})}. \tag{5.19}$$

Now we know $J = \mathrm{f}(y)$ and we can calculate the power losses in the conductor. The current in an element dy is Jb_c dy and the resistive losses are

$$P_{AC} = \int_0^{h_c} (Jb_c \mathrm{d}y)^2 \frac{l}{\sigma_c b_c \mathrm{d}y} = \frac{b_c l}{\sigma_c}\int_0^{h_c} J^2 \mathrm{d}y. \tag{5.20}$$

In complex form, we can write Equation (5.20) as

$$P_{AC} = \frac{b_c l}{\sigma_c}\int_0^{h_c} \underline{J}\,\underline{J}^* \mathrm{d}y, \tag{5.21}$$

where $\underline{J}^*$ is the complex conjugate of $\underline{J}$, and $|\underline{J}^*|$ and $|\underline{J}|$ are the effective values.

When a direct current having the same effective value I as the alternating current is flowing in the conductor, the losses are

$$P_{DC} = R_{DC} I^2 = \frac{l}{\sigma_c b_c h_c} I^2. \tag{5.22}$$

The factor by which the DC resistive losses have to be multiplied to get the AC resistive losses is the resistance factor k_R. It is also the ratio of the AC and DC resistances of the

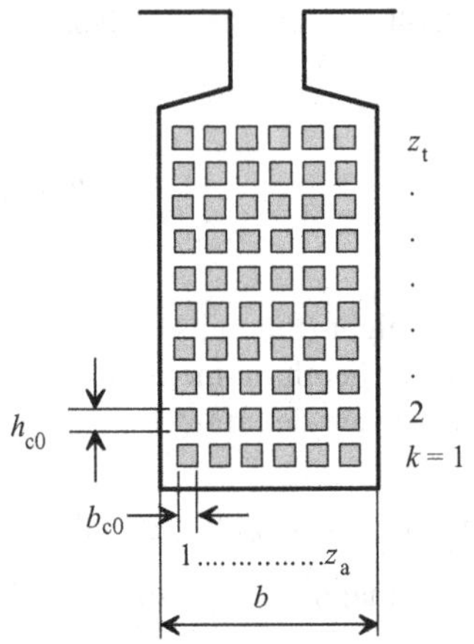

Figure 5.2 Determination of the reduced conductor height and the resistance factor of a winding with several conductors in the width and height directions of a slot with a uniform width in the conductor area. All the conductors are connected in series. The height and the width of the subconductors are h_{c0} and b_{c0}. There are z_t conductors on top of each other and z_a adjacent conductors.

conductor. According to Equations (5.20), (5.21) and (5.22), the resistance factor in the slot area is

$$k_{Ru} = \frac{R_{AC}}{R_{DC}} = \frac{P_{AC}}{P_{DC}} = \frac{b_c^2}{I^2}\int_0^{h_c} J^2 dy = \frac{b_c^2}{I^2}\int_0^{h_c} \underline{JJ^*} dy. \quad (5.23)$$

The integration in (5.23) is laborious but not very difficult. Here, it is passed over because there are solutions and applications readily available in the literature, for instance in Lipo (2007), Richter (1967), Stoll (1974), Vogt (1996) and Küpfmüller (1959). Some of them are given below.

Let us consider the case shown in Figure 5.2, where several conductors are placed in a slot. All the conductors are series connected. The height of the subconductors is h_{c0} and the width b_{c0}. There are z_t conductors on top of each other and z_a adjacent conductors.

The reduced conductor height ξ is calculated using Equation (5.18) by substituting for $b_c = z_a b_{c0}$ and for the conductor height, the height of an individual conductor h_{c0}, that is

$$\xi = \alpha h_{c0} = h_{c0}\sqrt{\frac{1}{2}\omega\mu_0\sigma_c\frac{z_a b_{c0}}{b}}. \quad (5.24)$$

Here we can see that dividing the conductor into adjacent subconductors does not affect the reduced conductor height.

The resistance factor of the kth layer is

$$k_{Rk} = \varphi(\xi) + k(k-1)\psi(\xi), \quad (5.25)$$

where the functions $\varphi(\xi)$ and $\psi(\xi)$ are

$$\varphi(\xi) = \xi \frac{\sinh 2\xi + \sin 2\xi}{\cosh 2\xi - \cos 2\xi} \tag{5.26}$$

and

$$\psi(\xi) = 2\xi \frac{\sinh \xi - \sin \xi}{\cosh \xi + \cos \xi}. \tag{5.27}$$

Equation (5.25) shows that the resistance factor is smallest on the bottom layer and largest on the top layer. This means that in the case of series-connected conductors, the bottommost conductors contribute less to the resistive losses than the topmost conductors.

The average resistance factor k_{Ru} over the slot is

$$k_{Ru} = \varphi(\xi) + \frac{z_t^2 - 1}{3}\psi(\xi), \tag{5.28}$$

where z_t is the number of conductor layers (Figure 5.2).

If $0 \le \xi \le 1$, a good approximation for the resistance factor is

$$k_{Ru} = 1 + \frac{z_t^2 - 0.2}{9}\xi^4. \tag{5.29}$$

The equations above are valid for rectangular conductors. The eddy current losses of round wires are 0.59 times the losses of rectangular wires, and therefore, for round wires, the approximate Equation (5.29) takes the form

$$k_{Ru} = 1 + 0.59\frac{z_t^2 - 0.2}{9}\xi^4. \tag{5.30}$$

According to Equations (5.29) and (5.30), the reduced conductor height ξ as well as the conductor height h_{c0} itself (Equation 5.18) strongly affect the resistance factor. With high conductors, also the resistance factor is high. Despite the fact that the resistance factor is proportional to the square of the number of conductors on top of each other, z_t, to reduce the resistance factor and eddy current losses in conductors, the conductors are divided into subconductors, which are connected together only at the beginning and at the end of the winding and transposed. Without transposition of the sub-conductors a divided conductor is fairly useless.

To get the full benefit of dividing the conductors, circulating currents in the conductors must be prevented. To prevent such circulating currents flowing through the connections at the beginning and at the end of the winding, every parallel-connected subconductor has to be linked, in average, exactly by the same amount of slot leakage flux. This is put into effect by transposing the subconductors ideally. If the number of subconductors is z_p, the transposition has to be done $z_p - 1$ times. For instance, if a winding consists of six coils and the conductors are divided into three subconductors, the transposition has to be done twice, which can be done at the end winding area after every two coils, Figure 5.3.

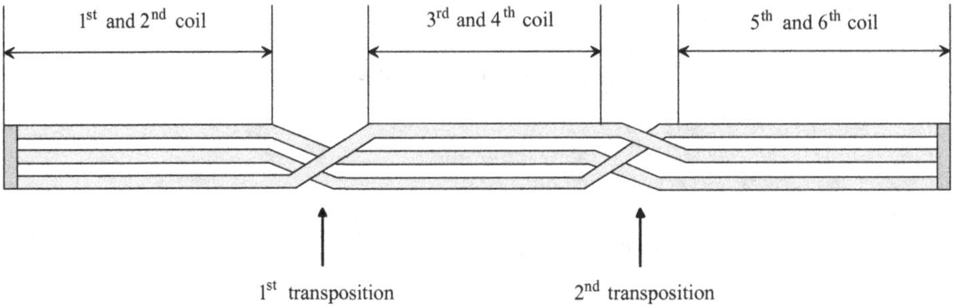

Figure 5.3 Transposition of three parallel-connected conductors is done twice in the end winding area after every two coils in a winding having six coils.

To clarify the meaning of transposition, let us consider Figure 5.4, where two conductors are connected in series and six conductors in parallel. If the subconductors were located in similar positions from slot to slot, in other words there should be no transposition of the conductors, the conductor height h_c to be substituted in Equation (5.18) is $h_c = z_p h_{c0}$, where z_p is the number of parallel-connected conductors (in Figure 5.4 $z_p = 6$) and h_{c0} the height of the subconductor. This means that dividing a conductor into subconductors but not transposing them equates to a solid conductor.

As the skin effect is remarkably smaller in the end winding area, dividing the conductors also helps slightly without transposing. The number of layers z_t in Equation (5.28) is, without transposition, the number of series-connected conductors (in Figure 5.4, $z_t = 2$). If the transposition is done so that every subconductor is located at an equal length in every position of the slot, that is in Figure 5.4 at every six positions, the conductor height in Equation (5.18) is $h_c = h_{c0}$ and the number of layers z_t in Equation (5.28) is the total number of conductors in the slot (in Figure 5.4, $z_t = 12$).

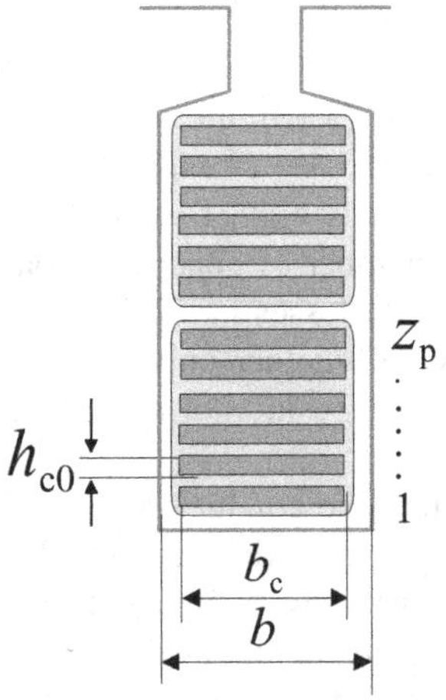

Figure 5.4 Two series-connected conductors ($z_Q = 2$) with six parallel subconductors on top of each other in a slot $z_p = 6$. (1) No transposition: $h_c = z_p h_{c0}$ and $z_t = 2$. (2) Full transposition: $h_c = h_{c0}$ and $z_t = 12$.

Example 5.1: Calculate the reduced conductor height ξ and the resistance factor k_{Ru} for the slot areas of the winding of Figure 5.4 for $h_{c0} = 2$ mm, $b_c = 10$ mm, $b = 14$ mm, $z_p = 6$ and the total number of conductors in the slot is 12, when (a) the winding is not transposed and (b) the winding is fully transposed. The frequency is 50 Hz and the conductor temperature 20 °C.

Solution: The specific conductivity of copper at 20 °C is 57 MS/m:

(a) $h_c = z_p h_{c0} = 6 \times 2$ mm $= 12$ mm and $z_t = 2$. According to Equations (5.18), (5.26), (5.27) and (5.28):

$$\xi = h_c\sqrt{\frac{1}{2}\omega\mu_0\sigma_c\frac{b_c}{b}} = 6\cdot 2\cdot 10^{-3}\sqrt{\frac{1}{2}2\pi\cdot 50\cdot 4\pi\cdot 10^{-7}\cdot 57\cdot 10^6\frac{10}{14}} = 1.076$$

$$\varphi(\xi) = \xi\frac{\sinh 2\xi + \sin 2\xi}{\cosh 2\xi - \cos 2\xi} = 1.076\frac{\sinh(2\cdot 1.076) + \sin(2\cdot 1.076)}{\cosh(2\cdot 1.076) - \cos(2\cdot 1.076)}$$

$$= 1.076\frac{4.241 + 0.836}{4.357 + 0.549} = 1.113$$

$$\psi(\xi) = 2\xi\frac{\sinh\xi - \sin\xi}{\cosh\xi + \cos\xi} = 2\cdot 1.076\frac{\sinh 1.076 - \sin 1.076}{\cosh 1.076 + \cos 1.076}$$

$$= 2.152\frac{1.296 - 0.880}{1.637 + 0.475} = 0.423$$

$$k_{Ru} = \varphi(\xi) + \frac{z_t^2 - 1}{3}\psi(\xi) = 1.113 + \frac{2^2 - 1}{3}0.423 = 1.54.$$

Without transposition the AC resistance is 54 % higher than the DC-resistance in the slot areas. This is caused by the circulating currents in the parallel strands.

(b) $h_c = h_{c0} = 2$ mm and $z_t = 12$:

$$\xi = 2\cdot 10^{-3}\sqrt{\frac{1}{2}2\pi\cdot 50\cdot 4\pi\cdot 10^{-7}\cdot 57\cdot 10^6\frac{10}{14}} = 0.179$$

$$\varphi(\xi) = 0.179\frac{\sinh(2\cdot 0.179) + \sin(2\cdot 0.179)}{\cosh(2\cdot 0.179) - \cos(2\cdot 0.179)} = 0.179\frac{0.366 + 0.351}{1.065 - 0.936} = 0.995$$

$$\psi(\xi) = 2\cdot 0.179\frac{\sinh 0.179 - \sin 0.179}{\cosh 0.179 + \cos 0.179} = 0.359\frac{0.180 - 0.178}{1.016 + 0.984} = 0.000345$$

$$k_{Ru} = \varphi(\xi) + \frac{z_t^2 - 1}{3}\psi(\xi) = 0.995 + \frac{12^2 - 1}{3}0.000345 = 1.01.$$

With ideal transposition the AC resistance, in practice, equals the DC-resistance. The calculation gave only 1 % increase in the AC resistance compared to the DC-resistance. The physical circulating currents are prevented in the transposed winding and the skin effect in the conductors themselves is low at these frequencies.

Transposition is, therefore, a very effective way to reduce the skin effect; in the example, the reduction is from 1.54 to 1.01.

Example 5.2: Calculate the reduced conductor height when the temperature of a copper winding is 50 °C and the frequency is 50 Hz.

Solution: The specific conductivity of copper at 50 °C is 50 MS/m and, according to Equation (5.17),

$$\alpha = \sqrt{\frac{1}{2}\omega\mu_0\sigma_c\frac{b_c}{b}} = \sqrt{\frac{1}{2}2\pi\cdot 50\cdot 4\pi\cdot 10^{-7}\cdot 50\cdot 10^6\frac{b_c}{b}} \approx 1.0\sqrt{\frac{b_c}{b}}\frac{1}{\mathrm{cm}}.$$

According to Equation (5.18),

$$\xi = \alpha h_c \approx \frac{h_c}{[\mathrm{cm}]}\sqrt{\frac{b_c}{\mathrm{b}}}. \tag{5.31}$$

Therefore, the reduced conductor height calculation for 50 Hz machines is simplified to Equation (5.31). If the insulation is thin and $b_c/b \approx 1$ the reduced conductor height corresponds to the height of the conductor in centimeters.

Example 5.3: Derive an equation for the resistance factor of a squirrel cage winding in the locked rotor state when the frequency is 50 Hz and the temperature of the winding 50 °C.

Solution: The number of conductors $z_t = 1$, the ratio $b_c/b = 1$ and the reduced conductor height ξ according to Example 5.2 is $\xi \approx h_c/[\mathrm{cm}]$. According to Equation (5.28), when $z_t = 1$

$$k_{Ru} = \varphi(\xi) + \frac{z_t^2 - 1}{3}\psi(\xi) = \varphi(\xi) = \xi\frac{\sinh 2\xi + \sin 2\xi}{\cosh 2\xi - \cos 2\xi}.$$

Usually, $h_c > 2$ cm, and therefore $\xi > 2$, in which case $\sinh 2\xi \gg \sin 2\xi$, $\cosh 2\xi \gg \cos 2\xi$ and $\sinh 2\xi \approx \cosh 2\xi$, and hence

$$k_{Ru} = \varphi(\xi) \approx \xi \approx h_c/[\mathrm{cm}].$$

The height of the conductor in cm in 50 Hz applications directly corresponds to the resistance factor. In a squirrel cage machine with rotor bars of e.g. 3 cm high, the rotor resistance is tripled at stall and the starting torque of the rotor significantly increased by the skin effect.

Example 5.4: Calculate the resistance factor for a rectangular squirrel cage bar with height $h_c = 50$ mm. The temperature of the bar is 50 °C and the frequency 50 Hz.

Solution: According to Example 5.3,

$$k_{Ru} \approx h_c/[\text{cm}] = 5.$$

More precisely,

$$k_{Ru} = \varphi(\xi) = \xi\frac{\sinh 2\xi + \sin 2\xi}{\cosh 2\xi - \cos 2\xi} = 5.0\frac{\sinh 10 + \sin 10}{\cosh 10 - \cos 10} = 5.0$$

so the approximate equation $k_{Ru} \approx h_c/[\text{cm}]$ gives as good a result as the accurate equation when the temperature is 50 °C, the frequency 50 Hz and the conductor is high.

The skin effect of the end windings is usually negligible. If the end windings are arranged as in Figure 5.5, the reduced conductor height ξ is again obtained from Equation (5.18) by substituting $b_c = z_a b_{c0}$ and $b = b_c + 1.2h$. Richter (1967) gives the following equation for the resistance factor of an end winding:

$$k_{Rw} \approx 1 + \frac{z_t^2 - 0.8}{36}\xi^4. \tag{5.32}$$

If the skin effect changes along a conductor, the conductor has to be divided into different sections for detailed analysis. For instance, the skin effect in slots is notably higher than in the end windings. The proportion of the end winding length can be high, even more than 50 % of the total winding length, and therefore the end winding resistance behavior, of course, has a significant effect on the total resistance and possible circulating currents. If the resistance

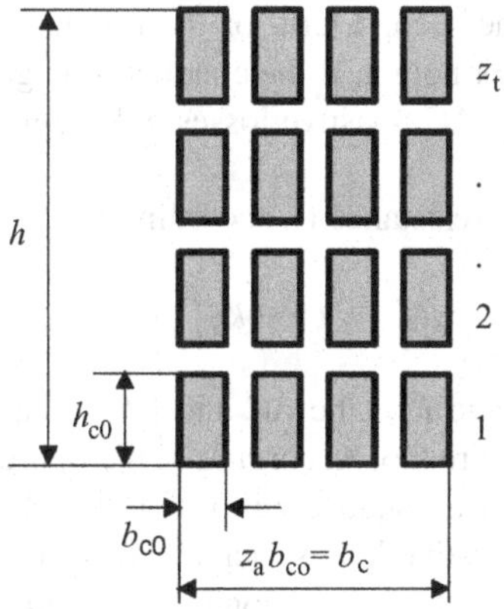

Figure 5.5 Imaginary arrangement of an end winding for the calculation of the reduced conductor height and the resistance factor.

factor in the slots is k_{Ru} as calculated above and in the end winding k_{Rw}, the resistance factor of a total winding in case of ideal transposition is

$$k_{\mathrm{R}} = \frac{k_{\mathrm{Ru}} R_{\mathrm{uDC}} + k_{\mathrm{Rw}} R_{\mathrm{wDC}}}{R_{\mathrm{uDC}} + R_{\mathrm{wDC}}} = k_{\mathrm{Ru}} \frac{R_{\mathrm{uDC}}}{R_{\mathrm{uDC}} + R_{\mathrm{wDC}}} + k_{\mathrm{Rw}} \frac{R_{\mathrm{wDC}}}{R_{\mathrm{uDC}} + R_{\mathrm{wDC}}}, \tag{5.33}$$

where R_{uDC} and R_{wDC} are the DC resistances of the slot part and winding end part of the winding, respectively.

Assuming the temperature to be the same everywhere in the conductor, the resistances are proportional to the conductor lengths. Denoting the average length of the coil turn as l_{av} and the equivalent length of the iron core as l', the resistance factor is

$$k_{\mathrm{R}} = k_{\mathrm{Ru}} \frac{2l'}{l_{\mathrm{av}}} + k_{\mathrm{Rw}} \frac{l_{\mathrm{av}} - 2l'}{l_{\mathrm{av}}}. \tag{5.34}$$

Usually, the skin effect in the end windings is negligible, thus $k_{\mathrm{Rw}} = 1$ and

$$k_{\mathrm{R}} = 1 + (k_{\mathrm{Ru}} - 1) \frac{2l'}{l_{\mathrm{av}}}. \tag{5.35}$$

If the winding is not ideally transposed there will be circulating currents in the parallel conductors and the resistance factor calculation becomes very complicated. Transposition cannot always be arranged ideally and then special measures have to be taken into use to define the resistance factor. For example, Hämäläinen *et al.* (2013) has suggested a method to calculate the circulating currents in case of nonideal transposition.

5.2.2 *Critical Conductor Height in Slot*

If the conductor height h_{c} in Figure 5.1 increases and the width b_{c} remains constant, the DC resistance of the conductor decreases in inverse proportion to the height. The reduced conductor height ξ (Equation 5.18) in the slot region increases linearly in proportion to the conductor height h_{c}, and the second term of the resistance factor (5.29) increases with the fourth power of the conductor height. If the conductor height is high, the resistance factor may have a very high value and AC resistive losses will be multiple compared with the DC resistive losses of the winding.

In the area $0 \le \xi \le 1$, the AC resistance is, according to Equation (5.29),

$$R_{\mathrm{AC}} \approx (1 + k\xi^4) R_{\mathrm{DC}}, \tag{5.36}$$

where k is a constant and R_{AC} and R_{DC} the AC and DC resistances. Equation (5.36) and its components are presented in Figure 5.6. According to the figure, there is a measure $h_{\mathrm{c}} = h_{\mathrm{c,cr}}$, with which the AC resistance of a conductor inserted in a slot has a minimum. Above this critical conductor height, the resistive losses increase with increasing conductor height. The resistance factor corresponding to the critical conductor height for copper and 50 Hz is

$$k_{\mathrm{Rcr}} \approx 1.33. \tag{5.37}$$

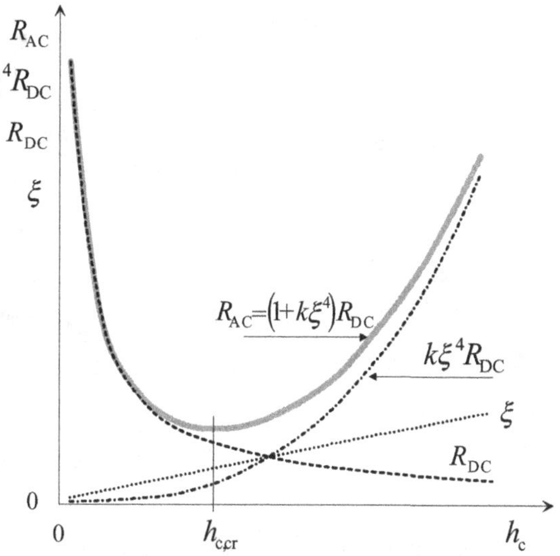

Figure 5.6 AC resistance R_{AC} has its minimum at the critical conductor height $h_{c,cr}$.

5.2.3 Methods to Limit the Skin Effect

The skin effect can be limited with the following methods:

1. The most effective method is to divide the conductors into subconductors and transpose them as shown in Section 5.2.1.
2. Instead of dividing the bulky conductors into subconductors and transposing them, we can use parallel paths, that is, parallel-connected poles and groups of poles. If the number of parallel paths is a, the series-connected turns must be a-fold in order to keep the induced voltage with the same air-gap flux density constant. This means that the conductor height is reduced in the ratio of $1/a$.
3. Using multi-thread, twisted conductors with perfect conductor transposition is also a very effective way to limit the skin effect. An example of such an arrangement is the Roebel bar or Litz wire shown in Figure 5.7.

In a Roebel bar, a constant shift takes place between the layers. This way, the sums of partial fluxes of the integration paths causing eddy currents are removed. The skin effect is thus effectively minimized. However, some skin effect occurs in the subconductors. One way to diminish this skin effect is to employ subconductors with as small a cross-sectional area as possible. Roebel bars are common in applications in which there is only a single effective turn per coil. Usually, the subconductors of a Roebel bar are connected together at the end winding areas, where the skin effect is remarkably lower than in the slots.

Litz wire may also be used. Litz wires are offered as tape-insulated rectangular conductors in a wide range of dimensions either with insulated or noninsulated strands. The strands are collected in bundles which are then twisted around a center bundle. Therefore, the transposition is not ideal but is still very effective. It seems to be very effective to use such Litz wires in

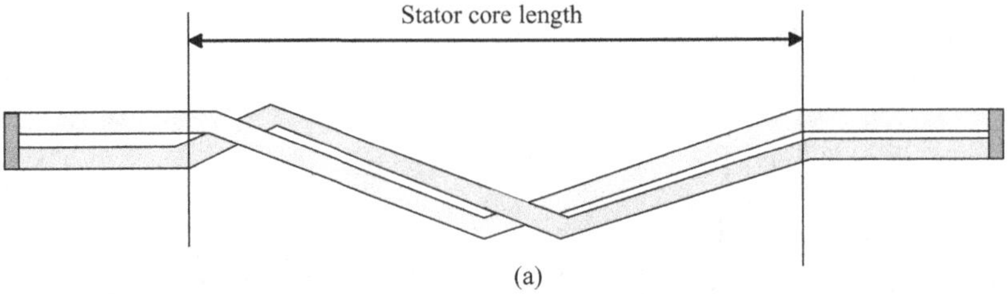

(a)

(b) Roebel bar from top

(c) Roebel bar from one side

(d) Litz wire with multiple either insulated or non-insulated strands typically with 0.5–1 mm diameter each

Figure 5.7 Transposing: (a) Simplified presentation of two adjacent subconductors. (b) and (c) Photographs illustrating the construction of a Roebel bar. A Roebel bar has minimal skin effect in a slot because of its perfect conductor transposition. (d) Photograph illustrating a Litz wire with several strands. (Photo by Juha Haikola, LUT.)

low-voltage high-power machines having high conductors. Even the version without strand insulation offers an effective alternative to traditional windings at frequencies lower than ca. 100 Hz (Hämäläinen *et al.* 2013).

5.2.4 *Inductance Factor*

The skin effect depends on the leakage flux of a conductor, which also changes the leakage inductance L_σ of the conductor. The change in the leakage inductance is notable, for instance at the beginning of starting up squirrel cage motors. A one-sided skin effect takes place in the rotor bars, and the current density increases towards the slot opening. The leakage flux in the conductor area crowds towards the slot opening, which in turn leads to a reduced leakage flux linkage, and thus to a reduced slot leakage inductance L_σ, as given in Section 4.3.3. If we denote the leakage inductance with direct current $L_{\sigma\mathrm{DC}}$, we may write for the inductance factor

$$k_\mathrm{L} = \frac{L_\sigma}{L_{\sigma\mathrm{DC}}}. \tag{5.38}$$

The inductance factor can be calculated from Equations (4.90)–(4.93).

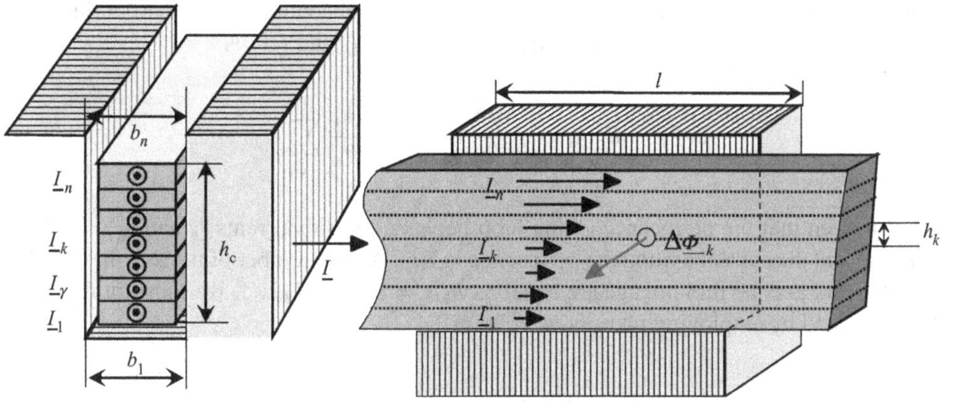

Figure 5.8 Division of solid conductors into several layers or subconductors in order to define the skin effect. The current of a conductor is positioned in the middle of imaginary subconductors, arrows 1 to n. The width of the slot varies from b_1 to b_n. The length of the iron core is l.

5.2.5 Calculation of Skin Effect in Slots Using Circuit Analysis

Besides the method presented in Section 5.2.1, the skin effect can be analyzed by applying circuit theory. This method is very suitable for rotor bars of arbitrary cross-sectional shape, for example for double-cage rotors. In this method, a solid conductor is divided into n imaginary layers or real subconductors as depicted in Figure 5.8. The height of the conductor is h_c and the height of the kth subconductors is h_k. The width of the slot varies over b_1 to b_n, and also the conductor width may vary. This usually happens when the skin effect of a squirrel cage die-cast bar is evaluated. The length of the iron core is l. The conductor is assumed to be of infinite length, and thus the division of the conductor does not have any influence on the shape of the current density distribution. Further, it is assumed that no skin effect occurs inside a subconductor.

The currents in subconductors k and $k+1$ are $\underline{I}_k$ and $\underline{I}_{k+1}$. It is assumed that the currents flow along the center lines of the subconductors. In the steady state, the voltage induced in the kth conductor by the slot leakage flux is

$$\underline{E}_k = -\mathrm{j}\omega\Delta\underline{\Phi}_k = R_k\underline{I}_k - R_{k+1}\underline{I}_{k+1}, \tag{5.39}$$

where $\Delta\Phi_k$ is the leakage flux flowing between the kth and $(k+1)$th subconductors (actually, between currents $\underline{I}_k$ and $\underline{I}_{k+1}$). The induced voltage creates a circulating current limited by the resistances $\underline{R}_k$ and $\underline{R}_{k+1}$ of the layers as it is indicated in (5.39)

The flux density B_k in the subconductor k between the currents $\underline{I}_k$ and $\underline{I}_{k+1}$ depends on the current linkage Θ_k calculated from the bottom of the slot to the subconductor k

$$B_k = \mu_0\frac{\Theta_k}{b_k} = \mu_0\frac{1}{b_k}\sum_{\gamma=1}^{k} i_\gamma, \tag{5.40}$$

where b_k is the width of the slot at the position of the kth subconductor. The partial flux $\Delta\Phi_k$ through the area ($l{\cdot}h_k$) limited by the paths of the currents $\underline{I}_k$ and $\underline{I}_{k+1}$ is

$$\Delta\Phi_k = B_k l h_k = \mu_0 \frac{\Theta_k}{b_k} l h_k = \mu_0 \frac{l h_k}{b_k} \sum_{\gamma=1}^{k} i_\gamma. \tag{5.41}$$

Here it is seen that the permeance of the loop between the subcurrents $\underline{I}_k$ and $\underline{I}_{k+1}$ is $\Lambda_k = \mu_0(lh_k/b_k)$ and the current exiting the permeance is Σi_γ. As the number of turns in the "coil" is $N = 1$ the flux and the flux linkage are equal, $\Delta\Phi_k N = \Delta\Psi_k = L_k \Sigma i_\gamma$. By substituting (5.41) into (5.39), and by employing phasors, we obtain

$$\underline{E}_k = -\mathrm{j}\omega\mu_0 \frac{l h_k}{b_k} \sum_{\gamma=1}^{k} \underline{I}_\gamma = R_k \underline{I}_k - R_{k+1} \underline{I}_{k+1}. \tag{5.42}$$

Next, the current is solved:

$$\underline{I}_{k+1} = \frac{R_k}{R_{k+1}} \underline{I}_k + \frac{\mathrm{j}\omega\mu_0}{R_{k+1}} \frac{l h_k}{b_k} \sum_{\gamma=1}^{k} \underline{I}_\gamma = \frac{R_k}{R_{k+1}} \underline{I}_k + \mathrm{j} \frac{\omega L_k}{R_{k+1}} \sum_{\gamma=1}^{k} \underline{I}_\gamma. \tag{5.43}$$

The flux linkage $\Delta\Psi_k$ of the one-turn loop between the paths of the subcurrents $\underline{I}_k$ and $\underline{I}_{k+1}$ forms an inductance L_k determined by

$$L_k = \Lambda_k N^2 = \mu_0 \frac{l h_k}{b_k} 1^2 = \mu_0 \frac{l h_k}{b_k}. \tag{5.44}$$

This inductance is a partial inductance linked with the area ($l{\cdot}h_k$) limited by the loop formed by the currents $\underline{I}_k$ and $\underline{I}_{k+1}$.

If an initial value for the current $\underline{I}_1$ is known, we obtain from Equation (5.43) the current of the next subconductor $\underline{I}_2$

$$\underline{I}_2 = \frac{R_1}{R_2} \underline{I}_1 + \mathrm{j} \frac{\omega L_1}{R_2} \underline{I}_1. \tag{5.45}$$

If the initial value for the current $\underline{I}_1$ is not known, we can choose an arbitrary value for it, for instance 1 A.

The current of the next subconductor $\underline{I}_3$ is obtained:

$$\underline{I}_3 = \frac{R_2}{R_3} \underline{I}_2 + \mathrm{j} \frac{\omega L_2}{R_3} (\underline{I}_1 + \underline{I}_2) \tag{5.46}$$

and generally

$$\underline{I}_{k+1} = \frac{R_k}{R_{k+1}} \underline{I}_k + \mathrm{j} \frac{\omega L_k}{R_{k+1}} (\underline{I}_1 + \underline{I}_2 + \cdots + \underline{I}_k). \tag{5.47}$$

This way, all the subcurrents of the bar are determined current by current. Now, the total current of the bar is obtained:

$$\underline{I} = \sum_{\gamma=1}^{n} \underline{I}_{\gamma}. \tag{5.48}$$

Normally, a total current I of a bar is given, and thus the subcurrents have to be iterated until the sum of the currents is equal to the total current of the bar.

If the width of the slot is constant and the height of each subconductor is selected equal, h_k, all the resistances R_k of the subconductors are equal as well as the inductances L_k. In that case, in Equation (5.43), the ratio of resistances $R_k/R_{k+1} = 1$, and correspondingly the ratio $\omega L_k/R_{k+1} = c =$ constant, and we may write

$$\underline{I}_{k+1} = \underline{I}_k + \mathrm{j}c(\underline{I}_1 + \underline{I}_2 + \cdots + \underline{I}_k). \tag{5.49}$$

Example 5.5: Consider a rectangular copper armature conductor of width $b_c = 15$ mm in a rectangular slot of width $b = 19$ mm. The height of the conductor is divided into seven subconductors with height $h_{k,} = 4$ mm. The length of the stator stack is 700 mm and the frequency of the machine is 75 Hz. The temperature of the copper is 100 °C. Calculate the skin effect in the subconductors and draw a phasor diagram of the currents and voltages when 1200 A is flowing in the bar.

Solution: At room temperature (20 °C) $\rho_{\mathrm{Cu}} = 1/(57 \times 10^6\ \mathrm{S/m}) = 1.75 \times 10^{-8}\ \Omega\,\mathrm{m}$. The temperature coefficient of resistivity for copper is $\alpha_{\mathrm{Cu}} = 3.81 \times 10^{-3}/\mathrm{K}$. At 100 °C the resistivity of copper is

$$\rho_{\mathrm{Cu,100^\circ C}} = \rho_{\mathrm{Cu,\,20^\circ C}}(1 + \Delta T\alpha_{\mathrm{Cu}}) = 1.75 \cdot 10^{-8}\ \Omega\mathrm{m}\,(1 + 80\,\mathrm{K} \cdot 3.81 \cdot 10^{-3}/\mathrm{K})$$
$$= 2.28 \cdot 10^{-8}\ \Omega\mathrm{m}$$

All the resistances R_{k} of the subconductors and the inductances L_{k} are now equal. The DC resistance of a subconductor is

$$R_{\mathrm{DC},k} = \frac{\rho l}{S_{\mathrm{c}}} = \frac{2.28 \cdot 10^{-8}\ \Omega\mathrm{m}\ 0.7\ \mathrm{m}}{0.015\ \mathrm{m} \cdot 0.004\ \mathrm{m}} = 0.266\ \mathrm{m}\Omega.$$

The inductance of a subconductor is

$$L_k = \mu_0 \frac{l h_k}{b} = 4\pi \cdot 10^{-7} \frac{\mathrm{Vs}}{\mathrm{Am}} \frac{0.7\,\mathrm{m} \cdot 0.004\,\mathrm{m}}{0.019\ \mathrm{m}} = 0.185\ \ \mu\mathrm{H}$$

The ratio of all resistances $R_k/R_{k+1} = 1$, and the ratio

$$c = \frac{\omega L_{\mathrm{k}}}{R_{\mathrm{k}}} = \frac{2\pi 75 \frac{1}{\mathrm{s}} \cdot 0.185\ \mu\mathrm{H}}{0.266\ \mathrm{m}\Omega} = 0.327$$

We are now ready to utilize Equation (5.49) to calculate the currents in the seven subconductors. If we fix the phase angle of the lowest bar to zero, we get the following division for the currents:

	Current, I [A]	Per unit current	Phase angle [°]	Current density J [A/mm^2]
Conductor 1	78.33	0.066	0	1.30
Conductor 2	82.41	0.069	18.1	1.37
Conductor 3	103.91	0.087	47.6	1.73
Conductor 4	155.28	0.130	76.4	2.58
Conductor 5	241.46	0.202	101.0	4.02
Conductor 6	370.80	0.310	123.9	6.17
Conductor 7	561.74	0.470	146.3	9.37
Whole bar	1200	1.0	112.4	Average 2.86

Figure 5.9 illustrates the sum of current phasors corresponding to Equation (5.49) for the bar of Example 5.5. The figure shows clearly how the phase angle of the current changes radically and how most of the current crowds in those parts of the bar which are located close to the slot opening. Since the skin effect constantly increases when approaching the slot opening, the situation is called a single-sided skin effect. A current solution reached with RMS values does not give exact information about the flux density distribution, since it should be defined by the instantaneous values of the current. The RMS values are, however, suitable for solving the loss distribution in the slot.

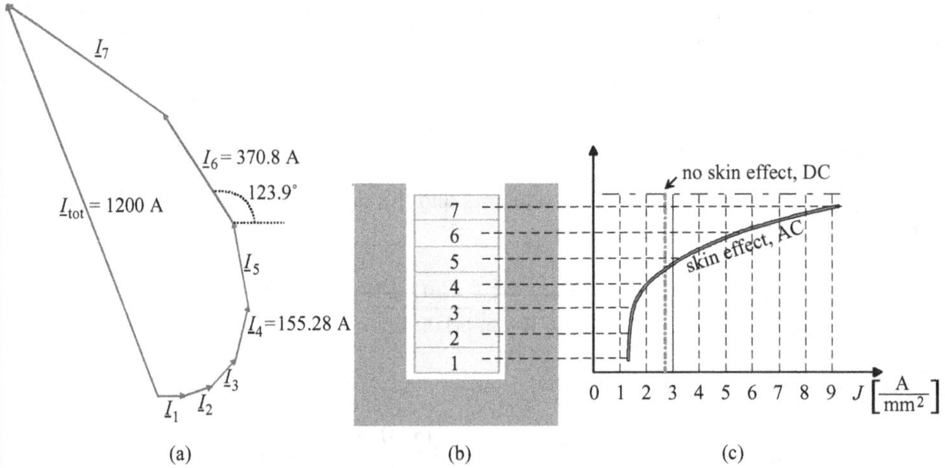

Figure 5.9 Skin effect according to Example 5.5: (a) phasor diagram of the currents of a conductor divided into seven subconductors (4 mm × 15 mm) with $c = 0.327$; (b) the bar with seven subconductors, (c) current density. With a direct current, no skin effect occurs, and the current density J is evenly distributed in the cross-sectional area of a conductor. With an alternating current at 75 Hz, the skin effect causes a large increase in the current density when approaching the slot opening.

We will now further examine the voltage behavior of the subconductors in a slot. The resistive voltage drop of the kth subconductor is

$$\underline{U}_{rk} = R_k \underline{I}_k. \tag{5.50}$$

The voltage drop of the $(k + 1)$th subconductor is calculated accordingly. The result is substituted in Equation (5.42), and we obtain

$$\underline{U}_{rk} - \underline{U}_{r(k+1)} = -j\omega\mu_0 \frac{l h_k}{b_k} \sum_{\gamma=1}^{k} \underline{I}_\gamma, \tag{5.51}$$

which leads to

$$\underline{U}_{r(k+1)} = \underline{U}_{rk} + j\omega L_k \sum_{\gamma=1}^{k} \underline{I}_\gamma = R_{k+1} \underline{I}_{k+1}. \tag{5.52}$$

On the surface of the conductor, we may write for the AC impedance of the bar

$$\underline{U}_{AC} = \underline{U}_{rn} + j\omega L_n \underline{I}. \tag{5.53}$$

The topmost inductance L_n differs from L_k (5.44) because the height of the last loop is only half of the height h_k between subcurrents $\underline{I}_k$ and $\underline{I}_{k+1}$. Now, as there are no more subconductors, the loop height linked by the last flux component is $h_k/2$. Therefore, also for a rectangular slot with similar rectangular subconductors, the inductance of the topmost conductor is

$$L_n = L_k/2. \tag{5.54}$$

The voltage over the topmost bar consists of the resistive voltage drop over the bar and the inductive voltage drop caused by all the currents of the subconductors

$$\underline{U}_{AC} = R_n \underline{I}_n + j\omega L_n \sum_{\gamma=1}^{n} \underline{I}_\gamma. \tag{5.55}$$

The voltage $\underline{U}_{AC}$ is the same over the last subconductor on the top of the bar as the whole conductor: $\underline{U}_{AC} = R_n \underline{I}_n + j\omega L_n \underline{I}_{tot}$ The total current $\underline{I}_{tot}$ meets the total effective resistance R_{AC} and inductance L_{AC} : $\underline{U}_{AC} = (R_{AC} + j\omega L_{AC})\underline{I}_{tot}$. Therefore, a comparison of the equations enables us to get the AC resistance with respect to the skin effect: $(R_{AC} + j\omega L_{AC})\underline{I}_{tot} = (R_n \underline{I}_n + j\omega L_n \underline{I}_{tot})$ from which we obtain

$$(R_{AC} + j\omega L_{AC}) = R_n \frac{\underline{I}_n}{\underline{I}_{tot}} + j\omega L_n. \tag{5.56}$$

Then the total effective AC resistance of the conductor with respect to the skin effect is the real right-hand side of Equation (5.56) and the reactance is its imaginary part.

Example 5.6: Calculate the resistive and reactive voltages and the AC impedance of the seven subconductor bars in Example 5.5.

Solution: From Example 5.5 $R_k = 0.266\ \text{m}\Omega$, $L_k = 0.185\ \mu\text{H}$, $\text{j}\omega L_k = \text{j}87.2\ \mu\Omega$, $L_n = 0.0925\ \mu\text{H}$, $\text{j}\omega L_n = \text{j}43.6\ \mu\Omega$:

	Current, I_k [A]	Phase [°]	$\underline{U}_{rk} = R_k \underline{I}_k [V]$	$\text{j}\omega L_k \sum_{\gamma=1}^{k} \underline{I}_\gamma [\text{V}]$	Phase [°]	$I_k^2 R_k [W]$
Conductor 1	78.33	0	0.0208	0.006 83	90	1.62
Conductor 2	82.41	18.1	0.0218	0.0138	99.3	1.80
Conductor 3	103.91	47.6	0.0277	0.0216	114.3	2.88
Conductor 4	155.28	76.4	0.0412	0.0318	144.9	6.40
Conductor 5	241.46	101.0	0.0641	0.0467	156.2	15.5
Conductor 6	370.80	123.9	0.0984	0.0696	179.3	36.4
Conductor 7	561.74	146.3	0.149	0.0525	202.4	83.7
Whole bar	1200	112.4				148.3

$$\underline{U}_{\text{AC}} = \underline{U}_{r7} + \text{j}\omega L_n \underline{I} = 0.149\,\text{V} \cdot \cos(146.3° - 112.4°) + \text{j}0.149\,\text{V}\ \sin(146.3° - 112.4°) + \text{j}0.0525\text{V}$$

$$\underline{U}_{\text{AC}} = (0.124 + \text{j}0.1355)\text{V}$$

with the total current phase of 112.4° as a reference direction.

Finally, for the AC impedance of the total bar at 75 Hz we get

$$(R_{\text{AC}} + \text{j}\omega L_{\text{AC}}) = \frac{\underline{U}_{\text{AC}}}{\underline{I}_{\text{tot}}} = \frac{(0.124 + \text{j}0.1355)\text{V}}{1200\text{A}} = 0.103\,\text{m}\Omega + \text{j}0.113\,\text{m}\Omega.$$

To check:

$$(R_{\text{AC}} + \text{j}\omega L_{\text{AC}}) = R_n \frac{\underline{I}_n}{\underline{I}_{\text{tot}}} + \text{j}\omega L_n$$

$$(R_{\text{AC}} + \text{j}\omega L_{\text{AC}}) = R_7 \frac{\underline{I}_7}{\underline{I}_{\text{tot}}} + \text{j}\omega L_7 = 266\ \mu\Omega \frac{561.74\,\text{A}|\underline{146.3°}}{1200\,\text{A}|\underline{112.4°}} + \text{j}\frac{87.2}{2}\ \mu\Omega,$$

$$(R_{\text{AC}} + \text{j}\omega L_{\text{AC}}) = 124.5\ \mu\Omega|\underline{33.9°} + 43.6\ \mu\Omega|\underline{90°},$$

$$(R_{\text{AC}} + \text{j}\omega L_{\text{AC}}) = 103\ \mu\Omega + \text{j}69.4\ \mu\Omega + \text{j}43.6\ \mu\Omega = 103\ \mu\Omega + \text{j}113\ \mu\Omega,$$

which is the same result as above.

Figure 5.10 illustrates the currents and the voltages over the subconductors of the conductor in Figure 5.9.

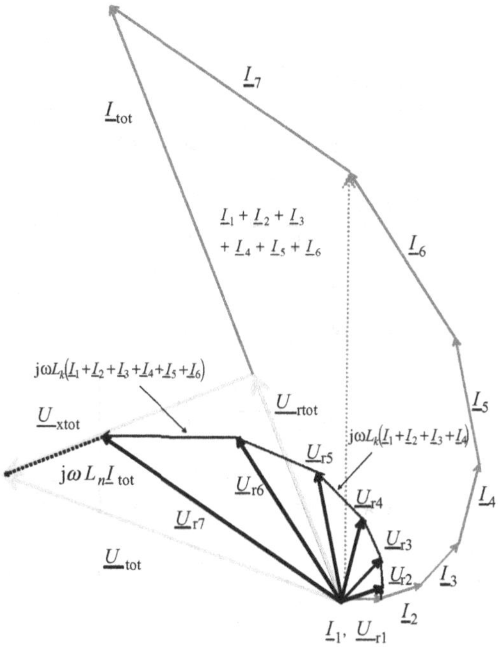

Figure 5.10 Skin effect of a seven-subconductor bar corresponding to Examples 5.5 and 5.6. The diagram is drawn as follows. The current $\underline{I}_1$ in the resistance R_1 creates a voltage difference $\underline{U}_{r1}$ parallel to $\underline{I}_1$. The reactive voltage $\underline{U}_{x1}$ is perpendicular to $\underline{I}_1$. The current $\underline{I}_2$ in the resistance R_2 creates a voltage $\underline{U}_{r2}$ parallel to $\underline{I}_2$. The reactive voltage $\underline{U}_{x2}$ is perpendicular to $\underline{I}_1 + \underline{I}_2$, and so on. The reactive voltage $\underline{U}_{xtot}$ is perpendicular to $\underline{I}_{tot}$.

Based on Equations (5.50)–(5.55), the skin effect of a conductor divided into subconductors can be illustrated by the equivalent circuit in Figure 5.11. The figure shows that the voltage defined by Equation (5.55) occurs in the terminals of the equivalent circuit, through which the total current I of the bar flows. The voltage $\underline{U}_{AC}$ can thus be called a terminal voltage of a conductor divided into subconductors, or it can also be called the voltage of a solid conductor divided into imaginary sections, and $R_{AC} + j\omega L_{AC}$ is the AC impedance of the conductor.

The above given explanation could also be given in the following form: The equivalent circuit of Figure 5.11 consists of T-equivalent circuits. In each of the T-circuits the inductances are $L_k/2$. These inductances are calculated corresponding to the leakage fluxes traveling through of an area limited by the center-line of the conductor and the conductor edge $l \cdot h_k/2$. When two adjacent T-circuits are connected together the total inductance between resistances R_k and R_{k+1} is $2 \cdot L_k/2 = L_k$ as shown in Figure 5.11. As the last T-equivalent circuit does no more connect to a following T-circuit the last inductance of the equivalent circuit is $L_k/2$.

When a bar of an armature winding is divided into insulated subconductors that travel transposed from slot to slot to avoid the skin effect, the previous equations may be utilized to analyze the inductance of individual subconductors. When a series-connected individual subconductor is transposed to all possible positions in different slots, the average inductances

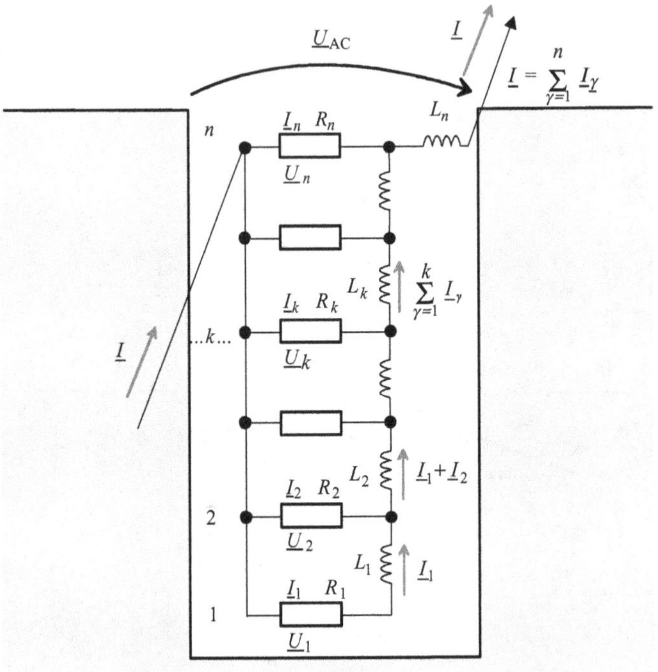

Figure 5.11 Equivalent circuit of a conductor in a slot divided into subconductors of varying width, corresponding to the phasor diagram of Figure 5.10. U_{AC} describes the voltage over the conductor at alternating current. In Figure 5.10, the voltage U_{tot} corresponds to this voltage.

of all subconductors become equal and the distribution of currents is equal between subconductors. When a perfect transposition is achieved, all the subconductor currents remain the same, which makes evaluation of the inductance of a subconductor easier. If the width of the slot b is constant and all the subconductors have the same height, we may, instead of summing the currents, multiply the inductance by the position number. We modify Equation (5.44) to get a new inductance for insulated parallel subconductors

$$L_k = k\mu_0 \frac{lh_k}{b_k}. \tag{5.57}$$

Now, the equivalent circuit of the connected subconductors simplifies to an equivalent circuit of the insulated, ideally transposed system in Figure 5.12 that represents the equivalent circuit of one slot. The inductance of a conductor in a phase winding is $L = L_k + 2L_k + \cdots + nL_k$, which is equal to all the conductors of the slot, so there is no skin effect present except for the skin effect of individual subconductors.

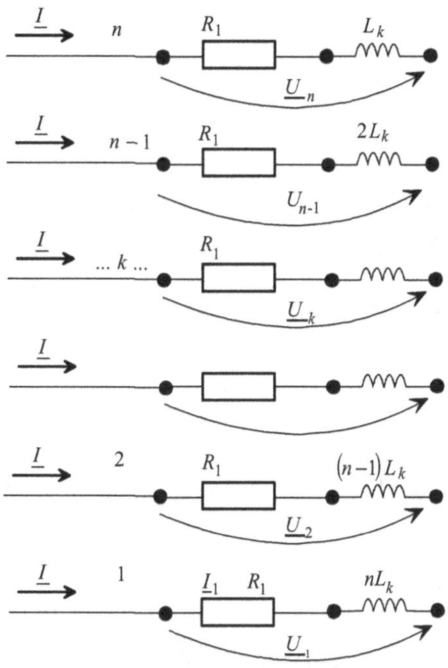

Figure 5.12 Equivalent circuit of a slot with subconductors that are transposed in winding ends so that there is no large skin effect present and all the subconductors carry the same current. There is no common AC voltage for such conductors over a single slot. The effect of the flux created by other conductors is taken into account by multiplying the inductance of each subconductor.

5.2.6 *Double-Sided Skin Effect*

In slots carrying conductors on top of each other, a double-sided skin effect is possible if the windings are not fully transposed. A similar effect may be observed in double-squirrel-cage rotors. We study a simple case in which there are two conductors on top of each other in a slot. The uppermost bar experiences, in addition to the flux created by it, the flux created by the lower bar, see Figure 5.13.

The behavior of currents may be studied in the same way as in the previous section for the single-sided skin effect. Here both the upper and the lower bars are divided into real or imaginary subconductors. The upper bar that is divided into imaginary subconductors experiences a flux created by a lower conductor. This makes the investigation of the skin effect of the upper bar somewhat more complicated than in the previous case.

The currents of the upper subconductors k have to be divided into two imaginary sections: $\underline{I}'_k$, caused by the flux of the underlying conductor; and $\underline{I}''_k$, which would be the current of the subconductor k if it were alone in the slot. That is,

$$\underline{I}_k = \underline{I}''_k + \underline{I}'_k. \tag{5.58}$$

Thus, a phasor diagram similar to the diagram in Figure 5.10 is valid as such for the subcurrents $\underline{I}''_k$. Now, we have to define the subcurrents $\underline{I}'_k$. A simple calculation method can

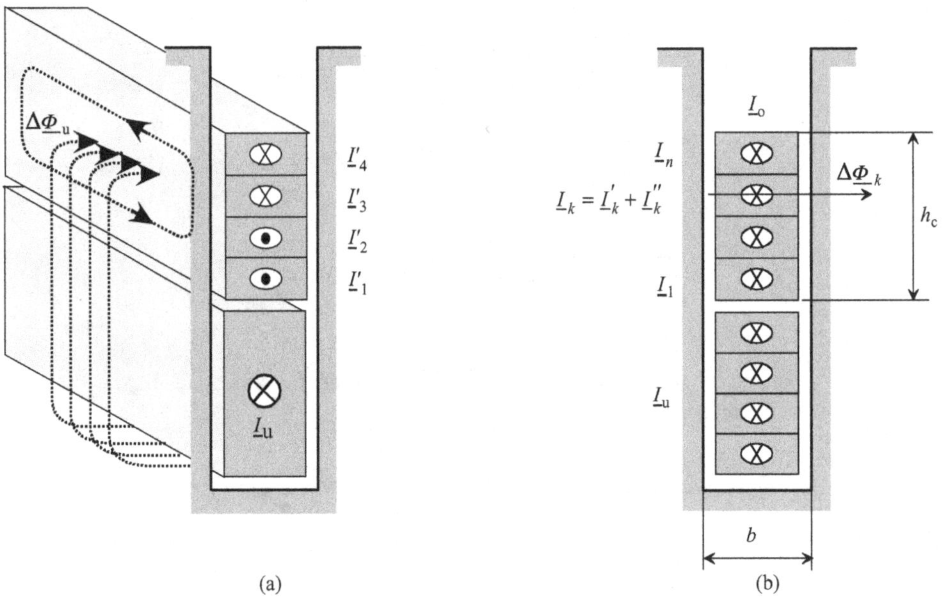

Figure 5.13 Calculation of the skin effect in a bar located above another current-carrying bar in a slot. (a) The flux created by the bottom conductor current alone and the corresponding induced currents $\underline{I}'_k$ opposing the flux created by the bottom bar. (b) Cut-away view of a slot; $\underline{I}_o$ is the total current in the upper conductor and $\underline{I}_u$ that in the bottom conductor.

be found in this case. The currents of the lower bar $\underline{I}_u$ create a time-varying magnetic flux density in the area of the upper bar. The eddy current pattern created by this flux density has to be symmetrical with respect to the center line of the upper conductor. Consequently, it is advisable to divide the upper bar into an even number of subconductors. The subconductors of the upper bar carry eddy currents that, according to Lenz's law, attempt to cancel the flux created by the lower bar current.

Therefore, the eddy current components at equal distances below and above the center of the conductor have to be equal in magnitude and in opposite directions:

$$\underline{I}'_k = -\underline{I}'_{n-k+1}. \tag{5.59}$$

In Figure 5.13, the currents $\underline{I}'_1$ and $\underline{I}'_2$ travel towards the observer, and the currents $\underline{I}'_3$ and $\underline{I}'_4$ run in the same direction as the sum of currents of the lower conductor. Hence, we get two current loops opposing the flux components $\Delta\underline{\Phi}_u$.

The flux created by the current $\underline{I}_u$ of the lower bar $\Delta\underline{\Phi}_u$ induces a current $\underline{I}'_{k+1}$ in the upper subconductor $k+1$. This current $\underline{I}'_{k+1}$ can be calculated in the $(k+1)$th imaginary subconductor analogously according to Equation (5.43)

$$\underline{I}'_{k+1} = \frac{R_k}{R_{k+1}}\underline{I}'_k + \mathrm{j}\frac{\omega L_k}{R_{k+1}}\left(\sum_{\gamma=1}^{k}\underline{I}'_\gamma + \underline{I}_u\right). \tag{5.60}$$

If we select $R_k/R_{k+1} = 1$, and $\omega L_k/R_{k+1} = c$, we get

$$\underline{I}'_{k+1} = \underline{I}'_k + \mathrm{j}c(\underline{I}'_1 + \underline{I}'_2 + \cdots + \underline{I}'_k + \underline{I}_\mathrm{u}). \tag{5.61}$$

With an even n, the subconductors $k = n/2$ and $k + 1 = n/2 + 1$ constitute the center of the conductor. Here, we may write for the currents, according to Equations (5.60) and (5.61),

$$\underline{I}'_{n/2+1} = -\underline{I}'_{n/2} = \underline{I}'_{n/2} + \mathrm{j}c\left(\sum_{\gamma=1}^{n/2} \underline{I}'_\gamma + \underline{I}_\mathrm{u}\right) = \underline{I}'_{n/2} + \mathrm{j}c\underline{I}_\mu. \tag{5.62}$$

In the brackets, we have the sum of the currents of the lower conductor and the imaginary currents in the lower half of the upper conductor caused by the lower conductor $\underline{I}_\mathrm{u}$. We solve Equation (5.62) for the subconductor $k = n/2$ lying just below the center line of the upper bar

$$\underline{I}'_{n/2} = -\frac{\mathrm{j}c\underline{I}_\mu}{2} = -\frac{\mathrm{j}c}{2}\left(\sum_{\gamma=1}^{n/2} \underline{I}'_\gamma + \underline{I}_\mathrm{u}\right). \tag{5.63}$$

Again, we have to find an iterative solution. It seems easiest to make an initial guess for $\underline{I}_\mu$ after which we are able to determine the currents of the subconductors. We now write for the other lower subconductors of the upper bar, according to Equation (5.60),

$$\underline{I}'_{n/2-1} = \underline{I}'_{n/2} - \mathrm{j}c\left(\sum_{\gamma=1}^{n/2-1} \underline{I}'_\gamma + \underline{I}_\mathrm{u}\right) = \underline{I}'_{n/2} - \mathrm{j}c(\underline{I}_\mu - \underline{I}'_{n/2}), \tag{5.64}$$

and so on, until we get for the lowest subconductor of the upper bar

$$\underline{I}'_1 = \underline{I}'_2 - \mathrm{j}c(\underline{I}'_1 + \underline{I}_\mathrm{u}) = \underline{I}'_2 - \mathrm{j}c\left(\underline{I}_\mu - \sum_{\gamma=1}^{n/2} \underline{I}'_\gamma\right). \tag{5.65}$$

As the currents from $\underline{I}'_1$ to $\underline{I}'_{n/2}$ have now been defined, we also know the currents from $\underline{I}'_{n/2+1}$ to $\underline{I}'_n$ based on Equation (5.59). With Equation (5.62), the sum of the currents of the lower conductor can also be defined as

$$\underline{I}_\mathrm{u} = \underline{I}_\mu - \sum_{\gamma=1}^{n/2} \underline{I}'_\gamma. \tag{5.66}$$

Example 5.7: Consider a double-layer winding with rectangular copper conductors divided into four subconductors of width $b_\mathrm{c} = 15$ mm in a rectangular slot of width $b = 19$ mm. The height of the upper bars of 16 mm is divided into four subconductors with a height $h_k = 4$ mm. The length of the stator stack is 700 mm and the frequency of the machine is 50 Hz. The copper temperature is 100 °C. Calculate the skin effect in the subconductors if no transposing of the windings is done, when 1000 A in the same phase is flowing in both of the conductors. The skin effect inside individual subconductors can be neglected.

Solution: All the resistances R_k of the subconductors and the inductances L_k are now equal. The DC resistance of a subconductor is $R_{DC,k} = 0.266\,\mathrm{m\Omega}$. The inductance of a subconductor is

$$L_k = \mu_0 \frac{l h_k}{b} = 4\pi \cdot 10^{-7} \frac{\mathrm{Vs}}{\mathrm{Am}} \frac{0.7\,\mathrm{m} \cdot 0.004\,\mathrm{m}}{0.019\,\mathrm{m}} = 0.185\,\mu\mathrm{H}$$

The ratio of all resistances $R_k/R_{k+1} = 1$, and the ratio

$$c = \frac{\omega L_k}{R_k} = \frac{2\pi 50 \frac{1}{\mathrm{s}} \cdot 0.185\,\mu\mathrm{H}}{0.266\,\mathrm{m\Omega}} = 0.218$$

We must now solve first the eddy currents of the upper bar. With an even $n = 4$, the subconductors $k = n/2 = 2$ and $k + 1 = n/2 + 1 = 3$ are in the center of the bar. We solve Equations (5.59) and (5.62), (5.63) for the subconductors $k = 2$ and $k = 3$ lying just below and over the center line of the upper bar

$$\underline{I}'_2 = -\frac{\mathrm{j}0.218}{2}\left(\sum_{\gamma=1}^{2} \underline{I}'_\gamma + \underline{I}_\mathrm{u}\right) = -\mathrm{j}0.11\underline{I}_\mu = -\underline{I}'_3.$$

From (5.65) we get

$$\underline{I}'_1 = \underline{I}'_2 - \mathrm{j}c(\underline{I}'_1 + \underline{I}_\mathrm{u}) = \underline{I}'_2 - \mathrm{j}c\left(\underline{I}_\mu - \sum_{\gamma=1}^{n/2} \underline{I}'_\gamma\right)$$

$$\Rightarrow \underline{I}'_1 = \underline{I}'_2 - \mathrm{j}0.218(\underline{I}_\mu - \underline{I}'_1 - \underline{I}'_2) = -\underline{I}'_4$$

$$\Rightarrow \underline{I}'_1(1 - \mathrm{j}0.218) = \underline{I}'_2(1 + \mathrm{j}0.218) - \mathrm{j}0.218\underline{I}_\mu$$

$$\underline{I}'_1 = \frac{\underline{I}'_2(1 + \mathrm{j}0.218) - \mathrm{j}0.218\underline{I}_\mu}{(1 - \mathrm{j}0.218)} = -\underline{I}'_4$$

For the lower bar current we get $\underline{I}_\mathrm{u} = \underline{I}_\mu - \underline{I}'_1 - \underline{I}'_2$.

The currents solved after iteration are $\underline{I}_\mu = (995 + \mathrm{j}96)$ A:

	$\underline{I}'_k$[A]	Phase [°]	$\underline{I}''_{k,0}$[A]	Phase [°]	$I_{k,0}$[A]	Phase [°]	$I_{k,\mathrm{u}}$[A]	Phase [°]
1	321	−68	232	0	462	−40.2	232	0
2	110	−84.4	238	12.3	250	−13.6	238	12.3
3	110	−95.5	269	34.5	336	51.1	269	34.5
4	321	112	350	59.6	602	84.6	350	59.6

Figure 5.14 illustrates the phasor diagram of the currents of Example 5.7.

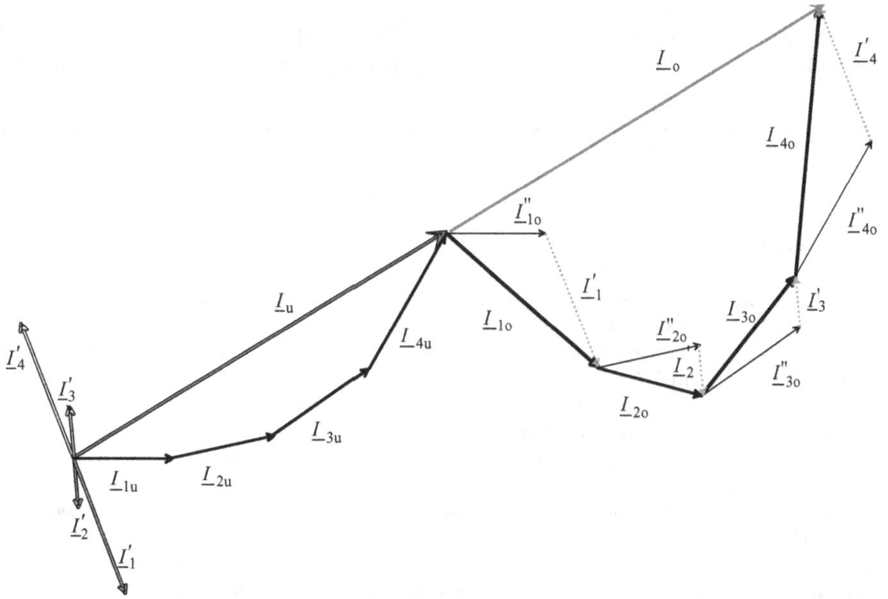

Figure 5.14 Phasor diagram of the subconductor currents in an induction motor rotor double cage at locked rotor state. The subcurrents of the upper bar are written $I_{k,0} = I''_{k,o} + I'_k$, where $I''_{k,o}$ are the eddy current components caused by the upper bar alone and I'_1, I'_2, I'_3, I'_4 the eddy current components caused by the lower bar current. The lower conductor currents behave similarly for a single bar in a slot.

Figure 5.15 depicts the RMS current density values of the subconductor currents. A symmetrical skin effect occurs in the upper bar as a result of the currents flowing under the observed conductor (J'_o). The resultant current density of the upper bar is J_o, in which the influence of the lower bar has been taken into account. The influence of the lower bar on the skin effect is significant. J_{DC} is the current density with direct current.

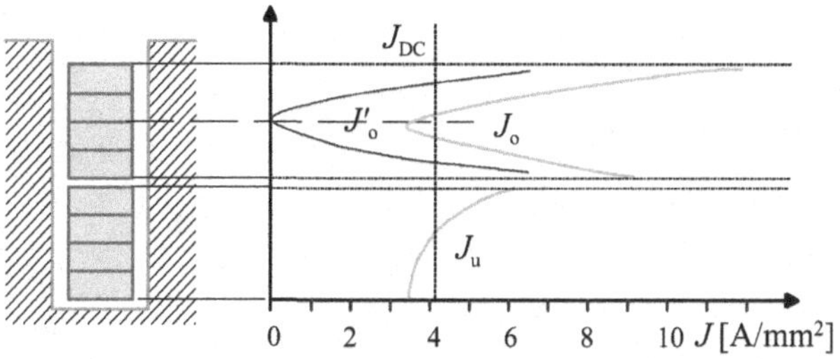

Figure 5.15 Skin effect of two solid conductors positioned on top of each other. The conductors have been divided into four imaginary subconductors. The upper conductor has a symmetrical current density J'_o with respect to the middle of the upper conductor. J'_o is created by the opposing current

densities caused by the current in the lower conductor. Taking into account the current in the upper conductor, the current density in the upper conductior J_o is not anymore symmetric. The current density in the lower conductor is J_u.

As we can see, the double-sided skin effect is remarkable in conductors with no transposition. In practice, such a winding arrangement is not acceptable except in the rotor slots or squirrel cage machines.

Bibliography

Hämäläinen, H., Pyrhönen, J., Nerg, J. AC resistance factor in one-layer form-wound winding used in rotating electrical machines. Accepted in *IEEE Transactions on Magnetics*, 2013.

Hämäläinen, H., Pyrhönen, J., Nerg, J., Talvitie, J. AC resistance factor of Litz-wire windings used in low-voltage generators. Accepted in *IEEE Transactions on Industrial Electronics*, 2013.

Küpfmüller, K. (1959) *Introduction to Theoretical Electrical Engineering (Einführung in die theoretische Elektrotechnik)*, 6th rev. edn, Springer Verlag, Berlin.

Lipo, T.A. (2007) *Introduction to AC Machine Design*, 3rd edn, Wisconsin Power Electronics Research Center, University of Wisconsin.

Richter, R. (1954) *Electrical Machines: Induction Machines (Elektrische Maschinen: Die Induktionsmaschinen)*, Vol. **IV**, 2nd edn, Birkhäuser Verlag, Basle and Stuttgart.

Richter, R. (1967) *Electrical Machines: General Calculation Elements. DC Machines (Elektrische Maschinen: Allgemeine Berechnungselemente. Die Gleichstrommaschinen)*, Vol. **I**, 3rd edn, Birkhäuser Verlag, Basle and Stuttgart.

Stoll, R. (1974) *The Analysis of Eddy Currents*, Clarendon Press, Oxford.

Vogt, K. (1996) *Design of Electrical Machines (Berechnung elektrischer Maschinen)*, Wiley-VCH Verlag GmbH, Weinheim.

6

Design Process of Rotating Electrical Machines

In the previous chapters, the general theory governing the design of an electrical machine was presented: Chapter 1 addressed the necessary fundamentals of electromagnetic theory, and Chapter 2 concentrated on winding arrangements. Chapter 3 described the behavior of the magnetic circuit. Chapter 4 discussed the inductances and, finally, Chapter 5 focused on the resistances of the windings. We should now be able to commence the discussion of the design process of the electrical machine. However, before entering the final design phase we have to study the eco-design principles of electrical machines. The target of the eco-design regulations is to reduce the life-cycle impacts of manufacturing and using motors.

6.1 Eco-Design Principles of Rotating Electrical Machines

Electric motors, which are marketed in the European Union area, have to fulfill the eco-design requirements for electric motors (Directive 2009/125/EC, Commission Regulation 640/2009). The aim of the requirements is to reduce the energy consumption and environmental impacts of the motors throughout the motor's life-cycle. Environmental aspects include material use, water use when applicable, polluting emissions, waste issues and recyclability. The energy consumption requirements cover minimum efficiency requirements for all kinds of three-phase 2–6 pole AC motors that are sold for direct-on-line operation, and whose rated voltage is less than or equal to 1000 V, and the rated power between 0.75 kW and 375 kW. The efficiency of motors is classified into three classes denoted by IE1–IE3. In 2013, new motors in the range of 7.5–375 kW must fulfill at least the requirements of the class IE2. From January 1, 2015 onwards, motors have to fulfill at least the requirements of the class IE3 or if they are supplied by a frequency converter at least the requirements of the class IE2. From January 1, 2017, motors from 0.75 kW to 7.5 kW are also subject to the regulation. The efficiency means the efficiency the motor reaches when supplied by a sinusoidal voltage, also in inverter fed drives. The efficiency classification is given in IEC standard 60034-30. The product range and classification in the new edition IEC 60034-30-1 (2014) is larger than the requirements given in the Commission Regulation 640/2009 (see Section 7.1).

Design of Rotating Electrical Machines, Second Edition. Juha Pyrhönen, Tapani Jokinen and Valéria Hrabovcová.

Table 6.1 Product information on motors that shall be displayed on technical documentation and free access websites of manufacturers of motors and manufactures of products in which motors are incorporated. The information referred to in points 1, 2 and 3 should be durably marked on or near the rating plate of the motor

1. Nominal efficiency at the full, 75 % and 50 % rated load and voltage
2. Efficiency level according to IE-classification
3. The year of manufacture
4. Manufacturer's name or trade mark, commercial registration number and place of manufacturer
5. Product's model number
6. Number of poles of the motor
7. The rated power output(s) or range of rated power output(s) [kW]
8. The rated input frequency(s) of the motor [Hz]
9. The rated voltage(s) or range of rated voltage [V]
10. The rated speed(s) or range of rated speed [min^{-1}]
11. Information relevant for disassembly, recycling or disposal at end-of-life
12. Information on the range of operating conditions for which the motor is specially designed: (i) altitudes above sea-level; (ii) ambient air temperature, including for motors with air cooling; (iii) water coolant temperature at the inlet to the product; (iv) maximum operating temperature; (v) potentially explosive atmospheres

Commission Regulation 640/2009 also gives decrees for the product information that manufacturers have to give (Table 6.1).

Life-cycle thinking is essential in eco-design. The life-cycle of a motor begins from raw materials used in the motor and ends to reuse and disposal of material (Figure 6.1). The designer of a product is responsible for which material he or she is using but not responsible on how the material is produced. The designer has to take into account e.g. if the material or the solution planned brings with it any noxious effects during the life-cycle of the product and, therefore, should avoid such a solution.

6.2 Design Process of a Rotating Electrical Machine

The flow chart in Figure 6.2 illustrates the main items of a design process of an electrical machine. It can be directly applied to asynchronous motors, but it is also applicable to the design of other machine types with minor changes. In the following chapters, the 15 actions presented in Figure 6.2 will be explained in detail.

6.2.1 Starting Values

The design of a rotating electrical machine can be commenced by defining certain basic characteristics, the most important of which are:

- Machine type (synchronous, asynchronous, DC, switched or synchronous reluctance machine, etc.).
- Type of construction (external pole, internal pole, axial flux, radial flux machine, etc.).

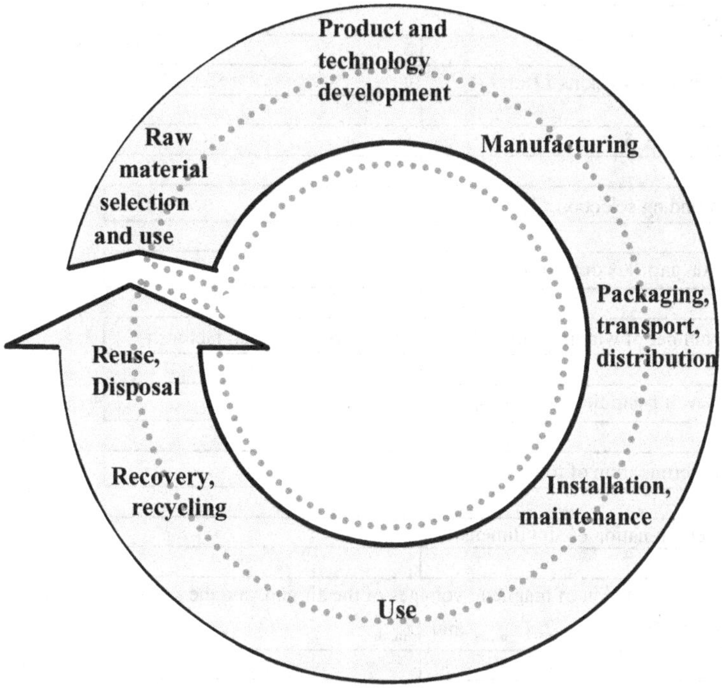

Figure 6.1 Product's life-cycle starts from the raw material selection and ends to material disposal and reuse. A designer is responsible for the choice of materials, construction solutions, maintenance solutions, etc. but only partly. For example, a designer is responsible for which material he or she is using but is not responsible on how the material is produced. The area of which the designer is responsible for is symbolically illustrated with the dashed line.

- Rated power:
  - For electric motors, the shaft output power P_N in W is given.
  - For synchronous motors, also the power factor ($\cos\varphi$ overexcited) is given.
  - For induction and DC generators, the electric output power P_N in W is given. Induction generators take reactive power from the network according to their power factor. The reactive power must usually be compensated by capacitor banks.
  - For synchronous generators, the apparent output power S_N is given in VA. The power factor (typically $\cos\varphi = 0.8$ overexcited) is also given.
- Rated rotational speed n_N of the machine or rated angular speed Ω_N.
- Number of pole pairs p of the machine (with frequency converter drives, this is also a subject of optimization).
- Rated frequency f_N of the machine (with frequency converter drives this is also a subject of optimization).
- Rated voltage U_N of the machine.
- Number of phases m of the machine (with frequency converter drives this can be also a subject of optimization).
- Intended duty cycle in direct-on-line drive (S1–S9) or variable speed drive.

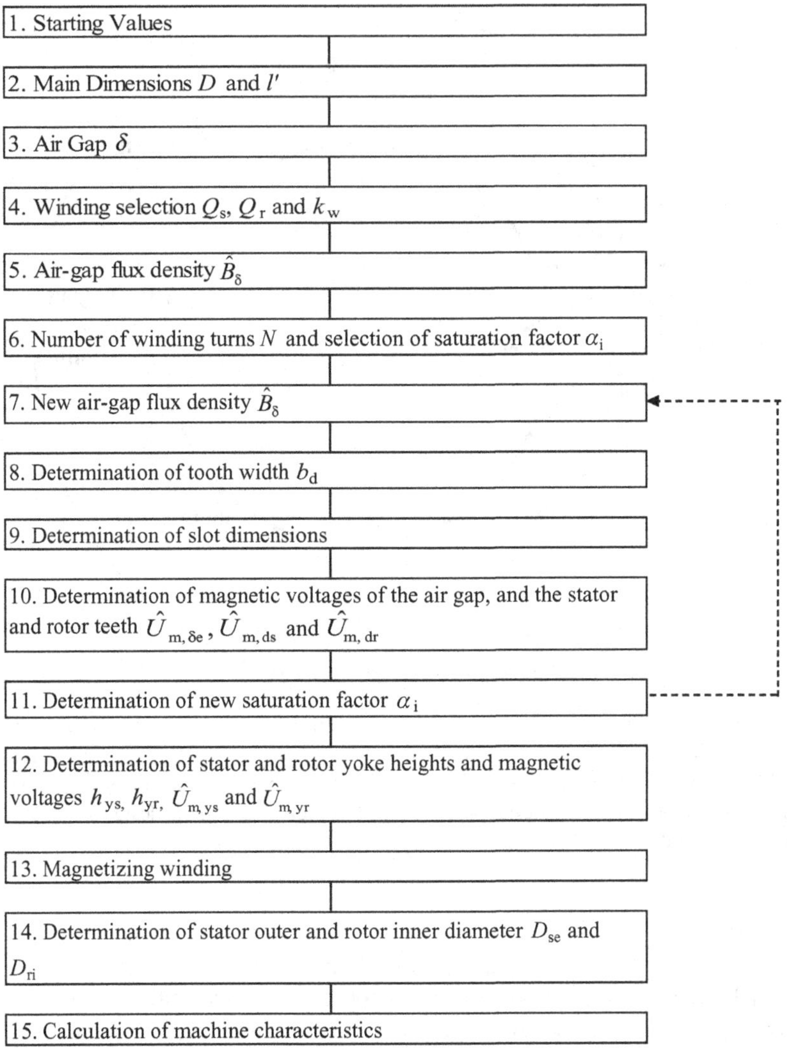

Figure 6.2 Design process of a rotating electrical machine in brief. This chart is originally intended for induction motor design but may also be applied to other rotating-field machine types. The saturation factor (α_i) behaves in a different way especially in rotor surface magnet permanent magnet machines. The effective relative magnet width may be used as an initial value for α_i in PMSMs with rotor surface magnets with a uniform thickness.

- Enclosure class and structure of the machine.
- Additional information, such as efficiency, required locked rotor torque, pull-up torque, peak torque, locked-rotor current, a speed-controlled drive etc.
- Standards applied in the machine design.
- Economic boundary conditions.
- Manufacturability and life-cycle assessment.

In machine design, there are a considerable number of free parameters. When aiming for an optimal solution, the task becomes extremely complicated unless the number of these free parameters is somehow limited. Many free parameters vary only slightly, and therefore, to simplify the task, these parameters can be assumed constant. The following ten parameters can be selected as free parameters:

- outer diameter of the stator stack (with the standard IEC frames, this parameter is often fixed to certain values);
- length of the stator stack;
- width of the stator slot;
- height of the stator slot;
- diameter of the air gap;
- air-gap length;
- peak value of the air-gap flux density;
- width of the rotor slot;
- height of the rotor slot;
- pole-pair number and frequency.

6.2.2 *Main Dimensions*

The actual machine design starts with the selection of the main dimensions of the machine. The term "main dimensions" refers to the air-gap diameter D_s measured at the stator bore (see Figures 3.1 and 3.2) and the equivalent core length l' (see Figure 3.10 and Equation 3.36). The equivalent length of the core takes into account the influence of the flux fringing at possible cooling ducts of the machine and also at the ends of the machine. In case of a permanent magnet machine the effective length of the machine must be observed with care and separately for the rotor and the stator excitations. As the permanent magnet material does not produce an axially continuous current linkage in a rotor with substacks the PM flux fringing must be taken into account. In case of rotor windings the current linkages will be continuous and similar fringing at the substack ends cannot be observed as in cases of noncontinuous permanent magnets.

In electrical machine design, there are certain empirically defined variation ranges of current and flux densities, which can be applied in the preliminary phase of the design. Tables 6.2 and 6.3 introduce some values of electromagnetic loadings for well-designed standard electrical machines.

The permitted loading levels are defined for a machine on the basis of the design of the insulation and the cooling of the machine. The values in the tables give some empirical information related to the selection of machine parameters. In principle, machine design is a rather complicated iteration process, in which the initial values are first selected for the dimensions of the machine. Next, the machine is designed electrically, and finally the cooling of the machine is computed. If the cooling of the machine is not efficient enough, the design has to be started from the beginning again by increasing the dimensions of the machine, by using better materials or selecting a more efficient cooling method. The material selection has a significant influence on both the losses and the thermal resistances. If a low-loss iron material and high-thermal-class insulation materials are selected, the output power of the machine can be improved without increasing its size.

Table 6.2 Permitted flux densities of the magnetic circuit for various standard electrical machines

	Flux density B[T]			
	Asynchronous machines	Salient-pole synchronous machines	Nonsalient-pole synchronous machines	DC machines
Air gap	0.7–0.9 ($\hat{B}_{\delta 1}$)	0.85–1.05 ($\hat{B}_{\delta 1}$)	0.8–1.05 ($\hat{B}_{\delta 1}$)	0.6–1.1 (B_{max})
Stator yoke	1.4–1.7 (… 2)	1.0–1.5	1.1–1.5	1.1–1.5
Tooth (apparent maximum value)	1.4–2.1 (stator) 1.5–2.2 (rotor)	1.6–2.0	1.5–2.0	1.6–2.0 (compensating winding) 1.8–2.2 (armature winding)
Rotor yoke	1–1.6 (… 1.9)	1.0–1.5	1.3–1.6	1.0–1.5
Pole core	–	1.3–1.8	1.1–1.7	1.2–1.7
Commutating poles	–	–	–	1.3

Table 6.3 Permitted RMS values for current densities J and linear current densities A for various electrical machines. Depending on the size of a permanent magnet machine, a synchronous machine, an asynchronous machine or a DC machine, suitably selected values can be used. Copper windings are generally assumed

			Non-salient pole synchronous machines			
			Indirect cooling			
	Asynchronous machines	Salient-pole synchronous machines or PMSM	Air	Hydrogen	Direct cooling	DC machines
A [kA/m]	30–65	35–65	30–80	90–110	150–200	25–65
	Stator winding	Armature winding	Armature winding	Armature winding	Armature winding Water cooling 7–10 Hydrogen cooling	Armature winding
J [A/mm^2]	3–8	4–6.5	3–5	4–6	6–13	4–9
	Copper rotor winding	Field winding Multi-layer	Field winding	Field winding		Pole winding
J [A/mm^2]	3–8	2–3.5	3–5	3–5		2–5.5
	Aluminium rotor winding	Field winding Single-layer				Compensating winding
J [A/mm^2]	3–6.5	2–4				3–4
			With direct water cooling also in field windings 13–18 A/mm^2 and 250–300 kA/m can be reached			

For permanent magnet machines we may select values according to Table 6.2 for synchronous machines.

When investigating the values of Table 6.3, it is worth noticing that when the dimensions of a slot are increased and indirect cooling used, lower values usually have to be selected than in the case of small slots. Hence, the lower values of J are for larger machines, and the highest values of J are suitable for small machines. Despite this, the lower values of A are valid for smaller machines and the higher ones for larger machines. If we are constructing a PMSM with concentrated fractional-slot nonoverlapping windings (tooth coils) with wide slots, the applicable values given in the table for pole windings are valid also in this case.

In Equation (1.115) we defined the tangential stress $\sigma_{F\tan}$ in the air gap. The local value for the tangential stress depends on the local linear current density $A(x)$ and the local flux density $B(x)$, $\sigma_{F\tan}(x) = A(x)B(x)$. If a sinusoidal air-gap flux density with the peak value $\hat{B}_\delta$ is assumed, and a sinusoidal linear current density with peak value $\hat{A}$ and RMS value A is applied, we obtain for the average tangential stress with spatial phase shift ζ between the fundamentals of the distributions of A and B.

$$\sigma_{F\tan} = \frac{\hat{A}\hat{B}_\delta \cos\zeta}{2} = \frac{A\hat{B}_\delta \cos\zeta}{\sqrt{2}}. \tag{6.1}$$

This tangential stress produces the torque of the machine when acting upon the rotor surface. Table 6.4 illustrates the guiding limit values for the tangential stress of the air gap calculated from Tables 6.2 and 6.3.

Table 6.4 Tangential stresses $\sigma_{F\tan}$ calculated from the values of Tables 6.2 and 6.3. There are three stress values, calculated with the lowest linear current density and flux density, with the average values and with the highest values. The flux density and linear current density distributions are assumed sinusoidal. For DC machines, a pole width coefficient of 2/3 is assumed. The cosζ of synchronous machines is assumed to be one, and for asynchronous machines, 0.8

	Totally enclosed asynchronous machines	Salient pole synchronous machines or PMSMs	Non-salient pole synchronous machines			DC machines
			Indirect cooling		Direct water cooling	
			Air	Hydrogen		
A[kA/m], RMS	30–65	35–65	30–80	90–110	150–200	25–65
Air-gap flux density $\hat{B}_{\delta 1}$[T]	0.7–0.9	0.85–1.05	0.8–1.05	0.8–1.05	0.8–1.05	0.6–1.1
Tangential stress $\sigma_{F\tan}$[Pa]						
minimum	12000*	21000*	17000*	51000*	85000*	12000*
average	21500*	33500*	36000*	65500*	114500*	29000*
maximum	33000*	48000*	59500*	81500*	148500*	47500*
	*cosζ = 0.8	*cosζ = 1	*cosζ = 1	*cosζ = 1	*cosζ = 1	*α_{DC} = 2/3

Typical tangential stress values give us a starting point for the design of an electrical machine. We may define the size of the rotor first by using a suitable tangential stress value on the rotor surface. If the rotor radius is r_r, the rotor equivalent length is l', the rotor surface facing the air gap is S_r, and the average tangential stress on the surface is $\sigma_{F\tan}$, we may write the torque T of the rotor simply as

$$\begin{aligned} T &= \sigma_{F\tan} r_r S_r \\ &= \sigma_{F\tan} r_r \left(2\pi r_r l'\right) \\ &= \sigma_{F\tan} 2\pi r_r^2 l' \\ &= \sigma_{F\tan} \pi \frac{D_r^2}{2} l' = 2\sigma_{F\tan} V_r. \end{aligned} \tag{6.2}$$

The rotor volume V_r to produce a certain torque can easily be estimated with Equation (6.2).

A similar basis for the design of the machine rotor size is the machine constant C of a well-designed electrical machine. The machine constant C expresses the magnitude of the "internal" apparent power S_i or the power P_i given by the rotor volume of the machine. The apparent power S_i for rotating-field machines rotating at a synchronous speed $n_{syn} = f/p$ is

$$S_i = mE_m I_s. \tag{6.3}$$

where E_m is induced emf over the magnetizing inductance L_m of the phase and I_s the stator phase current. The emf E_m is calculated in the following. The main flux penetrating a winding varies almost sinusoidally as a function of time

$$\Phi_m(t) = \hat{\Phi}_m \sin \omega t. \tag{6.4}$$

According to Faraday's induction law, the air-gap flux linkage $\Psi_m = N_s k_w \omega \hat{\Phi}_m$ (where N is the number of winding turns in a phase and k_w the winding factor) induces in the winding a voltage

$$e_m = -\frac{d\Psi_m}{dt} = -Nk_w \frac{d\Phi_m}{dt} = -Nk_w \omega \hat{\Phi}_m \cos \omega t. \tag{6.5}$$

The RMS value of the induced voltage is

$$E_m = \frac{1}{\sqrt{2}} \hat{e}_m = \frac{1}{\sqrt{2}} \omega k_w N \hat{\Phi}_m. \tag{6.6}$$

By substituting E_m into Equation (6.3) we obtain

$$S_i = m\frac{1}{\sqrt{2}} \omega \hat{\Psi}_m I_s = m\frac{1}{\sqrt{2}} \omega N k_w \hat{\Phi}_m I_s. \tag{6.7}$$

The maximum flux $\hat{\Phi}_{\mathrm{m}}$ penetrating a phase winding will be found by integrating the air-gap flux density $B_\delta(x)$ over the pole surface S_{p}

$$\hat{\Phi}_{\mathrm{m}} = \int_{S_{\mathrm{p}}} B_\delta(x)\mathrm{d}S_{\mathrm{p}}. \tag{6.8}$$

If the air-gap flux density has no variation with respect to the length of the machine l', the surface integral can be simplified as

$$\hat{\Phi}_{\mathrm{m}} = l'\tau_{\mathrm{p}}\alpha_{\mathrm{i}}\hat{B}_\delta. \tag{6.9}$$

The product $\alpha_{\mathrm{i}}\hat{B}_\delta$ represents the average value of the flux density of one pole in the air gap. In the case of a sinusoidal distribution $\alpha_{\mathrm{i}} = 2/\pi$. In other cases, a suitable value for α_{i} has to be found by integrating the flux density over the pole surface. For instance, if we have rotor surface magnets, the air-gap flux density created by the permanent magnets is usually nonsinusoidal. In such a case, the average value α_{i} for the flux density in the air gap can be defined from the effective relative magnet width $\alpha_{\mathrm{PM}} \approx w_{\mathrm{PM}}/\tau_{\mathrm{p}}$ (Figure 6.4).

The RMS value of the linear current density A of the stator may be defined with the slot pitch τ_{s} and the RMS value of slot current I_{u} (assuming that there are no parallel paths in the winding and the winding is a full-pitch one, $I_{\mathrm{u}} = I_{\mathrm{s}}z_{\mathrm{Q}}$). The number of slots in the stator is Q_{s}. Thus

$$A = \frac{I_{\mathrm{u}}}{\tau_{\mathrm{s}}}, \tag{6.10}$$

$$\tau_{\mathrm{s}} = \frac{\pi D}{Q_{\mathrm{s}}}. \tag{6.11}$$

Here D is, for generality, used instead of the stator inner or rotor outer diameter ($D \approx D_{\mathrm{r}} \approx D_{\mathrm{s}}$).

The number of conductors in the slot z_{Q} with all turns in series is

$$z_{\mathrm{Q}} = \frac{N}{pq} = \frac{N}{p\frac{Q_{\mathrm{s}}}{2pm}} = \frac{2Nm}{Q_{\mathrm{s}}}. \tag{6.12}$$

Now we may write for the linear current density

$$A = \frac{I_{\mathrm{u}}}{\tau_{\mathrm{s}}} = \frac{I_{\mathrm{u}}Q_{\mathrm{s}}}{\pi D} = \frac{I_{\mathrm{s}}z_{\mathrm{Q}}Q_{\mathrm{s}}}{\pi D} = \frac{2I_{\mathrm{s}}Nm}{\pi D}. \tag{6.13}$$

We solve the above for I_{s} and substitute I_{s} into Equation (6.7)

$$S_{\mathrm{i}} = m\frac{1}{\sqrt{2}}\omega N k_{\mathrm{w}}\hat{\Phi}_{\mathrm{m}}I_{\mathrm{s}} = m\frac{1}{\sqrt{2}}\omega N k_{\mathrm{w}}\hat{\Phi}_{\mathrm{m}}\frac{A\pi D}{2Nm} = \frac{1}{\sqrt{2}}\omega k_{\mathrm{w}}\hat{\Phi}_{\mathrm{m}}\frac{A\pi D}{2}. \tag{6.14}$$

We now substitute $\omega = 2p\pi n_{\text{syn}}$ and the peak flux from Equation (6.9) into Equation (6.14)

$$S_{\text{i}} = \frac{1}{\sqrt{2}} 2p\pi n_{\text{syn}} k_{\text{w}} \frac{2}{\pi} \frac{\pi D}{2p} \hat{B}_{\delta} l' \frac{A\pi D}{2}. \\ = \frac{\pi^2}{\sqrt{2}} n_{\text{syn}} k_{\text{w}} A \hat{B}_{\delta} l' D^2 \tag{6.15}$$

We may now rewrite this as

$$S_{\text{i}} = mE_{\text{m}}I_{\text{s}} = \frac{\pi^2}{\sqrt{2}} k_{\text{w}} A \hat{B}_{\delta} D^2 l' n_{\text{syn}}. \\ = CD^2 l' n_{\text{syn}} \tag{6.16}$$

The machine constant C can be written in the following form for rotating-field machines (synchronous and asynchronous machines), according to Equation (6.16)

$$C = \frac{\pi^2}{\sqrt{2}} k_{\text{w}} A \hat{B}_{\delta} = \frac{\pi^2}{2} k_{\text{w}} \hat{A} \hat{B}_{\delta}, \tag{6.17}$$

where $A = \hat{A}/\sqrt{2}$. l' is the equivalent length of the machine and A is the RMS value of the linear current density, which corresponds to the tangential magnetic field strength H_{tan} in the air gap, see Chapter 1.

In direct current machines, the air-gap flux density is not sinusoidal. Under the pole at no-load, we have the air-gap flux density $B_{\delta\,\text{max}}$. For DC machines, the internal power of which is defined $P_{\text{i}} = \pi^2 \alpha_{\text{DC}} A B_{\delta\,\text{max}} D^2 l' n_{\text{syn}} = C D^2 l' n_{\text{syn}}$, we may write

$$C = \pi^2 \alpha_{\text{DC}} A B_{\delta\,\text{max}}. \tag{6.18}$$

α_{DC} is the relative pole width for DC machines (typically about 2/3).

The dependence of the internal apparent power S_{i} on the mechanical power P_{mec} of a rotating-field machine is obtained from the power factor $\cos\varphi$ of the machine and the efficiency η (which have to be estimated at this stage). Now, the machine constant of mechanical power C_{mec} can be introduced:

$$P_{\text{mec}} = \eta mUI \cos\varphi = \eta \cos\varphi \frac{U}{E_{\text{m}}} S_{\text{i}} = C_{\text{mec}} D^2 l' n_{\text{syn}}. \tag{6.19}$$

In a DC machine, the internal power ($P_{\text{i}} = E_{\text{m}} I_{\text{a}}$) depends on the input power $P_{\text{in}} = U_{\text{a}} I_{\text{a}}$; consequently, based on their ratio, we may write

$$P_{\text{i}} = \frac{E_{\text{m}}}{U_{\text{a}}} \frac{I_{\text{a}}}{I_{\text{a}}} P_{\text{in}}. \tag{6.20}$$

where I_{a} is the armature current.

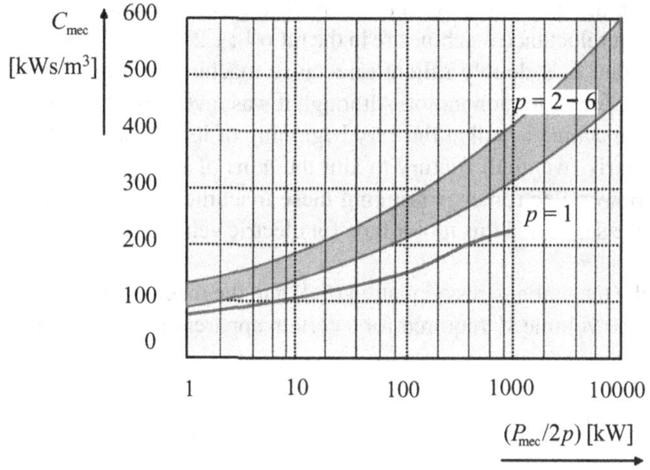

Figure 6.3 Machine constants of totally enclosed asynchronous and synchronous machines as a function of pole power.

The allowed linear current densities A and the air-gap flux densities B_δ of electrical machines are strongly dependent on the cooling methods, as shown in Tables 6.2 and 6.3. Highly effective cooling methods increase the permitted electromagnetic loadings to 1.5–2-fold, and correspondingly the weight of the machine drops by 30–50 %. In well-designed machines, the linear current density and the flux density depend on the machine size in such a way that they both increase as the size of the machine increases. As a result, the machine constant also depends on the size of the machine. Figure 6.3 illustrates the machine constant of induction and synchronous machines of different sizes per the pole power. Because of the low excitation losses in permanent magnet machines, the machine constants for PMSMs could be high. The values given in the Tables 6.3 and 6.4 are, however, also limitedly valid for PMSMs, because the permanent magnet rotor temperature must be limited to lower values than in wound rotors or die-cast rotors, especially when NdFeB magnets are used. Typically, 100 °C should not be exceeded with NdFeB magnets to minimize the magnet demagnetization risk under the demagnetizing armature reaction. SmCo magnets tolerate higher temperatures, but their remanence is remarkably lower (Table 3.3) than that of NdFeB magnets.

The machine constants of different machine types are not exactly equal. The definition of machine constants is based on empirical knowledge and the tradition of machine construction, and therefore even quite contradictory values for machine constants can be found in the literature. With a constant supply frequency, the rotor peripheral speed in rotating-field machines is proportional to the pole pitch. The permitted armature current and thereby the machine constant are functions of the pole pitch and the frequency of the machine. Since the pole pitch of a machine cannot be deduced from its rated values, the machine constant is often given as a function of pole power ($P_{mec}/2p$). The method can be justified by the fact that the ratio of the length and the pole pitch is fairly constant irrespective of the number of pole pairs.

The machine constant of doubly salient reluctance machines can with low outputs be notably higher than the machine constant of an induction machine. For instance, according to

Lawrenson (1992), the machine constants of an 11 kW DC machine, an induction machine, and a doubly salient reluctance machine are in the ratio 1 : 1.23 : 1.74. Thus, in this power class, the machine constant of a doubly salient reluctance machine is about 40 % higher than the machine constant of an induction motor. Although it was invented a long time ago, the doubly salient reluctance machine is still at the very beginning of its development. The machine was employed in the early twentieth century to aim the guns of British warships. Nowadays, the development of power electronics is bringing these machines into wider use. Lately, several manufacturers have suggested this motor type for electric vehicles probably because of its low material prices.

According to the tangential stress Equation (6.2) and the machine constant Equations (6.16) and (6.19), the rotor volume V_r required for a certain apparent power can be written as

$$V_r = \frac{\pi}{4}D^2 l' = \frac{T}{2\sigma_{F\,\tan}} = \frac{\pi}{4}\frac{S_i}{C n_{syn}} = \frac{\pi}{4}\frac{P_{mec}}{C_{mec} n_{syn}}. \tag{6.21}$$

From Equation (6.21) we obtain

$$D^2 l' = \frac{2}{\pi}\frac{T}{\sigma_{F\,\tan}} = \frac{S_i}{C n_{syn}} = \frac{P_{mec}}{C_{mec} n_{syn}}. \tag{6.22}$$

In standard machines, the ratio of the equivalent machine length and the air-gap diameter

$$\chi = \frac{l'}{D}. \tag{6.23}$$

varies within rather tight limits (Table 6.5).

Now, we have two Equations (6.22) and (6.23) to solve the air-gap diameter D and the equivalent machine length l'. If the machine has radial air-cooling channels, the stator is built of laminated stacks 40–80 mm long with 5–10 mm cooling channels in between. The equivalent length of the machine is then defined according to the rules given in Chapter 3. In the case of an axial flux machine, the diameter ratio of the machine decides the radial "length" of the machine. The best theoretical diameter ratio (the ratio of the inner diameter of the stator to the outer diameter, D_s/D_{se}) is approximately 0.6.

Table 6.5 Typical $\chi = l'/D$ ratios for different electrical machines

Asynchronous machines	Synchronous machines, $p > 1$	Synchronous machines, $p = 1$	DC machines
$\chi \approx \frac{\pi}{2p}\sqrt[3]{p}$	$\chi \approx \frac{\pi}{4p}\sqrt{p}$	$\chi = 1 - 3$	$\chi \approx \frac{0.8 - 1.6}{p}$

Example 6.1: Determine the main dimensions for a 30 kW, 690 V, 50 Hz, four-pole, three-phase squirrel cage induction motor.

Solution: $P_{mec}/2p = 30/4$ kW $= 7.5$ kW. From Figure 6.3 the machine constant $C_{mec} =$ 150 kWs/m^3. The ratio χ is according to Table 6.5

$$\chi = \frac{l'}{D} \approx \frac{\pi}{2p}\sqrt[3]{p} = \frac{\pi}{2\cdot 2}\sqrt[3]{2} = 0.9895.$$

Substituting $l' = \chi D$ in Equation (6.22) we obtain the air-gap diameter

$$D = \sqrt[3]{\frac{P_{mec}}{\chi C_{mec} n_{syn}}} = \sqrt[3]{\frac{30}{0.9895\cdot 150\cdot 50/2}}\,\mathrm{m} = 200.7\ \mathrm{mm}$$

and the effective length

$$l' = \chi D = 0.9895\cdot 200.7\ \mathrm{mm} = 198.6\ \mathrm{mm}.$$

Example 6.2: Calculate the tangential stress of the induction motor of Example 6.1. The rated speed is 1474 rpm.

Solution: The rated torque is

$$T = \frac{P}{2\pi n} = \frac{30000}{2\pi\cdot 1474/60}\,\mathrm{Nm} = 194.4\ \mathrm{Nm}$$

and the tangential stress according to Equation (6.2)

$$\sigma_{F\,\tan} = \frac{2T}{\pi D^2 l'} = \frac{2\cdot 194.4}{\pi\cdot 0.2007^2\cdot 0.1986}\,\mathrm{Pa} = 15.47\ \mathrm{kPa}.$$

The tangential stress is between the minimum and average values given in Table 6.4.

6.2.3 *Air Gap*

The length of the air gap of a machine has a significant influence on the characteristics of an electrical machine. In machines in which the magnetizing current is taken from the supply network, the length of the air gap is dimensioned to produce a minimum magnetizing current and, on the other hand, an optimal efficiency. In principle, a small air gap gives a low magnetizing current, while the eddy current losses of the rotor and stator surface increase because of permeance harmonics created by the open or semi-closed slots. A small air gap also increases the surface losses in the rotor caused by the current linkage harmonics of the stator. Although the air gap is of great significance, no theoretical optimum has been solved for its length, but usually empirical equations are employed instead in the definition of the length of the air gap. An air gap δ of a 50 Hz asynchronous machine can be calculated in meters as a function of power P with equations

$$\delta = \frac{0.2 + 0.01 \cdot P^{0.4}}{1000} \mathrm{m}, \text{ when } p = 1, \tag{6.24a}$$

$$\delta = \frac{0.18 + 0.006 \cdot P^{0.4}}{1000} \mathrm{m}, \text{ when } p > 1. \tag{6.24b}$$

P must be given in Watts. The smallest technically possible air gap is approximately 0.2 mm. In drives for extremely heavy duty, the air gap is increased by 60 %. In machines with an exceptionally large diameter, an air gap ratio of $\delta/D \geq 0.001$ has to be selected because of the manufacturing tolerances and the mechanical properties of the frame and the shaft of the machine.

In frequency converter drives, the air gap may be increased similarly as in heavy-duty drives (60 % increase) to get lower rotor surface losses. In large machines with prefabricated windings and open slots, the air-gap length must be selected high enough (60–100 % increase) to reduce pulsation losses.

If an asynchronous machine is designed for high speeds, to avoid excessive iron losses in the stator and rotor teeth, the air-gap length has to be increased considerably from the value obtained with Equation (6.24a or 6.24b) for a standard electric motor. If a high-speed machine is equipped with a solid rotor, the air gap has to be designed with special care, since the losses at the surface of a solid rotor decrease radically when the air gap is increased, whereas an increase in the magnetizing current in the stator leads into a notably smaller increase in the losses. A suitable value for the length of the air gap of an inverter fed, high-speed asynchronous machine (the peripheral speed of the rotor > 100 m/s), can be calculated e.g. using equation [Pat. U.S. 5,473,211]

$$\delta = 0.001\ \mathrm{m} + \frac{D_r}{0.07} + \frac{v}{400\ \mathrm{m}/s} \mathrm{m} \tag{6.25}$$

where D_r is the outer rotor diameter and v the peripheral speed of the rotor.

Example 6.3: What is a suitable air gap for the 30 kW, heavy-duty totally-enclosed industrial induction motor of Example 6.1? Calculate the core length of the stator, the stator inner and rotor outer diameters.

Solution: $\delta = 1.6 \cdot \frac{0.18 + 0.006 \cdot P^{0.4}}{1000} \mathrm{m} = 1.6 \cdot \frac{0.18 + 0.006 \cdot 30000^{0.4}}{1000} \mathrm{m}$
$= 0.88 \cdot 10^{-3} \mathrm{m} = 0.90\ \mathrm{mm}.$

The factor 1.6 is used because of the 60 % increase in a heavy duty machine. As there are no cooling channels the real physical length of the stator core is

$$l_s = l' - 2\delta = 198.6\ \mathrm{mm} - 2 \cdot 0.9\ \mathrm{mm} = 196.8\ \mathrm{mm} = 197\ \mathrm{mm}.$$

According to Example 6.1, the average air-gap diameter is 200.7 mm. We can choose for the stator inner diameter a round number $D_s = 202$ mm whereupon the rotor outer diameter is $D_r = D_s - 2 \cdot \delta = 202\ \mathrm{mm} - 2 \cdot 0.9\ \mathrm{mm} = 200.2\ \mathrm{mm}$.

In DC and synchronous machines, the air gap is basically defined by the permitted armature reaction. We have to ensure that the armature reaction (flux caused by the current-linkage of the armature) does not reduce the flux density excessively on one side of a magnetic pole. To meet this condition, the current linkage of the field winding has to be higher than the current linkage of the armature

$$\Theta_f \geq \Theta_a. \tag{6.26}$$

This condition may be rewritten for DC machines as (see Equation (2.132))

$$\frac{B_{\delta\max}}{\mu_0}\delta k_C \geq \frac{1}{2}\alpha_{DC}\tau_p A_a. \tag{6.27}$$

On the left of (6.27), we have the field winding current linkage expressed with the no-load air-gap flux density, and, on the right, the armature current linkage expressed by the armature linear current density A_a and the relative pole width α_{DC}.

The same condition for a synchronous machine is given as

$$\frac{\hat{B}_\delta}{\mu_0}\delta k_C \geq \frac{1}{2}\alpha_{SM}\tau_p A_a. \tag{6.28}$$

Then, the air gap of a DC machine must be

$$\delta \geq \frac{1}{2}\alpha_{DC}\mu_0\tau_p\frac{A_a}{k_C B_{\delta\max}} = \gamma\tau_p\frac{A_a}{B_{\delta\max}}, \tag{6.29}$$

and for synchronous machines

$$\delta \geq \frac{1}{2}\alpha_{SM}\mu_0\tau_p\frac{A_a}{k_C\hat{B}_\delta} = \gamma\tau_p\frac{A_a}{\hat{B}_\delta}, \tag{6.30}$$

where γ (Table 6.6) includes, according the type of the machines, the relative pole width of the pole shoe α_{DC} or α_{SM}, μ_0, k_C and a constant 1/2.

Table 6.6 Coefficient γ for the definition of the air gap of DC and synchronous machines

Salient-pole constant air-gap synchronous machines	$\gamma = 7.0 \times 10^{-7}$
Salient-pole synchronous machines, the air gap of which is shaped to produce a sinusoidal flux density distribution	$\gamma = 4.0 \times 10^{-7}$
Nonsalient-pole synchronous machines	$\gamma = 3.0 \times 10^{-7}$
DC machines without compensating winding and commutating poles	$\gamma = 5.0 \times 10^{-7}$
DC machines with commutating poles without compensating winding	$\gamma = 3.6 \times 10^{-7}$
Compensated DC machines (with commutating poles and compensating winding)	$\gamma = 2.2 \times 10^{-7}$

Example 6.4: A salient-pole synchronous machine with a no-load sinusoidal air-gap flux density has a stator linear current density of $A = 60$ kA/m, an air-gap flux density amplitude of 1 T and a pole pitch of 0.5 m. Find a suitable air gap for the machine.

Solution: The sinusoidal no-load air-gap flux density of the machine results from a suitable pole shape that produces the sinusoidal flux density. Hence, $\gamma = 4 \cdot 10^{-7}$ and we get

$$\delta_0 = \gamma \tau_{\mathrm{p}} \frac{A}{\hat{B}_\delta} = 4 \cdot 10^{-7} \cdot 0.5 \frac{60000}{1} \mathrm{m} = 0.012\,\mathrm{m}$$

Synchronous reluctance machines must have a high inductance ratio, which suggests selecting a small d-axis air gap. To be able to compete with the performance of asynchronous machines, synchronous reluctance motors may have to be equipped with smaller d-axis air gaps than those in induction motors. However, if the air gap is made very small, the surface loss, because of permeance harmonics, may increase remarkably if high-quality rotor laminations are not used.

In doubly salient reluctance machines, the aim is to construct as small an air gap as possible to achieve a high inductance ratio between the direct and quadrature position of the machine.

In PMSMs, the air-gap length is determined by mechanical constraints. It is similar to those values encountered in asynchronous machines and can be calculated from Equations (6.24) and (6.25). The synchronous inductance depends on the air-gap length. Since the magnet height (h_{PM} Figure 3.44) itself has a significant influence on the magnetic air gap of the machine, the synchronous inductance easily becomes low and the machine maximum torque high. However, in some cases, the height of the magnet and even the length of the physical air gap must be increased to get a smaller synchronous inductance. Generally, the physical air gap is made as small as possible to save the amount of material in the permanent magnet. This holds especially for low-speed, high-torque permanent magnet machines. In higher-speed machines, the air-gap harmonic content may cause very high losses in the permanent magnet material or in the ferromagnetic material under the permanent magnets, and in such cases the air gap must be increased to keep the magnet temperature low enough. In a PMSM, the determination of the air gap and the thickness of the magnets themselves is thus a demanding optimization task.

In rotor surface magnet machines, the magnetic air gap of the machine may be calculated as

$$\delta_{\mathrm{PM}} = h_{\mathrm{PM}}/\mu_{\mathrm{rPM}} + \delta_{\mathrm{e}} \tag{6.31}$$

where h_{PM} is the rotor surface permanent magnet height, μ_{rPM} the permanent magnet material relative permeability and δ_{e} is the equivalent air gap, the physical air gap corrected with the Carter factor ($\delta_{\mathrm{e}} = k_{\mathrm{C}}\delta$).

If the magnets are embedded, their effects in the reluctance of the magnetic circuit are somewhat complicated and have to be taken into account in the calculation of the effective air gap δ_{ef}. Numerical methods are often applied.

Also the behavior of the effective machine length l', especially in rotor surface magnet PMSMs is different compared to other machine types. When calculating the machine magnetizing inductance similar approach as earlier $l' \approx l + 2\delta$ can be used but to get a sufficient PM flux the rotor must be longer than the stator. It was recommended by Pyrhönen *et al.* (2010) that the rotor surface magnets' axial length must be, for example, $l_{rPM} \approx l_s + 2\delta$ to achieve a constant flux density level in the air gap.

In case of stator cooling ducts and, especially in cases of both stator and rotor cooling ducts the effective length of the machine has to be solved with great care – a finite element approach is suggested to evaluate the average flux density and therefore the effective length of a PMSM.

6.2.4 Winding Selection

The next step is to select a suitable winding for the stator. This is a decisive phase with respect to the final characteristics of the machine. A guiding principle is that a poly-phase winding produces the more sinusoidal current linkage, the more slots there are in the stator. The winding factors for an integer slot winding usually become lower for the harmonics as the number of slots increases (the current linkage produced by the winding approaches sinusoidal waveform). The limit value for this inference is the case in which there are no slots at all, but the winding is assembled straight into the air gap of the machine. Now, in principle, the current linkage created by the stator currents has a smooth distribution and, because of a long air gap, an optimum sinusoidal flux density distribution on the rotor surface. A slotless armature winding is typical in small and in high-speed permanent magnet synchronous machines. Leaving an asynchronous motor without slots would lead, as a result of a long air gap, into an excessive magnetizing current and therefore a poor power factor.

A large number of slots increases the number of coils and also the price of the machine. Table 6.7 introduces some recommendations for the slot pitch τ_u.

The lowest slot pitches occur in small machines, whereas the highest slot pitches are found in large machines. For instance, in a 4 kW, 3000 min-1 induction motor, the slot pitch τ_u of the stator is about 8.5 mm. On the other hand, the slot pitch τ_u of the stator in a 100 kVA four-pole synchronous machine is about 16 mm. In a low-speed 3.8 MW, 17 min^{-1} direct-driven wind generator, the slot pitch τ_u is about 38 mm ($D = 5.2$ m, $p = 72$). The largest slot pitches are suitable in direct water cooling. Bulky copper coils are difficult to cool if the machine is air cooled.

When the winding type and the number of slots are selected, as a result the winding factors of the machine are also defined. The most important winding factor is the winding factor k_{w1} of the fundamental. Simultaneously, attention must be paid to the winding factors of the winding harmonics. For instance, in a symmetrical three-phase machine with integer slot windings, the lowest harmful harmonics are the fifth and the seventh harmonics.

If the rotor winding is embedded in the slots, the rotor slot number has to differ from the stator slot number. Instruction for selection of the slot number of a cage winding is given in Chapter 7, Section 7.2.6.

Table 6.7 Recommended slot pitches for normal armature windings of different machine types

Machine type	Slot pitch, τ_u[mm]
Asynchronous machines and small PMSMs	7–45
Synchronous machines and large PMSMS	14–75
DC machines	10–30

Example 6.5: Select suitable stator and rotor windings for the motor in Examples 6.1–6.3 (30 kW, four-pole squirrel cage motor). Calculate the winding factors.

Solution: Integral slot windings are recommended for squirrel cage motors. Let us calculate the slot pitches (τ_{us}) and the slots numbers (Q_s) for three different numbers of slots per pole and phase (q_s).

q_s	3	4	5
$Q_s = 2pmq_s = 2 \cdot 2 \cdot 3 \cdot q_s$	36	48	60
$\tau_{us} = \pi D_s/Q_s = (\pi \cdot 202/Q_s)$ mm	17.6 mm	13.2 mm	10.6 mm

30 kW motor is between medium and small size motors. According to Table 6.7 a slot pitch of 13.2 mm should be the right choice and so we choose $q_s = 4$, $Q_s = 48$. Let us choose a two-layer winding for the stator and short-pitch it with two slot pitches, so the winding pitch $W = 5/6 \cdot \tau_p$ (the pole pitch measured as slot pitches $y_Q = Q_s/2p = 48/4 = 12$ slot pitches).

According to Equation (2.35) the stator winding factor for the fundamental is

$$k_{ws1} = k_{ps1}k_{ds1} = \sin\left(\frac{W}{\tau_p}\frac{\pi}{2}\right) \cdot \frac{\sin\left(\dfrac{\pi}{2m}\right)}{q\sin\left(\dfrac{\pi}{2mq}\right)} = \sin\left(\frac{5}{6}\frac{\pi}{2}\right) \cdot \frac{\sin\left(\dfrac{\pi}{2\cdot 3}\right)}{4\sin\left(\dfrac{\pi}{2\cdot 3\cdot 4}\right)}$$

$$= 0.9659 \cdot 0.9577 = 0.925.$$

According to Table 7.3, we can choose for the rotor slot number $Q_r = 44$ and skew the slots by one stator slot pitch (the skew $s_{sp} = 1$).

A squirrel cage winding can be considered as a Q_r-phase winding, where there is in every phase only one bar. Without skewing, the winding factor of this kind of winding is equal to one. With skewing, the winding factor of the rotor is equal to the skewing factor, which according to Equation (4.21) is for the fundamental

$$k_{sq1} = \frac{\sin\left(\dfrac{s}{\tau_p}\dfrac{\pi}{2}\right)}{\dfrac{s}{\tau_p}\dfrac{\pi}{2}} = \frac{\sin\left(\dfrac{\pi}{2}\dfrac{s_{sp}}{mq}\right)}{\dfrac{\pi}{2}\dfrac{s_{sp}}{mq}} = \frac{\sin\left(\dfrac{\pi}{2}\dfrac{1}{3\cdot 4}\right)}{\dfrac{\pi}{2}\dfrac{1}{3\cdot 4}} = 0.997.$$

6.2.5 *Air-Gap Flux Density*

Since the tangential stress or the machine constant has already been selected, the air-gap flux density $\hat{B}_\delta$ has to correlate with the selected machine constant. The initial value employed in the calculation can be selected according to Table 6.2. In permanent magnet machines, the air-gap flux density must be in a sensible relation to the remanent flux density of the permanent magnet material. From the economy point of view of this material, the maximum air-gap flux density with permanent magnets should be about half of the remanent flux density, namely 0.5–0.6 T. Such low values should, however, lead to a large machine, and hence remarkably higher values are also used in PMSMs.

Example 6.6: Choose the air-gap flux density for the 30 kW induction motor in Examples 6.1–6.3 and 6.5. Check that the chosen flux density and the tangential stress calculated in Example 6.2 give a proper linear current density. The cosine of the angle ς between the linear current density and the air-gap flux density is assumed to be 0.8.

Solution: According to Table 6.2 the air-gap flux density varies normally from 0.7 T to 0.9 T. Let us choose $\hat{B}_\delta = 0.8$ T. The linear current density is according to Equation (6.1)

$$A_s = \frac{\sqrt{2}\sigma F \tan}{\hat{B}_\delta \cos\varsigma} = \frac{\sqrt{2}\cdot 15470}{0.8\cdot 0.8}\text{A/m} = 34180\ \text{A/m}.$$

The value corresponds well to the values given in Table 6.4.

6.2.6 *The No-Load Flux of an Electrical Machine and the Number of Winding Turns*

As the main dimensions, the winding method and the air-gap density have been selected, the required number of winding turns N can be defined with a desired emf. The emf $E_m = \omega\Psi_m$ induced by the air-gap flux linkage ($\Psi_m = I_m L_m$) can first be estimated from the RMS value U_s of the fundamental terminal voltage; for induction motors it is $E_m \approx (0.93 - 0.98)\,U_s$ and for generators $E_m \approx (1.03 - 1.10)\,U_s$.

With synchronous machines and PMSMs, we first have to estimate the required E_f or E_{PM}, (cf. appropriate phasor diagrams in Chapter 7) which is crucial for the torque production of a synchronous machine. The stronger the armature reaction is, the higher the values of E_f or E_{PM} have to be selected in order to reach an adequate power factor at the rated point. In controlled PMSM drives the power factor has somewhat different meaning than in other rotating field machines. Especially, in case of rotor surface magnets, at lower than the rated speed, it is most efficient to control the motor in such a way that there is no d-axis current. And now, depending on the value of the synchronous inductance, the power factor will be significantly smaller than one. If, however, a high value of PMSM power factor at low speeds should be valued itself the amount of magnet material should be increased to have a higher E_{PM}, which is needed for a high power factor at the constant torque region. Later, when moving to field weakening of a PMSM the power factor will first reach unity and then, inevitably, become capacitive. This matter will be observed in details later in the design of a PMSM.

In synchronous machines with a strong armature reaction (large L_m), $E_f \approx (1.2 - 2)\,U_s$ and in permanent magnet synchronous machines, typically, the $E_{PM} \approx (0.9 - 1.1)U_s$. It must, however, be remembered that values at the higher end of $E_f \approx (1.2 - 2)\,U_s$ are fictitious values used in the machine linear analysis. The values are not possible at no load since the machine magnetic circuit should saturate heavily. Even in high-armature-reaction machines, the no-load maximum voltage seldom exceeds $E_f \approx (1.2 - 1.3)\,U_s$.

The number of the coil turns in series in a phase winding is solved from Equations (6.6) and (6.9)

$$N_s = \frac{\sqrt{2}E_m}{\omega k_w \hat{\Phi}_m} = \frac{\sqrt{2}E_m}{\omega k_w l' \tau_p \alpha_i \hat{B}_\delta}. \tag{6.32}$$

Note that the peak flux $\hat{\Phi}_m$ is calculated per single pole. In a symmetrical machine, the flux of each pole is of equal magnitude. The number of coil turns N can also, if required, be divided into several pole pairs. If there are for instance two pole pairs in the machine, they can be connected in parallel (the number of parallel paths is in such a case $a = 2$) if desired, the number of turns being now N in both pole pairs. If the pole pairs are connected in series instead ($a = 1$), the number of turns becomes $N_s/2$ for both pole pairs, the total number being N_s. If possible, it is often advisable to connect the coils in series, since in that case the possible asymmetries between the pole pairs do not cause circulating currents in the machine. However, parallel paths result in a larger number of slot conductors z_{Qs} which is favorable when trying to avoid skin effect and circulating currents in the windings.

The term α_i in Equation (6.32) is a coefficient showing the arithmetical average of the flux density of one pole, which takes the value $\alpha_i = 2/\pi$ in a sinusoidal flux density distribution, Figure 6.4.

In PMSMs, the magnets are often installed on the rotor surface. If the magnets have an equal thickness over the pole pitch, the flux density in the air gap is more or less rectangular, and the average flux may be defined based on the relative magnet width $\alpha_{PM} = w_{PM}/\tau_p$. In such a case, in Equation (6.32), the peak value of the air-gap flux must be calculated by integrating the air-gap flux density over the pole pitch. Rotor surface magnets may, of course, also be given a form that produces a sinusoidal flux density. Embedded magnets and a suitable pole shoe shape may also give a sinusoidal flux density in the air gap. In such cases, the average flux density is again close to $\alpha_i = 2/\pi$.

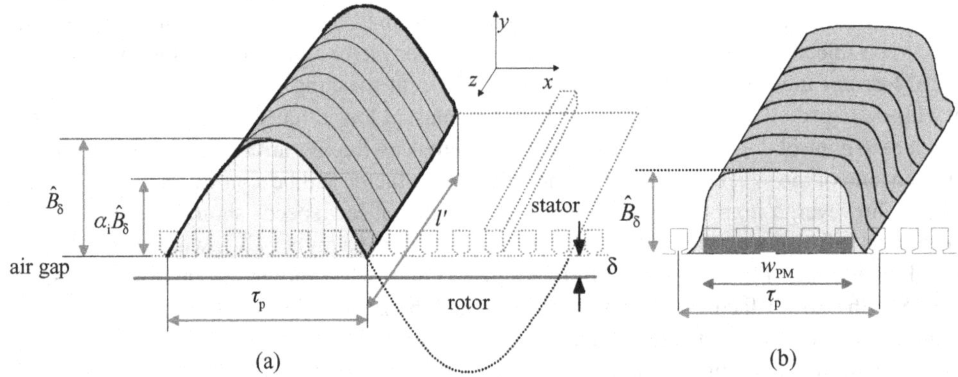

Figure 6.4 (a) Sinusoidal flux density distribution across a pole pitch τ_p ignoring the effect of slotting, peak value $\hat{B}_\delta$, average value $\alpha_i \hat{B}_\delta$. The cross-section of the flux density distribution is often somewhat flattened from the sinusoidal form because of saturation in the teeth. In the case of a sinusoidal distribution $\alpha_i = 2/\pi$ and in the case of a flattened distribution $\alpha_i > 2/\pi$. b) The air-gap flux behavior with a rectangular permanent magnet. The width of the magnets is w_{PM}.

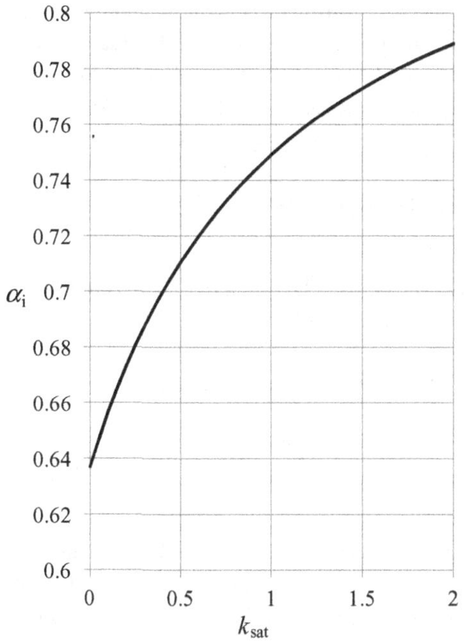

Figure 6.5 Effect of the saturation factor on the factor α_i of the arithmetical average of the flux density. The more the teeth of the machine are saturated, the higher the value of α_i becomes.

If the machine is dimensioned in such a way that the iron parts saturate at the peak value of the flux density, the flux density distribution is flattened. In ordinary network-supplied induction motors, both the stator and rotor teeth are saturated at the peak value of the flux density. This leads to a higher reluctance of these teeth when compared with other teeth, and thus α_i takes notably higher values than the value corresponding to a sinusoidal distribution. The factor α_i has to be iterated gradually to the correct value during the design process. The value $\alpha_i = 0.64$ of an unsaturated machine can be employed as an initial value, unless it is known at the very beginning of the design process that the aim is to design a strongly saturating machine, in which case a higher initial value can be selected. The theoretical maximum value of a maximally saturated machine is $\alpha_i = 1$. In practice, $\alpha_i = 0.77$ is not exceeded. In that case, the magnetic voltage $\hat{U}_{m,ds} + \hat{U}_{m,dr}$ in the stator and rotor teeth is already higher than the magnetic voltage $\hat{U}_{m,\delta}$ of the air gap. To simplify the machine design, the factor α_i can be determined beforehand for different phases of saturation. Figure 6.5 illustrates α_i as a function of saturation factor k_{sat}:

$$\alpha_i = \frac{1.24k_{sat} + 1}{1.42k_{sat} + 1.57}, \tag{6.33}$$

$$k_{sat} = \frac{\hat{U}_{m,ds} + \hat{U}_{m,dr}}{\hat{U}_{m,\delta}}. \tag{6.34}$$

This saturation factor simply takes the teeth into account because their possible saturation is the main reason for the flattening of the air-gap flux density distribution.

In case of semi-closed slots the slotting effect can be taken into account by the Carter factor. In case of open slots in tooth coil machines the Carter-factor-based study may no more be valid. This will be studied in Chapter 7. In case of rotor surface magnet machine the flux density distribution in a slotless case can be analyzed according to Zhu *et al.* (2007).

The rotor surface PM caused no-load air-gap flux density distribution in polar coordinates $B_{\mathrm{mr}}\ (r,\theta)$ in a slotless permanent magnet machine can be expressed for the radial component as a function of the radius r and angle θ as

$$B_{\mathrm{mr}}(r,\theta)=\sum_{n=1,3,5\ldots}^{\infty} K_{\mathrm{B}}(n)\, f_{\mathrm{Br}}(r)\cos(np\theta), \tag{6.35}$$

and for the tangential component as

$$B_{\mathrm{m}\theta}(r,\theta)=\sum_{n=1,3,5\ldots}^{\infty} K_{\mathrm{B}}(n)\, f_{\mathrm{B}\theta}(r)\sin(np\theta). \tag{6.36}$$

Factors $K_{\mathrm{B}}, f_{\mathrm{Br}}, f_{\mathrm{B\theta}}$, in Equations (6.35) and (6.36) depend on the pole pair number p, radial location within the air gap r, the stator inner radius r_{s} and the rotor surface magnet inner and outer radii r_{ryi}, r_{mr} and the magnet relative permeability μ_{r} in the following way, first for $np=1$

$$K_{\mathrm{B}}(n)=\frac{\mu_0 J_n}{2\mu_{\mathrm{r}}}\left\{\frac{A_{3n}\left(\frac{r_{\mathrm{r}}}{r_{\mathrm{s}}}\right)^2-A_{3n}\left(\frac{r_{\mathrm{ryi}}}{r_{\mathrm{s}}}\right)^2+\left(\frac{r_{\mathrm{ryi}}}{r_{\mathrm{s}}}\right)^2\ln\left(\frac{r_{\mathrm{r}}}{r_{\mathrm{s}}}\right)^2}{\frac{\mu_{\mathrm{r}}+1}{\mu_{\mathrm{r}}}\left[1-\left(\frac{r_{\mathrm{ryi}}}{r_{\mathrm{s}}}\right)^2\right]-\frac{\mu_{\mathrm{r}}-1}{\mu_{\mathrm{r}}}\left[\left(\frac{r_{\mathrm{r}}}{r_{\mathrm{s}}}\right)^2-\left(\frac{r_{\mathrm{ryi}}}{r_{\mathrm{r}}}\right)^2\right]}\right\}, \tag{6.37}$$

$$f_{\mathrm{Br}}(r)=1+\left(\frac{r_{\mathrm{s}}}{r}\right)^2, \tag{6.38}$$

$$f_{\mathrm{B}\theta}(r)=-1+\left(\frac{r_{\mathrm{s}}}{r}\right)^2, \tag{6.39}$$

and for $np\neq 1$

$$K_{\mathrm{B}}(n)=\frac{\mu_0 J_n}{2\mu_{\mathrm{r}}}\frac{np}{(np)^2-1}\left\{\frac{(A_{3n}-1)+2\left(\frac{r_{\mathrm{ryi}}}{r_{\mathrm{r}}}\right)^{np+1}-(A_{3n}+1)\left(\frac{r_{\mathrm{ryi}}}{r_{\mathrm{r}}}\right)^{2np}}{\frac{\mu_{\mathrm{r}}+1}{\mu_{\mathrm{r}}}\left[1-\left(\frac{r_{\mathrm{ryi}}}{r_{\mathrm{s}}}\right)^{2np}\right]-\frac{\mu_{\mathrm{r}}-1}{\mu_{\mathrm{r}}}\left[\left(\frac{r_{\mathrm{r}}}{r_{\mathrm{s}}}\right)^{2np}-\left(\frac{r_{\mathrm{ryi}}}{r_{\mathrm{m}}}\right)^{2np}\right]}\right\}, \tag{6.40}$$

$$f_{\mathrm{Br}}(r)=1+\left(\frac{r_{\mathrm{s}}}{r}\right)^{np-1}\left(\frac{r_{\mathrm{r}}}{r_{\mathrm{s}}}\right)^{np+1}+\left(\frac{r_{\mathrm{r}}}{r}\right)^{np+1}, \tag{6.41}$$

$$f_{\mathrm{B}\theta}(r)=-\left(\frac{r}{r_{\mathrm{s}}}\right)^{np-1}\left(\frac{r_{\mathrm{r}}}{r_{\mathrm{s}}}\right)^{np+1}+\left(\frac{r_{\mathrm{r}}}{r}\right)^{np+1}, \tag{6.42}$$

where J_n for radial polarization depends on the relative magnet width α_{PM} and the PM-material remanent flux density B_{r} as

$$J_n = 2\frac{B_{\mathrm{r}}}{\mu_0}\alpha_{\mathrm{PM}}\frac{\sin\left(\frac{n\pi\alpha_{\mathrm{PM}}}{2}\right)}{\frac{n\pi\alpha_{\mathrm{PM}}}{2}}, \tag{6.43}$$

and for parallel polarization as

$$J_n = \frac{B_{\mathrm{r}}}{\mu_0}\alpha_{\mathrm{PM}}\left(A_{1n} + A_{2n}\right) + np\frac{B_{\mathrm{r}}}{\mu_0}\alpha_{\mathrm{PM}}\left(A_{1n} + A_{2n}\right), \tag{6.44}$$

with

$$A_{1n} = \frac{\sin\left((np+1)\,\alpha_{\mathrm{PM}}\frac{\pi}{2p}\right)}{(np+1)\,\alpha_{\mathrm{PM}}\frac{\pi}{2p}}, \tag{6.45}$$

$$A_{2n} = 1, \text{ for } np = 1, \tag{6.46}$$

$$A_{2n} = \frac{\sin\left((np-1)\,\alpha_{\mathrm{PM}}\frac{\pi}{2p}\right)}{(np+1)\,\alpha_{\mathrm{PM}}\frac{\pi}{2p}} \text{ for } np \neq 1, \tag{6.47}$$

$$A_{3n} = \begin{Bmatrix} 2\frac{J_{\mathrm{r}1}}{J_1} - 1 \text{ for } np = 1 \\ \left(np - \frac{1}{np}\right)\frac{J_{\mathrm{r}n}}{J_n} + \frac{1}{np} \text{ for } np \neq 1 \end{Bmatrix}, \tag{6.48}$$

for parallel case, and

$$A_{3n} = \begin{Bmatrix} 1 \text{ for } np = 1 \\ np \text{ for } np \neq 1 \end{Bmatrix}, \tag{6.49}$$

for radial case.

In Equation (6.48) $J_{\mathrm{r}n}$ can be written

$$J_{\mathrm{r}n} = 2\frac{B_{\mathrm{r}}}{\mu_0}\alpha_{\mathrm{PM}}\frac{\sin\left(\frac{n\pi\alpha_{\mathrm{PM}}}{2}\right)}{\frac{n\pi\alpha_{\mathrm{PM}}}{2}}, \tag{6.50}$$

for radial, and

$$J_n = \frac{B_{\mathrm{r}}}{\mu_0}\alpha_{\mathrm{PM}}\left(A_{1n} + A_{2n}\right), \tag{6.51}$$

for parallel magnetization. J_{r1} is found by inserting 1 for n in (6.50) and (6.51). Finally the air-gap flux radial component $\hat{\Phi}_m$ for Equation (6.32) is found by integrating (6.35).

Next, we have to find a suitable integer closest to the previously calculated number of turns N. In a phase winding, there are N turns in series. A single coil turn is comprised of two conductors in slots, connected by the coil ends. In a phase winding, there are thus $2N$ conductors in series. With m phases in a machine, the number of conductors becomes $2mN$. There may be a number of a parallel paths in a winding, in which case the number of conductors is $2amN$. The number of conductors per slot becomes

$$z_Q = \frac{2am}{Q}N. \tag{6.52}$$

Here Q is the slot number of either the stator or the rotor (in a slip-ring asynchronous motor or a DC machine). z_Q has to be an integer or in case of a double layer winding an even integer. When rounding z_Q off to an integer, we have to pay attention to the appropriateness of the slot number Q and the number of parallel paths a to avoid too large a rounding-off. After rounding-off, a new number of turns N is calculated for the phase winding.

In some cases, especially in low-voltage, high-power machines, there may be a need to change the stator slot number, the number of parallel paths or even the main dimensions of the machine in order to find the appropriate number of conductors in a slot.

6.2.7 New Air-Gap Flux Density

The selected number of turns for the phase winding has an effect on the value $\hat{B}_\delta$ of the air-gap flux density. A new value is calculated with Equation (6.32).

Example 6.7: Find a suitable number of turns for the stator phase winding for the 30 kW motor in Examples 6.1–6.3, 6.5–6.6. Calculate the new air-gap maximum flux density.

Solution: The number of turns in series is according to Equation (6.32)

$$N_s = \frac{\sqrt{2}E_m}{\omega k_{w1} l' \tau_p \alpha_i \hat{B}_\delta}.$$

According to previous calculation and guidelines:

$E_m \approx 0.97 \cdot U_{s,ph} = 0.97 \cdot 690/\sqrt{3}$ V; $k_{w1} = 0.925$; $l' = l + 2 \cdot \delta = 197$ mm + $2 \cdot 0.9$ mm $= 198.8$ mm;

$\tau_p = \pi D_s/2p = \pi \cdot 202/4$ mm $= 158.7$ mm; $\alpha_i \approx 0.65$ (a small saturation assumed) and $\hat{B}_\delta = 0.8$ T.

$$N_s = \frac{\sqrt{2}E_m}{\omega k_{w} l' \tau_p \alpha_i \hat{B}_\delta} = \frac{\sqrt{2} \cdot 0.97 \cdot 690}{\sqrt{3} \cdot 2\pi \cdot 50 \cdot 0.925 \cdot 0.1988 \cdot 0.1587 \cdot 0.65 \cdot 0.8} = 114.63.$$

The number of conductors in a slot is

$$z_{Qs} = \frac{2am}{Q_s} N_s.$$

Choosing $a = 1$ parallel branches we obtain

$$z_{Qs} = \frac{2am}{Q_s} N_s = \frac{2 \cdot 1 \cdot 3}{48} 114.63 = 14.3$$

and if we choose $a = 2$ we obtain

$$z_{Q_s} = \frac{2am}{Q_s} N_s = \frac{2 \cdot 2 \cdot 3}{48} 114.63 = 28.6.$$

z_{Qs} has to be an even number as there are two layers in the stator winding. Let us choose $a = 2$ parallel branches, so $z_{Qs} = 28$ and the number of turns are

$$N_s = \frac{Q_s z_{Qs}}{2am} = \frac{48 \cdot 28}{2 \cdot 2 \cdot 3} = 112.$$

The new air-gap flux density is

$$\hat{B}_\delta = \frac{\sqrt{2} E_m}{\omega k_w l' \tau_p \alpha_i N_s} = \frac{\sqrt{2} \cdot 0.97 \cdot 690}{\sqrt{3} \cdot 2\pi \cdot 50 \cdot 0.925 \cdot 0.1988 \cdot 0.1587 \cdot 0.65 \cdot 112} \mathrm{T} = 0.819\mathrm{T}.$$

6.2.8 *Determination of Tooth Width*

The air-gap flux density $\hat{B}_\delta$ being determined, the stator and rotor teeth are dimensioned next. The flux densities of the stator and rotor teeth are chosen for normal machines according to the permitted values presented in Table 6.2. In high-frequency machines, it may be necessary to select values notably lower than the values presented in Table 6.2 to avoid excessive iron losses. When the apparent reference flux densities ($\hat{B}'_d$) are selected for the stator and rotor teeth, the tooth widths (b_d) are obtained from Equation (3.42)

$$b_d = \frac{l' \tau_u}{k_{Fe} (l - n_v b_v)} \frac{\hat{B}_\delta}{\hat{B}'_d} + 0.1 \text{ mm} \tag{6.53}$$

where l' is the equivalent stator (rotor) stack length, k_{Fe} the space factor of iron, l the total stator (rotor) length of the machine stack, and n_v and b_v the number of ventilating ducts and their width. The influence of punching on the crystal structure and permeability of iron has been taken into account by adding 0.1 mm to the width.

Example 6.8: Determine the stator and rotor tooth width for the 30 kW motor in Examples 6.1–6.3, 6.5–6.7 if there are no ventilating ducts.

Solution: According to Table 6.2, the flux density of a stator tooth can vary from 1.4 T to 2.1 T and the rotor tooth from 1.5 T to 2.2 T. Let us choose for the apparent flux densities $\hat{B}'_{ds}$ = 1.6 T and $\hat{B}'_{dr}$ = 1.6 T. We obtain for the stator tooth width according to Equation (6.53)

$$b_{ds} = \frac{l'\tau_{us}}{k_{Fe}(l - n_v b_v)} \frac{\hat{B}_\delta}{\hat{B}'_{ds}} + 0.1\ \text{mm} = \frac{0.1988 \cdot \dfrac{\pi \cdot 0.202}{48}}{0.97 \cdot 0.197} \frac{0.819}{1.6} \cdot 1000\ \text{mm} + 0.1\ \text{mm}$$
$$= 7.14\ \text{mm}$$

and for the rotor tooth

$$b_{dr} = \frac{l'\tau_{ur}}{k_{Fe}(l - n_v b_v)} \frac{\hat{B}_\delta}{\hat{B}'_{dr}} + 0.1\ \text{mm} = \frac{0.1988 \cdot \dfrac{\pi \cdot 0.2002}{44}}{0.97 \cdot 0.197} \frac{0.819}{1.6} \cdot 1000\ \text{mm} + 0.1\ \text{mm}$$
$$= 7.71\ \text{mm}.$$

6.2.9 Determination of Slot Dimensions

In order to determine the dimensions of the stator and rotor slots, we have first to estimate the stator and rotor currents. In synchronous and asynchronous motors, the stator current I_s is obtained with the shaft power P, the stator phase voltage $U_{s,ph}$, the efficiency η and the power factor $\cos\varphi$

$$I_s = \frac{P}{m\,\eta U_{s,ph} \cos\varphi}. \tag{6.54}$$

We have to estimate the efficiency η and, for induction motors, also the power factor utilising e.g. IEC standard 60034-30-1 (2014 forecasted) and Figure 7.24. For synchronous machines, the power factor is a design parameter.

The stator current of a generator is

$$I_s = \frac{P}{mU_{s,ph}\cos\varphi} = \frac{S}{mU_{s,ph}} \tag{6.55}$$

where P is the electric output power. For synchronous generators, the power factor $\cos\varphi$ is a design parameter, and for induction generators, it has to be estimated in this phase of the design process.

In an induction motor, the rotor current referred to the stator is approximately of the same magnitude of the real component of the stator current (since the magnetizing current flows only in the stator)

$$I'_r \approx I_s \cos\varphi. \tag{6.56}$$

The real rotor current is defined by the transformation ratio between the rotor and the stator. The current of the bar of the cage winding of an induction motor is written on the base of Equations (7.57), (7.54) and (6.52) as

$$I_r = K_{rs} I_r' \approx \frac{z_Q}{a} \frac{Q_s}{Q_r} I_s \cos\varphi. \tag{6.57}$$

The relationship between the rotor current and its equivalent in the stator winding (K_{rs}) will be given in (7.54).

The armature current of a DC motor is determined by an equation resembling Equation (6.54)

$$I_a \approx \frac{P}{U\eta}. \tag{6.58}$$

The magnetizing windings of a salient-pole synchronous machine and a DC machine are salient-pole windings and are thus not inserted in slots. The magnetizing winding of a nonsalient-pole synchronous machine is a slot winding, but the space required by it can be defined only when the total magnetizing current of the machine has been calculated.

When the stator and rotor currents have been solved, the areas of the conductors become S_c. The resistive losses of the windings are chiefly determined by the current densities J_s and J_r of the stator and rotor windings. In standard machines, the values of Table 6.3 can be employed. The number a of the parallel paths has also to be borne in mind:

$$S_{cs} = \frac{I_s}{a_s J_s}, \; S_{cr} = \frac{I_r}{a_r J_r}. \tag{6.59}$$

The areas S_{us} and S_{ur} of the stator and rotor slots are now obtained by taking the space factors $k_{Cu,s}$ and $k_{Cu,r}$ of the slot into account

$$S_{us} = \frac{z_{Qs} S_{cs}}{k_{Cu,s}}, \quad S_{ur} = \frac{z_{Qr} S_{cr}}{k_{Cu,r}} \tag{6.60}$$

The space factor k_{Cu} of the slot depends principally on the winding material, the voltage level and the winding type of the machine. The windings of small electrical machines are usually made of round wire. In that case, the space factor of an insulated wire in a free slot (with the area reserved for the slot insulation subtracted) varies, depending on the quality of winding assembly from 60 to 66 %.

The space factor value is, however, defined for noninsulated slots. Values $k_{Cu,s} \in (0.5, 0.6)$ are typical for low-voltage machines. The lower limit is for round enameled wires and the upper limit for ideal prefabricated rectangular windings in low-voltage machines. In high-voltage machines, the insulation takes more space and the slot space factor varies, $k_{Cu,s} \in (0.3, 0.45)$, the lower value being for round wires and the upper for rectangular wires.

The armature winding of large machines is usually constructed of preformed copper. When using square wire or a prefabricated, preformed winding, the space factor is somewhat better than the space factor of a round wire winding.

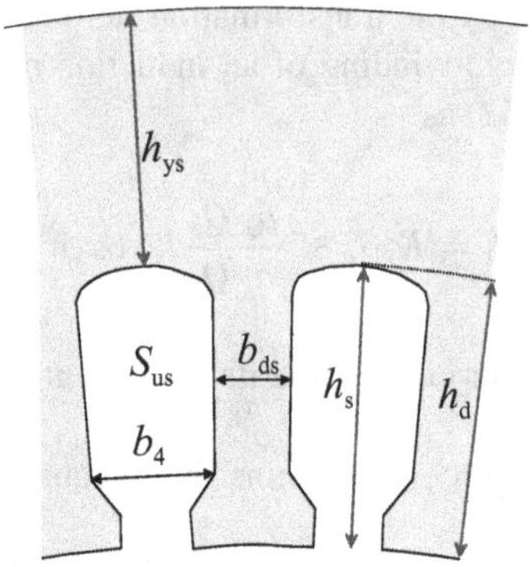

Figure 6.6 Stator tooth and two semi-closed stator slots and their main dimensions. h_{ys} is the stator yoke height. The slot depth equals the tooth height $h_s = h_d$.

If aluminum bars are die cast in the induction motor, the space factor becomes $k_{Cu,r} = 1$ (z_{Qr} is also now $z_{Qr} = 1$). If a cage winding is produced of copper bars by soldering, a clearance of approx. 0.4 mm in width and 1 mm in height has to be left in the rotor slot. This clearance also decreases the space factor.

In point 8, the widths b_{ds} (Figure 6.6) and b_{dr} of the stator and rotor teeth have been selected such that the selected permitted flux densities are located at respective teeth at the peak value of the flux density.

When the air-gap diameter and the teeth widths have been selected, the width of the slot is automatically known. The height h_s of the slot is always equal to the height of the tooth, and it is selected to reach the area S_{us} required for the winding and the insulations.

Example 6.9: Determine the stator and rotor slot dimensions for the 30 kW totally enclosed fan-cooled motor in Examples 6.1–6.3, 6.5–6.8.

Solution: Estimates for the efficiency $\eta = 0.936$ (efficiency class IE3), power factor $\cos\varphi = 0.87$ (a value little over the statistics given in Figure 7.24), the stator current I_s and the rotor bar current I_r and short-circuit ring current I_{ring} (Equations (6.54), (6.56) and (7.45)) are

$$I_s = \frac{P}{m\,\eta U_{s,ph}\cos\varphi} = \frac{\sqrt{3}\cdot 30000}{3\cdot 0.936\cdot 690\cdot 0.87}\,\mathrm{A} = 30.83\ \mathrm{A},$$

$$I_r = \frac{z_{Qs}}{a}\frac{Q_S}{Q_r}I_s\cos\varphi = \frac{28\cdot 48}{2\cdot 44}\cdot 30.83\cdot 0.87\ \mathrm{A} = 409.6\ \mathrm{A}\ \text{and}$$

$$I_{ring} = \frac{I_r}{2\cdot\sin(\pi p/Q_r)} = \frac{409.4}{2\cdot\sin(\pi\cdot 2/44)}\mathrm{A} = 1439.3\ \mathrm{A}.$$

Let us select for the stator and rotor current densities $J_s = 3.8$ A/mm^2, $J_r = 3.8$ A/mm^2 and $J_{ring} = 4$ A/mm^2 (Table 6.3) which are close to the lower limit because the motor is totally closed and the efficiency requirement is high. The stator conductor S_{cs}, rotor bar S_{cr} and the short-circuit ring S_{ring} cross-sectional areas are

$$S_{cs} = \frac{I_s}{aJ_s} = \frac{30.83}{2 \cdot 3.8} \text{mm}^2 = 4.06\ \text{mm}^2,\ S_{cr} = \frac{I_r}{J_r} = \frac{409.6}{3.8}\ \text{mm}^2 = 107.8\ \text{mm}^2 \text{and}$$

$$S_{cring} = \frac{I_{ring}}{J_{ring}} = \frac{1439.3}{4}\ \text{mm}^2 = 360\ \text{mm}^2.$$

To make stator coils easy to bend and to fabricate, the stator conductors are divided into five parallel wires of 1.0 mm of diameter. The final stator conductor area is

$$S_{cs} = 5 \cdot \frac{\pi \cdot 1.0^2}{4}\ \text{mm}^2 = 3.927\ \text{mm}^2.$$

The space factor inside the stator slot insulation is about $k_{Cus} = 0.62$ for a random wound coil. So the wound area of the slot should be

$$S_{Cus} = \frac{z_{Qs} S_{cs}}{k_{Cus}} = \frac{28 \cdot 3.927}{0.62} \text{mm}^2 = 177.3\ \text{mm}^2.$$

Let us choose the slot form presented in Figure 6.7 for the stator slot and following dimensions: $b_1 = 3.0$ mm, $h_1 = 0.5$ mm, $h_2 = 2.5$ mm, $h_3 = 2$ mm, $h_6 = 0.5$ mm and $h' = 0.5$ mm. The slot width and height are determined so that the tooth width is constant

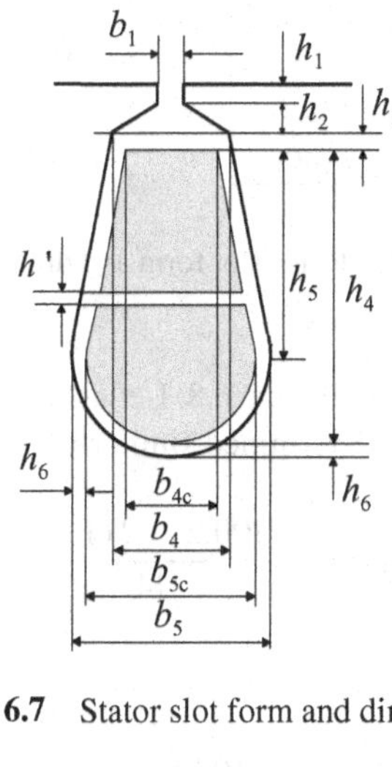

Figure 6.7 Stator slot form and dimensions.

and the wound area of the slot is $S_{\mathrm{Cus}} = 177.3\ \mathrm{mm}^2$. The dimensions of b_{4c}, b_{5c} and h_5 are solved from Equations

$$b_{4c} = \frac{\pi\left[D_s + 2(h_1 + h_2 + h_3)\right]}{Q_s} - 2h_6 - b_{ds} = \frac{\pi\left[202 + 2(0.5 + 2.5 + 2)\right]}{48}\mathrm{mm} - 2\cdot 0.5\ \mathrm{mm} - 7.14\ \mathrm{mm} = 5.74\ \mathrm{mm}$$

$$b_{5c} = b_{4c} + \frac{2\pi h_5}{Q_s}$$

$$S_{\mathrm{Cus}} = \frac{b_{4c} + b_{5c}}{2}(h_5 - h') + \frac{\pi}{8}b_{5c}^2 = 177.3\ \mathrm{mm}^2$$

and the slot width b_4

$$b_4 = \frac{\pi\left[D_s + 2(h_1 + h_2)\right]}{Q_s} - b_{ds} = \frac{\pi\left[202 + 2(0.5 + 2.5)\right]}{48}\ \mathrm{mm} - 7.14\ \mathrm{mm} = 6.5\ \mathrm{mm}.$$

For h_5 and b_{5c} we get the values $h_5 = 21.3$ mm and $b_{5c} = 8.5$ mm and further $b_5 = b_{5c} + 2h_6 = 9.5$ mm and $h_4 = h_5 + b_{5c}/2 = 25.5$ mm. The total height of the slot is $h_s = h_1 + h_2 + h_3 + h_4 + h_6 = 31.0$ mm. The dimensions are rounded to the nearest 0.1 mm and the final wound area of the stator slot is 176.3 mm^2.

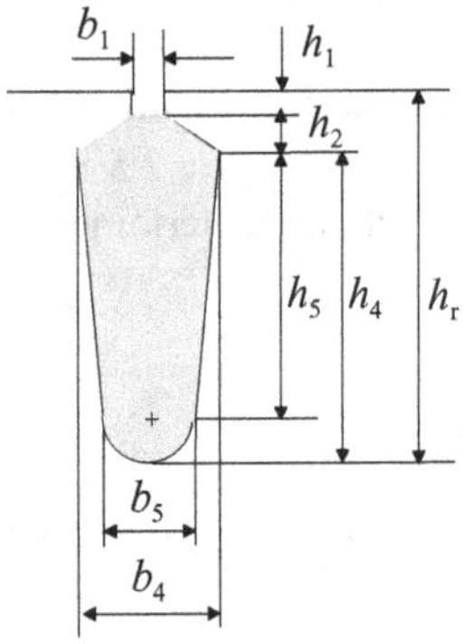

Figure 6.8 Rotor slot form and dimensions.

The rotor slot form is presented in Figure 6.8. Let us choose $b_1 = 3$ mm, $h_1 = 1$ mm and $h_2 = 2$ mm. Other dimensions are determined similarly as for the stator slot:

$$b_4 = \frac{\pi\left[D_r - 2(h_1 + h_2)\right]}{Q_r} - b_{dr} = \frac{\pi\left[200.2 - 2(1 + 2)\right]}{44}\mathrm{mm} - 7.71\ \mathrm{mm} = 6.16\ \mathrm{mm}$$

$$b_5 = b_4 - \frac{2\pi h_5}{Q_r}$$

$$S_{cr} = \frac{b_4 + b_5}{2}h_5 + \frac{\pi}{8}b_5^2 + \frac{b_1 + b_4}{2}h_2 = 107.8\ \mathrm{mm}^2.$$

The solution for h_5 and b_5 is $h_5 = 20.0$ mm and $b_5 = 3.3$ mm and further $h_4 = h_5 + b_5/2 = 21.65$ mm. The total slot height is $h_r = h_1 + h_2 + h_4 = 24.65$ mm. Because of rounding-offs, the final area for rotor bars is 108.0 mm^2.

6.2.10 Determination of the Magnetic Voltages of the Air Gap, and the Stator and Rotor Teeth

When the air-gap diameter, the air gap, the peak value of the air-gap flux density, and the dimensions of the stator and rotor slots of the machine are known, we may start to calculate the magnetic voltages over the air gap and the teeth. An exact definition of the magnetic voltages in these areas requires an analysis of the flux diagram in the respective areas. The manual solution of a flux diagram is a difficult task; however, with sufficient accuracy, the calculations can be made to estimate the line integral of the field strength $\boldsymbol{H}$ in this area

$$U_m = \int \boldsymbol{H} \cdot d\boldsymbol{l}. \tag{6.61}$$

The magnetic voltage U_m is calculated for each section individually. For instance, the flux diagram in Figure 6.9 is analyzed. Figure 6.10 shows the basically corresponding flux density and the field strength distribution at the middle of the tooth.

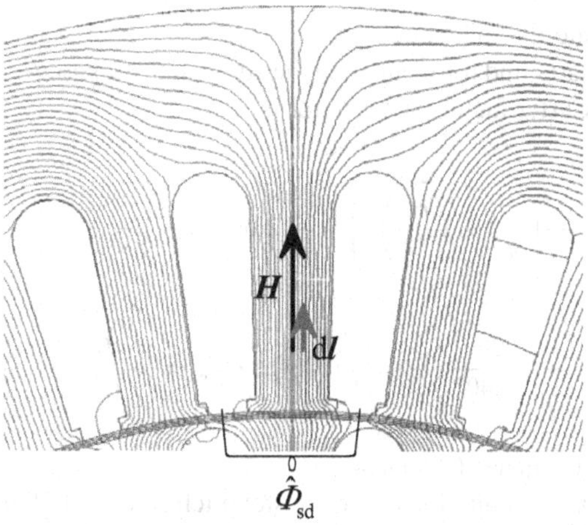

Figure 6.9 The apparent peak value $\hat{\phi}'_{sd}$ (see Equation (3.40), too) of the stator tooth flux is obtained by calculating the flux at the distance of a tooth pitch at the peak value of the flux density. We see that a part of the flux flows along the air space of the slot as the teeth saturate. In the analysis, the tooth in the middle in the figure occurs at the peak value of the air-gap flux density of the entire machine. In the middle of the tooth, the field strength $\boldsymbol{H}$ and the differential section d$\boldsymbol{l}$ of the integration path are parallel, and therefore the integration is easy.

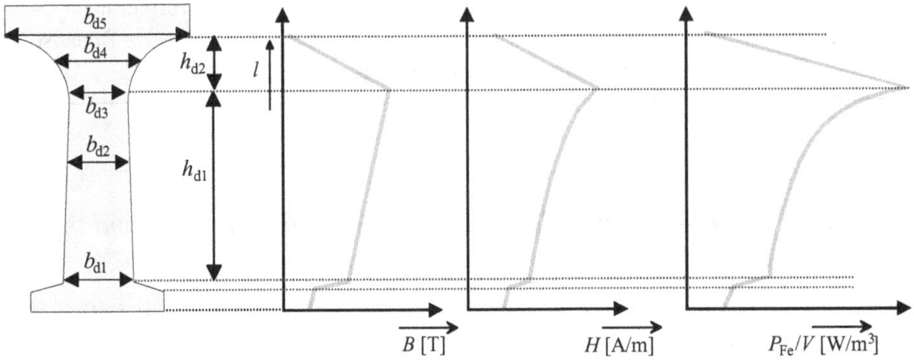

Figure 6.10 Principle of the flux density distribution B of a nonuniform stator tooth, the absolute value of the field strength H, and the iron loss density P_{Fe}/V on the integration path in the middle of the tooth in Figure 6.9. The curves are derived according to the stator tooth flux, the local tooth width, and the BH curve of the tooth material.

Example 6.10: Determine the magnetic voltages of the air gap, the stator and rotor teeth for the 30 kW motor in Examples 6.1–6.3, 6.5–6.9. Iron core material is M800-50A (Appendix A). The magnetization of the core material can be depicted by the functions

$$\hat{H} = 1835.2\hat{B}^5 - 6232.3\hat{B}^4 + 7806.7\hat{B}^3 - 4376.3\hat{B}^2 + 1227.6\hat{B}, \text{ if } \hat{B} < 1.5\text{ T and}$$
$$\hat{H} = 0.011637\mathrm{e}^{7.362\hat{B}}, \text{ if } \hat{B} \geq 1.5\text{ T}$$

Solution: To calculate the magnetic voltage of the air gap we have first to calculate the Carter factors of the stator and rotor and then the equivalent air gap. According to Equations (3.7b) and (3.8), the Carter factor of the stator is

$$\kappa_s = \frac{2}{\pi}\left(\arctan\frac{b_{1s}}{2\delta} - \frac{2\delta}{b_1}\ln\sqrt{1+\left(\frac{b_{1s}}{2\delta}\right)^2}\right) = \frac{2}{\pi}\left(\arctan\frac{3}{2\cdot 0.9} - \frac{2\cdot 0.9}{3}\ln\sqrt{1+\left(\frac{3}{2\cdot 0.9}\right)^2}\right)$$
$$= 0.4021$$

$$k_{Cs} = \frac{\tau_{us}}{\tau_{us} - \kappa b_{1s}} = \frac{13.22}{13.22 - 0.4021\cdot 3} = 1.100 \text{ where } \tau_{us} = \frac{\pi\cdot D_s}{Q_s} = \frac{\pi\cdot 202}{48}\text{ mm} = 13.22\text{ mm}$$

and $\delta_{es} = k_{Cs}\delta = 1.10\cdot 0.9\text{ mm} = 0.990\text{ mm}$.

Substituting in the above equations the rotor slot pitch ($\tau_{ur} = 14.29$ mm) and slot opening ($b_{1r} = 3.0$ mm) and for δ the above calculated δ_{es}, we obtain $\kappa_r = 0.3780$ and the Carter factor of the rotor $k_{Cr} = 1.086$ and finally the equivalent air gap $\delta_e = k_{Cr}\cdot\delta_{es} = 1.086\cdot 0.990$ mm $= 1.075$ mm. The magnetic voltage of the air gap is according to Equation (3.35)

$$\hat{U}_{m,\delta e} = \frac{\hat{B}_\delta}{\mu_0}\delta_e = \frac{0.819}{4\cdot\pi\cdot 10^{-7}}1.075\cdot 10^{-3}\text{ A} = 700.6\text{ A}$$

The magnetic voltage over the tooth can be calculated in three parts (Figure 6.10): the tooth tip (until b_{d1}), the straight part of the slot (b_{d1}, b_{d2}, b_{d3}) and the bottom half circle (b_{d3}, b_{d4}, b_{d5}). The integral (6.61) can be replaced by the Simpson rule, e.g. for the straight part:

$$\hat{U}_{m,dstraight} = h_{dl}\frac{\hat{H}_{d1} + 4\hat{H}_{d2} + \hat{H}_{d3}}{6} \quad (6.62)$$

where $\hat{H}_{d1}$, $\hat{H}_{d2}$ and $\hat{H}_{d3}$ are the magnetic field strengths after the tooth tip (the tooth width b_{d1}), in the middle of the tooth (the tooth width b_{d2}) and at the end of the straight part (the tooth width b_{d3}) respectively and h_{dl} the height of the straight part (Figure 6.10). The magnetic voltages over the tooth tip and the half circle part are calculated alike.

The magnetic field strengths are determined using Equations (3.42), (3.44) and (3.46) and the principle presented in Figure (3.14). Let us determine the field strength $\hat{H}_{d1}$. The apparent flux density is

$$\hat{B}'_{dl} = \frac{\Phi'_d}{S_{dl}} = \frac{l'\tau_{us}}{k_{Fe} l b_{dl}}\hat{B}_\delta = \frac{198.8 \cdot \dfrac{\pi \cdot 202}{48}}{0.97 \cdot 197.0 \cdot \left[\dfrac{\pi \cdot (202 + 2 \cdot 3)}{48} - 6.5 - 0.1\right]} 0.819\ \mathrm{T}$$
$$= 1.606\ \mathrm{T}.$$

The influence of punching on the crystal structure and permeability of iron has been taken into account by deducing 0.1 mm from the real tooth width.

The real flux density is according to Equations (3.46) and (3.44)

$$\hat{B}_{dl} = \hat{B}'_{dl} - \frac{S_{ul}}{S_{dl}}\mu_0 \hat{H}_{dl} = 1.606\,\mathrm{T} - 1.019 \cdot 4 \cdot \pi \cdot 10^{-7}\frac{\hat{H}_{dl}}{[\mathrm{A}]}\mathrm{T},$$

where $\dfrac{S_{ul}}{S_{dl}} = \dfrac{l'\tau_{ul}}{k_{Fe} l b_{dl}} - 1 = \dfrac{l'\pi[D_s + 2(h_{1s} + h_{2s})]/Q_s}{k_{Fe} l b_{dl}} - 1 = 2.0189 - 1 = 1.019.$

The intersection of the $\hat{B}_{dl}$-line and the magnetization curve of M800-50A gives for the real flux density the value $\hat{B}_{dl} = 1.569$ T and for the corresponding field strength the value $\hat{H}_{dl} = 1213$ A/m. Similarly, we obtain the field strengths $\hat{H}_{d2} = 1169$ A/m and $\hat{H}_{d3} = 1127$ A/m. The magnetic voltage of the straight part is

$$\hat{U}_{m,dstraight} = h_{dl}\frac{\hat{H}_{d1} + 4\hat{H}_{d2} + \hat{H}_{d3}}{6} = (0.002 + 000213)\frac{1213 + 4 \cdot 1169 + 1127}{6}$$
$$= 27.24\ \mathrm{A}$$

The magnetic voltage of the tooth tip part is small and can be neglected. For the half circle part we obtain the following flux densities $\hat{B}_{d3} = 1.559$ T, $\hat{B}_{d4} = 1.327$T and $\hat{B}_{d5} = 0.646$ T and the corresponding field strengths $\hat{H}_{d3} = 1227$ A/m, $\hat{H}_{d4} = 391$ A/m and $\hat{H}_{d5} = 192$ A/m. The magnetic voltage of the half circle part is

$$\hat{U}_{m,dcircle} = h_{d2} \cdot \frac{\hat{H}_{d3} + 4\hat{H}_{d4} + \hat{H}_{d5}}{6} = \frac{0.0095}{2} \cdot \frac{1227 + 4 \cdot 391 + 192}{6}\mathrm{A} = 2.36\ \mathrm{A}.$$

The height of the half circle $h_{d2} = b_5/2$, where b_5 is the width of the slot at the end of the straight part of the slot (Figure 6.7). The total magnetic voltage of the stator tooth is

$$\hat{U}_{m,ds} = \hat{U}_{m,dstraight} + \hat{U}_{m,dcircle} = 27.25\,A + 2.36\,A = 29.6\,A$$

Similarly, we obtain the rotor tooth magnetic voltage $\hat{U}_{m,dr} = 26.6\ A/m$.

6.2.11 Determination of New Saturation Factor

The factor α_i and the saturation factor k_{sat} were defined in Chapter 6.2.6. As we remember, the factor α_i was given as a function of the factor k_{sat}, Equation (6.34) and Figure 6.5. Now, we have to check up the saturation factor and determine a new α_i. If α_i does not correspond with sufficient accuracy to the factor selected in the initial phase of the calculation, the peak value $\hat{B}_\delta$ of the air-gap flux density has to be recalculated according to Equation (6.32) because N is now fixed. Simultaneously, the flux density values of the stator and rotor teeth have to be corrected, and new magnetic voltages have to be calculated for the teeth and the air gap. The factor α_i has to be iterated gradually to a correct value. For PMSMs with rotor surface magnets with a uniform thickness, this step is not valid.

Example 6.11: Define a new factor α_i for the 30 kW motor in Examples 6.1–6.3, 6.5–6.10

Solution: The saturation factor k_{sat} is according to Equation (6.34)

$$k_{sat} = \frac{\hat{U}_{m,ds} + \hat{U}_{m,dr}}{\hat{U}_{m,\delta}} = \frac{30.9 + 26.6}{700.6} = 0.082,$$

and the corresponding α_i is according to Equation (6.33) $\alpha_i = 0.65$. The value is the same as the previously predicted value 0.65.

6.2.12 Determination of Stator and Rotor Yoke Heights and Magnetic Voltages

The flux density maxima $\hat{B}_{ys}$ and $\hat{B}_{yr}$ of the stator and rotor yokes are selected according to Table 6.2. With the peak value of the flux of the machine, together with the flux density peaks $\hat{B}_{ys}$ and $\hat{B}_{yr}$, we are able to determine the heights h_{ys} and h_{yr} of the rotor and stator yokes from Equations (3.48) and (3.49). The stator and rotor yoke fluxes can be assumed to be equal in induction machines. In synchronous machines, the rotor yoke flux is greater than the stator yoke flux because of the leakage flux of the rotor poles and permanent magnets. In salient-pole synchronous machines, the rotor yoke flux (the flux at the foot of the pole) is according to Equation (3.47) $(1.1-1.3)\Phi_{\mathrm{m}}$. In permanent magnet machines with rotor surface mounted magnets, the rotor yoke flux is round $(1.1-1.2)\Phi_{\mathrm{m}}$. More exact figures we can obtain by solving the magnetic field e.g. with finite element method.

The magnetic voltages of the stator and rotor yokes are calculated from Equations (3.51) and (3.52).

Example 6.12: Determine the stator and rotor yoke heights and magnetic voltages of the 30 kW motor in Examples 6.1–6.3, 6.5–6.11.

Solution: According to Table 6.2, let us choose $\hat{B}_{\mathrm{ys}} = 1.4$ T and $\hat{B}_{\mathrm{yr}} = 1.4$ T. The yoke heights and magnetic voltages are

$$h_{\mathrm{ys}} = \frac{\hat{\Phi}_{\mathrm{m}}}{2k_{\mathrm{Fe}}l\hat{B}_{\mathrm{ys}}} = \frac{\alpha_{\mathrm{i}}\hat{B}_{\delta}\tau_{\mathrm{p}}l'}{2k_{\mathrm{Fe}}l\hat{B}_{\mathrm{ys}}} = \frac{0.65\cdot 0.819\cdot 0.1587\cdot 0.1988}{2\cdot 0.97\cdot 0.197\cdot 1.4}\,\mathrm{m} = 31.4\ \mathrm{mm}$$

$$h_{\mathrm{yr}} = \frac{\hat{\Phi}_{\mathrm{m}}}{2k_{\mathrm{Fe}}l\hat{B}_{\mathrm{yr}}} = 31.4\ \mathrm{mm}.$$

$$\hat{U}_{\mathrm{m,ys}} = c\hat{H}_{\mathrm{ys}}\tau_{\mathrm{ys}} = c\hat{H}_{\mathrm{ys}}\frac{\pi\left(D_{\mathrm{s}} + 2h_{\mathrm{s}} + h_{\mathrm{ys}}\right)}{2p}$$

$$= 0.27\cdot 500.9\cdot\frac{\pi\,(0.202 + 2\cdot 0.0310 + 0.0314)}{2\cdot 2}\,\mathrm{A} = 31.4\ \mathrm{A}$$

$$\hat{U}_{\mathrm{m,yr}} = c\hat{H}_{\mathrm{yr}}\tau_{\mathrm{yr}} = c\hat{H}_{\mathrm{yr}}\frac{\pi\left(D_{\mathrm{r}} - 2h_{\mathrm{r}} - h_{\mathrm{yr}}\right)}{2p}$$

$$= 0.27\cdot 500.9\cdot\frac{\pi\,(0.2002 - 2\cdot 0.02465 - 0.0314)}{2\cdot 2}\,\mathrm{A} = 12.7\mathrm{A}$$

where the stator slot height $h_{\mathrm{s}} = 0.0310$ m and the rotor slot height $h_{\mathrm{r}} = 0.02465$ m are calculated in Example 6.9. The coefficient c is obtained from Figure 3.17 and the field strengths $\hat{H}_{\mathrm{ys}}$ and $\hat{H}_{\mathrm{yr}}$ from Equations given in Example 6.10.

6.2.13 *Magnetizing Winding*

Now all the main dimensions of the machine have been determined, next we have to check the magnetic voltages required by different parts of the machine. The sum of the magnetic voltages has to be covered by the current linkage Θ produced by some (or multiple) of the windings or by permanent magnets. Magnetizing is accomplished by different methods in different machines. A DC machine is magnetized with a separate field winding or permanent magnets, whereas an induction machine is magnetized with the magnetizing current of the stator winding, or a synchronous machine with the current of the rotor field winding or by permanent magnets, and so on.

In this book, systematically, half of the magnetic circuit is calculated (e.g. half of the stator yoke length, one stator tooth, one air gap, one rotor tooth, half of the rotor yoke length). The magnetic voltage sum of half of the magnetic circuit has to be covered by the current linkage amplitude produced by the rotating-field winding or by the current linkage of one pole winding of an excitation winding or by one permanent magnet.

A stator rotating field produces the fundamental current linkage of amplitude $\hat{\Theta}_{s1} = mk_{w_1} N_s\sqrt{2}\, I_{s,mag}/(\pi p)$ according to Equation (2.15) modified for the fundamental component. The required magnetizing current $I_{s,mag}$ can be calculated by keeping the sum $\hat{U}_{m,tot}$ of the magnetic voltages of half magnetic circuit equal to the amplitude $\hat{U}_{m,tot} = \hat{\Theta}_{s1}$ (see Example 3.5 and Equations (3.59) and (3.60)). Now we obtain the RMS value of the stator magnetizing current

$$I_{s,mag} = \frac{\hat{U}_{m,tot}\pi p}{mk_{w1} N_s \sqrt{2}}. \tag{6.63}$$

For machines magnetized with pole windings, the calculation of the no-load magnetizing current is a simple task, since the current linkage of a single pole winding is $N_f I_f$. This, again, has to equal the sum of the magnetic voltages of half a magnetic circuit. When dimensioning a synchronous machine field winding, for instance, it must be remembered that the field winding must also be able of compensating for the armature reaction and, hence, the amount of copper must be large enough to carry a much larger current than at no load. Let the ratio of magnetizing current at load to magnetizing current at no-load to be k. When

$$N_f I_f = N_f J_f S_{cf} = k\hat{U}_{m,tot}$$

where N_f is the number of turns, J_f the field winding current density (Table 6.3) and S_{cf} the cross-sectional area of the conductor of a pole coil. If the space factor of a pole coil is k_{Cu}, the total cross-sectional area of a pole coil is

$$S_f = \frac{N_f S_{cf}}{k_{Cu}} = \frac{k\hat{U}_{m,tot}}{k_{Cu} J_f}. \tag{6.64}$$

This area is divided into a coil height and width according to the available space between the poles.

At this point, in case of a rotor surface magnet PMSM the height h_{PM} of the permanent magnet is calculated. The magnetic voltage sum of the magnetic circuit is

$$\hat{U}_{m,tot} = \Theta_{PM} = \frac{B_r}{\mu_{PM}} h_{PM} = \hat{U}_{m,\delta e} + \hat{U}_{m,ds} + \frac{B_{PM}}{\mu_{PM}} h_{PM} + \frac{\hat{U}_{m,ys}}{2} + \frac{\hat{U}_{m,yr}}{2} \tag{6.65}$$

from which

$$h_{PM} = \frac{\hat{U}_{m,\delta e} + \hat{U}_{m,ds} + \dfrac{\hat{U}_{m,ys}}{2} + \dfrac{\hat{U}_{m,yr}}{2}}{\dfrac{B_r - B_{PM}}{\mu_{PM}}}. \tag{6.66}$$

where $\mu_{PM} = \mu_{rPM}\mu_0$ is the permeability of the permanent magnets, μ_{rPM} the relative permeability of the permanent magnets ($\mu_{rPM} \approx 1.05$ for the neodymium-iron-boron magnets), B_r

the remanent flux density of the permanent magnets and B_{PM} the flux density in the permanent magnet for the operating point. B_{PM} is equal to the air-gap flux density $\hat{B}_\delta$ that has been chosen in Chapter 6.2.5 (see also Figure 6.4).

6.2.14 *Determination of Stator Outer and Rotor Inner Diameter*

When the air-gap diameter D_s, the slot heights h_s and h_r, and the yokes heights h_{ys} ja h_{yr} are known, we obtain the outer diameter D_{se} of the stator and the inner diameter D_{ri} of the machine; cf. Figures 3.1 and 3.2.

Example 6.13: Determine the stator outer and rotor inner diameter of the 30 kW motor in Examples 6.1–6.3, 6.5–6.12.

Solution:

$$D_{se} = D_s + 2(h_s + h_{ys}) = 202.0 + 2\,(31.0 + 31.4)\,\text{mm} = 326.8\ \text{mm}$$
$$D_{ri} = D_r - 2(h_r + h_{yr}) = 200.2 - 2\,(24.65 + 31.4)\,\text{mm} = 88.1\ \text{mm}.$$

6.2.15 *Calculation of Machine Characteristics*

Since the dimensions have been defined and the winding has been selected, the resistances and inductances of the machine are now calculated. The magnetizing inductance and the leakage inductances were discussed in Chapter 4 and the resistances in Chapter 5. With them, the equivalent circuit parameters of the machine per phase are obtained. Now the losses, efficiency, temperature rise and torques of the machine can be determined.

In Figure 6.10, the iron losses in the teeth can be solved by manual calculations. The frequency of the flux density alternations corresponds to the rated frequency of the machine. The specific power loss (W/kg) corresponding to this frequency and the peak flux density can be checked from the manufacturer's catalogue, and then a dissipation power density curve can be constructed corresponding to the flux density curve of Figure 6.10 for the determination of losses in a single tooth, when the weight of the tooth is known. Using the same principle, all the iron parts of the machine are analyzed. Since stresses and burrs occur in punched plates, and since the shaped parts on both sides of the air gap cause flux pulsation, there are notably higher losses in the teeth and also in other iron parts of the machine than the power loss calculated with the base frequency would suggest. Further, the power losses given by the manufacturer are presented for AC magnetizing, not for rotating magnetizing, which is the dominant form of motion of the field in the stator yoke. These facts have led to the utilization of empirical factors of Table 3.2.

The factors in Table 3.2 are used to correct the iron loss calculations in the most significant parts of the magnetic circuit of a machine. In machine design, special attention must be paid to the fact that the frequencies in all parts of the machine are not equal. In the stator of rotating-field machines, the base frequency is the input frequency f_s of the machine. However, in the teeth of the stator and the rotor, high-frequency flux components occur, which are based on the motion of the teeth with respect to each other. For instance, in the rotor of a synchronous machine, the base frequency is zero, but other pulsation losses occur on the rotor surface because of stator slotting.

Resistive losses are defined with the methods discussed in Chapter 9 by determining the resistances of the windings. Following the definition of the iron and resistive losses, the windage and friction losses and the additional losses of the machine can be defined according to the guidelines given in Chapter 9. Now the efficiency of the machine can be resolved. The thermal rise, on the other hand, finally determines the resistances of the machine, and thus also the resistive losses, and therefore the heat transmission calculations should still be carried out before the analysis of final losses of the machine.

Bibliography

Commission Regulation (EC) No 640/2009 of 22 July 2009 implementing Directive 2005/32/EC of the European Parliament and of the Council with regard to ecodesign requirements for electric motors.

Directive 2009/125/EC of the European Parliament and of the Council of 21 October 2009 establishing a framework for the setting of ecodesign requirements for energy-related products.

IEC 60034-30-1 (2014 forecasted) *Rotating Electrical Machines Part 30-1: Efficiency Classes of Line Operated AC Motors (IE-code).* International Electrotechnical Commission, Geneva.

Lawrenson, P.J. (1992) A brief status review of switched reluctance drives. *EPE Journal*, **2**(3), 133–44.

Pat. U.S. 5,473,211. *Asynchronous electric machine and rotor and stator for use in association therewith.* High Speed Tech Oy Ltd, Finland. (Antero Arkkio). Appl. 86,880, July 7, 1993. (Appl. in Finland July 7, 1992).

Pyrhönen, J., Ruuskanen, V., Nerg, J. Puranen, J., Jussila, H. (2010) Permanent magnet length effects in AC-machines, *IEEE Transactions on Magnetics*, **46** (10), 3783–9, ISSN 0018-9464.

Richter, R. (1954) *Electrical Machines: Induction Machines (Elektrische Maschinen: Die Induktionsmaschinen)*, Vol. **IV**, 2nd edn, Birkhäuser Verlag, Basle and Stuttgart.

Richter, R. (1963) *Electrical Machines: Synchronous Machines and Rotary Converters. (Elektrische Maschinen: Synchronmaschinen und Einankerumformer)*, Vol. **II**, 3rd edn, Birkhäuser Verlag, Basle and Stuttgart.

Richter, R. (1967) *Electrical Machines: General Calculation Elements. DC Machines. (Elektrische Maschinen: Allgemeine Berechnungselemente. Die Gleichstrommaschinen)*, Vol. **I**, 3rd edn, Birkhäuser Verlag, Basle and Stuttgart.

Vogt, K. (1996) *Design of Electrical Machines. (Berechnung elektrischer Maschinen)*: Wiley-VCH Verlag GmbH, Weinheim.

Zhu, Z.Q., Ishak, D., Howe, D., Chen, J. (2007) Unbalanced magnetic forces in permanent magnet brushless machines with diametrically asymmetric phase windings, *IEEE Transactions on Industry Applications*, **43**(6), 1544–53.

7

Properties of Rotating Electrical Machines

In this chapter we shall study the design and properties of the most important industrial electric machines, squirrel cage induction motors (IMs), different synchronous machines (traditional field-winding excited-synchronous machines (SMs), permanent magnet synchronous machines (PMSMs) and synchronous reluctance machines (SyRMs)), DC-machines and switched reluctance (SR) machines. The biggest attention is paid to IMs and SMs as they are the dominating machines in the industry. However, the importance of other machine types is also admitted.

Before entering into specific machine type properties, however, we will start by observing different loadings of the machines. First the relation of the machine speed and size and then mechanical, electrical and magnetic loadabilities are studied to find out the performance limits of different machine types.

7.1 Machine Size, Speed, Different Loadings and Efficiency

7.1.1 Machine Size and Speed

"Electro-mechanics" is the technology covering the design of electrical machines. The term itself indicates that in electrical machines there is an interaction between electromagnetism and mechanics. Actually, mechanics and material properties such as the yield strength dictate the dimensions and limiting speeds of the rotating parts in an electrical machine. As has been observed earlier the machine size is directly proportional to the torque of the machine. The torque is, of course, one of the most important starting points when designing a machine. However, in many cases the power of the machine is also an equally important starting point. The higher the speed of a machine the smaller torque it has to deliver to reach a certain power level. Therefore, a higher speed machine is small compared to a lower speed machine of the same power.

Design of Rotating Electrical Machines, Second Edition. Juha Pyrhönen, Tapani Jokinen and Valéria Hrabovcová.

According to Equation (6.21), the mass of the rotor m_r is proportional to the apparent power S_i and inversely proportional to the speed $n = f/p$

$$m_r \sim V_r = \frac{\pi}{4} \frac{pS_i}{Cf}. \tag{7.1}$$

With Equation (7.1) and Equations (6.1) and (6.17), we obtain for AC machines

$$m_r \sim \frac{pS_i}{fA\hat{B}_\delta} = \frac{pS_i}{f\sigma_{F\tan}}. \tag{7.2}$$

Thus, the higher the tangential stress and the supply frequency are, the lower is the mass of the rotor.

By selecting a small number of pole pairs p and a high supply frequency f, we obtain, in principle, a light machine for a certain output power. Since the output power of a machine depends on the torque T and the mechanical angular frequency Ω, as $P = \Omega T$, the torque of a high-speed machine is small when compared with a low-speed machine of equal output power. At higher speeds, however, the motor power density increases together with the loss density, and hence effective methods to reduce the losses and to improve the cooling of the machine have to be applied. With high frequencies, suitable means to reduce the stator winding skin effect have to be adopted so that there is no need to reduce significantly the linear current density. As the frequency rises, in order to maintain the air-gap flux density, better stator steel materials and more effective cooling methods have to be selected.

When analyzing the mass of 22 kW serial-produced machines, we may construct a curve according to Figure 7.1 as a function of the number of pole pairs. The straight line in the figure approximates the mass of the machine as a function of the number of pole pairs. We see that, in principle, the mass of the machine follows the equation

$$m = m_0 + Kp. \tag{7.3}$$

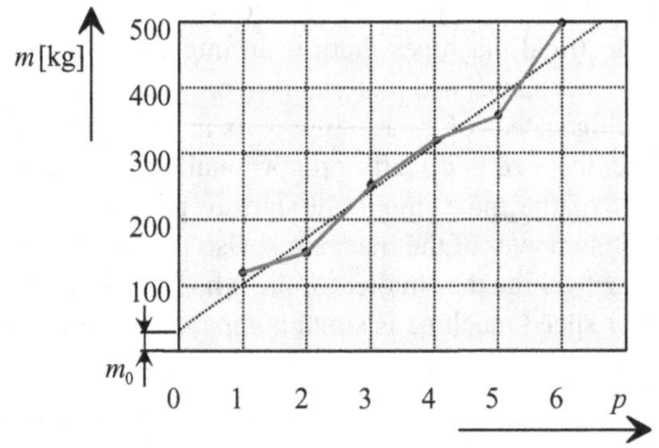

Figure 7.1 Mass of typical, totally enclosed, industrial 22 kW, 50 Hz, 400 V induction motors as a function of the number of pole pairs. The broken straight line is the trend line of the mass. It crosses the ordinate at $m_0 = 40$ kg.

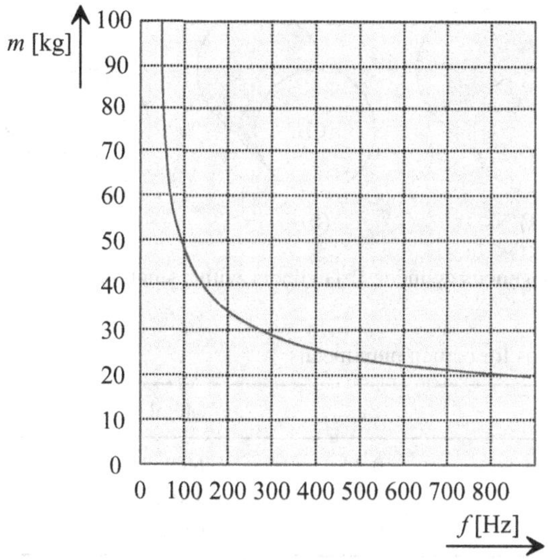

Figure 7.2 Mass of the active parts of a two-pole 22 kW induction motor as a function of frequency. The exponent $g \approx 0.8$ in Equation (7.4).

According to Equation (7.2), the mass of a machine is inversely proportional to the frequency, and thus the mass of the machine should be

$$m = m_0 + \frac{K}{f^g}, \tag{7.4}$$

where m_0, K and g are constants. Further, g is usually less than one, since the linear current density and the air-gap flux density cannot be considered constant as the frequency of the machine increases. As a result, we obtain Figure 7.2, which illustrates the approximate mass m of a two-pole induction machine as a function of frequency f.

7.1.2 Mechanical Loadability

In addition to temperature rise, the output power and maximum speed of a machine are restricted by the highest permissible mechanical stresses caused by the centrifugal force, natural frequencies, and by the highest permissible electrical and magnetic loadings. Analyzing the mechanical stresses in rotating parts is a challenging task and we do not try to cover the issue here in details. However, some very basic equations are given to get some kind of an image of the stresses affecting rotating parts.

When the speed or the size of a rotor is increased the limits of the material strength are easily reached. The highest stress σ_{mec} caused by the centrifugal force in the rotor is proportional to the square of the angular speed

$$\sigma_{mec} = C'\rho r_r^2 \Omega^2. \tag{7.5}$$

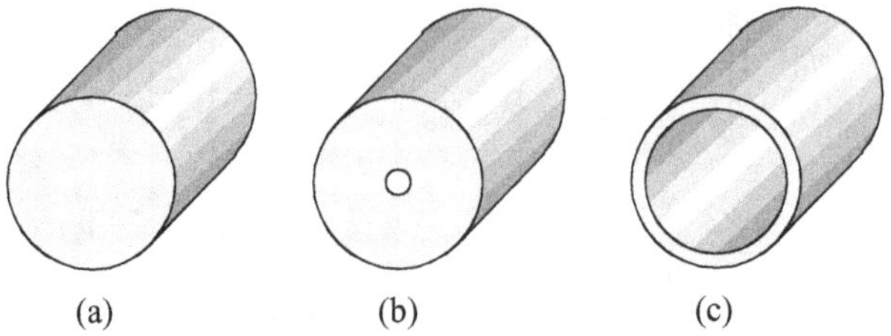

Figure 7.3 (a) Homogenous cylinder, (b) cylinder with a small bore, (c) thin (hollow) cylinder.

Table 7.1 Poisson's ratios for certain pure metals

Metal		ν	Metal		ν
Aluminum	Al	0.34	Nickel	Ni	0.30
Copper	Cu	0.34	Titanium	Ti	0.34
Iron	Fe	0.29	Cobalt	Co	0.31

Here

$C' = \dfrac{3+\nu}{8}$ for a smooth homogeneous cylinder (Figure 7.3a),
$C' = \dfrac{3+\nu}{4}$ for a cylinder with a small bore (Figure 7.3b),
$C' \approx 1$ for a thin hollow cylinder (Figure 7.3c),
r_r is the radius of the rotor,
Ω is the mechanical angular speed,
ρ is the density of the material,
ν is Poisson's ratio (i.e. the ratio of lateral contraction to longitudinal extension in the direction of stretching force).

When the maximal allowed mechanical stress of the rotor material is known, the equation can be used to determine the maximum allowable radius r_r of the rotor. Typically, a safety factor has to be used so as not to exceed the integrity of the rotor material.

Poisson's ratios vary slightly for different materials. Table 7.1 lists some ratios for pure materials.

Example 7.1: Calculate the maximum diameter for a smooth steel cylinder having a small bore. The cylinder is rotating at 15 000 min^{-1}. The yield strength for the material is 300 N/mm^2 = 300 MPa. The density of the steel is $\rho = 7860\ \text{kg/m}^3$.

Solution: Poisson's ratio for steel is 0.29. For a cylinder with a small bore

$$C' = \frac{3+\nu}{4} = \frac{3+0.29}{4} = 0.823.$$

The stress is calculated as $\sigma_{\mathrm{mec}} = C'\rho r_{\mathrm{r}}^2 \Omega^2$.

$$\sigma_{\mathrm{yield}} = \sigma_{\mathrm{mec}} = C'\rho r_{\mathrm{r}}^2 \Omega^2$$

$$\Leftrightarrow r_{\mathrm{r,max}} = \sqrt{\frac{\sigma_{\mathrm{yield}}}{C'\rho\Omega^2}} = \sqrt{\frac{300\,\mathrm{MPa}}{0.823 \cdot 7860\,\mathrm{kg/m^3} \cdot \frac{(2 \cdot (15\,000/60) \cdot \pi)^2}{\mathrm{s}^2}}} = 0.14\,\mathrm{m}.$$

As there must be some security in the yield stress, the radius of the rotor diameter has to be smaller than what was calculated above.

Equation (7.5) cannot be directly applied to the measurement of stresses in rotor laminations, for instance, because of the fairly complicated geometry of the laminations. However, some informative results may be found, since the stress is always highest at the center of a solid plate or at the inner surface of a plate with a center bore. At these points, the permitted proportion of the yield stress of the material should not be exceeded.

In the case of salient-pole rotors, in small machines, the pole cores should be fastened to the shaft for instance using screws. The screw fastenings have to be dimensioned with a sufficient security depending on the yield stress. If a screw fastening is out of the question, a dovetail joint may be applied.

Rotor constructions vary a lot and in addition to the above mentioned rotor core strength study, especially, the rotor windings or permanent magnets must be supported so that the centrifugal forces do not break the rotor. The centrifugal force F_{cf} exerted on an object with mass m rotating at certain peripheral velocity v at radius r is

$$F_{\mathrm{cf}} = \frac{mv^2}{r_{\mathrm{r}}} = mr_{\mathrm{r}}\Omega^2. \tag{7.6}$$

If, for example, permanent magnets are attached on a rotor surface the magnet retaining ring has to be dimensioned so that it can carry the centrifugal force F_{cf} exerted on the magnets. Similarly, the end windings of a synchronous machine field winding must be supported against the centrifugal force. The salient poles of low-speed synchronous machines are often fixed to the rotor core by steel bolts. The screws must, naturally, tolerate with suitable security factor the centrifugal force in the machine.

The length of the rotor is chiefly restricted by the critical angular speeds of the rotor. At a critical speed, the rotor has one of its mechanical resonances. There are several bending modes for the mechanics of each rotor. At the lowest critical speed, the rotor bends like a banana having two nodal points, Figure 7.4. At the second critical speed, the rotor bends into an S

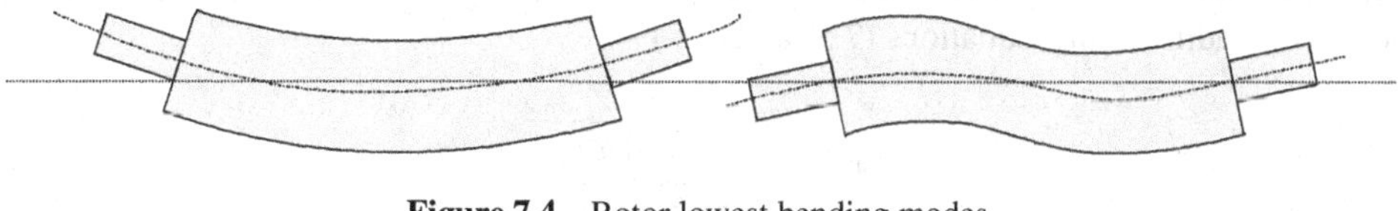

Figure 7.4 Rotor lowest bending modes.

shape with three nodal points, and so on. There are also torsional bending modes that may restrict the use of a rotor.

Usually, the ratio of the length of the machine to the air-gap diameter $\chi = l/D$ (recall Table 6.5) is selected for operation of the rotor below the first critical rotation speed. This cannot, however, be guaranteed. For instance, large turbo generators generally operate between different critical speeds. The maximum length of the rotor l_{max} that guarantees operation below the first critical speed is defined according to Wiart (1982)

$$l_{max}^2 = n^2 \frac{\pi^2}{k\Omega}\sqrt{\frac{EI}{\rho S}}, \tag{7.7}$$

where

S is the area of the cross-section of the cylinder (m^2),
E is the modulus of elasticity (Young's modulus) of the rotor material, typically 190–210 GPa for steel,
I is second moment of inertia of area (m^4), $I = \pi(D_{out}^4 - D_{in}^4)/64$ for a cylinder,
n is the order of critical rotation speed,
k is the safety factor (the ratio of the nth critical angular speed to rated angular speed),
ρ is the density of material.

Example 7.2: Calculate the maximum length with a safety factor $k = 1.5$ for a smooth solid-steel rotor operating under the first critical speed, when the rotor diameter is 0.15 m and the rotor speed is 20 000 min^{-1}.

Solution: $l_{max}^2 = n^2 \dfrac{\pi^2}{k\Omega}\sqrt{\dfrac{EI}{\rho S}}$

$$l_{max} = \sqrt{1^2 \frac{\pi^2}{1.5 \dfrac{20\,000}{60\,\mathrm{s}} 2\pi} \sqrt{\frac{200\,\mathrm{GPa} \cdot \dfrac{\pi \cdot 0.15^4}{64}\mathrm{m}^4}{7860 \dfrac{\mathrm{kg}}{\mathrm{m}^3} \dfrac{\pi \cdot 0.15^2}{4}\mathrm{m}^2}}} = 0.77\,\mathrm{m}.$$

In this theoretical case $l_{max}/r_r = 10$. In practice, the rotors have for instance slits and low-diameter shafts for bearings and so on, which reduces the l_{max}/r_r ratio.

If we employ the safety factor in the yield k_σ, we obtain for the ratio of the length of the rotor to the radius, using Equations (7.5) and (7.7)

$$\frac{l_{max}}{r_r} = n\pi\sqrt{\frac{k_\sigma}{k}}\sqrt[4]{\frac{C'E}{4\sigma_{mec}}}. \tag{7.8}$$

This equation yields the maximum length of the rotor after the permitted rotor radius has been defined with Equation (7.5). l_{max}/r_r is not a function of rotation speed. If a solid-steel rotor rotates below the first critical rotation speed, the ratio should usually be $l/r_r < 7$. In practice, the ratio is often $l/r_r \approx 5$.

The maximum stresses in a rotor caused by the centrifugal forces are, according to Equations (7.5) and (7.6), proportional to the square of the mechanical angular speed Ω. Thus, we can find fixed values for a maximum speeds and stresses. If the rotor dimensions (diameter, length, etc.) of the machine are assumed to be variable over the scale λ (lengths and diameters are proportional to λ, areas to λ^2, and volumes to λ^3), the maximum speed of the machine becomes inversely proportional to the scale λ:

$$n_{max} \sim \lambda^{-1}. \tag{7.9}$$

7.1.3 Electrical Loadability

The rotor of a higher speed machine can be made smaller as the torque is smaller with the same power. The losses of the machine are, however, not torque but power dependent and, therefore, as the machine speed increases, its power loss density tends to increase. Resistive losses P_{Cu} in a winding are proportional to the square of the current density J and to the mass of the conductors

$$P_{Cu} \sim J^2 m_{Cu} \sim J^2 \lambda^3. \tag{7.10}$$

The thermal resistance R_{th} between the conductors and the teeth is

$$R_{th} = \frac{d_i}{\lambda_i S_i}, \tag{7.11}$$

where d_i is the thickness of the slot insulation, λ_i the thermal conductivity of the insulation, and S_i the area of the slot wall. The thickness of the insulation d_i is constant independent of the machine size (it depends on the rated voltage) and hence

$$R_{th} \sim \frac{1}{\lambda^2}. \tag{7.12}$$

The temperature difference ΔT between the conductors and the teeth is

$$\Delta T = P_{Cu} R_{th} \sim J^2 \lambda. \tag{7.13}$$

Consequently, for a given temperature difference

$$J^2 \lambda = \text{constant} \tag{7.14}$$

and

$$J \sim \frac{1}{\sqrt{\lambda}}. \tag{7.15}$$

Thus, small machines tolerate higher current densities better than large machines.

The linear current density A in this case may be calculated as the total RMS current in a slot $JS_{\mathrm{Cu,u}}$ divided by the slot pitch τ_{u}

$$A = \frac{JS_{\mathrm{Cu,u}}}{\tau_{\mathrm{u}}} \sim \frac{\frac{1}{\sqrt{\lambda}}\lambda^2}{\lambda} = \sqrt{\lambda}. \tag{7.16}$$

The linear current density A and the current density J are dimensions of the electrical loading of a machine. As the current density in small machines can be higher than in large machines, the linear current density behaves in the opposite sense: the linear current density is typically higher in large machines than in small machines. The product of A and J

$$AJ \sim \sqrt{\lambda}\frac{1}{\sqrt{\lambda}} = 1 = \text{constant}, \tag{7.17}$$

and thus AJ is independent of the size of the machine; it depends only on the effectiveness of the cooling of the machine. For totally enclosed machines, AJ is smaller than for open-circuit cooling. In air-cooled machines, the product AJ ranges from 10×10^{10} $\mathrm{A^2/m^3}$ to 35×10^{10} $\mathrm{A^2/m^3}$, but in different machines with the same kind of cooling, the product is approximately the same independent of the machine size. In the case of direct water cooling, AJ attains essentially higher values.

Example 7.3: Using Table 6.3, calculate the AJ values for different machines.

Solution:

	Asynchronous machines	Salient-pole synchronous machines or PMSMs	Nonsalient-pole synchronous machines: Indirect cooling: Air	Nonsalient-pole synchronous machines: Indirect cooling: Hydrogen	Nonsalient-pole synchronous machines: Direct water cooling	DC machines
A [kA/m]	30–65	35–65	35–80	90–110	150–200	25–65
	Stator winding	Armature winding		Armature winding		Armature winding
J [$\mathrm{A/m^2}$]	$3–8 \times 10^6$	$4–6.5 \times 10^6$	$3–5 \times 10^6$	$4–6 \times 10^6$	$7–10 \times 10^6$	$4–9 \times 10^6$
AJ [$\mathrm{A^2/m^3}$]	9×10^{10} to 52×10^{10}	14×10^{10} to 42.25×10^{10}	10.5×10^{10} to 40×10^{10}	36×10^{10} to 66×10^{10}	105×10^{10} to 200×10^{10}	10×10^{10} to 58.5×10^{10}

The values for asynchronous, DC and air-cooled synchronous machines are similar. Hydrogen cooling or the direct water cooling method in larger synchronous machines, however, give remarkably higher values.

7.1.4 Magnetic Loadability

The air-gap flux density and the supply frequency of a machine determine the magnetic loading of the machine. Let us next study how the flux density varies as the speed of a machine increases with the increasing line frequency.

Iron losses P_{Fe} are approximately proportional to the square of flux density $\hat{B}_\delta$ and, with high frequencies, to the square of frequency f and to the volume of iron V_{Fe}

$$P_{Fe} \sim \hat{B}_\delta^2 f^2 V_{Fe} \tag{7.18}$$

The maximum rotation speed n_{max} is proportional to the frequency f, and taking into account Equation (7.9) and proportionality ($V \sim \lambda^3 \sim n_{max}^{-3}$), we obtain

$$P_{Fe} \sim \hat{B}_\delta^2 n_{max}^2 \frac{1}{n_{max}^3} = \hat{B}_\delta^2 \frac{1}{n_{max}}. \tag{7.19}$$

The temperature rise of a machine depends on the power loss per cooling area S. Assuming a constant temperature rise and taking into account Equation (7.9) and proportionality ($S \sim \lambda^2 \sim n_{max}^{-2}$), we obtain

$$\frac{P_{Fe}}{S} \sim \frac{\hat{B}_\delta^2}{n_{max} S} \sim \frac{\hat{B}_\delta^2}{n_{max}} n_{max}^2 = \hat{B}_\delta^2 n_{max} = \text{constant}. \tag{7.20}$$

Thus, the air-gap flux density depends on the speed:

$$\hat{B}_\delta \sim \frac{1}{\sqrt{n_{max}}}. \tag{7.21}$$

We now get the maximum available power P_{max} that can be obtained by increasing the speed of the machine. According to Equations (6.16), (7.15), (7.21), and (7.9)

$$P_{max} \sim A\hat{B}_\delta D^2 l' n_{max} \sim \frac{1}{\sqrt{n_{max}}} \frac{1}{\sqrt{n_{max}}} \frac{1}{n_{max}^2} \frac{1}{n_{max}} n_{max} = \frac{1}{n_{max}^3}. \tag{7.22}$$

The maximum power limit is inversely proportional to the cube of the speed. Figure 7.5 illustrates a study of different types of air-cooled machines and shows the rated power that can be reached at a required maximum speed or the maximum speed allowed for a given rated power. The lines in Figure 7.5 are based on conventional electrical and magnetic loading. The figure gives the average values for a large number of machines. Line e represents the limiting power of induction motors, the rotors of which are constructed of smooth solid-steel cylinders, which are coated with copper cylinders. The cylinder is fixed on the rotor core by explosion welding, which guarantees a perfect mechanical joint between copper and steel and, therefore, enables using copper on a higher speed rotor surface.

The limitations of materials decide the realistic upper limits for the power outputs of electrical machines. Present developments in large synchronous machines have led to notable outputs. Today, up to 1500 MW, 1500 min^{-1} synchronous machines are designed as generators of large nuclear power plants. For instance, the rotor dimensions of the nonsalient-pole synchronous generator in the Finnish Olkiluoto 1793 MW nuclear power plant are $D_r = 1.9$ m, $l_r = 7.8$ m ($l/r_r \approx 8.2$) with 300 m/s rotor surface speed.

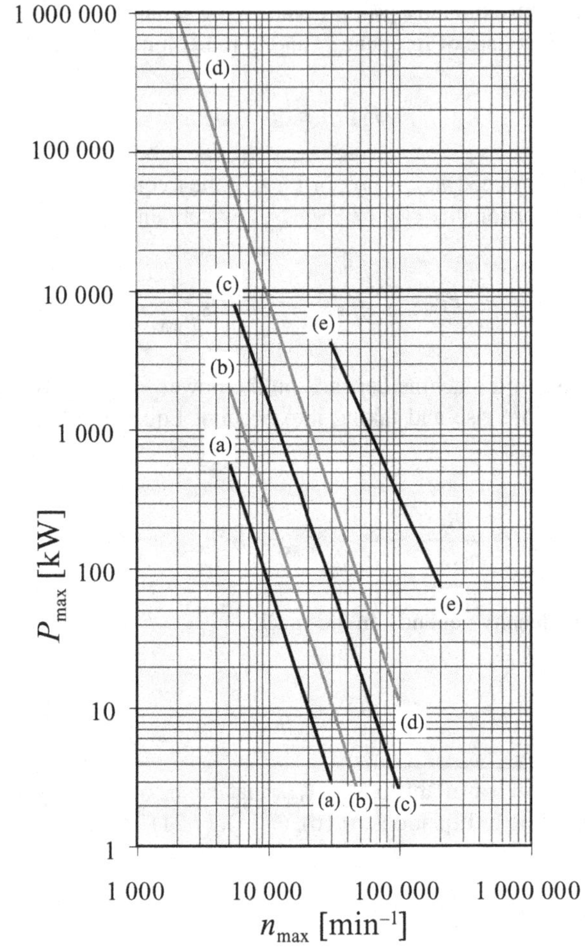

Figure 7.5 Maximum power and speed of air-cooled machines. (a) DC machines, rotor surface speed $\leq$ 110 m/s, (b) synchronous cylindrical-rotor machines with a laminated rotor, rotor surface speed $\leq$ 130 m/s, (c) induction motors with a laminated squirrel cage rotor, rotor surface speed $\leq$ 200 m/s, (d) (synchronous) machines with a solid rotor, rotor surface speed $\leq$ 400 m/s, (e) induction motors with a smooth solid, copper-coated rotor, rotor surface speed $\leq$ 550 m/s (a–d: Gutt 1988, e: Saari 1998).

7.1.5 Efficiency

One of the requirements set for the machine design is the efficiency. Manufacturers have to mark the efficiency class of an AC motor on the rating plate and brochures of the motor (see Section 6.1). The efficiency classes are defined in IEC standard 60034-30-1 (2014 forecasted). The standard deals with all kinds of single-speed electric motors that are rated for direct-on-line operation including starting at reduced voltage. The standard includes single- and three-phase induction motors as well as line-start permanent-magnet and synchronous reluctance motors which

- have a rated power from 0.12 kW to 1000 kW;
- have a rated voltage above 50 V up to 1 kV (so-called low voltage motors);
- have 2, 4, 6 or 8 poles (synchronous speed in a 50 Hz network 3000–750 rpm);
- are capable of continuous operation at their rated power;
- are marked with any ambient temperature within the range of −20 °C to +60 °C; the efficiency is, however, based on measurements at 25 °C ambient temperature;
- are marked with an altitude up to 4000 m above sea level; the efficiency is, however, based on a rating for altitudes up to 1000 m above sea level.

Geared motors, inverted-supplied motors and brake motors are covered by this standard if the motor can be tested without the losses of the gear, converter or brake. The tests are always performed on sinusoidal supply voltage.

The designation of the energy-efficiency class consists of the letters "IE" followed by a number representing the classification. The lowest class is 1 and the highest 5. The highest class IE5 is envisaged for future. As an example, the efficiency limit curves for the 50 Hz, four pole (synchronous speed 1500 min^{-1}) motors are shown in Figure 7.6.

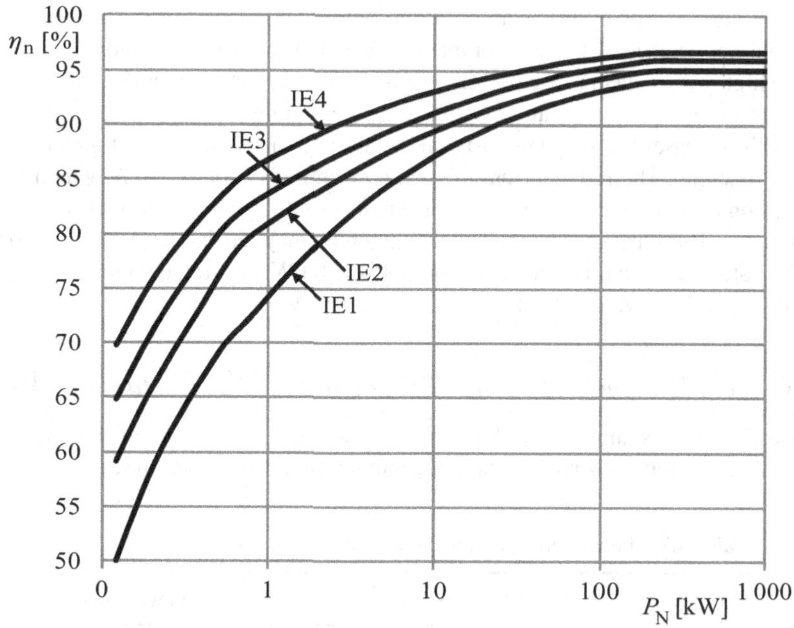

Figure 7.6 Nominal efficiency limits for 50 Hz, four pole motors. In the range from 200 kW to 1000 kW, the efficiency is constant and it is for the class IE1 94.0 %, for IE2 95.1 %, for IE3 96.0 % and for IE4 96.7 %.

7.2 Asynchronous Motor

An induction motor equipped with a squirrel cage rotor winding is the most common motor type applied in industry. According to its name "the asynchronous motor" needs a slip to create torque. The rotor of an asynchronous motor is rotating at the angular velocity Ω_r while the air gap flux travels at Ω_s. The difference is called the slip and it is often given as a per unit value $s = (\Omega_s - \Omega_r)/\Omega_s$. The slip makes the rotor bars experience a slowly altering flux at frequency $f_{slip} = sf_s$ that induces the rotor windings a voltage and as a result currents will start flowing in the rotor and creating torque.

Usually, the rated output powers follow a geometric series, in which the ratio of rated output powers is approximately $\sqrt[n]{10}$. The ordinal n of the root defines a power series that can be denoted "a series n." Typically, the series 5, the series 7 and the series 9 are employed. Different series are employed at different power ranges. The recommended power ranges are given in Table 7.2.

The output powers of induction motors do not strictly follow the recommendations. Usually, the rated powers of the low-voltage motors (50–1000 V) are the following: 0.18, 0.25, 0.37, 0.55, 0.75, 1.1, 1.5, 2.2, 3.0, 4.0, 5.5, 7.5, 11, 15, 18.5, 22, 30, 37, 45, 55, 75, 90, 110, 132, 160, 200, 250, 315, 400, 450, 500, 560, 630, and 710 kW. We also see that at the lower end of this power series, the series 7 is approximately followed. Machines above 710 kW are often constructed for high voltages, however, with an inverter drive; for instance, even a power of 5 MW and beyond can be reached with a 690 V voltage. In machines having several independent parallel winding paths supplied by several converter units there is, actually, no clear upper power limit.

Each rated voltage, however, has an optimum power range, the lower and upper limits of which should not be exceeded if a favorable construction is to be produced. If the voltage is excessively high with respect to the power, the conductors become rather thin, and the number of turns increases drastically. In an opposite case, the winding has to be made of preformed copper, which increases the winding costs. In the above series, 630 kW is, typically, the limit for a 400 V machine. The rated current of such a machine is about 1080 A, which is a value that can be considered a practical maximum for the rated current of a direct-on-line (DOL) induction motor. The starting current of such a machine may reach even 10 kA. According to IEC 60034-1 standard, the minimum rated output is 100 kW for a rated voltage $U_N = 1$–3 kV, 150 kW for $U_N = 3$–6 kV, and 800 kW for $U_N = 6$–11 kV.

7.2.1 *Current Linkage and Torque Production of an Asynchronous Machine*

A poly-phase current is supplied to the windings mounted in the slots of the machine. The slot currents can be replaced with sufficient accuracy by a slot's local linear current density

Table 7.2 Power series of induction motors

Series n	Power ratio $\sqrt[n]{10}$	Power range
5	1.58	< 1.1 kW
7	1.39	1.1 – 40 kW
9	1.29	> 40 kW

constant value A_u along the width of the slot opening b_1.

$$A_u = \frac{z_Q I}{b_1}, \text{ at the slot opening, elsewhere } A = 0. \tag{7.23}$$

Note that in some cases in this book (e.g. in Tables 6.3 and 6.4), for convenience, the linear current density is defined as an RMS value of the fundamental component A_1 of the linear current density. In some cases, in the literature, a step function $A_u = z_Q I/\tau_s$ is also used. The amplitude of the fundamental component of the current linkage is nevertheless the same. We may also assume that the slot opening width is infinitesimal. In such a case the slot linear current density consists of impulse functions at the slot openings.

Figure 7.7 illustrates how the linear current density $A(\alpha)$ of the machine with finite slot openings forms, in practice, a bar chart with bars having the slot opening width. Integrating $A(\alpha)$ we get the fairly sharp stepped (in case of infinitesimal slot openings a sharp stepped) current linkage curve $\Theta(\alpha)$. The current linkage waveform has an electrical phase shift of $\pi/2$ compared with the fundamental linear current density, Figure 7.7a. The linear current density distribution can be developed into a Fourier series. Each term ν in the series produces a corresponding flux density harmonic as a function of position. Figure 7.7b illustrates a linear current density A_1 of this type on the inner surface of the stator. An axial current $A_\nu r\mathrm{d}\alpha$ flows in an element of the equivalent rotor length l' of the cylinder

$$\mathrm{d}I_\nu = A_\nu r \frac{\mathrm{d}\alpha}{p} = A_\nu \frac{D_s}{2p} \mathrm{d}\alpha. \tag{7.24}$$

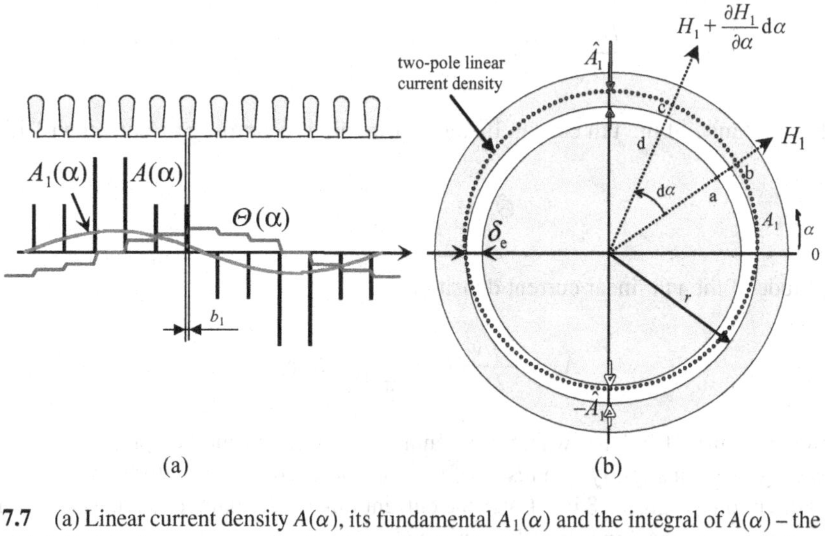

Figure 7.7 (a) Linear current density $A(\alpha)$, its fundamental $A_1(\alpha)$ and the integral of $A(\alpha)$ – the current linkage $\Theta(\alpha)$. (b) The linear current density A_1 created by the three-phase currents of a two-pole rotating-field stator with $Q = 12$, $q = 2$ and slot opening with b_1. The fundamental of the linear current density creates field strength in the air gap and a corresponding flux density. In this figure, the linear current density created by slot currents is illustrated as poles(a), the height of which is the slot current divided by the slot opening width. $A = z_Q I/b_1$.

Here $\mathrm{d}\alpha/p$ is used, since the angle α obtains values from 0 to $p{\cdot}2\pi$ as the rotor periphery is traversed.

When the effective air gap δ_{ef} is notably smaller than the radius of the rotor ($\delta_{\mathrm{ef}} \ll r_{\mathrm{r}}$), the air-gap field strength H_ν can be assumed constant in the direction of the radius r_{r}. According to the penetration law, when traveling along the path a-b-c-d-a of the length l', the rotor being currentless, we obtain

$$\oint H_\nu \mathrm{d}l = H_\nu \delta_{\mathrm{ef}} - \left(H_\nu + \frac{\partial H_\nu}{\partial \alpha} \mathrm{d}\alpha \right) \delta_{\mathrm{ef}} = A_\nu \frac{D_{\mathrm{s}}}{2p} \mathrm{d}\alpha. \tag{7.25}$$

Here, we obtain the linear current density

$$A_\nu = -\frac{2p\delta_{\mathrm{ef}}}{D_{\mathrm{s}}} \frac{\partial H_\nu}{\partial \alpha} = -\frac{2p\delta_{\mathrm{ef}}}{\mu_0 D_{\mathrm{s}}} \frac{\partial B_\nu}{\partial \alpha} = -\frac{2p}{D_{\mathrm{s}}} \frac{\partial \Theta_\nu}{\partial \alpha}. \tag{7.26}$$

D_{s} is the stator bore diameter and p is the number of pole pairs.

Choosing the coordinates so that at the time instant $t = 0$ the fundamental linear current density has its maximum at $\alpha = \pi/2$ (as in Figure 7.7b), the νth term in the Fourier series of the linear current density can be written as

$$A_\nu = \hat{A}_\nu \left| \underline{\omega t - \nu\alpha + \pi/2} \right. = \hat{A}_\nu e^{j(\omega t - \nu\alpha + \pi/2)}, \tag{7.27}$$

According to Equation (7.26), the current linkage is the integral of the linear current density:

$$\Theta_\nu = -\frac{D_{\mathrm{s}}}{2p} \int A_\nu \mathrm{d}\alpha = \frac{D_{\mathrm{s}}}{2p\nu} \hat{A}_\nu \left| \underline{\omega t - \nu\alpha} \right. = \hat{\Theta}_\nu \mathrm{e}^{\mathrm{j}(\omega t - \nu\alpha)} \tag{7.28}$$

Since the amplitude of the νth current linkage harmonic is according to Equation (2.15)

$$\hat{\Theta}_\nu = \frac{m}{\pi} \frac{k_{\mathrm{w}\nu} N_{\mathrm{s}}}{\nu p} \hat{i},$$

the amplitude of the νth linear current density is

$$\hat{A}_\nu = \frac{2p\nu}{D_{\mathrm{s}}} \hat{\Theta}_\nu = \frac{2}{\pi} \frac{m}{D_{\mathrm{s}}} k_{\mathrm{ws}\nu} N_{\mathrm{s}} \hat{i}. \tag{7.29}$$

The stepped current linkage waveform containing all the harmonics propagates in the air gap changing its form slightly as a result of the harmonics traveling at different speeds and in different directions. Figure 7.8 illustrates the current linkage fundamentals (harmonics filtered out) of 6- and 2-pole machines traveling in the air gaps.

The current linkage amplitude is exerted on half of the main magnetic circuit. The ratio of the air-gap flux density to the current linkage, i.e. the permeance per unit area, is called the specific permeance of the magnetic circuit. The specific permeance is a periodic function of time and space. The slotting, saturation of iron, eccentricity and saliency in salient-pole

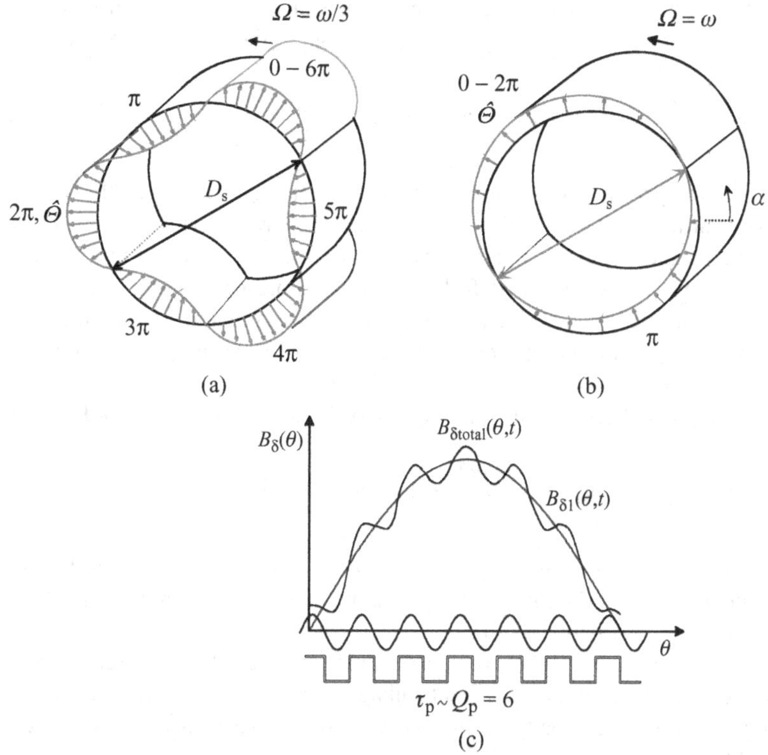

Figure 7.8 (a) Six-pole ($p = 3$) current linkage fundamental propagating at a physical angular speed $\Omega = \omega/3$, and (b) a two-pole ($p = 1$) fundamental, propagating at a speed $\Omega = \omega$. If the angular frequency ω of the input current is equal in both windings, the current linkage propagates in a six-pole winding locally at one-third speed when compared with the flux of the two-pole winding. Both distributions propagate one wavelength during one supply frequency period, which explains the propagation speed difference. The influence of the number of pole pairs on the rotation speed of a machine is based on this fact. (c) Relationship between the fundamental and the slot harmonics for $Q = 24$, $2p = 4$, $Q_p = 24/4 = 6$ waves for half wave of the fundamental harmonic.

machines are the origins of the permeance variations. The specific permeance Λ' is expressible as a Fourier series

$$\Lambda' = \Lambda'_0 + \sum_{\mu} \hat{\Lambda}'_{\mu} e^{j(\omega_\mu t - \mu\alpha + \varphi_\mu)}, \tag{7.30}$$

where μ is the ordinal of the specific permeance, Λ'_0 is the average of the specific permeance ($\Lambda'_0 = \mu_0/\delta_{ef}$) and $\hat{\Lambda}'_{\mu}$ the amplitude of the μth permeance harmonic. The single effective air gap δ_{ef} includes the influence of the slotting and iron saturation (cf. Chapter 3, Equation (3.56)). Naturally, similar functions can be written for the air gap permeance Λ when the observed area S is known.

The flux density is the product of the current linkage and the specific permeance. The current linkage creates flux density harmonics with the average of the specific permeance

$$B_{\delta\nu} = \Lambda_0 \hat{\Theta}_\nu = \frac{\mu_0}{\delta_{\text{ef}}} \hat{\Theta}_\nu = \hat{B}_{\text{m}\nu} \underline{|\omega t - \nu\alpha} = \hat{B}_{\delta\nu} \text{e}^{\text{j}(\omega t - \nu\alpha)} = \frac{\mu_0}{\delta_{\text{ef}}} \frac{m}{\pi} \frac{k_{\text{ws}\nu} N_{\text{s}}}{\nu p} \hat{i} \text{e}^{\text{j}(\omega t - \nu\alpha)}. \tag{7.31}$$

With the specific permeance harmonics, the current linkage creates further a large amount of flux density harmonics. Usually, the largest harmonics are due to the slotting. The maximum value of the permeance occurs at the center of the tooth and the minimum value at the center of the slot opening (Figure 7.8c). It follows that the pole pair numbers of the permeance harmonics are multiples of the slot number. As the fundamental of the sinusoidal current linkage having p pole pairs is multiplied by the sinusoidal permeance harmonic having cQ pole pairs we obtain flux density harmonics having $p \pm cQ$ pole pairs, where $c = 1, 2, 3, \ldots$ and Q the slot number. These harmonics are called slot harmonics. Their ordinals are

$$\nu_{\text{u}} = \frac{p \pm cQ}{p} = 1 \pm c\frac{Q}{p} = 1 \pm 2mqc. \tag{7.32}$$

The slot harmonics always occur in pairs. The pair of waves obtained from Equation (7.32) with $c = 1$ and with plus and minus sign is called the first slot harmonic. Correspondingly, the second slot harmonic is obtained with $c = 2$.

The current linkage Θ_ν and the average of the permeance Λ_0 also create slot harmonics since there are Q steps in the total current linkage wave. The slot harmonics created by the slot openings and by the stepped current linkage have a phase shift and we have to use the geometric addition to obtain the final slot harmonic (Jokinen 1972).

The harmonic (7.31) represents the flux penetrating the air gap, the peak value of which is

$$\hat{\Phi}_{\text{m}\nu} = \frac{D_{\text{s}} l'}{\nu p} \hat{B}_{\delta\nu} = \frac{\mu_0 m D_{\text{s}} l' k_{\text{ws}\nu}}{\pi p^2 \delta_{\text{ef}} \nu^2} N_{\text{s}} \hat{i}. \tag{7.33}$$

The peak value of the total air-gap flux is the sum

$$\hat{\Phi}_{\text{m}} = \sum_{\nu=1}^{\pm\infty} \hat{\Phi}_{\text{m}\nu} = \frac{\mu_0 m D_{\text{s}} l'}{\pi p^2 \delta_{\text{ef}}} N_{\text{s}} \hat{i} \sum_{\nu=1}^{\pm\infty} \frac{k_{ws\nu}}{V^2} \approx \frac{\mu_0 m D_{\text{s}} l' k_{w1}}{\pi p^2 \delta_{\text{ef}}} N_{\text{s}} \hat{i}. \tag{7.34}$$

The series converges so rapidly that it normally suffices to take only the fundamental into account in the calculation. The composition of the main flux of the fundamental and harmonics is illustrated by Equation (7.34). It is worth remembering that the harmonics analogously induce a voltage at the supply base frequency in the windings of the machine. This can be explained by the reciprocity: As the winding supplied by sinusoidal (fundamental) voltages and currents creates spatial harmonics in the air gap, spatial harmonics must be similarly capable of inducing corresponding fundamental sinusoidal voltages in the windings.

Harmonics cause several problems in the operation of AC machines, mainly in asynchronous ones: (a) Harmonics produce parasitic torques, asynchronous and/or synchronous ones. These torques can seriously affect the shape of the torque-speed curve of an asynchronous motor, see Sections 7.2.4 and 7.2.5. (b) They introduce vibration and noise in the machine. (c) They induce harmonics into the generated voltage. (d) They increase core losses because of high-frequency components of magnetic flux density.

We can reduce the harmonics by short-pitching the winding and by skewing the slots. Short-pitching influence especially the low order harmonics such as the fifth and the seventh harmonic (cf. Example 2.12). The winding factors of the slot harmonics are always the same as that of the fundamental, and therefore, by short-pitching the slot harmonics reduce similarly as the fundamental. We can reduce the effects of the slot harmonics by skewing the slots. This reduces efficiently the effects of slot harmonics and only slightly the fundamental (cf. Example 4.1).

Currents flow in both the stator and rotor windings of an asynchronous machine operating under load. Both currents create a current linkage of their own, and therefore the current linkage exerted on a half of a magnetic circuit is the sum of these two linkages

$$\Theta_{\mathrm{m}}(\alpha) = \Theta_{\mathrm{r}}(\alpha) + \Theta_{\mathrm{s}}(\alpha). \tag{7.35}$$

This sum creates the real magnetic flux density in the air gap. When the machine is linearized for the calculations, also the local flux densities can be superposed. Now we obtain for the air-gap flux density

$$B_{\delta}(\alpha) = B_{\mathrm{s}}'(\alpha) + B_{\mathrm{r}}'(\alpha). \tag{7.36}$$

The imaginary flux densities $B_{\mathrm{s}}'(\alpha)$ and $B_{\mathrm{r}}'(\alpha)$ presented here cannot be measured, yet they can be calculated as follows

$$B_{\mathrm{r}}'(\alpha) = \frac{\mu_0}{\delta_{\mathrm{ef}}}\Theta_{\mathrm{r}}(\alpha); \quad B_{\mathrm{s}}'(\alpha) = \frac{\mu_0}{\delta_{\mathrm{ef}}}\Theta_{\mathrm{s}}(\alpha). \tag{7.37}$$

Figure 7.9 illustrates a linear current density on the boundary of the rotor and air gap. At an arbitrary position α of the air gap, the differential current is written as

$$\mathrm{d}I(\alpha) = A(\alpha)\frac{r}{p}\mathrm{d}\alpha. \tag{7.38}$$

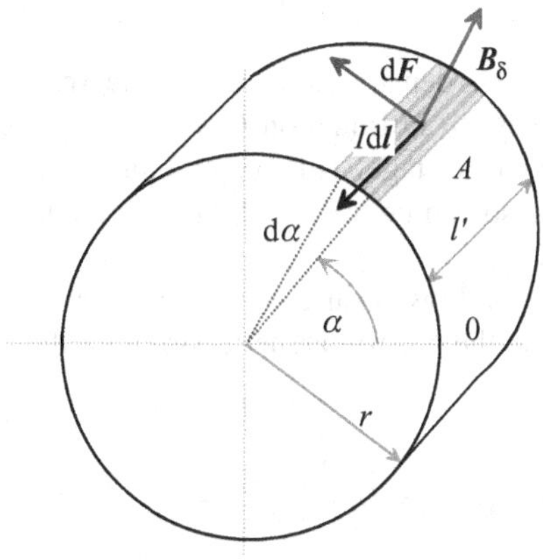

Figure 7.9 Definition of the torque T acting upon the rotor. The torque is defined with a peripheral force dF exerted on the linear current density element Idl.

The prevailing linear current density at the position α can be defined according to Equation (7.27). The air-gap flux density vector $\boldsymbol{B}_\delta$ and the rotor current $Id\boldsymbol{l}$ (where $\boldsymbol{l}$ is the unit vector in the rotor axis direction) are perpendicular to each other and, according to Lorentz force equation, they cause a peripheral force element d$\boldsymbol{F}$ parallel to the tangent to the cylinder

$$dF(\alpha) = l' B_\delta(\alpha) dI(\alpha) = \frac{D_s l'}{2p} A(\alpha) B_\delta(\alpha)\, d\alpha. \tag{7.39}$$

For simplicity, $D_s \approx D_r \approx D \approx 2r$ is used in the following.

Since the peripheral forces are tangential everywhere, their vector sum around the rotor is zero, yet they can be employed in the calculation of the torque. The peripheral force of a machine is solved by line integrating d$\boldsymbol{F}$ around the surface of the rotor over an angle $2\pi p$. Simultaneously, we obtain the electromagnetic torque of the machine by multiplying the obtained force by the diameter of the rotor radius ($r \approx D/2$)

$$T_{em} = \frac{D^2 l'}{4p} \int_0^{2\pi p} A(\alpha) B_\delta(\alpha) d\alpha. \tag{7.40}$$

This result is in line with the results found in section 1.5 where the tangential stress was first studied based on the Lorentz force.

The fundamental $\nu = 1$ of the linear current density of the rotor is solved by substituting the rotor current into Equations (7.27) and (7.29), and by taking into account the angular difference ζ_r of the linear current density of the rotor with respect to the rotor induced voltage

$$A_r = \hat{A}_r \underline{|s\omega t - \alpha - \zeta_r + \pi/2}\,, \tag{7.41}$$

where

$$\hat{A}_r = \frac{2}{\pi} \frac{m_r}{D} k_{wr} N_r \hat{i}_r. \tag{7.42}$$

Since the rotor linear current density A_r is written in the rotor coordinate system, its frequency is the slip frequency $s\omega$ in Equation (7.41). ζ_r represents the phase angle of the rotor current lagging the air-gap voltage. This phase angle is caused by the rotor impedance $R_r + sj\omega_s L_{r\sigma} = Z_r \underline{|\zeta_r}$ but it also can be used to define the angle between the distributions of A_r and B_δ which is $\pi/2 - \zeta_r$.

The real air-gap flux density follows Equation (7.31). When the elements are substituted in Equation (7.40), we may write the electromagnetic torque of the induction machine (see also Equation (6.2)) as

$$T_{em} = \pi D l \sigma_{tan} r = \pi D l \frac{\hat{A} \hat{B}_\delta \cos \zeta_r}{2} r = \frac{\pi}{4} D^2 l \hat{A} \hat{B}_\delta \cos \zeta_r. \tag{7.43}$$

In (7.43), the peak values of the fundamental distributions of $\hat{A}$ and $\hat{B}_\delta$ are used. That is why the result must be divided by two ($\sqrt{2}^2$). Note also that $\cos\zeta_r$ corresponds to the rotor power factor. In Equation (1.115), there is no angle involved, since instantaneous local values are used.

In principle, Equation (7.43) is the same equation we discussed previously with the definition of tangential stress in Section 1.5. This is a general torque equation based on the Lorentz force and derived from the flux and current distributions of the machine. Formally, the equation is valid for all machine types. The torque acts upon both the rotor and the stator with equal magnitude but of opposite direction. When Equation (7.43) is repeated for stator quantities, it may be applied to any rotating-field machine type. For example, if a synchronous machine is operating with an overlapping linear current density and flux density distributions, the electromagnetic torque will be $T_{em} = \pi D_s^2 l \hat{A}_s \hat{B}_\delta / 4$.

This equation shows that the torque of a machine is proportional to the volume of the rotor and the product $\hat{A}\hat{B}_\delta$. The maximum value that a machine may produce in continuous operation is determined by the temperature rise of the machine. The maximum of the fundamental component of the air-gap flux density in iron-cored machines is usually limited to the order of 1 T or slightly above. If the air-gap flux density curve is rectangular and the maximum flux density is 1 T, the fundamental peak value reaches $4/\pi$. On the other hand, the variation in the linear current density A is large and depends on the cooling of the machine.

7.2.2 *Impedance and Current Linkage of a Cage Winding*

Figure 7.10 illustrates a simplified cage winding (a) and phasors (b) of the bar and ring currents for the cage winding. The bars are numbered from 1 to Q_r. The bar and ring currents are denoted respectively, the arrows indicating the positive directions. If the resistance of a single bar is R_{bar} (the DC value of the resistance of the bar is obtained by the area S of the slot of the rotor, the length l of the rotor core, the skew angle α (Figure 4.10), and the specific conductivity σ of the conductor material $R_{bar} = l/(\cos\alpha \cdot \sigma S)$) and the inductance is L_{bar}, and

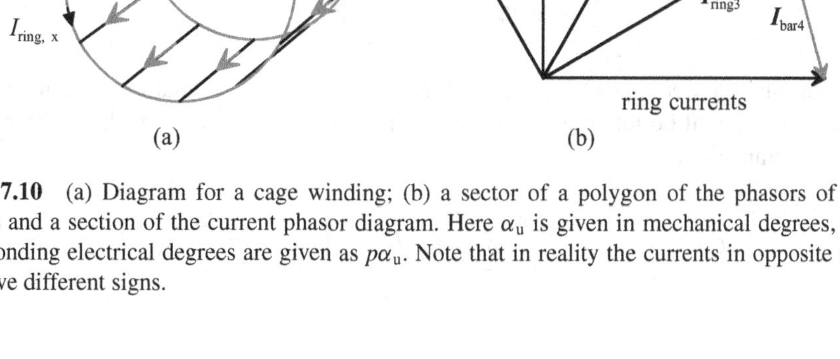

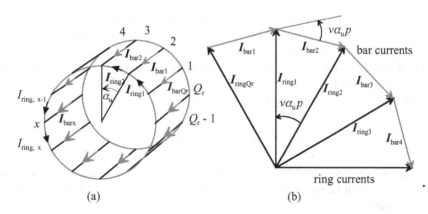

Figure 7.10 (a) Diagram for a cage winding; (b) a sector of a polygon of the phasors of the bar currents and a section of the current phasor diagram. Here α_u is given in mechanical degrees, and the corresponding electrical degrees are given as $p\alpha_u$. Note that in reality the currents in opposite sides of cage have different signs.

finally, R_{ring} and L_{ring} are the respective values of the ring section between two bars, then the resultant impedance of the complete winding can be calculated as follows.

A current phasor diagram for the rotor bars is constructed for the harmonic ν in such a way that the angular phase shift of the bar currents is $\nu\alpha_u p$. As a squirrel cage rotor can operate with different stator pole pair numbers p, we use α_u here as a mechanical angle, so then the corresponding electrical angle is $p\alpha_u$ (cf. Figure 7.10). Next, a polygon is constructed for the bar currents. A single sector of this polygon is illustrated by the phasors $\underline{I}_{bar1}$–$\underline{I}_{bar4}$ in Figure 7.10. The phasors from $\underline{I}_{ringQ_r}$ to $\underline{I}_{ring3}$ drawn from the center of the polygon to the angle points are the phasors of the ring currents. They follow Kirchhoff's first law at each connection point of the bar and the ring – see Figure 7.10a:

$$\underline{I}_{ring,x} = \underline{I}_{bar,x} + \underline{I}_{ring,x-1}. \tag{7.44}$$

The mutual phase angle between the ring currents is also $\nu\alpha_u p$.

Since the bar alone comprises the phase winding of the rotor ($N_r = 1/2$), the bar current is the phase current of the rotor. Now, the RMS value of the bar current induced by theνth flux density harmonic is generally denoted by $I_{bar\nu}$. Correspondingly, the RMS value of the ring current is now denoted by $I_{ring\nu}$. Thus, based on Figure 7.10b, we obtain

$$I_{ring\nu} = \frac{I_{bar\nu}}{2\sin\frac{\alpha_{u\nu}}{2}}; \quad \alpha_{u\nu} = \nu\frac{2\pi p}{Q_r}. \tag{7.45}$$

The currents create a resistive loss in the rotor

$$P_{Cu\nu} = Q_r\left(R_{bar\nu}I_{bar\nu}^2 + 2R_{ring}I_{ring\nu}^2\right) = Q_r I_{bar\nu}^2\left(R_{bar\nu} + \frac{R_{ring}}{2\sin^2\frac{\alpha_{u\nu}}{2}}\right), \tag{7.46}$$

where R_{ring} is the resistance of the part of the ring belonging to one bar. The resistance of a single rotor phase is higher than the mere bar resistance $R_{bar\nu}$ by an amount equal to the term of the second element in the large brackets. The phase inductance is calculated by the same principle. When we take into account the electric angular frequency supplying the machine, we obtain the equation for the phase impedance of the rotor of an induction machine for the air-gap flux harmonic ν for a locked rotor, $s = 1$

$$\underline{Z}_{r\nu} = R_{r\nu} + j\omega_s L_{r\nu} = Z_{r\nu}\left|\underline{\zeta_{r\nu}}\right. . \tag{7.47}$$

Note that the angular frequency ω_s of the stator is employed in the calculation of the impedance. The effect of slip will be taken into account later. Now we can write for the resistance and leakage inductance of the rotor

$$R_{r\nu} = R_{bar\nu} + \frac{R_{ring}}{2\sin^2\frac{\nu\pi p}{Q_r}}; \quad L_{r\nu} = L_{bar} + \frac{L_{ring}}{2\sin^2\frac{\nu\pi p}{Q_r}}. \tag{7.48}$$

In Equations (7.47) and (7.48) L_{bar} and L_{ring} are the leakage inductances of the bars and the ring sections. The leakage inductance of the end ring can be solved simply with Equation (4.101), which replaces the latter part of Equation (7.48). When analyzing the equations, we note that the value of the phase impedance of the rotor is a function of the ordinal ν of the inducing flux density harmonic. A cage rotor reacts only to such flux density harmonics created by the stator, the ordinals ν of which meet the condition

$$\nu p \neq cQ_r, \text{ where } c = 0, \pm 1 \pm 2, \pm 3, \ldots. \tag{7.49}$$

This explains the fact that for the harmonics, the ordinal of which is

$$\nu' = \frac{cQ_r}{p} \tag{7.50}$$

the impedance of the rotor phase is infinite. The pitch factor $k_{p\nu}$ of these harmonics is zero. The wavelength of the harmonic ν' is equal to the slot pitch of the rotor, or an integral part of it. In that case, each bar always has a flux density of equal magnitude. This creates an equal induced current linkage to each bar, and the emfs of a closed electric circuit compensate each other, and thus the voltages induced by the harmonic ν' do not generate currents. According to Equations (7.47) and (7.48), a cage winding can be replaced by such an equivalent cage, the impedance of the short-circuit rings of which is zero and the impedance of the bars is $\underline{Z}_{r\nu}$. The impedance of the cage winding or the resistance and inductance are usually transferred to the stator side. The case is analyzed next.

The currents in the bars of a single pole pitch of a cage winding are all of different phases. In a symmetrical m-phase system, the angles between the phases is $360°/m$. Thus, there are as many phases in the rotor as the number of rotor bars. If the number of bars is Q_r in the rotor, the phase number of the rotor is accordingly

$$m_r = Q_r. \tag{7.51}$$

Generally, there has to be at least two conductors in a coil turn at a distance of about 180° from each other. We may thus consider that a single rotor bar comprises half a turn, and we can write $N_r = 1/2$. To make the equivalent circuit analysis easy we have to refer the rotor quantities to the stator. The induction machine can be treated analogously with a two-winding transformer. For the fundamental the number of effective coil turns in a stator is $m_s k_{w1s} N_s$ and in the squirrel cage rotor $m_r k_{w1r} N_r$. Here $m_r = Q_r$, $k_{w1r} = 1$ and $N_r = 1/2$. If the slots of the stator and rotor are skewed with respect to each other, we also have to take the skewing factor k_{sq} into account. If a wound rotor is used analogous equations can be written using m_r, k_{w1r} and N_r for the wound rotor. When considering the current linkages of a harmonic ν, the rotor current $I'_{\nu r}$ referred to the stator and flowing in the stator winding has to produce an equal current linkage to the original rotor current when flowing in the rotor. Thus we may write

$$m_s N_s k_{w\nu s} I'_{\nu r} = m_r N_r k_{sq\nu r} k_{w\nu r} I_{\nu r}. \tag{7.52}$$

The quantities with the prime symbol are referred to the stator from the rotor. The transformation ratio for the harmonic ν from the rotor to the stator therefore becomes

$$K_{\mathrm{rs},\nu} = \frac{I_{\nu\mathrm{r}}}{I'_{\nu\mathrm{r}}} = \frac{m_\mathrm{s} k_{\mathrm{w}\nu\mathrm{s}} N_\mathrm{s}}{m_\mathrm{r} k_{\mathrm{w}\nu\mathrm{r}} k_{\mathrm{sq}\nu\mathrm{r}} N_\mathrm{r}}. \tag{7.53}$$

By applying the above to the cage winding and the fundamental component, we obtain

$$K_{\mathrm{rs},1} = \frac{m_\mathrm{s} k_{\mathrm{w1s}} N_\mathrm{s}}{m_\mathrm{r} k_{\mathrm{w1r}} k_{\mathrm{sq1r}} N_\mathrm{r}} = \frac{m_\mathrm{s} k_{\mathrm{w1s}} N_\mathrm{s}}{Q_\mathrm{r} \cdot 1 \cdot k_{\mathrm{sq1r}} \cdot 1/2} = \frac{2 m_\mathrm{s} k_{\mathrm{w1s}} N_\mathrm{s}}{Q_\mathrm{r} k_{\mathrm{sq1r}}}. \tag{7.54}$$

If R_r is the resistance of the rotor bar added with the proportion of the short-circuit rings, and I_r is the RMS value of the current of the rotor bar, we obtain, by writing the resistive loss of the rotor equal in both the stator and in the rotor

$$m_\mathrm{s} I_\mathrm{r}'^2 R'_\mathrm{r} = Q_\mathrm{r} I_\mathrm{r}^2 R_\mathrm{r}. \tag{7.55}$$

The phase resistance of the rotor referred to the stator can now be written as

$$R'_\mathrm{r} = \frac{Q_\mathrm{r} I_\mathrm{r}^2 R_\mathrm{r}}{m_\mathrm{s} I_\mathrm{r}'^2}. \tag{7.56}$$

Since

$$\frac{I_\mathrm{r}}{I'_\mathrm{r}} = K_\mathrm{rs}, \tag{7.57}$$

the resistance of the rotor referred to the stator now becomes

$$R'_\mathrm{r} = \frac{Q_\mathrm{r}}{m_\mathrm{s}} \left[\frac{I_\mathrm{r}}{I'_\mathrm{r}}\right]^2 R_\mathrm{r} = \frac{Q_\mathrm{r}}{m_\mathrm{s}} K_{\mathrm{rs},1}^2 R_\mathrm{r} = \frac{Q_\mathrm{r}}{m_\mathrm{s}} \left[\frac{2 m_\mathrm{s} k_{\mathrm{w1s}} N_\mathrm{s}}{Q_\mathrm{r} k_{\mathrm{sq1r}}}\right]^2 R_\mathrm{r} = \frac{4 m_\mathrm{s} \left(k_{\mathrm{w1s}} N_\mathrm{s}\right)^2}{Q_\mathrm{r} k_{\mathrm{sq1r}}^2} R_\mathrm{r}. \tag{7.58}$$

When it is necessary to refer the rotor resistance to the stator, in general, it has to be multiplied by the term

$$\rho_\nu = \frac{m_\mathrm{s}}{m_\mathrm{r}} \left(\frac{N_\mathrm{s} k_{\mathrm{w}\nu\mathrm{s}}}{N_\mathrm{r} k_{\mathrm{sq}\nu\mathrm{r}} k_{\mathrm{w}\nu\mathrm{r}}}\right)^2. \tag{7.59}$$

In a squirrel cage induction motor, this is written in the form

$$\rho_\nu = \frac{4 m_\mathrm{s}}{Q_\mathrm{r}} \left(\frac{N_\mathrm{s} k_{\mathrm{w}\nu\mathrm{s}}}{k_{\mathrm{sq}\nu\mathrm{r}}}\right)^2. \tag{7.60}$$

If there is no skewing, we write further

$$\rho_\nu = \frac{4 m_\mathrm{s}}{Q_\mathrm{r}} \left(N_\mathrm{s} k_{\mathrm{w}\nu\mathrm{s}}\right)^2. \tag{7.61}$$

The same referring factor is valid also in the referring of the inductances. It cannot, however, be obtained based on equal stator and rotor losses (I^2R), but on equal energy stored in the inductance ($^1/_2LI^2$). Hence we get

$$R'_{\nu r} = \rho_\nu R_{\nu r}; \quad L'_{\nu r} = \rho_\nu L_{\nu r}. \tag{7.62}$$

Here it is worth noting that in a rotating electrical machine, the referring deviates from the referring in a transformer by the fact that the impedance quantities are not referred directly with the square of the transformation ratio of the current, but we also have to take into account the ratio of the numbers of phases.

With low values of slip, the resistance R_r can be given as a DC value, but for instance at start-up, the rotor frequency is so high that the skin effect in the squirrel cage has to be taken into account. Also, in rapid transients, the rotor resistance deviates notably from its DC value.

In a cage winding, there are no coils, and there can be an odd number of bars in the winding. Therefore, the definition of the current linkage is not quite as straightforward as was the case with the coil windings discussed previously. First, a current linkage of a single bar is defined, and next the current linkages of all the bars are summed. The analysis can be carried out in a reference frame attached to the rotor, since we are now analyzing phenomena that occur only between the resulting air-gap flux density harmonic ν and the rotor cage. The cage winding itself does not form poles, but the number of pole pairs always settles to be the same as the stator harmonic influencing it. In the analysis of a cage winding, geometrical angles ϑ are employed as here. Now the electric angles are at the fundamental $p\vartheta$ and at harmonics $p\nu\vartheta$. Let us choose the instant $t = 0$, when the peak $\hat{B}_{\delta\nu}$ of the air-gap flux density of the harmonic ν occurs at the first bar. In a bar that is at an arbitrary position angle $\vartheta_x = xp\vartheta$ an emf is induced with a slip s_ν of a certain harmonic

$$e_{\nu x}(t) = \hat{e}_{\nu x} \cos(s_\nu \omega_s t - x\nu p\vartheta). \tag{7.63}$$

This harmonic emf is found by calculating the flux and its time derivative and is written with the absolute value and the phase angle as a complex number

$$e_{\nu x}(t) = \frac{s_\nu \omega_s}{\nu} \frac{\pi D_r l}{2p} \hat{B}_{\delta\nu} \left| \underline{s_\nu \omega_s t - x\nu p\vartheta} \right. . \tag{7.64}$$

The bar current $i_{\nu x}(t)$ is determined by dividing the emf by the equivalent bar impedance $\underline{Z}_{r\nu}(s)$. When the case is investigated in the rotor coordinate system, the slip becomes important with respect to the impedance. The imaginary part of the impedance changes as a function of the slip angular frequency, the phase angle of the rotor impedance for the νth harmonic being $\zeta_{r\nu}(s_\nu)$. Correspondingly, the angle of the current depends on the slip as $s_\nu \omega_s t$

$$i_{\nu x}(t) = \frac{s_\nu \omega_s}{2p\nu} \frac{\pi D l}{Z_{r\nu}} \hat{B}_{\delta\nu} \left| \underline{s_\nu \omega_s t - x\nu p\vartheta - \zeta_{r\nu}(s_\nu)} \right. . \tag{7.65}$$

The amplitude of this current corresponds to the rotor bar current needed later in (7.67). When considering a current linkage of a single bar, we have to find a suitable graph for the current

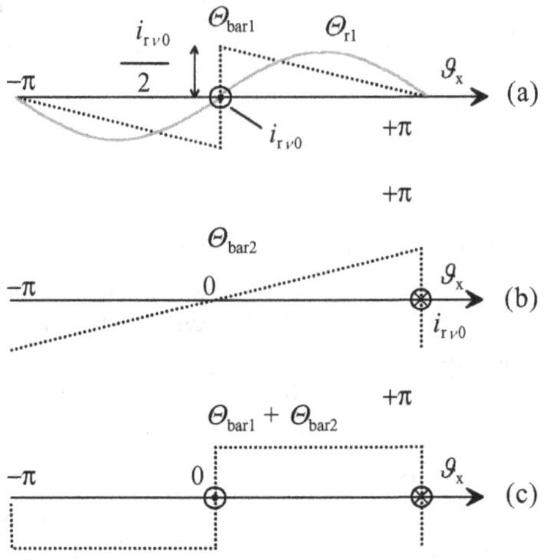

Figure 7.11 (a), (b) Current linkage created by a single rotor bar at $\vartheta_x = 0$ and $\vartheta_x = \pi$. (c) The current linkage created by two bars together.

linkage created by the current of the bar. With intuitive inference, we may construct a sawtooth wave illustrated in Figure 7.11a. Normally, a current loop is always required to produce a current linkage, as discussed with Figure 2.19. In that figure, a single current penetrates the system at two points, thus creating a closed loop. This kind of a loop is not unambiguously created for a cage winding, since in a cage winding the number of bars is not necessarily even. The current of a certain bar under investigation may be divided between several bars on the other side of the pole pitch. Therefore, the current linkage created by a single bar has to be observed in the analysis.

The intuitive deduction can be complemented by investigating Figures 7.11a–c. Now a second, separate bar is mounted at a point $\vartheta_x = \pi$ (i.e. at a distance of a pole pitch from the first bar). The current of this bar is opposite to the current of the bar in the position $\vartheta_x = 0$. As a result, a graph 7.11b that is opposite to the current linkage graph of 7.11a is created, and half of the graph is shifted a full pole pitch left in the case of two poles. By combining the graphs in Figures 7.11a and b, we obtain a familiar graph for the current linkage of a single loop that corresponds to Figure 2.19. Therefore, with intuitive deduction, we may draw the graph of a single bar that corresponds to Figure 7.11a. The figure also illustrates the fundamental Θ_{r1} of the current linkage of one bar $\Theta_{bar}(\vartheta_x)$. We may now write the current linkage for the bar at $x = 0$. Its current linkage is thus

$$\Theta_{bar,0} = \frac{\hat{i}_{\nu 0}}{2}\frac{(\pi - \vartheta_x)}{\pi}, \quad \vartheta_x \ni [0, 2\pi]. \tag{7.66}$$

The current linkage of an arbitrary bar at x is obtained by substituting the current of the original bar $i_{\nu 0}$ with the current $i_{\nu x}$, Equation (7.65),

$$\Theta_{bar,x} = \frac{\hat{i}_{r\nu}}{2}\frac{(\pi - \vartheta_x + \nu x p\vartheta)}{\pi}, \quad \vartheta_x \ni [0, 2\pi]. \tag{7.67}$$

The change in the sign results from the fact that the temporal phase shift $-\nu xp\vartheta$ corresponds to the local position angle $+\nu xp\vartheta$. The function is continuous only in the range 0–2π. When the function $\Theta_{\text{bar,x}}$ of the current linkage of the rotor is developed into Fourier series, its term ν_r becomes

$$\Theta_{\nu x \nu r} = \frac{\hat{i}_{r\nu}}{2\pi\nu_r}\underline{\left|s_\nu\omega_s t - \nu_r\vartheta + \beta_{r\nu} - (\nu p - \nu_r)x\vartheta\right.} = \frac{\hat{i}_{r\nu}}{2\pi\nu_r}e^{j(s_\nu\omega_s t - \nu_r\vartheta + \beta_{r\nu} - (\nu p - \nu_r)x\vartheta)}, \quad (7.68)$$

where $\beta_{r\nu} = -\frac{\pi}{2} - \zeta_{r\nu}(s_\nu)$.

Only the last term of the exponent of the constant e (Napier's constant, also known as Euler's number) depends on the ordinal number x of the bar. The sum of the Fourier series taking all the bars into account is

$$\sum_{x=0}^{Q_r} e^{-j(\nu p - \nu_r)x\vartheta} = \frac{1 - e^{-j(\nu p - \nu_r)Q_r\vartheta}}{1 - e^{-j(\nu p - \nu_r)\vartheta}}. \quad (7.69)$$

Since $(\nu p - \nu_r)$ is an integer and $Q_r\vartheta = 2\pi$, the numerator is always zero. The sum obtains values other than zero only when

$$\nu p - \nu_r = cQ_r, \quad (7.70)$$

where $c = 0, \pm 1, \pm 2, \pm 3, \ldots$. The harmonic ν of the resultant flux density of the air gap can create rotor harmonics ν_r that meet the condition (7.70). Equation (7.69) then takes the limit value Q_r. The ordinals ν and ν_r can be either positive or negative. The cage winding thus creates current linkages

$$\Theta_{\nu r} = \hat{\Theta}_{\nu r}\underline{\left|s_\nu\omega_s t - \nu_r\vartheta + \beta_{r\nu}\right.}, \quad (7.71)$$

where

$$\hat{\Theta}_{\nu r} = \frac{Q_r}{2\pi\nu_r}\hat{i}_{r\nu} = \frac{s_\nu\omega_s}{4\pi\nu p\nu_r}\frac{Dl}{Z_{r\nu}}\hat{B}_{\delta\nu}. \quad (7.72)$$

The corresponding linear current density of the rotor is

$$A_{\nu r} = \hat{A}_{\nu r}\underline{\left|s_\nu\omega_s t - \nu_r\vartheta - \zeta_{r\nu}\right.}, \quad (7.73)$$

where

$$\hat{A}_{\nu r} = \frac{Q_r}{\pi D}\hat{i}_{r\nu}. \quad (7.74)$$

$\zeta_{r\nu}$ is the phase angle of the rotor impedance for the harmonic ν.

7.2.3 Characteristics of an Induction Machine

The characteristics of an induction motor highly depend on the fulfillment of Equations (7.49) and (7.70). When the number of bars Q_r is finite, $c = 0$ always meets the condition (7.49). Each flux-density-producing air-gap harmonic ν that meets the condition (7.49) can induce a vast number of rotor current linkage harmonics with ordinals ν_r (7.68).

Now, the equivalent circuit of an asynchronous motor per phase, the quantities of which are calculated in the machine design, is worth recollecting. Figure 7.12 illustrates a single-phase equivalent circuit of an ordinary induction motor per phase, a simplified equivalent circuit and a phasor diagram.

In Figure 7.12, the stator is supplied with the voltage U_s. The stator resistance R_s is the resistance of the stator winding at operating frequency and in operating temperature, U_s' is the stator voltage after the resistive voltage drop is subtracted, $L_{s\sigma}$ is the leakage inductance of the stator, L_m is the magnetizing inductance of the machine at the rated voltage, R_{Fe} is the resistance describing the losses of the iron circuit of the machine, $L_{r\sigma}'$ is the rotor leakage inductance of the machine referred to the stator, and R_r' is the rotor resistance referred to the stator. s is the slip of the rotor. The term $R_r'(1-s)/s$ describes the electromechanical power produced by the machine (cf. Equation 7.85). A part of the mechanical power is consumed in the friction and windage losses of the machine. Ψ_s is the stator flux linkage, which includes

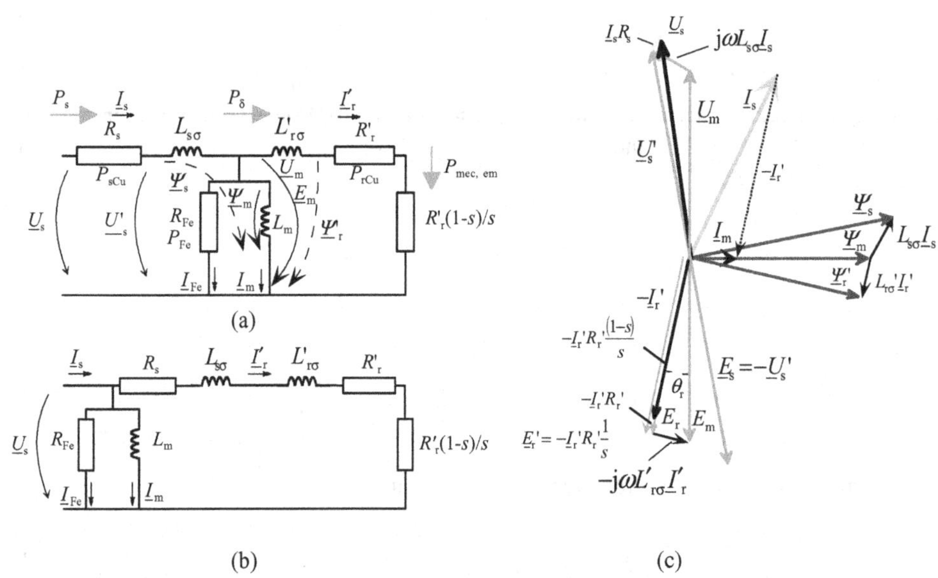

Figure 7.12 (a) Steady-state equivalent circuit of an asynchronous machine per phase. (b) A simplified equivalent circuit of the machine, the parameters of which are calculated in the machine design. (c) A phasor diagram of an asynchronous machine (ignoring the small iron loss current). The input stator power is P_s. In the stator resistance, a resistive loss P_{sCu} takes place. The iron loss is the loss P_{Fe} created in the magnetic circuit. The air-gap power P_δ flows across the air gap to the rotor. In the rotor, some resistive losses P_{rCu} take place. The power $P_{mec, em}$ is the electromechanical power of the machine. When friction and windage losses are subtracted, the output power P of the machine on the shaft can be found.

the air-gap flux linkage Ψ_m and the stator leakage flux $\Psi_{s\sigma}$. Correspondingly, Ψ_r is the flux linkage of the rotor, which includes the air-gap flux linkage Ψ_m and the rotor leakage flux $\Psi_{r\sigma}$. The stator voltage U'_s creates the stator flux linkage, from which, in turn, a back emf E_s is derived. A voltage E_m is induced over L_m by the air-gap flux linkage Ψ_m. This voltage is consumed completely in the rotor apparent resistance R'_r/s and in the rotor leakage reactance. The iron loss current I_{Fe} is small and is therefore not illustrated in the phasor diagram.

Stator power P_s is fed to the motor. Some power is consumed in the stator resistance and in the iron loss resistance. An air-gap power P_δ crosses the air gap. In the rotor, a part of the air-gap power is lost in the rotor resistance R'_r, and a part is converted into mechanical power in $R'_r(1-s)/s$.

The fundamental flux density created by the stator and rotor currents causes a varying flux, which in turn induces a voltage and a current acting against the flux variation. At no load, the rotor rotates approximately with a synchronous speed, and the rotor frequency and the currents approach zero. If the rotor is loaded, its speed decreases and the relative speed with respect to the fundamental flux propagating in the air gap increases. Now the emf induced in the rotor increases. Also the inductive reactance of the rotor increases as the rotor frequency increases. The peripheral force created by the fundamental flux density of the air gap and the rotor torque reaches its maximum at a certain slip.

If required, the rotor of an asynchronous machine can be driven by an above synchronous speed. In that case, the rotor currents create a torque opposing the accelerating torque and the machine acts as a generator, the slip being negative.

Let us assume that the fundamental $\nu = 1$ of the resultant air-gap flux density follows Equation (7.31). As the rotor rotates with a slip s with respect to the fundamental air-gap wave, an emf is induced in the phase winding of the rotor. The peak value of the emf induced in rotor bars depends on the slip

$$\hat{e}_r(s) = s\hat{e}_{rk} \left| \underline{s\omega_s t - \pi/2} \right. . \tag{7.75}$$

In this equation, there occurs a peak value of the emf induced to the rotor at the slip $s = 1$:

$$\hat{e}_{rk} = \omega_s \hat{\Psi}_r = \omega_s \frac{2}{\pi} \frac{\pi D}{2p} \hat{B}_\delta l k_{wr} N_r. \tag{7.76}$$

The impedance of the rotor circuit depends on the slip angular frequency

$$\underline{Z}_r(s) = R_r + js\omega_s L_{r\sigma} = Z_r(s) \left| \underline{\zeta_r(s)} \right. . \tag{7.77}$$

The peak value for the rotor current phasor now becomes

$$\hat{\underline{i}}_r(s) = \frac{s\hat{e}_{rk}(s)}{\underline{Z}_r(s)} = \hat{i}_r(s) \left| \underline{s\omega_s t - \zeta_r(s) - \pi/2} \right. . \tag{7.78}$$

The amplitude is

$$\hat{i}_{\mathrm{r}}(s) = \frac{s\hat{e}_{\mathrm{rk}}}{Z_{\mathrm{r}}(s)} = \frac{s\hat{e}_{\mathrm{rk}}}{\sqrt{R_{\mathrm{r}}^2 + s^2(\omega_{\mathrm{s}} L_{\mathrm{r}\sigma})^2}}. \tag{7.79}$$

In the case of a slip-ring machine, additional impedances can be attached to the rotor circuit. Now the total impedance of the circuit has to be substituted in the equation. The current $\hat{i}_{\mathrm{r}}(s)$ represents the linear current density of the rotor surface, the amplitude of which can be calculated from the induced voltage and the rotor impedance

$$\hat{A}_{\mathrm{r}}(s) = \frac{2p}{\pi} s\omega_{\mathrm{s}} \frac{m_{\mathrm{r}} l k_{\mathrm{wr}}^2 \left(\dfrac{N_{\mathrm{r}}}{p}\right)^2}{\sqrt{R_{\mathrm{r}}^2 + s^2(\omega_{\mathrm{s}} L_{\mathrm{r}\sigma})^2}} \hat{B}_{\delta}. \tag{7.80}$$

When the above is substituted in the general torque Equation (7.43), and we note that

$$\cos \zeta_{\mathrm{r}}(s) = \frac{R_{\mathrm{r}}}{\sqrt{R_{\mathrm{r}}^2 + s^2(\omega_{\mathrm{s}} L_{\mathrm{r}\sigma})^2}}, \tag{7.81}$$

which corresponds to the power factor of the rotor impedance, we obtain the electromagnetic torque

$$T_{\mathrm{em}}(s) = \frac{p m_{\mathrm{r}}}{\omega_{\mathrm{s}}} \frac{s R_{\mathrm{r}}}{R_{\mathrm{r}}^2 + s^2(\omega_{\mathrm{s}} L_{\mathrm{r}\sigma})^2} E_{\mathrm{rk}}^2. \tag{7.82}$$

Here E_{rk} is the RMS value of the emf of the phase winding of the rotor when the machine is held at stall, $E_{\mathrm{rk}} = \hat{e}_{\mathrm{rk}}/\sqrt{2}$. Also, the power distribution leads to the same equation. Thus, the RMS value of the current in the rotor circuit of the equivalent circuit with the slip s is

$$I_{\mathrm{r}}(s) = \frac{sE_{\mathrm{rk}}}{\sqrt{R_{\mathrm{r}}^2 + (s\omega_{\mathrm{s}} L_{\mathrm{r}\sigma})^2}} = \frac{E_{\mathrm{rk}}}{\sqrt{\left(\dfrac{R_{\mathrm{r}}}{s}\right)^2 + (\omega_{\mathrm{s}} L_{\mathrm{r}\sigma})^2}}. \tag{7.83}$$

Transferring Equation (7.83) to the stator side we get

$$I_{\mathrm{r}}'(s) = \frac{E_{\mathrm{rk}}'}{\sqrt{\left(\dfrac{R_{\mathrm{r}}'}{s}\right)^2 + (\omega_{\mathrm{s}} L_{\mathrm{r}\sigma}')^2}}. \tag{7.84}$$

The rotor circuit of the equivalent circuit in Figure 7.12 follows this equation. In the figure, the rotor resistance has been divided into two parts, the sum of which is R_2'/s. Correspondingly, the active power P_δ crossing the air gap is divided into the resistive loss P_{rCu} of the rotor and the mechanical power $P_{mec,em}$

$$P_\delta = R'_r I'^2_r + \frac{1-s}{s} R'_r I'^2_r = \frac{R'_r}{s} I'^2_r = P_{rCu} + P_{mec,\,em}, \tag{7.85}$$

$$\frac{P_{rCu}}{P_{mec,\,em}} = \frac{s}{1-s}. \tag{7.86}$$

In motor drive, the air-gap power P_δ is the power transmitted from the stator via the air gap to the rotor. Of this power, the proportion P_{rCu} is consumed in the resistive losses of the rotor, and the rest is electromechanical power $P_{mec,\,em}$. The mechanical power P_{mec} is obtained from the shaft, when friction and windage losses are subtracted from the electromechanical power. We can write Equations (7.85) and (7.86) in the form

$$P_{rCu} = s P_\delta; \quad P_{mec,em} = (1-s) P_\delta \tag{7.87}$$

$$T_{em}(s) = \frac{P_{mec,em}}{\Omega} = \frac{p}{(1-s)\,\omega_s} P_{mec,em} = \frac{p}{s\omega_s} P_{rCu} = \frac{p}{\omega_s} P_\delta \tag{7.88}$$

where Ω is the actual angular rotor speed.

The torque can thus be solved with the resistive loss power of the rotor. The torque is always (including when the machine is at stall) proportional to the air-gap power P_δ: From Equation (7.58) and (7.54) it can be derived:

$$T_{em} = \frac{m_r E_{rk} I_r \cos\varsigma_r}{\omega_s/p} = \frac{P_\delta}{\omega_s/p} = \frac{m_r E^2_{rk}}{\omega_s/p} \frac{R_r/s}{\left(\dfrac{R_r}{s}\right)^2 + (\omega_s L_{r\sigma})^2} = \frac{p m_r E^2_{rk}}{\omega_s} \frac{s R_r}{R^2_r + (s\omega_s L_{r\sigma})^2},$$

$$\approx \frac{p m_s U^2_s}{\omega_s} \frac{R'_r/s}{\left(R_s + \dfrac{R'_r}{s}\right)^2 + (\omega_s L_k)^2}. \tag{7.89}$$

The approximation in Equation (7.89) is obtained from the simplified equivalent circuit on Figure 7.12b by substituting $I_r = sE_{rk}/Z_r$ and $\cos\varsigma_r = R_r/Z_r$ and, based on the simplified equivalent circuit, by assuming that the air-gap voltage is equal to the terminal voltage of the machine. Furthermore, we employ the short-circuit inductance $L_k \approx L_{s\sigma} + L'_{r\sigma}$. Neglecting the effects of R_s, the highest value of T_{em}, the pull-out torque T_b, can be found from the per unit slip

$$s_b = \pm \frac{R'_r}{\omega_s L_{s\sigma} + \omega_s L'_{r\sigma}} = \frac{R'_r}{\omega_s L_k}, \tag{7.90}$$

$$T_b = \pm \frac{mp}{2\omega^2_s} \frac{U^2_s}{L_k}. \tag{7.91}$$

Here we remember that $L_k \approx L_{s\sigma} + L'_{r\sigma} \approx 2L'_{r\sigma}$. We see that the peak torque is inversely proportional to the short-circuit inductance of the machine. If the per unit value of L_k is 0.2

for instance, the maximum torque is about $5T_n$. Contrary to the slip value of the peak torque, the maximum torque is independent of the rotor resistance. By substituting Equation (7.90) in Equation (7.81) we obtain

$$\cos\zeta_r(s) = \frac{1}{\sqrt{\left(\frac{s}{s_b}\right)^2 + 1}} \tag{7.92}$$

The term takes the value $\cos\zeta_r(s) = \frac{1}{\sqrt{2}}$ at peak torque slip.

Dividing the torque T_{em} (7.89) by the maximum torque T_b (7.91) and neglecting the stator resistance ($R_s = 0$) and using Equation (7.90), we obtain

$$\frac{T_{em}}{T_b} = \frac{2}{\frac{s}{s_b} + \frac{s_b}{s}}. \tag{7.93}$$

The calculation of torque introduced above can thus be simplified with the equivalent circuit to the definition of the power of the resistance. With the single-phase equivalent circuit, the load variations can chiefly be observed in the changes in the slip.

Somewhat more accurate results can be obtained for an asynchronous machine by again employing a simplified equivalent circuit, but applying a reduced voltage in the calculation of the rotor current of the machine

$$I'_r = \frac{U_s\left(1 - \frac{L_{s\sigma}}{L_m}\right)}{\sqrt{\left(R_s + R'_r/s\right)^2 + \left(\omega_s L_{s\sigma} + \omega_s L'_{r\sigma}\right)^2}}. \tag{7.94}$$

Now, the generated electromechanical torque is

$$T_{em} = \frac{3\left[U_s\left(1 - \frac{L_{s\sigma}}{L_m}\right)\right]^2 \frac{R'_r}{s}}{\frac{\omega_s}{p}\left[\left(R_s + R'_r/s\right)^2 + \left(\omega_s L_{s\sigma} + \omega_s L'_{r\sigma}\right)^2\right]}. \tag{7.95}$$

The starting torque produced by the fundamental is obtained by setting $s = 1$ in Equation (7.89). However, the rotor resistance is higher at large slips because of the skin effect, and the resistance must be defined at every slip value before calculating the torque. The pull-out torque is solved by derivation with respect to R'_r/s, after which we can write for the slip of the maximum torque, if R_s is taken into account

$$s_b = \pm\frac{R'_r}{\sqrt{(R_s)^2 + \left(\omega_s L_{s\sigma} + \omega_s L'_{r\sigma}\right)^2}}, \tag{7.96}$$

where the plus sign is for motoring, and the minus sign for generating operation. The corresponding torque for motoring is

$$T_b = \frac{3\left[U_s\left(1-\frac{L_{s\sigma}}{L_m}\right)\right]^2}{2\frac{\omega_s}{p}\left[R_s+\sqrt{R_s^2+\left(\omega_s L_{s\sigma}+\omega_s L'_{r\sigma}\right)^2}\right]}, \tag{7.97}$$

and for generating

$$T_b = -\frac{3\left[U_s\left(1+\frac{L_{s\sigma}}{L_m}\right)\right]^2}{2\frac{\omega_s}{p}\left[R_s-\sqrt{R_s^2+\left(\omega_s L_{s\sigma}+\omega_s L'_{r\sigma}\right)^2}\right]}. \tag{7.98}$$

7.2.4 Equivalent Circuit Taking Asynchronous Torques and Harmonics into Account

The parameters of a cage winding are investigated next. When the number of bars Q_r of the rotor is finite and the factor $c = 0$, the condition (7.70) is always met. Each resulting air-gap flux density harmonic ν that meets the condition (7.49) can induce a large number of rotor harmonics ν_r (7.68). In a case where $c = 0$, we investigate the harmonic ν_r, the number of pole pairs of which is the same as the inducing air-gap harmonic ν. The torque equation for the pair of harmonics ν and ν_r is derived in the same way as for the torque of the fundamental (7.82). First we obtain two equations

$$T_\nu(s_\nu) = \frac{\nu p}{\omega_s} Q_r \frac{s_\nu R_{r\nu}}{R_{r\nu}^2 + s_\nu^2\omega_s^2 L_{r\sigma\nu}^2} E_{rk\nu}^2. \tag{7.99}$$

From which similarly as in Equation (7.76)

$$T_\nu(s_\nu) = \frac{s_\nu\omega_s}{8\nu p} Q_r \frac{D^2 l^2 R_{r\nu}}{R_{r\nu}^2 + s_\nu^2\omega_s^2 L_{r\sigma\nu}^2} \hat{B}_{m\nu}^2. \tag{7.100}$$

Here $E_{rk\nu}$ is the RMS value of the emf induced by the harmonic ν at the slip $s_\nu = 1$. We may now infer that each harmonic meeting the conditions behaves accordingly. Torque is a continuous function of slip that becomes zero at the slip $s_\nu = 0$, hence the term "asynchronous torque."

When the machine is running at a fundamental slip s_1, the slip of the rotor with respect to the νth stator harmonic is written as

$$s_\nu = 1 - \nu(1 - s_1). \tag{7.101}$$

The angular frequency of the νth harmonic in the rotor is thus

$$\omega_{\nu r} = \omega_s(1 - \nu(1 - s_1)). \tag{7.102}$$

Based on Equation (7.101), by setting the slip of the harmonic ν zero, we obtain the zero slip of the torque of the harmonic with respect to the fundamental

$$s_1(s_\nu = 0) = \frac{\nu - 1}{\nu}. \tag{7.103}$$

The harmonics rotate by the speed

$$n_{\mathrm{syn}\nu} = \frac{n_{\mathrm{syn1}}}{\nu}. \tag{7.104}$$

The ordinal of the harmonic ν has to be inserted into Equations (7.103) and (7.104) with its sign.

Figure 7.13 illustrates the torque curve of an asynchronous machine. At the slip of the fundamental $s_1 = 6/7 \approx 0.86$, there is a zero point of the torque produced by the seventh harmonic ($\nu = 7$). The seventh harmonic is discussed because its synchronous speed is the first synchronous speed after the fundamental synchronous speed at positive slips and positive speeds. For instance the fifth harmonic has its synchronous speed at a negative rotation speed of the rotor $n_{\mathrm{syn5}} = -n_{\mathrm{syn1}}/5$ and $s_1 = 6/5 = 1.2$. The peak values of the harmonic torques are

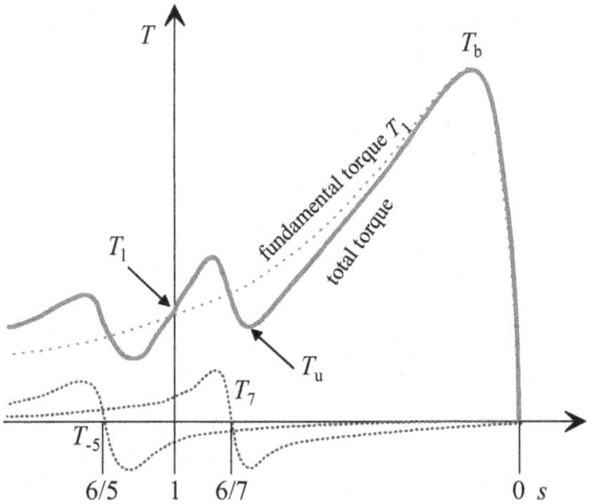

Figure 7.13 Total torque of an induction motor with a cage winding, and the torque of the fifth and seventh harmonics as a function of slip s. We see that both the fifth and the seventh harmonic reduce the torque of the machine at speeds higher than the synchronous speed of the seventh harmonic (slip 6/7). The negative peak of the seventh harmonic torque makes the total torque lower than the locked rotor torque T_l. T_u, is the pull-up torque and T_b, the peak torque. The synchronous speeds of the fifth and seventh harmonics are at 6/5 and 6/7 of the slip of the fundamental.

located approximately at slips $s_{\nu b} = \pm R'_{r\nu}/\omega_s L_{k\nu}$ (7.90). According to Equation (7.101), the negative peak torque at the base slip is written as

$$s_1(s_{\nu b}) \approx 1 - \frac{R'_{r\nu} + \omega_s L_{k\nu}}{\nu \omega_s L_{k\nu}}. \tag{7.105}$$

where $s_{\nu b}$ has the negative sign.

Equation (7.103) shows that the harmonic-wave-generated torques of a three-phase motor are high at high values of slip, and therefore they may impede start-up of the motor. The start-up and drive characteristics can be improved by various structural means in machine construction. According to (7.90), the peak torque of an asynchronous machine shifts in the direction of a higher slip as the rotor resistance increases. On the other hand, rotor losses can be minimized by employing a rotor with minimum resistive losses. A high starting resistance and a low operational resistance are achieved by designing the rotor such that the skin effect increases the low DC resistance of the rotor bars at high slips. The rotor resistance can be adjusted by shaping the rotor bars. In some machines, good start-up and operation characteristics are achieved by employing a double cage or deep rotor slots, Figure 7.14.

In Figure 7.14a, the cross-sectional area of the outer bar of the double cage is small and therefore has a high resistance. The leakage inductance of the outer bar is low, while the leakage inductance of the inner bar is high and the resistance low. Therefore, at start-up, at the line frequency (e.g. 50 Hz), the current density displacement takes place in the direction of the air gap because of the skin effect, the inner bar is not effective and the outer bar acts as the chief current carrier and yields a good torque at lower start-up current because of the resulting high rotor resistance. The inner bar, because of its high inductance, carries operational current only at low frequencies, that is, when the machine is running at a low slip. In that case, the low resistance of the bar leads to a small slip. In a deep-slot rotor (Figure 7.14b), similar phenomena arise when the machine turns from the start-up to continuous running. The resistance of the cages in the figure therefore changes thus as a function of the rotor frequency. The resistance can be analyzed at different frequencies for instance with the methods presented for skin effect in Chapter 5.

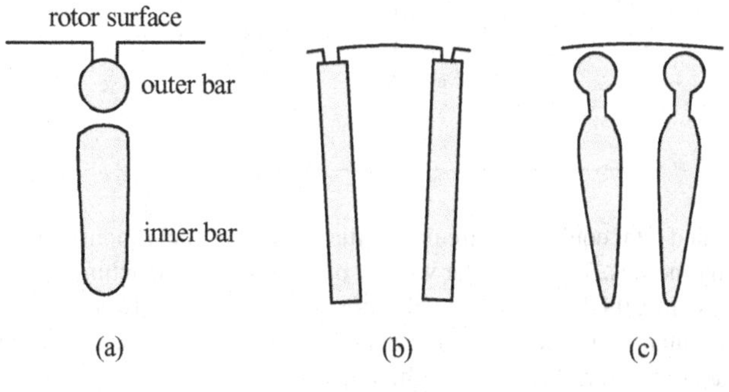

Figure 7.14 Shapes of slots and rotor bars for (a) a double cage, (b) a deep slot, (c) a typical cast-aluminum double cage rotor slot. The slot opening is closed to ease the squirrel cage die-cast process (no separate mould is required).

The rotor resistance of a high slip can be increased also by employing a ferromagnetic material as the conductive material of the cage winding. In ferromagnetic materials, the skin effect is extremely strong at high frequencies, when the penetration depth is small. An example of the influences of a high rotor resistance is a completely solid-steel rotor, the starting torque of which is relatively high (see Figure 2.57). In practice, ferromagnetic material could be applied in the cage winding of an induction motor, for instance by constructing the rotor bars of copper and by soldering them to thick steel rings at the rotor ends. Now the end rings saturate heavily during the start-up, and the rotor resistance is high. One problem in the application of iron as the conducting material of a cage rotor is that the resistivity of iron is high (for pure iron about 9.6 μΩcm, structural steels about 20–30 μΩcm, and for ferromagnetic stainless steels 40–120 μΩcm) when compared with aluminum (2.8 μΩcm) and copper (1.7 μΩcm), and therefore a large conductor area is required to achieve adequate operational characteristics. Additional problems arise in joining the iron to another conductor material, which is not an easy task. Copper and iron can be soldered with silver, whereas joining iron and aluminum is fairly difficult. Copper can also be welded to the iron by electron beam welding.

The effects of harmonics are usually investigated in asynchronous machines with the equivalent circuit illustrated in Figure 7.15. In the circuit, each harmonic frequency forms an

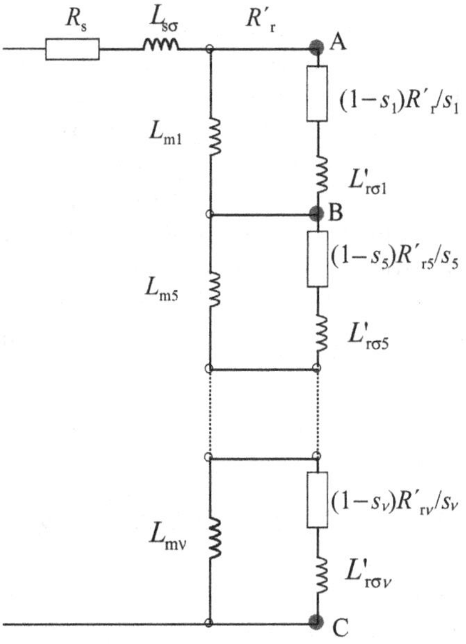

Figure 7.15 Simplified total equivalent circuit of a three-phase asynchronous motor, also involving the influence of spatial harmonics. Typically, the voltage of the harmonic machines ($\nu = -5, +7 \ldots \pm\infty$) remains rather low. Usually, 98–99 % of the total stator voltage occurs between the points AB. Thus, the sum voltage of the harmonic machines is only about 1–2 %. However, harmonics are able to produce quite high torques, and therefore they have to be taken into account in machine design. The starting properties of squirrel cage machines in particular are affected by the harmonics. In many cases related to the use of induction motors, the harmonic machines (−5, +7, −11 ...) are replaced with a single leakage inductance, or they are neglected altogether.

individual electrical machine, which is connected in series with all the other harmonic machines. The connection in series is justified, since the emf of the phase winding is the sum of emfs induced by different harmonics. The machine acts like a group of machines assembled on the same axis, the windings of the machines being in series. Each machine represents a number of pole pairs νp. The sum voltage is divided between the machines in the ratio of the impedances. Phasor calculation has to be employed in the analysis of the equivalent circuit. We may state that the torque caused by the harmonic ν (7.100) can be calculated according to Equation (7.87) from the resistive loss of the resistance $R'_{\mathrm{r}\nu}$ (see Equation (7.85) for the fundamental harmonic) of the equivalent circuit. The stator winding of a three-phase machine produces the harmonics presented in Table 2.2 (see Example 2.11).

The resistance and inductance of a cage winding for a harmonic ν were defined earlier (see Equations (7.47) and (7.48)) but are repeated here for the sake of convenience:

$$R_{\mathrm{r}\nu} = R_{\mathrm{bar}} + \frac{R_{\mathrm{ring}}}{2\sin^2\dfrac{\alpha_\nu}{2}}, \tag{7.106}$$

$$L_{\mathrm{r}\sigma\nu} = L_{\mathrm{bar}} + \frac{L_{\mathrm{ring}}}{2\sin^2\dfrac{\alpha_\nu}{2}}, \tag{7.107}$$

where $R_{\mathrm{bar}\nu}$ is the resistance of the rotor bar for the νth harmonic and R_{ring} is part of the segment of a short-circuit ring belonging to one bar on the both ends Inductance parameters behave analogically. α_ν describes the rotor phase angle for the νth harmonic

$$\alpha_\nu = \nu\frac{2\pi p}{Q_{\mathrm{r}}}. \tag{7.108}$$

Figure 7.16 illustrates the definition of α_1 for a motor, for which $p = 1$ and $Q_{\mathrm{r}} = 28$. α_ν is the νth multiple of α_1.

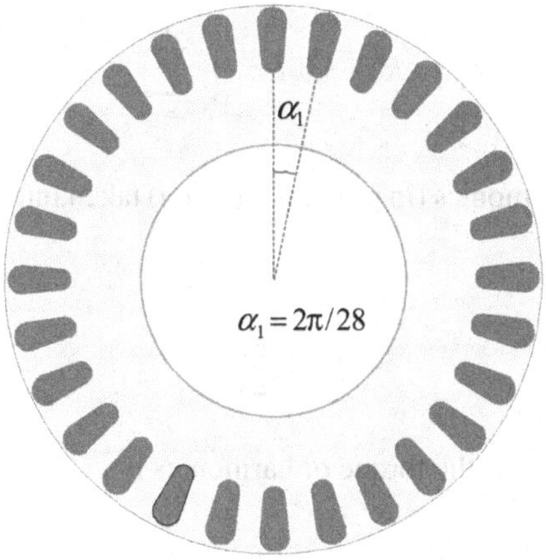

Figure 7.16 Cross-section of a 28-slot rotor ($Q_{\mathrm{r}} = 28$) and the definition of the angle α_1 in the case where the number of pole pairs is $p = 1$. α_1 is given in electrical radians.

When analyzing the phenomena in a simplified form, we take only the influence of the equivalent air gap into account (in other words, the machine is assumed to remain linear). Now we may calculate the stator magnetizing inductance of the νth harmonic with the magnetizing inductance L_{m1} defined for the fundamental

$$L_{m\nu} = L_{m1}\frac{1}{\nu^2}\left(\frac{k_{ws\nu}}{k_{ws1}}\right)^2. \tag{7.109}$$

In reality, the permeance of the magnetic circuit of a machine is different for different frequencies, and therefore Equation (7.109) is not completely valid, but suffices as a good approximation in the machine analysis. The rotor impedance of the νth harmonic is a function of slip

$$\underline{Z}'_{r\nu} = \frac{R'_{r\nu}}{s_\nu} + j\omega_s L'_{r\sigma\nu}, \tag{7.110}$$

where $s_\nu = 1 - \nu(1 - s_1)$ according to Equation (7.101).

The total impedance that describes the effects of the νth harmonic is a parallel connection of the rotor circuit and the magnetizing circuit. The impedances of the cage rotor have to be referred to the stator before connecting in parallel (Equations (7.59) and (7.60))

$$\underline{Z}'_{r\nu} = \frac{m_s}{m_r}\left(\frac{N_s k_{w\nu s}}{N_r k_{sq\nu r} k_{w\nu r}}\right)^2 \underline{Z}_{r\nu} = \frac{4m_s}{Q_r}\left(\frac{N_s k_{w\nu s}}{k_{sq\nu r}}\right)^2 \underline{Z}_{r\nu}.$$

$$\underline{Z}'_\nu = \frac{\underline{Z}'_{r\nu} j\omega_s L_{m\nu}}{\underline{Z}'_{r\nu} + j\omega_s L_{m\nu}}. \tag{7.111}$$

Now it is possible to determine the impedance of the total equivalent circuit of Figure 7.15

$$\underline{Z}_e = R_s + j\omega_s L_{s\sigma} + \sum_{\nu=-n}^{n} \underline{Z}'_\nu, \tag{7.112}$$

where n is the number of harmonics (in principle infinite) taken into account in the calculation.

The stator current $\underline{I}_s$ is

$$\underline{I}_s = \frac{\underline{U}_s}{\underline{Z}_e}. \tag{7.113}$$

The rotor current that creates the torque of harmonics is

$$\underline{I}'_{r\nu} = \frac{\underline{Z}_e}{\dfrac{R'_{r\nu}}{s_\nu} + j\omega_s L'_{r\sigma\nu}} \underline{I}_s. \tag{7.114}$$

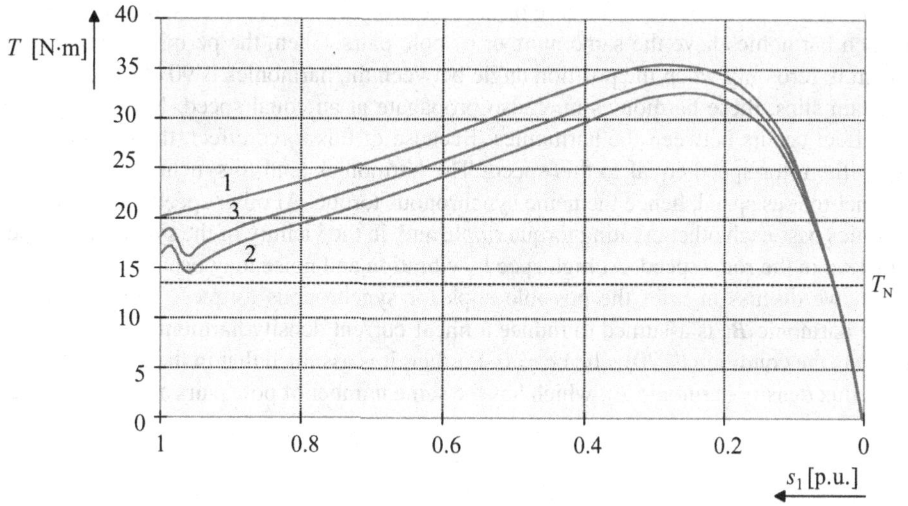

Figure 7.17 Torque curves of a two-pole three-phase, 4 kW induction motor as a function of slip s_1. The harmonic torques have been taken into account in the calculation of the curves. 1, static torque of the fundamental; 2, total torque with a full-pitch winding ($W/\tau_p = 1$, $q = 6$); 3, total torque with a short-pitched double-layer winding ($W/\tau_p = 5/6$, $q = 6$). A high number of slots per pole and phase lead to a situation in which the influence of the fifth and the seventh harmonic is almost insignificant (not visible at all in the figure). Therefore, the torque saddle is located in the vicinity of the slip $s = 1$ caused mostly by the 19th harmonic. T_N gives the rated torque of the machine

The torque of the νth harmonic as a function of slip is written as

$$T_\nu(s) = \frac{m_s p}{\omega_{s1}} \nu \frac{R'_{s\nu}}{s_\nu} I'^2_{r\nu}, \tag{7.115}$$

and the total torque

$$T(s) = \sum_{\nu=1}^{\infty} T_\nu(s). \tag{7.116}$$

This equation is valid in the steady state. The time constants of electrical machines are nevertheless so small that during slow acceleration, for instance, the machine follows its static torque curve rather strictly. Figure 7.17 illustrates the influence of harmonic torques on the total torque of the machine.

7.2.5 Synchronous Torques

Next, we investigate harmonic pairs that meet the condition (7.70) when $c \neq 0$. In that case, we have harmonics ν and ν_r that are alien to each other, in other words, they are not induced by each other as in the case of asynchronous torque but now the rotor harmonics ν_r seem to have their own excitation as in the case of the synchronous machines. It is, however, possible that such harmonics have the same number of pole pairs. Then, the peripheral force caused by them is zero only when the position angle between the harmonics is 90 electrical degrees. At certain slips, these harmonics may also propagate at an equal speed. Now, a permanent force effect occurs between the harmonics. Because of this force effect, the harmonics tend to keep the rotor speed equal to this speed. The harmonics tend to synchronize the rotor at this synchronous speed, hence the name synchronous torque. At other speeds of the rotor, the harmonics pass each other creating torque ripple and, in the vicinity of the synchronous speed, fluctuation in the rotor speed. At high speeds, vibration and noise may occur.

Next, we discuss in brief the possible slips for synchronous torques. The resultant flux density harmonic B_ν is assumed to induce a linear current density harmonic $A_{\nu r}$ in the rotor. $A_{\nu r}$ meets the condition (7.70) when $c \neq 0$. Further, it is assumed that in the air gap, there is a certain flux density harmonic B_μ which has the same number of pole pairs as the harmonic ν_r:

$$p\mu = \nu_r. \tag{7.117}$$

The geometrical speed of the rotor harmonic ν_r with respect to the rotor is

$$\Omega_{vr} = \frac{s_v \omega_s}{p \nu_r}. \tag{7.118}$$

When we add here the rotor's own speed with respect to the stator rotating field, we obtain the speed of the harmonic with respect to the stator

$$\Omega_{vrs} = \left[\frac{s_v}{\nu_r} + \frac{1 - s_1}{p}\right] \omega_s. \tag{7.119}$$

Now we substitute the slip of the harmonic (Equation (7.101)) and condition (7.70), $\nu p - \nu_r = cQ_r$, where $c = 0, \pm 1, \pm 2, \pm 3, \ldots$, and eliminate the ordinal ν_r. This yields

$$\Omega_{vrs} = \left[1 - \frac{cQ_r}{p}(1 - s_1)\right] \frac{\omega_s}{\nu_r}. \tag{7.120}$$

Here s_1 is, as always, the fundamental slip. The speed of the synchronizing harmonic with respect to the stator is $\Omega_{\mu s} = \omega_s/(\mu p)$. By setting these two equal, it means $\Omega_{\mu s}$ and (7.120), we may solve the fundamental slip at which the harmonic field has its synchronous speed

$$s_{syn} = 1 - \frac{p}{cQ_r}\left(1 - \frac{\nu_r}{\mu p}\right). \tag{7.121}$$

Based on condition (7.117), the absolute value of the second term in the brackets is always one. Depending on the sign, the slip thus obtains two values

$$\begin{aligned} \pm|\nu_r| &= \pm|\mu|p \rightarrow s_{syn1} = 1, \\ \pm|\nu_r| &= \mp|\mu|p \rightarrow s_{syn2} = 1 - \frac{2p}{cQ_r}. \end{aligned} \tag{7.122}$$

The former shows that the start-up of an induction motor is impeded by a synchronous torque at a slip $s_{syn1} = 1$. The second synchronous slip depends on the sign of c. If $c > 0$, this point occurs in the motor or generator range, otherwise in the braking range. The angular speed of the machine at the slip s_{syn2} is

$$\Omega_r = (1 - s_{syn2})\Omega_s = \frac{2p\Omega_s}{cQ_r} = \frac{2\omega_s}{cQ_r}. \tag{7.123}$$

The magnitude of the synchronous torque can be calculated from Equation (7.43) by substituting the real values of the harmonics B_μ, $A_{\nu r}$ with a synchronous slip. In the torque curve of the machine, the synchronous torques are indicated by peaks at the synchronous slip. The magnitude of the harmonic torques depends highly on the ratio of the slot numbers of the stator and the rotor. The torques can be reduced by skewing the rotor slots with respect to the stator. In that case, the slot torques resulting from the slotting are damped. In the design of an induction motor, special attention has to be paid to the elimination of the harmonic torques.

7.2.6 Selection of the Slot Number of a Cage Winding

In order to avoid the disturbance caused by asynchronous and synchronous torques and the vibration hampering the operation of an induction motor, the slot number Q_r of the rotor has to be selected with special care. To reduce the asynchronous harmonic torques, the slot number of the rotor has to be small. It is generally recommended that

$$Q_r < 1.25Q_s. \tag{7.124}$$

Next, only machines with three-phase integral slot stator windings are investigated. To limit the synchronous torques when the motor is at stall, the slot number of the rotor has to meet the condition

$$Q_r \neq 6pg, \tag{7.125}$$

where g may be any positive integer. To avoid synchronous torques created by slot harmonics, the selection of the slot number has to meet the following conditions:

$$Q_r \neq Q_s; \quad Q_r \neq {}^1\!/\!_2 Q_s; \quad Q_r \neq 2Q_s. \tag{7.126}$$

To avoid synchronous torques during running, the following inequality has to be in force:

$$Q_r \neq 6pg \pm 2p. \tag{7.127}$$

Here g is again any positive integer. The plus sign holds for positive rotation speeds and the minus sign is valid for negative rotation speeds.

To avoid dangerous slot harmonics, the following inequalities have to be in force:

$$\begin{aligned} Q_r &\neq Q_s \pm 2p, \\ Q_r &\neq 2Q_s \pm 2p, \\ Q_r &\neq Q_s \pm p, \\ Q_r &\neq \frac{Q_s}{2} \pm p. \end{aligned} \tag{7.128}$$

Also in these conditions, the plus sign holds for the positive rotation speeds and the minus sign is valid for negative rotation speeds.

To avoid mechanical vibrations, the following inequalities have to be in force:

$$\begin{aligned} Q_r &\neq 6pg \pm 1 \\ Q_r &\neq 6pg \pm 2p \pm 1 \\ Q_r &\neq 6pg \pm 2p \mp 1. \end{aligned} \tag{7.129}$$

The above avoidable slot numbers hold basically when there is no rotor skewing. If the rotor slots are skewed with respect to the stator, the harmonic torques and vibrations are more or less weakened. For choosing the slot number of the rotor, the following equation is given. It depends on the slots per pole and phase q_s of the stator and the number of pole pairs p, when the rotor bars are skewed for the amount of one stator slot pitch

$$Q_r = (6q_s + 4)\, p = Q_s + 4p, \tag{7.130}$$

where Q_s is the number of the stator slots.

Tables 7.3 and 7.4 illustrate examples of the selection of the rotor slot number for a rotor with straight slots for the pole pairs $p = 1$, 2 and 3 with different numbers of stator slots. A ❍ symbol indicates the slot numbers that predict particularly harmful synchronous torques when the motor is held at stall. A + sign indicates the slot numbers that predict particularly harmful synchronous torques at positive rotation speeds. A – sign shows those slot numbers that predict particularly harmful synchronous torques at negative rotation speeds (in counter-current braking). A × symbol indicates pairs of slot numbers with harmful mechanical vibrations.

Tables 7.3 and 7.4 show that there are only a few possible combinations. For instance, for a two-pole machine there are in practice no suitable stator–rotor slot number combinations for general use. The slot number of the rotor has to be selected individually in such a way that disturbances are minimized.

If no strict limits are set for the start-up noise, we may apply odd rotor bar numbers that, according to the tables, produce mechanical vibrations (×). When $Q_s = 24$, the slot numbers $Q_r = 19$, 27 and 29 may be noisy at the start-up. Correspondingly, the slot number pairs $Q_s = 36$ and $Q_r = 31$ and also $Q_s = 48$ and $Q_r = 43$ may be noisy. Even slot numbers usually provide a more silent running. If the motor is not designed for counter-current braking (s is always < 1), for instance, we may select a rotor slot number at a – sign, the permitted slot numbers for a 24-slot stator then being $Q_r = 10$, 16, 22, 28 and 34. For a 36- or 48-slot stator we may also select $Q_r = 40$, 46, 52 and 58, and for a 48-slot stator further $Q_r = 64$.

Table 7.3 Selection of the slot number. Only the combinations without a symbol are safe choices. There are some disadvantages with all the other combinations. −, harmful torques in counter-current braking; +, harmful torques at positive speeds; ×, harmful mechanical vibrations; ○, harmful synchronous torques at a standstill

Q_s	Tens of the slot number of the rotor	Number of pole pairs $p = 1$ Ones of the rotor slot number Q_r										Number of pole pairs $p = 2$ Ones of the rotor slot number Q_r									
		0	1	2	3	4	5	6	7	8	9	0	1	2	3	4	5	6	7	8	9
	1	−	×	○	×	+	×	−	×	○	×	−	×	○	×	+	×	+	×		×
24	2	+	×	−	×	○	×	+	×	−	×	−	×	−	×	○	×	+	×	+	×
	3	○	×	+	×	−	×	○	×	+	×		×	−	×		×	○	×		×
	1	−	×	○	×	+	×	−	×	○	×		×	○	×		×	∓	×	○	×
	2	+	×	−	×	○	×	+	×	−	×	±	×		×	○	×		×	+	×
36	3	○	×	+	×	−	×	○	×	+	×		×	−	×	−	×	○	×	+	×
	4	−	×	○	×	+	×	−	×	○	×	+	×		×	−	×		×	○	×
	5	+	×	−	×	○	×	+	×	−	×		×	+	×		×	−	×		×
	1	−	×	○	×	+	×	−	×	○	×		×	○	×		×	+	×		×
	2	+	×	−	×	○	×	+	×	−	×	−	×	−	×	○	×	+	×	+	×
48	3	○	×	+	×	−	×	○	×	+	×		×	−	×		×	○	×		×
	4	−	×	○	×	+	×	−	×	○	×	+	×		×	−	×	−	×	○	×
	5	+	×	−	×	○	×	+	×	−	×	+	×	+	×		×	−	×		×
	6	○	×	+	×	−	×	○	×	+	×	○	×		×	+	×		×	−	×

Source: Adapted from Richter (1954).

When $Q_s = 24$, the slot numbers $Q_r = 12$, 18 and 24 in a $p = 1$ machine have to be avoided at all instances, since they produce so high synchronous torques at stall that the machines will not start up. $Q_r = 26$ produces intolerably high synchronous torques at positive rotation speeds. Further, if the motor is also running at slips $s > 1$, the slot number $Q_r = 22$ cannot be selected under any circumstances. An example of a two-pole industrial machine is a 4 kW, 3000 min^{-1} motor produced by ABB, which is equipped with a combination of slot numbers 36/28 and also the skewing of rotor bars.

Table 7.5 illustrates the recommended numbers of slot pairs when using a skewing of one to two stator slot pitches. The numbers for rotor slots are given in the order of superiority, the best first.

7.2.7 Construction of an Induction Motor

The IEC has developed international standards for the mounting dimensions of electrical machines and defined the respective codes. The machines with equal codes are physically interchangeable. When the shaft height of the machine is 400 mm at the maximum, the IEC code for foot-mounted electrical machines comprises the frame code and the diameter of the free shaft end. The frame code consists of the shaft height and the letter code S, M, or L, which

Table 7.4 Selection of the rotor slot number. (Only the combinations without any symbol are safe choices. There are some disadvantages with all the other combinations)

Q_s	Tens of the slot number of the rotor	Number of pole pairs $p = 3$ Ones of the rotor slot number Q_r									
		0	1	2	3	4	5	6	7	8	9
	1		×	−	×		−		×	○	×
	2		+		×	+	×				×
36	3	−	×		−		×	○	×		+
	4		×	+	×				×	−	×
	5				×	○	×				×
	1		×	−	×				×	○	×
	2				×	○	×		○		×
	3	±	×				×	○	×		
54	4		×	+	×				×	−	×
	5		−		×	○	×		+		×
	6	+	×				×	−	×		
	7		×	○	×				×	+	×
	1		×	−	×				×	○	×
	2				×	+	×				×
	3	−	×		−		×	○	×		+
	4		×	+	×				×	−	×
72	5				×	○	×				×
	6	+	×				×	−	×		−
	7		×	○	×		+		×	+	×
	8				×	−	×				×
	9	○	×				×	+	×		

Source: Adapted from Richter (1954).

expresses the length of the frame. The code 112 M 28 thus designates a machine for which the height of the center point of the shaft, measured from the level defined by the lower surface of the feet, is 112 mm; the stator stack of the machine belongs to the medium-length class; and the diameter of the free shaft is 28 mm. If the frame code does not involve a letter code, the dimensions are separated with dash, for instance 80−19. Table 7.6 contains some IEC codes for small machines.

If a foot-mounted machine is also equipped with a connection flange, the code of the flange is given, for instance 112M 28 FF215. Those machines suitable for only flange mounting are designated according to the shaft diameter and the flange, for instance 28 FF215. IEC 60034−8 (2007) defines the terminal markings and the direction of rotation for a rotating machine and the dependency between these two when the machine is connected to the network. Figure 7.18 illustrates the terminal markings of the windings of ordinary three-phase rotating-field machines. In addition to these markings, the markings of the magnetizing windings of a synchronous machine are F1–F2.

Table 7.5 Most advantageous slot numbers for rotors with slots skewed for 1–2 stator slot pitches

p	Q_s	Q_r
1	24	28, 16, 22
	36	24, 28, 48, 16
	48	40, 52
	60	48
2	36	24, 40, 42, 60, 30, 44
	48	60, 84, 56, 44
	60	72, 48, 84, 44
3	36	42, 48, 54, 30
	54	72, 88, 48
	72	96, 90, 84, 54
4	36	48
	48	72, 60
	72	96, 84

Source: Adapted from Richter (1954).

The IEC 60034-7 standard defines the shaft ends of a rotating machine with letter codes D (Drive end) and N (Non-Drive end). The rotation direction of the motor shaft is observed either clockwise or counterclockwise at the D end. The internal coupling of three-phase electrical machines is carried out in such a way that when the network phases L1, L2 and L3 are connected to the terminals U_1, V_1, and W_1, respectively, the machine rotates clockwise when observed standing in front of the machine D end (see the illustration of Table 7.6).

The most common asynchronous machines are enclosed squirrel cage winding induction motors of enclosure class IP 55. As typical examples of these machines, Figures 7.19a and b illustrate the profiles of two totally enclosed fan-cooled induction motors of different sizes manufactured by ABB.

7.2.8 Cooling and Duty Types

In electrical machines, part of the supplied energy is always converted into heat. For instance, when the efficiency of a 4 kW standard induction machine at a rated power is 85 %, the following percentages of the energy supplied to the terminals of the machine turn into heat: 6.9 % in the stator resistive losses, 1.9 % in the stator iron losses, 0.5 % in the additional losses and 4.7 % in the rotor losses. The rest of the supplied energy is converted to mechanical energy but about 1 % is still lost in the mechanical losses. The heat generated in the machine has to be led to the medium surrounding the machine. The cooling methods of the electrical machines are defined in IEC 60034–6 and the enclosure classes in IEC 60034–5. The class of enclosure depends on the cooling method. For instance, the enclosure class IP 44 designates good mechanical and moisture protection that are not compatible with the cooling method IC 01, since it requires an open machine. Table 7.7 gives the most common IC classes.

The duty types of electrical machines are designated as S1–S9 and described according to IEC 60034–1 (2004).

Table 7.6 Mechanical dimensions of electrical machines according to the IEC codes

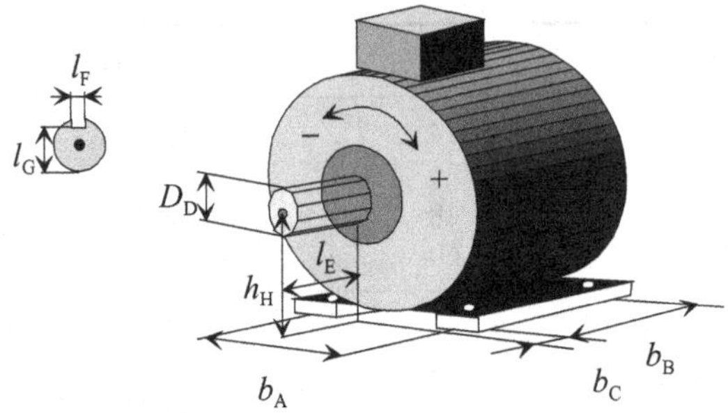

IEC Code	b_A [mm]	b_B [mm]	b_C [mm]	D_D [mm]	l_E [mm]	l_F [mm]	l_G [mm]	h_H [mm]	Fixing screws
71-14	112	90	45	14	30	5	11	71	M6
80-19	125	100	50	19	40	6	15.5	80	M8
90S24	140	100	56	24	50	8	20	90	M8
90L24	140	125	56	24	50	8	20	90	M8
100L28	160	140	63	28	60	8	24	100	M10
112M28	190	140	70	28	60	8	24	112	M10
132S38	216	140	89	38	80	10	33	132	M10
132M38	216	178	89	38	80	10	33	132	M10
160M42	254	210	108	42	110	12	37	160	M12
160L42	254	254	108	42	110	12	37	160	M12
180M48	279	241	121	48	110	14	42.5	180	M12
180L48	279	279	121	48	110	14	42.5	180	M12
200M55	318	267	133	55	110	16	49	200	M16
200L55	318	305	133	55	110	16	49	200	M16
225SM, $p = 1$	356	286	149	55	110,	16,	48,	225	M18
$p > 1$				60	140	18	53		
250SM, $p = 1$	406	311	168	60	140	18	53,	250	$\phi 24$*
$p > 1$				65			58		
280SM, $p = 1$	457	368	190	65	140	18	58,	250	$\phi 24$*
$p > 1$				75			67.5		
315SM, $p = 1$	508	406	216	65,	140,	18,	58,	315	$\phi 30$*
$p > 1$				80	170	22	71		
315ML, $p = 1$	508	457	216	65,	140,	18,	58,	315	$\phi 30$*
$p > 1$				90	170	25	81		
355S, $p = 1$	610	500	254	70,	140,	20,	62.5,	355	$\phi 35$*
$p > 1$				100	210	28	90		
355SM, $p = 1$	610	500	254	70,	140,	20,	62.5,	355	$\phi 35$*
$p > 1$				100	210	28	90		
35ML, $p = 1$	610	560	254	70,	140,	20,	62.5,	355	$\phi 35$*
$p > 1$				100	210	28	90		
400M, $p = 1$	686	630	280	70,	140,	20,	62.5,	400	$\phi 35$*
$p > 1$				100	210	28	90		
400LK, $p = 1$	686	710	280	80,	170,	22,	71,	400	$\phi 35$*
$p > 1$				100	210	28	90		

* Fixing hole diameter.

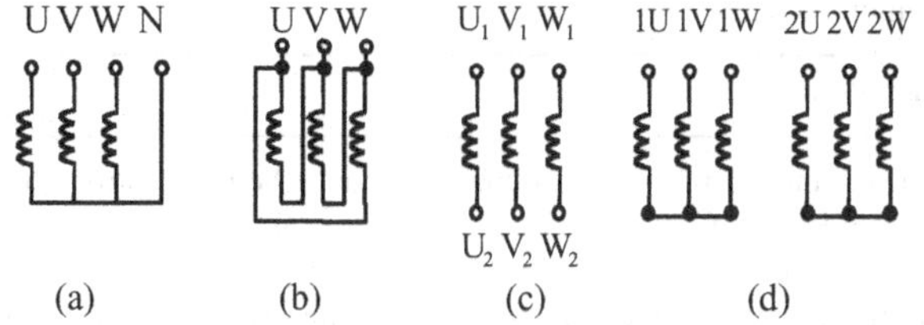

Figure 7.18 Connections and terminal markings of three-phase machines. (a) The machine is connected in star. The neutral point N is not necessarily available. (b) The machine is connected in delta. (c) The ends of the phase windings (U_1–U_2) of the machine are connected to a terminal board. If there are intermediate taps in the windings, they are marked U_3–U_4. The connection of the machine in star or in delta takes place either in the terminal box or completely outside the machine. (d) The terminal markings of a two-speed machine. The higher front number designates the winding of the higher rotation speed of the machine.

7.2.8.1 Duty Type S1 – Continuous Running Duty

This deals with operation at constant load maintained for sufficient time to allow the machine to reach thermal equilibrium. A machine for this kind of operation is stamped with the abbreviation S1. This is the most common duty-type stamping in industrial machines irrespective of the real use of the motor.

7.2.8.2 Duty Type S2 – Short-Time Duty

This covers operation at constant load for a given time, less than that required to reach thermal equilibrium. Each operation period is followed by a time at rest and de-energized, with a sufficient duration to re-establish the temperature of the surrounding air. For machines of short-time duty, the recommended durations of the duty are 10, 30, 60 and 90 minutes. The appropriate abbreviation is S2, followed by an indication of the duration of the duty, for

Figure 7.19 (a) ABB aluminum motor, shaft height 132 mm. 5.5 kW, 3000 min^{-1}. (b) ABB cast iron motor, shaft height 280 mm, 75 kW, 3000 min^{-1}. Reproduced by permission of ABB Oy.

Table 7.7 Most common IC classes of electrical machines

Code	Definition
IC 00	The coolant surrounding the machine cools the inner parts of the machine. The ventilating effect of the rotor is insignificant. The coolant is transferred by free convection.
IC 01	As IC 00, but there is an integral fan mounted on the shaft or the rotor to circulate the coolant. This is a common cooling method of open induction motors.
IC 03	A method similar to IC 01, but with a separate motor mounted blower having same power source with the machine to be cooled.
IC 06	A method similar to IC 01, but the coolant is circulated with a separate motor-mounted blower with a different power source. There can also be a single extensive blower system supplying the coolant for several machines.
IC 11	The coolant enters the machine via a ventilating duct and passes freely to the surrounding environment. The circulation of the coolant is carried out with a motor- or shaft-mounted blower.
IC 31	The rotating machine is inlet- and outlet-pipe-ventilated. The circulation of the coolant is carried out with a motor- or shaft-mounted blower.
IC 00 41	Totally enclosed internal circulation of the coolant by convection and cooling through the frame with no separate blower.
IC 01 41	As IC 00 41, but the frame-surface cooling takes place with a separate shaft-mounted blower causing the circulation of the coolant. This is a cooling method of ordinary enclosed induction motors.
IC 01 51	Totally enclosed internal cooling by convection. The heat is transferred through an internal air-to-air heat exchanger to the surrounding medium, which is circulated by a shaft-mounted blower.
IC 01 61	As IC 01 51, but the heat exchanger is mounted on the machine.
IC W37 A71	Totally enclosed internal cooling by convection. The heat is transferred through an internal water-to-air heat exchanger to the cooling water, which is circulated either by supply pressure or an auxiliary pump.
IC W37 A81	As IC W37 A71, but the heat exchanger is mounted on the machine.

instance stamping S2 60 min. Stampings for short periods of use, for instance S2 10 min, are common in motors integrated in cheap tools such as small grinders.

7.2.8.3 Duty Type S3 – Intermittent Periodic Duty

This concerns a sequence of identical duty cycles, each including a time of operation at constant load and a time at rest and de-energized. Thermal equilibrium is not reached during a duty cycle. Starting currents do not significantly affect the temperature rise. The cyclic duration factor is 15, 25, 40 or 60 % of the 10 min duration of the duty. The appropriate abbreviation is S3, followed by the cyclic duration factor, for instance the stamping S3 25 %.

7.2.8.4 Duty Type S4 – Intermittent Periodic Duty with Starting

This covers a sequence of identical duty cycles, each cycle including a significant starting time, a time of operation at constant load and a time at rest and de-energized. Thermal equilibrium is not reached during a duty cycle. The motor stops by naturally decelerating, and thus it is not thermally stressed. The appropriate abbreviation for the stamping is S4, followed by the cyclic duration factor, the number of cycles in an hour (c/h), the moment of inertia of the motor (J_M), the moment of inertia of the load (J_{ext}) referred to the motor shaft, and the permitted average counter torque T_v during a change of speed given by means of the rated torque. For example, the stamping S4 – 15 % – 120 c/h – $J_M = 0.1$ kg m^2 – $J_{ext} = 0.1$ kg m^2 – $T_v = 0.5\ T_N$.

7.2.8.5 Duty Type S5 – Intermittent Periodic Duty with Electric Braking

This covers a sequence of identical duty cycles, each cycle consisting of a starting time, a time of operation at constant load, a time of braking and a time at rest and de-energized. Thermal equilibrium is not reached during a duty cycle. In this duty type, the motor is decelerated with electric braking, for instance counter-current braking. The appropriate abbreviation for stamping is S5, followed by the cyclic duration factor, the number of cycles per hour (c/h), the moment of inertia of the motor J_M, the moment of inertia of the load J_{ext}, and the permitted counter torque T_v. For example, the stamping S5 – 60 % – 120 c/h – $J_M = 1.62$ kg m^2 – J_{ext} = 3.2 kg m^2 – $T_v = 0.35\ T_N$.

7.2.8.6 Duty Type S6 – Continuous Operation Periodic Duty

This covers a sequence of identical duty cycles, each cycle consisting of a time of operation at constant load and a time of operation at no load. Thermal equilibrium is not reached during a duty cycle. The cyclic duration factor is 15, 25, 40 or 60 % and the duration of the duty is 10 min. The stamping for instance is S6 60 %.

7.2.8.7 Duty Type S7 – Continuous Operation Periodic Duty with Electric Braking

This concerns a sequence of identical duty cycles, each cycle consisting of a starting time, a time of operation at constant load and a time of electric braking. The motor is decelerated by counter-current braking. Thermal equilibrium is not reached during a duty cycle. The appropriate abbreviation is S7, followed by the moment of inertia of the motor, the moment of inertia of the load and the permitted counter torque (cf. S4). The stamping for instance is S7 – 500 c/h – $J_M = 0.06$ kg m^2 – $T_v = 0.25\ T_N$.

7.2.8.8 Duty Type S8 – Continuous Operation Periodic Duty with Related Load/Speed Changes

This covers a sequence of identical duty cycles, each cycle consisting of a time of operation at constant load corresponding to a predetermined speed of rotation, followed by one or more times of operation at other constant loads corresponding to different speeds of rotation (carried out, for example, by means of a change in the number of poles in the case of induction motors). There is no time at rest and de-energized. Thermal equilibrium is not reached during a duty cycle. The appropriate abbreviation is S8, followed by the moment of inertia of the motor, the moment of inertia of the load and the number of duty cycles in an hour. Also a permitted counter torque and the cyclic duration factor have to be given. Stamping for instance is

S8 – $J_M = 2.3$ kg m^2 – $J_{ext} = 35$ kg m^2
30 c/h – $T_v = T_N$ – 24 kW – 740 r/min – 30 %
30 c/h – $T_v = 0,5\ T_N$ – 60 kW – 1460 r/min – 30 %
30 c/h – $T_v = 0,5\ T_N$ – 45 kW – 980 r/min – 40 %.

The load and combinations of rotation speeds are stamped in the order in which they occur in the duty.

7.2.9 Examples of the Parameters of Three-Phase Industrial Induction Motors

In the following, typical values of equivalent circuit parameters of commercial induction motors are given. Figure 7.20 illustrates the average behavior of the per unit magnetizing inductances of certain induction motors as a function of power. The motors have been produced by ABB.

A second significant parameter in the equivalent circuit of the machine is the flux leakage of the stator. Figure 7.21 illustrates the per unit leakage inductances of the same machines manufactured by ABB as in the previous figure. The values for the flux leakage are calculated here by employing the catalogue values of the machine, for instance for the pull-out torque.

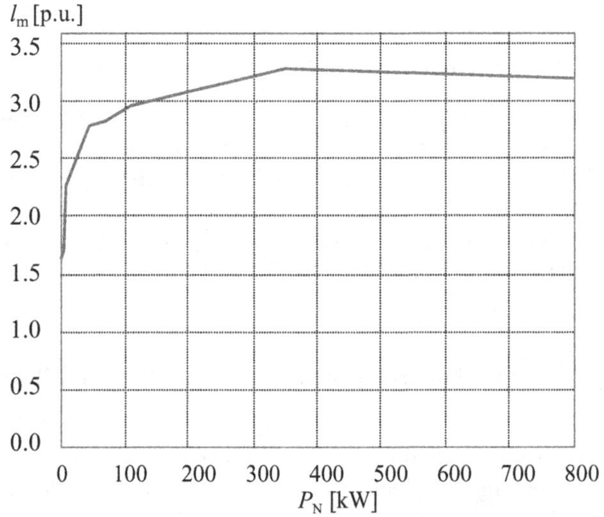

Figure 7.20 Average behavior of the per unit magnetizing inductances of four-pole induction motors as a function of shaft output power. The curve is composed of the values of 1.1, 2.2, 5.5, 11, 45, 75, 110, 355 and 710 kW machines. Reproduced by permission of Markku Niemelä.

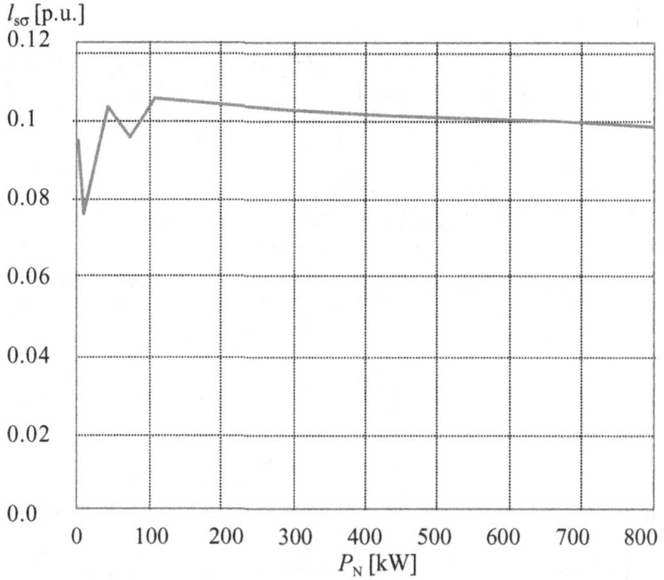

Figure 7.21 Average behavior of the per unit leakage inductances of four-pole induction motors as a function of shaft output power. The curve is composed of the values of 1.1, 2.2, 5.5, 11, 45, 75, 110, 355 and 710 kW machines. Reproduced by permission of Markku Niemelä.

The rotor flux leakage per unit inductance at rated load is typically of the same magnitude as the stator flux leakage. For manufacturing reasons, closed rotor slots are often used in die-cast rotors, which makes the rotor leakage behavior somewhat complicated. At low slips, the rotor leakage inductance is high, but at larger loads the leakage flux path saturates and similar per unit values are found as for the stator.

The behavior of the stator resistance is a significant factor in the machines. Figure 7.22 illustrates the behavior of the stator resistance of induction motors as a function of shaft output power. The figure also reveals how difficult it is to compare small and large machines, unless the resistance is calculated as a per unit value.

Figure 7.22 shows that in small machines, the resistance is a very significant component. Already at rated current there is an 11 % resistive loss in the stator resistance of a 1.1 kW machine. Therefore, the efficiency cannot reach very high values either. With a full logarithmic scale, both the absolute and per unit resistances seem to behave linearly.

The rotor resistances are typically of the same magnitude as the stator resistance in induction motors. The resistive loss of the rotor is proportional to the slip s and takes place in the resistance R_r of the rotor. The magnitude of the rotor resistive loss is $P_{Cur} = I_r^2 R_r$. In small machines, the rotor resistance is proportionally smaller than the stator resistance. This is indicated in Figure 7.23, where the different losses of the machine are illustrated as their proportional shares. A small power factor in a small machine leads to the fact that the stator loss in small machines is relatively high.

At a power of just 90 kW, the resistive losses of the rotor and stator approach each other. Bearing in mind that the stator current is slightly higher than the rotor current referred to the

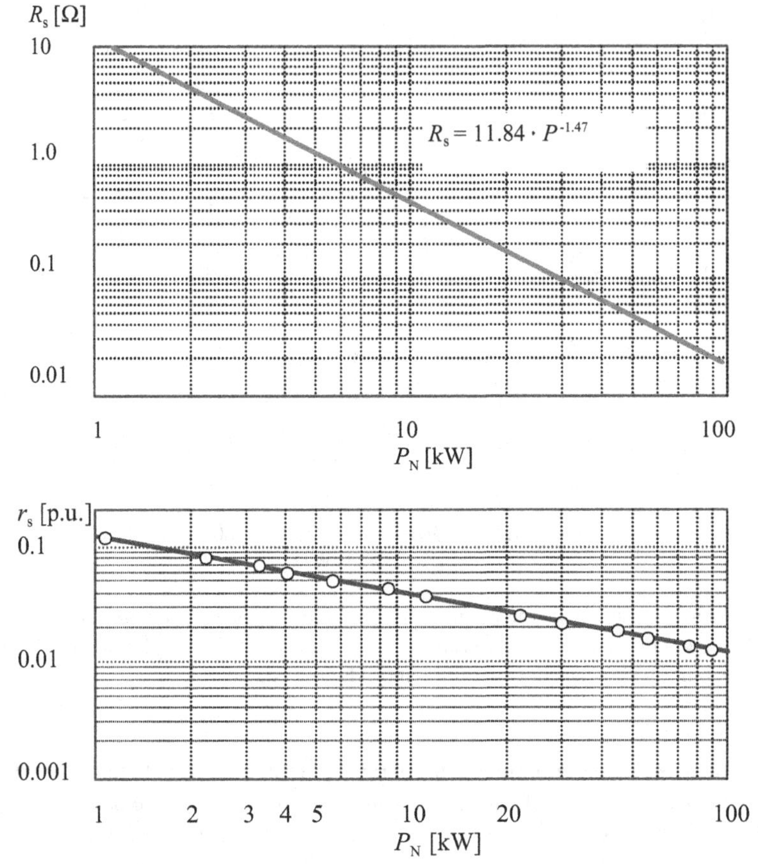

Figure 7.22 Stator resistances of typical modern, totally enclosed industrial induction motors from ABB as a function of shaft output power, given as absolute and per unit values. Reproduced by permission of Jorma Haataja.

stator side $I_r' \approx I_s \cos \varphi$, we may infer from the loss distribution that the stator-referred rotor resistances are typically of the same magnitude as stator resistances.

Figure 7.24 illustrates how the power factor of the induction motors of ABB's M3000 series behaves as a function of power, the number of pole pairs being the parameter.

The increase in the number of pole pairs notably reduces the power factors of the induction motors. This is due to the fact that the magnetizing inductance of a rotating field machine is inversely proportional to the square of the number of pole pairs. As the magnetizing inductance decreases, the machine consumes more reactive current, and consequently the power factor is lower.

7.2.10 *Asynchronous Generator*

If desired, an asynchronous machine is applicable also as a generator. When connected to the network, it has to be rotated by a power engine at a speed higher than the synchronous speed.

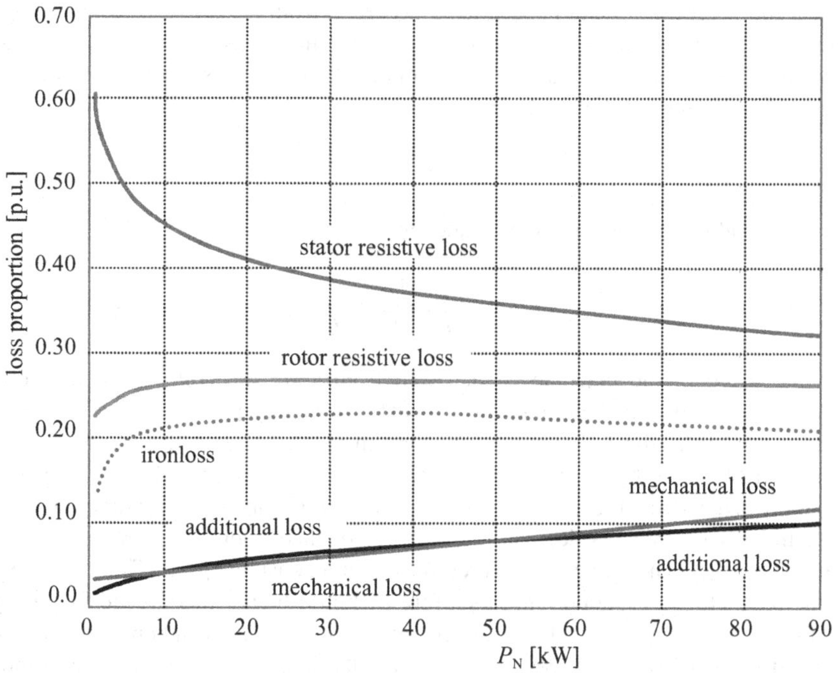

Figure 7.23 Average loss distributions of induction motors at powers below 100 kW. The proportion of the losses is referred to the machine losses at the rated operating point. Adapted from Auinger (1997).

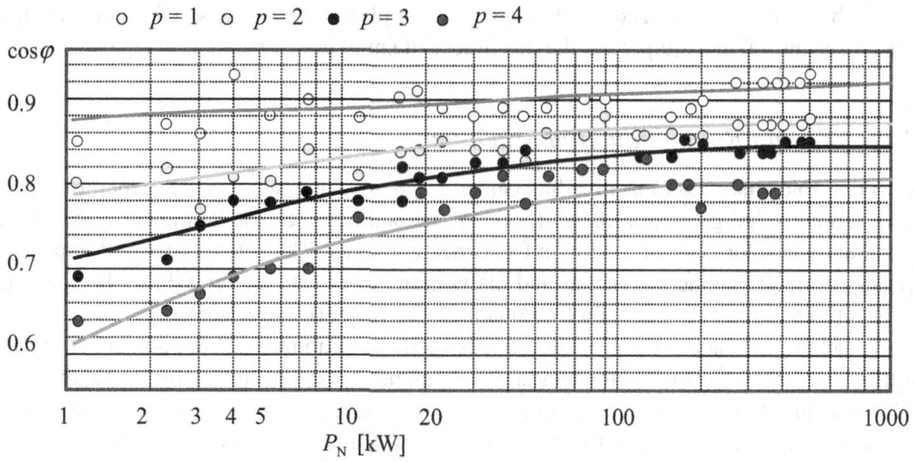

Figure 7.24 Power factors of totally enclosed induction motors as a function of power, with the number of pole pairs as a parameter (ABB M3000).

An asynchronous generator takes its magnetizing current from the network and, usually, has to be compensated by a capacitor bank. Under certain circumstances, an asynchronous machine is also used as a capacitor-excited island generator. The capacitors have to be in resonance with the total inductance of the machine,

$$|X_{\mathrm{C}}| = |X_{\mathrm{m}} + X_{\mathrm{s}\sigma}|. \tag{7.131}$$

The corresponding resonant angular frequency is

$$\omega = \frac{1}{\sqrt{(L_{\mathrm{m}} + L_{\mathrm{s}\sigma})C}}. \tag{7.132}$$

When the speed of the machine is increased from zero, the build up of the generator voltage requires a remanence of the rotor or an external energy pulse supplied to the system. The current flowing in the *LC* circuit of the system is very low until the resonant angular frequency is approached. In the vicinity of the resonance, the current of the machine increases rapidly and it magnetizes itself. The voltage is rapidly built up. As the saturation increases, the magnetizing inductance is decreased and the resonance frequency is increased, and the terminal voltage of the machine remains reasonable at a constant speed. The load affects the size of the magnetizing capacitor, and therefore it should be adjusted as the frequency and the load fluctuate. However, asynchronous generators intended for island operation are usually dimensioned in such a way that their magnetic circuits saturate strongly as soon as the rated voltage is exceeded at the rated speed. In that case the capacitance of a capacitor can be increased quite high without changing the terminal voltage. Thus we are able to construct an asynchronous generator for island operation, the terminal voltage of which remains almost constant as the load varies.

Asynchronous generators equipped with squirrel cage rotors are fairly common. Due to their reliability, they are utilized in unattended power plants. Since an asynchronous machine needs no synchronizing equipment, the machinery of small power plants is kept cost effective.

7.2.11 Wound Rotor Induction Machine

The wound rotor induction machine has gained popularity as a wind power generator. In the case of a doubly-fed generator, the stator of the induction generator is connected directly to the grid and the rotor is controlled by a frequency converter. The rotor distributed winding is internally normally connected in star and the winding terminals are brought to the slip rings and from there to the terminals of a four quadrant converter (Figure 7.25).

The machine can be driven in asynchronous mode with about ±30 % slip. This indicates that the power driven via the rotor winding can be about 30 % of the stator power, therefore resulting in an efficient drive and a limited power converter. Six pole generators are popular in the megawatt power range. With generator stator winding connected to a 50 Hz network at 1000 min^{-1} speed the generator works as a synchronous generator while the rotor has to be supplied by a DC-current.

When the turbine power is high the generator runs at super-synchronous speeds (higher than the synchronous speed) and the rotor converter has to operate as an active rectifier and the network converter then feeds in the slip power of the doubly-fed generator to the grid. At

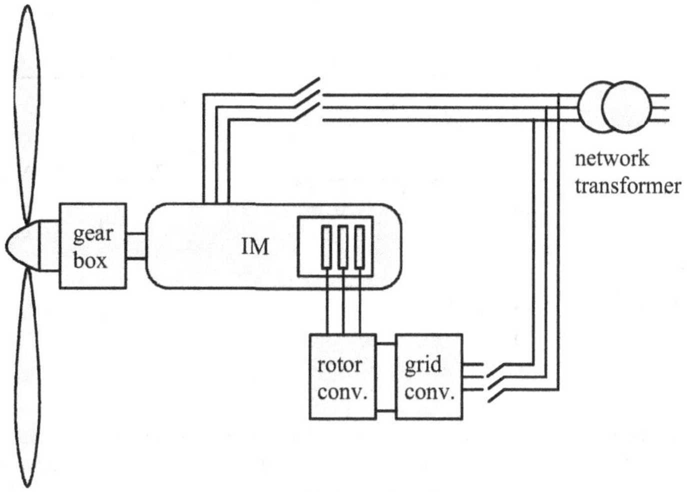

Figure 7.25 Doubly-fed induction generator drive for wind power generation.

subsynchronous (lower than the synchronous) speed the rotor converter has to actively supply the rotor with a three-phase current. This makes the subsynchronous generator mode possible.

The design of a doubly fed generator follows the induction machine design rules. However, the rotor voltage must be matched according to the maximum allowable slip of the machine and the converter ratings. Contrary to normal squirrel cage motors the rotor of a doubly-fed generator will also have significant iron loss while operating at a high slip. At 30 % slip the rotor frequency in a 50 Hz network will be 15 Hz.

7.2.12 Asynchronous Motor Supplied with Single-Phase Current

A winding supplied with single-phase current creates a flux density which does not rotate, but pulsates and can be illustrated as phasor pairs rotating in opposite directions in the air gap. Next, the operation of a machine is analyzed with respect to a fundamental pair of phasors $\nu = \pm 1$. Of these $\nu = +1$ is considered the actual positive-sequence field. In that case, the equivalent circuit of a single-phase machine appears as illustrated in Figure 7.15 with the following adjustments. Since there is only one stator winding in the machine the magnetizing reactances X_{sm1} and X_{sm5} etc. have to be replaced with halves of the main reactance $X_{\text{pD}} = \omega L_{\text{p}}$ (see Chapter 4). The leakage reactance of the rotor at a slip $s = 1$ is equal for the positive- and negative-phase-sequence field. The only difference lies in the resistances of the rotor circuit. The positive-sequence resistance is 0.5 R'_{r}/s, and the negative-sequence resistance is 0.5 $R'_{\text{r}}/(2-s)$ (see Figure 7.26). Phasor calculation is the easiest method to solve the impedances, and thus the positive and negative-sequence impedances of the rotor can be written in the form

$$\begin{aligned} \underline{Z}'_{\text{r,ps}} &= \frac{0.5R'_{\text{r}}}{s} + \text{j}0.5X'_{\text{r}\sigma}, \\ \underline{Z}'_{\text{r,ns}} &= \frac{0.5R'_{\text{r}}}{2-s} + \text{j}0.5X'_{\text{r}\sigma}. \end{aligned} \tag{7.133}$$

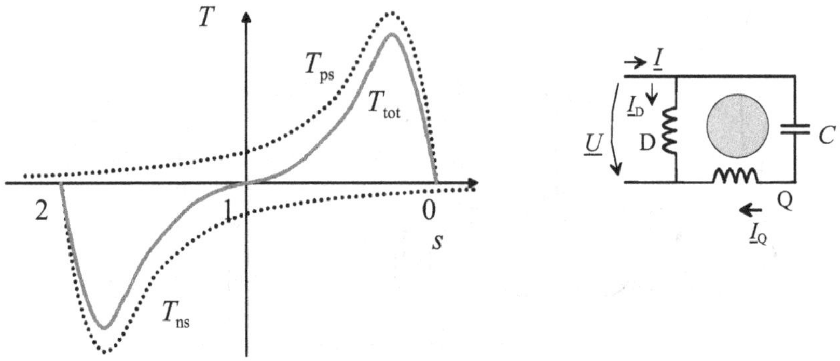

(a) (b)

Figure 7.26 (a) Static torque curves of a single-phase induction motor and the sum curve of the torque as a function of slip. T_{ps} is the positive-sequence torque and T_{ns} the negative-sequence torque. T_{tot} is the sum of these components. At zero speed, the sum of torques is zero ($T_{tot} = 0$). (b) Model of a stator winding of a single-phase supplied capacitor motor.

Now, the positive and negative-sequence impedances are obtained

$$\underline{Z}_{ps} = \frac{j0.5X_{pD}\underline{Z}'_{r,ps}}{j0.5X_{pD} + \underline{Z}'_{r,ps}},$$
$$\underline{Z}_{ns} = \frac{j0.5X_{pD}\underline{Z}'_{r,ns}}{j0.5X_{pD} + \underline{Z}'_{r,ns}}. \tag{7.134}$$

When the stator impedance $\underline{Z}_s = R_s + jX_{s\sigma}$ is added to the sum of these impedances, we obtain the total impedance $\underline{Z}$ of the machine, which is connected to the terminal voltage. Then we obtain the current of the machine. The emf components of the machine are $\underline{E}_{ps} = \underline{Z}_{ps}\underline{I}_{ps}$ and $\underline{E}_{ns} = \underline{Z}_{ns}\underline{I}_{ns}$. They can be employed in the calculation of the corresponding components of the rotor currents

$$\underline{I}'_{r,ps} = \frac{\underline{Z}_{ps}\underline{I}_{ps}}{\underline{Z}'_{r,ps}},$$
$$\underline{I}'_{r,ns} = \frac{\underline{Z}_{ns}\underline{I}_{ps}}{\underline{Z}'_{r,ns}}. \tag{7.135}$$

The resistive losses in the rotor resistances created by these components are

$$P_{rCu,ps} = I'^2_{r,ps}\frac{0.5R'_r}{s},$$
$$P_{rCu,ns} = I'^2_{r,ns}\frac{0.5R'_r}{2-s}. \tag{7.136}$$

These resistive loss powers both represent opposite torques. The resultant torque at a slip s is (see Equation (7.87))

$$T = \frac{p}{\omega}\left(\frac{P_{\mathrm{rCu,ps}}}{s} - \frac{P_{\mathrm{rCu,ns}}}{2-s}\right). \tag{7.137}$$

When the machine is at stall $s = 1$, the torques are equal, and therefore the machine does not start up without assistance. At synchronous slip $s = 0$, the positive-sequence impedance is $\underline{Z}_{\mathrm{ps}} = \mathrm{j}0.5X_{\mathrm{p}}$ and the negative-sequence impedance $|\underline{Z}_{\mathrm{ns}}| \ll |\underline{Z}_{\mathrm{ps}}|$, and thus $E_{\mathrm{ps}} \gg E_{\mathrm{ns}}$. $0.5X_{\mathrm{p}}$ $(0.5\omega L_{\mathrm{p}})$ is used instead of X_{m} because of the single-phase arrangement. When both the emf components are induced in the same coil, it means that at low values of slip the torque of the counter-rotating field is low. It nearly disappears and the machine operates almost like a symmetrically supplied poly-phase machine. As the resultant emf has to be almost the magnitude of the terminal voltage, correspondingly the positive-sequence field is so high that it alone induces the required emf. The amplitude of the negative-sequence field is damped almost to zero at low values of slip. Figure 7.26a illustrates the static torque curves of the machine.

This motor type is actually of no use, since a starting torque is lacking. Therefore, there are some starting techniques commonly used to start these motors. An improvement of the torque-speed characteristic can be achieved by an auxiliary winding with a capacitor in series. Such an arrangement will be analyzed below. The single-phase motors are in practice two-phase motors supplied with a single-phase voltage. Figure 7.26b illustrates such a machine. There is a two-phase winding in the stator of the machine. The magnetic axes of the coils D and Q are located at a distance of 90 electrical degrees. There is an additional impedance connected in series with the coil Q, usually a capacitance C, which causes a certain temporal phase shift between the currents of the windings, when the machine is supplied with a single phase. Usually, there is a cage winding in the rotor. There is usually also a different number of turns in the stator windings of a capacitor motor and $N_{\mathrm{Q}} > N_{\mathrm{D}}$. This asymmetry complicates the analysis of the machine's characteristics considerably.

The voltage equations of the main (D) and auxiliary (Q) windings can be written in the form (Matsch and Morgan 1987)

$$\begin{aligned} \underline{U}_{\mathrm{D}} &= (R_{\mathrm{sD}} + \mathrm{j}X_{\mathrm{s}\sigma\mathrm{D}})\underline{I}_{\mathrm{D}} + \underline{E}_{\mathrm{2D}} \\ \underline{U}_{\mathrm{A}} &= (R_{\mathrm{sQ}} + \mathrm{j}X_{\mathrm{s}\sigma\mathrm{Q}} + R_{\mathrm{C}} + \mathrm{j}X_{\mathrm{C}})\underline{I}_{\mathrm{Q}} + \underline{E}_{\mathrm{2Q}}. \end{aligned} \tag{7.138}$$

$R_{\mathrm{sD}} + \mathrm{j}X_{\mathrm{s}\sigma\mathrm{D}} = \underline{Z}_{\mathrm{1D}}$ and $R_{\mathrm{sQ}} + \mathrm{j}X_{\mathrm{s}\sigma\mathrm{Q}} = \underline{Z}_{\mathrm{1Q}}$ are the stator impedances of the main and auxiliary windings. $\underline{E}_{\mathrm{2D}}$ includes, in addition to the two rotating flux components of the main winding, the effects of the two rotating components of the auxiliary winding. $R_{\mathrm{C}} + \mathrm{j}X_{\mathrm{C}}$ is the impedance of the capacitor. The direction of rotation of a capacitor motor is from the auxiliary winding to the main winding, since the capacitor shifts the current of the auxiliary winding temporally ahead of the current of the main winding. When the main winding is open, the voltage equations are written in the form

$$\begin{aligned} \underline{U}_{\mathrm{D}} &= \underline{E}_{\mathrm{2D}} \\ \underline{U}_{\mathrm{A}} &= (R_{\mathrm{sQ}} + \mathrm{j}X_{\mathrm{s}\sigma\mathrm{Q}} + R_{\mathrm{C}} + \mathrm{j}X_{\mathrm{C}})\underline{I}_{\mathrm{Q}} + \underline{E}_{\mathrm{2Q}} \end{aligned} \tag{7.139}$$

where

$$\underline{E}_{2Q} = \underline{E}_{psQ} + \underline{E}_{nsQ} \tag{7.140}$$

is the sum of the positive-sequence component (ps) and the negative-sequence component (ns) of the emf. Since the main winding is located at a distance of 90 electrical degrees from the auxiliary winding, the voltage induced by the positive-sequence flux component of the auxiliary winding has to be 90 electrical degrees lagging, and the voltage induced by the negative-sequence flux component has to be 90 electrical degrees leading. Thus

$$\underline{E}_{2D} = -j\frac{\underline{E}_{psQ}}{K} + j\frac{\underline{E}_{nsQ}}{K}, \tag{7.141}$$

where K (see (7.148) is the ratio of the effective number of coil turns of the main winding to the effective number of coil turns of the auxiliary winding. When both windings are connected to the network, the effect of all flux components has to be taken into account, and the voltage equations are thus written in the form

$$\begin{aligned} \underline{U}_D &= \underline{Z}_{1D}\underline{I}_D + \underline{E}_{psD} - j\frac{\underline{E}_{psQ}}{K} + j\frac{\underline{E}_{nsQ}}{K} + E_{nsD}, \\ \underline{U}_A &= (\underline{Z}_{1Q} + \underline{Z}_C)\underline{I}_Q + \underline{E}_{psQ} + jK\underline{E}_{psD} + \underline{E}_{nsQ} - jK\underline{E}_{nsD}. \end{aligned} \tag{7.142}$$

Figure 7.27 illustrates the equivalent circuit of a capacitor motor constructed based on Equation (7.142). Also this equivalent circuit is based on the equivalent circuit of Figure 7.15, which has been modified to take into account the voltage components created by different

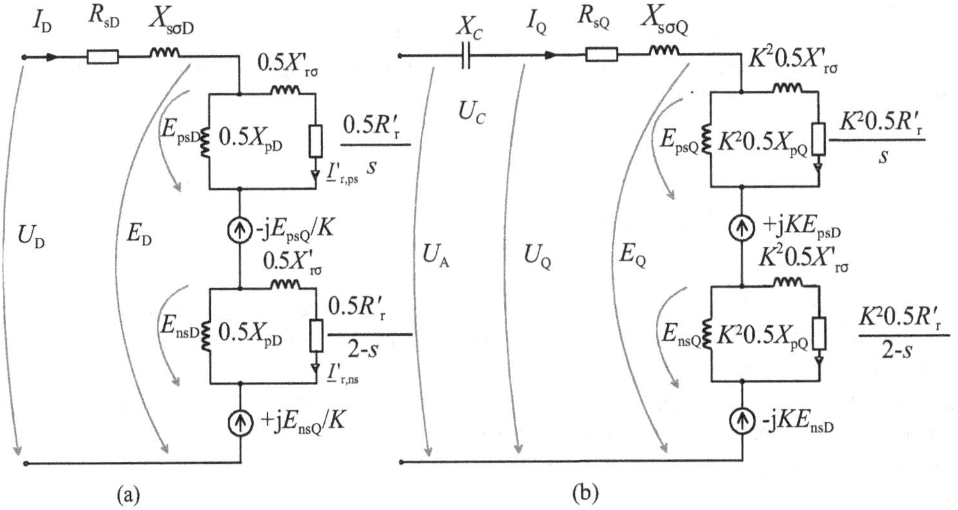

Figure 7.27 Equivalent circuits of the main and auxiliary windings of a capacitor motor. (a) Main winding, (b) auxiliary winding.

axes. Both the rotor quantities of the equivalent circuits of both phases are referred to the number of turns of the main winding.

With the equivalent circuit of Figure 7.27, we may simplify the voltage equations in the form

$$\begin{aligned} \underline{U}_{\mathrm{D}} &= (\underline{Z}_{1\mathrm{D}} + \underline{Z}_{\mathrm{ps}} + \underline{Z}_{\mathrm{ns}})\underline{I}_{\mathrm{D}} - \frac{\mathrm{j}(\underline{Z}_{\mathrm{ps}} - \underline{Z}_{\mathrm{ns}})}{K}\underline{I}_{\mathrm{Q}} \\ \underline{U}_{\mathrm{A}} &= \mathrm{j}K(\underline{Z}_{\mathrm{ps}} - \underline{Z}_{\mathrm{ns}})\underline{I}_{\mathrm{D}} + [\underline{Z}_{\mathrm{C}} + \underline{Z}_{1\mathrm{Q}} + K^2(\underline{Z}_{\mathrm{ps}} + \underline{Z}_{\mathrm{ns}})]\underline{I}_{\mathrm{Q}}. \end{aligned} \tag{7.143}$$

Since the phases of a capacitor motor are usually connected in parallel, we may write

$$\underline{U}_{\mathrm{D}} = \underline{U}_{\mathrm{A}} = \underline{U}. \tag{7.144}$$

The total current supplied in the motor is

$$\underline{I} = \underline{I}_{\mathrm{D}} + \underline{I}_{\mathrm{Q}}. \tag{7.145}$$

The difference of the positive- and negative-sequence power components of both windings is with current complex conjugate values

$$P_{\delta\mathrm{ps}} - P_{\delta\mathrm{ns}} = \mathrm{Re}[(\underline{E}_{\mathrm{psD}} - \underline{E}_{\mathrm{nsD}})\underline{I}_{\mathrm{D}}^{*} + \mathrm{j}(\underline{E}_{\mathrm{psQ}} - \underline{E}_{\mathrm{nsQ}})K\underline{I}_{\mathrm{Q}}^{*}]. \tag{7.146}$$

The torque of the machine is now written as

$$T_{\mathrm{em}} = \frac{p(P_{\delta\mathrm{ps}} - P_{\delta\mathrm{ns}})}{\omega_{\mathrm{s}}}. \tag{7.147}$$

A symmetry should prevail between the phases at the rated operation point of a capacitor machine. The negative-sequence field disappears at a symmetry point where the current linkages of the D- and Q-windings are equal. Such a point may be found at a certain slip. Now the emfs of the windings are phase shifted by 90 electrical degrees, and they are proportional to the numbers of coil turns

$$K = \frac{E_{\mathrm{D}}}{E_{\mathrm{Q}}} = \frac{k_{\mathrm{wD}}N_{\mathrm{D}}}{k_{\mathrm{wQ}}N_{\mathrm{Q}}} \approx \frac{U_{\mathrm{D}}}{U_{\mathrm{Q}}}. \tag{7.148}$$

The voltage of the capacitor is 90 electrical degrees behind the current $\underline{I}_{\mathrm{Q}}$ being thus in the direction of $\underline{I}_{\mathrm{D}}$. On the other hand, the sum $\underline{U}_{\mathrm{Q}} + \underline{U}_{\mathrm{C}} = \underline{U}_{\mathrm{D}}$. A phasor diagram can now be drawn according to Figure 7.28. According to the phasor diagram and Equation (7.148)

$$\tan\varphi_{\mathrm{D}} = \frac{U_{\mathrm{Q}}}{U_{\mathrm{D}}} = \frac{k_{\mathrm{wQ}}N_{\mathrm{Q}}}{k_{\mathrm{wD}}N_{\mathrm{D}}} = \frac{1}{K}, \tag{7.149}$$

$$U_{\mathrm{C}} = \frac{I_{\mathrm{Q}}}{\omega C} = \frac{U_{\mathrm{D}}}{\cos\varphi_{\mathrm{D}}}. \tag{7.150}$$

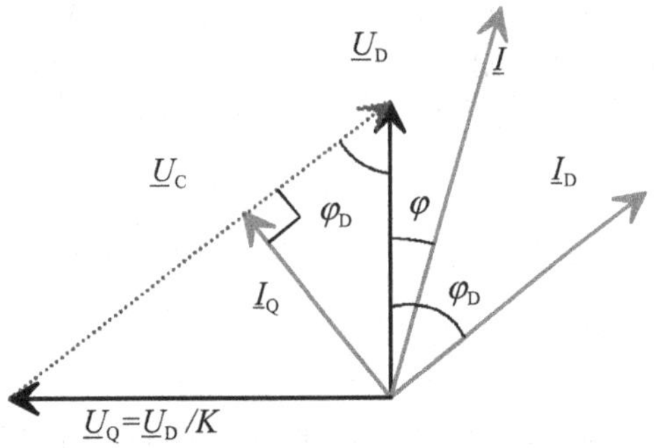

Figure 7.28 Phasor diagram of a capacitor motor at a symmetry point. As $N_D < N_Q$ and $I_D > I_Q$ the current linkages at this point have equal amplitudes and the machine operates with a rotating field.

Since at the symmetry point, the current linkages of the windings are also equal

$$k_{wD}N_DI_D = k_{wQ}N_QI_Q, \tag{7.151}$$

the symmetry requires a capacitance

$$C = \frac{I_D \cos^2 \varphi_D}{\omega U_D \sin \varphi_D}. \tag{7.152}$$

The phasor diagram shows that the voltage of the capacitor is notably higher than the rated voltage of the machine. Therefore, a high voltage rating is required for the capacitor.

The rated point of the machine is usually selected as the symmetry point. The torque requires a certain resistive loss of the rotor P_{rCu} (7.136), half of which is provided by the phase winding D to the rotor. The phase current of the rotor is calculated from this loss by employing the rotor resistance (7.48). When it is referred to the winding D, we obtain the current $I_D \approx I_r'$ (7.52).

Instead of a capacitor, we may also employ a resistance or an inductive coil. However, they do not produce as good starting properties as the capacitor does. In shaded-pole motors, the salient-pole section of the stator is surrounded by a strong copper short-circuit ring, where the induced voltage creates a current temporally delaying the flux of this shaded pole, thus producing the phase shift required for starting. Because of the great losses, this method is applicable only in very small machines.

7.3 Synchronous Machines

Synchronous machines constitute an entire family of electrical machines. The main types of the electrical machine family are (1) separately excited synchronous machines (SMs), (2) synchronous reluctance machines (SyRMs) and (3) permanent magnet synchronous machines (PMSMs). Separately excited machines are either salient or nonsalient-pole machines. In these machines, there is usually a three-phase winding in the stator, whereas the rotor is equipped with a single-phase field winding supplied with direct current. A synchronous machine and a permanent magnet machine differ essentially from an asynchronous machine by the fact that their resulting current linkage component created by the rotor is generated either by field winding direct current or by permanent magnets which, in steady state, work independent of the stator. In asynchronous motors, the rotor currents are induced by the air-gap flux as a slip occurs. In synchronous machines, the air-gap flux change caused by the stator current linkage, that is the armature reaction, is not compensated spontaneously, but the rotor field winding current has to be adjusted if necessary. If a synchronous machine carries a damper winding, its behavior during transients is very similar to the behavior of an asynchronous machine squirrel cage under slip. Since the current linkage of permanent magnets cannot normally be altered, some problems occur in the use of permanent magnet machines. Further, a special feature of permanent magnet machines is that, unlike in machines excited by field winding current, the source of the current linkage itself, the low-permeability permanent magnet material, belongs to the magnetic circuit. Therefore, the magnetizing inductances of permanent magnet machines are relatively low when compared with ordinary synchronous machines. Thus, correspondingly, the armature reaction also remains lower than the armature reaction of separately excited synchronous machines. Table 7.8 lists synchronous machines of different types.

The design of a synchronous machine proceeds in the same way as the design of an asynchronous machine. Section 3.1.2 discusses the different air gaps of a synchronous machine, which together with the windings determine the inductances of synchronous machines. When considering the stationary operation of the machine, the load angle equation forms a significant design criterion.

In the case of rotor-surface-mounted permanent magnets, a permanent magnet machine is, in principle, of the nonsalient-pole type. The magnetic air gap is long and the inductances are low. In the case of embedded magnets instead, the rotor is a salient-pole construction, the inductance of the direct axis being typically lower than the inductance of the quadrature axis ($L_q > L_d$). It is, however, also possible to create a PMSM with $L_d > L_q$ which is, however, quite seldom but has been suggested by some researchers.

Table 7.8 Family of synchronous machines

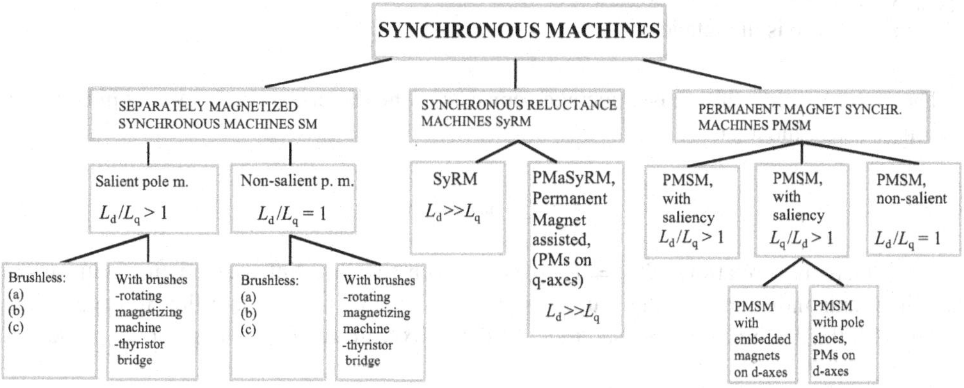

The different brushless excitation systems (a), (b) and (c) are discussed briefly in Section 7.3.9.

In the case of a synchronous reluctance machine, the objective is to create a maximum inductance difference between the direct and quadrature axes. The goal is reached by employing a minimum direct axis and a maximum quadrature air gap. Embedding magnets on the q-axis to reduce the q-axis armature reaction creates a permanent-magnet-assisted SyRM whose properties, especially the power factor, is better than in ordinary SyRM.

7.3.1 Inductances of a Synchronous Machine in Synchronous Operation and in Transients

The inductances chiefly decide the characteristics of synchronous machines. The air gaps of a synchronous machine may vary in direct and quadrature directions, which also leads to different stator inductances in the direct and quadrature directions, L_d and L_q, respectively. These inductances are comprised of the direct and quadrature magnetizing inductance and the leakage inductance $L_{s\sigma}$.

A salient-pole synchronous machine, connected to a three-phase network rotating synchronously but with no rotor field winding current (a synchronous reluctance machine) takes its magnetizing current from the network. A symmetrical three-phase current linkage creates a rotating flux with constant amplitude in the air gap. In the natural position (corresponding to the minimum of the reluctance), the direct axis (d-axis) of the salient-pole rotor is aligned with the axis of the rotating current linkage of the stator. Unlike a nonsalient-pole machine, the salient-pole machine keeps on running at a synchronous speed, and when operating as a motor or as a generator, produces a low torque even without rotor field winding current. A machine of this kind, when running at no-load, takes a current from the network

$$\underline{I}_d = \frac{\underline{U}_{s,ph}}{j\omega_s L_d}. \tag{7.153}$$

Here

$\underline{I}_d$ is the direct-axis current of the stator winding
$\underline{U}_{s,ph}$ is the phase voltage of the stator,
L_d is the direct-axis inductance.

The direct-axis synchronous inductance consists of the direct-axis magnetizing inductance and the leakage inductance

$$L_d = L_{md} + L_{s\sigma}. \tag{7.154}$$

Since there is a relation $d\Psi_{md} = L_{md}\,di_d$ between the flux linkage and the inductance, the direct-axis inductance of the stator can be calculated using the main flux linkage of the machine. If the machine does not saturate, the direct-axis air-gap flux linkage of the phase is

$$\hat{\Psi}_{md} = k_{ws1} N_s \frac{2}{\pi} \tau_p l' \hat{B}_{\delta d}. \tag{7.155}$$

In a saturating machine, another average flux density value of α_i than $2/\pi$ is used. On the other hand, the magnetic flux density of the air gap can be defined with the current linkage of the phase; in other words, the current linkage $\hat{\Theta}_{sd}$ of the stator creates in the effective direct-axis air gap δ_{def} (see Equation (3.56)) a flux linkage

$$\hat{\Psi}_{md} = k_{ws1} N_s \frac{2}{\pi} \frac{\mu_0 \hat{\Theta}_{sd}}{\delta_{def}} \tau_p l'. \tag{7.156}$$

The effective air gap δ_{def} takes into account the slotting and magnetic voltages in iron parts of the magnetic circuit.

The amplitude of the direct-axis current linkage of a single-phase winding, when the current is on the direct axis, is written as

$$\hat{\Theta}_{sd} = \frac{4}{\pi} \frac{k_{ws1} N_s}{2p} \sqrt{2} I_s. \tag{7.157}$$

Substitution yields for the main flux linkage

$$\hat{\Psi}_{md} = k_{ws1} N_s \frac{2}{\pi} \frac{\mu_0}{\delta_{def}} \frac{4}{\pi} \frac{k_{ws1} N_s}{2p} \tau_p l' \sqrt{2} I_s, \tag{7.158}$$

$$\hat{\Psi}_{md} = \frac{2}{\pi} \mu_0 \frac{1}{2p} \frac{4}{\pi} \frac{\tau_p}{\delta_{def}} l' (k_{ws1} N_s)^2 \sqrt{2} I_s. \tag{7.159}$$

When the peak value of the air-gap flux linkage is divided by the peak value of the stator current, we obtain the main inductance L_{pd} of a single stator phase in the d-direction

$$L_{pd} = \frac{2}{\pi} \mu_0 \frac{1}{2p} \frac{4}{\pi} \frac{\tau_p}{\delta_{def}} l' (k_{ws1} N_s)^2. \tag{7.160}$$

In a multiple-phase machine, other windings also affect the flux, and hence the magnetizing inductance L_{md} of the direct axis of an m-phase machine can be solved by multiplying the main inductance L_{pd} by $m/2$

$$L_{md} = \frac{m}{2} \frac{2}{\pi} \mu_0 \frac{1}{2p} \frac{4}{\pi} \frac{\tau_p}{\delta_{def}} l' (k_{ws1} N_s)^2 = \mu_0 \frac{2m\tau_p}{p\pi^2 \delta_{def}} l' (k_{ws1} N_s)^2 = \mu_0 \alpha_i \frac{m\tau_p}{\pi p \delta_{def}} l' (k_{ws1} N_s)^2 \tag{7.161}$$

where α_i is the saturation factor (6.33). For a sinusoidal flux density distribution $\alpha_i = 2/\pi$. This equation corresponds to the general equation of the magnetizing inductance for a rotating-field machine written in Chapter 4, and is applicable as such to different machine types, providing that the equivalent air gap and the saturation factor have been determined correctly. If the machine operates at a constant frequency, for instance as a power plant generator drive, reactances can be employed. The magnetizing reactance X_{md} corresponds to the magnetizing inductance of Equation (7.161) in the d-axis direction

$$X_{md} = \omega L_{md}. \tag{7.162}$$

A corresponding expression may be written for $X_{mq} = \omega L_{mq}$. If the machine is magnetized with the stator rotating-field winding alone and the machine is running in synchronism at no load with the reluctance torque, we obtain the current of the stator

$$\underline{I}_s = \frac{\underline{U}_s}{j\omega_s L_d}. \tag{7.163}$$

In practice, the no-load current is almost purely on the d-axis and Equation (7.163) gives the d-axis current.

If a nonexcited synchronous machine or a synchronous reluctance machine can be made to run 90 electrical degrees away from its natural direct-axis position, the largest effective air gap δ_{qef} of the machine is constantly at the peak of the rotating current linkage. To magnetize this large air gap, a notably higher current is now required in the stator

$$\underline{I}_q = \frac{\underline{U}_s}{j\omega_s L_q}. \tag{7.164}$$

The quadrature-axis synchronous inductance is in analogy to Equation (7.154)

$$L_q = L_{mq} + L_{s\sigma}. \tag{7.165}$$

L_{mq} is calculated analogically with L_{md} by employing δ_{qef} as the air gap instead of δ_{def}. Correspondingly, the leakage inductance of the stator is calculated with the methods discussed in Chapter 4. The flux leakage on the direct and quadrature axes is usually assumed equal, although also the length of the air gap affects the flux leakage. The flux leakage of the direct axis is somewhat lower than the flux leakage of the quadrature axis, since the current-carrying d-axis magnetizing conductors are located in practice on the quadrature axis, where the air gap is notably longer than on the direct axis.

The characteristics of a synchronous machine are strongly influenced by the quality of the damping of the machine. Previously, the core principles of the dimensioning of the damper winding were briefly discussed in the winding section in Section 2.18. The guidelines are based mainly on empirical knowledge. Damper windings resemble the rotor windings of induction motors, and the design principles of cage windings can be applied to also in the design of damper windings. The damper windings of salient-pole machines are usually mounted in the slots of the pole shoe. The slot pitch of the rotor has to be selected to deviate 10–15 % from the slot pitch of the stator to avoid the harmful effects of flux harmonics, such as noise. If there is skewing in the damper winding (usually for an amount of single stator slot pitch), the same slot pitch can be applied in both the stator and the rotor. The bars of the damper winding are connected with short-circuit rings. If the pole shoes are solid, they can, as the solid rotor of a nonsalient-pole machine, act as a damper winding, as long as the ends of the pole shoes are connected with firm short-circuit rings. A separate damper winding is seldom used in nonsalient-pole machines. However, conductors assembled under the keys or the slot keys themselves can be employed as bars of the damper winding.

In generators, one of the functions of damper windings is to damp counter-rotating fields created by unbalanced load currents. Therefore in such a case, a minimum resistance is selected for the damper windings. The cross-sectional area of the damper bars is usually selected to be 20–30 % of the cross-sectional area of the armature winding. The material of the bars is usually copper. In single-phase generators, cross-sectional areas above 30 % are employed. The area of the short-circuit rings is selected about 30–50 % of the cross-sectional area of the damper bars per pole.

In synchronous motors, the damper bars have to damp for instance the fluctuation of the rotation speed caused by pulsating torque loads, and ensure an optimal starting torque as an asynchronous machine. Thus, to increase the rotor resistance, brass damper bars are used. If copper bars are used, their cross-sectional area is selected to be only 10 % of the cross-sectional area of the copper of the armature winding.

In permanent magnet synchronous machines, in axial flux machines in particular, the damper winding is easily constructed by assembling a suitable aluminum plate on the rotor surface, on top of the magnets. The above principles can be applied to in dimensioning of the aluminum plate. The cross-sectional area of the aluminum plate can be dimensioned to be about 15 % of the total cross-sectional area of the stator copper material.

During transients, the inductances of a synchronous machine obtain first the subtransient values L''_d and L''_q, and then the transient value L'_d, which can be dimensioned precisely by measuring or by employing time-stepping numerical calculation methods, cf. Figure 7.30 below. The magnitude of these inductances is basically influenced by the characteristics of the damper winding and the field winding. The time constant τ''_d0 of the machine is usually relatively small when compared with the time constant τ'_d0. The solid-rotor frame of a nonsalient-pole machine forms a corresponding path for the eddy currents. At the beginning of the short circuit, as the stator current increases rapidly, there is an abrupt change in the current linkage of the stator, and the current linkage immediately tries to alter the main flux of the machine. The damper winding reacts strongly and resists the change by forcing the flux created by the stator winding of the machine to leakage paths in the vicinity of the air gap. This makes the initial subtransient inductances L''_d and L''_q low when compared with the synchronous inductance L_d. Correspondingly, the field winding of the machine resists the change during the transient. When calculating the short-circuit currents for this period, we employ the transient inductance L'_d and the time constant τ'_d, by which the short-circuit currents are damped to the values of the continuous state, calculated with the synchronous inductances L_d and L_q.

As a result of the magnetic anisotropy of a synchronous machine, such a machine has to be analyzed separately in the direct and quadrature directions in a reference frame attached to the rotor. Figure 7.29 illustrates these equivalent circuits of a synchronous machine according to the space vector theory. The theory is not discussed in detail here; however, the presented equivalent circuits are useful also from the machine design point of view.

We can derive different equivalent circuits for synchronous machines from these above circuits to illustrate different states, Figure 7.30:

(a) at steady state, there is a direct-axis synchronous inductance $L_\mathrm{d} = L_\mathrm{md} + L_\mathrm{s\sigma}$ dominating together with
(b) the quadrature-axis synchronous inductance $L_\mathrm{q} = L_\mathrm{mq} + L_\mathrm{s\sigma}$;
(c) the direct-axis transient inductance L'_d is the sum of the stator leakage inductance and the parallel connection of the direct-axis magnetizing inductance and the field winding leakage inductance

$$L'_\mathrm{d} = L_\mathrm{s\sigma} + \frac{L_\mathrm{md} L_\mathrm{f\sigma}}{L_\mathrm{md} + L_\mathrm{f\sigma}};$$

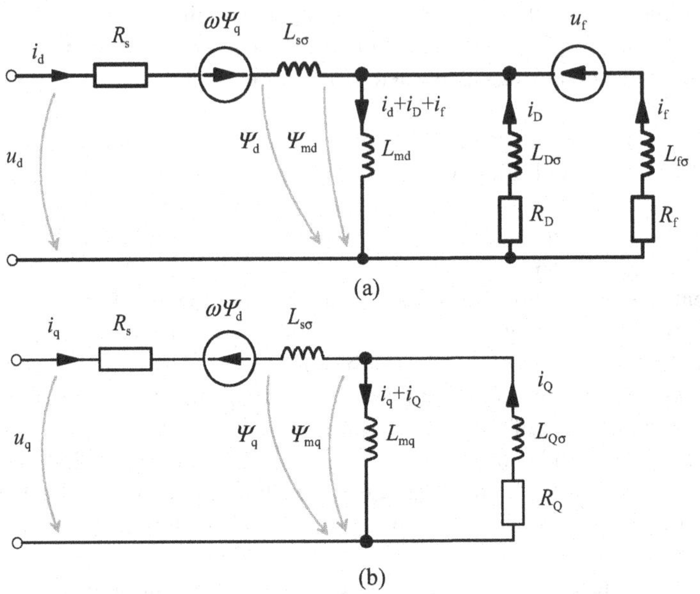

Figure 7.29 (a) Equivalent circuit of a synchronous machine in the d-direction according to the space vector theory. i_d and u_d are the d-axis components of the stator current and voltage. Ψ_d and Ψ_q are the direct- and quadrature-axis components of the stator flux linkage. i_D is the direct-axis current of the damper winding. i_f is the current of the field winding. R_s is the stator resistance, R_D is the direct-axis resistance of the damper winding and R_f is the resistance of the field winding. $L_{s\sigma}$ is the leakage inductance of the stator, L_{md} is the direct-axis magnetizing inductance, $L_{D\sigma}$ is the direct-axis leakage inductance of the damper winding, and $L_{f\sigma}$ is the leakage inductance of the field winding. u_f is the supply voltage of the field winding referred to the stator. (b) Equivalent circuit of a synchronous machine in the q-direction according to the space vector theory. i_q and u_q are the q-axis components of the stator current and voltage. Ψ_d and Ψ_q are the direct- and quadrature-axis components of the stator flux linkage. i_Q is the quadrature-axis current of the damper winding. R_s is the stator resistance, R_Q is the resistance of the quadrature axis damper winding. $L_{s\sigma}$ is the leakage inductance of the stator, L_{mq} is the quadrature-axis magnetizing inductance, $L_{Q\sigma}$ is the quadrature-axis leakage inductance of the damper winding all referred to the stator.

(d) the direct-axis subtransient inductance L''_d is the sum of the stator leakage inductance and the parallel connection of the direct-axis magnetizing inductance, the damper winding direct-axis leakage inductance and the field winding leakage inductance

$$L''_d = L_{s\sigma} + \frac{L_{md}\dfrac{L_{D\sigma}L_{f\sigma}}{L_{D\sigma}+L_{f\sigma}}}{L_{md} + \dfrac{L_{D\sigma}L_{f\sigma}}{L_{D\sigma}+L_{f\sigma}}};$$

(e) the quadrature axis has no field winding, and therefore the quadrature-axis subtransient inductance is

$$L''_q = L_{s\sigma} + \frac{L_{mq}L_{Q\sigma}}{L_{mq}+L_{Q\sigma}}.$$

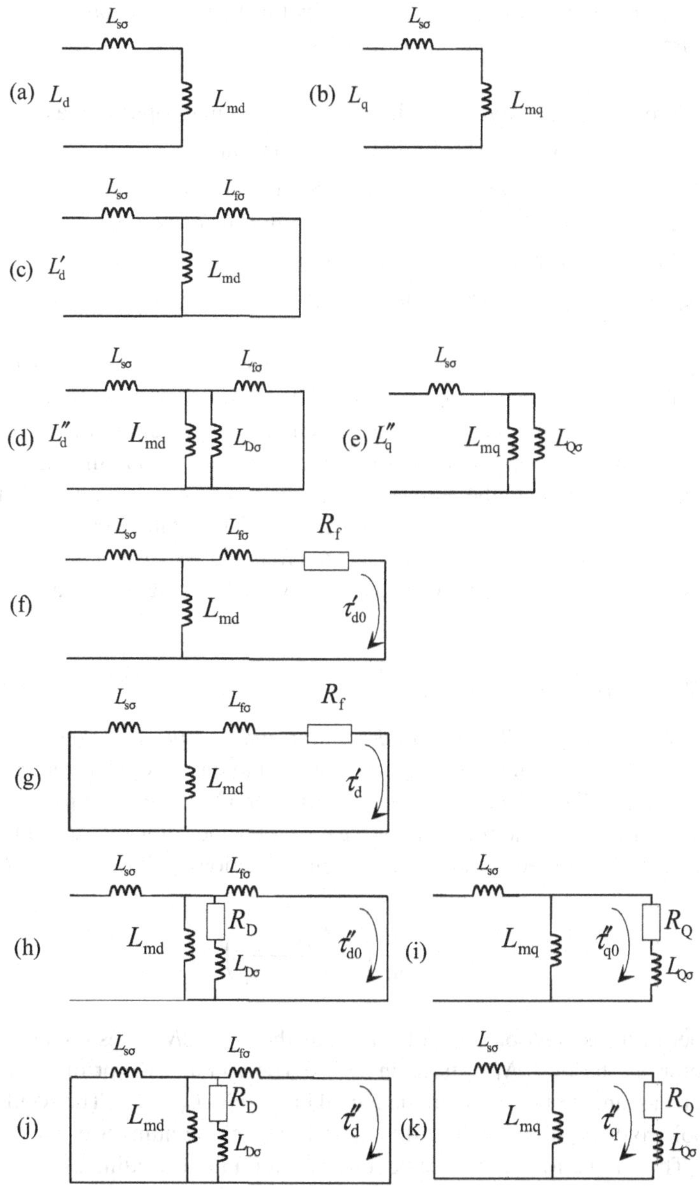

Figure 7.30 Equivalent circuits of transients that can be derived from the steady-state equivalent circuits of a synchronous machine, and some parameters related to these circuits.

In the figure, we also find the equivalent circuits for the time constants τ''_{d0}, τ''_{d}, τ'_{d0}, τ'_{d}, τ''_{q0} and τ''_{q}. The time constants are denoted as follows:

(f) τ'_{d0} is the direct-axis transient open-circuit (stator terminals open) time constant;
(g) τ'_{d} is the direct-axis transient short-circuit time constant;
(h) τ''_{d0} is the direct-axis subtransient time open-circuit constant;
(i) τ''_{q0} is the quadrature-axis subtransient open-circuit time constant;
(j) τ''_{d} is the direct-axis subtransient short-circuit time constant;
(k) τ''_{q} is the quadrature-axis subtransient short-circuit time constant.

A machine designer should be capable of giving values for all the above-listed parameters; in principle, this is quite a straightforward task and may be performed by using the methods discussed in this book. In practice, however, it is a very demanding task since the magnetic conditions vary from one operating point to another. This means that the inductances and the resistances are not constants in the equivalent circuits. As an example, it could be mentioned that even the stator leakage inductance saturates when large stator currents are present. The best results for the above-mentioned parameters are found by experiments or by transient finite element analysis (FEA). If the machine is large, only the FEA method is available.

7.3.1.1 Referred Rotor Parameters

Resistance of Field Winding Referred to the Stator and its Flux Leakage

First, we analyze the field winding parameters referred to the stator. The DC resistance of the field winding is R_{fDC}. The effective number of turns for a field winding per pole is N_f. There may be a_r parallel paths in the rotor. If we wish to refer the rotor resistance to the stator, in general it has to be multiplied with the term given in Section 7.2.2 (Equation (7.59)).

$$\rho_\nu = \frac{m_s}{m_r}\left(\frac{N_s k_{w\nu s}}{N_r k_{sq\nu r} k_{w\nu r}}\right)^2$$

The number of turns is N_f on one pole and thus there are $2N_f$ turns on a pole pair, and, in the whole rotor, a total of $2pN_f/a_r$ turns in series. The winding factor of a salient-pole field winding can be assumed one for the fundamental harmonic, $k_{w1f} = 1$. The winding factor of a nonsalient-pole rotor must be calculated, $k_{w1f} < 1$. The phase number of the field winding is one, $m_r = 1$. Thus, Equation (7.59) can be rewritten for a field winding as

$$\rho_1 = \frac{m_s}{1}\left(\frac{N_s k_{w1s}}{2pk_{w1f}N_f/a_r}\right)^2. \quad (7.166)$$

The resistance of the field winding of the rotor referred to the stator becomes thus

$$R'_f = m_s\left(\frac{N_s k_{w1s}}{2pk_{w1f}N_f/a_r}\right)^2 R_{fDC}. \quad (7.167)$$

Correspondingly, the leakage inductance of the field winding can be referred to the stator

$$L'_{\mathrm{f}\sigma} = m_{\mathrm{s}} \left(\frac{N_{\mathrm{s}} k_{\mathrm{w1s}}}{2p k_{\mathrm{w1f}} N_{\mathrm{f}} / a_{\mathrm{r}}} \right)^2 L_{\mathrm{f}\sigma}. \tag{7.168}$$

Applying Equation (4.57) to the calculation of the actual leakage inductance $L_{\mathrm{f1}\sigma}$ of one salient pole, we obtain $L_{\mathrm{f1}\sigma} = \mu_0 l' N_{\mathrm{f}}^2 \lambda_{\mathrm{p}}$, where λ_{p} is the permeance factor between the poles. As there are $2p/a_{\mathrm{r}}$ coils in series, the leakage inductance of the complete salient-pole rotor is

$$L_{\mathrm{f}\sigma} = \mu_0 l' \frac{2p}{a_{\mathrm{r}}} N_{\mathrm{f}}^2 \lambda_{\mathrm{p}}. \tag{7.169}$$

The leakage inductance of a nonsalient-pole rotor is calculated accordingly by using the methods introduced for slot windings.

Damper Winding

Next, the resistance and the leakage inductance of a damper winding are investigated. The simplest damper winding is a solid-rotor construction, in which eddy currents can be induced. The calculation of such a solid body is a challenging task, which cannot be solved analytically. However, if we assume the rotor material is linear, we are able to derive equations analytically for the material. Figure 7.31 illustrates the observation between the stator winding and the surface current.

Now, Ampère's law $\oint \boldsymbol{H} \cdot \mathrm{d}\boldsymbol{l} = \int_S \boldsymbol{J} \cdot \mathrm{d}\boldsymbol{S}$ is applied to the area between the stator and rotor surfaces along the integration path of Figure 7.31 for the determination of the impedance of the solid material of the rotor, observed from the stator. When the rotor surface moves at a speed that is different from the current linkage harmonic of the stator, currents are induced in the solid rotor. A field strength $\boldsymbol{H}_{\mathrm{r}}$ caused by the damper currents acts upon the rotor surface. The corresponding additional current of the stator is detectable as an additional current linkage of the stator. Now, the fundamental component Θ_1 of the current linkage of the stator and

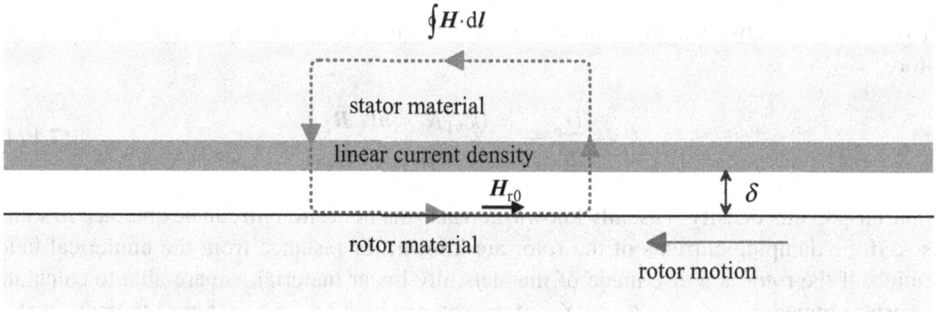

Figure 7.31 Application of Ampère's law between the surface currents and the stator currents. Note that boldface $\boldsymbol{l}$ indicates the general integration route, not the length of the machine. The field strength $\boldsymbol{H}_{\mathrm{r0}}$ is measured at the very surface of the solid rotor surface.

the tangential field strength $\hat{H}_r$ (time-harmonic field strength vector amplitude) of the rotor surface, assumed sinusoidal, are written equal to each other

$$\hat{\Theta}_1 = \frac{m}{2}\frac{4}{\pi}\frac{k_{w1}N_s}{p}\sqrt{2}I_r' = \int_0^{\tau_p} \hat{H}_{r0}e^{jax}dx, \quad a = \frac{\pi}{\tau_p}. \tag{7.170}$$

In (7.170), the variable x runs on the rotor surface along one pole pitch $x \in \lfloor 0, \tau_p \rfloor$

We may now write for the rotor current referred to the stator

$$\underline{I}_r' = \frac{\sqrt{2}j\pi p}{amk_{w1}N_s}\hat{\boldsymbol{H}}_{r0}. \tag{7.171}$$

$\hat{\boldsymbol{H}}_{r0}$ is the amplitude of the time-harmonic field strength represented as a complex vector on the rotor surface containing information of the phase angle.

The amplitude of the complex harmonic field strength $\hat{\boldsymbol{H}}_r$ on the rotor surface corresponds to the currents of the rotor. When the concept of surface current is employed, we may state that the magnetic field at the interface equals to the time-harmonic complex surface current $\hat{\boldsymbol{J}}_S$ with a normal vector direction

$$\hat{\boldsymbol{J}}_{Sr} = \boldsymbol{n} \times \hat{\boldsymbol{H}}_{r0}. \tag{7.172}$$

$\boldsymbol{n}$ is the normal unit vector of the surface, pointing away from the metal. The surface current is thus in the direction of the rotor axis when the field strength is tangential.

To be able to define the rotor "damper winding" impedance, we have to solve the corresponding induced voltage. Let us denote by $\hat{\boldsymbol{B}}_\delta$ the amplitude of the air-gap flux density on the rotor surface in the complex form. First, the flux is integrated, and next the flux linkage and the voltage referred to the stator are calculated

$$\underline{U}_r' = -j\omega_s\frac{k_{w1}N_s}{\sqrt{2}}\int_0^{\tau_p}\hat{\boldsymbol{B}}_\delta e^{jax}l'dx = j\omega_s\frac{2k_{w1}N_s l'}{\sqrt{2}a}\hat{\boldsymbol{B}}_\delta \tag{7.173}$$

where l' is the effective length of the machine. We obtain the rotor impedance referred to the stator

$$\underline{Z}_r' = \frac{\underline{U}_r'}{\underline{I}_r'} = \frac{\omega_s(k_{w1}N_s)^2ml'}{\pi p}\frac{\hat{\boldsymbol{B}}_\delta}{\hat{\boldsymbol{H}}_{r0}}. \tag{7.174}$$

The air-gap flux density is usually known, and also the field strength can be obtained in some cases, if the damping currents of the rotor are known, for instance from the numerical field solution. If the rotor is a disc made of magnetically linear material, we are able to calculate the surface impedance $\underline{Z}_{Sr} = R_{Sr} + jX_{Sr}$ of the rotor (which has to be referred to the stator)

$$\underline{Z}_{Sr} = \frac{\hat{\boldsymbol{E}}_{r0}(0,t)}{\hat{\boldsymbol{J}}_{Sr}(t)} = \frac{\hat{\boldsymbol{E}}_{r0}(0,t)}{\boldsymbol{n}\times\hat{\boldsymbol{H}}_{r0}(t)} = \frac{1+j}{\delta\sigma} = (1+j)\sqrt{\frac{\omega_r\mu}{2\sigma}} = R_{Sr} + jX_{Sr} \tag{7.175}$$

where δ is the penetration depth. The above can be applied for instance to the evaluation of the characteristics of a damper winding produced of aluminum sheet, if the sheet is thick compared with the penetration depth $\delta = \sqrt{2/\omega_r \mu \sigma}$. This is valid in particular for the harmonics. The fundamental component, however, usually travels through the aluminum sheet. The simplest way to evaluate such a plate is to divide the plate into fictional bars and to apply Equations (7.52)–(7.62) in analyzing the "squirrel cage." The angular frequency in the rotor with respect to the harmonic ν at the slip s_1 of the fundamental wave is

$$\omega_{\nu r} = \omega_s(1 - \nu(1 - s_1)). \tag{7.176}$$

The above rotor surface impedance can be referred to the stator quantity

$$\underline{Z}'_r = \frac{\omega_s (k_{w1} N_s)^2 ml}{\pi p} \underline{Z}_{Sr}. \tag{7.177}$$

If the rotor is made of a magnetically nonlinear material, an accurate analytic solution cannot be obtained, but the field solution of the rotor has to be carried out numerically or semi-analytically, for example with the multilayer transfer matrix method (discussed in Pyrhönen 1991). After the field solution, the average complex Poynting's vector is calculated on the rotor surfaces using the complex time-harmonic amplitudes of the E- and H-fields

$$\underline{S}_r = \frac{\hat{\boldsymbol{E}}_{r0} \cdot \hat{\boldsymbol{H}}^*_{r0}}{2}. \tag{7.178}$$

The value of Poynting's vector is integrated over all the rotor surfaces, and thus we obtain the apparent power $\underline{S}_r$ of the rotor. Now the referred impedance of the rotor is obtained with the peak value $\hat{U}_m$ of the air-gap magnetic voltage

$$\underline{Z}'_r = \frac{\hat{U}^2_m m}{2\underline{S}^*}. \tag{7.179}$$

If the damper winding is constructed as a cage winding, its resistance and leakage inductance in d and q directions referred to the stator are (Schuisky 1950)

$$R'_{Dd} = \frac{r'_{damp}}{\zeta_d}; \qquad R'_{Dq} = \frac{r'_{damp}}{\zeta_q}, \tag{7.180}$$

$$L'_{Dd} = \frac{l'_{damp}}{\zeta_d}; \qquad L'_{Dq} = \frac{l'_{damp}}{\zeta_q}, \tag{7.181}$$

where

$$r'_{damp} = 2\frac{m_s}{\sigma_{Db}} \frac{(k_{ws1} N_s)^2}{p} \left(\frac{b_p l_D}{\tau_p Q_{Dp} S_D} + \frac{D_{Dr}}{\pi p S_{Dr}} \frac{\sigma_{Db}}{\sigma_{Dr}} \right), \tag{7.182}$$

$$l'_{damp} = 2\mu_0 m_s l' \frac{(k_{ws1} N_s)^2}{p} \frac{b_p}{\tau_p Q_{Dp}} \left(\lambda_D + 0.131 \frac{b_p D_s}{\tau_p p Q_{Dp} k_C \delta_{de}} \right). \tag{7.183}$$

where m_s is the phase number of the stator, σ_{Db} is the conductivity of the damper bar material, σ_{Dr} is the conductivity of the material of the short-circuit ring, b_p the width of the pole shoe, τ_p the pole pitch, Q_{Dp} is the number of damper bars per pole, S_D is the cross-sectional area of the damper bar, l_D is the length of the damper bar, S_{Dr} is the area of the short-circuit ring, D_{Dr} is the average diameter of the short-circuit ring, δ_d the air-gap length in the middle of the pole, k_C the Carter factor of the stator and damper bar slotting and λ_D the slot permeance factor of the damper bar slot (see Section 4.3.3).

The leakage inductances of the damper winding include the slot leakage and air-gap leakage inductances. The so called harmonic factors ζ_d and ζ_q are calculated for $f_r > 5$ Hz

$$\zeta_d = \frac{b_p}{\tau_p} + \frac{1}{\pi}\sin\left(\pi\frac{b_p}{\tau_p}\right) - \frac{2}{\pi}\sqrt{\sigma_D}\cos\left(\pi\frac{b_p}{\tau_p}\right) - \frac{2}{\pi}\sqrt{\sigma_D} \tag{7.184}$$

$$\zeta_q = \frac{b_p}{\tau_p} + \frac{1}{\pi}\sin\left(\pi\frac{b_p}{\tau_p}\right) + \frac{4}{\pi}$$
$$\frac{\left[\sqrt{\sigma_D}\sin\left(\frac{\pi b_p}{2\tau_p}\right) + \cos\left(\frac{\pi b_p}{2\tau_p}\right)\right]\left\{\left[1 - \sqrt{\sigma_D}\left(1 - \frac{\delta_{def}}{\delta_{qef}}\right)\right]\cos\left(\frac{\pi b_p}{2\tau_p}\right) - \frac{\pi}{2}\left(1 - \frac{b_p}{\tau_p}\right)\sin\left(\frac{\pi b_p}{2\tau_p}\right)\right\}}{\sqrt{\sigma_D}\frac{\delta_{qef}}{\delta_{def}} + \frac{\pi}{2}\left(1 - \frac{b_p}{\tau_p}\right)}, \tag{7.185}$$

and for $f_r < 2$ Hz

$$\zeta_d = \frac{b_p}{\tau_p} - \frac{1}{\pi}\sin\left(\pi\frac{b_p}{\tau_p}\right) \tag{7.186}$$

$$\zeta_q = \frac{b_p}{\tau_p} - \frac{1}{\pi}\left(1 - 2\frac{\delta_{def}}{\delta_{qef}}\right)\sin\left(\pi\frac{b_p}{\tau_p}\right). \tag{7.187}$$

In the range $f_r = 2$–5 Hz, we do not have any simple equation for the harmonic factors. In Equations (7.184)–(7.187), σ_D is the leakage factor of the damper winding, and δ_{def} and δ_{qef} are the effective air-gap lengths in d- and q-directions.

In Figure 7.32, the harmonic factors are presented as the function of b_p/τ_p (pole shoe width/pole pitch). We may see that the leakage factor σ_D of the damper winding and the effective air-gap ratio $\delta_{def}/\delta_{qef}$ influence only a little the harmonic factors.

7.3.2 Loaded Synchronous Machine and Load Angle Equation

A synchronous machine can operate either as a motor or as a generator. A synchronous machine may also operate either as over- or underexcited. The terms "overexcitation" and "underexcitation" seem to have two different definitions in the literature. For instance, Richter (1963) has defined overexcitation in two ways:

1. In an overexcited synchronous machine, the field winding current is larger than at no load at the rated voltage.
2. An overexcited machine has a field winding current large enough to be capable of supplying inductive load magnetizing current.

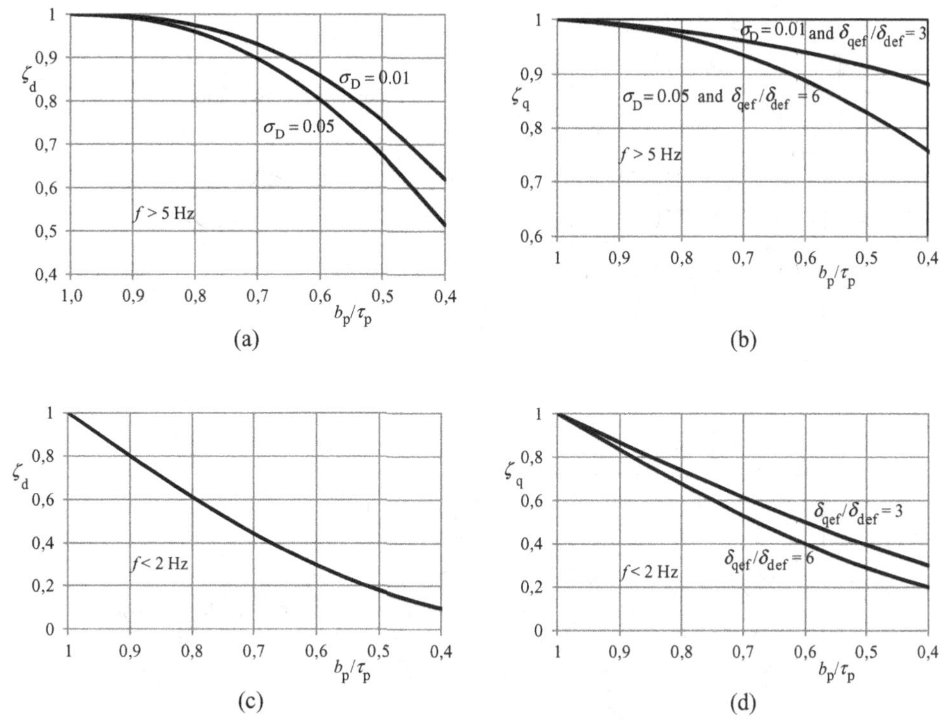

Figure 7.32 The harmonic factors according to Equations (7.184)–(7.187). (a) and (b) for frequencies higher than 5 Hz and (c) and (d) for frequencies lower than 2 Hz.

In practice, according to Definition I, a synchronous machine operates almost always overexcited. This definition is, therefore, not very useful.

The second definition necessitates a remarkably larger field winding current before overexcitation is reached. According to the second definition, a machine may be underexcited while operating overexcited according to Definition I. This may be seen for instance in Figure 7.39b, where the stator current has a d-axis component opposing the field winding current. According to Definition I, the machine operates overexcited, whereas according to Definition II, the machine operates underexcited.

In the following, however, we will use Definition II, since it is very practical and may be used in various contexts. Hence, an overexcited synchronous machine supplies the reactive current required by the inductive loads to the network. The overexcited machine appears as a capacitor to the network, and correspondingly, when underexcited, it takes the required additional magnetizing energy from the network appearing thus as a coil to the network. Normally, the synchronous machines run overexcited in the network, $\cos\varphi = 0.7–0.8$. The machines seldom operate underexcited, but for instance in a situation where a large generator supplies a long transmission line with a power below the natural power of the line, the voltage tends to increase towards the end of the line, and the machine itself may then run underexcited. Figure 7.33 illustrates the effect of various load currents on the magnetization of the generator. This generator is operating alone in an island.

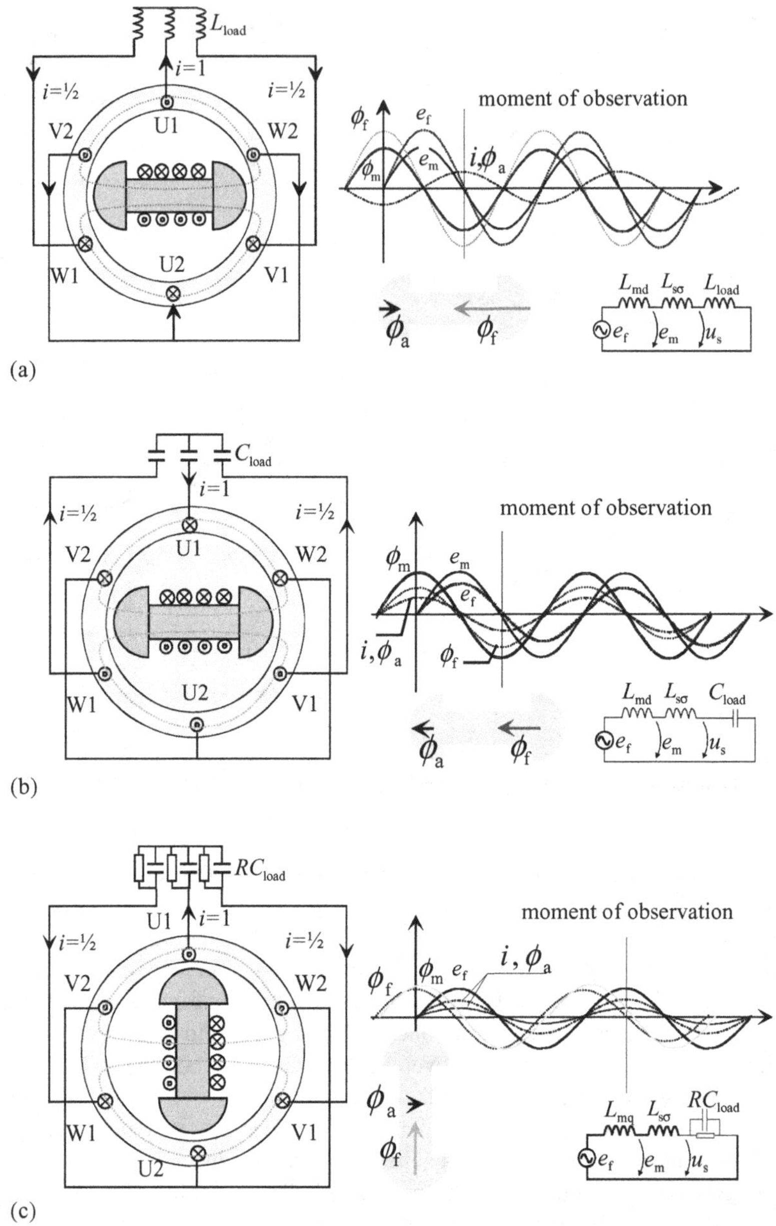

Figure 7.33 (a) Effect of an inductive load on the magnetization of a generator. The armature reaction ϕ_{a} of the stator created by the inductive load current i tends to weaken the flux ϕ_{f} generated by the field winding. The air-gap flux ϕ_{m} is a sum of ϕ_{f} and ϕ_{a}. The armature reaction weakens the flux produced by the magnetization of the field winding. The d-axis equivalent circuit with L_{md} is used. (b) Effect of a capacitive load current i on the flux of a synchronous generator. The armature reaction ϕ_{a} strengthens the flux ϕ_{f} produced by the field winding. (c) Effect of internally resistive load current on the flux of a synchronous machine. The armature flux ϕ_{a} created by the resistive current i is normal to the flux ϕ_{f} created by the field winding. Note that now the inductance in the equivalent circuit is changed to L_{mq} as the q-axis of the machine is magnetized.

In each case, there is a peak value $\hat{i} = 1$ of the current flowing in the generator winding U1–U2 at the moment of observation. Since in a symmetrical three-phase system the sum of phase currents is zero, there flows a negative current $i = -1/2$ in the windings V1–V2 and W1–W2. Now, in Figure 7.33a we see that when the inductive current is 90° behind the emf $e_f(t)$ induced by the flux linkage, the flux ϕ_a that is created by the armature current and that is at the same phase as the load current has a phase shift of 180° with the flux ϕ_f created by the actual field winding. The flux ϕ_a thus weakens the flux ϕ_f. If we wish to keep the emf of the machine at a value corresponding to the nominal voltage, the DC field winding current of the generator has to be increased from the no- load value to compensate for the armature reaction caused by the inductive load current; the generator is then operating overexcited.

Figure 7.33b illustrates a generator supplying a capacitive load. The moment of observation is selected at the negative peak value of the load current in the winding U1–U2. Now the armature winding creates a flux with the capacitive current, this flux being parallel with the flux ϕ_f created by the magnetizing current. Thus, the DC field winding current of the generator can be reduced from the no-load value while the terminal voltage remains constant.

Figure 7.33c investigates an armature reaction caused by an internally resistive current. An internally resistive load current is a current that is resistive with respect to the emf. The emf and the load current are temporally in the same phase. Since the internal reactance of the generator is inductive, to reach the situation described above we have to connect the capacitors in parallel with the load resistances. The absolute values of the reactances of these capacitors have to be equal to the absolute value of the reactance of the generator. At the moment of observation, the current of the winding U1–U2 is at maximum. Now the current linkage of the stator winding creates a flux Φ_q, which is transverse to the flux of field winding Φ_f. A large air gap in a salient-pole machine restricts the flux created by the quadrature current linkage. The situation of Figure 7.33c is illustrated as a phasor diagram in Figure 7.34.

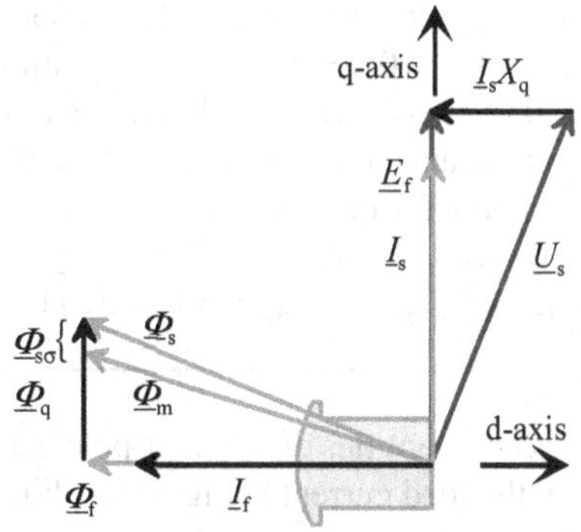

Figure 7.34 RMS value phasors corresponding to the instantaneous values of the generator of Figure 7.33c according to generator logic. Φ_f is the air-gap flux created by the field winding that induces the emf E_f of the machine, U_s is the terminal voltage, I_s is the phase current, Φ_q is the quadrature flux of the armature reaction, which includes also the effect of the flux leakage, Φ_s is the total flux of the stator, $I_s X_q$ is the inductive voltage drop created by the internally resistive load current at the quadrature reactance. An internally resistive load has no DC component. The "generator logic" discussed in Chapter 1 is followed here.

Example 7.4: A synchronous machine has the following per unit quantities: $l_{s\sigma} = 0.1$, $l_{md} = 1.5$, $l_{mq} = 0.6$, $r_s \approx 0$. Calculate the per unit field winding current needed to maintain $\psi_s = 1$ at no load, or when the machine is loaded with an inductive current $i_L = 0.5$, a capacitive current $i_C = 0.5$ or an internally resistive current $i_R = 0.5$.

Solution: At no load, the field winding current must alone produce the stator flux linkage absolute value of $\psi_{sd} = 1$, and hence $i_f = \psi_{sd}/l_{md} = \psi_{md}/l_{md} = 1/1.5 = 0.67$. We can see that the magnetizing inductance size defines how much field winding current is needed. In this case, the d-axis magnetizing inductance l_m is fairly large, and the low no-load i_f is fairly small. If l_m is small, the armature reaction is also small, but more i_f is needed.

During an inductive load, the armature reaction is opposing the field winding current linkage, and hence we define the current as negative. The armature reaction is now $\Delta\psi_{sd} = l_{md} \cdot i_L = 1.5 \times (-0.5) = -0.75$. The d-axis stray flux linkage is $\Delta\psi_{s\sigma d} = l_{s\sigma} \times i_L = 0.1 \times (-0.5) = -0.05$. The d-axis stator flux linkage is found as $\psi_{sd} = l_{s\sigma} \times (i_L) + l_{md}(i_L + i_f)$. The field winding current must produce an extra flux component to compensate for the armature reaction and the stray flux linkage. As the stator flux linkage must remain at its initial unity value, we have $1 = l_{s\sigma} \times i_L + l_{md}(i_L + i_f)$.

Now we get

$$i_f = (1 - l_{s\sigma} \cdot i_L - l_{md} i_L)/l_{md}, \quad i_f = [1 - 0.1 \cdot (-0.5) - 1.5 \cdot (-0.5)]/1.5 = 1.2.$$

During a capacitive load, the armature reaction is $+\, l_{md} \times i_C = +1.5 \times 0.5 = +0.75$. The stator stray flux linkage is $+\, l_{s\sigma} \times i_C = +0.1 \times 0.5 = +0.05$. Both of these values are now increasing the stator d-axis flux linkage, which means that the field winding current has to be decreased by $\Delta i_f = -0.537$. Now $i_f = [1 - 0.1 \times (+0.5) - 1.5 \times (+0.5)]/1.5 = 0.133$.

Hitherto, all the flux and current phenomena have to take place on the d-axis. During an internally resistive load, however, the armature reaction and the leakage flux linkage are according to Figure 7.34 on the quadrature axis. The absolute value of the quadrature armature reaction is $l_{mq} \times i_R = 0.6 \times 0.5 = 0.3$. The quadrature axis stray flux linkage is again $l_{s\sigma} \times i_R = 0.1 \times 0.5 = 0.05$. Together these flux linkages change the quadrature flux linkage by $\Delta\psi_{sq} = 0.35$. On the d-axis, there is only the field winding current and the flux linkage created by it. We write now for the stator flux linkage

$$\psi_s = \sqrt{\psi_{sd}^2 + \psi_{sq}^2} = \sqrt{(i_f l_{md})^2 + (i_R l_{s\sigma} + i_R l_{mq})^2} = 1 = \sqrt{(i_f 1.5)^2 + 0.35^2} \rightarrow i_f = 0.624.$$

Figure 7.35 illustrates the phasor quantities of a salient-pole machine when the power factor of an external load is inductive, the load current being thus behind the terminal voltage.

Assuming the voltage $\underline{I}_s R_s$ over the resistance R_s of the stator winding to be zero (the underlining denotes a phasor, while a symbol without underlining denotes an absolute value), we may now employ the synchronous inductances L_d and L_q ($L_d = L_{md} + L_{s\sigma}$, $L_q = L_{mq} + L_{s\sigma}$). All the quantities in the equations are phase quantities assuming a star connection:

$$\underline{E}_f - \underline{I}_d j\omega_s L_d = (\underline{U}_s \cos\delta) e^{j\delta}, \tag{7.188}$$

$$\underline{I}_q j\omega_s L_q = \underline{U}_s \sin\delta e^{+j(\frac{\pi}{2}+\delta)} = j\underline{U}_s \sin\delta e^{j\delta}. \tag{7.189}$$

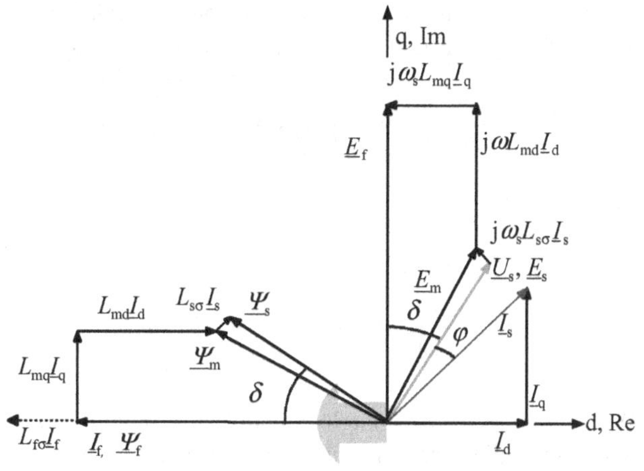

Figure 7.35 Phasor diagram of the steady state of a salient-pole generator with an inductive load current $\underline{I}_s$ (an overexcited generator). $\underline{I}_s$ is divided into two components. I_d is the direct-axis current, since it creates a parallel but an opposite flux with the flux linkage $\Psi_f = L_m i_f$ of the DC field winding current. In fact there also exists a total field winding flux linkage including also $L_{f\sigma} i_f$ but it is not usually shown in the diagrams. I_q is the quadrature-axis current, since it creates a quadrature flux. The quadrature axis current I_q is also called an internal active current, since it is in the same phase with the emf E_f. The angle δ, which is the angle between the E_f and the terminal voltage U_s, is the internal load angle of the machine. The figure is constructed with per unit values, although the notations are in principle real values. In the figure $l_{md} = 1$, $l_{mq} = 0.7$ and $l_{s\sigma} = 0.1$, $u_s = 1$ and $i_s = 1$. Note that the quantities in the phasor diagram are phase quantities. Thus, for example U_s is now the phase voltage.

Equations (7.188) and (7.189) yield for the currents $\underline{I}_d$ and $\underline{I}_q$

$$\underline{I}_d = \frac{-\underline{U}_s \cos\delta e^{j\delta} + \underline{E}_f}{j\omega_s L_d}, \tag{7.190}$$

$$\underline{I}_q = \frac{\underline{U}_s \sin\delta e^{j\delta}}{\omega_s L_q}. \tag{7.191}$$

Since $\underline{I}_s = \underline{I}_d + \underline{I}_q$, we obtain

$$\underline{I}_s = \underline{U}_s \frac{-\cos\delta e^{j\delta}}{j\omega_s L_d} + \underline{U}_s \frac{\sin\delta e^{j\delta}}{\omega_s L_q} + \frac{\underline{E}_f}{j\omega_s L_d}. \tag{7.192}$$

Using the relations $\cos\delta = (e^{j\delta} + e^{-j\delta})/2$ and $j\sin\delta = (e^{j\delta} - e^{-j\delta})/2$, Equation (7.192) reduces to

$$\begin{aligned}\underline{I}_s &= \frac{\underline{U}_s}{2j\omega_s L_d}(-e^{j\delta} - e^{-j\delta})e^{j\delta} + \frac{\underline{U}_s}{2\omega_s L_q}(-j)(e^{j\delta} - e^{-j\delta})e^{j\delta} + \frac{\underline{E}_f}{j\omega_s L_d} \\ &= j\frac{\underline{U}_s}{2\omega_s L_d}(e^{j2\delta} + 1) - j\frac{\underline{U}_s}{2\omega_s L_q}(e^{j2\delta} - 1) - j\frac{\underline{E}_f}{\omega_s L_d} \\ &= j\frac{\underline{U}_s}{2\omega_s}\left(\frac{1}{L_d} + \frac{1}{L_q}\right) + j\frac{\underline{U}_s}{2\omega_s}\left(\frac{1}{L_d} - \frac{1}{L_q}\right)e^{j2\delta} - j\frac{\underline{E}_f}{\omega_s L_d}.\end{aligned} \tag{7.193}$$

Next, the power $P = 3\ U_s I_s \cos\varphi$ is calculated with the phase quantities. Choosing $\underline{U}_s$ as the real axis and substituting for $\underline{E}_f = E_f(\cos\delta + j\sin\delta)$, we get for the real component of $\underline{I}_s$

$$\mathrm{Re}I_s = -\frac{U_s}{2\omega_s}\left(\frac{1}{L_d} - \frac{1}{L_q}\right)\sin 2\delta + \frac{E_f}{\omega_s L_d}\sin\delta, \tag{7.194}$$

and for the power

$$\begin{aligned} P &= 3U_s I_s \cos\varphi \\ &= 3\left(\frac{U_s\, E_f}{\omega_s L_d}\sin\delta + U_s^2 \frac{L_d - L_q}{2\omega_s L_d\, L_q}\sin 2\delta\right). \end{aligned} \tag{7.195}$$

Equation (7.195) is known as the load angle equation of a salient-pole machine. Equation (7.195) shows that with a nonsalient-pole machine, for which $L_d = L_q$, the equation is simplified as the latter term becomes zero. If the voltages of the equation are written as line-to-line voltages, the factor 3 in front of the parenthetical expression is omitted. When the mechanical power is zero and the losses are neglected, the reactive power is obtained with Equation (7.193) by substituting $\delta = 0$

$$Q = 3\left(\frac{U_s^2 - U_s\, E_f}{\omega_s L_d}\right). \tag{7.196}$$

Figure 7.36 illustrates the graphs of the load angle equation as a function of load angle. The figure shows the notable effect of the dimensioning of the magnetic circuit on the operation

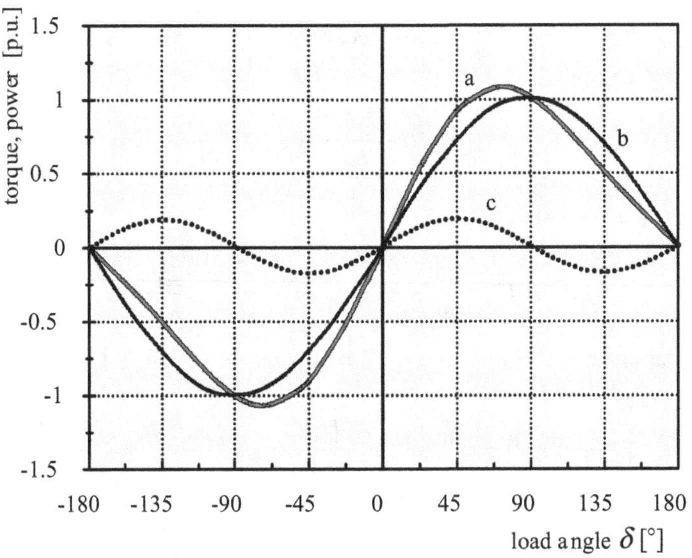

Figure 7.36 Graph of the load angle equation of a salient-pole machine (a) and a nonsalient-pole machine (b). The curve (c) illustrates the reluctance torque of a salient-pole machine.

of the machine. When the synchronous inductances of the machine diverge in the direct and quadrature direction, a load angle curve of a salient-pole machine is created. The load angle curve of a nonsalient-pole machine comprises a single sinus term only. The highest power is reached with a nonsalient-pole machine at a load angle $\delta = 90^\circ$. The maximum torque and power of a salient-pole machine are reached at a load angle less than 90°. One of the core tasks of a designer is to design a synchronous machine in such a way that the required inductance values are reached, since these values have a considerable influence on the characteristics of the machine.

Figure 7.36 also illustrates the principle of a synchronous reluctance machine. In Equation (7.195), the latter term, based on the magnitude difference between the direct and quadrature inductances, describes the load angle equation of a synchronous reluctance machine. Such a machine is thus in principle a salient-pole machine without a field winding. In the case of a synchronous reluctance machine, the target is to maximize the difference between the direct and quadrature inductance so that the curve (c) of Figure 7.36 produces a maximum peak torque.

7.3.3 RMS Value Phasor Diagrams of a Synchronous Machine

There are two basic logical systems in the phasor diagrams: a "generator logic" and a "motor logic," see Chapter 1. In the generator logic, the flux created by the rotor excitation induces the stator emf, and then, the stator current is created as a result of the induced emf; therefore, it is approximately parallel to it (within ±90°). In the motor logic, the supplied stator terminal voltage is first integrated with respect to time to obtain the stator flux linkage $\Psi_s(t) \approx \int u_s(t)\mathrm{d}t$; the emf is considered a counter-rotating electromotive force $e_s(t) = -(\mathrm{d}\Psi_s(t)/\mathrm{d}t)$. The current is considered to be created as an influence of the terminal voltage, and therefore, in the motor drive, its phasor is approximately in the same direction as the voltage.

Next, phasor diagrams of various synchronous machines with different load conditions are investigated. The phasor diagram gives a clear picture of the phase shifts between different waveforms of the machines. However, the RMS value phasor diagram is valid only for sinusoidal quantities at steady state. With phasor diagrams corresponding to the RMS value diagrams, we are able also to investigate dynamic states, which, however, are outside the scope of this chapter.

7.3.3.1 Nonsalient Synchronous Machine

A nonsalient synchronous machine is nearly isotropic, since the uniformity of the air gap is disrupted only by the slotting of the stator and the rotor. The rotor slots are usually not located regularly on the periphery of the rotor to produce as sinusoidal a flux density as possible with the rotor current linkage, as it was discussed in Figure 2.3. If a uniform slot pitch is selected, normally two-thirds of the rotor surface is equipped with slots and windings. Therefore a slight magnetic anisotropy occurs also in nonsalient-pole machines. In these machines, the stator resistance R_s causes a voltage drop. The magnetizing reactance X_m obtained from the magnetizing inductance L_m and the leakage reactance $X_{s\sigma}$ obtained from the leakage inductance $L_{s\sigma}$ of the stator phase winding are impedances, over which a voltage can be considered to take place. In reality, the flux linkages created in these inductances reduce the total flux

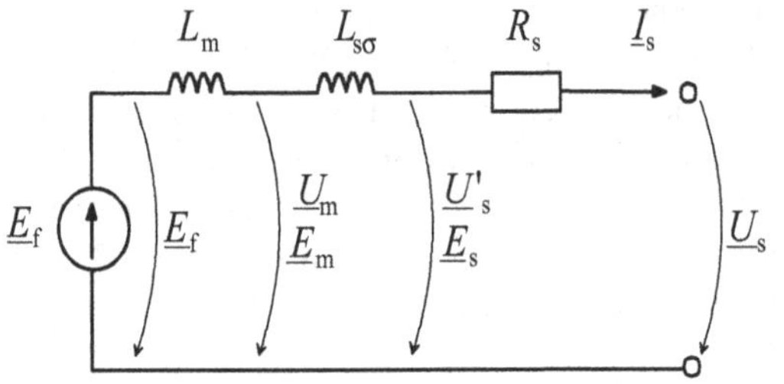

Figure 7.37 RMS value equivalent circuit of a nonsalient-pole generator in synchronous operation.

linkage of the machine so that the voltages induced in the machine remain lower than the imaginary emf induced by the rotor flux linkage. When the equivalent circuits are presented, it is usually advisable to interpret the phenomena arising in the inductances as voltages. In the phasor diagrams, we employ a direct-axis synchronous inductance L_d, which is the sum of the magnetizing inductance L_{md} and the leakage inductance $L_{s\sigma}$. Figure 7.37 illustrates a suitable equivalent circuit for a nonsalient-pole machine. This is a RMS value equivalent circuit corresponding to the equivalent circuit of the d-axis of a previously discussed phasor equivalent circuit.

The emf $\underline{E}_f$ induced by the flux linkage $\underline{\Psi}_f$ created by the rotor field winding is the driving force in the equivalent circuit. The terminal voltage is $\underline{U}_s = \underline{E}_f - \underline{I}_s R_s - \underline{I}_s j\omega_s L_{s\sigma} - \underline{I}_s j\omega_s L_m$. The magnetizing flux linkage $\underline{\Psi}_f$ is 90° ahead of the emf $\underline{E}_f$. In the stator circuit, a current $\underline{I}_s$ is created by the emf. This current in turn creates an armature flux linkage $\underline{\Psi}_a = L_m \underline{I}_s + L_{s\sigma}\underline{I}_s$. The armature flux linkage, that is the armature reaction, is thus in the same phase as the current. The resulting total flux linkage, or the stator flux linkage, is $\underline{\Psi}_s = L_m \underline{I}_s + L_{s\sigma}\underline{I}_s + \underline{\Psi}_f$.

7.3.3.2 Phasor Diagram of a Nonsalient-Pole Synchronous Generator

A synchronous generator may be connected either to an active network containing other machines that maintain the grid voltage, or to a passive island load where the generator alone is responsible for the voltage. In a network drive, we assume that the network determines the terminal voltage of the generator and the field winding current of the generator determines the reactive power balance of the generator. In an island drive, the load determines the power factor of the generator and the field winding current is used to adjust the terminal voltage of the generator. In principle, a phasor diagram construction is independent of the operating mode of a generator. In this presentation, however, the phasor diagrams are drawn with a fixed terminal voltage. In synchronous generator drives, such a case may be reached both in island and rigid network operation.

As a synchronous machine operates as a generator, we assume naturally that the emf $\underline{E}_f$ produces the current $\underline{I}_s$. Therefore, in the phasor diagrams of a synchronous generator, the active components of the voltage $\underline{U}_s$ and the current $\underline{I}_s$ are illustrated as being approximately parallel. In the diagram, there occurs also the voltage $\underline{U}'_s$, which is the terminal voltage

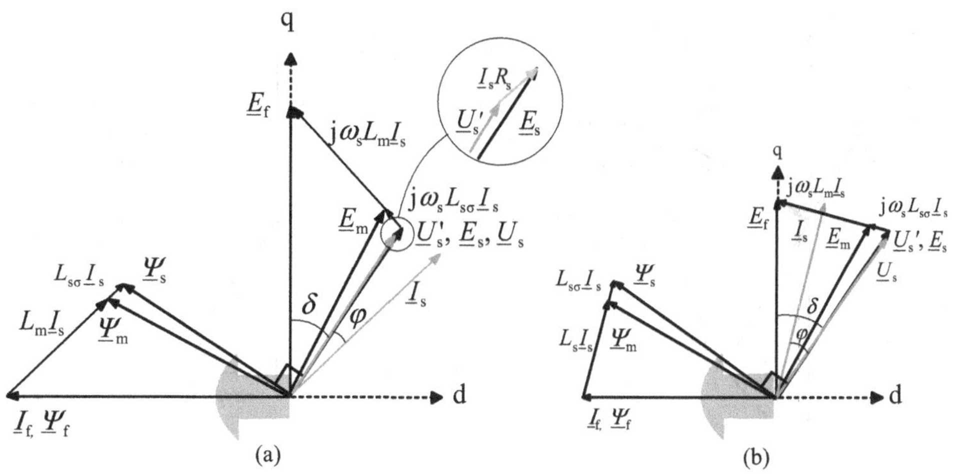

Figure 7.38 Phasor diagrams of a nonsalient-pole synchronous generator operating at constant terminal voltage $\underline{U}_{s,pu} = 1$ according to generator logic (a) overexcited (lagging power factor, delivers inductive power to the network) and (b) underexcited (leading power factor, absorbs inductive power from the network). A salient-pole shoe, despite the nonsaliency of the machine, is depicted to show the direction of the rotor excitation.

reduced with the resistive voltage drop $\underline{I}_s R_s$ of the stator. The voltage $\underline{U}'_s$ is employed in the determination of the direction of the stator flux linkage $\underline{\Psi}_s$. The field-winding-created flux linkage $\underline{\Psi}_f$ is illustrated 90° ahead of the emf $\underline{E}_f$, and the armature reaction plus leakage $L_m\underline{I}_s + L_{s\sigma}\underline{I}_s$ parallel to the stator current. For the geometric sum of the magnetizing flux linkage $\underline{\Psi}_f$ and the armature reaction flux linkage plus leakage ($\underline{\Psi}_a = L_m\underline{I}_s + L_{s\sigma}\underline{I}_s$), we obtain the total flux linkage $\underline{\Psi}_s$, which is 90° ahead of the voltage $\underline{U}'_s$. Figure 7.38 illustrates the RMS value phasor diagrams of a nonsalient-pole generator (a) overexcited and (b) underexcited.

In Figure 7.38a, the current of a nonsalient-pole synchronous generator is lagging the voltage, and therefore the machine operates overexcited supplying inductive reactive power to the network (the phase shift of the external load is inductive). An overexcited machine is easily recognizable from the magnitude of the emf E_f, the absolute value of which for an overexcited machine is significantly higher than the terminal voltage U_s. Note that the load angle remains notably lower in the case of an overexcited machine when compared with an underexcited machine. Overexcited synchronous machines are used in many industrial plants to provide the magnetizing power, that is inductive reactive power for the asynchronous machines and other inductive loads in the plant. Figure 7.38b shows that the phase shift angle φ of the synchronous generator is below 90° and capacitive. We may sometimes read that the generator, when operating underexcited, supplies the network with capacitive reactive power. This, however, is a misleading conception, implying that the capacitor would consume capacitive reactive power (and thus, in terms of dual thinking, would produce inductive reactive power). As we know, a capacitor is not assembled to consume capacitive reactive power, but to compensate for the effects of inductive loads in the network. A capacitor is an electrical energy storage,

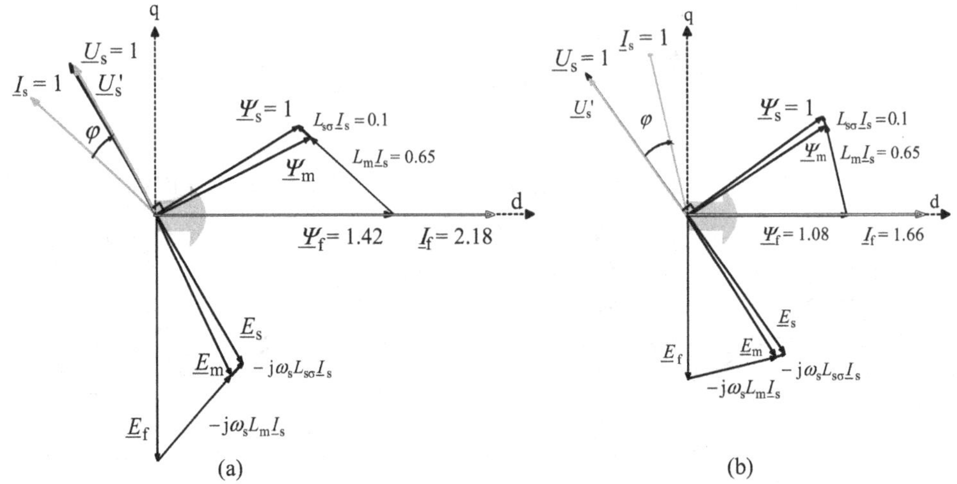

Figure 7.39 Phasor diagrams of a nonsalient-pole synchronous motor with $i_s = 1$, $u_s = 1$, $l_m = 0.65$ and $l_{s\sigma} = 0.1$ At no-load $\psi_f = \psi_m = \psi_s = 1$ and $i_f = \psi_f / l_m = 1.53$. At load (a) overexcited, $\psi_f = 1.42$ and $i_f = \psi_f / l_m = 1.42/0.65 = 2.18$ and (b) underexcited $\psi_f = 1.08$, $i_f = 1.08/0.65 = 1.66$ Despite the nonsaliency of the motor, the pole shoes are illustrated to show the directions of the d- and q-axes. The synchronous inductance in the case is relatively small, $l_s = 0.75$.

in which for instance the energy stored in magnetic circuits can be transferred and released over time. In the case of a synchronous machine, the most unambiguous way to describe the situation is to speak, according to Definition II in 7.3.2, about over- or underexcited machines, and thus to avoid misunderstandings.

7.3.3.3 Nonsalient-Pole Synchronous Motor

Figure 7.39a illustrates a phasor diagram of an overexcited nonsalient-pole motor. The direct armature flux is, as a result of overexcitation, partly opposite to the field winding flux, and the total flux corresponding to the terminal voltage $\underline{U}'_s$ of the machine tends to decrease from its no-load value. Since the network voltage decides the terminal voltage $\underline{U}_s$ of the motor, and in the real machine $\underline{U}'_s$ and $\underline{U}_s$ are approximately equal because of the small stator resistance, the field winding current of the machine has to be adjusted in order to keep the reactive power conditions unchanged. In the figure, the imaginary emf $\underline{E}_f$ is higher than the terminal voltage $\underline{U}_s$.

For a machine disconnected from the network and running at no load, the terminal voltage $\underline{U}_s$, $\underline{U}'_s$ and the emf $\underline{E}_f$ are equal. If a synchronous motor is employed in the compensation of inductive loads (the reactive current equals the reactive current of the overexcited synchronous generator), the motor has to be overexcited.

Figure 7.39b illustrates a phasor diagram of the corresponding underexcited synchronous motor. The current still has a small negative direct component that resists the field winding magnetization of the rotor. However, we see that the armature flux enforces the effect of the magnetizing flux when compared with the previous case of overmagnetizing; $\underline{\Psi}_s$ is higher than $\underline{\Psi}_m$. This is necessary, since the terminal voltage is defined by the supplying network, and the voltage $\underline{U}'_s$ induced by the total flux has to remain close enough to the terminal voltage $\underline{U}_s$.

Figures 7.39a and b illustrate the motor logic. When a voltage is connected to the stator winding, the stator flux linkage is created. We know that the flux linkage can be given as an integral of the voltage, $\Psi_s(t) = \int (u_s(t) - i_s(t)R)dt$. This flux linkage is about 90 electrical degrees behind the voltage, when the motor operates in a stationary state with sinusoidal quantities. Even if there existed a flux linkage created by the rotor current in the machine before the stator voltage was connected, after the transience, the flux linkage of the machine settles to correspond to the integral. The real stator flux linkage $\underline{\Psi}_s$ corresponds to the terminal voltage $\underline{U}'_s$. Note that this flux includes, in addition to the air-gap flux, the influence of the stator flux leakage. Therefore, the flux related to this flux linkage cannot be directly measured, although it induces the emf $\underline{E}_s$ of the stator, which is opposite to the voltage $\underline{U}'_s$ that created the flux. The current of the machine is defined in this logic by the small sum voltage of $\underline{U}_s$ and $\underline{E}_s$, which creates a current in the stator resistance R_s of the machine.

It is worth emphasizing that in reality, only the air-gap flux $\underline{\Phi}_m$ can be measured. The air-gap flux linkage $\underline{\Psi}_m$ corresponds to this flux. When the leakage flux linkage $\underline{\Psi}_{s\sigma} = L_{s\sigma}\underline{I}_s$ is added to the air-gap flux linkage, we obtain the sum flux linkage, that is the stator flux linkage $\underline{\Psi}_s$, which is created as an effect of all the currents in the machine. The logic is based on the principle according to which the total flux linkage of the machine is the superposition of the magnetizing flux linkage $\underline{\Psi}_f$ (the time derivative of which creates the imaginary emf) the armature flux linkage $\underline{\Psi}_a$ and leakage flux linkage $\underline{\Psi}_{s\sigma}$, all of which create an induction of their own ($\underline{\Psi}_f$ vs. $\underline{E}_f$, $\underline{\Psi}_a = L_m\underline{I}_s$ vs. $-j\omega_s L_m \underline{I}_s$, and $\underline{\Psi}_{s\sigma} = +L_{s\sigma}\underline{I}_s$ vs. $-j\omega_s L_{s\sigma}\underline{I}_s$). In reality, only a single total flux linkage occurs in the machine, namely the stator flux linkage $\underline{\Psi}_s$, and in the armature windings, only the electromotive voltage $\underline{E}_s$ is induced as a result of the total flux linkage. In a loaded machine, the emf $\underline{E}_f$ is only imaginary, since there exists no corresponding flux linkage component $\underline{\Psi}_f$ in reality, as the armature reaction $\underline{\Psi}_a$ and the leakage $\underline{\Psi}_{s\sigma}$ have reduced the real flux linkage of the machine to the sum flux linkage $\underline{\Psi}_s$.

If the machine is running at no load (as a generator), disconnected from the network and excited with the rotor field winding current, no armature current occurs in the machine, and thus $\underline{E}_f$ can be measured from the terminals of the machine ($\underline{\Psi}_f = \underline{\Psi}_m = \underline{\Psi}_s$). $\underline{\Psi}_f$ values as high as illustrated with a loaded machine in the phasor diagrams can never be measured from a real machine, since it should be heavily saturated at no load with a field winding current corresponding to the state of rated loading.

7.3.3.4 Salient-Pole Synchronous Machine

Because of the rotor structure of a salient-pole machine, the air gaps are of different in the direct and quadrature directions. As a result of this magnetic anisotropy, the equivalent circuits of synchronous inductances illustrating the inductances of a salient-pole machine are given separately for direct and quadrature directions in Figure 7.30. The direct synchronous inductance is according to Figure 7.30a $L_d = L_{md} + L_{s\sigma}$, where L_{md} is the direct-axis magnetizing inductance and $L_{s\sigma}$ is the leakage inductance of the stator. The quadrature-axis synchronous

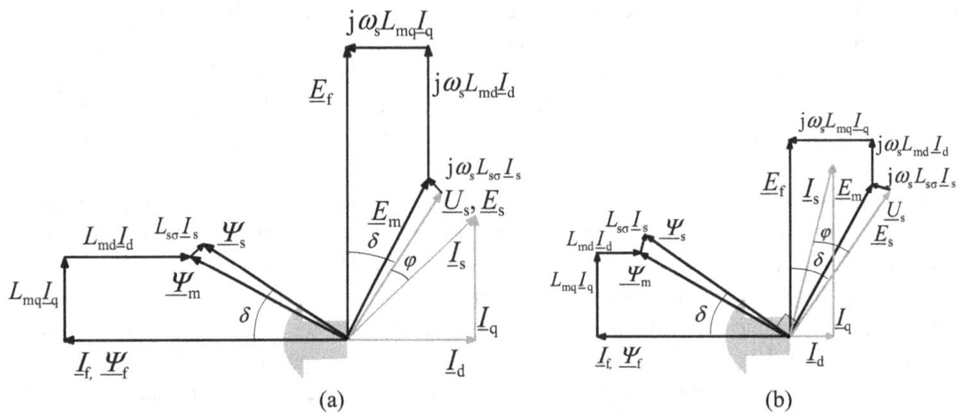

Figure 7.40 Phasor diagram of a salient-pole generator with $i_s = 1$, $u_s = 1$, $l_{md} = 1$, $l_{mq} = 0.65$, $l_{s\sigma} = 0.1$ and $r_s = 0$ (a) overexcited and (b) underexcited.

inductance is correspondingly $L_q = L_{mq} + L_{s\sigma}$, where L_{mq} is the quadrature-axis magnetizing inductance.

The stator current $\underline{I}_s$ of a salient-pole machine is comprised of two components, namely the direct-axis current $\underline{I}_d$ and the quadrature axis current $\underline{I}_q$, and is called the internal active power current of the machine, since it is in the same phase as the emf $\underline{E}_f$ of the machine. The load current creates the armature flux linkage, which is divided into direct and quadrature components. The direct-axis armature flux linkage $\underline{\Psi}_d$ is at the angle of 180° to the magnetizing flux linkage $\underline{\Psi}_f$, and the quadrature-axis flux linkage $\underline{\Psi}_q$ is at the angle of 90°. The stator flux linkage $\underline{\Psi}_s$ is the geometric sum of the direct- and quadrature-axes armature flux linkages, leakage flux linkage and the flux linkage of the DC excitation $\underline{\Psi}_f$. According to the generator logic, the stator flux linkage is 90° ahead the terminal voltage $\underline{U}_s$, cf. Figure 7.40. The terminal voltage is $\underline{U}_s = \underline{E}_f - \underline{I}_s R_s - j\underline{I}_d \omega_s \underline{L}_{md} - j\underline{I}_q \omega_s \underline{L}_{mq} - j\underline{I}_s \omega_s \underline{L}_{s\sigma}$ (cf. the salient-pole machine phasor diagrams in the following).

7.3.3.5 Salient-Pole Synchronous Generator

With respect to the reactive power of the network, an overexcited synchronous machine acts like a capacitor. In the phasor diagram of a synchronous generator, the stator current $\underline{I}_s$ is behind the voltage $\underline{U}_s$. Figure 7.40 illustrates the phasor diagrams of an overexcited and an underexcited salient-pole synchronous generator according to the generator logic. An underexcited generator operates like a coil with respect to the reactive power of the network, and thus takes the required extra magnetizing from the network. The stator current $\underline{I}_s$ is now ahead of the voltage $\underline{U}_s$. This may seem confusing, but the terminal voltage and current phasors in the case of a generator describe the current of the generator load. When the load is inductive, the current lags the voltage, and when the load is capacitive, the current leads the voltage.

By removing the rotor field winding current, a salient-pole machine is turned into a synchronous reluctance machine. A synchronous reluctance machine can naturally operate either

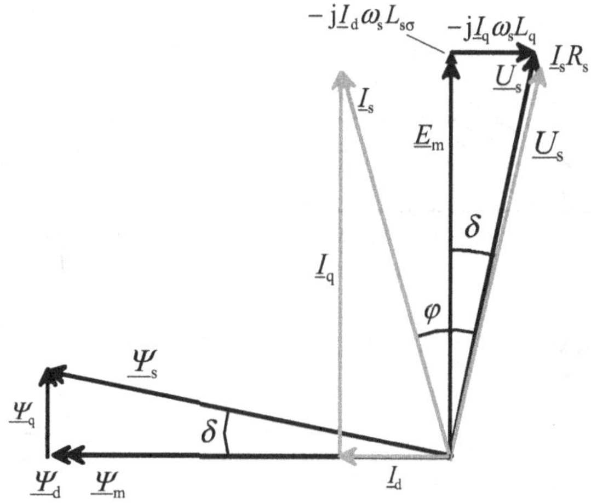

Figure 7.41 Synchronous reluctance generator phasor diagram drawn according to the generator logic with $i_s = 1$, $u_s = 1$, $l_{md} = 3.44$, $l_{mq} = 0.14$ and $l_{s\sigma} = 0.1$. $L_d/L_q = 15$ is as high as possible, and $E_f = 0$. The stator current must have a component I_d that magnetizes the machine d-axis.

as a motor or as a generator. Figure 7.41 illustrates the operation of a synchronous reluctance machine with a phasor diagram. There is no magnetizing current, and thus the respective flux linkage is lacking in the machine. The flux linkage components have to be created by the stator currents if this generator is working direct on line with constant voltage and frequency. If not, there must be a set of capacitors brought to the terminals as noticed in the section of an induction generator.

7.3.3.6 Salient-Pole Synchronous Motor

Figure 7.42a illustrates a phasor diagram of an overexcited salient-pole motor. The angle δ' between the emf $\underline{E}_f$ and the air-gap emf $\underline{E}_m$ or the field winding flux linkage $\underline{\Psi}_f$ and the air-gap flux linkage $\underline{\Psi}_m$ is the internal load angle. Correspondingly, the angle δ between the emf $\underline{E}_f$ and the emf $\underline{E}_s$ or the field winding flux linkage $\underline{\Psi}_f$ and the stator flux linkage $\underline{\Psi}_s$ is the load angle. The higher the angles, the more torque the synchronous machine yields as long as the pull-out point is not exceeded. Here $-j\omega L_{mq}\underline{I}_q$ is the voltage over the quadrature-axis reactance and $-j\omega L_{md}\underline{I}_d$ the voltage over the direct-axis reactance. $\underline{I}_s R_s$ is the voltage over the resistance of the stator winding. The direct-axis armature flux reduces the relative length of the field winding flux. This is necessary to keep the voltage $\underline{U}_s'$ in proper proportion with respect to the terminal voltage, since a strong network or a converter chiefly determines the terminal voltage $\underline{U}_s$. The voltage $\underline{U}_s'$ defines the stator flux linkage. The currents of the machine have to adjust themselves so that they also produce the same stator flux linkage. According to the equivalent circuits of Figure 7.29 and the phasor diagrams, we can understand that in the motor mode, the stator flux linkage is formed by integrating the voltage $\underline{\Psi}_s = \int (U_{sd} - I_d R_s + j(U_{sq} - I_q R_s))dt$ and the currents have to adapt themselves to the flux linkage determined by the voltage so

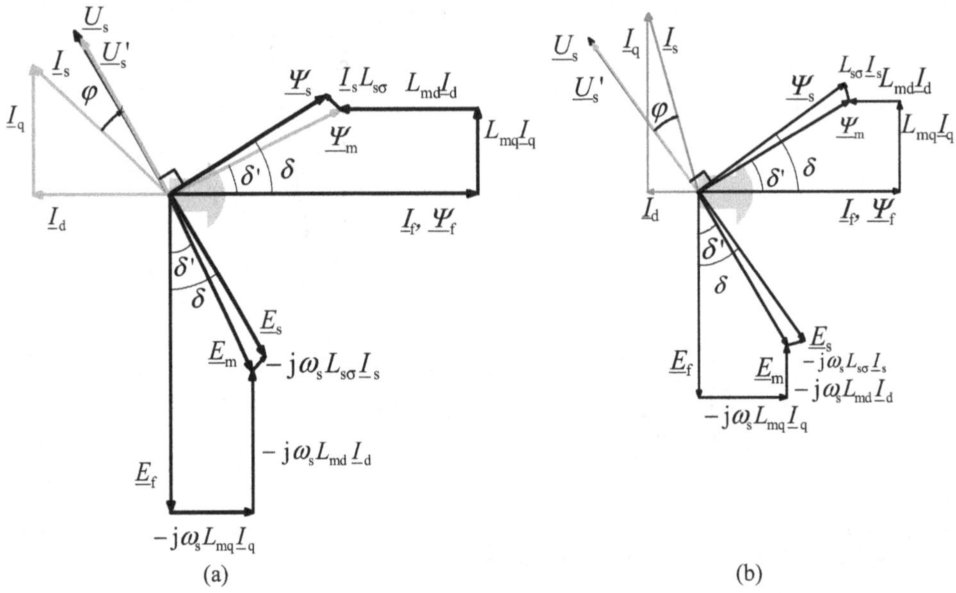

Figure 7.42 (a) Overexcited salient-pole motor with $i_s = 1$, $u_s = 1$, $l_{md} = 1$, $l_{mq} = 0.65$, and $l_{s\sigma} = 0.1$. The direct-axis armature flux considerably reduces the flux of the DC magnetizing. (b) Phasor diagram of an underexcited salient-pole motor. The direct-axis current here is also still slightly negative and reduces the flux linkage created by the field winding magnetizing.

that exactly the same stator flux linkage is obtained as $\underline{\Psi}_s = I_d L_{s\sigma} + (I_d + I_D + I_f)L_{md} + \mathrm{j}[I_q L_{s\sigma} + (I_q + I_Q)L_{mq}]$.

Figures 7.42a and b show that usually the armature flux opposes the effect of the magnetizing flux. This is valid for both over- and underexcited machines according to Definition II.

Example 7.5: In case of small machines the load angle equation ignoring the stator resistance can be far too erroneous. Derive the current and load angle equations for nonsalient-pole and salient-pole motor and generator taking the stator resistive voltage drop into account. The iron loss current can be neglected as it is normally small in all cases:

Solution: Referring to the phasor diagrams give above the generator stator phase voltage equation can be written as

$$U_s = U_s \cos\delta + \mathrm{j}U_s \sin\delta \tag{E1}$$

(E1) divided in d and q components gives

$$u_q = U_s \cos\delta' = E_f - R_s i_q - \mathrm{j}\omega_s L_d i_d, \tag{E2}$$

$$u_d = \mathrm{j}U_s \sin\delta' = R_s i_d - \mathrm{j}\omega_s L_q i_q. \tag{E3}$$

Corresponding currents are solved from (E2) and (E3)

$$i_q = \frac{-U_s \cos\delta' + E_f - j\omega_s L_d i_d}{R_s}, \tag{E4}$$

$$i_d = \frac{jU_s \sin\delta' + j\omega_s L_q i_q}{R_s}. \tag{E5}$$

By inserting (E4) in (E5) and vice versa we get the current components for a generator

$$i_q = \frac{R_s E_f + U_s(\omega_s L_d \sin\delta' - R_s \cos\delta')}{R_s^2 + \omega_s^2 L_q L_d}, \tag{E6}$$

$$i_d = \frac{\omega_s L_q E_f - U_s(\omega_s L_q \cos\delta' + R_s \sin\delta')}{R_s^2 + \omega_s^2 L_q L_d}. \tag{E7}$$

Corresponding voltage equations for a motor

$$u_q = U_s \cos\delta' = R_s i_q + E_f + j\omega_s L_d i_d, \tag{E8}$$

$$u_d = jU_s \sin\delta' = R_s i_d + j\omega_s L_q i_q. \tag{E9}$$

Currents are solved from (E8) and (E9)

$$i_q = \frac{U_s \cos\delta' - E_f - j\omega_s L_d i_d}{R_s}, \tag{E10}$$

$$i_d = \frac{jU_s \sin\delta' - j\omega_s L_q i_q}{R_s}. \tag{E11}$$

And further

$$i_q = \frac{-R_s E_f + U_s(\omega_s L_d \sin\delta' + R_s \cos\delta')}{R_s^2 + \omega_s^2 L_q L_d}, \tag{E12}$$

$$i_d = \frac{\omega_s L_q E_f + U_s(-\omega_s L_q \cos\delta' + R_s \sin\delta')}{R_s^2 + \omega_s^2 L_q L_d}. \tag{E13}$$

The electromagnetic power P_{em} of a machine with no saliency is

$$P_{em} = m i_q E_f \tag{E14}$$

The electromagnetic power of a nonsalient-pole generator output power taking stator resistive voltage drop into account is

$$P_{em,gen} = \frac{m E_f (R_s E_f + U_s(\omega_s L_d \sin\delta' - R_s \cos\delta'))}{R_s^2 + \omega_s^2 L_q L_d}. \tag{E15}$$

Correspondingly the input electromagnetic power of a nonsalient-pole motor is

$$P_{\mathrm{em,mot}} = \frac{mE_{\mathrm{f}}(-R_{\mathrm{s}}E_{\mathrm{f}} + U_{\mathrm{s}}(\omega_{\mathrm{s}}L_{\mathrm{d}}\sin\delta' + R_{\mathrm{s}}\cos\delta'))}{R_{\mathrm{s}}^2 + \omega_{\mathrm{s}}^2 L_{\mathrm{q}}L_{\mathrm{d}}}. \quad \text{(E16)}$$

(E15) and (E16) give the electromagnetic powers of the generator and of the motor which are

$$P_{\mathrm{em,gen}} = P_{\mathrm{out,gen}} + P_{\mathrm{Cu}}, \quad \text{(E17)}$$

and correspondingly for a motor

$$P_{\mathrm{out,mot}} = P_{\mathrm{em,mot}} - P_{\mathrm{Cu}} \quad \text{(E18)}$$

The output and input powers can now be written

$$P_{\mathrm{out.gen}} = \frac{-mU_{\mathrm{s}}^2 R_{\mathrm{s}}}{R_{\mathrm{s}}^2 + \omega_{\mathrm{s}}^2 L_{\mathrm{d}}L_{\mathrm{q}}} + \frac{mU_{\mathrm{s}}E_{\mathrm{f}}(R_{\mathrm{s}}\cos\delta' + \omega_{\mathrm{s}}L_{\mathrm{q}}\sin\delta')}{R_{\mathrm{s}}^2 + \omega_{\mathrm{s}}^2 L_{\mathrm{d}}L_{\mathrm{q}}}, \quad \text{(E19)}$$

$$P_{\mathrm{in,mot}} = \frac{mU_{\mathrm{s}}^2 R_{\mathrm{s}}}{R_{\mathrm{s}}^2 + \omega_{\mathrm{s}}^2 L_{\mathrm{d}}L_{\mathrm{q}}} + \frac{mU_{\mathrm{s}}E_{\mathrm{f}}(-R_{\mathrm{s}}\cos\delta' + \omega_{\mathrm{s}}L_{\mathrm{q}}\sin\delta')}{R_{\mathrm{s}}^2 + \omega_{\mathrm{s}}^2 L_{\mathrm{d}}L_{\mathrm{q}}}, \quad \text{(E20)}$$

in which the first term represents Joule losses P_{Cu}.
With $R_{\mathrm{s}} = 0$ both (E19) and (E20) become for nonsalient-pole motor and generator

$$P_{\mathrm{in,mot}} = P_{\mathrm{out,gen}} = P = \frac{mU_{\mathrm{s}}E_{\mathrm{f}}}{\omega_{\mathrm{s}}L_{\mathrm{d}}}\sin\delta. \quad \text{(E21)}$$

With salient-pole machines the equations become more complicated because of the reluctance torque. We get the electro-magnetic powers for a generator and a motor as:

$$P_{\mathrm{out,gen}} = \frac{-mU_{\mathrm{s}}^2 R_{\mathrm{s}}}{R_{\mathrm{s}}^2 + \omega_{\mathrm{s}}^2 L_{\mathrm{d}}L_{\mathrm{q}}} + \frac{mU_{\mathrm{s}}E_{\mathrm{f}}(R_{\mathrm{s}}\cos\delta' + \omega_{\mathrm{s}}L_{\mathrm{q}}\sin\delta')}{R_{\mathrm{s}}^2 + \omega_{\mathrm{s}}^2 L_{\mathrm{d}}L_{\mathrm{q}}} + \frac{mU_{\mathrm{s}}^2(\omega_{\mathrm{s}}L_{\mathrm{d}} - \omega_{\mathrm{s}}L_{\mathrm{q}})}{2(R_{\mathrm{s}}^2 + \omega_{\mathrm{s}}^2 L_{\mathrm{d}}L_{\mathrm{q}})}\sin 2\delta', \quad \text{(E22)}$$

$$P_{\mathrm{in,mot}} = \frac{mU_{\mathrm{s}}^2 R_{\mathrm{s}}}{R_{\mathrm{s}}^2 + \omega_{\mathrm{s}}^2 L_{\mathrm{d}}L_{\mathrm{q}}} + \frac{mU_{\mathrm{s}}E_{\mathrm{f}}(-R_{\mathrm{s}}\cos\delta' + \omega_{\mathrm{s}}L_{\mathrm{q}}\sin\delta')}{R_{\mathrm{s}}^2 + \omega_{\mathrm{s}}^2 L_{\mathrm{d}}L_{\mathrm{q}}} + \frac{mU_{\mathrm{s}}^2(\omega_{\mathrm{s}}L_{\mathrm{d}} - \omega_{\mathrm{s}}L_{\mathrm{q}})}{2(R_{\mathrm{s}}^2 + \omega_{\mathrm{s}}^2 L_{\mathrm{d}}L_{\mathrm{q}})}\sin 2\delta'. \quad \text{(E23)}$$

Again, with $R_{\mathrm{s}} = 0$ both of these reduce to the familiar load angle equation

$$P_{\mathrm{out,gen}} = P_{\mathrm{in,mot}} = P = \frac{mU_{\mathrm{s}}E_{\mathrm{f}}}{\omega_{\mathrm{s}}L_{\mathrm{d}}}\sin\delta + \frac{mU_{\mathrm{s}}^2(\omega_{\mathrm{s}}L_{\mathrm{d}} - \omega_{\mathrm{s}}L_{\mathrm{q}})}{2(\omega_{\mathrm{s}}^2 L_{\mathrm{d}}L_{\mathrm{q}})}\sin 2\delta. \quad \text{(E24)}$$

The same equations can be used for PMSMs and SyRMs by replacing $E_{\mathrm{f}} = E_{\mathrm{PM}}$ for PMSMs and $E_{\mathrm{f}} = 0$ for SyRMs.

7.3.4 No-Load Curve and Short-Circuit Test

To determine the characteristics of a synchronous machine, an extensive series of tests has to be carried out. The methods are standardized (e.g. in the European standard IEC 60034-4). Of these methods, only the two simplest are discussed here.

If a constant voltage source has a constant inductance, the corresponding impedance can be determined as a ratio of the voltage of the open circuit and the corresponding short-circuit current. This can be applied also to synchronous generators. When the inductance is a function of the no-load voltage, as is the case in the saturating iron circuit, a no-load curve in generating mode is required. The so-called unsaturated synchronous inductance is, however, constant, since it is determined by the air gap of the machine.

To determine the no-load curve, the machine is rotated at rated speed, and the terminal voltage is measured as a function of field winding current. In the sustained short-circuit test, the machine is rotated at rated speed, the terminals are short-circuited, and the phase currents are measured as a function of field winding current. The results of these tests are illustrated in Figure 7.43.

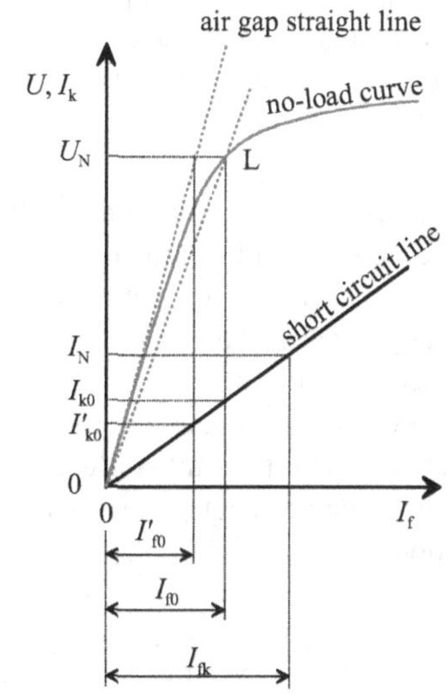

Figure 7.43 No-load curve of a synchronous machine and the short-circuit straight line. The figure also illustrates the air-gap straight line, which the no-load curve would follow if the permeability of iron were infinite. The machine can also be linearized with a straight line passing through point L corresponding to the rated voltage of the machine U_N. The figure also illustrates the behavior characteristic of synchronous machines: the short-circuit current can be lower than the rated current in a permanent short circuit. This is due to the direct-axis synchronous inductance. A synchronous machine produces temporarily a high short-circuit current only at the initial state of short circuit. In no-load conditions, the field winding current I_{f0} corresponds to the rated stator voltage (in a machine with infinite iron permeability, the no-load field winding current should be replaced by I'_{f0}). If the stator is short-circuited, the short circuit stator current is I_{k0} at the same I_{f0}. In short circuit, the field winding current I_{fk} corresponds to the rated stator current. Based on the ratio of these two currents, the referring factor from stator to rotor is defined: $g = I_{f0}/I_{k0}$. This referring factor is of particular importance in power electronic synchronous motor drives, where the field winding current takes part in the vector control.

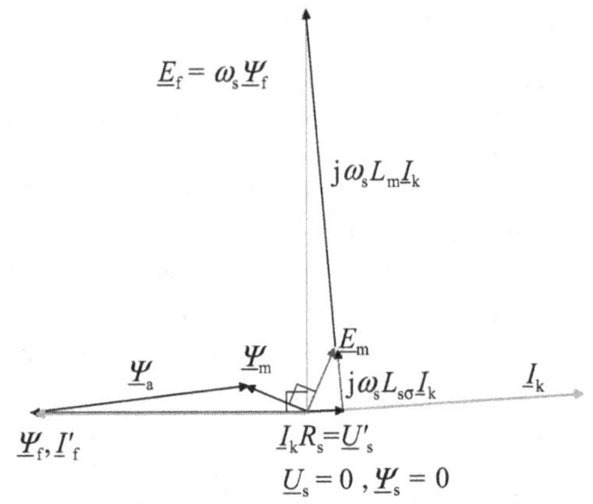

Figure 7.44 Phasor diagram of a short-circuit nonsalient-pole generator at steady-state. The flux linkage Ψ_a of the armature reaction nearly compensates the flux linkage Ψ_f of the magnetizing current, leaving a low flux linkage Ψ_m in the air gap, which creates a voltage that is consumed as an effect of I_k in the stator resistance R_s and in the leakage inductance $L_{s\sigma}$. Note that in the short-circuit test, the current linkages (as I'_f and I_k are about the same size when referred to the stator side as is the case here) of the stator and rotor are approximately equal but opposite. Thus, it is possible to approximate the referring factor of the rotor current with the short-circuit test.

The machine is excited in these tests with a rotor field winding current I_f. As the machine rotates at constant speed, a no-load curve can be created for the machine running at no load. Typically, the curve saturates immediately after the rated voltage, since the iron parts of the magnetic circuit begin to saturate. However, the curve is linearized in two ways: the air-gap straight line gives only the effect of the air gap, whereas the straight line that passes through the point L includes also the effect of the iron linearized to this exact point.

The phasor diagram of a steady-state short circuit is illustrated in Figure 7.44.

The unsaturated synchronous impedance is determined with the air-gap straight line. With the notations of Figure 7.43 we obtain

$$Z'_d = \frac{U_N}{\sqrt{3} I'_{k0}}, \tag{7.197}$$

where U_N is the rated line-to-line voltage of the machine. However, the most cost-effective solution is usually reached when the machine is saturated with its rated voltage. The corresponding synchronous impedance is

$$Z_d = \frac{U_N}{\sqrt{3} I_{k0}}, \tag{7.198}$$

which corresponds to the impedance of a linearized machine (the straight line OL). To determine the synchronous inductance, the stator resistance is measured and the impedance is divided into its components

$$L_d = \frac{\sqrt{Z_d^2 - R_s^2}}{\omega_s} \approx \frac{Z_d}{\omega_s}. \tag{7.199}$$

The short-circuit ratio (k_k) is the ratio of the magnetizing current required to produce the rated voltage at the rated rotation speed I_{f0} to the magnetizing current that is required to produce the rated current in a short-circuit test. The same ratio is valid also for the stator currents I_{k0} and I_N. With the notations of Figure 7.43, we can write

$$k_k = \frac{I_{f0}}{I_{fk}} = \frac{I_{k0}}{I_N}. \tag{7.200}$$

The short-circuit test is regarded as an indicator of the physical size of the machine. If we double the air gap of the previous machine (Figure 7.44), and the influence of the iron in the magnetic voltage is neglected, a double-field-winding current I_{f0} is required compared with the previous no-load test. Correspondingly, the synchronous inductance L_d would be cut into half, and thus only half of the original flux linkage would be required to produce the short-circuit current corresponding to the rated current. Thus, the field winding current corresponding to this short-circuit current would have remained unaltered. The size of the field winding has to be increased to create a flux sufficient to produce the voltage. This in turn increases the size of the machine.

7.3.5 Asynchronous Drive

Among synchronous machines, synchronous motors in particular are usually equipped with a proper damper winding. Above all, a damper winding is necessary in a machine that is supplied from a network. The present vector-controlled motor drives, synchronous motor drives included, operate very well without a damper winding. In these drives, a damper winding can even be harmful, if the switching harmonics of the frequency converter heat the damper winding excessively.

In a machine supplied from a network, the function of the damper winding is, in addition to keep the machine in synchronous running, to produce a sufficient starting torque. The motors are often of salient-pole type, and therefore the magnetic circuit and also the damper winding of the rotor are anisotropic. Next, the starting characteristics of a salient-pole machine are briefly discussed.

In a synchronous motor asynchronous drive, the saliency and the partial cage winding together create a pulsating flux bound to the rotor itself. This pulsating flux can be described as a sum of positive- and negative-sequence flux components both bound up with the rotor. The electrical angular speed of the negative-sequence flux component ω_{ns} of a counter-rotating field depends on the slip s of the machine as

$$\omega_{ns} = \omega_s(1 - 2s). \tag{7.201}$$

At synchronism ($s = 0$, $\Omega_{r,pu} = 1$), there is no pulsating field, hence no counter-rotating field at all in the rotor, and its speed, in principle, is the stator synchronous speed. When the slip s increases, for instance at $s = 0.25$ ($s = 0.25$, $\Omega_{r,pu} = 0.75$), $\omega_{ns} = \omega_s(1 - 1/2) = \omega_s/2$ with respect to the stator. At $s = 0.5$ ($s = 0.5$, $\Omega_{r,pu} = 0.5$) $\omega_{ns} = \omega_s(1 - 1) = 0$ with respect to the stator. At $s = 0.75$ ($s = 0.75$, $\Omega_{r,pu} = 0.25$), $\omega_{ns} = \omega_s(1 - 1.5) = -\omega_s/2$ with respect to the stator. Hence, the rotation direction of the rotor counter-rotating field changes with respect

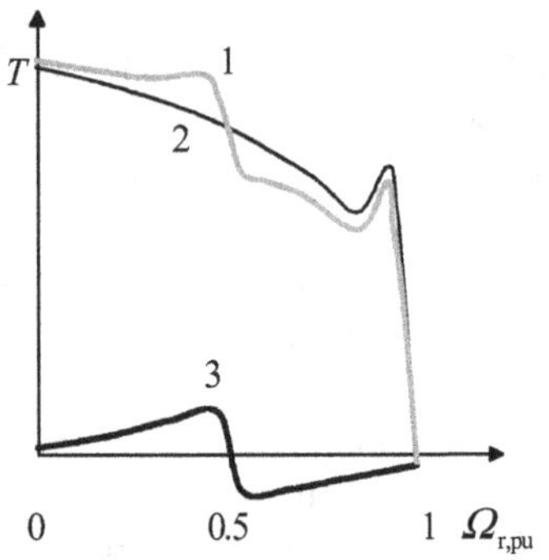

Figure 7.45 Asynchronous torque components of a synchronous motor at start-up, and the total torque as a function of angular speed. 1, total torque; 2, asynchronous torque produced by the stator synchronously rotating flux component; 3, torque of the counter-rotating field of the rotor. The curves are plotted as a function of mechanical angle per unit frequency.

to the stator at the slip $s = 0.5$. Simultaneously, the torque produced by the field changes its direction. Figure 7.45 depicts the torque components of a salient-pole machine during start-up, and the generated total torque.

7.3.5.1 Effect of Saliency

Because of saliency, the inductances of the phase windings of a synchronous motor depend on the position of the rotor. If we compare the magnetizing inductance $L_{m\delta}$ of a nonsalient-pole machine with the magnetizing inductance of a salient-pole machine, we may state that the direct- and quadrature-axis magnetizing inductances of a salient-pole machine are

$$L_{md} = k_d L_{m\delta}, \tag{7.202}$$

$$L_{mq} = k_q L_{m\delta}, \tag{7.203}$$

where $k_d \approx 0.85$ and $k_q \approx 0.35$ in typical machines. Thus, Schuisky (1950) shows that the current I_{ns} produced by the counter-rotating field is at the beginning of start-up

$$\frac{I_{ns}}{I_s} = \frac{(k_d^2 - k_q^2)\underline{Z}_1}{(k_d + k_q)^2(\underline{Z}_1 + \underline{Z}_2')}, \tag{7.204}$$

where

$$\underline{Z}_1 = (R_s + jX_{s\sigma}), \tag{7.205}$$

$$\underline{Z}_2' = \left(\frac{R_D'}{s} + jX_{D\sigma}'\right). \tag{7.206}$$

If, in addition to the typical values of k_d and k_q, $Z_1 = 3Z_2'$ in start-up, we obtain the ratio for the starting torques

$$\frac{T_{ns}}{T_s} = \left(\frac{I_{ns}}{I_s}\right)^2 \frac{Z_2'}{Z_1} \approx 0.31^2 \frac{1}{3} = 0.032. \tag{7.207}$$

The torque of a counter-rotating field is so small that it can be neglected, at least at these starting values. At start-up, the field winding has to be short-circuited in order to avoid damagingly high voltage stresses in the winding.

7.3.5.2 Partial Cage Winding

Since the machine is a salient-pole one and therefore anisotropic, the cage winding is not isotropic either, but there are bars missing when compared with the cage winding of an induction machine. Further anisotropy is caused by the single-phase field winding, which is normally short-circuited during the start-up to damp high overvoltages.

The missing bars cause distortion in the linear current density of the cage winding. There is more current flowing in the bars closest to the pole edges, that is in the outermost bars, compared with the case of a complete winding, Figure 7.46.

In an ideal, imaginary winding, there would be $1/\alpha$ more bars than in a real winding. Because of this lack of bars, the rotor winding produces a lower fundamental than an ideal winding. This can be taken into account by determining new, lower winding factors of the direct and quadrature axes for the rotor. The winding factor of an ideal winding is one, $k_w = 1$. Schuisky (1950) has determined the harmonic factors ζ_d and ζ_q for salient-pole cage windings, Equations (7.184)–(7.187). These factors describe the distributions of the currents of a partial cage winding. With the harmonic factors, we may also define the new winding factors for the cage winding of the rotor in the direct and quadrature directions

$$k_w = \sqrt{\zeta}. \tag{7.208}$$

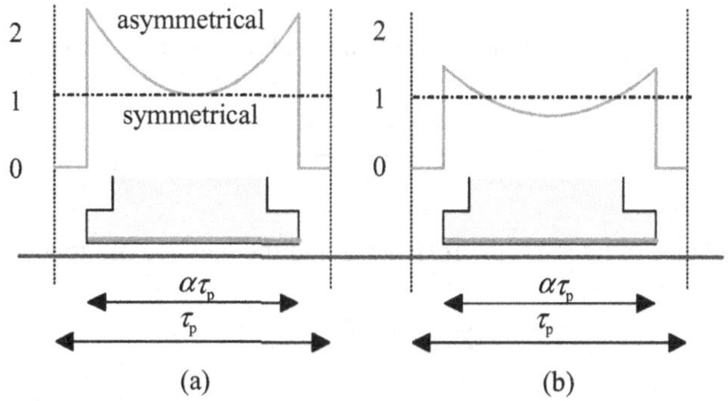

Figure 7.46 Distortion of the linear current density of a partial cage winding (a) at high frequencies (at the beginning of start-up) and (b) at low frequencies. The level of reference is the level of the linear current density of a complete winding. α describes the relative width of the pole shoe. Adapted from Schuisky (1950).

Harmonic factors are not constant, for instance, with respect to frequency (see Figure 7.32).

For synchronous motors, the leakage factor of a damper winding usually approaches the value $\sigma_D = X_{D\sigma}/X_m = 0.05$. The air-gap ratio $\delta_{qef}/\delta_{def}$ is a fictional value that is not directly proportional to real air gaps. For normal synchronous machines ($\alpha = 0.7$), this ratio is usually, for instance because of the saturation, of the scale $\delta_{qef}/\delta_{def} = 6$. For common induction machines, we obtain $\delta_{qef}/\delta_{def} = 1$ even if some of the rotor bars were damaged.

During an asynchronous start-up, we may employ Equations (7.184) and (7.185) or Figures 7.32a and b. For ordinary synchronous motors, we obtain $\zeta_d = 0.9$ and $\zeta_q = 0.95$ on average. Near synchronous running, the average values $\zeta_d = 0.45$ and $\zeta_q = 0.8$ can be employed.

The starting current and the torque can now be approximated with an equivalent direct and quadrature circuit. In the equivalent circuit of a synchronous machine, we have the magnetizing inductance, the damper branch and the short-circuited field winding in parallel. Next, admittances are solved for the direct- and quadrature-axis impedances connected in parallel. When calculating referred rotor quantities, we may employ the above harmonic factors that yield the winding factors of the rotor damping:

$$\underline{Y}_{rd} = -\frac{j}{X_{md}} + \frac{R'_f/s - jX'_f}{(R'_f/s)^2 + (X_f)^2} + \frac{R'_{Dd}/s - jX'_{Dd}}{(R'_{Dd}/s)^2 + (X'_{Dd})^2}, \tag{7.209}$$

$$\underline{Y}_{rq} = -\frac{j}{X_{mq}} + \frac{R'_{Dq}/s - jX'_{Dq}}{(R'_{Dq}/s)^2 + (X'_{Dq})^2}. \tag{7.210}$$

The corresponding impedances of parallel connections are

$$\underline{Z}_{rd} = \frac{1}{\underline{Y}_{rd}}, \tag{7.211}$$

$$\underline{Z}_{rq} = \frac{1}{\underline{Y}_{rq}}. \tag{7.212}$$

Next, during the start-up, we employ the average value of d- and q-axes

$$\underline{Z}_r \approx \frac{\underline{Z}_{rd} + \underline{Z}_{rq}}{2}. \tag{7.213}$$

The stator impedance created by the resistance and the leakage inductance is written as

$$\underline{Z}_s = R_s + jX_{s\sigma}. \tag{7.214}$$

The starting current is now approximately

$$\underline{I}_{s,start} \approx \frac{\underline{U}_s}{\underline{Z}_s + \underline{Z}_r}, \tag{7.215}$$

and the corresponding starting torque is approximately

$$T_{s,start} \approx \frac{pI_{s,start}^2 \mathrm{Re}(\underline{Z}_r)\eta_N \cos\varphi_N}{\omega_s}. \tag{7.216}$$

We employ the rated efficiency η_N and the power factor $\cos\varphi_N$ in the equation.

7.3.6 Asymmetric-Load-Caused Damper Currents

According to Equation (2.15), the rated three-phase current I_N of the stator creates a fundamental current linkage

$$\hat{\Theta}_{s1} = \frac{3}{\pi}\frac{k_{w1}N_s}{p}\sqrt{2}I_N \tag{7.217}$$

The fundamental component of the flux density caused by the armature reaction in the d-axis air gap is

$$\hat{B}_{\delta 1} = \frac{\mu_0}{\frac{4}{\pi}k_C\delta_{def}}\hat{\Theta}_{s1}. \tag{7.218}$$

Here the effective d-axis air-gap length δ_{def} defined in Chapter 3 is used. Let us assume that there are two phases with a rated current, and one phase is currentless. We have a pulsating flux that may be described with positive- and negative-sequence components. In this case, the counter-rotating flux component amplitude is 50 % of the positive-sequence amplitude

$$\hat{B}_{\delta,ns} = 0.5\hat{B}_{\delta,ps}. \tag{7.219}$$

This flux density induces a voltage at a double supply frequency in the rotor damper bars. This voltage causes a current opposing the asymmetric armature reaction in the rotor damping.

The peripheral speed v_{ns} of the flux propagating in opposite direction is

$$v_{ns} = 2\cdot 2\cdot \tau_p f_{s1}. \tag{7.220}$$

A flux propagating at this speed induces in a bar of the length l_D a voltage alternating at a double stator frequency, the RMS value of which is

$$U_{ns} = \frac{l_D v_{ns}\hat{B}_{s,ns}}{\sqrt{2}}. \tag{7.221}$$

The slot leakage reactance of a single rotor bar at a double frequency is

$$X_{ns,D} = 2\cdot\pi\cdot 2f_{s1}\mu_0 l_D\lambda_D. \tag{7.222}$$

For instance, the permeance factor of a round damper bar is

$$\lambda_{\mathrm{D}} = 0.47 + 0.066\frac{b_4}{b_1} + \frac{h_1}{b_1},$$

see Figure 4.15. The current of the damper bar is now

$$\underline{I}_{\mathrm{ns}} = \frac{l_{\mathrm{D}} v_{\mathrm{ns}} \hat{B}_{\delta,\mathrm{ns}}}{\sqrt{2}(R_{\mathrm{D}} + \mathrm{j} \cdot 2 \cdot \pi \cdot 2 f_{\mathrm{s1}} \mu_0 l_{\mathrm{D}} \lambda_{\mathrm{D}})}. \tag{7.223}$$

The power created by the counter-rotating field in the damper bar is

$$P_{\mathrm{ns}} = I_{\mathrm{ns}}^2 R_{\mathrm{D}}. \tag{7.224}$$

This power constantly heats the damper bars of a machine at asymmetric load, and therefore a sufficient heat transfer from the bars has to be ensured.

Frequency converter supply is a demanding case also with respect to the damper bars of a synchronous machine. The amplitude ripple of the air-gap flux density at the switching frequency similarly induces currents of switching frequency to the damper bars. These currents may in some cases even damage the damper winding. Therefore, the use of damper winding in a frequency converter drive should be carefully considered. A frequency converter drive is able to control a synchronous machine even without damping.

7.3.7 *Shift of Damper Bar Slotting from the Symmetry Axis of the Pole*

To reduce the slot harmonics in the generator voltages, and the disturbances in the motor start-ups, the damper bar slotting can be shifted with the following method canceling the slot harmonics. Figure 7.47 illustrates the shift χ of a damper bar slotting of a single pole pair.

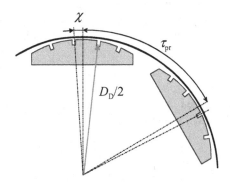

Figure 7.47 Shift of the damper bar slotting of the rotor in order to reduce slot harmonics and disturbances during start-up. The pole pitch on the pole surface τ_{pr}, the diameter at which the damper bars are positioned is D_{D}, and the shift χ. The shift is done on the pole 1 to clockwise and on the pole 2 to counterclockwise.

First, we investigate a factor c, which should approach zero. The factor is defined as a multiple of two geometrical factors k_p and k_χ as

$$c = k_p \cdot k_\chi = \frac{\sin(p\nu_{us}\pi)}{p\sin(\nu_{us}\pi)} \cdot \sin\left(\nu_{us}\frac{\pi}{2} + \nu_{us}\frac{\chi}{\tau_{pr}}\pi\right). \tag{7.225}$$

where ν_{us} is the ordinal number of stator slot harmonics

$$\nu_{us} = 1 + k\frac{Q_s}{p} = 1 + k \cdot 2mq_s = 1 + k \cdot 6q_s; \qquad k = \pm 1, \pm 2, \ldots. \tag{7.226}$$

In the shift of the damper bar slotting, we may present separate cases for integral and fractional slot windings of the stator.

(1) q_s (of the stator winding) is an integer. Thus $k_p = 1$, and therefore k_χ has to be zero. For the first slot harmonics $k = 1$ and

$$k_\chi = \sin\left((1 \pm 6q_s)\frac{\pi}{2} + (1 \pm 6q_s)\frac{\chi}{\tau_{pr}}\pi\right)$$

The ordinal number $1 \pm 6q_s$ is odd and k_χ is zero if

$$(1 \pm 6q_s)\frac{\chi}{\tau_{pr}} = \frac{1}{2},$$

from which

$$\chi = \frac{\tau_{pr}}{2\left(1 \pm \dfrac{Q_s}{p}\right)} = \frac{\pi D_D}{4(p \pm Q_s)}, \tag{7.227}$$

where

$$\tau_{pr} = \frac{\pi D_D}{2p} \tag{7.228}$$

(2) q_s is a fraction, the denominator of which is two. Again, $k_p = 1$, and thus k_χ has to be zero. Now, $1 \pm 6q_s$ is even and we can choose $\chi = 0$ for the shift.

(3) q_s is a fraction, the denominator being four. Now, when $k = \pm 1$ the ordinal number $1 \pm k6q_s$ is a fractional number and $k_p = 0$, and when $k = \pm 2$ the ordinal number $1 \pm k6q_s$ is an integer and $k_p \neq 0$. Thus the shift can be made by employing the second-order $k = \pm 2$. The magnitude of the shift is selected such that $k_\chi = 0$

$$k_\chi = \sin\left(\left(1 \pm 2\frac{Q_s}{p}\right)\frac{\chi}{\tau_{pr}}\pi\right) = 0 \Rightarrow \chi = \frac{\pi D_D}{2(p \pm 2Q_s)}. \tag{7.229}$$

Other fractions are not favored.

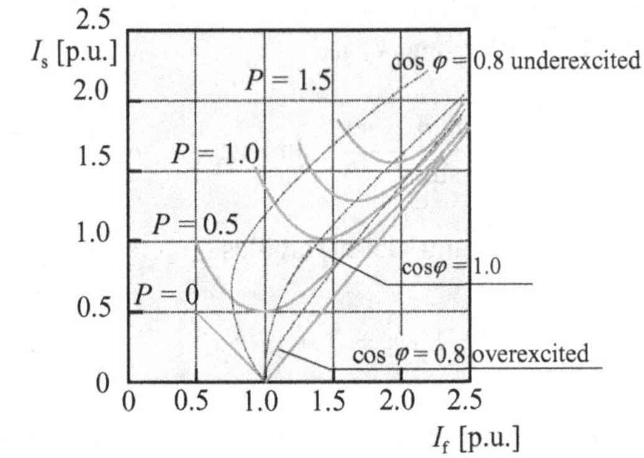

Figure 7.48 V curve of a synchronous machine. The current axes and the power parameters are presented in per unit values.

7.3.8 V Curve of a Synchronous Machine

The behavior of the armature current of a synchronous machine as a function of field winding current when the voltage, frequency and active power are kept as constants is called the V curve because of its form. The solid curves of Figure 7.48 indicate the V curves of an unsaturated synchronous machine, when the synchronous inductance as a per unit value is $l_{d,pu} = 1.0$. The dotted lines indicate constant $\cos\varphi$ values. Usually, synchronous motors operate overexcited producing a magnetizing current for the induction motors operating parallel to them. A synchronous motor can also be employed as a synchronous capacitor, in which case the motor runs at no load and overexcited, which means the right side of the curve $P = 0$ is valid. This way, by controlling the field winding current, it is possible to construct a "stepless controllable capacitor."

7.3.9 Excitation Methods of a Synchronous Machine

In principle, there are two methods to magnetize a separately excited machine (see also Table 7.8):

1. The field winding current is conducted via slip rings to the rotor. The current is supplied either by a thyristor bridge or a rotating exciting DC generator. The DC generator may be mounted on the same axis with the synchronous machine, and in that case the mechanical power needed for excitation is obtained from the turbine shaft. A disadvantage of this method is that the field winding current passes through both the commutator of the DC machine and the slip rings of the AC machine, and therefore two sets of brushes are required. These brushes carry high currents and require constant maintenance. When the field winding current is brought via slip rings to the rotor, and the excitation circuit is dimensioned to allow a considerable voltage reserve, we are able to construct a quite fast-responding excitation control despite the high inductance of the field winding.

2. By using brushless excitation. There are several different brushless excitation systems. Some main principles will be discussed here.

(a) There are two or even three synchronous machines on a common shaft: (i) the main synchronous machine; (ii) an outer pole synchronous generator having its field windings in the stator and the armature on the rotor supplying the field winding current for the main machine. Machine (ii) is excited either with a network supply or with a permanent magnet generator rotating on the same shaft. A thyristor bridge is needed to control the current. A permanent magnet generator excitation system is totally independent of external exciting electric power if only the prime mover can rotate the main machine and the auxiliary machines.

There is also a rotating diode bridge that rectifies the current supplied by the main excitation machine. This system is totally independent of auxiliary exciting electric power. Another and simpler version of this excitation system is the one in which the outer pole synchronous machine (ii) has one permanent-magnet-excited pole pair which suffices to excite the main machine so that its voltage may be used to magnetize the excitation machine.

(b) There is a rotationally symmetrical axial transformer at the end of the shaft having its magnetic circuit divided into a rotating and stationary part and a diode rectifier and being supplied with a stationary switched-mode AC power supply.

(c) There are two rotating machines on a common shaft: (i) the main machine and (ii) a wound rotor induction generator. The field winding current is rectified by a rotating diode bridge. The stationary part of the wound-rotor induction generator is supplied by a controllable three-phase voltage. This method is applicable for converter-fed motors that have to produce full torque also at zero speed. In the control, a three-phase triac controller is often used.

Figure 7.49 illustrates brushless generator excitation constructions. In Figure 7.49a, an outer pole synchronous generator is mounted on the shaft of the synchronous machine. This machine gets its excitation from a network supply. The current produced by the generator is rectified by diodes rotating along the machine. A brushless synchronous machine is quite maintenance free. On the other hand, the slow control of the field winding current of the machine is a disadvantage of this construction.

With regard to motors, a brushless magnetizing can be arranged also with a rotating-field winding in the magnetizing machine. This winding is supplied with a frequency converter when the main machine is not running. Now the main machine can be excited also at zero speed.

7.3.10 Permanent Magnet Synchronous Machines

What has been written above about synchronous machines is in many cases valid also for permanent magnet machines. For example, the phasor diagrams are similar except that in PMSMs, the magnets create a constant Ψ_{PM} instead of controllable Ψ_{f}. The Ψ_{PM} induces an emf E_{PM} instead of E_{f}. Permanent magnet excitation makes it possible to design very high-efficiency machines. As there are, in principle, no losses in the excitation, the efficiency should inherently be high. Unfortunately, many permanent magnet materials are conductive, and hence Joule losses take place in them. The permanent magnet excitation cannot be changed, which gives some boundary conditions for the machine design. As the magnets have a very low permeability (ideally $\mu_{\mathrm{rPM}} = 1$), the magnetizing inductances of permanent magnet machines

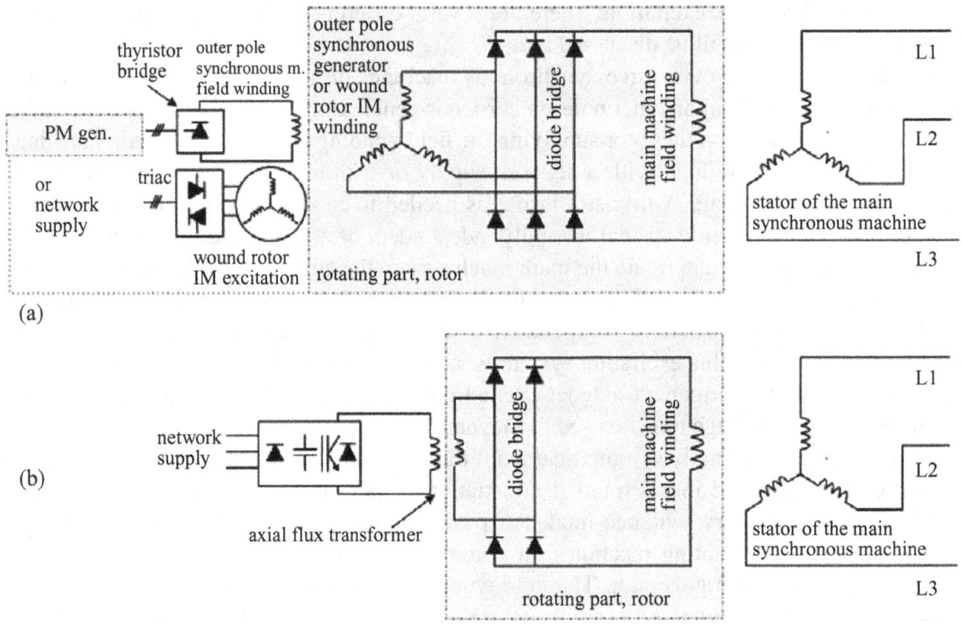

Figure 7.49 Excitation of a brushless synchronous generator. (a) Two machines have been mounted on a common shaft. The machines are either the main machine and an outer pole synchronous generator for excitation, or the main machine and a wound-rotor induction generator for the main machine excitation. In addition to the field winding of the main machine, the armature winding of the outer pole synchronous machine or the rotating field winding of the induction machine has been mounted on the rotor to produce the required magnetizing power for the main machine. The outer pole generator excitation is obtained either from network or from a permanent magnet generator rotating on the same shaft: The excitation is controlled with a thyristor bridge. The wound-rotor induction motor configuration may also operate at zero speed, when the stator of the magnetizing is supplied, for example with a three-phase triac. (b) An axial flux transformer with an air gap supplies the excitation energy needed.

become usually low. In permanent magnet machines the synchronous inductance per unit value must normally be lower than one (1 pu) to produce enough peak torque at the nominal voltage and speed. This is because the maximum torque is inversely proportional to the synchronous inductance. The load angle in Equation (7.195) is valid for a PMSM in the form

$$P = 3\left(\frac{U_s E_{PM}}{\omega_s L_d}\sin\delta + U_s^2 \frac{L_d - L_q}{2\omega_s L_d L_q}\sin 2\delta\right), \tag{7.230}$$

where we write E_{PM} instead of E_f. If by the permanent-magnet-induced emf pu value $e_{PM} = 1$ and the supply voltage $u_s = 1$ in case of nonsalient-pole PMSM we have to select $l_s = 1/1.6 = 0.625$ to achieve peak power P_{max} and torque T_{max} per unit value of 1.6, which is normally required by standards. In permanent magnet motors, the per unit e_{PM} is typically selected to vary between 0.9 and 1, and in generators $e_{PM} = 1.1$ is typically selected. The designer must remember that the remanence of magnets usually decreases as the temperature increases.

Permanent magnet motors have gained popularity in low-speed high-torque drives, which suit this machine type very well. The magnetizing inductance in section 4.1 was found to be inversely proportional to the square of the pole pair number p, $L_m = mD_\delta \mu_0 l'(k_{ws1}N_s)^2/(\pi p^2 \delta_e)$. This makes for instance multiple-pole low-speed induction motor characteristics poor, because the low magnetizing inductance causes a poor power factor for an induction machine. A PMSM does not suffer this as the magnetizing is arranged by permanent magnets. As the NdFeB and SmCo permanent magnets are quite well conductive, low-speed applications are easy also for rotor surface magnets because, at low speeds, the losses in the magnets remain low. In higher-speed applications, special care to avoid losses in permanent magnet material must be taken. In some cases bulky high remanence ($B_r \sim 0.4$ T) Ferrite magnets can be used as, in practice, there are no eddy current losses in them.

The characteristics of a permanent magnet machine are largely determined based on the rotor construction. Figure 7.50 illustrates different permanent magnet rotor constructions. If the magnets are mounted on the rotor surface, the rotor is in principle nonsalient, since the relative permeability of neodymium-iron-boron magnets is approximately one ($\mu_r = 1.04$–1.05; see Table 3.3). Magnets embedded in the rotor construction produce almost without exception a machine for which the quadrature-axis synchronous inductance is higher than the direct-axis synchronous inductance. Also, the pole shoe construction produces a similar inductance ratio. Such machines do produce also some reluctance torque according to (7.230).

Embedding the magnetic material completely inside the rotor structure wastes a significant proportion, typically one quarter, of the flux produced by the magnet. The flux is consumed in leakage components of the rotor, Figure 7.51. On the other hand, the embedded magnets are protected both mechanically and magnetically. In embedded mounting, we may also employ two magnets per pole (Figure 7.50f), in which case it is possible to reach a rather high air-gap

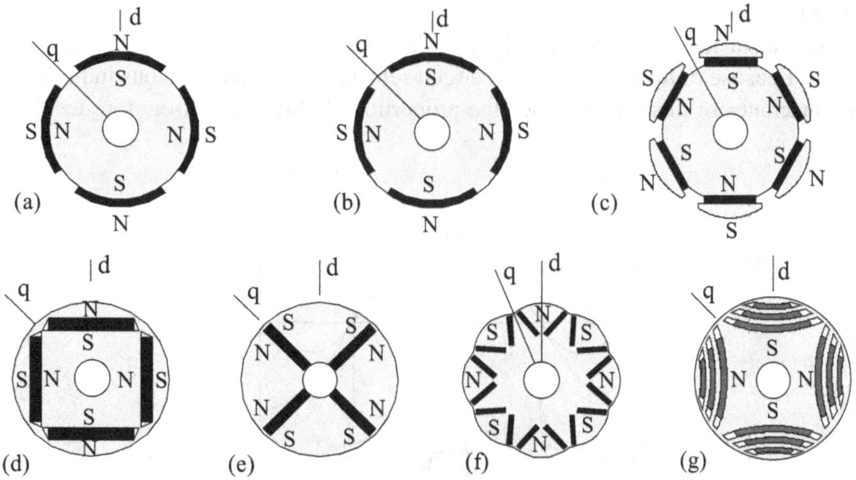

Figure 7.50 Rotors of permanent magnet machines. (a) Rotor surface mounted magnets, (b) magnets embedded in the surface, (c) pole shoe rotor, (d) tangentially embedded magnets, (e) radially embedded magnets, (f) two magnets per pole in V position, (g) a synchronous reluctance rotor equipped with permanent magnets. Reproduced by permission of Tanja Hedberg, based on Morimoto, Sanada and Taniguchi (1994).

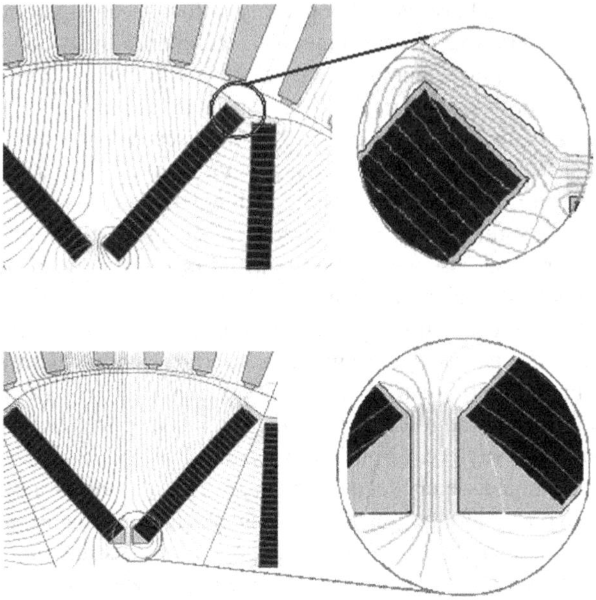

Figure 7.51 Air-gap flux of a V-magnet machine and the leakage fluxes of the magnet. In the bottom illustration, the leakage flux is reduced by constructing a flux barrier of air. Reproduced by permission of Tanja Hedberg.

flux density in no-load conditions. If there is iron on the rotor surface of a permanent magnet machine, a considerable armature reaction occurs, somewhat weakening the characteristics of the machine, Figure 7.52.

The permanent magnet material is best utilized in rotor-surface-magnet machines (Figure 7.50a). Because of the high magnetic circuit reluctance, the synchronous inductances are low and machines of this type produce the proportionally highest pull-out torque. However,

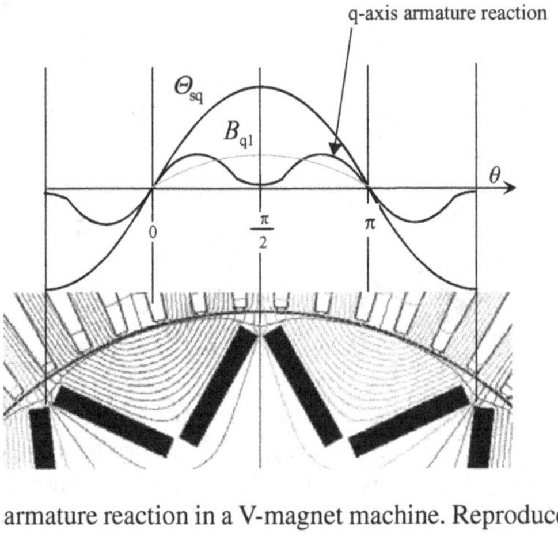

Figure 7.52 Quadrature armature reaction in a V-magnet machine. Reproduced by permission of Tanja Hedberg.

rotor surface magnets are liable to mechanical and magnetic stresses and also eddy current losses. In some cases, even the NdFeB magnets may be demagnetized. Figure 7.53 compares the air-gap densities of a V-magnet machine and a surface-magnet machine in an integer slot winding machine. Figure 7.53b also shows how the flux density is nearly constant in the area facing the magnet and zero elsewhere. Hence, the average flux of the permanent magnet machine may be calculated simply by using the relative magnet width ($\alpha_{PM} = w_{PM}/\tau_p$). Because of the rough flux plot, the figure does not reveal the surface magnet leakage flux at all. Typically, however, about 5–20 % of the flux of a surface magnet is lost in the leakage in the magnet edge areas.

In particular, in the case of a rotor surface magnet machine, special attention has to be paid to minimizing the cogging torque. The relative width of a permanent magnet has a significant influence on the quality of the torque of the machine. The quality of the torque has been investigated for instance by Heikkilä (2002), Kurronen (2003) and Salminen (2004).

Permanent magnet machines are mostly applied in frequency converter drives, where no damping is necessary. A damper winding is, however, necessary in network drives. The machine constructions of Figure 7.50c–f offer the best possibilities of embedding also a damper winding in the poles, but also a rotor surface-magnet machine may be equipped with a damper; however, to obtain enough conducting surface, some compromises have to be made. A thin aluminum plate for instance on top of the magnets is not enough, because, for example, 30 % (see Chapter 2 for damper winding principles) of the stator copper amount has to be reached in the damper.

If the machine E_{PM} is high as it usually has to be in generators, a line start as a motor is not possible despite the presence of a damper. Hence, generators may not achieve a line start property. This is explained by the fact that as a permanent magnet machine starts as an induction motor in a network supply, the permanent magnet flux linkage generates an asynchronous E_{PM}, which is short-circuited via the network impedance. A high current and a high braking torque are created in this short-circuit process, but the asynchronous torque created by the damper winding is too weak to accelerate the machine to the synchronous speed. Such a machine has to be synchronized to the network as a traditional synchronous machine. In motors, E_{PM} may be lower ($e_{PM} \ll 1$), and a line start-up may be possible with a correctly designed damper winding.

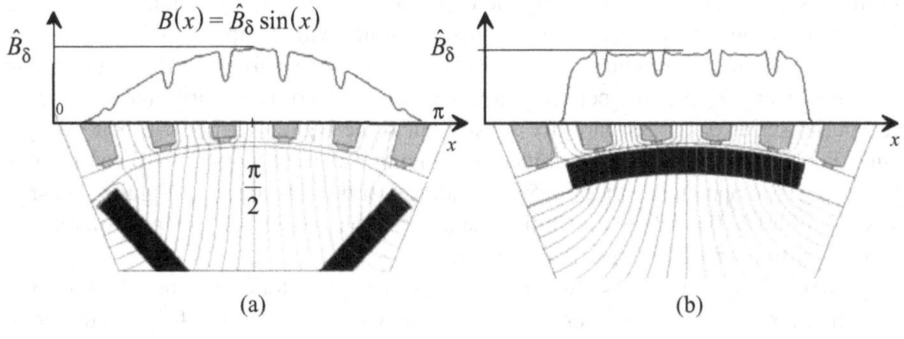

Figure 7.53 Comparison of the pole constructions and no-load air-gap densities of (a) a V-magnet machine with variable air gap and (b) a surface-magnet machine with constant air gap. Reproduced by permission of Tanja Hedberg.

Under certain circumstances, synchronous reluctance machines are equipped with auxiliary magnets, Figure 7.50g. In that case, the function of the magnets is to improve the power factor of the machine. It is somewhat difficult to define where is the boundary between the PMSM with embedded magnets with L_q typically larger than L_d producing positive reluctance torque with negative direct-axis and positive quadrature axis currents or a PM assisted SynRM where the $L_d \gg L_q$. To confuse the situation even more e.g. Moncada *et al.* (2008) have presented a PMSM with $L_d > L_q$.

What was stated above about PMSMs is mainly valid for normal integer slot winding arrangements. Using fractional-slot concentrated-non-overlapping i.e. tooth-coil windings in permanent magnet machines has became popular because of the easy manufacturing of the tooth coils, especially, in the case of open slots. This machine type's properties do not suffer from using a large amount of poles, and therefore, it is often used as a "high-torque" machine which produces a large torque per volume as the machine yokes can be built small and as the end windings do not need large spaces.

Using tooth coils is impractical in other AC machine types but PMSMs because practical nonconductivity of the rotor is needed to avoid excessive rotor Joule losses. Special care must, therefore, be taken when using the well conductive NdFeB- or SmCo-magnets in tooth-coil machines. It might even be also possible to design tooth-coil synchronous machines with field windings and laminated rotors without damping. Significant voltages should be induced in the field winding coils and they should, therefore, be connected in a way that the sum of these voltages is minimized.

Some of the machine design principles presented in earlier chapters have to be slightly altered in case of tooth-coil machines. The biggest principal change is that a tooth-coil machine does not necessarily operate with the fundamental which the stator winding creates but with a harmonic which is called the operating harmonic. Actually, such principles become valid when starting to use any type of fractional slot windings – also when $q > 0.5$. In case of distributed fractional slot windings ($q > 0.5$), however, we traditionally, think that the machine's operating harmonic is the fundamental similarly as in case of integer slot machines even though the winding produces subharmonics whose wavelength is longer than the wavelength of the fundamental. Such thinking may be clever as the amplitudes of the subharmonics stay low in traditional fractional slot windings. Anyway, the subharmonics should be treated as leakage and be taken into account, especially in calculating the air-gap leakage inductance. The subharmonics also contribute to the machine losses if the rotor has any conductivity, which is the case in induction machines and synchronous machines with damper winding.

However, in tooth-coil windings with $q \leq 0.5$ it is clever to start using different thinking. The machine operates with an operating harmonic which is not necessarily the first member of the Fourier series of the air-gap flux density but some higher harmonic produced by the winding. For example, the base machine type having 12 stator slots and 10 rotor poles – the 12/10-machine operates with the -5th harmonic, not the fundamental. Figure 7.54 shows the flux plot of a 12/10 machine illustrating clearly that there is a weak two-pole fundamental and a strong fifth operating harmonic in the machine.

The 12/10 machine with $q = 0.4$ happens to be a special case of tooth-coil machines: It creates the same harmonic orders as an integer slot winding, $+1, -5, +7, -11, +13\ldots$ but operates with the -5th which can also be called the synchronous harmonic. The same stator equipped with an eight-pole rotor – the 12/8-machine with $q = 0.5$ – however, is a multiple of a 3/2-base machine operating with the fundamental. When defining the tooth-coil machine principles it

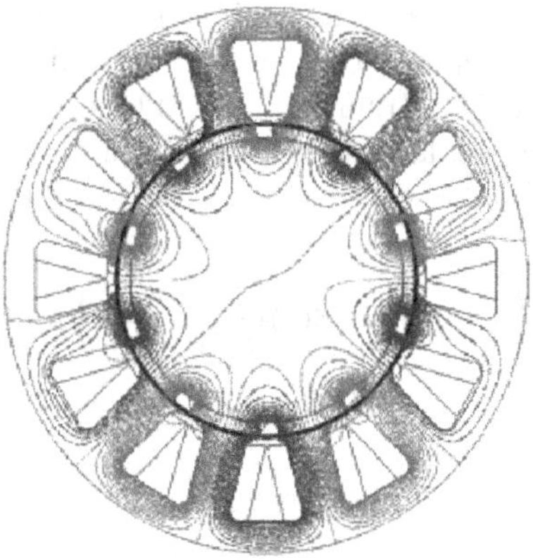

Figure 7.54 Flux paths of a 12-slot 10-pole radial flux machine. The machine clearly produces ten poles for the operating harmonic but there is also a two-pole flux line traveling from southwest to northeast. Figure reproduced by the permission of Hanne Jussila.

is important first to find the base machine similarly as in traditional winding machines. The base machine is found by dividing the fraction (e.g. 12/10 or 12/8) by the greatest common divisor of the numerator and denumerator resulting in even number of rotor poles (e.g. 1 or 4 in the cases of 12/10 or 12/8). The operating harmonic then is the base machine rotor pole number divided by two: In case of a 12/10 machine the fifth harmonic and in case of the 3/2 machine the fundamental. The machine rotates at the speed defined by the number of the rotor pole pairs and the supply frequency. For instance a 12/10 machine rotates at 600 min^{-1} with 50 Hz supply.

As a further example case an 18/16 permanent magnet machine with $q = 3/8$ can be studied. The base machine is a 9/8-machine which, again, produces a small two-pole fundamental and whose operating harmonic is the 4th spatial current linkage harmonic. Except the operating harmonic all other air-gap harmonics can be regarded as leakage components. Also the fundamental is a leakage component in this case.

7.3.10.1 Effect of Slotting on the Air-Gap Magnetic Field Distribution of Tooth-Coil PMSMs

In Chapter 3 we studied the effect of slotting in integer slot winding machines. Normally, the equivalent air-gap length δ_e calculated by the Carter factor k_C takes the slotting into account in machine design.

In tooth-coil machines, however, the air-gap flux density is badly distorted as the pole pitch is short and the slot openings are relatively wide. In Chapter 6, we analyzed the flux density distribution of rotor surface magnets in case of slotless air gap. In tooth-coil machines, open and possibly wide slots are often used because of manufacturing purposes which makes the flux density analysis more difficult than with integer slot winding machines.

It is, however, with reasonable accuracy, possible to analyze the air-gap flux density by multiplying the slotless air-gap flux density distribution by a relative permeance function introduced in Chapter 3 (Equation (3.13)) used for example by Zhu and Howe (1993). In the following we shall take also the curvature into account because the slot openings can be relatively wide compared with the slot openings of traditional machines. Zhu and Howe reformulate the Heller Hamata (1977) function (3.13) to a permeance function $\lambda(\alpha, r)$. The permeance function is defined similarly as in Chapter 3 at the center-axis of a stator slot and it depends on the radial position as

$$\lambda(\alpha, r) = \begin{cases} \Lambda_0 \left(1 - \beta(r) - \beta(r)\cos\dfrac{\pi}{0.8\alpha_0}\alpha\right) \text{ for } 0 \leq \alpha \leq 0.8\alpha_0 \\ \Lambda_0 \text{ for } 0.8\alpha_0 \leq \alpha \leq \alpha_t/2 \end{cases} \tag{7.231}$$

with

$$\Lambda_0 = \mu_0/\delta_{\text{ef}},\ \alpha_0 = b_1/r_{\text{s}},\ \alpha_{\text{t}} = \tau_{\text{u}}/r_{\text{s}}\ (r_{\text{s}} \text{ is the stator active surface radius}) \tag{7.232}$$

and is based on conformal mapping (Zhu and Howe 1993)

$$\beta(r) = \frac{1}{2}\left[1 - \frac{1}{\sqrt{1 + \left(\dfrac{b_1}{2\delta'_{\text{ef}}}\right)^2 (1 + \upsilon^2)}}\right], \tag{7.233}$$

where υ has to be iterated from

$$y\frac{\pi}{b_1} = \frac{1}{2}\ln\left(\frac{\sqrt{a^2 + \upsilon^2} + \upsilon}{\sqrt{a^2 + \upsilon^2} - \upsilon}\right) + \frac{2\delta'_{\text{ef}}}{b_1}\arctan\left(\frac{2\delta'_{\text{ef}}}{b_1}\frac{\upsilon}{\sqrt{a^2 + \upsilon^2}}\right), \tag{7.234}$$

and

$$a^2 = 1 + \left(\frac{2\delta'_{\text{ef}}}{b_1}\right)^2. \tag{7.235}$$

With stator active surface radius r_{s} and effective air-gap length $\delta'_{\text{ef}} = \delta + \dfrac{h_{\text{PM}}}{\mu_{\text{r}}}$ it can be written as a function of the radius r

$$y = \begin{cases} r - r_{\text{s}} + \delta'_{\text{ef}} = r - r_{\text{s}} + \delta + \dfrac{h_{\text{PM}}}{\mu_{\text{r}}} & \text{for inner rotor and} \\ r_{\text{s}} + \delta'_{\text{ef}} - r = r_{\text{s}} + \delta + \dfrac{h_{\text{PM}}}{\mu_{\text{r}}} - r & \text{for outer rotor} \end{cases}, \tag{7.236}$$

Now, open circuit air-gap flux density is just multiplied with the permeance variation function to get the flux density distribution which takes stator slotting into account. Figure 7.55 describes the result of finding the radial flux density distribution in a 12/10 machine.

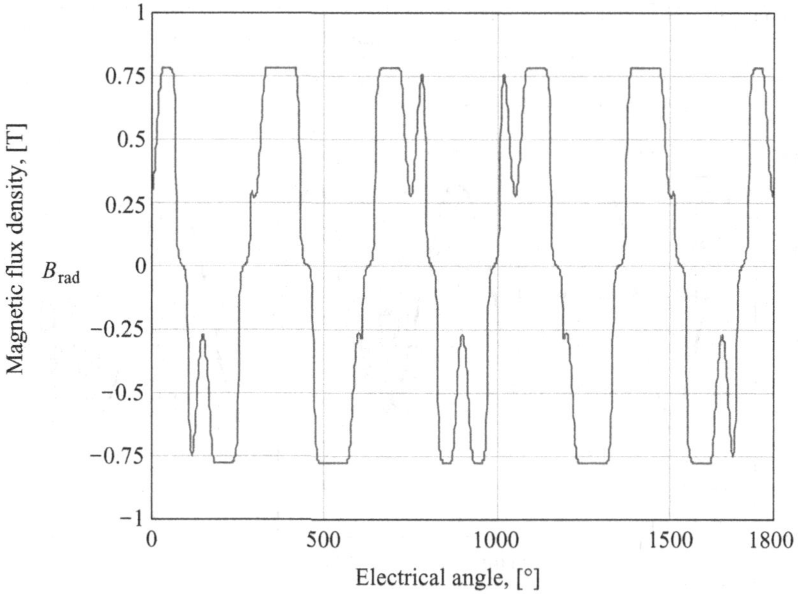

Figure 7.55 Analytic calculation result of the air-gap flux density of a radial flux 12/10 PMSM.

The operating harmonic flux density amplitude is found by analyzing the result of Figure 7.55 and writing its Fourier series. The synchronous harmonic flux density amplitude is then integrated to the peak value of the main flux of the machine similarly as in traditional machines.

7.3.10.2 Inductance Calculation in Tooth-Coil Machines

The synchronous inductance L_s of a nonskewed PMSM is calculated normally as a sum of partial inductances as

$$L_s = L_m + L_\sigma = L_m + L_{ew} + L_u + L_d + L_\delta, \tag{7.237}$$

where, L_m is the magnetizing inductance; L_{ew} is the end winding leakage inductance; L_u is the slot leakage inductance; L_d is the tooth tip leakage inductance and L_δ is the air-gap leakage inductance.

The main inductance L_{sp} of a single phase winding is calculated similarly as in Chapter 4, Section 4.1. The air-gap flux density of the operating harmonic B_{sp} (created by single phase stator current linkages) is, naturally, distributed sinusoidally. The peak value of Ψ_{sp} (the flux linkage created by the operating harmonic of the stator single-phase current linkage) can, therefore, be expressed as

$$\hat{\Psi}_{sp} = k_{wp} N_s \frac{2}{\pi} \tau_p l' \hat{B}_{sp}, \tag{7.238}$$

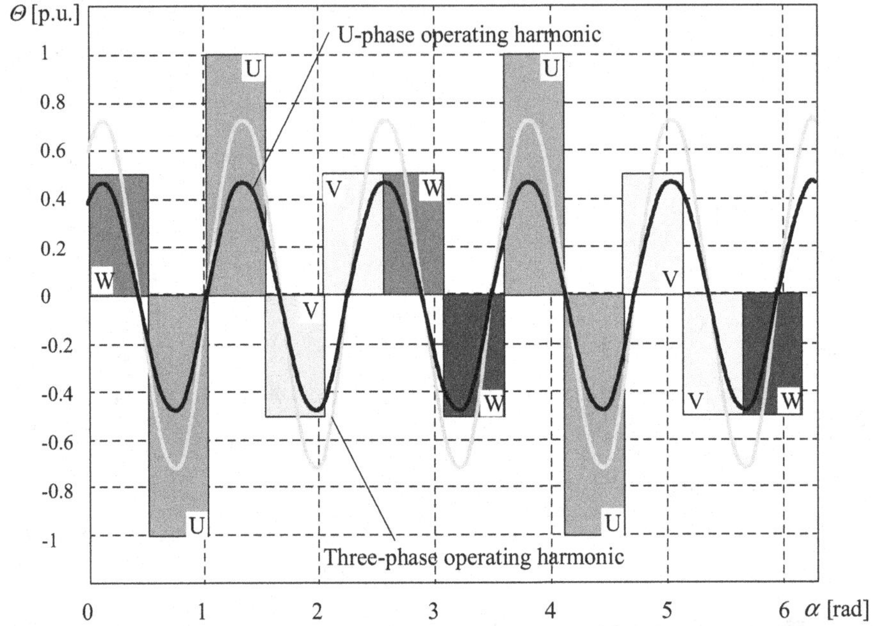

Figure 7.56 Stator-winding-created current linkage waveforms of a 12/10 Tooth Coil (TC)-PMSM when phase U carries its peak current ($i_U(t) = 1$, $i_V(t) = -0.5$ and $i_w(t) = -0.5$). The slot openingas are assumed infinitesimas and therefore the current linkages (integrated from linear current desities) behave as rectangles. Even though the mutual inductance seems negligible also phases V and W contribute to the operating harmonic in the tooth coil machine resulting in a similar magnetizing inductance as in integer slot machines. Fig. reproduced by the permission of Pavel Ponomarev.

where τ_p is the pole pitch; l' is the effective stator core length; N_s is the number of phase turns in series; k_{wp} is the winding factor of the operating harmonic and $2/\pi$ the arithmetic per-unit average value of a sinusoidal half wave.

The air-gap flux density of the operating harmonic B_{sp} produced by the tooth-coil winding's single phase stator current linkage Θ_{sp} (see Figure 7.56) is

$$\hat{B}_{sp} = \mu_0 \frac{\hat{\Theta}_{sp}}{\delta_{ef}} = \frac{\mu_0}{\delta_{ef}} \frac{4q}{\pi} \frac{z_Q}{c} k_{wp} \hat{I}_s, \tag{7.239}$$

where δ_{ef} is the effective air gap; q is the number of stator slots per pole per phase; z_Q is the number of stator conductors in a slot; $c = 1$ for single-layer windings and $c = 2$ for double-layer windings.

Only double-layer windings are studied further. For double-layer windings (7.239) can be rewritten as

$$\hat{B}_{sp} = \frac{\mu_0}{\delta_{ef}} \frac{4q}{\pi} \frac{m}{Q_s} k_{wp} N_s \hat{I}_s, \tag{7.240}$$

where m is the number of phases and Q_s is the number of stator slots. Combining (7.238), (7.239) and (7.240) the phase inductance (main inductance) can be found

$$L_{sp} = k_{wp}N_s \frac{2}{\pi}\tau_p l' \frac{\mu_0}{\delta_{ef}} \frac{4q}{\pi} \frac{m}{Q_s} k_{wp}N_s, \tag{7.241}$$

$$L_{sp} = \frac{2}{\pi}\tau_p l' \frac{\mu_0}{\delta_{ef}} \frac{4q}{\pi} \frac{m}{Q_s}(k_{wp}N_s)^2. \tag{7.242}$$

The magnetizing inductance L_m of an integer-slot rotating-field winding is found by multiplying (7.242) by a factor $m/2$ as the resultant current linkage wave is produced by all the m phases of the machine together.

$$L_m = \frac{m}{2}L_{sp} = \tau_p l' \frac{\mu_0}{\delta_{ef}} \frac{4q}{Q_s}\left(\frac{m}{\pi}k_{wp}N_s\right)^2 \tag{7.243}$$

Even though the mutual inductance between phases in tooth-coil machines can be completely different – even nonexistent – compared with the mutual inductance integer-slot winding machines where $L_{mn} = -{}^1/_2 L_p$ always, the operating harmonic magnitude is anyway contributed by all three windings and the magnetizing inductance of tooth-coil machines must be calculated analogously to integer slot machines.

7.3.10.3 Mutual Coupling Coefficient in Tooth-Coil Machine Phases

The main-flux linked mutual inductance in tooth-coil windings behaves in a more complicated way than with integer slot windings. In a 12/10 machine, there are no V- or W-phase current linkages which could create additional magnetic fluxes through the first phase winding U, see Figure 7.57. For example 12/14 and 24/20 machines behave similarly.

There are also other types of tooth coil PMSMs where the interaction between current linkages of different phases is more complex. For example, the current linkage for an 18/16 machine is presented in Figure 7.58.

In Figure 7.58 in some sectors the current linkages of different phases strengthen each other, but in other sectors they weaken each other. The mutual coupling coefficient m_c can be calculated as

$$m_c = \frac{\displaystyle\int_0^{2\pi} \Theta_U \Theta_V d\alpha}{\displaystyle\int_0^{2\pi} \Theta_U \Theta_U d\alpha}, \tag{7.244}$$

where Θ_U and Θ_V are the current linkages of corresponding phases.

For integer slot sinusoidally distributed windings the mutual coupling between two phases is $m_c = -0.5$. For an 18/16 machine the main flux mutual coupling coefficient is $m_c = -0.0385$ and for a 12/10 machine $m_c = 0$. Main flux mutual coupling coefficients m_c for different tooth-coil windings are shown in Table 7.9.

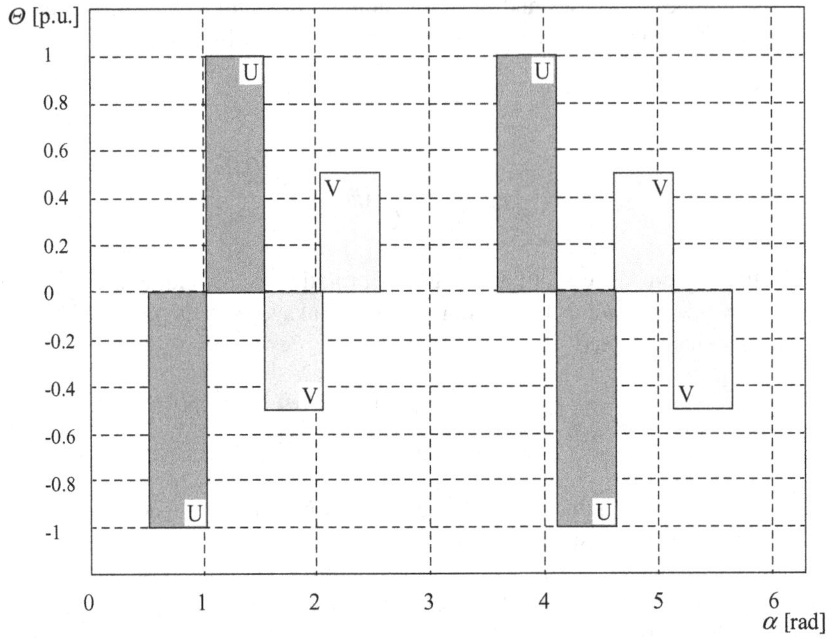

Figure 7.57 Simplified current linkage waveforms of two phases of a 12/10 tooth coil PMSM with $i_U(t) = 1$, $i_V(t) = -0.5$. The current linkages are nowhere overlapping with each other. Fig. reproduced by the permission of Pavel Ponomarev.

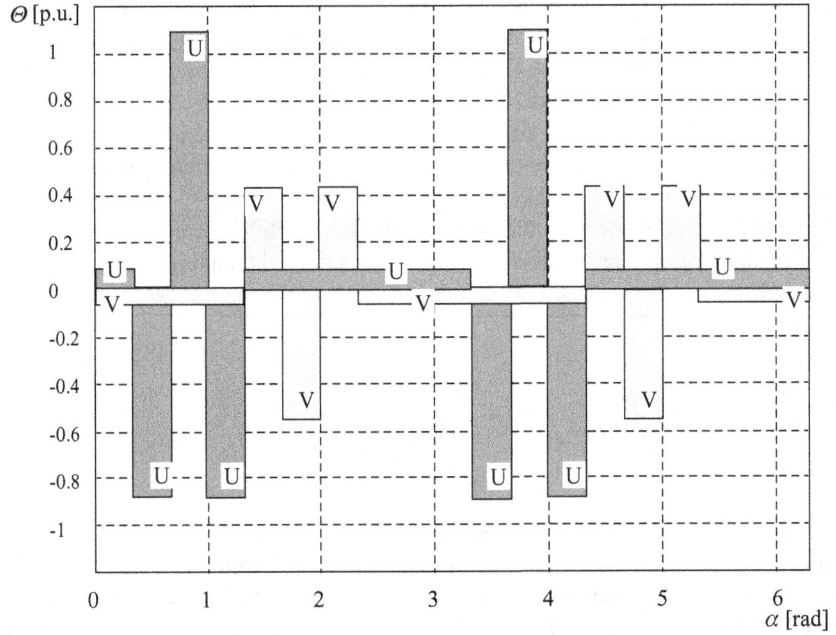

Figure 7.58 Simplified current linkage waveforms of two phases of an 18/16 tooth coil PMSM. The current linkages are overlapping. Fig. reproduced by the permission of Pavel Ponomarev.

Table 7.9 Harmonic air-gap leakage factors of tooth-coil PMSMs

Q_s \ $2p$		4	6	8	10	12	14	16	18	20
6	q	1 / 2		1 / 4	1 / 5		1 / 7	1 / 8		1 / 10
	k_{wp}	0.866		0.866	0.5		0.5	0.866		0.866
	m_c	−0.5		−0.5	0		0	−0.5		−0.5
	σ_δ	**0.46**		**4.8**	**26**		**53**	**22**		**36**
9	q	3/4	1 / 2	3 / 8	3 / 10	1 / 4	3 / 14	3 / 16		3 / 20
	k_{wp}		0.866	0.945	0.945	0.866	0.617	0.328		0.328
	m_c		−0.5	−0.039	−0.039	−0.5	−0.039	−0.039		−0.039
	σ_δ		**0.46**	1.2	2.4	**4.8**	15	71		112
12	q	1	2/3	1 / 2	2 / 5		2 / 7	1 / 4		1 / 5
	k_{wp}			0.866	0.933		0.933	0.866		0.5
	m_c			−0.5	0		0	−0.5		0
	σ_δ			**0.46**	**0.96**		**2.9**	**4.8**		**26**
15	q	1 1/4	5/6	5/8	1 / 2		5 / 14	5 / 16		1 / 4
	k_{wp}				0.866		0.951	0.951		0.866
	m_c				−0.5		−0.013	−0.013		−0.5
	σ_δ				**0.46**		1.4	2.1		**4.8**
18	q	1 1/2	1	3/4	3/5	1 / 2	3 / 7	3 / 8		3 / 10
	k_{wp}					0.866	0.902	0.945		0.945
	m_c					−0.5	0	−0.039		−0.039
	σ_δ					**0.46**	**0.83**	**1.2**		**2.4**
21	q	1 3/4	1 1/6	7/8	7/10	7/12	1 / 2	7 / 16		7 / 20
	k_{wp}						0.866	0.890		0.953
	m_c						−0.5	−0.007		−0.007
	σ_δ						**0.46**	0.8		1.5
24	q	2	1 1/3	1	4/5	2/3	4/7	1 / 2		2 / 5
	k_{wp}							0.866		0.933
	m_c							−0.5		0
	σ_δ							**0.46**		**0.96**
27	q	2 3/4	1 1/2	1 3/8	9/10	3/4	9/14	9/16	1 / 2	9 / 20
	k_{wp}								0.866	0.877
	m_c								−0.5	−0.004
	σ_δ								**0.46**	0.75

integral product of number of phases, not applicable.

magnetic pull in the base machine is unbalanced (the base machine can be multiplied if used in practice).

Table reproduced by permission of Pavel Ponomarev.

Table 7.9 takes into account only the mutual coupling via the air-gap flux. There are, however, other magnetic couplings between phases – coupling via the slot leakage flux and coupling via the end winding leakage flux. The end winding leakage flux mutual coupling is relatively weak for a tooth coil PMSM as the end windings are compact and different phases are usually coupled just through the air. The slot leakage mutual coupling, however, appears strong when coil sides of different phases are located in the same slot.

7.3.10.4 Air-Gap Harmonic Leakage Inductance

A tooth-coil winding produces a very nonsinusoidal current linkage waveform containing a big proportion of the harmonics resulting in a large air-gap leakage inductance (see Chapter 4, Section 4.3.2). In distributed winding machines the air-gap leakage inductance can be almost negligible but in tooth-coil machines this leakage component plays a major role and to a large extent defines the machine synchronous inductance, and therefore the machine properties.

The air-gap harmonic leakage inductance L_δ can be defined, in principal similarly as earlier in Chapter 4 but the harmonic air-gap leakage factor σ_δ is now defined slightly differently as

$$\sigma_\delta = \sum_{\substack{\nu=1 \\ \nu\neq 3,6,9\ldots \\ \nu\neq p}}^{\nu=+\infty} \left(p\frac{k_{\mathrm{w}\nu}}{\nu k_{\mathrm{wp}}}\right)^2, \tag{7.245}$$

where $k_{\mathrm{w}\nu}$ is the winding factor of the νth harmonic. The term $\nu = p$ represents the operating harmonic, and thus, the magnetizing inductance L_m component. Note, that elsewhere in this book ν means the order of the harmonic but here it is the pole pair number of the harmonic. The pole pair numbers divisible by three are excluded also in tooth-coil machines with three phases.

Table 7.9 shows the calculated air-gap harmonic leakage factors for different tooth-coil PMSMs. The numbers of slots per pole and phase q, winding factors of operating harmonics k_{wp} and mutual coupling coefficients m_c are given as well.

The behavior of air-gap harmonic leakage factors calculated for integer slot three-phase machines and the corresponding values for tooth-coil windings are presented in Figure 7.59.

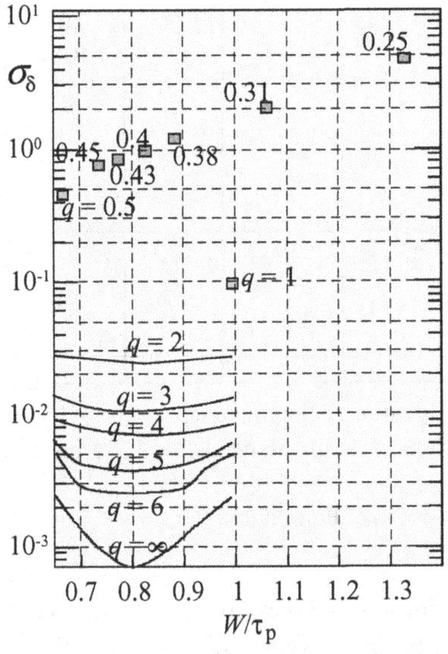

Figure 7.59 Air-gap harmonic leakage factors σ_δ for integer-slot and tooth-coil windings versus coil span W/τ_p for integer-slot windings. With tooth-coil windings $W = \tau_\mathrm{u}$. The smaller the number of slots per pole and phase q is, the bigger is the air-gap leakage.

The lower q a three-phase winding has the higher will be the air-gap harmonic leakage factor σ_δ.

As can be seen, tooth-coil machines can have an air-gap leakage inductance which has a very big impact on the machine properties and is even tens of times the magnetizing inductance, making such combinations impractical. One of the smallest numbers of slots per pole and phase, in practice, $q = 0.25$ results in $L_\delta = 4.8L_\mathrm{m}$. A machine designer can select the tooth-coil machine topology more or less based on the air-gap leakage inductance. If a low synchronous inductance is desired, topologies with low air-gap harmonic leakage coefficients should be selected. However, in some cases, e.g. in traction applications, a high leakage might be desired to get a long field weakening range and then a topology with a higher air-gap leakage might be clever.

Alberti *et al.* (2010) and Dajaku *et al.* (2012) suggested subharmonic flux barriers which can decrease the subharmonic component of the air-gap harmonic leakage inductance L_δ and reduce losses. Machines with $m_\mathrm{c} = 0$ can be built with separate stator phase modules. Such modules break the two-pole flux path in the stator thus resulting in a clever suppression of the sub harmonics. For example a 12/10 machine can be built with six stator-yoke-breaking flux barriers to almost totally cancel the fundamental in the machine. Also the adverse effects of the fundamental in a 12/10 machine will be cancelled simultaneously.

7.3.10.5 End Winding Leakag Inductance

Tooth coils have the shortest possible end windings producing small leakage inductance and very small mutual leakage inductance. Figure 7.60 illustrates the end winding leakage flux. The end winding sections of tooth-coil PMSMs can be considered as halves of a solenoid. Therefore, the leakage inductance of tooth-coil windings can be calculated using expressions for an air-cored solenoid inductance

$$L_\mathrm{solenoid} = \mu_0 \mu_\mathrm{env} \frac{\left(\frac{z_Q}{c}\right) S}{h_4}, \tag{7.246}$$

where h_4 is the height of the solenoid; S is the cross-section area of the air core; μ_env is the relative permeability of the environment. If the end turns (Figure 4.20) are made as half-circles ($l_\mathrm{ew} = w_\mathrm{ew}/2$), the total inductance of the end windings of a double layer winding can be expressed as

$$L_\mathrm{ew} = \frac{Q_\mathrm{s}}{m} L_\mathrm{solenoid} = \frac{Q_\mathrm{s}}{m} \mu_0 \mu_\mathrm{env} \frac{\left(\frac{z_Q}{2}\right)^2 \pi (l_\mathrm{ew})^2}{h_4}. \tag{7.247}$$

Because of the presence of iron parts in the vicinity (iron frame, end laminations of the stator stack) the relative permeability μ_env can be chosen from the range of 1.2–2 depending on materials and how compactly the end regions of the machine are assembled. There is also some mutual coupling between the phases due to the end-winding mutual inductance but this coupling is very weak and, therefore, can be neglected without making a big mistake.

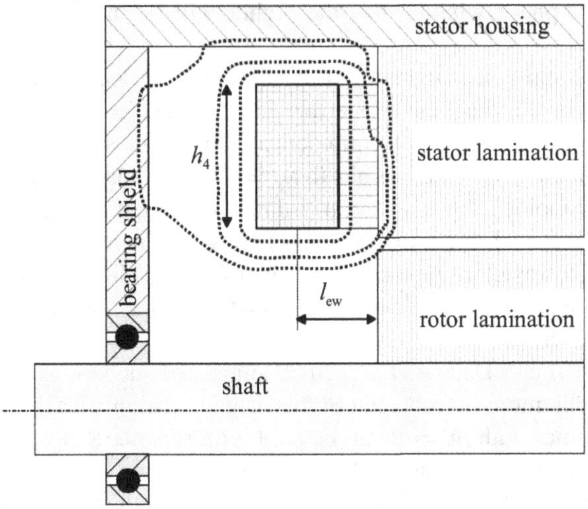

Figure 7.60 End-winding leakage flux paths. Replacing the end windings by an air cored solenoid with average height h_4 and average length l_{ew}.

7.3.10.6 Slot Leakage Inductance

The slot leakage inductance L_u is calculated as advised in Ch 4. It can be kept minimal with shallow and wide slots. To achieve a high torque density the aspect ratio (height/width) of the slot should be as small as possible – even close to one which can be feasible in tooth coil machines.

7.3.10.7 Tooth-Tip Leakage Inductance

The tooth tip leakage inductance is calculated as advised in Ch 4. However, the wide slot openings can cause that the permeance factor λ_d given in (4.96) is no more valid. In case of big b_1/δ (wide open slots) the toot tip leakage L_d can be even negative. Figure 7.61 illustrates the behavior of the slot and toot tip leakages in cases of open or semi-closed slots.

The permeance factor can be calculated according to Voldek (1974) as

$$\lambda_d = \frac{1}{2\pi}\left[\ln\left(\frac{\delta^2}{b_1^2} + \frac{1}{4}\right) + 4\frac{\delta}{b_1}\text{atan}\frac{b_1}{2\delta}\right]. \tag{7.248}$$

The dependency of λ_d on b_1/δ is given in Figure 7.62. With small slot openings λ_d has a significant positive value but with extremely wide slot openings λ_d becomes slightly negative.

For rotor surface magnet machines the air-gap length δ should include the physical air gap and also the height of the magnets divided by the relative permeability h_{PM}/μ_{rPM} of the permanent magnet material. For an internal permanent magnet configuration only the physical air gap should be used in the tooth tip leakage calculation.

To minimize the tooth tip leakage L_d, open slots and nonmagnetic wedges should be used. This, however, also affects the magnetizing inductance by increasing the effective air gap for rotor surface permanent magnet machines.

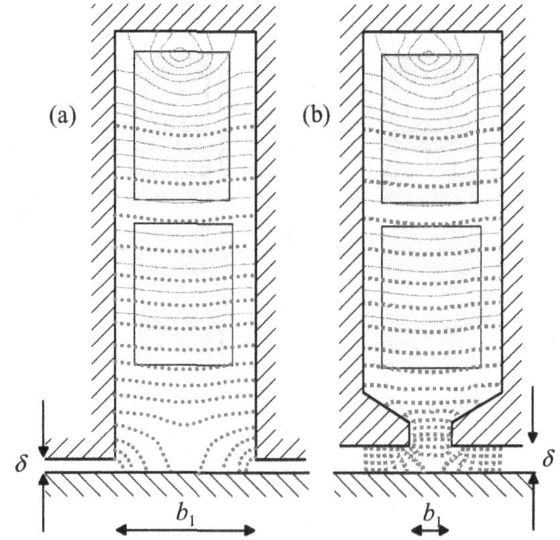

Figure 7.61 Slot and tooth tip leakage flux lines. (a) open slot, (b) semi-closed slot with slot opening b_1 and air gap δ. Modified from Voldek (1974).

7.3.10.8 Equivalent Length of a PMSM

What was stated in Chapter 3 about the equivalent length l' of electrical machines is not really valid for a permanent magnet machine. In Equation (3.36) it was stated that $l' \approx l + 2\delta$ which can be used for a machine having no sub-stacks when the stator and rotor stacks are of equal length. In case of a rotor surface magnet PMSM, the stator current linkage meets a very long air gap as the permanent magnet itself behaves almost like an air gap and h_{PM}/μ_r must be included in the air-gap length

$$\delta_{PM} = \delta + h_{PM}/\mu_r. \tag{7.249}$$

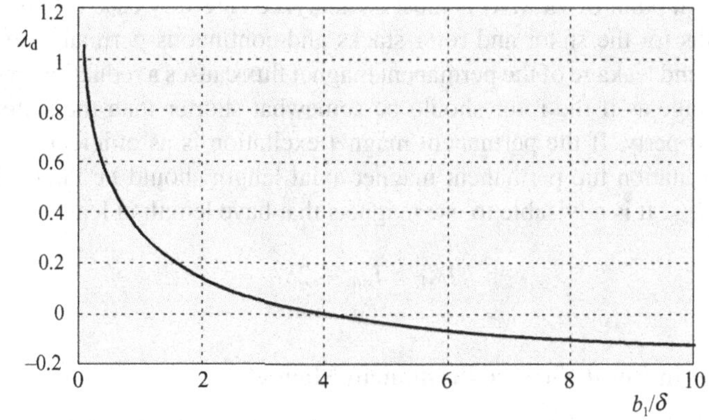

Figure 7.62 Permeance factor of the tooth-tip leakage.

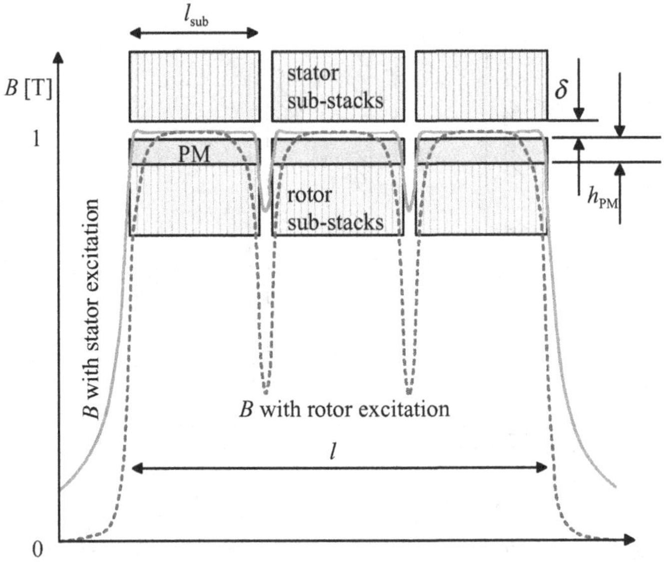

Figure 7.63 Permanent magnet machine length effects when the rotor surface magnets have equal axial lengths with stator substacks.

Therefore, the equivalent length of a PMSM when calculating the stator inductance is

$$l'_{PM} \approx l + 2\delta + h_{PM}/\mu_r. \tag{7.250}$$

This equivalent length, however, is not at all valid when calculating the rotor excitation effects in the stator. In machines having continuous current linkages, the machine length seen from the rotor winding or the stator winding are equivalent. In case of a PMSM, the current linkage caused by the rotor permanent magnet material is not necessarily continuous. If there are free spaces between magnets, there will be permanent magnet flux leakage and the Ψ_{PM} and therefore also E_{PM} induced in the stator will decrease and the machine looks shorter from the rotor excitation point of view. This must be analyzed carefully case by case but in case of equivalent lengths for the stator and rotor stacks and continuous permanent magnets on the rotor surface the end leakage of the permanent magnet flux causes a reduction in the permanent magnet flux linkage as if the rotor should be somewhat shorter than the stator. Figure 7.63 illustrates this property. If the permanent magnet excitation is as efficient as corresponding field winding excitation the permanent magnet axial length should be larger than the stator substack length l_{sub}. It is advisable to use magnets that have length at least

$$l_{PM} \approx l_{sub} + 2\delta. \tag{7.251}$$

7.3.10.9 Losses in Rotor Surface Permanent Magnets

Air-gap harmonics cause eddy current losses in conducting permanent magnets on the rotor surface. Accurate calculation of the losses is difficult and 3D finite element analysis is recommended. However, analytic approaches can be used to evaluate the losses with indicative results. The following method based on Carter factor was described by Pyrhönen *et al.* (2012). As magnet materials are linear and fairly well conducting, the eddy currents in the magnets follow the surface impedance phase angle of 45°. The method assumes that an eddy current path can be described with its resistance taking the depth of penetration in the permanent magnets into account and that there is no eddy-currents-caused reaction in the air-gap flux density.

The flux density variation under a slot pitch may be written as a Fourier series with $\alpha \in [0, 2\pi/Q_s]$, the slot being in the middle

$$B_\delta(\alpha) = B_{av}\left[1 - \sum_{k=1}^{\infty}(-1)^k \beta k_C a_{1k}\cos(kQ_s\alpha)\right], \tag{7.252}$$

where

$$a_{1k} = \frac{2\sin\left(k\pi\dfrac{b_1'}{\tau_u}\right)}{k\pi\left[1-\left(k\dfrac{b_1'}{\tau_u}\right)^2\right]},\quad b_1' = \gamma\frac{\delta}{\beta},\quad \gamma = \frac{\left(\dfrac{b_1}{\delta}\right)^2}{5+\dfrac{b_1}{\delta}} \tag{7.253}$$

and β according to (3.10a).

In (7.253), b_1' describes the total width of the area where the slot opening has an impact on the rotor surface flux density (Figure 3.5). The harmonic (with ordinal k) flux density amplitudes caused by the permeance variation are

$$\hat{B}_k = B_{av}\beta k_C \frac{2\sin\left(k\pi\dfrac{b_1'}{\tau_u}\right)}{k\pi\left[1-\left(k\dfrac{b_1'}{\tau_u}\right)^2\right]}. \tag{7.254}$$

Each of these amplitudes has its own pole pitch

$$\tau_k = \tau_u/(2k). \tag{7.255}$$

In the case of wide slot openings and narrow teeth there will be mainly the fundamental slot frequency amplitude present. However, in the case of narrow slot openings, significantly higher frequencies will be present. Figure 7.64 illustrates a comparison of the relative permeance behavior and the harmonic contents on the rotor surface in two different slot opening cases.

As the rotor rotates, the flux density dips described above will travel on the magnet surfaces. The frequencies and angular frequencies of the flux density variations in the magnets caused by the stator slotting and the rotor rotation are

$$f_{PM,k} = kn_{syn}Q_s,\quad \omega_{PM,k} = 2\pi\, f_{PM,k}. \tag{7.256}$$

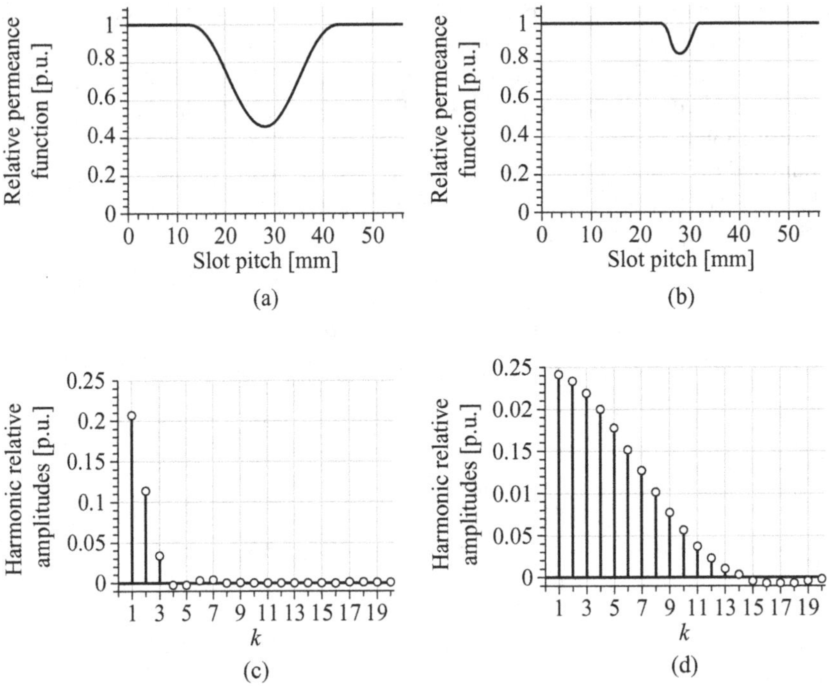

Figure 7.64 Relative flux density dips under a wide slot opening (a) and a narrow slot opening (b) on the permanent magnet surface of the machine used as an example in the experimental part. The harmonic relative amplitude spectra of these two different cases differ significantly from each other (c and d). The former (c) mainly contains a fundamental (amplitude 0.21 p.u.) with the wavelength of one slot pitch and the latter (d) contains several significant permeance harmonics, however, with much lower amplitudes (fundamental amplitude 0.024 p.u.).

The relative permeability of sintered neodymium iron boron magnets, which are the most important permanent magnets at the moment, is typically $\mu_{\mathrm{PM}} = 1.05$. The resistivity of sintered neodymium magnets varies in the range of 100–200 μΩcm. The depth of penetration in the permanent magnet material is

$$\delta_{\mathrm{PM}} = \sqrt{\frac{2\rho_{\mathrm{PM}}}{\omega_{\mathrm{PM}}\mu_0\mu_{\mathrm{r,PM}}}}. \tag{7.257}$$

Even though the resistivity of the permanent magnet material is relatively low the depth of penetration can be larger than the thickness of the magnets, and hence, the flux density variation penetrates through the whole magnet in some practical applications. This emphasizes the fact that in open slot machines the material under the permanent magnets should be nonconducting. In practice, laminated NdFeB or SmCo magnets have to be used or magnets have to be embedded inside laminations to avoid excessive eddy current losses in them.

The harmonics traveling on the rotor surface cause eddy current densities $J_{PM,k}(z)$. In the depth z in the permanent magnets, a flux density varies with respect to the depth of penetration δ_{PM} with the current density on the surface of PM $J_{0,PM,k}$ as

$$J_{PM,k} = J_{0,PM,k} e^{-z/\delta_{PM,k}}. \tag{7.258}$$

The eddy-current-caused loss density is proportional to the current density squared

$$\frac{P_{PM,k}}{V} \hat{=} J_{PM,k}^2(z) = J_{0,PM,k}^2 e^{-2z/\delta_{PM,k}}. \tag{7.259}$$

The total loss caused by the k-harmonic of the eddy current density is

$$\begin{aligned} P_{PM,k,tot} &= \int_V \frac{P_{PM,k}}{V} dV \\ &\hat{=} \int_0^{\infty} J_{0,PM,k}^2 e^{-2z/\delta_{PM,k}} dz = J_{0,PM,k}^2 \frac{\delta_{PM,k}}{2}. \end{aligned} \tag{7.260}$$

Therefore, we can calculate the average power loss in the magnet using eddy current paths that penetrate to half of the depth of penetration for each of the harmonics. If the magnet, however, is thinner than the depth of penetration ($h_{PM} < \delta_{PM}$), the resistance of the eddy current path should be calculated with an eddy current depth giving the average value for the eddy current loss. The model assumes that the eddy current distributions have an exponential dependence. The magnet height h_{PM} and the skin depth δ_{PM} are taken into account

$$\begin{aligned} P_{hPM,k,tot} &\hat{=} \int_0^{h_{PM}} J_{0,PM,k}^2 e^{-2z/\delta_{PM,k}} dz \\ &= J_{0,PM,k}^2 \frac{\delta_{PM,k}}{2} (1 - e^{-2h_{PM,k}/\delta_{PM,k}}). \end{aligned} \tag{7.261}$$

Now, if for instance the slot opening is $b_1 = 0.018$ m, $\tau_u = 0.056$ m, $\delta = 0.002$ m, $h_{PM} =$ 8 mm, $Q_s = 12$ and the rotational speed is 2400 min^{-1} we get for $\delta_{PM,1} = 27$ mm. The depth of penetration of the permeance harmonic fundamental is, therefore, significantly greater than the magnet depth. We now have to calculate the eddy current path resistance taking (7.261) into account.

If a single, sintered permanent magnet piece is wide compared with the harmonic pole pitch τ_k, we can calculate the losses caused by the amplitudes of sinusoidal flux density variations traveling on the rotor surface based on Figure 7.65 in which approximate eddy-current paths are shown. The directions of eddy-currents as indicated by the arrows correspond to a from-left-to-right movement of the permanent magnets across the stator slots.

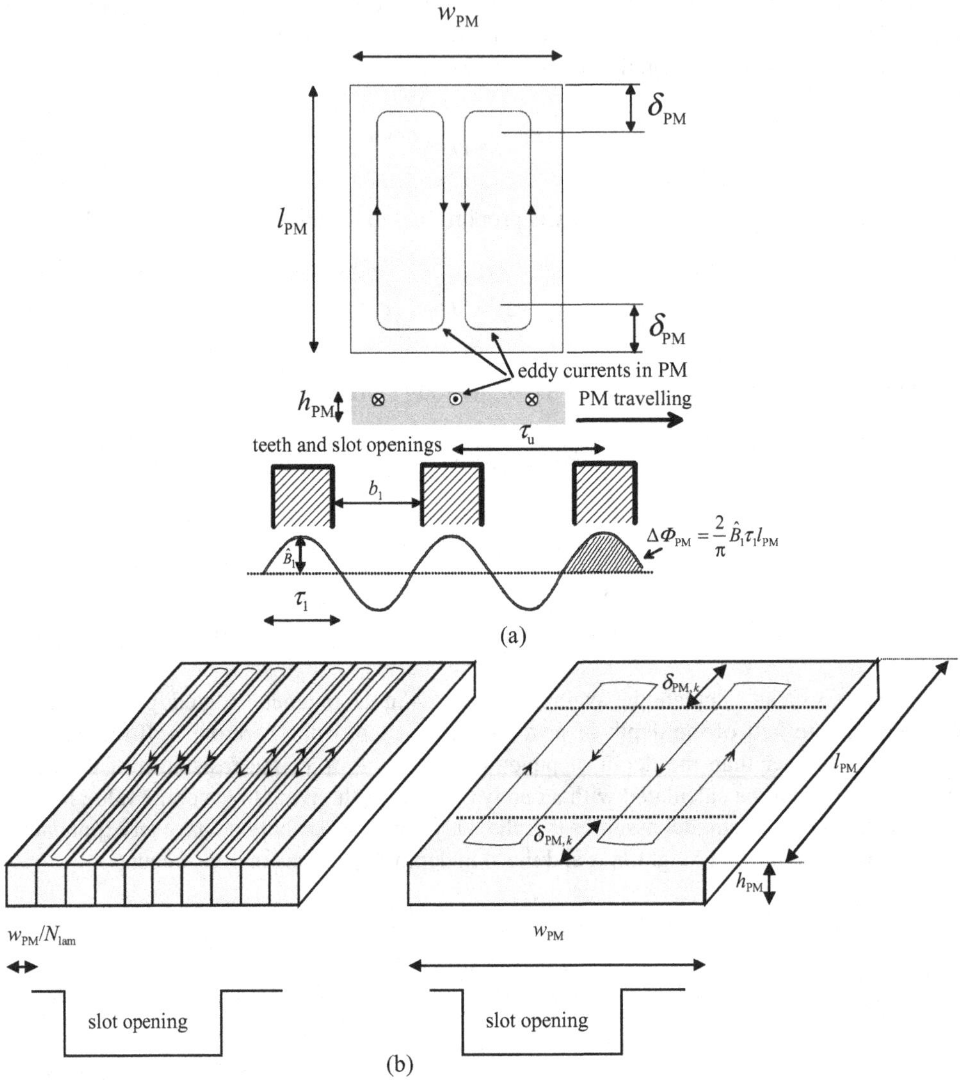

Figure 7.65 (a) Eddy current paths in a wide magnet piece caused by permeance variations, (b) illustration of a laminated and a bulky magnet.

Equations given above are applied to define the flux density variation under the slot opening as a Fourier series. As we have defined the amplitudes of the permeance-variation-caused flux densities, we can calculate the flux of each wave traveling on the rotor surface.

$$\hat{\Phi}_{PM,k} = \frac{2}{\pi}\hat{B}_k \tau_k l_{PM} \tag{7.262}$$

The emf induced in the permanent magnet while traveling under a flux is

$$\hat{e}_{\mathrm{PM},k} = \omega_k \hat{\Phi}_{\mathrm{PM},k}. \tag{7.263}$$

The following task is to define the eddy current path resistance in the permanent magnet. We observe first a wide magnet. If the magnet piece is longer than the depth of penetration ($l_{\mathrm{PM}} > \delta_{\mathrm{PM},k}$) in the material at the slot permeance harmonic angular frequency, the depth of penetration at the end areas is used to calculate approximately the eddy current path resistance, Figure 7.65.

In the case of wide magnet pieces, the cross-sectional surface ($w_{\mathrm{ax}} \times h$) of the axial route carrying the eddy currents on the path around the area traveling under the slot is according to (7.261)

$$S_{\mathrm{PM,ax},k} = w_{\mathrm{ax}} \times h \approx \frac{\tau_k}{2} \times \frac{\delta_{\mathrm{PM},k}}{2}(1 - e^{-2h_{\mathrm{PM}}/\delta_{\mathrm{PM},k}}). \tag{7.264}$$

Both axial and tangential path resistances must be taken into account. At the ends of the wide magnet, the path is assumed to be as wide as the depth of penetration in the magnet $\delta_{\mathrm{PM},k}$.

$$S_{\mathrm{PM,tan},k} = w_{\mathrm{tan}} \times h \approx \delta_{\mathrm{PM},k} \times \frac{\delta_{\mathrm{PM},k}}{2}(1 - e^{-2h_{\mathrm{PM}}/\delta_{\mathrm{PM},k}}). \tag{7.265}$$

The length of the path is defined next, and it can be done according to Figure 7.65. We are observing sinusoidal excitation, and therefore, the eddy current densities are sinusoidally distributed on average. We have to increase the path apparent resistances by $\pi/2$ (inverse of the average of sine wave). The resistances for the eddy current paths in a wide magnet are then given as

$$R_{\mathrm{PM,Ft},k} = \frac{\pi}{2}\left(\frac{2(l - \delta_{\mathrm{PM},k})}{S_{\mathrm{PM,ax},k}} + \frac{2\tau_k}{S_{\mathrm{PM,tan},k}}\right)\rho_{\mathrm{PM}}. \tag{7.266}$$

The RMS eddy current in the magnet (or magnet lamination) is now

$$I_{\mathrm{PM},k} = \frac{\hat{e}_{\mathrm{PM},k}/\sqrt{2}}{R_{\mathrm{PM},k}\sqrt{2}}. \tag{7.267}$$

The resistance of the magnet is multiplied by $\sqrt{2}$ to take also the reactance of the path into account in the path impedance. The phase shifts φ_k in the eddy currents will be 45° according to the linear surface impedance theory. The average eddy current power in a magnet lamination under one harmonic pole pitch is then

$$P'_{\mathrm{PM,Ft},k} = E_{\mathrm{PM},k} I_{\mathrm{PM},k} \cos\varphi_k. \tag{7.268}$$

There will be w_{PM}/τ_k harmonic current paths in a magnet and in the motor totally $2p$ magnets, and hence, the total eddy current loss in a wide magnet caused by the slot openings per harmonic is

$$P_{PM,Ft,tot,k} = 2p\frac{w_{PM}}{\tau_k}P'_{PM,Ft,k}. \tag{7.269}$$

Often wide bulky magnets cannot, however, be used at all because of the too large losses in them, and in such cases the magnet width w_{PM} has to be divided into N_{lam} laminations.

In the cases where the magnet is laminated so that an individual permanent magnet lamination width is significantly narrower than the harmonic pole pitch ($w_{PM}/N_{lam} \ll \tau_k$), we can use an approach based on the traditional eddy current loss equation for laminations, see Chapter 3, Section 3.6.2.

$$P_{PM,Ft,lam,k} = \frac{V_{PM}\pi^2 f_{PM,k}^2 w_{PM}^2 \hat{B}_k^2}{6\rho_{PM}}. \tag{7.270}$$

The transition from the above presented approximate calculation to using (7.270) cannot be done straightforwardly as the pole pitches of several permeance harmonics can be several millimeters and (7.270) should give too large a value for the loss if it is used for laminations being several mm wide. Therefore, we need to develop a modified version of the above presented method. In case of $\tau_k/n < w_{PM}/N_{lam} < \tau_k$, where $n \in (3, 4, 5\ldots)$ the method must be selected so that it gives at the boundary (7.270) the same loss value as the modified equations below; so we must modify the flux (7.262), the resistance (7.266) and the cross-sectional surfaces (7.265) and (7.264) to give appropriate values for narrow magnet laminations:

$$\hat{\Phi}_{PM,k} \approx \hat{B}_k\frac{w_{PM}}{N_{lam}}l_{PM}. \tag{7.271}$$

The resistive path in the magnet will be different as the harmonic pole pitch does not now define the width of the path and does not let us to have sinusoidal eddy current distribution. Also, the narrow lamination forces the eddy currents to run towards the ends of the magnet laminations, therefore

$$R_{PM,Ft,k} \approx 2\left(\frac{l}{S'_{PM,ax,k}} + \frac{w_{PM}/N_{lam}}{S'_{PM,tan,k}}\right)\rho_{PM}, \tag{7.272}$$

where

$$S'_{PM,ax,k} \approx \frac{w_{PM}}{2N_{lam}} \times \frac{\delta_{PM,k}}{2}(1 - e^{-2h_{PM}/\delta_{PM,k}}). \tag{7.273}$$

The axial currents run until the end of narrow laminations, and therefore, we take the tangential path width the same as the lamination width and the width gets the same value as (7.273)

$$S'_{PM,ax,k} \approx S'_{PM,tan,k}. \tag{7.274}$$

7.3.10.10 Winding Harmonics Causing Eddy-Current Losses

A tooth coil winding machine produces an especially large amount of winding harmonics. In such a machine the rotor will experience a series of asynchronous flux frequencies. The stator-harmonics-caused eddy-current losses in the magnets can be estimated similarly as the eddy-current losses caused by the permeance harmonics. The stator current linkages represented by equivalent current sheets on the surface of the magnets can be calculated as

$$\hat{\Theta}_{s\nu} = \frac{m}{2}\frac{4}{\pi}\frac{k_{w\nu}}{p\nu}\frac{N_s}{2}\frac{1}{2}\sqrt{2}I_s = \frac{mk_{w\nu}N_s}{\pi p\nu}\hat{i}_s. \tag{7.275}$$

In a smooth air gap, the stator current linkages produce amplitudes for the harmonics according to the winding factors as

$$\hat{B}_{s\nu} = \mu_0\frac{\hat{\Theta}_{s\nu}}{\delta_{ef}}. \tag{7.276}$$

Each of the significant harmonics now have to be analyzed from the PM eddy current loss point of view.

Example 7.6: The winding factors of a 12/10 machine are $k_{w1} = 0.067$, $k_{w-5} = 0.933$, $k_{w7} = 0.933$, $k_{w-11} = 0.067$, $k_{w13} = 0.067$, $k_{w-17} = 0.933$ etc. In case of a 12/10 machine operating at the -5th spatial harmonic, define the angular frequencies seen by the rotor surface permanent magnets.

Solution: The mechanical angular speed $\Omega = -\omega_s/5$. The stator fundamental travels at speed ω_s in the positive direction while the permanent magnets travel with the synchronous speed $-\omega_s/5$ in the negative direction. The angular velocity caused by the fundamental on the PM surface is $1\,{}^1\!/_5\omega_s$. The seventh harmonic travels at speed $+\omega_s/7$ in the positive direction. Its pole pitch is one-seventh of the pole pitch of the fundamental and therefore the angular velocity seen by the rotor surface based on this harmonic is $({}^1\!/_5 + {}^1\!/_7) \times 7 = ({}^{12}\!/_{35}) \times 7 = {}^{84}\!/_{35} = 2\,{}^2\!/_5$. The rest of the harmonics have the following angular per-unit velocities.

Spatial harmonic order ν	Rotor surface $\omega_{PM,\,p.u.}$
1	$1\,{}^1\!/_5$
7	$2\,{}^2\!/_5$
−11	$1\,{}^1\!/_5$
13	$3\,{}^3\!/_5$
−17	$2\,{}^2\!/_5$
19	$4\,{}^4\!/_5$

The method described above for permeance harmonics can be applied also to the stator harmonics. It is important to notice that now several of the stator harmonics, depending on the number of laminations, can have a pole pitch much wider than the permanent magnet pieces.

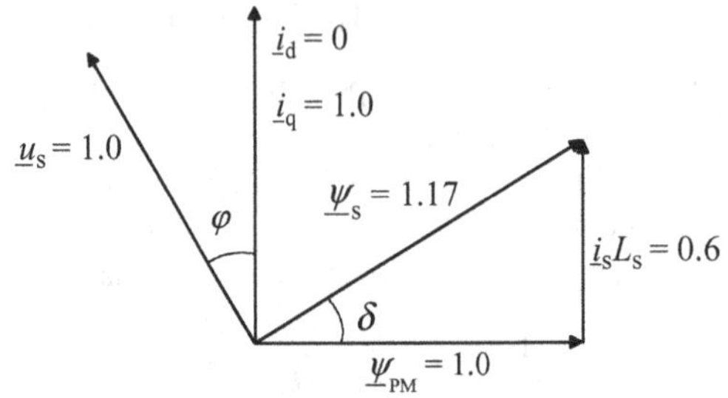

Figure 7.66 $i_d = 0$ controlled machine phasor diagram at the machine's field weakening limit. The power factor angle $\varphi \approx 31^\circ$ and the load angle $\delta \approx 31^\circ$ will be the same with $i_d = 0$.

For example the fundamental pole pitch covers in a case of a 12/10-machine five rotor magnet poles. In (Pyrhönen *et al.*, 2012) a good agreement with the measured PM losses of a 37 kW PMSM was found.

7.3.10.11 PMSM Properties and Power Factor with Rotor Surface Magnet PMSM

A brief analysis of the power factor on nonsalient-pole PMSMs will be given. A nonsalient-pole PMSM produces torque only with the q-axis current. Therefore it is normally driven with $i_d = 0$ control. We will study a PMSM with rotor surface magnets $\psi_{PM} = 1$ and $L_s = 0.6$. $L_s = 0.6$ because such a machine, with rated voltage supply, has a suitable peak torque capability of c. 1.6 pu. Generally $i_d = 0$ is used in the control as far as possible. Figure 7.66 illustrates the machine with its rated voltage and current at the highest possible speed of $\omega_s = 0.86$ which is 86 % of the base speed $\omega_s = 1.0$ with $i_d = 0$.

With $i_d = 0$ control the machine with $\psi_{PM} = 1$ and $L_s = 0.6$ reaches its maximum voltage $u_s = 1$ at speed at $\omega_s = 0.86$ of the no-load speed $\omega_s = 1.0$ as the stator flux linkage – in a nonsaturable machine – has grown to $\psi_s = 1.17$ because of the armature reaction and the leakage flux. The converter rated apparent power $S = 1$ is also reached at this point. Below this speed there is no interest in increasing the power factor. It should only increase the motor current which should be used in demagnetizing the d-axis flux linkage. Also the torque should decrease if the current is not increased at the same time.

If for some reason there is a need to increase the power factor value to $\cos\varphi = 1$ there are two alternatives, (1) either the supply voltage has to be made lower, or (2) the permanent magnet flux linkage has to be increased.

If instead $i_d = 0$ and $\cos\varphi = 0.86$, $i_d < 0$ and $\cos\varphi = 1$ is desired at the original operating speed $\omega_s = 0.86$ we should reduce the supply voltage to a value $u_s = 0.8 \times 0.86 = 0.69$ to reach $\psi_s = 0.8$. Now, the torque of the machine with $i_s = 1$ is reduced to $T = 0.8$ (instead of earlier $T = 1$ with $i_d = 0$) and the machine power to $P = 0.69$. In principle, a corresponding converter with $u = 0.69$ and $i_s = 1$ could be used. This is, however, not possible as the no-load voltage of the motor should still be 1 unless it is always driven with a substantial negative i_d current. Figure 7.67 illustrates the phasor diagram at unity power factor.

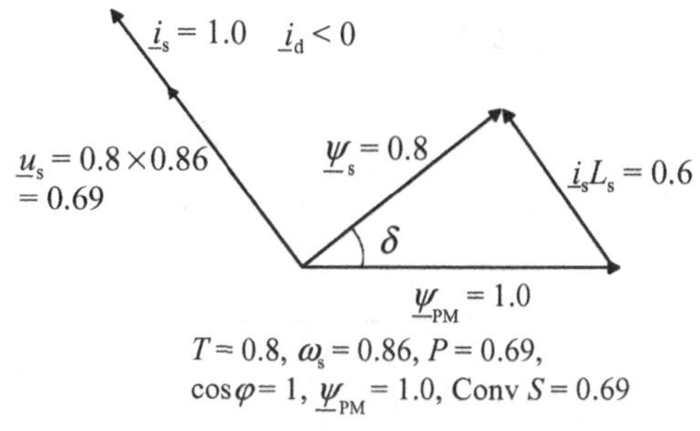

Figure 7.67 Unity power factor at 86 % of the rated speed. The stator flux linkage, the torque and the power are reduced as the voltage of the machine must be reduced to achieve $\cos\varphi = 1$.

If $\cos\varphi = 1$, $\omega_s = 1$, $T = 1$ and $P = 1$ are desired the machine should be equipped with more magnets to get $\psi_{PM} = 1.15$. Now, however, the no-load speed of the machine is reduced to $\omega_s = 0.87$, or the machine must be operated at $i_d = -0.15/0.6 = -0.25$ at no load to reduce the stator flux linkage to $\psi_s = 1.0$ (see Figure 7.68).

Normally, a rotor surface magnet machine is driven with $i_d = 0$ as long as possible and then the amount of negative i_d is gradually increased to reach the no-load speed under load. Figure 7.69 illustrates the machine operation at the rated speed, voltage and current with $\psi_{PM} = 1$.

When the motor is further accelerated the torque gets lower but the speed increases faster and a point with $\cos\varphi = 1$, $T = 0.8$ and $P = 1$ will be reached (see Figure 7.70). Therefore, the highest power of the motor will be reached when the power factor reaches unity. This is, however achieved by demagnetizing the magnet and by increasing the speed. The situation resembles the situation of Figure 7.67. To achieve a unity power factor there is a need to demagnetize the PM flux linkage.

The speed of the machine can further be increased with a stronger demagnetizing d-axis current. Figure 7.71 illustrates the machine behavior at twice the rated speed.

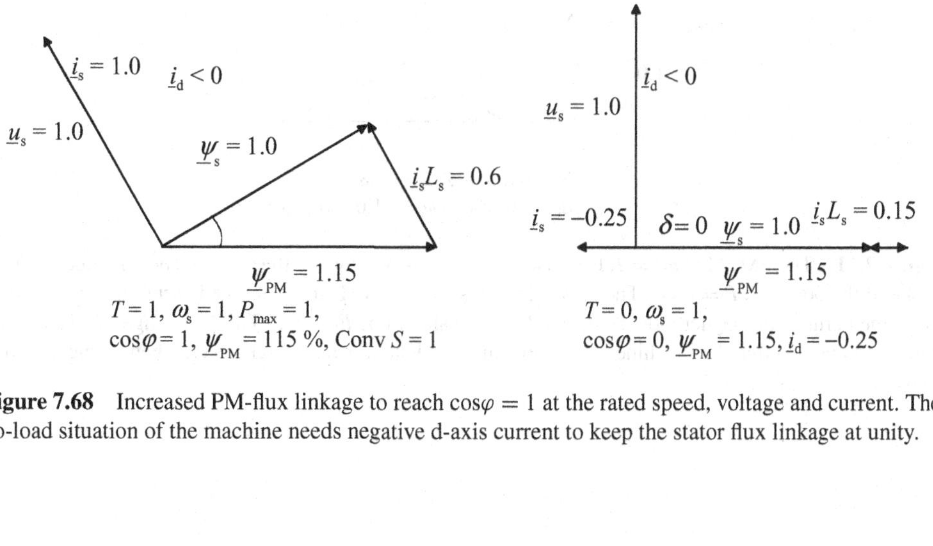

Figure 7.68 Increased PM-flux linkage to reach $\cos\varphi = 1$ at the rated speed, voltage and current. The no-load situation of the machine needs negative d-axis current to keep the stator flux linkage at unity.

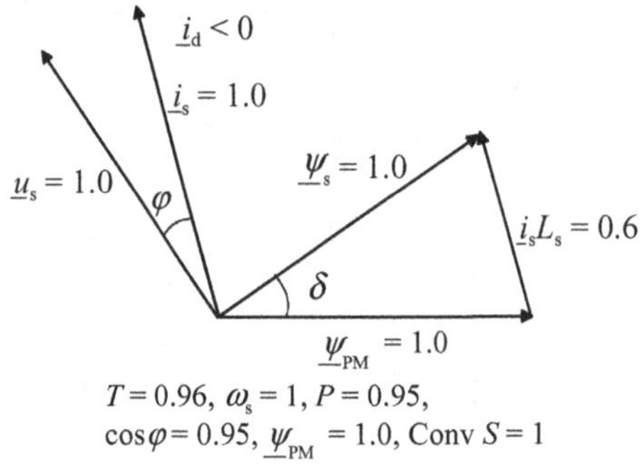

Figure 7.69 Motor operation at the rated voltage, current and speed $\omega_s = 1$. The power of $P = 0.95$ will be reached with $\cos\varphi = 0.96$,

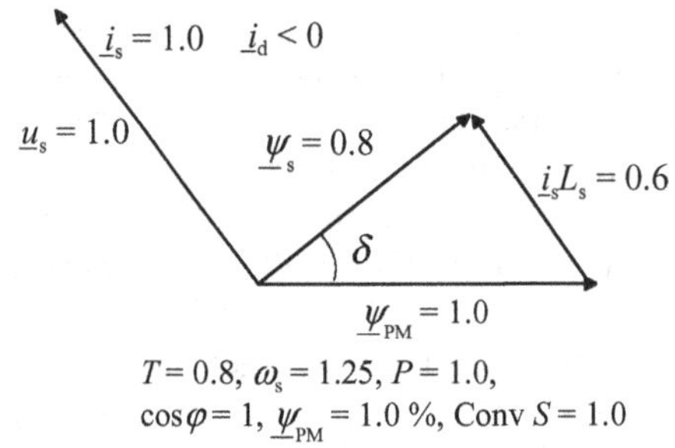

Figure 7.70 Phasor diagram at $\omega_s = 1.25$. The power factor reaches unity at this speed and power. Also the highest power of the machine is found here, $P = 1$. The stator flux linkage, however, is reduced to $\psi_s = 0.8$.

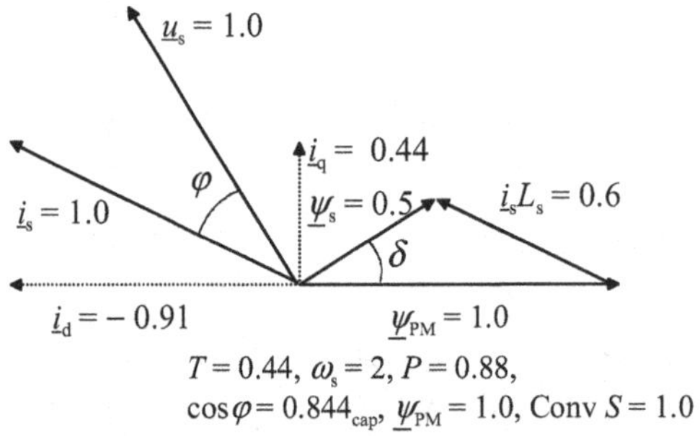

Figure 7.71 The PMSM at $\omega_s = 2$. Because of the low q-axis current the torque is now reduced to $T = 0.44$ and the power to $P = 0.88$. The power factor is capacitive. If the machine is running freely at this speed the permanent magnet induced no-load voltage should be $E_{PM} = 2$. This is a dangerous value and it must be ensured that the machine can never run at no load at such a high speed when connected to a converter.

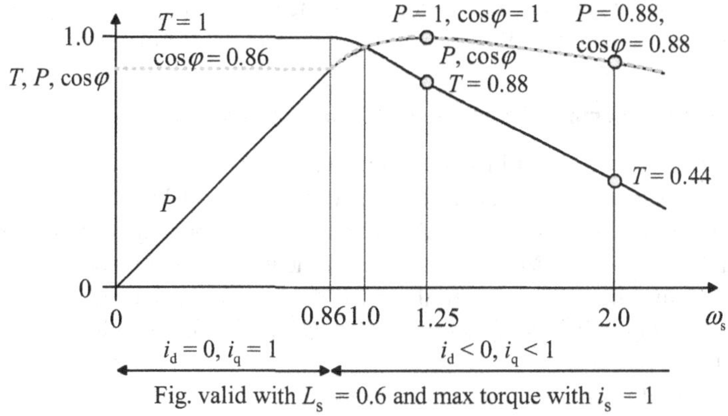

Figure 7.72 Behavior of torque, power and power factor at different speeds of PMSM drive with $L_s = 0.6$.

It has been proven that the $i_d = 0$ control is the most efficient as far as it can be used and there is no need to worry about the power factor. Figure 7.72 also illustrates how the drive moves from $i_d = 0$ to $i_d < 0$ when the base speed $\omega_s = 1$ is reached.

In case of $L_s = 0.6$, $u_s = 1$, $i_s = 1$ a right-angled triangle $\psi_{PM} = 1$, $L_s i_s = 0.6$, $\psi_s = 0.8$ will be formed at $\omega_s = 1.25$. The torque will be $T = 0.88$ and the power at this point will be $P = 1$. The torque is already getting lower but as the speed increases the power and the power factor of the drive increase. After this point the power factor turns capacitive and the torque and the power start to decrease, Figure 7.72.

At first the power factor stays constant as far as $i_d = 0$ is followed. After that it follows the per unit power.

When a motor is driven with lower power and $i_d = 0$ the power factor will be higher at low speeds and the $i_d = 0$ area increases. For example, if $i_q = 0.5$ Figure 7.66 will change to Figure 7.73.

In case of embedded magnets and $L_d \neq L_q$ there are several possible strategies to drive the machine. Utilizing the reluctance torque will lead in machine with $L_q > L_d$ to a situation

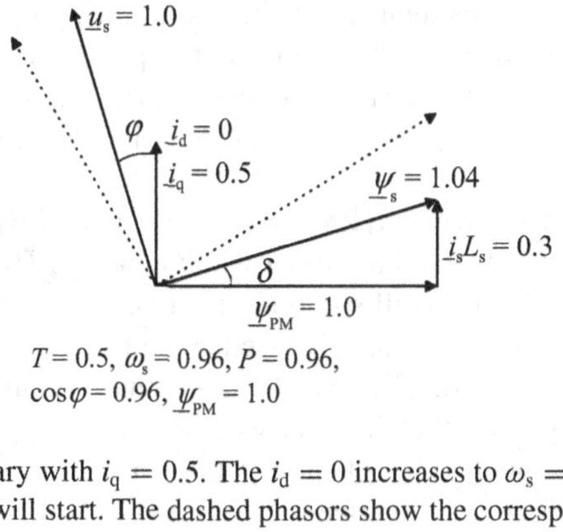

Figure 7.73 $i_d = 0$ boundary with $i_q = 0.5$. The $i_d = 0$ increases to $\omega_s = 0.96$ – instead of $\omega_s = 0.86$ with $i_q = 1$ – before $i_d < 0$ will start. The dashed phasors show the corresponding phasors of Fig 7.66.

where the highest torque is reached by using a small negative d-axis and a large q axis current. Optimal driving of a salient-pole PMSM will not, however, be studied here further.

7.3.10.12 Brushless Permanent Magnet DC-machine (BLDC)

The BLDC has been very popular in automation and is also used widely in low power traction applications. In principle, the difference between a BLDC and a PMSM is clear but in practice these machines can be very similar. A BLDC is a permanent magnet (AC) machine with rectangular back emf while a PMSM is an AC machine with sinusoidal back emf. In principle there is a significant difference between them but in practice these machine types are easily confused with each other. It is common that a BLDC has sinusoidal voltage but is still called a BLDC based on its control.

The easiest way of designing a BLDC is to use rectangular flux density producing rotor surface magnets and a distributed three-phase single layer winding with two slots per pole and phase ($q = 2$). With such an approach the machine back emf stays in principle rectangular.

Controlling of these machines is, however, totally different, in principle. A BLDC is driven with DC-current pulses (positive and negative pulses looking like rectangular AC) and the control is, in principle, a DC-machine control. These pulses are supplied to the armature based on rotor position encoder so that the current pulses are synchronized with the back emf rectangular pulses and, as a result, smooth torque is achieved.

A PMSM needs sinusoidal currents and vector control to produce smooth torque. It can, naturally, also operate direct on line if equipped with a damper winding.

7.3.11 Synchronous Reluctance Machines

The synchronous reluctance machine (SyRM) is the simplest of synchronous machines. The operation of a (SyRM) is based on the difference between the inductances of the direct and quadrature axes. The power of a synchronous reluctance machine is obtained from the load angle equation (see also (7.195) or (7.230)).

$$P = 3U_s^2 \frac{L_d - L_q}{2\omega_s L_d L_q} \sin 2\delta = \frac{3U_s^2}{2\omega_s} \left(\frac{1 - \dfrac{L_q}{L_d}}{L_q} \right) \sin 2\delta. \tag{7.277}$$

We can see in (7.277) that the smaller the quadrature-axis synchronous inductance and the higher the direct-axis inductance, the higher is the power and torque at a certain load angle. In practice, the limiting value for L_q is the stator leakage inductance $L_{s\sigma}$; consequently, $L_q > L_{s\sigma}$. It is the task of the designer to maximize the inductance ratio L_d/L_q to achieve a good performance.

In practice, a high-inductance-ratio SyRM may be driven only with a frequency converter irrespective of whether the rotor carries a damper winding or not. The high rotor saliency ratio results in the rotor not starting at its full speed in direct network drive but instead remains at half speed. This is due to the phenomenon discussed previously in conjunction with the *partial* damper windings (cf. Figure 7.45). The rotor-bound counter-rotating field produces such a high torque at high saliency ratios that the rotor will not start.

Figure 7.74 illustrates various kinds of the rotors of a synchronous reluctance machine. In Figure 7.74a and b, the rotors are converted from the rotor of an induction motor. This

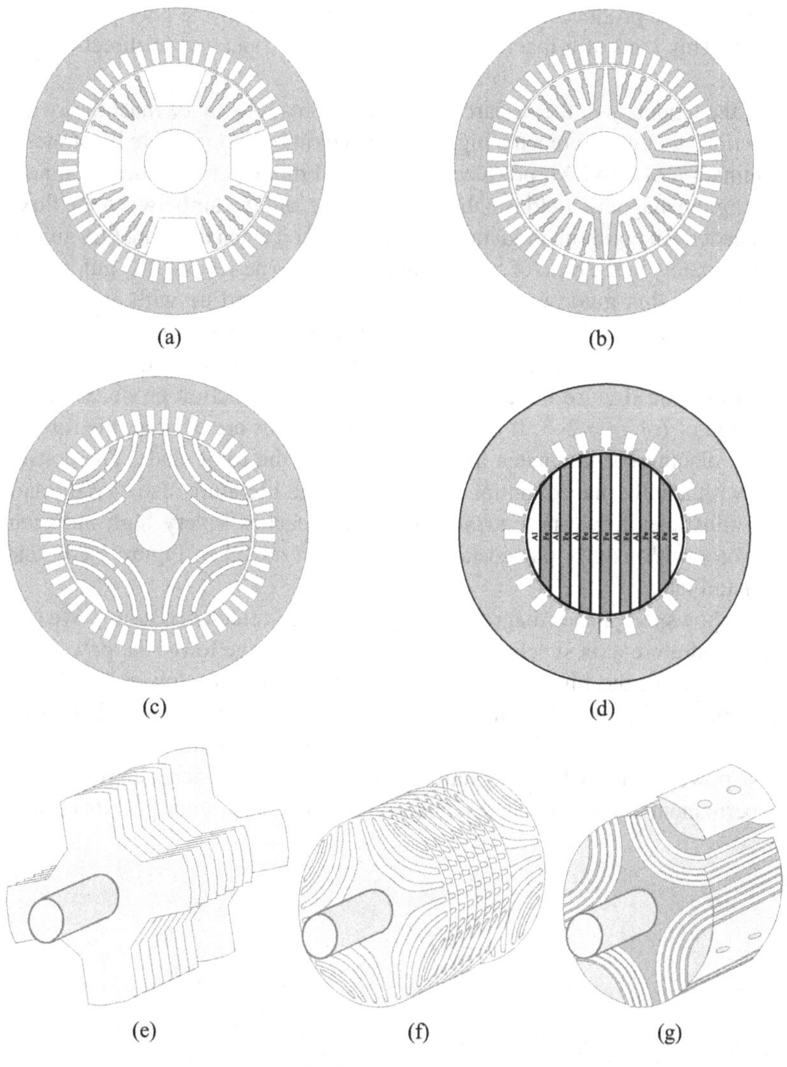

Figure 7.74 Various kinds of synchronous reluctance machines (SyRM): (a), (b) Rotors of early SyRMs converted from the rotor of an induction machine. Inductance ratio L_d/L_q is typically of the order 2–3. (d) Stator and rotor laminations of a two-pole synchronous reluctance machine with axially placed rotor laminations. The rotor consists of rectangular iron and aluminum laminations of different sizes. The iron sheets act a flux guides and aluminum sheets as flux barriers. This type of a rotor creates a very small quadrature-axis inductance, yet the direct-axis inductance may be large. Very large inductance ratios may be obtained. The construction, however, is difficult to manufacture and the rotor iron losses tend to be high. In the figure, aluminum and steel sheets are diffusion welded one after another. Instead of aluminum, stainless steel or some nonconducting composite can also be used. Note that in the figure there is no bore for a shaft. The shaft ends can be attached e.g. with friction welding. (c) Four-pole machine stator and rotor laminations with cut barriers. The rotor sheet has been punched from a single sheet, and therefore, to keep the shape of the construction, several bridges have been left between the flux barriers and the flux guides. The bridges notably reduce the inductance ratio. With the presented structure, and by applying a small air gap, an inductance ratio of $L_d/L_q = 10$ may be reached. (e) 3D view of a four pole rotor with radially placed laminations with low inductance ratio. (f) 3D view of the version in (c). Laminations are with barriers. (g) 3D view of a four pole rotor with axially placed laminations.

manufacturing method produces a relatively low inductance ratio, a poor power factor and a poor efficiency. This kind of a machine may, however, be operated in direct-on-line (DOL) applications.

Previously, the tendency in the research on synchronous reluctance machines was to create rotor constructions that would reach a high inductance ratio. Figure 7.74c illustrates the stator and rotor lamination of a SyRM, produced from round thin electrical sheet laminations. The steel conducting the magnetic field is called a flux guide and the air between the flux guides is called a flux barrier. An optimum inductance ratio is reached when there are flux guides and flux barriers approximately in the ratio of 50:50. There must also be several flux guides and barriers per pole. If just one flux guide is used, see Fig. 7.74b, we end up with an inductance ratio typical of traditional synchronous machines. In such a case, the flux guide forms a construction resembling the pole shoe construction. A wide pole shoe adjacent to the air gap guarantees a good route to the quadrature flux, and this kind of a construction produces an inductance ratio in the range of $L_d/L_q = 2–3$. If higher inductance ratios are desired, the rotor should be constructed by placing the laminates in the direction of the rotor shaft. The construction is called "rotor with axially placed laminations" or in some literature also "axially laminated". Figure 7.74d illustrates a two-pole arrangement that reaches a very high inductance ratio. A four pole rotor with axially placed laminations is presented in Figure 7.74g and will be analyzed in detail later.

Figure 7.75 shows the phasor diagram of a synchronous reluctance motor. We can see that the lower the quadrature-axis synchronous inductance L_q is, the lower the power angle δ will be. On the other hand, the higher the direct-axis inductance, the lower the current needed to magnetize the d-axis, and the better the power factor of the machine. It is also obvious that the quadrature axis armature reaction should remain low. Otherwise it is impossible to reach a good power factor when the stator flux linkage turns increasingly away from the q-axis. The permanent magnet assisted synchronous reluctance motor gets its basic idea from

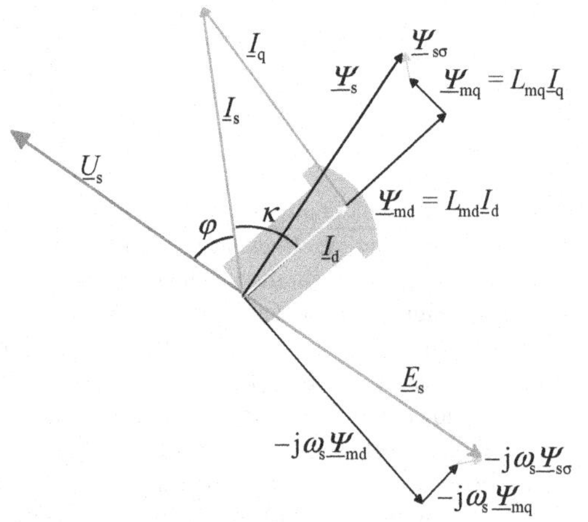

Figure 7.75 Phasor diagram of a synchronous reluctance machine. In the figure, the per unit values are $l_{md} = 2$ and $l_{mq} = 0.20$, $l_{s\sigma} = 0.1$ and therefore the inductance ratio $L_d/L_q = 2.1/0.3 = 7$. The power factor $\cos\varphi \approx 0.66$.

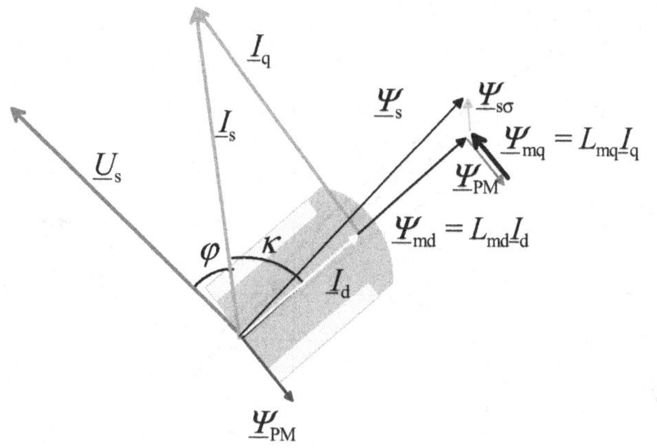

Figure 7.76 Phasor diagram of the same SyRM as in Figure 7.75 but equipped with armature reaction compensating permanent magnets producing a flux linkage of $\Psi_{PM} = 0.25$. The power factor has increased significantly to $\cos\varphi \approx 0.8$.

this fact: Permanent magnets are used to compensate the quatrature axis armature reaction. Figure 7.76 illustrates the effect of the permanent magnet flux in compensating the q-axis armature reaction.

Figure 7.77 illustrates the effect of the inductance ratio on the torque production capability of a SyRM and the current angle. As the figure shows, the synchronous reluctance motor never

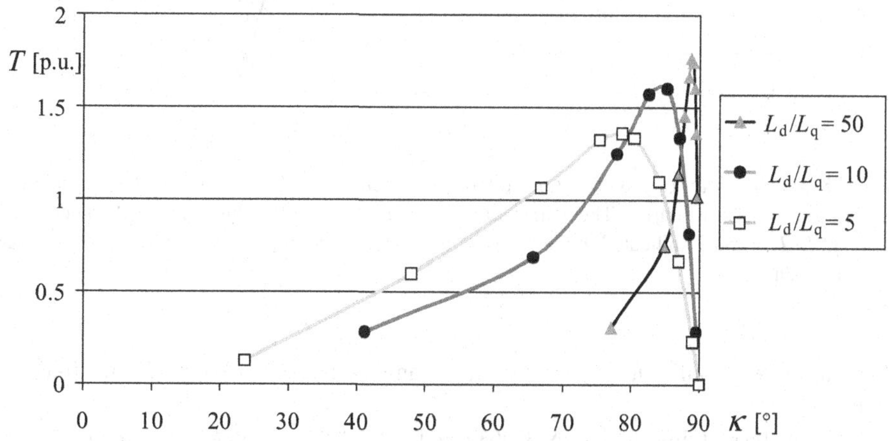

Figure 7.77 Effect of the inductance ratio on the torque production capability of a synchronous reluctance machine and the current angle κ. In the calculation, the rated current of the corresponding stator of an induction machine is employed. The values are calculated for a 30 kW, four-pole, 50 Hz machine. $L_d/L_q = 50$ is indicated in the figure for academic interest. Reproduced by permission of Jorma Haataja.

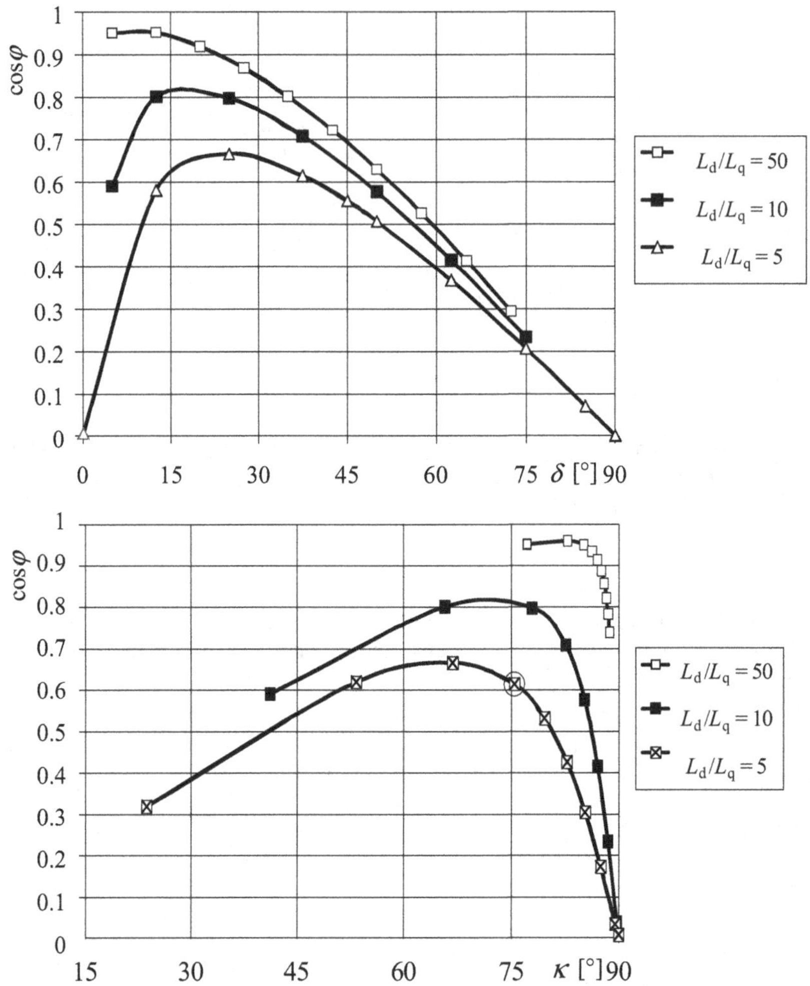

Figure 7.78 Power factor of a synchronous reluctance machine as a function of load angle and current angle at various inductance ratios. The values are calculated for a 30 kW, four-pole, 50 Hz machine. The impractical, $L_d/L_q = 50$ is indicated in the figure only for academic interest. Reproduced by permission of Jorma Haataja.

produces a very high pull-out torque. The current angle κ is measured between the d-axis and the stator current.

A high inductance ratio in a SyRM leads to a high current angle, and thus to a small load angle and a good power factor, Figure 7.78.

Figure 7.79 depicts the behavior of the power factor of the motor as a function of the shaft output power (motor power). The figure clearly indicates the importance of a high inductance ratio for obtaining good machine characteristics. In practice, an inductance ratio of 20 is within the bounds of possibility.

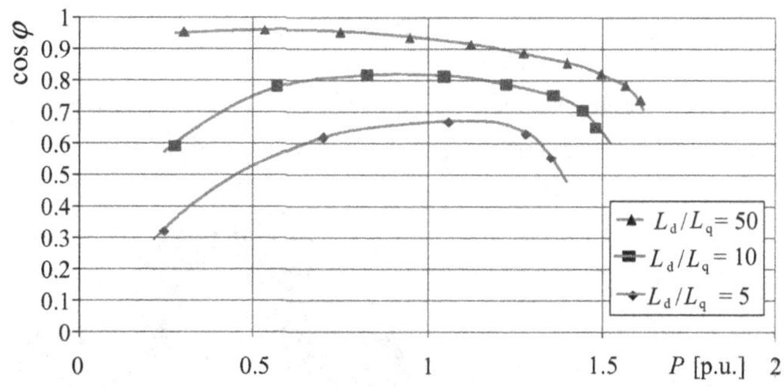

Figure 7.79 Power factor of a synchronous reluctance machine as a function of shaft output power at rated speed. The values are calculated for a 30 kW, four-pole, 50 Hz machine. $L_d/L_q = 50$ is indicated in the figure for academic interest. In practice, such an inductance ratio may not be reached. Reproduced by permission of Jorma Haataja.

7.3.11.1 Rotor with Axially Placed Laminations

This kind of a rotor arrangement is called also axially laminated rotor. Figure 7.74d illustrates a two pole motor and in Figures 7.74g, 7.80 and 7.82 there are drafts of a possible four-pole rotor arrangement. The laminations are bent under the angle α_{lam} and fixed on a nonmagnetic spider forming the mechanical core of the rotor.

An analytical approach to the analysis of such a rotor is based on the magnetic flux distribution in the d- and q-axes. The magnetic conductivity along the magnetic circuit is investigated in two basic positions as

the stator current linkage is on the d-axis: $\hat{\Theta}_{s1d}$, see Figure 7.81a and
the stator current linkage is on the q-axis: $\hat{\Theta}_{s1q}$, see Figure 7.81b.

The performance of the motor is determined by comparing the magnetizing inductances in d- and q-axes (L_{md}, L_{mq}) with the magnetizing inductance $L_{m\delta}$ of a cylindrical rotor with non-salient poles. Saliency factors k_d and k_q are introduced based on Equations (7.202) and (7.203). These factors are investigated in details because they describe the rotor geometry effects on the performance. The saliency factors are defined in Figure 7.81, where the waveforms of the fundamental harmonic of the stator current linkage $\Theta_{s1d,q}$ with their magnitudes $\hat{\Theta}_{s1d,q}$, real waveforms of the magnetic flux density, their envelope profiles, fundamental harmonics $\hat{B}_{\delta d1}$, $\hat{B}_{\delta q1}$, and the span of the laminations thickness τ on one pole pitch τ_p, are illustrated.

The saliency factors k_d and k_q are defined as the ratio of the $\hat{B}_{\delta d1}$, $\hat{B}_{\delta q1}$, and $\hat{B}_{\delta 1}$ of the nonsalient-pole rotor:

$$k_d = \frac{\hat{B}_{\delta d1}}{\hat{B}_{\delta 1}} = \frac{4\delta k_{Cs}}{\tau_p} \int_0^{\tau_p/2} \frac{1}{\delta_d(x)} \cos^2 \frac{\pi}{\tau_p} x dx \tag{7.278}$$

$$k_q = \frac{\hat{B}_{\delta q1}}{\hat{B}_{\delta 1}} = \frac{4\delta k_{Cs}}{\tau_p} \int_0^{\tau_p/2} \frac{1}{\delta_q(x)} \sin^2 \frac{\pi}{\tau_p} x dx \tag{7.279}$$

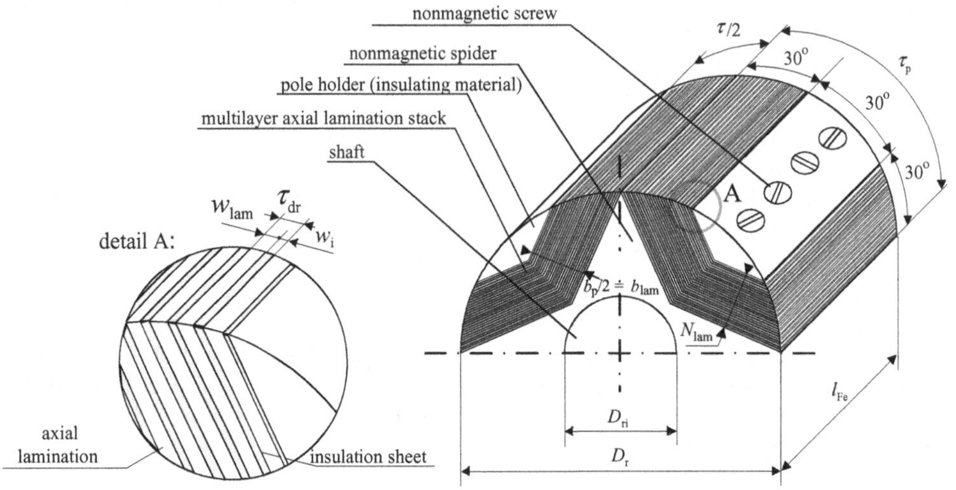

Figure 7.80 Sketch of four-pole rotor with axially placed laminations.

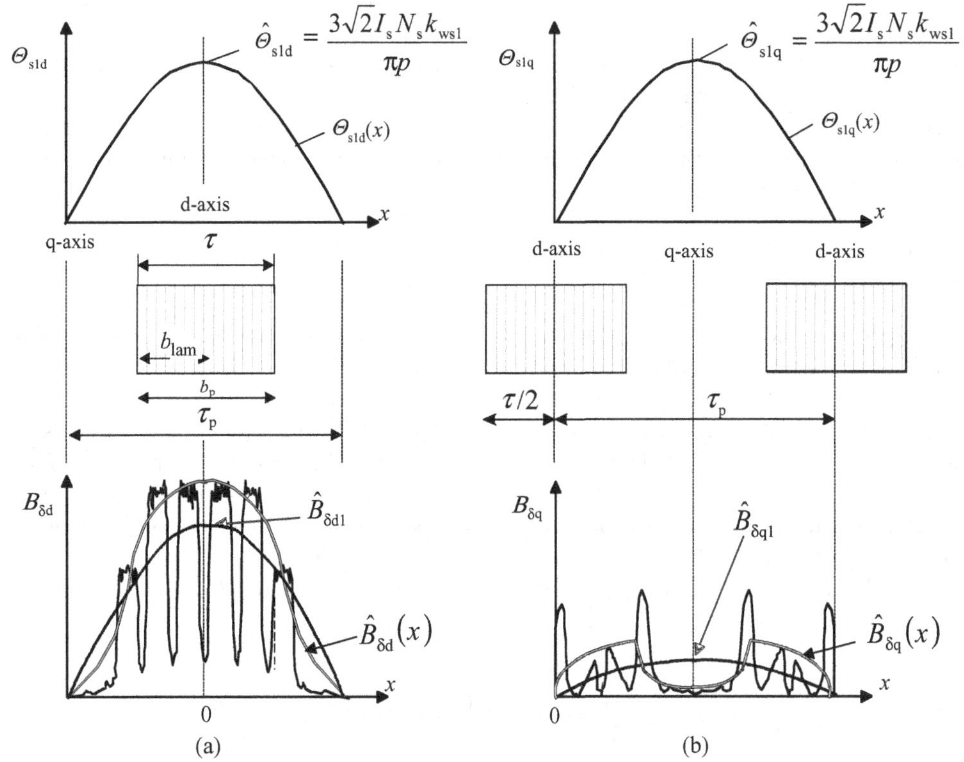

Figure 7.81 Waveforms of the current linkage and corresponding magnetic flux densities in the d- and q-axes, including real flux density curves. The real flux density curves $B_{\delta d,q}(x)$ are very distorted because of the permeance variations on both sides of the air gap.

where

$$\hat{B}_{\delta 1} = \frac{\mu_0}{\delta k_C} \frac{3}{\pi} \sqrt{2} I_s \frac{N_s k_{ws1}}{p}. \tag{7.280}$$

In (7.278)–(7.280) the air gap δ is the physical air gap on the d-axis. To get proper saliency factors, the following aspects must be taken into account: At first, in d-axis, the saturation of laminations must be introduced by means of a saturation factor k_{dsat} (see Equation (6.34)). Saturation in the q-axis can be neglected in this kind of a rotor with no saturating iron bridges on the q-axis. After the leakage inductance is added, the synchronous inductance in the d-axis is as follows:

$$L_d = \frac{L_{md}}{(1 + k_{dsat})} + L_\sigma = \frac{k_d L_{m\delta}}{(1 + k_{dsat})} + L_\sigma \tag{7.281}$$

where k_d is calculated from (7.278) and L_{md} is calculated for a nonsaturating machine.

Secondly, in the q-axis only part of magnetic flux lines cross perpendicularly the laminations, their insulations and the nonmagnetic spider. Another part of the flux lines are attracted along the laminations to the air gap and a stator tooth, coming back to the rotor, oscillating between the stator and rotor parts. Therefore, the factor k_{qpar} is introduced to take these parallel paths into account. This factor is calculated according (7.279). If $\delta_{qpar}(x)$, see (7.286), is introduced instead of $\delta_q(x)$ the synchronous inductance on the q-axis is

$$L_q = L_{mq} + L_{\sigma s} = k_{qpar} L_{m\delta} + L_\sigma \tag{7.282}$$

Air-gap profile functions $\delta_{d,q}(x)$ are very important in this calculation. The profile functions describe the paths of the magnetic flux lines in nonmagnetic materials, i.e. in a fictive enlarged air gap (see expressions below). Based on Figure 7.82, the following expressions may be derived:

in d–axis:

$$\begin{aligned}
\delta_d(x) &= \\
&= \delta k_C \quad \text{for } 0 \le x \le \tau/2 \\
&= \delta k_C + \left(D_r \sin\frac{\beta_{xd}}{2}\right)\left[\alpha_{lam} - \frac{\beta_{xd}}{2} - \arccos\frac{r}{D_r \sin(\beta_{xd}/2)}\right] \quad \text{for } \frac{\tau}{2} \le x \le \frac{\tau}{2} - r \\
&= \delta k_C + \left(D_r \sin\frac{\beta_{xd}}{2}\right)\left(\frac{\pi}{2} - \frac{\beta_{xd}}{2}\right) \quad \text{for } \frac{\tau}{2} - r \le x \le \frac{\tau_p}{2}
\end{aligned} \tag{7.283}$$

in q–axis for the route across insulations and rotor spider:

$$\begin{aligned}
\delta_{qi}(x) &= \delta k_C + D_r\left[a_i \sin\frac{\alpha_x}{2} \sin\gamma_x + \alpha_{sp} \sin\frac{\alpha_x}{2} \cos\gamma_x\right] \quad \text{for } 0 \le x \le \tau/2 \\
&= \delta k_C + a_i b_{lam} + \left(D_r \sin\frac{\beta_{xq}}{2}\right)\left[\frac{\pi}{2} - \gamma - \alpha_{sp} - \frac{\beta_{xq}}{2}\right] \\
&\quad + D_r\left(\sin\frac{\gamma}{2}\cos\gamma_t + \sin\frac{\beta_{xq}}{2}\right)\alpha_{sp} \quad \text{for } \frac{\tau}{2} \le x \le \frac{\tau_p}{2}
\end{aligned} \tag{7.284}$$

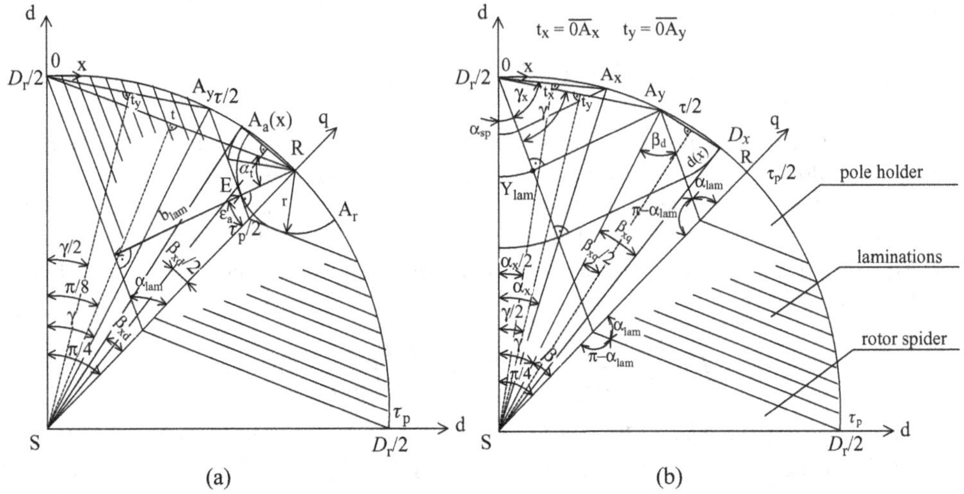

Figure 7.82 Magnetic paths of a four pole SyRM with axially placed laminations on the rotor if the magnitude of the stator current linkage is in (a) d – axis and (b) q – axis.

in q–axis for the way along the laminations:

$$\delta_{\text{qlam}}(x) = \begin{cases} 2k_C\delta + D_r a_i \sin\dfrac{\alpha_x}{2}\sin\gamma_x & 0 \le x \le \dfrac{\tau}{2} \\ 2k_C\delta + a_i b_{\text{lam}} + D_r\left(\sin\dfrac{\beta_{xq}}{2}\right)\left(\dfrac{\pi}{2} - \dfrac{\beta_{xq}}{2} - \alpha_{\text{sp}} - \gamma\right) & \dfrac{\tau}{2} < x \le \dfrac{\tau_p}{2} \end{cases} \tag{7.285}$$

and finally taking both parallel routes into account:

$$\delta_{\text{qpar}}(x) = \frac{\delta_{\text{qi}}(x)\delta_{\text{qlam}}(x)}{\delta_{\text{qi}}(x) + \delta_{\text{qlam}}(x)}. \tag{7.286}$$

Although Equations (7.283)–(7.286) are derived for a four pole rotor, it is possible to use them for any $2p$-pole machine and also for two pole machine, if the following expressions for the angles are introduced:

$$\begin{aligned} \beta_{xd} &= \frac{\pi}{2p}\frac{\tau_p - 2x}{\tau_p}, \alpha_x = \frac{\pi x}{p\tau_p}, a_i = \frac{w_i}{w_i + w_{\text{lam}}}, \beta_{xq} = \frac{\pi}{2p}\frac{2x - \tau}{\tau_p}, \gamma_x = \frac{\pi}{2} - \frac{\alpha_x}{2} - \alpha_{\text{sp}}, \\ r &= D_r \sin\frac{\pi}{4p}\cos\left(\alpha_{\text{lam}} - \frac{\pi}{4p}\right) - b_{\text{lam}}, \alpha_{\text{sp}} = \alpha_{\text{lam}} - \frac{\pi}{2p}, \gamma = \frac{\pi}{2p}\frac{\tau}{\tau_p}, \\ b_{\text{lam}} &= D_r \sin\frac{\gamma}{2}\sin\left(\frac{\pi}{2} - \alpha_{\text{sp}} - \frac{\gamma}{2}\right), \\ \beta &= \frac{\pi}{2p}\left(1 - \frac{\tau}{\tau_p}\right), \gamma_t = \frac{\pi}{2} - \frac{\gamma}{2} - \alpha_{\text{sp}}. \end{aligned} \tag{7.287}$$

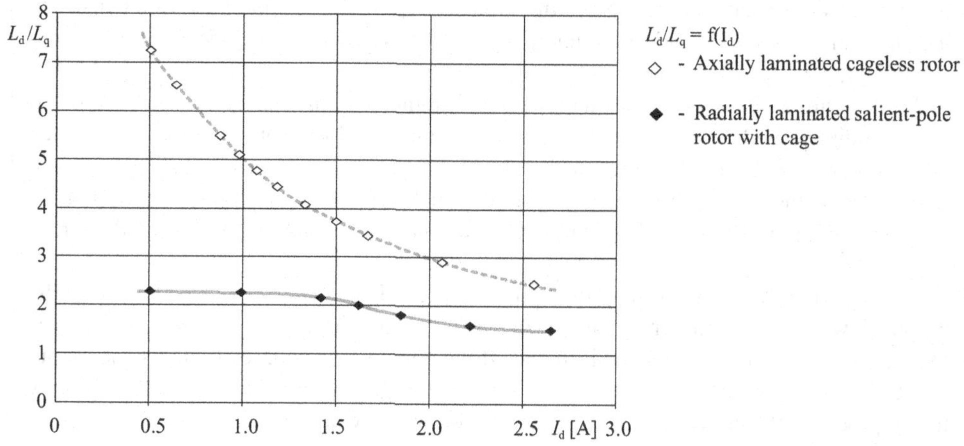

Figure 7.83 Saliency ratio $\zeta = L_d/L_q$, versus current if $I_d = I_q$, gained in Example 7.8.

From Equations (7.281)–(7.282) it is seen that L_d/L_q is not identical with the k_d/k_q ratio. However, the higher the k_d/k_q ratio is, the higher will be also the inductance ratio L_d/L_q. Therefore, the use of this analytical approach is recommended for fast calculation of k_d and k_q and their ratio k_d/k_q, to predict the SyRM performance and to make a choice for the rotor configuration. The final value of L_d/L_q can be calculated with FEM or analytically. Usually L_d/L_q is about 40–50 % of the k_d/k_q ratio.

The actual machine design starts with the stator in the same way as the design of an asynchronous motor. For the rotor design, the most important parameter is the magnetic flux penetrating the air gap. This magnetic flux must be accepted by the magnetic flux density in laminations B_{lam} on the area given by the thickness of the pole span τ and the length of the motor. Because of iron losses which are proportional to the square of the B_{lam}, the value of the B_{lam} must be selected carefully. The other parameters N_{lam}, w_{lam}, w_i, τ and τ/τ_p are chosen on the base of the graphs given in Figure 7.84. They have been constructed according to Equations (7.283)–(7.287) for a concrete motor (see Example 7.7).

Example 7.7: A nameplate of the original synchronous reluctance motor with radially placed laminations on the rotor (see Figure 7.74a) is as follows: 400 W, 4 poles, 380 V, Y, 2 A, 50 Hz, 1500 min^{-1}. Further data is gained based on an inverse design calculation: The number of stator turns $N_s = 408$, the stator fundamental winding factor $k_{w1} = 0.959$, the air gap length $\delta = 0.2$ mm, the air-gap flux density 0.848 T, the magnetic flux per pole 2.416 mWb, $D_r = 77.6$ mm. Design a new rotor with axially placed laminations using the same stator (see Figure 7.74g).

Solution: For rotor design, electric steel sheets M700-50A, 0.5 mm can be used. No additional insulation layers are added, and therefore, the insulation thickness is $w_i = 0.076$ mm (two layers of normal electrical steel insulations). The ratio $\tau/\tau_p = 2/3$ is selected based on Figure 7.84, the number of rotor laminations is $N_{lam} = 26$, the magnetic flux density in the laminations is 1.27 T and the saliency factor ratio is $k_{dsat}/k_{qpar} = 8.45$. Finally the ratio of the $L_d/L_q = 4.07$.

A rotor with axially placed laminations was designed and produced. The newly created motor equipped with this rotor and using the original SyRM stator improved its saliency ratio (Figure 7.83) and also the other parameters under full load (Table 7.10). At no load the new motor has, however, increased iron losses. This is typical for the axially placed laminations that suffer from the permeance harmonics more than the radially placed laminations.

Table 7.10 shows that the new motor has improved mainly the power factor and the output power, if the comparison is based on the same stator current. The efficiency is increased in spite of the increased iron loss. If the comparison is based on the same output power 393 W (approximately the rated value of the original motor), it is seen, that the new motor produces this power at 1.5 A, while the original one at 2.0 A.

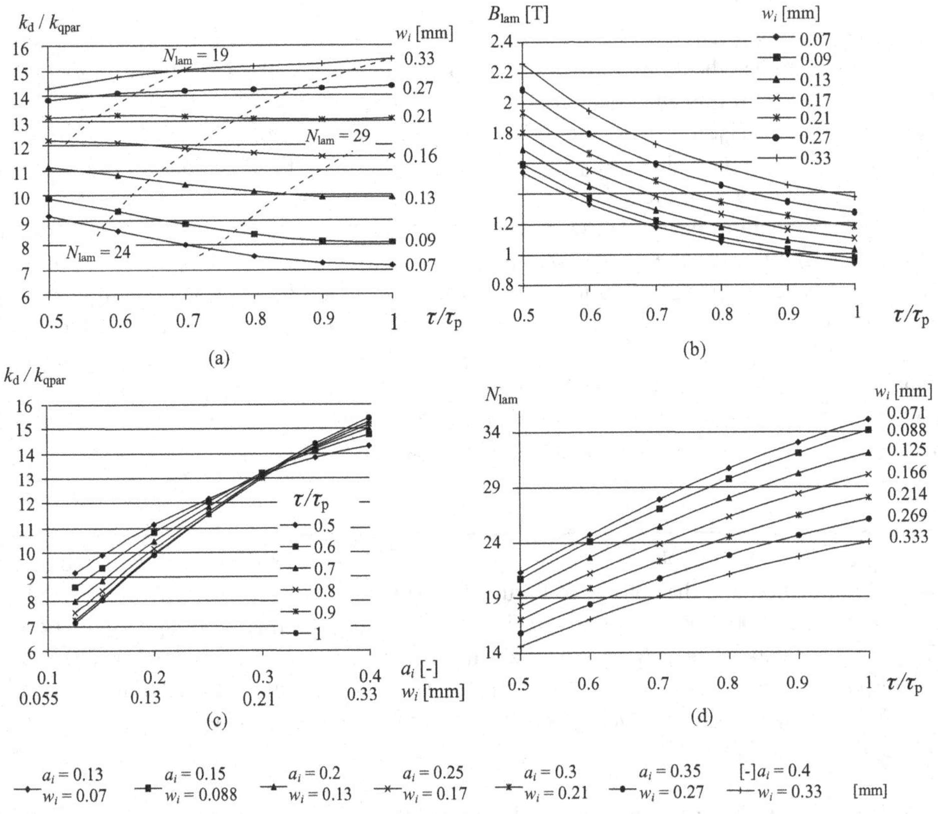

Figure 7.84 Dependence of the k_d / k_{qpar} of 4-pole SynRM on design parameters, $\delta = 0.2$ mm, $D_r = 77.6$ mm, $\alpha_{lam} = 70°$, $w_{lam} = 0.5$ mm, $a_i = w_i / (w_i + w_{lam})$.

Table 7.10 SyRM comparison of the radially and axially laminated rotor at full load with equal currents and equal output powers.

$U_{ph} = 220$ V	Original radially laminated rotor	New axially laminated rotor	
Stator current [A]	2A	2 A	1.5 A
apparent input power [VA]	1320	1320	999
input active power [W]	585	960	584
power factor	0.44	0.709	0.585
output power [W]	393	706	393
efficiency [%]	67.0	73.6	67.2

The other favorite rotor arrangement of the SyRM is a rotor with cut barriers, as it is shown in Figure 7.74c, f. Its main advantage in comparison with the axially placed laminations is that the sheet of such barrier rotor is more similar to the rotor sheet of the induction motor and the sheets are placed in the same way as in the case of induction motors. Therefore, the manufacturing technology is better elaborated.

The investigation of a flux barrier rotor SyRM performance is made similarly as with an axially laminated rotor. In Figure 7.85, there are the plottings of the d- or q-axis magnetic fields computed by FEM. The figure helps to understand the magnetic flux line profiles, and on this basis to create the expressions for the air-gap profile functions, as it was explained above with the axially placed rotor laminations.

The suitable shapes of the barriers, their number and the whole rotor structure can be optimized to ensure the optimal torque capability and the minimum torque ripple, to minimize

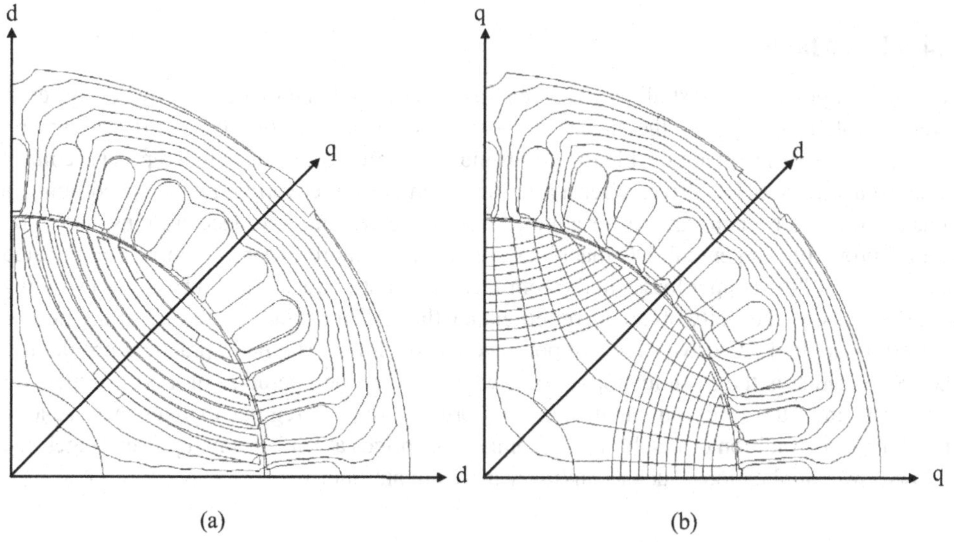

Figure 7.85 FEM analysis results of the magnetic field distribution in four-pole SyRM with barrier rotor as $\hat{\Theta}_{s1}$ is (a) on the d-axis and (b) on the q-axis.

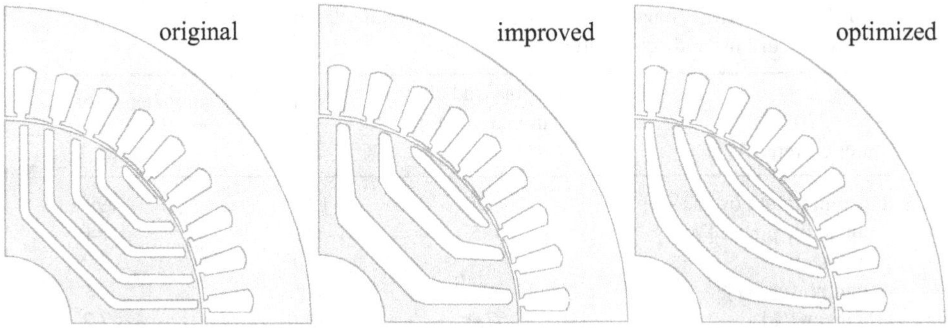

Figure 7.86 Different designs of the SyRM, initial, improved and optimized arrangement (modified from Moghaddam 2011). In practical machines radial ribs trough the innermost flux barriers are needed to support higher speed operation (not shown in the figure).

all secondary effects such as rotor iron losses, vibration and noise and to affect all electrical parameters for the maximum efficiency. The q-axis flux must be blocked as well as possible which results in a low L_q and simultaneously the flux in the d-axis must flow very smoothly which results in maximum of L_d. One possibility to achieve this is to align barrier shapes with the shapes of natural flux lines in a round lamination without barriers. A detailed analysis of such an approach can be read e.g. in (Moghaddam 2011).

In Figure 7.86, three different designs (initial, improved and optimized) are shown. The four-pole structure shows the best performance at constant torque condition for a wide speed range in terms of efficiency, power factor and temperature rise. The optimized rotor structure with flux barriers shows comparable results in terms of the torque capability, expected efficiency, power factor, winding temperature rise with induction motor.

7.4 DC Machines

Although in practice almost all electrical energy is produced and distributed by AC systems, a remarkable proportion of the electrical energy is consumed in the form of direct current. DC motors are still well applicable to various industrial processes that demand precise control of speed and torque. DC drives offer a rotation speed control covering a wide speed range, a constant torque or a constant power control, rapid accelerations and decelerations, and good control properties in general. DC machines are, however, more expensive, more complicated and, because of the copper losses in the complicated armature reaction compensation winding arrangements, also usually offer a lower efficiency than AC machines. In the era of frequency-converter-supplied AC drives, their importance has declined. The main practical problem of the DC machine is the commutator service, which makes the operation of a drive expensive in process industry. The commutator and the brushes require regular maintenance. A long manufacturing tradition offers, however, competitive prices at least in small power ranges and in low-power applications, such as auxiliary drives in automotives.

7.4.1 Configuration of DC Machines

Operation of a DC machine is based chiefly on the cooperation of two windings, namely a rotating armature winding and a stationary field winding. The armature winding is embedded

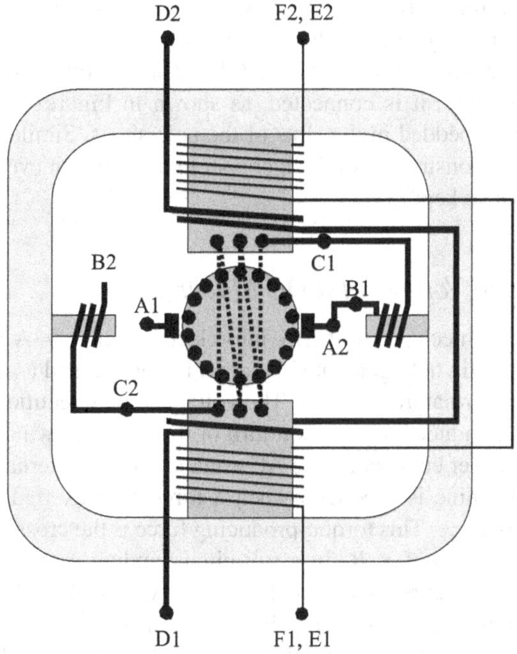

Figure 7.87 Terminal notations of the windings of a DC machine and the connection of the armature, commutating pole and the compensating windings. A1–A2, armature winding; B1–B2, commutating pole winding; C1–C2, commutating pole and compensating winding; F1–F2, separate field winding; E1–E2, shunt field winding; D1–D2, series field winding.

in slots on the outer periphery of the electrical sheet core and, although the flux variations in the iron core of the field winding are not significant, the pole shoes are also made of electrical sheet. The pole shoes are fastened to the yoke that closes the magnetic circuit.

In order to operate properly, several other windings are required as well in practical DC applications, as shown in Figure 7.87. The function of shunt and series windings in DC machines is to produce the main magnetic field in the machine, or the main field. Therefore, these windings together are called field windings. The shunt winding E1–E2 is connected in parallel with the armature, whereas a separate field current is led to the separate field winding F1–F2. The resistance of the windings E1–E2 and F1–F2 is high because of the thin wire and a high number of coil turns. The series winding D1–D2 is connected in series with the armature A1–A2, and thus it magnetizes the machine with the load current. Because of the high armature current, there are few coil turns in the winding, and its cross-sectional area is large. The windings of a DC machine were discussed in Chapter 2.

An alternating voltage is induced in the armature winding A1–A2. The induced AC voltage has to be rectified in the generator and, in the case of a motor, the supplying DC voltage has to be inverted to AC. The DC-to-AC inversion and the AC-to-DC rectification are mechanically carried out, together with brushes, by a commutator comprised of copper strips.

The function of the commutating pole winding B1–B2 is to ensure nonsparking commutation. In this function, the winding partly compensates the armature field created by the armature current. The commutating pole winding is connected in series with the armature, but it is fixed on the stator on the geometrically neutral axis.

The function of the compensation winding C1–C2 is to compensate the effects of the armature reaction. Therefore, it is connected, as shown in Figure 7.87, in series with the armature winding, and embedded in the slots of the pole shoes. Similarly as the other series windings, this winding is constructed of thick copper in order to achieve a low resistance. The number of coil turns is also kept low.

7.4.2 Operation and Voltage of a DC Machine

A DC generator converts mechanical energy into electrical energy. A power engine rotates the armature in the magnetic field generated by a field winding, and as a result an alternating voltage is induced in the armature winding. The voltage is now rectified by the commutator. An alternating voltage is induced in each coil turn of the armature winding. The commutator acts as a mechanical rectifier connecting the AC armature to the external DC circuit.

The torque of the machine is produced when a force $\boldsymbol{F}$ is exerted to the rotor winding according to the Lorentz force. This torque-producing force is the cross-product of the current $\boldsymbol{I}$ and the flux density $\boldsymbol{B}$: $\boldsymbol{F} = \boldsymbol{I} \times \boldsymbol{B}$. In a full-pitch winding extending across a full pole pitch, when one coil side occurs at the middle of the pole N, the other coil side occurs at the adjacent pole S. Thus in a two-pole machine, the coil sides are located on the opposite sides of the rotor. Now, if there are a number of $2p$ poles in the machine, the coil span is $2\pi/2p$ rad. In DC machines, a completely homogeneous field distribution cannot be achieved, but the flux density varies, as illustrated in Figure 7.88a, over a distance of two pole pitches. Figure 7.88b illustrates the voltage induced in a corresponding full-pitch winding, rectified by a commutator. The voltage is uniform with the flux density. Next, a full-pitch winding inserted in two slots is investigated. The number of coil turns is N, and the winding positioned at θ in Figure 7.88a is rotating at an electric angular speed ω.

The flux linkage of the armature in a distance of the differential angle $\theta = \mathrm{d}\omega t$ is

$$\mathrm{d}\Psi = NB(\omega t)\mathrm{d}S, \tag{7.288}$$

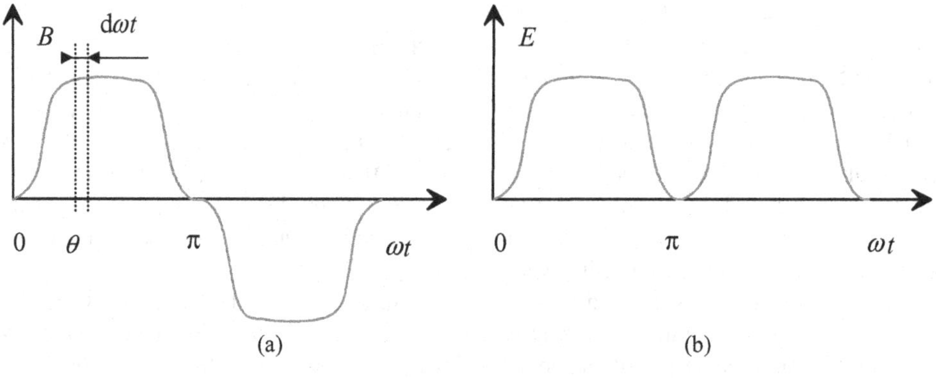

Figure 7.88 (a) Flux density waveform in a distance of two pole pitches and (b) rectified voltage in no-load operation.

where $dS = (D/2)ld\omega t$; that is, the area of the path of the flux, when D is the diameter and l is the length. The equation can now be written as

$$d\Psi = \frac{NDlB(\omega t)\,d\omega t}{2}. \tag{7.289}$$

The flux linkage is obtained by integration

$$\Psi = \frac{NDl}{2}\int_{\theta}^{\theta+\pi} B(\omega t)d\omega t. \tag{7.290}$$

Since the flux density is symmetrical ($B(\theta + \pi) = -B(\theta)$), it can be presented as a series of odd harmonics

$$B(\omega t) = B_1 \sin \omega t + B_3 \sin 3\omega t + \cdots + B_n \sin n\omega t. \tag{7.291}$$

By substituting this in Equation (7.290) we obtain the result of the integration

$$\Psi = NDl\left(B_1 \cos\theta + \frac{1}{3}B_3 \cos 3\theta + \cdots + \frac{1}{n}B_n \cos n\theta\right). \tag{7.292}$$

The voltage induced in the coil is

$$e = -\frac{d\Psi}{dt}, \tag{7.293}$$

and by substituting (7.292) into the above equation we obtain

$$e = \omega NDl(B_1 \sin\theta + B_3 \sin 3\theta + \cdots + B_n \sin n\theta). \tag{7.294}$$

This shows that the voltage follows the curvature form of the flux density, when ω is constant. We obtain a rectified voltage as presented in Figure 7.88b. As there are several coils instead of a single coil in practical applications, being connected via a commutator, the coils together produce a practically rippleless voltage. Figure 7.89 depicts the curvature form of the voltage, created by two coils located at a distance of $\pi/2$.

Based on Figures 7.88 and 7.89, it is obvious that the relative amplitude of the voltage ripple of two coils is half of the case of a single coil. Simultaneously, the frequency is doubled. When we increase the number of coils, we reach a rather even voltage. The voltage of several coils in series equals to the number coils multiplied with the average voltage of a single coil. The average voltage of a single coil is

$$\overline{E} = \frac{1}{\pi}\int_{0}^{\pi} e d\omega t. \tag{7.295}$$

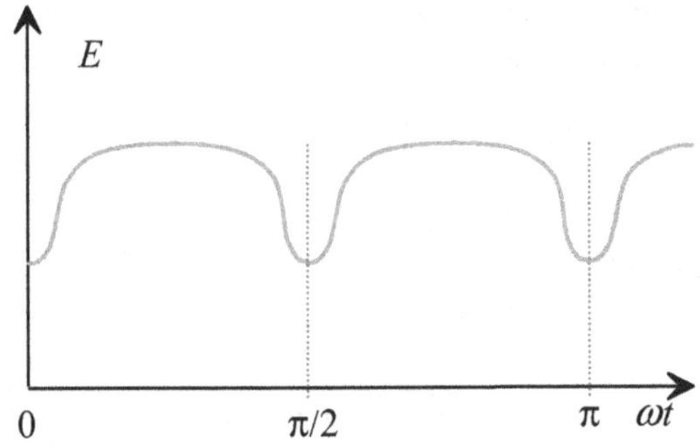

Figure 7.89 Common voltage of two rotor coils located electrically at a distance of $\pi/2$.

By substituting (7.294) in (7.295) we obtain, based on the (7.291)

$$\overline{E} = \frac{\omega}{\pi} N \int_0^{\pi} B(\omega t)\, Dl\mathrm{d}\omega t, \tag{7.296}$$

where $Dl\mathrm{d}\omega t = 2\mathrm{d}S$, and thus

$$\overline{E} = \frac{2\omega N}{\pi} \int_0^{\pi} B(\omega t)\mathrm{d}S. \tag{7.297}$$

The integral term in (7.297) yields the flux per pole Φ_p, and thus the voltage induced in the full-pitch coil is

$$\overline{E} = \frac{2\, \omega N \Phi_\mathrm{p}}{\pi}. \tag{7.298}$$

Equation (7.298) is valid both for two-pole machines or machines with multiple poles, bearing in mind that ω is electric angular speed. If n is the rotational speed, the mechanical angular speed Ω is

$$\Omega = 2\pi n. \tag{7.299}$$

When the number of poles is $2p$, we obtain the electric angular speed

$$\omega = p\Omega. \tag{7.300}$$

In general, we obtain for the average voltage of a coil with N turns

$$\overline{E} = 4pN\, \Phi_\mathrm{p} n. \tag{7.301}$$

The previous discussion has concerned a single full-pitch coil, that is a coil that covers a single slot per pole, and is of the width π in electrical degrees. However, Equation (7.301) holds with sufficient accuracy also for non-full-pitch windings common in DC machines. The coil span is usually more than two-thirds of the pole pitch, and the maximum flux density is nearly equal to the full-pitch winding. If now

$2a$ is the number of parallel current paths of the armature winding,
p is the number of pole pairs,
N_a is the number of turns in the armature winding,
n is the rotation speed and
Φ_p is the flux per pole,

then the number of coil turns of the armature between the brushes is $N_a/2a$ and the induced armature voltage E_a is approximately (irrespective of the coil span)

$$E_a \approx \frac{4pN_a n\Phi_p}{2a} = \frac{2pN_a \Omega \Phi_p}{\pi 2a} = \frac{p}{a}\frac{z}{2\pi}\Phi_p \Omega, \tag{7.302}$$

where z is the number of all armature conductors $z = 2N_a$. Based on the above discussion, we may now present the simplified equations for the DC machine. The emf depends on the magnetic flux per pole, the rotation speed n and the machine-related constant k_E

$$E_a = k_E n\Phi_p. \tag{7.303}$$

where $k_E = pz/a$.

The emf can also be written as

$$E_a = \frac{p}{a}\frac{z}{2\pi}\Phi_p \Omega = k_{DC}\Phi_p \Omega \tag{7.304}$$

where

$$k_{DC} = \frac{p}{a}\frac{z}{2\pi} \tag{7.305}$$

The electric power P_a of the armature of the machine is a product of the emf E and the armature current I_a

$$P_a = E_a I_a. \tag{7.306}$$

The torque corresponding to this power is

$$T_e = \frac{P_a}{\Omega} = k_{DC}\Phi_p I_a. \tag{7.307}$$

When we substitute (7.303) and (7.306), we obtain

$$T_e = \frac{k_E n\Phi_p I_a}{\Omega}. \tag{7.308}$$

7.4.3 Armature Reaction of a DC machine and Machine Design

Although a detailed description of the design of a DC machine is beyond the scope of this material, some design principles are still worth mentioning. The methods presented previously are applicable to the design of a DC machine with certain adjustments. One of the most important special features of a DC machine is the armature reaction and, in particular, its compensation.

According to IEC, the armature reaction is the current linkage set up by the currents in the armature winding or, in a wider sense, the resulting change in the air-gap flux. Since the brushes are on the quadrature axis, the armature current produces the armature reaction also in the quadrature direction, that is, transversal to the field-winding-generated flux. Figure 7.90 depicts the armature reaction in the air gap of a noncompensated DC machine.

As a result of the armature reaction, the air-gap flux density is distorted from the no-load value, and the flux density at the quadrature axes no longer remains zero. This is extremely harmful from the commutating point of view. To enhance the commutating procedure, the quadrature axis flux density has to be brought almost to zero (see Chapter 7.4.4). Figure 7.91 illustrates the air-gap flux density behavior in a noncompensated DC machine.

The armature reaction has a remarkable influence on the current commutating and special attention has to be paid to the design of successful commutation. Commutation is most successful in machines in which the quadrature axis flux density is brought to a value with which it is able to compensate the voltage induced in the commutating coil followed from its leakage inductance and current time variation during the commutation process and to achieve

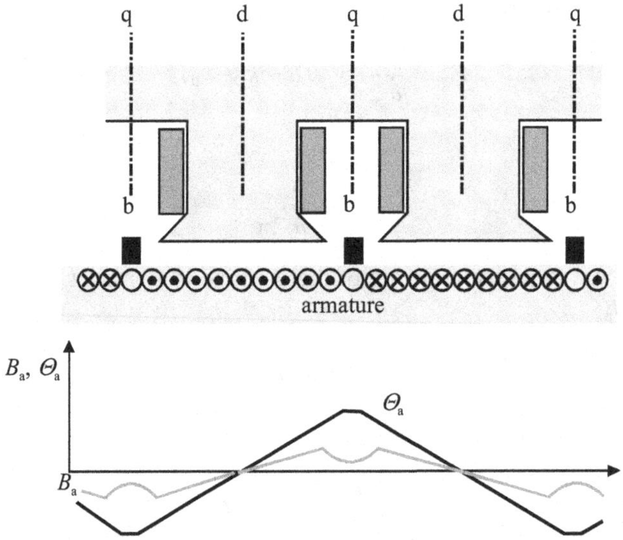

Figure 7.90 Armature reaction of a DC machine. The direct (d) and the quadrature (q) axes and the brushes (b) of the armature circuit are indicated in the figure. Θ_a is the armature current linkage and B_a the corresponding air-gap flux component, differing by shape from Θ_a mainly in the q-axis area, where the air gap and hence the magnetic reluctance are very high.

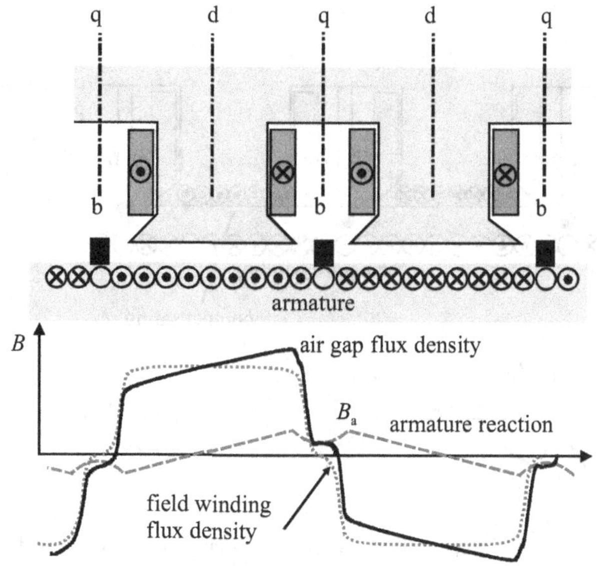

Figure 7.91 Resulting air-gap flux density as a sum of the field winding flux density and the armature reaction. As a result of the armature reaction, the flux densities at the quadrature axes are not zeros. This is harmful for the commutation of the machine.

so called "ideal commutation" (see Figure 7.94). This can be reached with a compensating and commutating pole winding with the right number of turns (see Chapter 2.16). The operation of the compensating and the commutating pole windings is illustrated in Figure 7.92.

The focus of the design of the magnetic circuit and the windings of the machine has to be on the dimensioning of the commutating poles, their windings, and the compensating winding. The machine is equipped with a turning brush rocker to ensure nonsparking commutation. Dimensioning of the compensating windings was discussed in brief in Chapter 2, Sections 2.15 and 2.16. Since it is not possible to remove completely the quadrature-axis flux density, a commutating pole winding is employed in addition to the compensating winding. In low-power machines, the disadvantages of commutating are usually compensated only with a commutating pole winding.

If we want to supply a DC series machine with an AC voltage supply (known as a universal motor), the commutating arrangements get even more complicated. As an exception to normal DC drive, we have to construct the entire magnetic circuit from laminated electrical sheet. These small applications are still employed in various high-speed consumer electronics such as drills, angle grinders, vacuum cleaners and the like.

7.4.4 Commutation

The current flowing in the armature winding is alternating current, whereas the current of the external circuit flowing through brushes is direct current. The commutator segment acts as

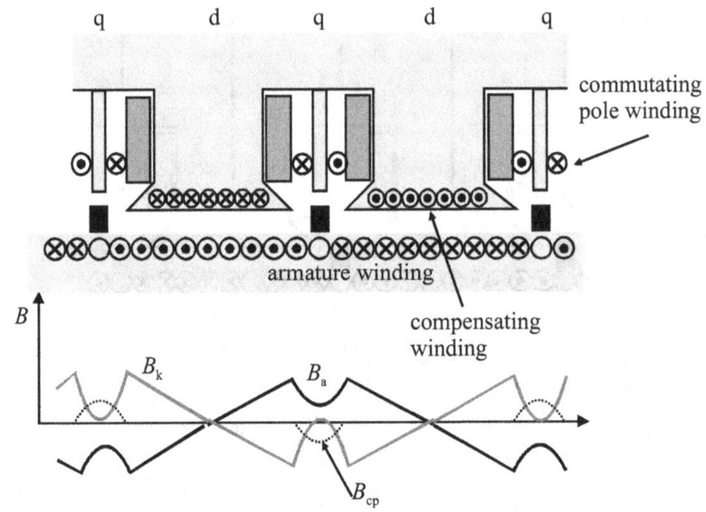

Figure 7.92 Operation of the compensating and the commutating pole windings. The armature reaction flux density B_a is totally compensated by the compensating flux density B_k and the commutating pole flux density B_{cp}.

an end point of the coil. When the commutator segment passes under a brush, the direction of current flow is always changed in the coil. The time interval required for this switching is known as the commutation period. Commutation is said to be linear when the current of the armature winding changes its polarity at constant speed, as shown in Figure 7.93. Linear commutation is the simplest way of to analyze commutation. Figure 7.93 illustrates the four steps of commutating. In the figure, the armature windings are presented simplified as coils connected to parallel commutator segments. There is a current I flowing per brush. When the commutator segment reaches the leading edge of the brush, the current of the coil is $I/2$ and, correspondingly, when the segment is passing the trailing edge of the brush, the current of the coil has attained the value $-I/2$.

Commutation is a fairly complicated process, and in reality it is not a completely linear process. Resulting from the change of current direction, a voltage is created by the leakage fluxes of the armature winding in the coil that is short-circuited by the brush. This voltage must be compensated either by turning the brush rocker or by commutating pole windings, or the inversion of the current flow direction must be retarded, and undercommutation occurs. As a result, the current decreases at an accelerated rate towards the end of the commutation period. This in turn causes an increase in the current density at the trailing part of the brush. If the current reversal is not completed when the commutator segment leaves the brush, sparking occurs at the trailing end of the brush. An excessive sparking burns the brushes and destroys the surface of the commutator. If the brush rocker is turned too much, or the influence of the commutating poles is excessive, an overcommutation occurs, and the direction of the current flow changes too rapidly. In the current reversal, the opposite current may increase too high, and the current density at the leading part of the brush is increased. Figure 7.94 illustrates the changes in the short-circuit current of a coil at different states of commutation.

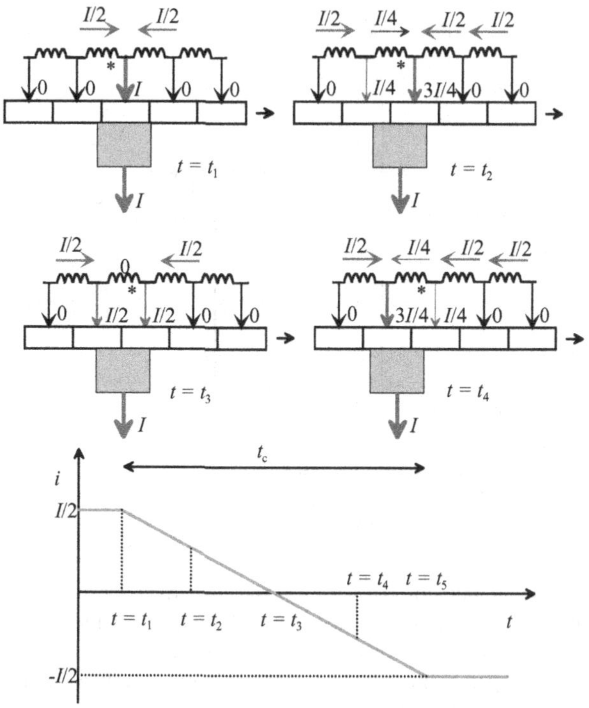

Figure 7.93 Phases of simplified linear commutation of an armature winding (indicated by an asterisk *) at the commutator and the current reversal during linear commutation in the armature winding. *tc* is the commutation period in the armature winding.

In addition to the voltage induced in the short-circuited armature coil, the commutation is also affected by the contact with brushes. In general, commutation can be stated to take place in four steps. In commutation, a current flows in the commutator segment via point-like areas. When currentless, these areas are covered with a thin oxide layer. Therefore, when a brush comes into contact with the tip of a commutator segment, there is no current flowing in the first, so-called blocking phase, since there is quite a high blocking voltage u_{bl} between the brush and the commutator segment, Figure 7.95a.

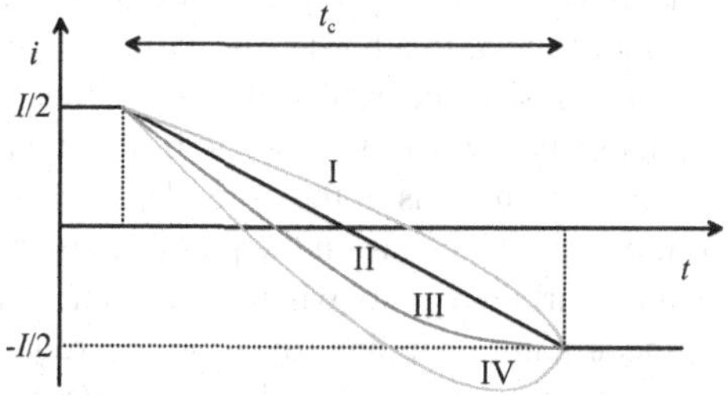

Figure 7.94 Current of an armature winding coil short-circuited by commutation: I, undercommutation; II, linear commutation; III, ideal commutation where at *tc*, d*i*/d*t* = 0; IV, overcommutation.

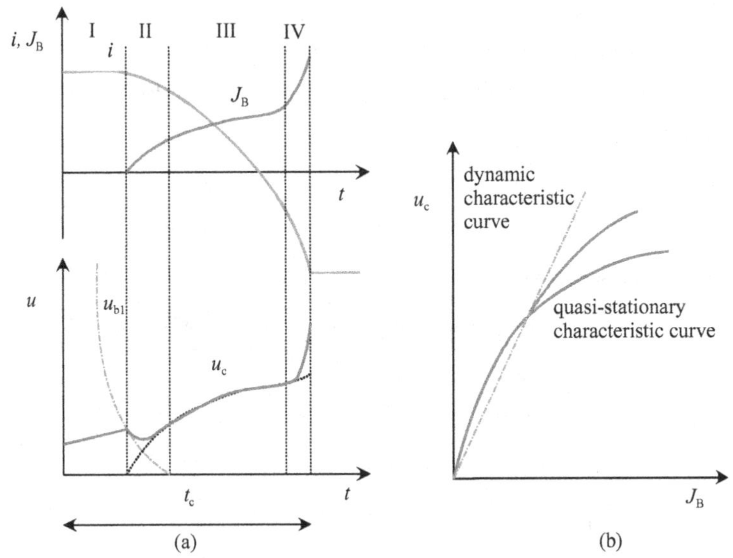

Figure 7.95 (a) Phases of commutation. The current i and the current density in the brush J_B. I, blocking phase; II, contact formation; III, quasi-stationary phase; IV, end phase. In the voltage graph, the solid line u_c depicts the real function of commutation voltage, and u_{bl} is the blocking voltage of the oxide layer. The dotted line indicates the time function of the voltage corresponding to the quasi-stationary characteristic curve. (b) The characteristic curve of the brush contact, with the commutation voltage as a function of current density. The dynamic characteristic curve is approached during the high-speed change of the current density at the end phase of the commutation.

Only when the contact voltage between the brush and the segment exceeds the reverse voltage that sinks rapidly because of the simultaneously occurring contact points does the actual contact formation begin between the commutator segment and the brush. The contact points gradually turn conductive as the voltage strength of the oxide layer decreases. The process takes place before the brush completely covers the commutator segment. In this phase, to create a current, a raised voltage level is required contrary to the next, quasi-stationary phase. In this phase, the quasi-stationary characteristic curve $u_c = f(J_B)$ is valid (Figure 7.95b) and gives the contact voltage as a function of current density J_B. The final phase of the commutation is characterized by rapid changes in current density, which the contact formation process cannot follow. In a borderline case, an excessive change in current density leads to a constant contact resistance, represented by a linear characteristic curve that approaches the dynamic characteristic curve of Figure 7.95b. The dynamic characteristic curve is valid when the change in current density is of the order $dJ_B/dt > 10^5$ A/(cm^{2}s).

The final phase of the commutation depends on the possible sparking between the brushes and the commutator segments. Sparking occurs when the current at the end of commutation deviates from the current of the armature winding (over- or undercommutation), and the commutator segment in question shifts away from below the brush, in which case the commutation has to be completed via the air gap. An arc is struck between the commutator segment and the brush. The arc disappears when the current has reached the correct value.

In general, brushes consist of carbon, graphite and organic materials. In low-voltage devices, such as in the 12 V starting motor of a car engine, the resistance between the brushes and the commutator has to be kept low. Therefore, carbon-coated graphite brushes are employed.

A thin layer of copper oxide, covered with graphite, occurs on the surface of the commutator (patina). This layer affects the operating life of the brush and the wear of the commutator. Graphite acts as a lubricant, and the resistance of the layer is high enough to restrict the short-circuit current and to create resistance commutation.

7.5 Doubly Salient Reluctance Machine

A doubly salient reluctance machine is an electrical machine that operates with intelligently controlled power electronics. The machine is also called a switched reluctance machine (SR machine) because it cannot operate without power electronic switches. The development of semiconductor power switches and digital technology has opened up new opportunities for controlled electric drives. The development of electric drives can be seen for instance in the increased application of frequency-converter-supplied induction machines among controlled electric drives.

Although simpler induction motor alternatives have been found to replace DC machines, neither the dynamic performance nor the power density have improved with the controlled electric drives. In particular, with low rotation speeds or at standstill, both drive types have problems in producing satisfactory torques. This is clearly visible for instance in machine automation, where high torques are produced by hydraulic motors. A doubly salient, power-electronics-controlled reluctance machine may in some cases provide improvement in the torque production at low rotation speeds.

A reluctance machine has already been employed for a long time in stepper motor drives, in which continuous, rippleless torque control has not been necessary. Only the development of power electronics and control systems has enabled the application of a reluctance machine in the power range of a few hundred kilowatts, while stepper motor applications have so far been limited to a few hundred watts at maximum.

The basic construction of a switched reluctance motor was introduced as early as in 1838 by Robert Davidson in Scotland, but the machine was not utilizable before the development of power electronics and suitable components. The motor construction and the theory of control were quite well reported by the end of the 1970s, and development has been continuous ever since, particularly in the field of control technology.

7.5.1 Operating Principle of a Doubly Salient Reluctance Machine

The English term for a switched reluctance motor originates from the fact that it always requires a controlled power source, that is an inverter. In the German literature, the term "doubly salient reluctance machine" occurs repeatedly (*Reluktanzmaschine mit beidseitig ausgeprägten Polen*). The latter term describes the geometry of the machine well, as shown in Figure 7.96.

The operating principle of a switched reluctance machine is based on the power effects of the magnetic circuit that tend to minimize the reluctance of the magnetic circuit. If current is led to phase A of the machine in Figure 7.96, the rotor tries to turn counterclockwise so

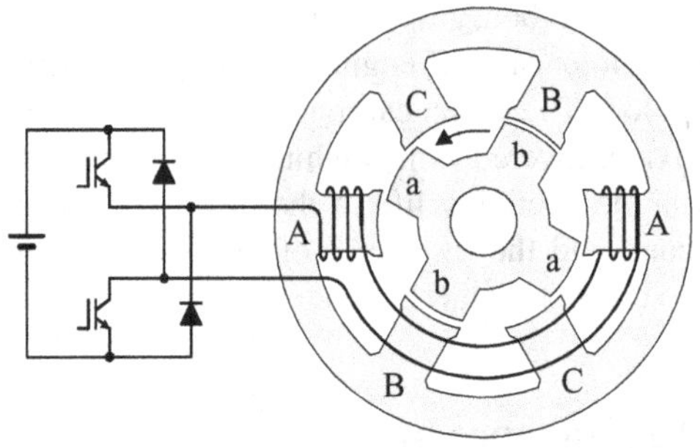

Figure 7.96 A 6 : 4-pole (six stator poles and four rotor poles), three-phase, doubly salient reluctance machine and the semiconductor switches controlling a single phase of the machine. Rotor poles have just passed their unaligned position with respect to stator phase A and the poles (a) will be in the aligned position with the stator poles A–A when the rotor moves counterclockwise towards the excited phase A–A.

that the reluctance of the magnetic circuit of phase A would reach the minimum and also the energy of the magnetic circuit would thus be minimized.

When the energy minimum of phase B has been reached, the magnetic forces try to keep the rotor in a position in which the energy minimum of the magnetic circuit is preserved. Now the magnetic energy has to be removed from phase B to make the machine rotate again. Correspondingly, current supply is started to phase A at an instant when the poles of the rotor and the stator are about to overlap. By connecting the currents to different phases in turn at the correct instants and with suitable magnitudes, a high, almost smooth sum torque can be reached over a wide rotation speed range. Also, the loading of the rotor at zero speed is easier than with traditional electrical machines.

7.5.2 *Torque of an SR Machine*

The torque calculation of an SR machine is based on solving the flux diagram and the determination of Maxwell's stress tensor. If we have the Ψi graphs of the machine, we can employ the principle of virtual work in the torque calculation, as discussed in Chapter 1.

The calculation of the torque is problematic because of the dependence of the inductance of the magnetic circuit on both the rotor angle and the stator current. The dependence results from the excessive saturation of the magnetic circuit at certain rotor positions and with high stator currents, Figure 7.97.

The instantaneous electromagnetic torque T_{em} is the ratio of the change of mechanical energy to the rotation angle. It can be written in the form

$$T_{em} = \frac{dW_{mec}}{d\gamma} = i\frac{\partial\Psi}{\partial\gamma} - \frac{\partial W_{em}}{\partial\gamma}. \tag{7.309}$$

The torque equation can be simplified by replacing the electromagnetic energy W_{em} with magnetic coenergy W'. The magnetic coenergy is determined as

$$W' = \int_0^i \Psi di. \tag{7.310}$$

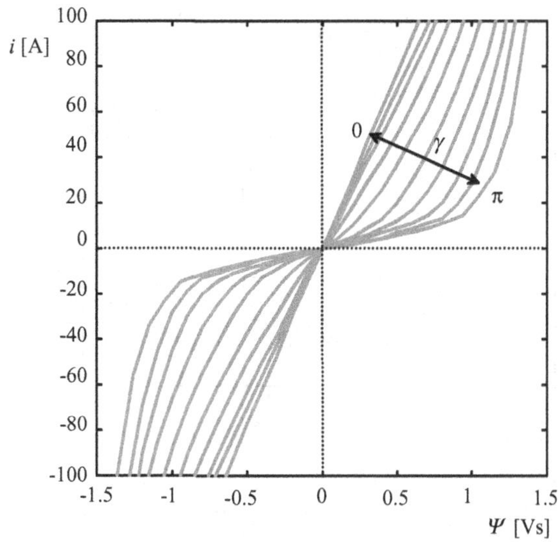

Figure 7.97 Current of a 15 kW, 1500 min^{-1} SR machine as a function of flux linkage, the rotor position angle γ being a parameter. When the poles are in the aligned position, the iron circuit makes the machine very nonlinear. When the poles are in the unaligned position, the large air gap makes the dependence quite linear. $\gamma = \pi$ corresponds to the aligned rotor position and $\gamma = 0$ the unaligned position.

The geometric interpretation for magnetic energy and coenergy is the area between the magnetizing curve and the i-axis presented in an $i\Psi$ plane, Figure 1.15:

$$W_{\text{em}} + W' = i\Psi. \tag{7.311}$$

By deriving W' with respect to the angle γ, the derivative of the coenergy can be written as

$$T_{\text{em}} = \frac{\partial W'}{\partial \gamma} = i\frac{\partial \Psi}{\partial \gamma} - \frac{\partial W_{\text{em}}}{\partial \gamma}. \tag{7.312}$$

By comparing the result with Equation (7.309) for the torque, we can see that the torque of a reluctance machine is equal to the change in the magnetic coenergy per angular change.

7.5.3 Operation of an SR Machine

The operation of an SR machine is governed in the linear region by the constant variation of the inductance as a function of the rotation angle of the machine. Furthermore, the machine saturates at the pole edges, and therefore the definition of the inductances is difficult in different rotor positions and with different currents. Saturation is clearly visible for instance in the dependence between the flux linkage and the current, illustrated in Figures 7.97 and 7.99. The higher the inductance difference between the aligned and unaligned positions, the higher the average torque of the machine. As stated previously, the instantaneous torque of the machine can be calculated from the change in coenergy as a function of rotation angle.

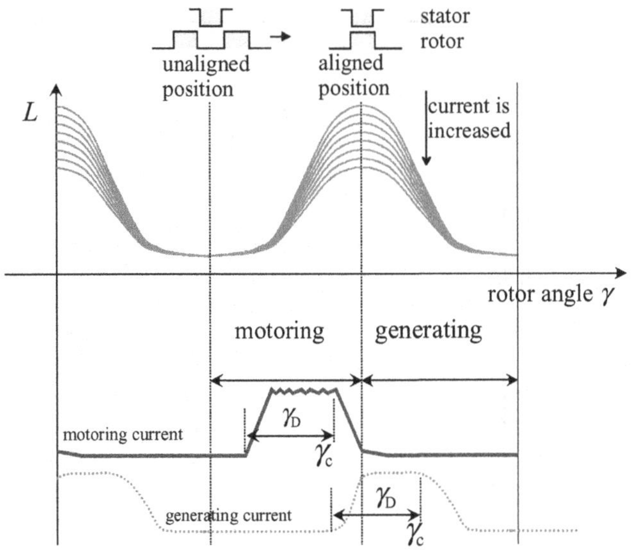

Figure 7.98 Inductance of a saturating reluctance motor as a function of rotor angle with current as a parameter, and the current pulses of a motor drive and a generator drive, when the voltage of the intermediate circuit remains constant. In motor or generator drive, the phase is commutated with the angle γ_c. The total conduction angle of the supplying positive voltage pulse has a length of γ_D (Switch ON). The current of the motor does not quite reach zero before the aligned position is reached. Now a slight braking takes place as the machine shifts from the aligned position. In generating, the machine is excited at the aligned position and the current is commutated well before the unaligned position.

In motoring action, the current of the machine is often kept constant with a switched-mode power supply, Figure 7.98.

The energy supplied from the voltage source cannot be completely converted into mechanical work, but a part of it is returned to the current source at the end of each working stroke. Figure 7.99 illustrates the behavior of the flux linkage and the current during a single stroke. The required energies are also indicated in the figure. The figure is constructed assuming that the rotation angular speed Ω of the reluctance machine remains constant.

When the supply voltage U_d (converter DC link voltage) and the phase resistance are constant, the speed is constant and the resistive voltage drop remains low, there is a linear increase in flux linkage Ψ as an integral of voltage U_d after the switching of the voltage:

$$\Psi(t) = \int (U_d - Ri)\,dt = \frac{1}{\Omega}\int (U_d - Ri)\,d\gamma.$$

The increase in current i is also initially linear because the inductance L remains low and almost constant in the vicinity of the unaligned position.

When the poles approach the aligned position, the inductance increases rapidly, and the resulting back emf restricts the current. This phase is illustrated as the period 0–C (Figure 7.99(a)). With the rotor angle γ_C at point C, the phase in question is commutated. Now the energy brought into the system is $W_{mt} + W_{fc}$ (the bright + shaded areas). Here W_{fc} is

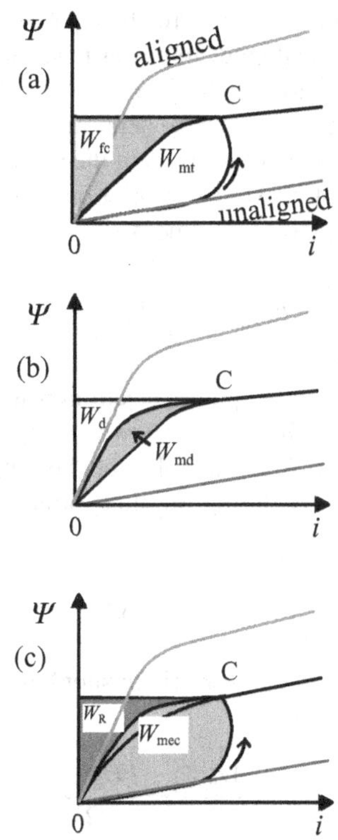

Figure 7.99 (a) Transistor conduction period, (b) diode conduction period (see also Figure 7.96), (c) energy conversion loop.

the energy stored in the magnetic field and W_{mt} is the energy converted into mechanical work when the transistor is conducting (see Figure 7.105a). In this phase, the mechanical work is approximately equal to the energy stored in the magnetic circuit. After the commutation, the polarity of the voltage is changed and the energy W_d is returned through the diode to the voltage source (see Figure 7.105c), and the remaining energy W_{md} is the obtained mechanical work in the interval C–0, Figure 7.99b. During the complete working stroke, the mechanical work is thus $W_{mec} = W_{mt} + W_{md}$ and the energy returning to the voltage source is $W_R = W_d$. The complete working stroke is illustrated in Figure 7.99c. According to the example in the figure, the proportion of mechanical energy of the total energy is about 65 %. The rest of the originally supplied energy is "reactive energy" of the reluctance machine that is stored either in the electric field of the capacitor of the intermediate circuit (DC link) or in the magnetic field of the magnetic circuit of the machine.

Usually, an energy ratio Γ is determined for an SR machine. This ratio expresses the energy that can be converted into mechanical energy during the energy conversion loop

$$\Gamma = \frac{W_{mec}}{W_{mec} + W_R} = \frac{W_{mec}}{W_{el}}, \tag{7.313}$$

where $W_{mec} + W_R$ is the apparent power delivered by the power electronic circuit. The energy ratio is to some degree a quantity analogous to power factor in AC machines. In the example of Figure 7.99 the energy ratio has a value of approximately $\Gamma = 0.65$.

The average torque of an SR machine can be determined when the number of strokes per revolution is known. In one revolution all poles N_r of the rotor must be worked on by all stator phases, and therefore the number of strokes per revolution is mN_r. The average electromagnetic torque over one revolution thus obtains the value

$$T_{em\ av} = \frac{mN_r}{2\pi} W_{mec}. \tag{7.314}$$

The input energy W_{el} supplied by the power electronics to the machine can be expressed as a fraction k of the product $i_c \Psi_c$ ($W_{el} = k\Psi_c i_c$), where Ψ_c is the value of the flux linkage at the instant of commutation and i_c is the value of the current at that respective instant. If the flux linkage increases linearly during the flux formation period 0–C as illustrated in Figure 7.99, then

$$\Psi_c = U_d \gamma / \Omega. \tag{7.315}$$

Here γ is the angle during which the power bridge supplies power to the machine. We now obtain

$$W_{el} = \frac{W_{mec}}{\Gamma} = \frac{kU_d \gamma i_c}{\Omega}, \tag{7.316}$$

and since i_c is the peak value of the current yielded by the power electronics, the required apparent power S_m processing ability of the output stage in an m-phase system is

$$S_m = mU_d i_c = \frac{mW_{mec}\Omega}{\Gamma k \gamma} = \frac{2\pi T \Omega}{N_r \Gamma k \gamma} \tag{7.317}$$

The product of the torque and angular speed corresponds to the air-gap power P_δ, and the product $N_r\gamma$ is constant, the maximum value of which is about $\pi/2$ at the base rotation speed of the machine. The power-processing capability of the power bridge thus has to be

$$S_m = \frac{4P_\delta}{k\Gamma} \tag{7.318}$$

The required power is therefore independent of the phase number and the number of poles, and it is inversely proportional to the energy conversion ratio Γ and the fraction k. Both Γ and k depend greatly on the static magnetizing curves of the machine, and on the curves of the aligned and unaligned position in particular. These curves, however, are in practice highly dependent on the pole number N_r, which thus has a strong indirect effect on the dimensioning of the power stage of the machine. When the power processing capability of the power stage is compared with the shaft output power, we can determine, assuming that $k = 0.7$ and $\Gamma = 0.6$, that $S_m/P_\delta \approx 10$. This value is typical of SR motor drives. Inverter power stages of the same scale are required in induction motor drives also.

7.5.4 Basic Terminology, Phase Number and Dimensioning of an SR Machine

The geometry of SR motors is typically either symmetric or asymmetric. The stator and rotor poles of a regular machine are symmetric with respect to their center lines, and the poles are located at even distances with respect to the peripheral angle. High-power machines are usually symmetrical. Asymmetrical rotors are employed mainly in single- and two-phase machines to produce starting torque.

An absolute torque zone is defined for an SR machine, referring to the angle through which a single phase of the machine may produce torque. In a regular machine, the maximum angle is π/N_r.

An effective torque zone refers to the angle through which the machine can produce an effective torque comparable with its rated torque. The effective torque zone corresponds in practice to a smaller one of the stator and rotor pole arcs. For instance, in Figure 7.100 $\beta_s = 30°$ and $\beta_r = 32°$, and thus the effective torque zone equals the stator pole angle $\beta_s = 30°$.

The stroke angle ε of the machine is defined by the number of strokes during a full circle

$$\varepsilon = \frac{2\pi}{mN_r} = \frac{360°}{mN_r} \tag{7.319}$$

The absolute overlap ratio ρ_A is the ratio of the absolute torque zone and the stroke angle

$$\rho_A = \frac{\pi/N_r}{\varepsilon} = \frac{\pi/N_r}{2\pi/(mN_r)} = \frac{m}{2} \tag{7.320}$$

In a symmetrical motor, this value has to be at least one in order to produce torque at all position angles of the rotor. In practice, the absolute overlap ratio has to be greater than one, since a single phase alone is not capable of producing smooth torque at the distance of the

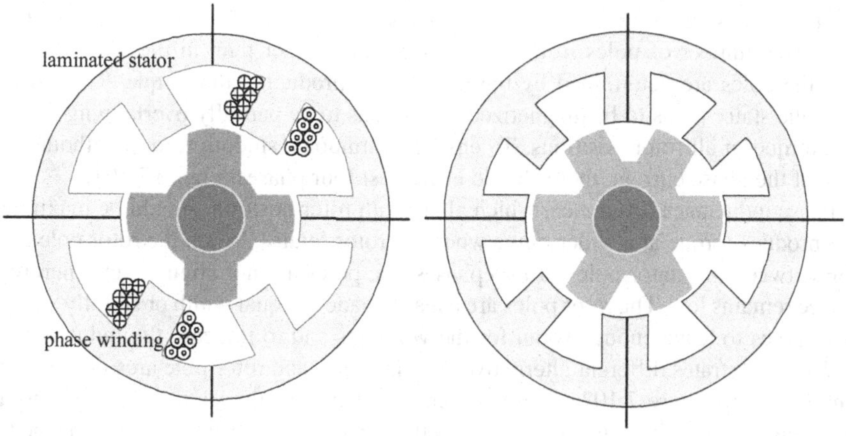

Figure 7.100 Three-phase 6 : 4 ($\beta_s = 30°$, $\beta_r = 32°$) and a four-phase 8 : 6 doubly salient reluctance motor.

whole absolute torque zone. Therefore, the effective overlap ratio ρ_E is defined as the ratio of the effective torque zone and the stroke angle

$$\rho_E = \frac{\min(\beta_s,\ \beta_r)}{\varepsilon}. \tag{7.321}$$

For instance, in Figure 7.73, $\beta_s = 30°$, $\varepsilon = 360°/3 \times 4 = 30°$ and thus we obtain $\rho_E = 30°/30° = 1$. At least the value of one has to be reached with the geometry of the machine in order to attain good starting torque with all rotor positions when only one phase is conducting.

In the construction of an efficient SR motor, usually the pole numbers presented in Figure 7.100 are employed. The ratio of the stator and rotor pole numbers is thus either 6 : 4 or 8 : 6. These machines are three- or four-phase ones respectively. However, other topologies are also possible. An increase in the pole number improves the operational accuracy of the motor and the quality of the torque, but simultaneously the structure and control of the converter switches get more complicated.

The poles of this motor type are typically long and narrow. As a result of the shape of the poles, the magnetic flux of the machine is smaller, and therefore more coil turns are required than in an AC motor operating at the same voltage. The rotor is usually long and thin in order to reduce the moment of inertia. The air gap should be as short as possible to reach a maximum average torque at small rotor volume. The length of the air gap is usually selected to be about 0.5–1 % of the rotor diameter.

There are a few basic rules governing the selection of the number of poles and phases and the angles between the poles. The ratio of the rotor speed n and the base switching frequency can be determined when we assume that the poles connected to the same phase are at the opposite sides of the stator. Now, as the rotor pole passes the phase, a torque pulse is produced. The base frequency f_1 is thus

$$f_1 = nN_r, \tag{7.322}$$

where n is the rotation speed and N_r is the number of poles in the rotor.

For a three-phase 6 : 4 motor, the stroke angle is $\varepsilon = 30°$ and for a four-phase 8 : 6 motor $\varepsilon = 15°$. The number of poles in the stator is usually higher than in the rotor. The angles between the poles are determined by the mechanism producing the torque. The pair of the rotor and the stator poles to be magnetized always has to be partially overlapping in order to produce torque in all rotor positions. To ensure continuous, smooth torque without difficult profiling of the phase current, there should be at least four phases, Figure 7.101.

The phase inductance difference, which alters with rotor position, should be maximized in order to produce torque at a wider range when the rotor rotates. When the rotor pole is in the position between the stator poles of two phases, the poles are not aligned, and therefore the inductance remains low. The rotor poles are usually made of equal width or slightly wider than the stator poles to leave enough room for the windings and to increase the inductance ratio. Figure 7.102 illustrates different alternatives for the stator and rotor pole arcs of the machine.

In the motor of Figure 7.102a, there is plenty of room for the stator windings and also a high inductance ratio. Thus, the efficiency and the power density are high, whereas the torque ripple is higher than in cases b and c of Figure 7.102. Also, the starting torque of the motor is low. In the stator of the motor in Figure 7.102b, there is enough room for the coils, but the

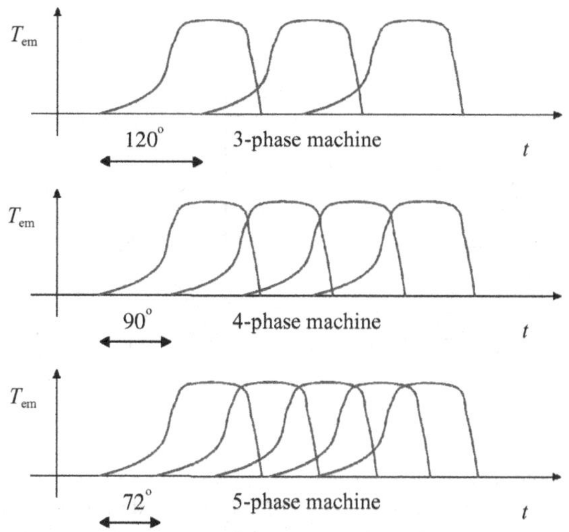

Figure 7.101 Torque pulses produced by the phases of a three-, four- and five-phase reluctance machine with single pulse control. To achieve smooth torque, the phases of the machine have to be controlled in such a way that in the commutating phase, two phases together produce the total torque. With three-phase machines, to smooth the total torque, quite high currents have to be employed in the commutation area, since, as shown above, at the intersections of the torque curves of different phases the sum of the torques does not reach the level of the peak torque of single phases. In four- and five-phase machines, the production of smooth torque is easier than with the three-phase machine, due to the torque reserve at commutation.

minimum inductance of the machine remains high in the unaligned position, since the poles are constantly aligned to some extent. In Figure 7.102c, the inductance of the motor is far too high in the unaligned position, and there is not enough room for the windings either. Miller (1993) presents the different alternatives for pole arcs in graphical form. In the graphical presentation, the feasible pole arcs are given in the form of a triangle ABC, Figure 7.103.

An optimum value can be solved for the pole size. This optimum value yields the maximum inductance ratio and simultaneously the maximum average torque. In addition, several other factors affecting the operation of the machine have to be taken into account, such as the torque

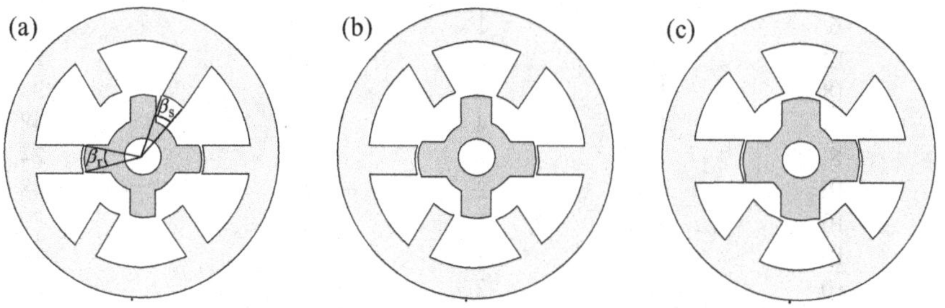

Figure 7.102 Examples of different pole arc combinations for a three-phase machine with a topology of 6 : 4 with different values of βs and βr.

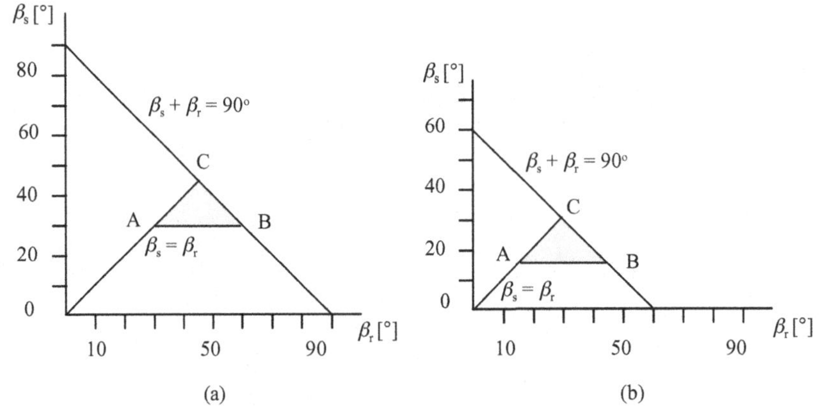

Figure 7.103 Feasible pole arcs for the stator and rotor poles of an SR motor; (a) 6 : 4 three-phase motor; (b) 8 : 6 four-phase motor. Angles of the triangles ABC correspond in principle to the combinations of Figures 7.102a, b and c. The pole arcs should lie within the triangular boundaries. Adapted from Miller (1993).

ripple, starting torque and the effects of saturation, and therefore no general solution can be attained.

Table 7.11 introduces various SR motor configurations for reluctance machines with 1–5 phases. Although several other combinations are possible e.g. several combinations where the number of rotor poles is larger than the number of stator poles, the table presents only those combinations most likely to be met in practice.

Table 7.11 Selection of feasible pole numbers for reluctance machines with 1–5 phases. m is the phase number, N is the number of poles, μ is the number of pole pairs operating simultaneously per phase and ε is the stroke angle

m	N_s	N_r	μ	ε/°	Strokes/revolution
1	2	2	1	180	2[a]
2	4	2	1	90	4[b]
3	6	4	1	30	12
3	6	8	1	15	24
3	6	10	1	7.5	36
3	12	8	2	15	24
3	18	12	3	10	36
3	24	16	4	7.5	48
4	8	6	1	15	24
4	16	12	2	7.5	48
5	10	6	1	12	30
5	10	8	1	9	40
5	10	8	2	18	20

[a]Needs assistance for starting.
[b]An asymmetric rotor is needed to create starting torque.

7.5.5 Control Systems of an SR Motor

SR motors always require an individual control system, the performance of which decides the overall characteristics of the machine drive. SR machines may not be operated directly on line. That is why power electronic controllers are briefly discussed here. The control system comprises switched-mode converters and the control and measuring circuits controlling them. The torque of a reluctance motor does not depend on the direction of the phase current, and therefore unidirectional switches can be employed in the converter. An advantage of direct current is the reduction of hysteresis losses.

In the circuits of Figure 7.104, the switches can be either power FETs (Field-Effect Transistors), or IGBTs (Insulated-Gate Bipolar Transistors). The phases are independent of each other, unlike in AC inverters for instance. The chopper switch is well protected in fault situations, since the phase winding of the motor is in series with both switches. The control circuit of the switches can be made far simpler when compared with an AC inverter. In the circuit of Figure 7.104b, fewer switches are required than in the circuit of Figure 7.104a without any significant decline in applicability. At high speeds, the energy of the magnetic field cannot be discharged fast enough with the circuit of Figure 7.104b, since the polarity of the voltage of the phase winding cannot be changed. Now a braking torque is created, and the losses increase rapidly.

The inductance varies with rotor position and therefore, with a fixed switching frequency, also the current ripple changes. This can be prevented by applying variable switching frequency, in which case the switching frequency varies according to the angle between the rotor and the

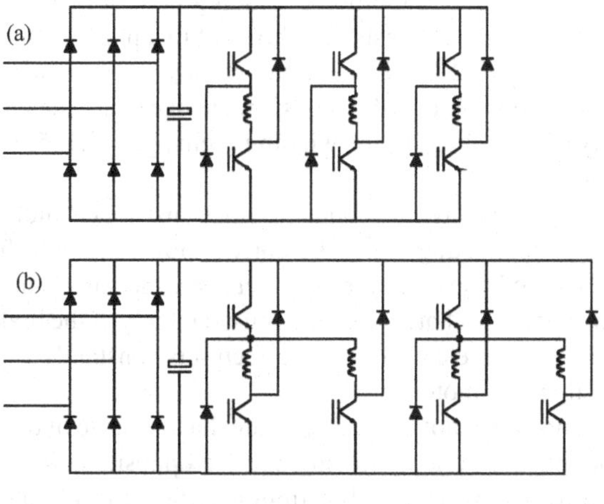

Figure 7.104 Converters operating a three- and four-phase SR motor. (a) Three-phase machine: each phase has a separate chopper leg comprising two switches, which can be controlled independent of the other phases. (b) In the control circuit of a four-phase machine, the transistor of the upper branch is shared by two phases. However, when the phases are selected appropriately, it is possible to control the machine in all the possible modes of Figure 7.105. When a transistor, located in the upper leg of a three-phase machine, is employed jointly for all three phases, the transistor of the upper leg has to be nearly always conducting, in which case some of the controllability is lost. In principle, the upper transistor controls the value of the current and the lower transistors guide the current to different phases.

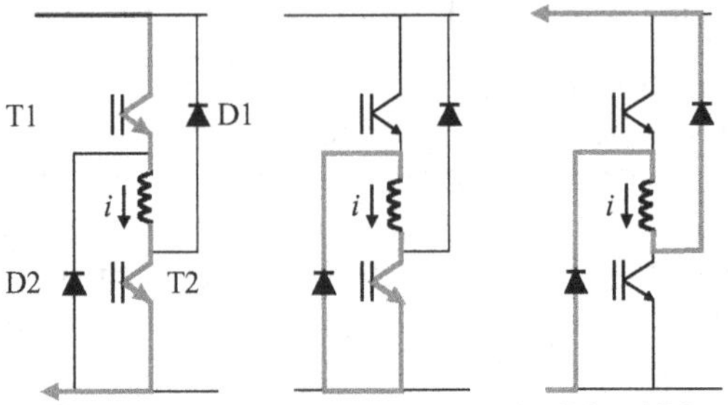

Figure 7.105 Application of power electronics to control the current of a single SR machine phase: (a) positive voltage, (b) zero voltage, (c) negative voltage in the coil.

stator. The switching frequency is usually selected above 10 kHz, as in other motor drives, in order to reduce noise. With high power values, low switching frequencies are selected because of the slowness of large switches; therefore, large and slow motors are usually quite noisy.

At low rotation speeds, the switches are usually controlled so that one transistor of the branch is employed for commutation and the other to control the current. At high rotation speeds, both transistors are constantly conductive, since the emf restricts the maximum value of the current, in which case the current curvature depends on the motor characteristics. Figure 7.105 illustrates the current flow during the conduction period. When the transistors are conducting, the current passes through them and the phase winding (7.105a). When the current reaches the upper limit, the upper chopping transistor is brought to a nonconducting state, the current of the winding passes through T2 and D2, and the voltage over the winding is close to zero (7.105b). At the commutation point, both transistors change state, the current transfers to the diodes, and the polarity of the voltage of the winding is changed (7.105c). This control method is called soft chopping.

The second alternative is always to control both transistors simultaneously. When the transistors are in the nonconducting state, the current flows through the diodes to the DC voltage source. In this method, the current ripple increases and simultaneously also the torque ripple and the noise increase. This hard chopping is used mainly in the braking of the rotor, that is in the generator drive. Nevertheless, some manufacturers constantly apply hard chopping in the control systems of their SR motors.

For the proper operation of the motor, it is important that the pulse of the phase current occurs at an instant when the inductance is increasing and the opposite poles of the rotor approach the stator poles of the respective phase. The timing of the current and the duration decide the torque, efficiency and other performance characteristics. There is no clear relationship between the phase current and the torque in a reluctance motor, as there is for instance in DC motors. Therefore, the production of a smooth torque requires highly intelligent control capable of detecting the rotor position and controlling the control as desired.

To improve the reliability of the motor and to simplify the construction, the target has been to develop an indirect position control without any devices attached to the rotor (position sensorless control). It has been shown that the rotor position can be determined with sufficient precision by analyzing the current and voltage curves of the motor. The varying inductance of the motor can be defined from these curves. Another, corresponding method is an active method in which the rotor angle is determined by supplying an instantaneous voltage pulse to a currentless phase.

7.5.6 Future Scenarios for SR Machines

Traditionally, reluctance machines have been controlled in the same way as stepper motors by supplying constant voltage and current pulses to the stator at a frequency defined by the rotor angle. This results in a considerable torque ripple, which has made reluctance machines unsuitable for several applications. Recently, various control methods have been introduced to bring the torque ripple to the level of traditional electrical machines. The methods are based on the exact measurement of the reluctance machine to analyze the magnetic properties of the machine. The control utilizes these documented measurement results. Therefore, the control has to be customized for each machine type individually.

The advantages of a reluctance machine when compared with traditional electric drives are for instance the following:

- No winding is required in the rotor; the rotor construction is simple and easy to manufacture.
- The moment of inertia of the rotor is low, a fact that improves the dynamics of a controlled electric drive.
- The stator winding is easy to construct, and the losses of an end winding are lower than in a corresponding induction machine.
- Most of the losses occur in the stator, and therefore cooling of the motor is easier, and a higher loadability is achieved.
- The large free spaces in the rotor enable efficient ventilation throughout the machine.
- The torque of the machine is independent of the direction of current, thus giving more degrees of freedom in the inverter and control solutions.
- The machine can produce a very high torque also at small rotation speeds and with a steady rotor at low current.
- The machine constant of an SR machine is higher than the machine constant of an induction motor.
- The torque is independent of the direction of the phase current, and therefore, in certain applications, it is possible to reduce the number of power switch components.
- In the event of a failure, the voltage of an open circuit and the short-circuit current are low.
- In reluctance motor applications, the power electronic circuits do not have a shoot-through path, which facilitates the implementation of the control system.
- Extremely high rotation speeds are possible. This, together with a high machine constant, makes the machine type interesting in aviation applications as a jet engine starter–generator.

A disadvantage of an SR motor is the discontinuous torque that causes vibration in the configuration and also acoustic noise. Over a low-speed range, the torque ripple can be restricted to 5–10 % or less, which is comparable to induction motor drives. Over a high-speed range, the restriction of the torque ripple is impossible in practice. This is not a problem thanks to mechanical filtering. At present, the best drives produce a very low degree of torque ripple at low speeds. The fact is that the smoothest torque is required only at low speeds, when the loads are most vulnerable to the harmful effects of torque ripple. In small motors, the noise can be damped at high speeds by selecting a switching frequency above the range of audibility.

In the torque control, the power is taken from the DC link in a pulsating manner, and therefore efficient filtering is necessary. In this sense, the drive does not differ considerably from the inverter drive of an induction motor. A small air gap advantageous to the operation of the motor increases the production costs. A small air gap is required to maximize the inductance ratio.

Despite notable advantages, the reluctance motor has so far been restricted in its application by the problems in smooth torque production over a sufficiently wide rotation speed range. In order to solve these problems, the operating principle of an SR machine requires new inverter and control solutions. On the other hand, the present processor technology and power electronics allow control algorithms of complicated electric drives also.

The design of an SR machine depends to a large extent on field calculation. In this calculation process, the shape of the magnetic circuit of the machine and the inductances at different rotor positions are determined. This is manually a demanding task because of the local saturation at salient-pole tips, which is typical of the operation of an SR machine. Because of the saturation, it is difficult to employ an orthogonal field diagram. Therefore, field calculation software is required in the design of an SR machine. The task becomes easier as the field solutions can usually be made at steady state excluding the problems caused by eddy currents. However, the use of an SR machine is so far so limited that no extensive calculation instructions can be found in the literature.

Speed control has become more common in pump and blower drives, thus improving their efficiency. In drives of this kind, SR motors are not yet competitive with frequency-converter-driven induction motors, for instance, because of their complicated control. The torque of an SR motor is very high at low rotation speeds, and therefore the motor type may become more popular in applications in which a high starting torque is required. Permanent magnet AC motors, however, seem to be more popular even in this application area.

The applicability of SR motors to electric tools has also been investigated. In these applications, the size and the torque properties of reluctance motors are most beneficial. A disadvantage of the universal current motor, which is commonly used at present, is wear of the mechanical commutator and electromagnetic disturbances caused by the commutation. These problems could be solved with an SR drive. Due to its durability and other favorable qualities, an SR motor is also a suitable power source for electric vehicles.

Bibliography

Alberti, L., Fornasiero, E., Bianchi, N. (2010) Impact of the rotor yoke geometry on rotor losses in permanent magnet machines. *Energy Conversion Congress and Exposition (ECCE), 2010 IEEE Proceedings*, Sep. 12–16, 2010, pp. 3486–92.

Auinger, H. (1997) Considerations about the determination and designation of the efficiency of electrical machines. In eds A. De Almeida, P. Bertoldi and W. Leonhard (eds), *Energy Efficiency Improvements in Electric Motors and Drives*. Springer-Verlag, Berlin, pp. 284–304.

Dajaku, G., Gerling, D. (2012) A novel 12-teeth/10-poles PM machine with flux barriers in stator yoke. *XXth International Conference on Electrical Machines (ICEM), 2012*, Sep. 2–5, 2012, pp. 36–40.

Gieras, F., Wing, M. (1997) *Permanent Magnet Motor Technology – Design and Applications*. Marcel Dekker, New York.

Gutt, H.-J. (1988) Development of small very high speed AC drives and considerations about their upper speed/output limits. *Proceedings of Conference on High Speed Technology, August 21–24, 1988. Lappeenranta, Finland*, pp. 199–216.

Haataja, J. (2003) *A comparative performance study of four-pole induction motors and synchronous reluctance motors in variable speed drives*. Dissertation, Acta Universitatis Lappeenrantaensis 153. Lappeenranta University of Technology, https://oa.doria.fi/.

Heikkilä, T. (2002) *Permanent magnet synchronous motor for industrial inverter applications –analysis and design*. Dissertation, Acta Universitatis Lappeenrantaensis 134. Lappeenranta University of Technology, https://oa.doria.fi/.

Heller, B. and Hamata, V. (1977) *Harmonic Field Effects in Induction Machines*. Elsevier Scientific, Amsterdam

Hendershot Jr, J.R., Miller, T.J.E. (1994) *Design of Brushless Permanent-Magnet Motors*. Magna Physics Publishing and Clarendon Press, Hillsboro, OH and Oxford.

Hrabovcová, V., Pyrhönen, J., Haataja, J. (2005) *Reluctance Synchronous Motor and its Performances*. Lappeenranta University of Technology, Finland.

Hrabovcová, V., Rafajdus, P., Hudák, P., Franko, M, Mihok, J. (2002) Design method of reluctance synchronous motor with axially laminated rotor. *Proceedings of the Conference EPE – PEMC 2002*, Dubrovník-Cavlat, P.1.

IEC 60034-1 (2004) *Rotating Electrical Machines Part 1: Rating and Performance*. International Electrotechnical Commission Geneva.

IEC 60034-4 (1985) *Rotating Electrical Machines Part 4: Methods for Determining Synchronous Machine Quantities from Tests*. International Electrotechnical Commission, Geneva.

IEC 60034-5 (2006) *Rotating Electrical Machines Part 5: Degrees of Protection Provided by the Integral Design of Rotating Electrical Machines (IP code) – Classification*. International Electrotechnical Commission, Geneva.

IEC 60034-6 (1991) *Rotating Electrical Machines Part 6: Methods of Cooling (IC Code)*. International Electrotechnical Commission, Geneva.

IEC 60034-7 (2001) *Rotating Electrical Machines Part 7: Classification of Types of Construction, Mounting Arrangements and Terminal Box Position (IM Code)*. International Electrotechnical Commission, Geneva.

IEC 60034-8 (2007) *Rotating Electrical Machines Part 8: Terminal Markings and Direction of Rotation*. International Electrotechnical Commission, Geneva.

IEC 60034-30-1 (2014 foregasted) *Rotating Electrical Machines Part 30-1: Efficiency Classes of Line Operated AC Motors (IE-code)*. International Electrotechnical Commission, Geneva.

Jokinen, T. (1979) *Design of a rotating electrical machine. (Pyörivän sähkökoneen suunnitteleminen)*, Lecture notes. Laboratory of Electromechanics. Helsinki University of Technology.

Jokinen, T. (1972) *Utilization of harmonics for self-excitation of a synchronous generator by placing an auxiliary winding in the rotor*. Dissertation, Acta Polytechnica Scandinavica, Electrical Engineering Series 32, Helsinki University of Technology. Available at http://lib.tkk.fi/Diss/197X/isbn9512260778/.

Jokinen, T. and Luomi, J. (1988) High-speed electrical machines. *Conference on High Speed Technology*, Aug. 21–24, 1988, Lappeenranta, Finland, pp. 175–85.

Kovács, K.P. and Rácz, I. (1959) *Transient Phenomena in AC Machines (Transiente Vorgänge in Vechselstrommaschinen)*. Verlag der Ungarischen Akademie der Wissenschaften, Budapest.

Kurronen, P. (2003) *Torque vibration model of axial-flux surface-mounted permanent magnet synchronous machine*, Dissertation, Acta Universitatis Lappeenrantaensis 154, Lappeenranta University of Technology.

Lawrenson, P.J. (1992) A brief status review of switched reluctance drives, *EPE Journal*, **2**(3): 133–44.

Marchenoir, A. (1983) High speed heavyweights take on turbines. *Electrical Review*, **212**(4): 31–3.

Matsch, L.D. and Morgan J.D. (1987) *Electromagnetic and Electromechanical Machines*. 3rd edn, John Wiley & Sons, Inc., New York.

Miller, T.J.E. (1993) *Switched Reluctance Motors and Their Controls*. Magna Physics Publishing and Clarendon Press, OH and Oxford.

Moghaddam, R.R. (2011) *Synchronous reluctance machine (SynRM) in variable speed drives (VSD) application*. Dissertation, Royal Institute of Technology, Stockholm.

Morimoto, S., Sanada, Y.T., and Taniguchi, K. (1994) Optimum machine parameters and design of inverter-driven synchronous motors for wide constant power operation. Industry Applications Society Annual Meeting, Oct. 2–6, 1994. *Conference Record of the IEEE* **1**: 177–82.

Moncada, R.H., Tapia, J.A. and Jahns, T.M. (2008) Saliency analysis of PM machines with flux weakening capability. *Proceedings of the 18th International Conference on Electrical Machines*, 2008. ICEM.

Niemelä, M. (2005) *Motor parameters*, unpublished.

Pyrhönen, J. (1991) *The High-Speed induction motor: calculating the effects of solid-rotor material on machine characteristics*. Dissertation, Electrical Engineering series EL 68. The Finnish Academy of Technology, Acta Polytechnica Scandinavica, Helsinki, https://oa.doria.fi/.

Pyrhönen, J. (1992) *Magnetic materials. (Magneettiset materiaalit.)* En B-74. Department of Electrical Engineering. Lappeenranta University of Technology. In Finnish.

Pyrhönen, J. Jussila, H. Alexandrova, Y. Rafajdus, P. Nerg, J. (2012) Harmonic loss calculation in Rotor Surface Magnets – New Analytic Approach. *IEEE Transactions on Magnetics*, **48**(8): 2358–66.

Pyökäri, T. (1971) *Electrical Machine Theory (Sähkökoneoppi)*. Weilin+Göös, Espoo.

Richter, R. (1954) *Electrical Machines: (Induction Machines. (Elektrische Maschinen: Die Induktionsmaschinen.)*, Vol. **IV**, 2nd edn. Birkhäuser Verlag, Basle and Stuttgart.

Richter, R. (1963) *Electrical Machines: Synchronous Machines and Synchronous Inverters (Elektrische Maschinen: Synchronmaschinen und Einankerumformer)*, Vol. **II**, 3rd edn. Birkhäuser Verlag, Basel and Stuttgart.

Richter, R. (1967) *Electrical Machines: General Calculation Elements. DC Machines (Elektrische Maschinen: Allgemeine Berechnungselemente. Die Gleichstrommaschinen)*, Vol. **I**, 3rd edn. Birkhäuser Verlag, Basel and Stuttgart.

Saari, J. (1998) *Thermal analysis of high-speed induction machines*, Dissertation, Acta Polytechnica Scandinavica, Electrical Engineering Series No. 90. Helsinki University of Technology.

Salminen, P. (2004) *Fractional slot permanent magnet synchronous motor for low speed applications*. Dissertation, Acta Universitatis Lappeenrantaensis 198, Lappeenranta University of Technology.

Schuisky, W. (1950) Self-starting of a synchronous motor. (Selbstanlauf eines Synchronmotors.) *Archiv für Elektrotechnik*, **39**(10): 657–67.

Vas, P. (1992) *Electrical Machines and Drives: a Space-Vector Theory Approach*. Clarendon Press, Oxford.

Vogel, J. (1977) *Fundamentals of Electric Drive Technology with Calculation Examples (Grundlagen der Elektrischen Antriebstechnik mit Berechnungsbeispielen)*. Dr. Alfred Hütig Verlag, Heidelberg and Basle.

Vogt, K. (1996) *Design of Electrical Machines (Berechnung elektrischer Maschinen)*. Wiley-VCH Verlag GmbH, Weinheim.

Voldek, A. I. (1974) Electrical machines [in Russian], *Energy*, 1974.

Wiart, A. (1982) New high-speed high-power machines with converter power supply. *Motorcon Proceedings*, Sep. 1982, pp. 641–6.

Zhu, Z. Q. and Howe, D. (1993) Instantaneous magnetic field distribution in brushless permanent magnet dc motors, Part III: Effect of stator slotting. *IEEE Transactions on Magnetics*. **29**(1): 143–51.

8

Insulation of Electrical Machines

Here, an insulator refers to a nonconducting material or an insulating material with a very low conductivity. An insulation system comprises insulating materials and insulation distances. The main function of insulation is to separate components of different electric potentials or of different electric circuits. Further, insulators improve the strength of the winding structures; they also have to act as heat conductors between the winding and the surroundings, and they have to protect the winding from external stresses such as dirt, moisture and chemicals.

In electrical machines, there are typically insulation distances of three types. First, a clearance in air is simply an air gap between the objects, the insulating strength of which is determined by the distance between the objects, their shape and the state of the air. A clearance in air can be defined by either insulated or noninsulated surfaces. In certain cases, the medium of the insulation distance, which also simultaneously acts as a coolant, can be a gas other than air. In the case of a homogeneous electric field, the breakdown voltage depends on the width of the air gap according to the Paschen curve, Figure 8.1.

The Paschen law essentially states that the breakdown characteristics of a gap are a function of the product of the gas pressure and the gap length. Figure 8.1 shows a special case with constant pressure. Air in an insulation construction is often prone to partial discharges. Partial discharges take place inside the insulation construction, especially in places where materials with different permittivities are connected in parallel. This takes place for instance in an insulation having air bubbles in the insulating resin. The dielectric strength of the air bubble is far smaller than the resin itself, and small partial discharges take place inside the air bubble. The sensitivity for partial discharges can be estimated using the Paschen curve. If there is 500 V over a 10 μm air bubble ($E = 50$ MV/m), partial discharges will take place. The insulation starts to deteriorate in these places and, finally, partial discharges may destroy the insulation if the materials selected do not tolerate partial discharges. Mica is used in high-voltage insulations since it tolerates partial discharges without deteriorating. The present-day switching power supplies may cause partial discharges also in low-voltage machines, especially in the first turn of the winding because of the nonuniform division of the fast-rising voltage in the winding.

Second, an insulation distance created by solid insulation is chiefly composed of solid insulators, in which the electric field is not significantly oriented along the interfaces of the

Design of Rotating Electrical Machines, Second Edition. Juha Pyrhönen, Tapani Jokinen and Valéria Hrabovcová.

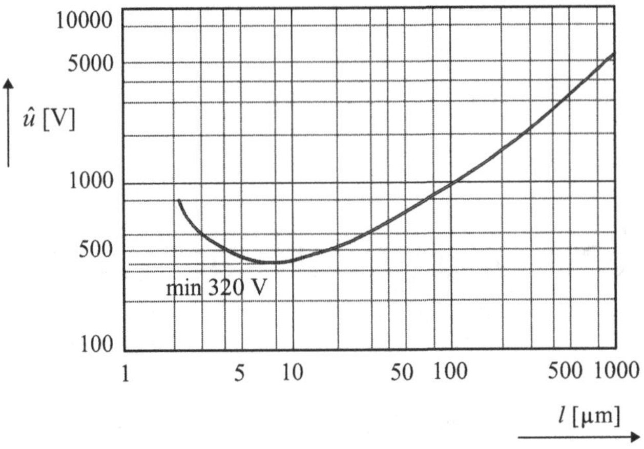

Figure 8.1 Paschen curve illustrating the voltage strength of an insulation distance formed from a homogeneous air gap at a pressure of 101 kPa. l is the gap spacing between the electrodes.

insulators. In this case, the insulating strength is determined by the thickness of the insulation and by the relative permittivity of the insulating material.

Third, a creepage distance is an insulation distance in which a bare live part is connected to a conductive or insulated component in another electric potential, such as the earthed frame of the machine. We have a creepage insulation distance also when a live part is only weakly insulated. Surface discharges may occur, or a flashover in the creepage distance, if the effective electric field has a component parallel to the surface in question. Typically, such a point occurs in an electrical machine on the surface of slot insulation just outside the iron core.

An open creepage distance may gather dirt or moisture that may produce creepage currents and discharges on the surface of the insulation. In machines below 1000 V, the creepage currents are chiefly a risk factor. With high voltages, surface discharges may also damage the insulation. The track resistance of a material describes the ability of the material to resist the formation of a carbonized track when a high voltage is applied to it. The track resistance of an insulating material is defined by the Comparative Tracking Index (CTI). As the track resistance is defined by the voltage (which will cause failure by tracking), the unit of CTI is the volt. The test method is defined for instance in the IEC 600112 standard.

In an electric circuit, there always has to be insulation, which naturally takes up some room. This is a fact that has to be taken into account in the dimensioning of the magnetic circuit and the winding. Real insulation is always slightly conductive, and it can be damaged by electrical, thermal, mechanical or ambient stress, or by chemical attack. Ambient stresses include for instance humidity, abrasive particles in cooling air, dirt and radiation. An insulating material has to have a sufficient voltage resistance to avoid flashovers in the voltage test or when exposed for instance to overvoltages. The estimation of stresses to which the insulation may be exposed during its service life and the dimensioning of the insulation based on this analysis together are called insulation coordination.

The conductivity of the insulation and the dielectric losses should remain low during operation. The insulation has to be thermally resistant to short-term overloads during operation and also to cumulative ageing caused by the stresses mentioned above. Although at present there are hundreds or even thousands of suitable alternatives for insulating materials, the most common insulating materials in electrical machines are easily listed: mica, polyester films, aramid paper and epoxy or polyester resins. Insulation materials somewhat less common in electrical machines are materials made of polyester fiber (Dacron, Terylene, Diolen, Mylar, etc.), polyimide films (Kapton) and silicon resins employed in the impregnation.

8.1 Insulation of Rotating Electrical Machines

Insulation can be roughly divided into two main categories: groundwall and conductor insulation. The function of groundwall insulation is to separate those components that may not be in galvanic contact with each other; for example, groundwall insulation galvanically separates the coil from the iron core of the machine. Conductor insulation separates the wires and turns of a coil. Usually, standards are not as high for conductor insulation as they are for groundwall insulation, and therefore the former is usually notably thinner than the latter. Consequently, a conductor or turn-to-turn insulation is usually, and in small machines in particular, a varnishing on the wire.

The main types of insulation are:

- slot insulation and the slot closer;
- phase-to-phase insulation in the slot and in the coil end;
- pole winding insulation;
- insulation of the connection leads and the terminals;
- impregnating varnish and resin; and
- surface varnish and protective paint.

The outermost layer of the slot insulator has to have a good mechanical strength, since sharp edges may occur in the slot. A polyester film, for instance, is a suitable material for slot insulation. If two insulating materials are employed in the slot, the inner layer is usually selected to be aramid paper since it has better thermal resistance and impregnation properties than polyester film. Aramid papers efficiently absorb the resin, and the resin is well fixed to the surface compared with polyester surfaces that are glossy and nonporous. In phase-to-phase spacings, flexible, clothlike insulation materials are selected for the end windings. When a high voltage strength is required, mica is employed, and when a high mechanical strength is needed, thermoplastics reinforced with fiberglass are used.

Mica is an inorganic natural substance occurring commonly in bedrock. Micas belong to the monoclinic system, and they are composed of thin flakes of silicate tetrahedra which are elastic and transparent. These silicate sheets are composed of interconnected six-membered rings, which form the typical pseudohexagonal symmetry structure of micas.

For the past hundred years mica has been a significant component in the insulation of high-voltage machines, due chiefly to its excellent partial discharge strength. Chemically, mica is composed of potassium, aluminum silicate or some other closely related mineral. The crystals of mica comprise layers of thin flakes or sheets that can be easily separated from each other. This crystal structure enables the flakes to be split into thin strips that are flexible and thus suitable as an insulation material for electrical machines.

Table 8.1 Thermal classes of insulating materials. Adapted from standards IEC 60085, IEC 60034-1

Thermal class	Previous designation	Hot spot allowance [°C]	Permitted design temperature rise [K], when the ambient temperature is 40 [°C]	Permitted average winding temperature determined by resistance measurement [°C]
90	Y	90		
105	A	105	60	
120	E	120	75	
130	B	130	80	120
155	F	155	100	140
180	H	180	125	165
200		200		
220		220		
250		250		

The thermal endurance of mica is very high. (See also thermal classes presented in Table 8.1.) At the lowest, the mica qualities start to lose their crystal water at a temperature of 500 °C, although some qualities endure even above 1100 °C. For electrical machines, these values are more than adequate, since the highest permitted temperatures for the machine parts are usually about 200 °C at the maximum. Mica has excellent chemical resistance; it is resistant to water, alkalis, various acids and common solvents. Only sulfuric acid and phosphoric acid dissolve mica. However, mica does not resist oil, because oil penetrates between its flakes separating them from each other.

The dielectric strength of mica is high, the dielectric losses are low and the surface resistance is high. Creepage currents do not damage mica, and it resists the effects of partial discharges far better than the best organic insulators. Therefore, mica is almost an indispensable material in high-voltage electrical machines, in which there are always some partial discharges. Partial discharges are difficult to handle. Usually in machines with a rated voltage above 4 kV, some partial discharges occur during operation. Mica tolerates this. It is, however, possible that even in large low-voltage machines, there may be some partial discharges in the converter supply. This is because the steep-edged voltage pulses are not evenly distributed in the winding turns, but may stress heavily the first turn of the winding. In such a case, the electric field strength may be large enough to cause partial discharges, and if there is no mica present the insulation will fail sooner or later.

In the mica insulator, the flakes of mica are bound with a suitable binding agent. Further, some layers of a suitable auxiliary substance are required, such as glass fabric or polyester foil to improve the tensile strength of the insulation. Nowadays, mica is used mainly as a paper in the insulation of electrical machines. Mica paper is an insulator composed of extremely small flakes of mica and produced in the same way as paper, hence the name. Thus, despite the term "paper," the material does not include any cellulose fibers. Natural mica is crushed either mechanically or with heat into small flakes that are glued with resin into a flexible, paper-like material. The properties of mica insulators are presented in Table 8.2.

Table 8.2 Characteristics of mica insulators

Characteristic	Unit	Commutator micanite	Moulding micanite	Mica folium	Glass–mica tape	Epoxy glass mica paper tape
Mica content	%	95–98	80–90	40–50	40	45–55
Binder content	%	2–5	10–20	25–40	18–22	35–45
Content of the supporting material		—	—	20	40	15
Compression strength	N/mm^2	110–170	—	—	—	—
Tensile strength	N/mm^2	—	—	30–50	40–80	80–120
Compression	%	2–6	—[a]	—	—	—
Continuous operating temperature, binder:	°C					
Shellac		F155[b]	F[b]	B130	B130	F155
Alkyds, epoxy		H180[c]	H[c]	F155	F155	
Silicone		—	—	H180	H180	
Voltage strength (1 min, 50 Hz)	kV/mm	25	20	20	16–20	20–30

[a]In the production process, mica flakes glide with respect to each other. When the binder has set, the compression is 4–8 %.
[b]In the commutators of insulation class F machines.
[c]In the commutators of insulation class H machines.
Source: Adapted from Paloniemi and Keskinen (1996).

Insulating films constitute a rather diverse group of insulators (Table 8.3). The films are usually duroplastics, the thermal resistance of which is restricted by the melting temperature and rapid ageing far below this temperature.

Polyester film is produced for various purposes at nominal thicknesses from 6 μm to 0.4 mm. PETP (polyethyleneterephthalate) membrane is a tough polyester film, which is used as such or as a laminated insulator as slot insulation and a phase coil insulation in small- and medium-size machines. Figure 8.2 illustrates the chemical composition of a Mylar polyester film.

Aromatic polyamides, or aramid fibers (trade name Nomex), have better heat-resistance qualities than the previous ones, but they are not quite as ductile as PETP films. Therefore, aromatic polyamides are often used as laminated insulators together with PETP or polyimide films. Figure 8.3 illustrates the chemical composition of aramid fibers.

Polyester film has the best mechanical strength of the insulating foils available; its yield limit and tensile strength approach the values of soft copper. Films are employed as such in the slot insulation of small machines, and in the slots of large machines, together with fiber insulation (aramid or polyester paper). The melting temperature of polyester film is approx. 250 °C. However, polyester has an inadequate resistivity to moisture at high temperatures. In such cases, aramid fibers are used alone.

Polyimide film endures constant temperatures of even 220 °C, and its instantaneous thermal resistance is as high as 400 °C. Further, the mechanical properties of the film are

Table 8.3 Characteristics of insulating foils

Characteristic	Unit	Polyester PETP	Polyimide	Polysulphone PS
Tensile strength	N/mm^2	140–160	180	90
Elongation at break	%	75	70	25
Modulus of elasticity	N/mm^2	3900	3000	2500
Density	g/cm^3	1.38	1.42	1.37
Continuous operation temperature	°C	130	220	180
Instantaneous thermal resistance	°C	190	400	210
Softening point	°C	80–210	530	235
Melting point	°C	250	Does not melt	
Shrinkage at 150 °C	%	3	—	—
Burning		Slow	Does not burn	Does not keep up burning
Moisture absorbency	% by weight	0.5	3	1.1
Chemical strength	Graded 0–4			
Acids		2	3	3
Alkalis		1	0	3
Organic solvents		4	4	1
Specific resistivity	Ω cm	10^{19}	10^{18}	5×10^{16}
Voltage strength	kV/mm	150	280	175
Trade names		Mylar Melinex Hostaphan	Kapton	Folacron PES

Source: Adapted from Paloniemi and Keskinen (1996).

relatively good, and its breakdown voltage is high. It is not affected by organic solvents. The thicknesses of the film vary usually between 0.01 and 0.12 mm. Polyimide insulation is rather expensive, but as a thin insulator it leaves more space for instance for the windings, thus being an advantageous alternative in certain cases. Figure 8.4 illustrates the polyimide construction.

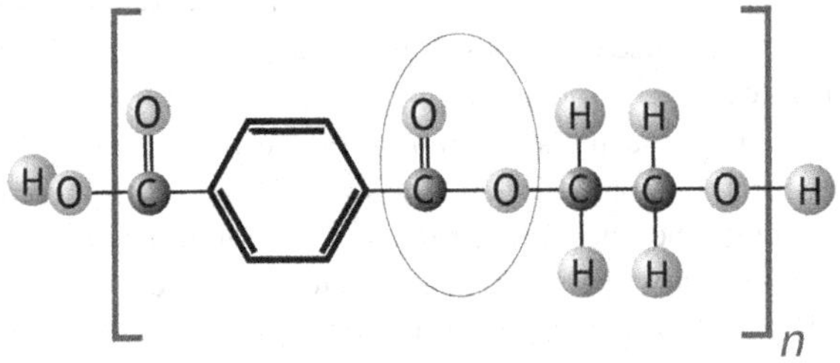

Figure 8.2 Mylar polyester construction. The material contains hydrogen (H), oxygen (O), carbon (C) and benzene hexagons (C_6H_6). The encircled part is an ester. The ester group is prone to hydrolysis, hence the material does not tolerate heat and moisture simultaneously.

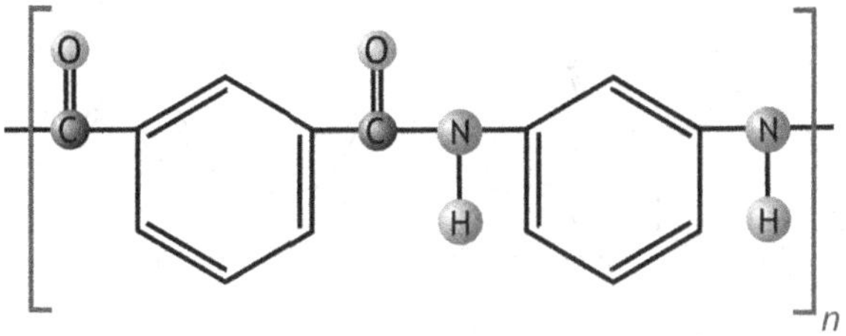

Figure 8.3 Nomex aramid fiber chemical structure. The material contains hydrogen (H), oxygen (O), carbon (C), nitrogen (N) and benzene hexagons (C_6H_6). This simple construction tolerates different stresses well, even heat and moisture simultaneously.

Insulation of conductors is the most demanding task in the insulation construction of an electrical machine, since the insulation is located closest to the hot copper wire and, moreover, it is the thinnest insulation component. Conductor insulation is often varnish-like thermoplastic. There are both crystalline and amorphous regions in the varnish. The crystallinity of the material improves thermal resistance, it forms an impermeable sealing against solvents, and, further, it improves the mechanical properties of the insulator. The amorphous property, on the other hand, makes the insulator flexible. The most common conductor varnishes used in electrical machines are ester–imides or amide–imides. Figure 8.5 shows the polyamide–imide polymer chain. They are thus related both to polyester and polyimide films. The highest permitted temperatures for ester–imides are of the order of 180–200 °C.

Conductor insulation can be formed from two or more different types of varnish. Two different varnishes are employed in order to achieve better thermal resistance, better mechanical properties and improved cost efficiency than with a single material. In practice, the varnish coating comprises several layers also in the case of a single varnishing material. In the coating, the bare conductor is heat treated and coated with varnish. Next, the solvent is vaporized by heating the wire in the oven, and the wire is coated with a new layer of varnish. The process is repeated 4–12 times. The wires are divided into three grades according to the thickness of the coating: "single" (Grade 1), "double" (Grade 2) and "triple" (Grade 3). In reality, as

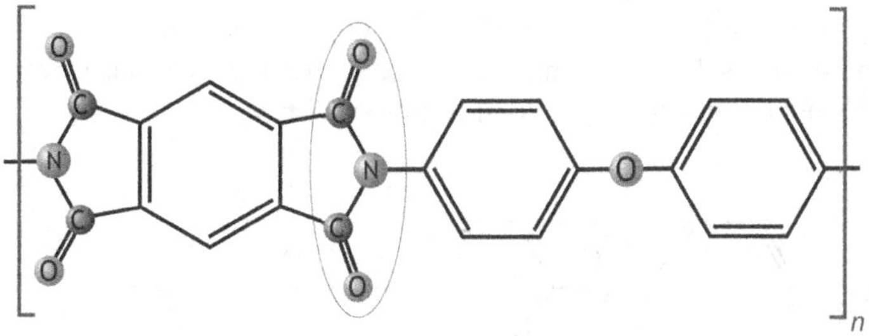

Figure 8.4 Kapton polyimide. The circled NC_2O_2 functional groups are the imide groups. The material contains oxygen (O), carbon (C), nitrogen (N) and benzene hexagons (C_6H_6). The imide groups are prone to hydrolysis, and the material does not tolerate heat and moisture simultaneously.

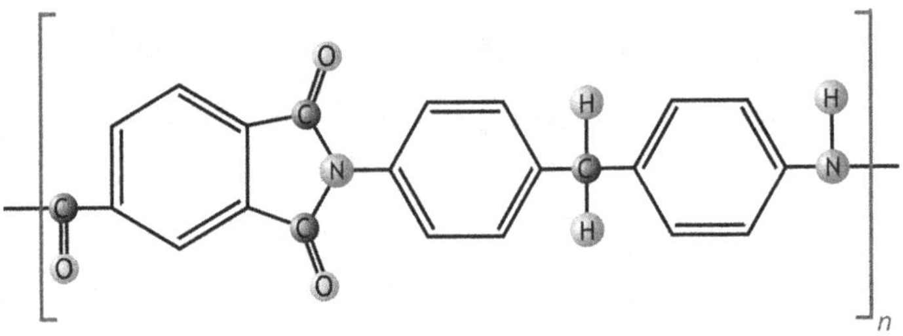

Figure 8.5 Polyamide–imide polymer. The material contains hydrogen (H), oxygen (O), carbon (C), nitrogen (N) and benzene hexagons (C_6H_6). The imide groups are prone to hydrolysis.

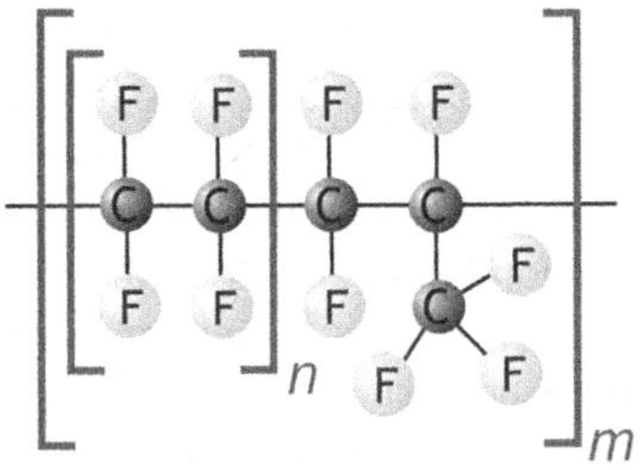

Figure 8.6 Chemical structure of FEP fluoropolymer. The structure, which contains two carbons and four fluorines and repeats *n* times, is the chemical structure of a PTFE fluoropolymer.

previously mentioned, there are always several layers of varnish. Furthermore, the thickness of the varnishing of each grade is always proportional to the diameter of the wire.

Polyimide films and aramid papers can also be used as conductor insulation. They are wrapped on a wire-like tape. They can be considered for special applications, but they are too expensive for common machines. For extremely harsh environments, fluoropolymer (Teflon) insulation can be used. It is extruded on the wire in the manufacturing process. Fluoropolymers offer good chemical resistivity and excellent resistivity to moisture even at high temperatures. However, their breakdown voltage can be one-fourth of that of polyester–imides and polyamide–imides, which has to be taken into account when designing an insulation system. Figure 8.6 illustrates the chemical composition of Teflon.

PEEK (polyether-etherketone) polymer is a very good conductor insulation material in demanding environments. PEEK is reported to have an excellent resistance even to hydrolysis up to 250 °C. Figure 8.7 illustrates the composition of PEEK.

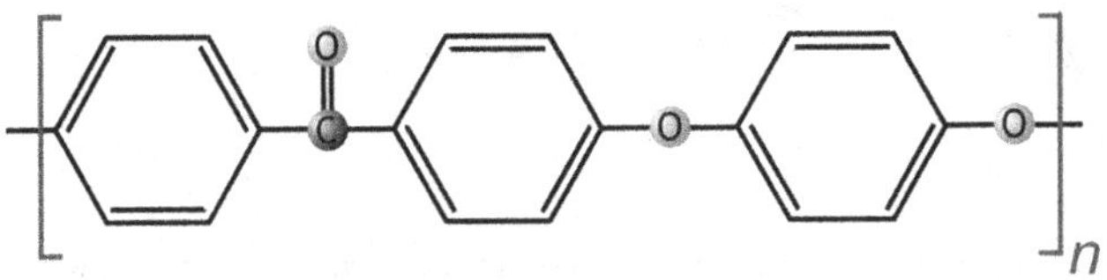

Figure 8.7 Chemical structure of PEEK.

8.2 Impregnation Varnishes and Resins

The function of an impregnating varnish or resin is to reinforce mechanically the winding, to protect it from moisture, dirt and chemicals, and to improve its thermal conductivity. On the other hand, excessive varnishing decreases thermal conductivity, for instance in the coil ends. An impregnating varnish comprises a base component (linear polymers), monomer (a cross-linking agent), solvents and possible oils. Depending on the composition, the varnishes are divided into oil-based and polyester coatings. The setting of oil-based varnishes requires oxygen, and therefore they should not be used in compact and thick windings. The electrical properties of oil-based varnishes are good, but their mechanical strength is rather poor. Polyester varnishes are nowadays the most common varnishing materials, Figure 8.8. They are either single- or two-component varnishes, the hardening of which usually requires heat treatment. In the curing process, the solvent is vaporized, monomer from one end attaches to the reactive part of one base component and from the other end to the reactive part of the other base component. This is called cross-linking, because the polymer chain of the monomer sets perpendicular to the polymer chain of the base component. This way, a complex three-dimensional polymer structure is formed. A plastic created this way is a thermoplastic.

In the *impregnating varnishes*, approximately half of the volume is an evaporable solvent, which is replaced by air when the varnish hardens. Therefore, alkyd- or polyester-based varnishes including solvents have mostly been replaced by impregnating resins that are polyester or epoxy based; they do not contain solvents and are chemically hardening impregnants. In common machines, polyester resins are used, because they have a low price and are easy to handle. In a polyester resin, the base component and the monomer are quite similar substances in nature and viscosity, and mix easily. Moreover, it is possible to use convenient mixing ratios, such as one to one. Epoxy resins are very reliable and have the composition illustrated in Figure 8.9.

A disadvantage of epoxy resins is their high price when compared with polyester resins. However, epoxy resins are usually employed in machines above 250 kW because of their good mechanical strength, adhesiveness and low shrinkability. In harsh environments, epoxy resins are favored. They have good resistivity to chemicals, moisture and radiation. On the other hand, polyester resins have better resistivity to oils, such as transformer oil. In special cases, as in traction motors involving very high temperatures (over 200 °C), silicone resins can be used. They offer excellent thermal properties, but have a poor mechanical strength.

Surface varnish improves the surface quality of the insulation. Surface varnish forms an impermeable coating that is easy to clean and that improves the track resistance of the insulation. Table 8.4 introduces the properties of insulation varnishes, resins and surface varnishes.

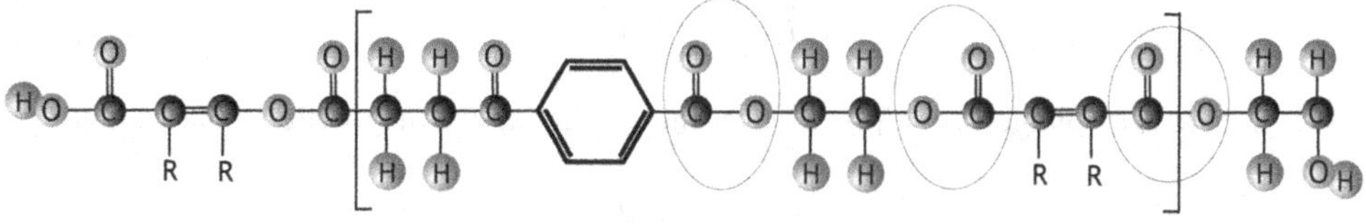

Figure 8.8 Polyester resin composition. The material contains hydrogen (H), oxygen (O), carbon (C), benzene hexagons (C_6H_6) and reactive parts (R). The encircled parts are esters. The monomer binds itself to the reactive parts indicated by R when the material is cured to a thermoplastic.

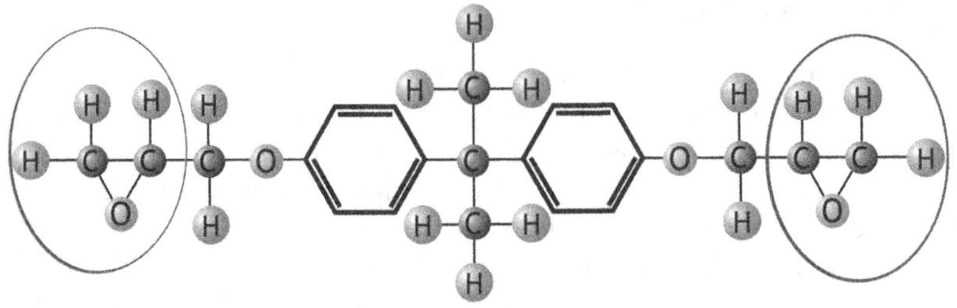

Figure 8.9 Chemical composition of basic epoxy resin. Epoxide functional groups are circled. These groups are characteristic for every epoxy, but the polymer chain in between them can vary.

The impregnation of the insulation has an influence on the track resistance of the insulation. Since the insulation is composed of several different components, one of the components is often an air gap. Air gaps may also be left in undesired places, for instance in a slot, when the impregnation has not been complete or there are bubbles in the varnish. This has to be taken into account in the dimensioning of the insulation and the selection of the impregnation method. The electric field strength E in a single material of thickness d is

$$E = \frac{U}{d}. \tag{8.1}$$

The electric field density D through the insulation must be constant in a homogeneous insulation and equal throughout the insulation construction made of several layers:

$$E \cdot \varepsilon = D. \tag{8.2}$$

When two different insulation materials 1 and 2 (thicknesses d_1 and d_2) are in series in the same outer field, we have the corresponding field strengths E_1 and E_2 and voltages U_1 and U_2 according to Equations (8.1) and (8.2):

$$\frac{E_1}{E_2} = \frac{\varepsilon_2}{\varepsilon_1} \tag{8.3}$$

and

$$\frac{U_1}{U_2} = \frac{d_1}{d_2} \cdot \frac{\varepsilon_2}{\varepsilon_1}. \tag{8.4}$$

When there is a voltage $U = U_1 + U_2$ over the insulation, we may write for the voltages

$$U_1 = U \cdot \frac{\dfrac{d_1}{\varepsilon_1}}{\dfrac{d_1}{\varepsilon_1} + \dfrac{d_2}{\varepsilon_2}} \tag{8.5}$$

Table 8.4 Characteristics of insulating varnishes and impregnating resins

Characteristic, Grades 0–4	Impregnating varnishes		Impregnating resins			Coating (surface) varnishes	
Continuous operating temperature [°C]	155	180	155	180	130	155	180
Mechanical strength at operating temperature	3	1	3	4	2	2	1
Flexibility	3	2	2	2	4	3	2
Moisture resistance	3	3	3	4	3	4	4
Chemical strength	3	3	3	4	3	4	3
Track resistance	3	3	3	2	3	2	3
Typical materials	Alkyd polyester	Silicone epoxy	Polyester alkyd Epoxy	Polyester Epoxy	Alkyd Polyurethane	Alkyd Polyurethane	Alkyd Silicone Epoxy

Source: Adapted from Paloniemi and Keskinen (1996).

and

$$U_2 = U \cdot \frac{\frac{d_2}{\varepsilon_2}}{\frac{d_1}{\varepsilon_1} + \frac{d_2}{\varepsilon_2}}. \tag{8.6}$$

Example 8.1: The thickness of insulation is 4 mm and its permittivity $\varepsilon_1 = 5$. Inside the insulation, there is a 0.25 mm air gap with $\varepsilon_2 = 1$. The voltage over the insulation is $U = 12$ kV. How is the voltage divided between the insulation and the air gap, and how high is the field strength in the air gap?

Solution: With Equations (8.5) and (8.6), we obtain the voltages $U_1 = 9.14$ kV and $U_2 =$ 2.86 kV. The electric field strength in the air gap is $E_2 = 2.86$ kV/0.25 mm $= 11.44$ kV/mm, which may exceed the breakthrough field strength. Even a small air gap in the insulation may cause partial discharges that may age the insulation. Even though the partial discharge in the air space remaining inside the insulation did not result in an immediate flashover in the complete insulation, it might, through local overheating and wear, rapidly cause a breakdown in the complete insulation construction.

We may take two epoxy-coated polyester films as an example: they are relatively heat resistant when not in contact with air. The dielectric strength of this kind of construction can be even dozens of times higher compared with the voltage at which the air gap between the films starts to spark, thus wearing the coating and insulating foils. However, exceeding this inception voltage of sparking drastically cuts the service life of the insulation, whereas the breakdown voltage has almost no influence on the service life of the insulation.

Therefore, nowadays such impregnation methods are selected that ensure adequate penetration of the impregnant and an impermeable construction. A possible method is provided for instance by vacuum pressure impregnation (VPI) technology. This method is employed at both low and high voltages in the temperature classes 155–220 (cf. Table 8.1).

For operation under special conditions, the windings can be cast in plastic. An advantage of the method is that the winding becomes completely waterproof. Further, the mechanical strength of the coil is improved. A disadvantage is the high price of the insulation. The plastics used in the method are either polyester or epoxy plastics. The method is seldom used in industrial motor applications.

Since chemicals are always released from the insulation during operation, it has to be ensured that these chemicals are not harmful to other insulating materials. The dissolved oxygen molecules creating ozone in partial discharges also rapidly weaken several polymers. The impregnating varnish in particular has to be compatible with other insulators. The manufacturers recommend testing insulating material combinations to ensure compatibility.

8.3 Dimensioning of an Insulation

The mechanical, electrical and thermal stresses that the insulation is exposed to are contributing factors to the dimensioning of an electrical machine. These stresses degrade the properties of the insulation. Further, in the selection of the insulating material, stresses caused by the operating conditions, such as accumulating dirt, chemicals, oils, moisture and radiation, have to be taken into account.

The compression strength of the insulation is usually higher than the tensile strength, and therefore, in the design, we should aim at insulations that are exposed to compression rather than tension. For instance, fiberglass bindings can be employed to receive the tensile stresses exerted on the insulation (cf. binding of the rotor of a DC machine). The end windings are also exposed to shearing stresses. Thus, the insulation has to be dimensioned according to these stresses. Although rigidity is also required of the insulation, under certain circumstances it has to be flexible too. Usually, and in low-voltage machines in particular, it suffices that the insulation adapts to the deformation caused by the thermal expansion of the copper wire.

An insulation construction is often employed to support the winding. Therefore, the structure has to withstand vibration and electrodynamic forces, such as starting and short-circuit currents. The more securely the coil end is supported, the higher the natural frequency gets. The target is to raise the natural frequency of the insulation construction to a level above the frequency range of the electrodynamic forces. The most important natural frequency to be avoided is the frequency that is double the supply frequency. This force is created as a result of the magnetic flux and the winding current. The direction of this force is illustrated in Figure 8.10.

In addition to the normal operating voltage, the insulation construction of an electrical machine also has to withstand temporary overvoltages at operating frequency, switching overvoltages and exposed overvoltages. These short-term overvoltages have to be taken into account particularly in the dimensioning of the air gap, since an insulation composed of solid insulators endures rather well the quite high temporary voltage stresses. In practice, a high-voltage insulation is usually designed with an effective field strength of 2–3 kV/mm in the insulation. An approximate value for the minimum thickness of the insulation layer is obtained by applying Equation (8.1)

$$d = U/E_{max}, \tag{8.7}$$

where d is the thickness of the insulating material, U is the voltage over the insulation and E_{max} is the highest allowable field strength in the material concerned.

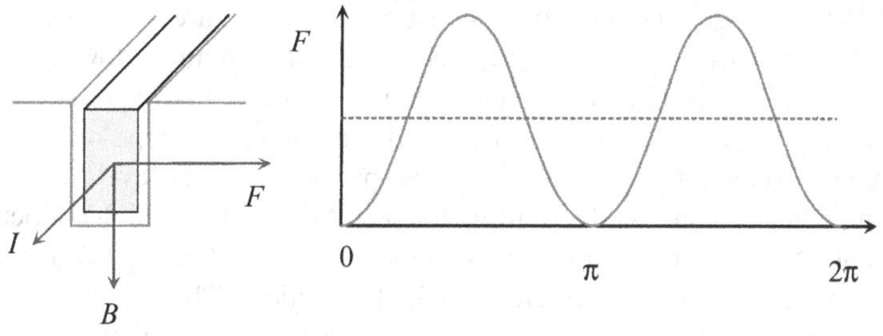

Figure 8.10 Lorentz force caused by the flux in the conductor in the slot on the periphery of the machine.

If the insulation is composed of several layers, the thicknesses of these layers can be estimated from the following:

$$U = E_1 d_1 + E_2 d_2 = D\left(\frac{d_1}{\varepsilon_1} + \frac{d_2}{\varepsilon_2}\right). \tag{8.8}$$

According to the IEC standard for rotating electrical machines (IEC 60034), the voltage test should be carried out with an AC voltage at a frequency of 50 or 60 Hz. With low-voltage three-phase motors, the test voltage for motors below 1 kW has to be

$$U_{\text{test}} = 2U_{\text{N}} + 500\,\text{V}. \tag{8.9}$$

For motors above 1 kW, the test voltage has to be

$$U_{\text{test}} = 2U_{\text{N}} + 1000\,\text{V (but at least 1500 V)}, \tag{8.10}$$

where U_{N} is the rated line-to-line voltage and U_{test} is the test voltage. The test period is 5 s for used motors, the power of which is less than or equal to 5 kW, and for new motors, 1 min.

For DC machines, the insulation level is tested between the winding and the frame with a 50 Hz AC voltage of 1.5 kV, when the rated voltage is from 50 to 380 V, and 2.5 kV when the rated voltage is from 380 to 1000 V. It is worth noting that if the test period is 1 min, the voltage must not exceed 50 % of the test voltage. After this period, the test voltage can be raised to the peak voltage in 10 s.

High-voltage machines also have to withstand impulse waves. The impulse wave tolerance level should be at least

$$\hat{U}_{\text{sj}} = 4U_{\text{N}} + 5\,\text{kV}. \tag{8.11}$$

Here $\hat{U}_{\text{sj}}$ is the peak value of the impulse voltage. Due to the risk of failure, a new, complete machine is not tested with this voltage; however, separate coils can be tested in the laboratory.

The impulse voltage is adjusted to have a rise time of 1.2 μs and a duration of 50 μs. Equation (8.11) yields the peak value for the overvoltage for high-voltage machines. This peak value is the basis for the dimensioning of the turn insulation. Further, in addition to the overvoltage strength, in the dimensioning of the insulation it has to be borne in mind that the electric field has an ageing effect on the construction. To ensure long-term durability, the partial discharge level of the insulation construction should remain as low as possible. This can be tested by a tan δ measurement. The tan δ value increases when the test voltage is raised, since the amount of partial discharges is increased. Therefore the slope of the tan δ voltage curve can be considered an indirect indicator of the partial discharge level.

Insulating materials are classified according to their ability to resist high temperatures without failures. Table 8.1 shows the temperatures according to the IEC standard, and the previous, although still commonly used, thermal classes with letter codes. The hot-spot allowance gives the highest permitted temperature that the warmest part of the insulation may reach. The temperature rise allowance indicates the highest permitted temperature rise of the winding at rated load.

The most common thermal class in electrical machines is 155 (F). The classes 130 (B) and 180 (H) are also of common occurrence.

The ageing of the insulation puts a limit on its long-term thermal resistance, that is on its temperature rise allowance. When evaluating the long-term thermal resistance of a single insulator, the concept of temperature index is employed. The temperature index is the maximum temperature at which the insulator can be operated to yield an average life of 20 000 h, or 2.3 years. This is a very short lifetime; in practice, the lifetime is longer than 2.3 years, because it is assumed in the preparation of the insulation classification standards that in reality the winding temperature does not remain continuously at the allowed upper limit. It is assumed that the machine runs intermittently at partial load, the ambient temperature seldom reaches its upper allowed limit, and there is some nonoperating time, too. Manufacturers often design the temperature rise for instance for the class 130 (B), but then use an insulation system belonging to the class 155 (F). This underclassification leads to a longer expected insulation life. It is also worth noting that the temperature index of the insulation belonging for instance to the class 155 has to be at least 155. When the thermal class of the insulation is determined, the temperature index is rounded down to the nearest thermal class. If a machine is running 24 h a day all year round (a situation typical in power plants), the temperature rise of the machine windings should be designed below the permitted value given in Table 8.1.

Short-term thermal resistance refers to thermal stresses, the duration of which is a few hours at maximum. During this stress, the insulator may melt, or bubbles may occur, or it may shrink or become charred. The insulation should not be damaged in any of these ways if the temperature is moderately exceeded in any situation under normal operating conditions. In Table 8.1, temperature rise refers to a permitted temperature rise in a winding at rated load. This temperature rise does not cause premature ageing of the insulator. An excessive temperature fluctuation may cause the development of brittleness and cracks in the insulator. It has to be borne in mind that when there are several ageing factors, such as temperature and moisture, at the same time, the critical temperature will be lower and must be approximated individually for each case. In certain operating situations, frost resistance may also decide the selection of the insulating material.

Thermal ageing is commonly estimated by the Arrhenius equation for the reaction rate

$$k = \eta\, e^{-E_a/RT} \tag{8.12}$$

where η is an experimental pre-exponential constant, E_a is the activation energy, R is the gas constant and T is the absolute temperature.

Practical insulator constructions have shown that a temperature rise of 8–10 K cuts the expected life expectancy in half.

8.4 Electrical Reactions Ageing Insulation

Partial discharge is an electrical breakdown in an air spacing, at least one side of which is confined to an insulator. Thus, a partial discharge does not immediately result in a breakdown of the complete insulation. The energy released in the partial discharge is relatively small, corresponding to the energy of a small capacitance discharged in this air spacing plus the discharging energy of a capacitance of the insulation section that is in series with this capacitance. A single partial discharge is not harmful to the insulation. It can only break some bonds in the polymer structure of the insulator. When this activity continues, the air spacing will expand and unite with other air spacings. In this way, long channels, or trees, are formed inside the insulator. The walls of the trees are at least partly conducting. When the tree is long enough, reaching nearly all the way from the stator core to the conductor, a breakdown will occur. Consequently, repeated partial discharges may deteriorate the insulation and lead to a failure. Only mica tolerates continuous partial discharges, and therefore it is frequently used in high-voltage insulation.

If there is a notably high creepage current flowing in the creepage distance, the insulation may gradually become damaged. As an influence of the current, the moist surface layer dries. Drying is unevenly distributed. When the creepage current disrupts, sparking occurs in the air in the vicinity of the insulator surface. In high-voltage machines, the voltage against the ground is so high that the breakdown strength of the air in the corner of the sheet stack may be exceeded, and the discharge occurring in the air spacing proceeds along the surface of the insulation until the field parallel to the surface becomes too low and the discharge extinguishes. Here, the discharge energy is far lower than in the sparking of creepage currents; however, repeated surface discharges may damage the insulation.

When the voltage at the surface discharge exceeds the gliding discharge voltage limit, the energy of the discharges is suddenly increased and the discharge starts to glide along the surface of the insulation. The discharges may glide even a distance of dozens of centimeters, and as the voltage rises further, they eventually reach a weaker insulation at the end of the coil, or a noninsulated point, and a breakdown occurs. Both gliding discharges and surface discharges can be decreased for instance by employing a layer containing silicon carbide in the insulation. The conductivity of silicon carbide is highly dependent on the electric field, and it increases in proportion to the fifth power of the field strength. As a result, the field strength peaks are substantially balanced out on the surface of the insulation.

Mathematical modeling of electrical ageing is a challenging task, since at least the discharge inception voltage creates a clear discontinuation point. Usually, ageing caused by electrical stress is described with a power equation that expresses the lifetime of the insulation, t, as a function of electric field strength E,

$$t = kE^{-n}, \tag{8.13}$$

in which case the breakdown period has been shown to follow approximately the Weibull exponential distribution.

In reality, estimation of the lifetime of an insulation construction is always a statistical problem that requires a set of tests; even then, the influence of several simultaneous stresses has to be taken into account as well as possible. This is of particular significance when fast results are required in accelerated ageing tests.

8.5 Practical Insulation Constructions

In practice, the most important factor influencing the structure of the insulation is the rated voltage of the machine. Based on the rated voltage, electrical machines can be divided into high-voltage and low-voltage machines. A low-voltage machine is a machine the rated voltage

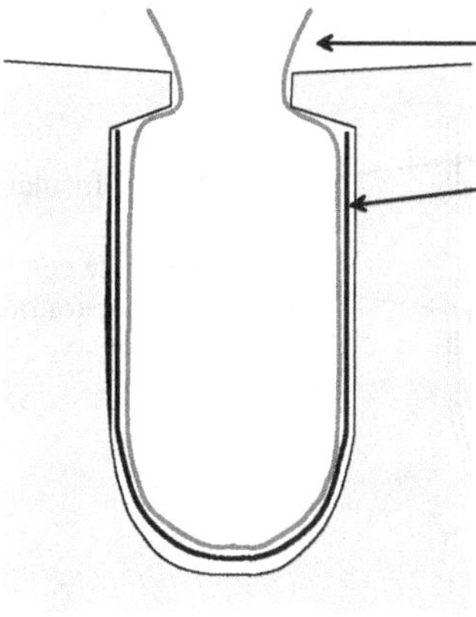

Figure 8.11 Insulation of a stator slot of a low-voltage AC machine. A distance of 0.2–0.3 mm from the slot wall has to be reserved for the insulation. In other words, the slot insulation is finished before placing the coil in the slot. The coils are fed in the slots wire by wire or in groups of a few wires.

of which is below 1 kV. In low-voltage machines, the voltage strength of the insulation is not usually a decisive design criterion. If the insulation construction is mechanically strong enough to endure the assembly, and the creepage currents of the insulation construction are in control, the voltage strength is sufficient enough. Nonsparking during operation has yet to be ensured, since the present, most common organic materials do not resist sparking. In high-voltage machines instead, the strength is a decisive factor in the design in the presence of partial discharges. Next, some practical examples of insulation constructions will be discussed.

8.5.1 Slot Insulations of Low-Voltage Machines

The slot insulation usually consists of two layers, namely an outer and an inner liner. Figure 8.11 illustrates the positioning of the liners in the slot. The inner insulation forms a funnel-shaped guide that facilitates the wire assembly. The stator slot is finally closed at the slot opening with a slot key or a slot wedge.

If the winding is constructed such that there will be more than one coil side in a single slot, an additional insulation (a slot separator) is required between the coil sides. This is not usually problematic in machines below 1 kV. The voltage between different coil sides is of the magnitude of the line-to-line voltage at maximum. Figure 8.12 illustrates a slot with two coil sides closed with a slot key. The insulation has to extend outside the slot at the ends. For instance, the overhang has to be 5 mm at minimum when the rated voltage is 500 V at maximum. This overhang is important, since a local electric field maximum is created at the position where the slot ends, creating creepage currents.

Nowadays there are durable materials available, the voltage strength of which is high enough (e.g. polyester films) and which can be used to construct a single-layer slot insulation. In that case, the coils are mounted in the slots with auxiliary guides or by employing extra-wide insulation films that lead the wires in the slot without damaging them. After the coils are mounted in the slot, the extra guides are removed or the wide insulation films cut and the slot is closed with a lid.

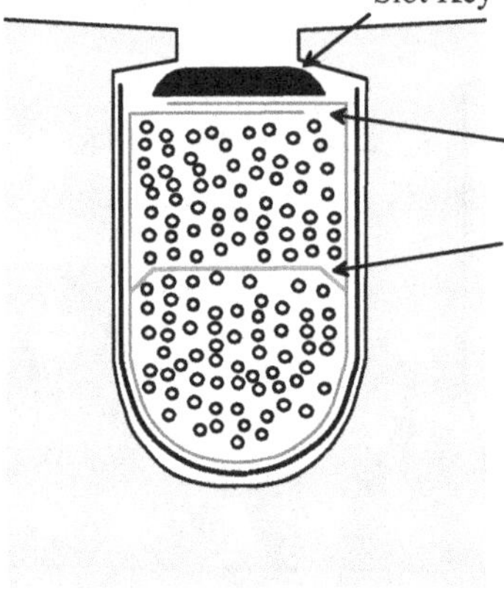

Figure 8.12 Key-closed slot of a double-layer winding arrangement. The winding is made of round enameled wires. The winding is constructed by winding the wires on a coil former neglecting the order of the wires. When the coil is fed into the stator slots, the order of the wires may get mixed even further, and therefore the winding is called a random-wound winding. In the dimensioning of the cross-sectional area of the slot, it has to be borne in mind that the highest possible space factor k_{Cu} of the free space for a round wire winding is in practice, its wire insulators included, about 0.66 (cf. Equation (6.60); the theoretical limit is $\pi/4 = 0.785$). Usually, the space factor varies between 0.6 and 0.66. Further, when the room required by the slot insulation is taken into account, we can see that less than half of the total area of the slot can be filled with actual wire copper. The insulation is thus of great significance in the determination of the resistive losses of the machine. In small machines, the proportion of resistive losses is relatively high, and therefore, when considering the efficiency of the machine, the slot filling deserves special attention.

8.5.2 *Coil End Insulations of Low-Voltage Machines*

In the previous chapter, phase coil insulations in a slot were discussed. Also at the coil ends, the different phases have to be separated. Pieces of insulating cloth are employed as a phase coil insulator. The pieces are stuffed into place after assembly of the windings, the end windings are tied with glass fiber bands, and finally the whole insulation system is hardened with an impregnant.

In addition to the previously mentioned types of insulation, connection leads also have to be insulated at lead-ins by using an insulating sleeve of suitable thickness. Finally, before impregnation, the coils are tied with bands. Special attention has to be paid to the reliability of the fastening of the connection wires and leads to ensure that the winding cannot move and thus wear mechanically.

8.5.3 *Pole Winding Insulations*

In DC and synchronous machines, DC windings are employed. These windings are called pole windings or field windings. Concentrated windings are also used in fractional slot PMSMs. These windings are very similar to the field windings of DC or synchronous machines. The poles are often wound with preformed copper, which is typically of rectangular shape, although round-shaped coils are also employed. In both cases, varnishing is the most common insulation method. The insulated winding is assembled from inside around an iron core. Normally, the dimensioning of the pole winding is not problematic, and therefore the thickness of the insulation does not set any limits on the construction. Commutating poles of DC machines are wound separately, like the coils. First, a layer of insulation is assembled and then the winding is wound on it.

8.5.4 Low-Voltage Machine Impregnation

Low-voltage machines can be impregnated with the dip and bake method by simply soaking the motor into the resin (dip) and then curing the resin in an oven (bake). This is a traditional method and has been in use since the early days of electrical machines. However, the method is still widely used because it is easy and cheap and requires no heavy or expensive facilities, such as vacuum pumps. Vacuum pressure impregnation (VPI) is used in special machines with a purpose-designed insulation, such as those to be driven by inverter drives. The method will be discussed in detail in the next section. For small machines, the trickle impregnation method can be applied. In this method, the stator is attached to a rotating, sloped bench. Current is applied to the windings to preheat them. The preheated resin is trickled onto the windings from the upper end of the machine, hence the name. Simultaneously, the bench begins to rotate slowly. The resin flows through the stator slots and fills them. When the resin begins to drop out from the lower end winding of the machine, it is turned upside down and the same procedure is repeated. When the desired impregnation is achieved, the resin is cured in an oven.

8.5.5 Insulation of High-Voltage Machines

In the insulation of high-voltage machines, partial discharges may prove a significant cause of the ageing of the insulation. High-voltage machines can be wound with round wire, when the rated voltage is 3 kV or below. At voltages above this, prefabricated coils made of preformed copper are nearly always selected. Figure 8.13 illustrates a cross-section of a slot of a preformed copper winding, known as a form-wound winding. Usually, from 6 kV onwards, there is also a conductive corona protection between the insulator and the stator stack in the slot. The function of this layer is to prevent sparking in the voids between the insulator and the sheet stack. Semiconductive and conductive materials can also be employed as slot filling. The most common insulating material of the coil in high-voltage machines is a mica tape wound in layers around the conductors. Due to its excellent partial discharge resistance and performance properties, mica has maintained its dominant position as the insulator in high-voltage machines for more than a century. In the insulation of high-voltage machines, the amount of voids in the impregnation has to be kept to a minimum. Therefore, only the VPI and resin-rich (RR) methods are usually applied.

In the VPI method, the object to be impregnated is placed in a vacuum chamber, which is closed, and a vacuum is created to a pressure of typically about 1 millibar. Next, resin that has been pretreated (viscosity checked, curing agent added, cooled) and degassed in a separate container is pumped into the vacuum chamber through a heat exchanger where the resin is preheated to about 70 °C until the object to be impregnated is completely covered

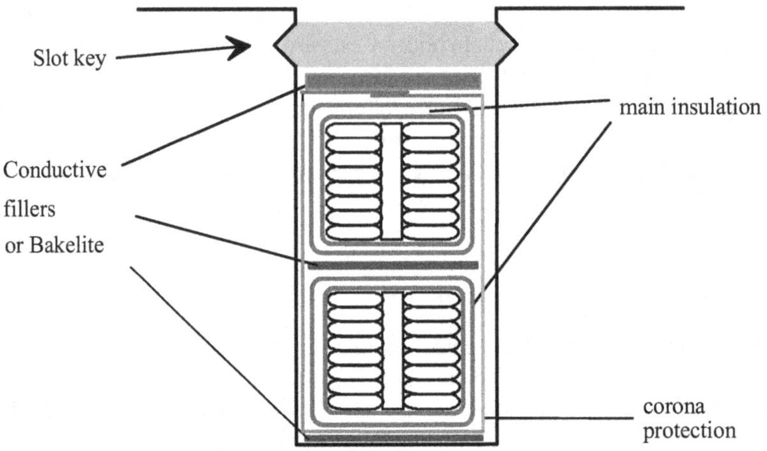

Figure 8.13 Insulation construction of a stator slot in a high-voltage machine. The figure shows that the proportion of the insulation is notably higher than in the slot of the low-voltage machine in Figure 8.12. On the other hand, preformed copper has been employed in the slot, and therefore the space factor of the copper becomes relatively high.

with warmed-up resin. The preheating of the resin is very important, because it significantly decreases the viscosity, and thus the resin can more easily penetrate the slots and fill them completely.

The vacuum is then released and the chamber is allowed to pressurize from 3 to 5 bar for a period of several hours. Finally, the resin is pumped through the heat exchanger back to the cool container. The cooling in the storage container is important to extend the lifetime of the unhardened resin. The object to be impregnated is then placed in the oven, where the resin is hardened. The VPI method is particularly well suited for preformed copper windings, the insulation thicknesses of which can be dimensioned precisely. Thus the thickness of the resin layers can also be controlled.

In the RR method, almost all the selected insulators and other materials are preimpregnated. The binding agent is usually epoxy resin, which is in a precured state, in which the resin is solid but malleable, the insulator thus being easy to process. The coils are insulated with several turns of mica tape wrapped around the coil from end to end. This is the main insulation of the coil. Finally, the epoxy resin of the insulation is hardened at a high temperature (about 160 °C) and high pressure. Usually, insulation of this type meets the requirements of thermal class 155 (F).

The VPI method differs from the RR method by the fact that in the former, the insulating materials used are porous, and do not contain a considerable amount of binding agents. When the insulation is mounted in place, its air spacings are carefully impregnated with an impregnating resin. When applying the VPI method, with accurate dimensioning it is possible to decrease the amount of fillers required in the key fitting. The finished insulation is in every respect equal in quality irrespective of the selected method.

As shown in Figure 8.13, in the RR method in particular, filler strips can be employed. One purpose of these strips is to ensure that the slot key evenly wedges the bar. Sometimes even flexible filler strips are selected to avoid loosening the coils. This loosening can also be avoided by applying a method in which the winding is aged by exposing it to mechanical force and heat. This method is called hot prewedging. It ensures a high and permanent compression force of the key fitting.

The insulation of the end windings is constructed in the same way as in the slots, only without filler strips. A conductive material is not used, unlike in the slot section. Nevertheless, at the point where the stator stack ends, semiconductive coating materials can be employed on the surface of the conductor. The most common stress grading materials are a semiconductive silicon carbide tape or paint.

8.6 Condition Monitoring of Insulation

The ageing process reduces the insulation capacity of the insulating materials. To ensure reliable performance of an electrical machine, the insulation has to be monitored regularly to anticipate a possible loss in insulation capacity. The most common monitoring methods are the measurement of insulation resistance and the tan δ measurement; however, in large machines, various on-line monitoring methods are common nowadays.

The insulation resistance of a winding consists of a surface resistance and a volume resistance. Usually, the requirements for the volume resistance are relatively low in electrical machines, since in these machines, far higher dielectric losses can be tolerated than for instance in the case of capacitors. The values of the volume resistance do not usually depend on the ambient humidity, unless the insulation is hygroscopic. The insulation resistance of a new low-voltage machine is typically between 5000 and 10 000 MΩ. This corresponds to a specific resistance of approximately 10^{14} Ω m. However, a low-voltage machine is still in a quite acceptable condition when its insulation resistance has decreased to 1 MΩ, corresponding to a specific resistance of approximately 10^{10} Ωm. With regard to dielectric losses, even a notably lower specific resistance, maybe about 10^7 Ωm, could be permitted. Hence, the ideal value of the insulation resistance is negligible in the dimensioning of the thickness of the insulation.

Surface resistance is highly dependent on the accumulating dirt and moisture on the surface of the insulation. Dirt accumulates on the insulated surfaces that act as creepage distances. This dirt is often conductive, but also hygroscopic. The insulation resistance of a dirty and damp winding can drop below the above-mentioned level of 1 MΩ, thus endangering the durability of the winding.

A creepage distance of electrical machine insulation is created for instance across the end of the slot insulation from the copper conductor to the stator stack. The effective length of the creepage distance exceeds its physical length because of the insulation of the conductors and the impregnation of the winding. Nevertheless, there may be occasional voids in the insulation and impregnation of the winding. The most critical creepage distances are open insulating distances, for instance on the commutator of a DC machine, which often gathers carbon dust during operation.

The measurement of insulation resistance is the most common method in the condition monitoring of electrical machines. The measurement is carried out with a relatively low (500–1500 V) DC voltage. The method is quite easy and fast, and does not require any expensive special equipment. However, no far-reaching conclusions on the internal state of the insulation can be drawn from these measurement results. Insulation resistance measurements can nevertheless be recommended particularly as a commissioning test and in routine condition check-ups. Temperature has a significant influence on the results of insulation resistance measurements. Unlike the resistance of conductors, the resistance of insulating materials decreases as the temperature rises.

Surge comparison testing is applied in the monitoring of the integrity and voltage strength of a turn insulation. The surge comparison test instrument creates a closed oscillating circuit between the voltage source and the tested winding, resulting in an oscillating voltage between the coil ends. The test result is displayed on the instrument's oscilloscope screen. If the two windings to be compared are identical, the reflected images are identical and appear as a single trace. With this test method, the interturn fault, earth–fault and impedance difference can be detected as a deviation of the frequency, amplitude and the damping period of the oscillation.

The measurement of the dielectric loss factor ($\tan\delta$) of an insulator is an advantageous and common measuring method that is especially suitable for large machines. The $\tan\delta$ curve is usually drawn as a function of voltage, the voltage being raised at least to the highest operating voltage of the machine. When evaluating the test results, special attention can be paid both to the magnitude of the $\tan\delta$ value and to its increase as the voltage is increased. Figure 8.14a illustrates the shapes of a $\tan\delta(U)$ curve with some examples. Curve 1 represents an ideal case. In practice, for a healthy insulator, $\tan\delta$ is almost constant as a function of voltage, as shown by curve 2. The shape of the curve also gives some information about the factors causing an increase in the value of $\tan\delta$. A $\tan\delta$ curve such as curve 4 may indicate excessive ageing of the insulation. On the other hand, curve 3 illustrates a situation in which $\tan\delta$ values start to increase abruptly, when a certain voltage stress is reached. This is a sign of the occurrence of remarkable partial discharges in the insulation. Since the discharges are local, the actual partial discharge measurement is usually a far more sensitive indicator of partial discharges than the method described above.

A $\tan\delta$ value as such expresses the relative magnitude of dielectric losses. An excessive dielectric loss may result in local or complete overheating of the insulation.

When carrying out $\tan\delta$ measurements, it has to be borne in mind that the temperature of the insulation may influence the test results. In some cases, it may prove advisable to take the measurements at several temperatures. In many insulation constructions, $\Delta\tan\delta = 10\,‰$ is considered the highest permitted value.

The measurement of the dielectric loss angle (δ) is carried out as a bridge measurement. The objective of the method is to determine the ratio of dielectric losses to the capacitive reactive power in the insulation, Figure 8.15. The circuit for measuring is illustrated in Figure 8.16.

Partial discharge measurement is applied to determine partial discharges occurring in the winding insulation. With these measurements, it is possible to anticipate insulation failures, their nature and approximate location. A notable advantage of the method is the applicability of the results in the evaluation of the insulation condition. Partial discharges are measured with special measuring equipment consisting of a transformer, a capacitor, a measuring impedance Z and a partial discharge measuring instrument, Figure 8.17. The measurement is carried out in phases against the stator frame and the other phases connected to it. Disturbances during the measurement have to be minimized by careful selection of the components of the measuring circuit and by selecting a measuring frequency that minimizes radio disturbances.

As a result of measurement, we obtain the values of partial discharge as a function of voltage. Figure 8.18 illustrates two printouts, example (a) including the measurement results

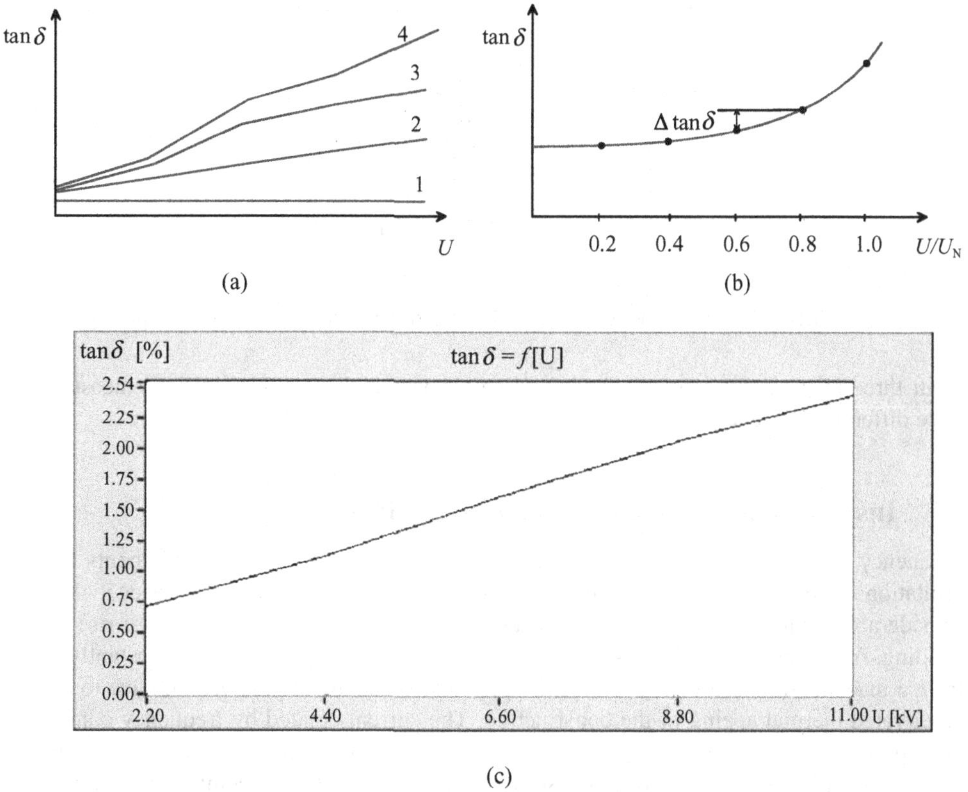

Figure 8.14 (a) Different tan $\delta(U)$ curves: 1, an ideal curve; 2, a typical curve of a healthy insulation; 3, extra partial discharges; 4, the behavior of aged insulation. (b) Measurement in steps of $0.2U_N$, where U_N is the rated voltage of the machine. The slope is defined as Δ tan $\delta/(0.2U_N)$. (c) The test result of a real machine with a healthy insulation corresponding to curve 2 in Figure 8.14a. Reproduced by permission of ABB Oy.

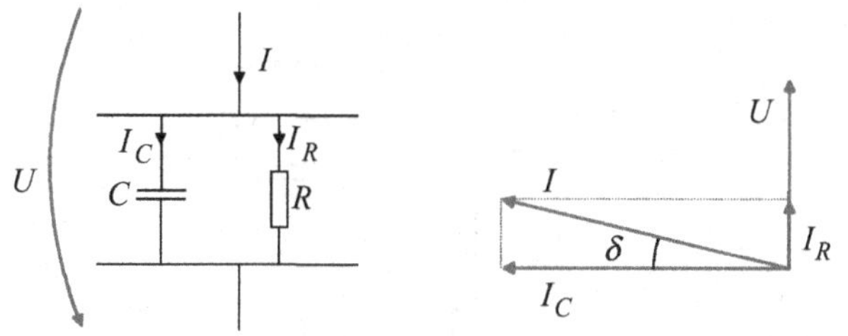

Figure 8.15 Equivalent circuit of an insulation construction and calculation of the loss angle. An ideal insulation forms a capacitance. The dielectric losses of a practical insulation are seen as resistive current real power in the insulation. The ageing of the insulation is seen as increasing resistive current and hence increasing tan δ.

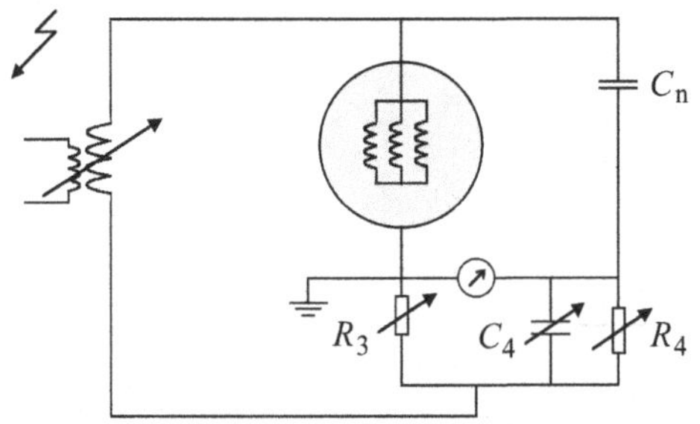

Figure 8.16 tan δ measurement circuit with Schering bridge. C_n is a normal capacitor.

in all three phases, and example (b) including the values for partial discharge measured at three different frequencies.

8.7 Insulation in Frequency Converter Drives

Frequency converter drives that are widely used nowadays set special requirements for the insulation constructions of electric motors. The voltage of a frequency converter includes a considerable number of harmonics that cause additional losses and temperature rise in the winding. As a result of this thermal rise, the maximum rated power cannot be constantly taken from a machine dimensioned for network operation, or, as an alternative, we have to accept accelerated thermal ageing of the construction. The current created by frequency converters that apply a high switching frequency is so sinusoidal that the efficiency of the machine is reduced by only about 1 %, and in this sense the loadability of the machine is not weakened significantly in a frequency converter drive. When applying a very low switching frequency (<1 kHz), the rated power of the machine has typically to be decreased by about 5 %.

The voltage of a frequency converter consists of pulses, the peak value of which may deviate notably from the peak value of normal sinusoidal voltage. Further, the rate of voltage rise can be so high that the pulse behaves like a surge (impulse) wave in the winding, being nonlinearly distributed in the winding and causing a voltage increase by reflection.

With frequencies notably higher than the normal operating frequency, the motor supply cable can no longer be described by concentrated cable parameters, but distributed parameters

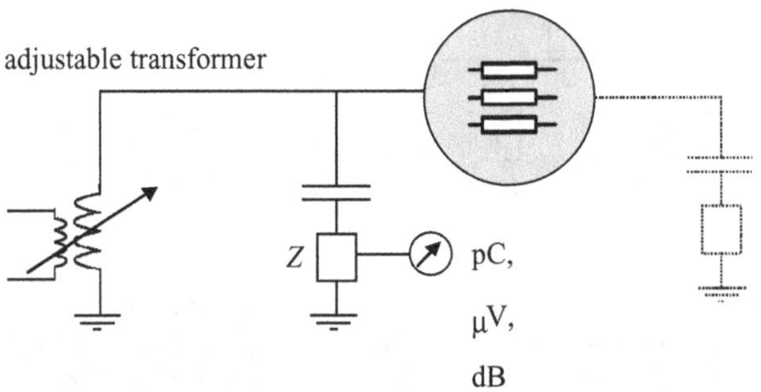

Figure 8.17 Measuring circuit of partial discharge.

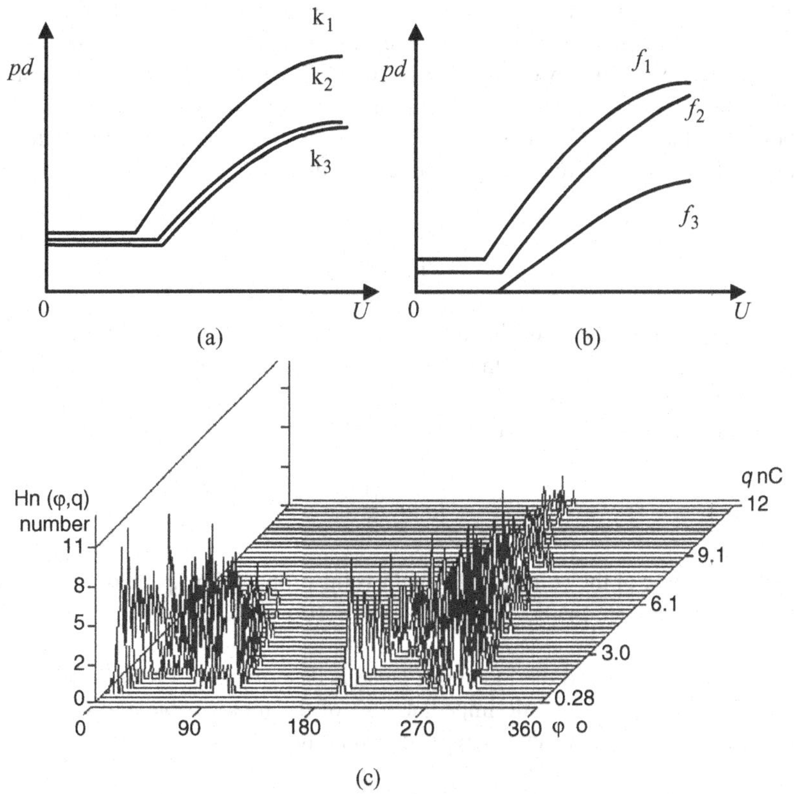

Figure 8.18 Printout of the partial discharge measurement. *pd* is the partial discharge (typically in nC), the level of partial discharge. k_1, k_2 and k_3 illustrate different windings in different insulation conditions, and f_1, f_2 and f_3 different measuring frequencies. The *pd* values start to increase when the voltage is increasing. The lower the *pd* values, the better the insulation. The graph (c) illustrates the measurement result of a real healthy machine using a Haefely Trench Tettex partial discharge measurement unit. The measurement time is 2 minutes and the measurement voltage 8 kV, 50 Hz. The figure indicates the amount of partial discharges as a function of apparent charge in nC and temporal phase angle of the sinusoidal test voltage. Note that there are more partial discharges during the negative phase of the voltage. Note also that the $H_n(\varphi, q)$ symbol on the vertical axis indicates the number of partial discharges having different amounts of electric charge (q in nC) and phase angle (φ in degrees). Reproduced by permission of ABB Oy.

have to be employed. Now the resistance, inductance, leakage conductance and the capacitance of the cable are considered to be evenly distributed along the total cable length. The inductance and the capacitance per unit length determine the characteristic impedance Z_0 individually for each cable

$$Z_0 = \sqrt{\frac{l}{c}}, \tag{8.14}$$

where c is the capacitance per unit of length and l is the inductance per unit of length.

The structure and the insulation of the cable determine the magnitude of the characteristic impedance Z_0, which is thus independent of the cable length. The value for the characteristic impedance of a cable is typically of the order of 100 Ω. The pulse velocity in the cable depends on the materials of the cable. The term "material" refers here to the media surrounding the cable, not the conductor itself. The maximum velocity is the speed of light that can be reached in a vacuum. The pulse velocity v can be defined by the characteristic values of the cable

$$v = \frac{C}{\sqrt{\mu_r \varepsilon_r}} = \frac{1}{\sqrt{lc}}. \tag{8.15}$$

The characteristic impedance of the machine always differs notably from the characteristic impedance of the cable, reflections thus being inevitable. An impulse wave propagates in a motor cable typically at a speed of 150 m/μs. The propagation distance of a wave required by a perfect reflection, that is the critical cable length l_{cr}, is thus

$$l_{cr} = \frac{t_r v}{2}. \tag{8.16}$$

where t_r is the rise time of the voltage pulse and v is the propagation speed of the voltage pulse.

Assuming for a wave that $v = 150$ m/μs and $t_r = 100$ ns (IGBT), the critical cable length becomes 7.5 m.

The ratio of the reflected to the incoming pulse is described by a reflection factor ρ. This factor depends on the characteristic impedance Z_0 of the motor cable and the characteristic impedance Z_M of the motor (winding) experienced by the wave

$$\rho = \frac{Z_M - Z_0}{Z_M + Z_0}. \tag{8.17}$$

As shown by the equation, the value of the reflection factor varies over $0 \le \rho \le 1$, when $Z_M \ge Z_0$. However, the calculation of the characteristic impedance of the motor is nearly impossible. Nevertheless, by measuring reflections at the machine terminals, it is possible to estimate the scale of the characteristic impedance of the machine. Since $Z_0 = \sqrt{l/c}$, we may conclude that the characteristic impedance of the motor is high because of the high inductivity of the motor.

The foundation of insulation design of electrical machines lies in the IEC 60034 standards that are based on long experience of insulation design. Further, when the stresses during operation of the machine, such as heat and operating conditions, are also taken into account, the insulation construction of the machine can be designed. Usually, in addition to mechanical and nonsparking requirements, it is not necessary to pay any special attention to the electrical security of the insulation construction of low-voltage machines. In high-voltage machines, however, it is usually necessary to design some protection to reduce partial discharges.

The tendency seems to be that controlled drives are becoming increasingly popular among low-voltage machines. Now, as the supply cable is of considerable length, and the rated voltage is for instance 690 V, the voltage of the machine may, as a result of the reflecting wave, increase so high that partial discharges occur in the winding between the turns belonging to the same coil. As the switches in power electronics become constantly faster, the pulse rise times get shorter. As a result, the compatibility of the drive, the cable and the machine gain further significance together with the proper 360° concentric conductor earthing of the cable.

Table 8.5 Characteristics of fiber insulators

Characteristic	Unit	Cotton fibers	Polyester fibers	Glass fibers	Aramid paper
Tensile strength	N/mm^2	250–500	500–600	1000–2000	1250
Elongation at break	%	6–10	20–25	1.5	17
Modulus of elasticity	N/mm^2	5000		70 000	—
Tear strength		Moderate	Good	Good	Good
Continuous operating temperature	°C	105[a]	155	130–200[b]	210
Instantaneous thermal resistance	°C	150	190	>600	300
Softening point	°C	—	210	670	Does not soften
Melting point	°C	—	260	850	Does not melt
Thermal conductivity	W/m K	0.07–0.14		0.99	0.1
Burning		Burn	Slow burning	Do not burn	Does not burn
Moisture absorbance	% by weight	10	0.4	—	7–9
Chemical strength	Graded 0–4				
Acids		1	2	4	3
Alkalis		2	1	3	3
Organic solvents		4	4	4	4
Voltage strength	kV/mm	—	—	—	20[c]

[a]Impregnated.
[b]Depends on the impregnant.
[c]1 min, 50 Hz test.
Source: Adapted from Paloniemi and Keskinen (1996).

Furthermore, the cable has to be as short as possible. Utilization of various low-pass filters also becomes more common for reasons other than the limitations set by the insulation alone.

In order to create a single partial discharge pulse, there must be several thousand fast voltage pulses, each having a rise time below 200 ns. Hence, there must be tens of thousands of pulses in 1 second. The number of pulses can be decreased with proper control of the machine; this is nevertheless not always possible. The designer of the machine also has some tools to reduce partial discharge activity. It is possible to select a partial-discharge-resistant conductor for the windings. These conductors include mica or some metal oxide, which can tolerate partial discharges. Also, additional insulation can be applied on different turns. Furthermore, the use of the VPI method considerably decreases the number of air spacings in the slot (Tables 8.2–8.5).

Bibliography

IEC 60034 (various dates) *Rotating Electrical Machines*. International Electrotechnical Commission, Geneva.

IEC 60034-1 (2004) *Rotating Electrical Machines. Part 1: Rating and Performance*. International Electrotechnical Commission, Geneva.

IEC 60085 (2007) *Electrical Insulation – Thermal Evaluation and Designation*. International Electrotechnical Commission, Geneva.

IEC 600112 (1979) Method for Determining the Comparative Tracking Index of Solid Insulating Materials Under Moist Conditions. International Electrotechnical Commission, Geneva.

Nousiainen, K. (1991) *Fundamentals of High-Voltage Engineering. (Suurjännitetekniikan perusteet)*, Study material 144. Tampere University of Technology, Tampere.

Paloniemi, P. and Keskinen, E. (1996) *Insulations of electrical machines. (Sähkökoneiden eristykset)*, Lecture notes. Helsinki University of Technology, Espoo.

Walker, J.H. (1981) *Large Synchronous Machines*. Clarendon Press, Oxford.

9

Losses and Heat Transfer

Heat transfer always occurs when there is a temperature difference in a system. The temperature difference evens out naturally as heat transfers from the higher temperature to the lower according to the second law of thermodynamics.

In electrical machines, the design of heat transfer is of equal importance as the electromagnetic design of the machine, because the temperature rise of the machine eventually determines the maximum output power with which the machine is allowed to be constantly loaded. As a matter of fact, accurate management of heat and fluid transfer in an electrical machine is a more difficult and complicated issue than the conventional electromagnetic design of an electrical machine. However, as shown previously in this material, problems related to heat transfer can to some degree be avoided by utilizing empirical knowledge of the machine constants available. When creating completely new constructions, empirical knowledge is not enough, and thorough modeling of the heat transfer is required. Finally, prototyping and measurements verify the successfulness of the design.

The problem of temperature rise is twofold: first, in most motors, adequate heat removal is ensured by convection in air, conduction through the fastening surfaces of the machine and radiation to ambient. In machines with a high power density, direct cooling methods can also be applied. Sometimes even the winding of the machine is made of copper pipe, through which the coolant flows during operation of the machine. The heat transfer of electrical machines can be analyzed adequately with a fairly simple equation for heat and fluid transfer. The most important factor in thermal design is, however, the temperature of ambient fluid, as it determines the maximum temperature rise with the heat tolerance of the insulation.

Second, in addition to the question of heat removal, the distribution of heat in different parts of the machine also has to be considered. This is a problem of heat diffusion, which is a complicated three-dimensional problem involving numerous elements such as the question of heat transfer from the conductors over the insulation to the stator frame. It should be borne in mind that the various empirical equations are to be employed with caution. The distribution of heat in the machine can be calculated when the distribution of losses in different parts of the machine and the heat removal power are exactly known. In transients, the heat is distributed completely differently than in the stationary state. For instance, it is possible to overload the motor considerably for a short period of time by storing the excess heat in the heat capacity of the machine.

Design of Rotating Electrical Machines, Second Edition. Juha Pyrhönen, Tapani Jokinen and Valéria Hrabovcová.

The lifetime of insulation can be estimated by statistical methods only. However, over a wide temperature range, the lifetime shortens exponentially with the temperature rise Θ of the machine. A rise of 10 K cuts the lifetime of the insulation by as much as 50 %. The machines may withstand temporary, often-repeated high temperatures depending on the duration and height of the temperature peak. A similar shortening of the lifetime applies also to the bearings of the motor, in which heat-resistant grease can be employed. In critical drives, oil mist lubrication can be used, in which case the oil is cooled elsewhere and then fed to the bearings. Even ball bearings can be used at elevated speeds if their effective cooling is ensured, for instance by oil lubrication.

The temperature rise of the winding of an electrical machine increases the resistance of the winding. A temperature rise of 50 K above ambient (20 °C) increases the resistance by 20 % and a temperature rise of 135 K by 53 %. If the current of the machine remains unchanged, the resistive losses increase accordingly. The average temperature of the winding is usually determined by the measurement of the resistance of the winding. At hot spots, the temperature may be 10–20 K above the average.

9.1 Losses

Power losses in electrical machines are composed of the following elements:

- resistive losses in stator and rotor conductors;
- iron losses in the magnetic circuit;
- additional losses;
- mechanical losses.

Resistive losses in conductors are sometimes called Joule losses or copper losses, and therefore the subscript Cu is used in the following for resistive losses.

Figure 9.1 illustrates the power balance of a typical totally enclosed 4 kW IE3 induction motor. Here 12 % of the electrical energy is converted into heat at the rated power of the machine. The proportion of the resistive losses in this case is high: 62.5 % of the total losses and 7.5 % of the rated power. The proportion of iron losses remains low, 1.5 % of the input power, although the iron circuits are usually strictly dimensioned. Mechanical losses are also low, 1 % of the input power. The temperature rise of the machine in Figure 9.1 is dimensioned according to the thermal class 130 (B) but the insulation belongs to the class 155 (F) (see Table 8.1), in order to prolong the lifetime of the insulation.

9.1.1 Resistive Losses

Resistive losses in a winding with m phases and current I are

$$P_{\mathrm{Cu}} = mI^2 R_{\mathrm{AC}}, \tag{9.1}$$

where R_{AC} is the AC resistance of the phase winding. The AC resistance is

$$R_{\mathrm{AC}} = k_{\mathrm{R}} \frac{N l_{\mathrm{av}}}{\sigma S_{\mathrm{c}}}, \tag{9.2}$$

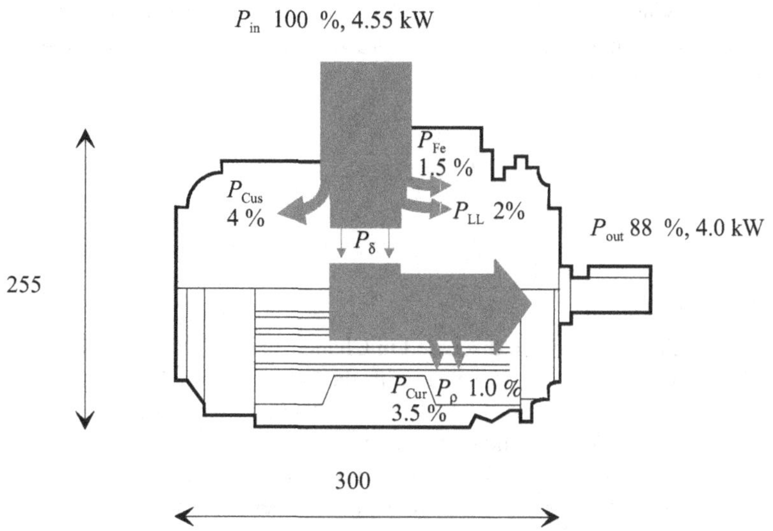

Figure 9.1 Sankey diagram of a 4 kW two-pole induction motor. P_{Fe}, iron losses; P_{Cus}, resistive losses of the stator; P_{ad}, additional losses; P_δ, air-gap power; P_{Cur}, resistive losses of the rotor; P_ρ, friction losses. The losses (550 W in total) have to be removed from the machine at an acceptable temperature difference to the ambient.

where k_R is the resistance factor Equation (5.34), N the number of turns, l_{av} the average length of a turn, S_c the cross-sectional area of the conductor and σ the specific conductivity of the conductor.

The mass of the conductors is

$$m_{Cu} = \rho N l_{av} S_c, \tag{9.3}$$

where ρ is the density of the conductor.

Solving Nl_{av} from Equation (9.3), substituting it into Equation (9.2) and substituting Equation (9.2) into Equation (9.1) yields for the resistive losses

$$P_{Cu} = mI^2 R_{AC} = \frac{k_R}{\rho\sigma}\frac{I^2}{S_c^2} m_{Cu} = \frac{k_R}{\rho\sigma} J^2 m_{Cu}, \tag{9.4}$$

where $J = I/S_c$ is the current density in conductors.

In temperature rise calculations, the resistive losses (9.1) or (9.4) have to be divided into the end winding losses and winding losses in slots as presented in Chapter 5. In efficiency calculation, the resistive losses are calculated using the DC resistance of the winding. The skin effect is taken into account in additional losses (Chapter 9.1.3).

In commutator and slip-ring machines, there are losses in brushes. The resistive losses in brushes are very small because the current density in brushes is low, about 0.1 A/mm^2, but the contact voltage between brushes and commutator may produce significant losses. The contact voltage drop varies depending on the type of brushes, the applied brush pressure and the brush current, being typically $U_{contact} = 0.5–1.5$ V for carbon and graphite brushes and $U_{contact} = 0.2–0.5$ V for metalline brushes. The brush losses P_B with brush current I_B per brush pair are

$$P_B = 2I_B U_{contact}. \tag{9.5}$$

9.1.2 *Iron Losses*

Losses in the iron circuit have been discussed in Chapter 3. In practice, Equation (3.78) gives good enough results if we have experience with the correction coefficients needed in using Equation (3.78). The coefficients are given in Table 3.2, but depending on the construction and dimensioning of the machine, and the manufacturing system, the coefficients may differ from the values in Table 3.2. The coefficients are determined by statistical means by analyzing the no-load tests of a large group of the same kinds of machines.

9.1.3 *Additional Losses*

Additional losses (called also stray-load losses) are defined as the difference between the total losses and the sum of stator and rotor resistive losses, stator and rotor iron losses, and mechanical losses, all measured or calculated from measured results according to IEC 60034-2-1, "Standard methods for determining losses and efficiency from tests." Additional losses are caused by several different phenomena. Some phenomena are easy to model and calculate but some are very difficult to calculate accurately. According to IEC 60034-2-1 standard, the resistive losses are calculated using the DC resistance of the winding, and therefore, the additional losses include losses caused by the skin effect in conductors, which can be calculated by the help of the resistance factor k_R (Equation (5.34)). The iron losses are determined from the no-load test, and consequently they include the additional losses at no load, e.g. eddy current losses that air-gap harmonics create on the rotor surface, stator and rotor tooth tips and windings, iron losses in press plates at the ends of the stator core, iron losses in the frame and end shields (if the winding ends are too close to the frame or end shields). Many of these loss components are small because the no-load current is small. During calculation, these loss components are evaluated empirically by using the correction coefficients $k_{Fe,n}$ in Equation (3.78) and in Table 3.2.

The additional losses defined in the beginning of this chapter are the losses, which the load current and its spatial harmonics cause in windings, laminations, frame and other construction parts and which are not taken into account in calculating resistive and iron losses, therefore they are called additional *load* losses in IEC 60034-2-1. These losses do not, however, take into account losses caused by possible time harmonics in the supply voltage.

Measurement methods to determine the additional load losses in sinusoidal supply are given in the IEC 60034-2-1. The measured additional load losses are used when the efficiency of a motor is calculated indirectly from the loss measurements. If additional loss

Table 9.1 Additional losses as a percentage of input power in electrical machines

Machine type	Additional losses of input power
Squirrel cage motor	0.3–2 % (sometimes up to 5 %)
Slip-ring asynchronous machine	0.5 %
Salient-pole synchronous machine	0.1–0.2 %
Nonsalient-pole synchronous machine	0.05–0.15 %
DC machine without compensating winding	1 %
DC machine with compensating winding	0.5 %

tests are not done, the additional load losses of an induction motor are assumed to be (IEC 60034-2-1, 2007)

$$\begin{aligned} P_{\mathrm{LL}} &= 0.025 P_{\mathrm{in},} \text{ for } P_{\mathrm{out}} \leq 1 \text{ kW} \\ P_{\mathrm{LL}} &= \left[0.025 - 0.005 \log_{10}\left(\frac{P_{\mathrm{out}}}{1 \text{ kW}}\right)\right] P_{\mathrm{in},} \text{ for } 1 \text{ kW} < P_{\mathrm{out}} < 10\,000 \text{ kW} \\ P_{\mathrm{LL}} &= 0.005 P_{\mathrm{in},} \text{ for } P_{\mathrm{out}} \geq 10\,000 \text{ kW}. \end{aligned} \tag{9.6}$$

where P_{in} is the input power and P_{out} the output power of the motor.

Typical additional loss values in different machine types are given in Table 9.1 (Lipo 2007; Schuisky 1960).

Additional load losses are proportional to the square of the stator current (I_{s}) minus the square of no-load current (I_0) and to the power of 1.5 of the frequency (f), that is

$$P_{\mathrm{LL}} \sim (I_{\mathrm{s}}^2 - I_0^2) f^{1.5}. \tag{9.7}$$

If additional losses are measured or known for one pair of the current and frequency, they can be determined for another pair of current and frequency using Equation (9.7).

9.1.4 Mechanical Losses

Mechanical losses are a consequence of bearing friction losses, windage losses of rotating rotor, and ventilator losses. Bearing losses depend on the shaft speed, bearing type, properties of the lubricant and the load on the bearing. Bearing manufacturers give guidelines for calculating bearing losses. According to SKF (2013), bearing friction losses are

$$P_{\rho,\mathrm{bearing}} = 0.5\ \Omega \mu F\ D_{\mathrm{bearing}}, \tag{9.8}$$

where Ω is the angular frequency of the shaft supported by a bearing, μ the friction coefficient (typically 0.08–0.20 for steel-on steel sliding contact surface combination), F the bearing load and D_{bearing} the inner diameter of the bearing.

Windage losses become more and more significant with increasing machine speed. These losses are a consequence of the friction between the rotating surfaces and the surrounding gas, usually air. The windage losses of the rotor are divided into two parts, to the losses in the air gap ($P_{\rho w1}$) and to the losses at the ends of the rotor ($P_{\rho w2}$). The air-gap part of the rotor can be modeled as a rotating cylinder in an enclosure. Saari (1995) gives an equation for the power associated with the resisting drag torque of the rotating cylinder:

$$P_{\rho w1} = \frac{1}{32} k C_M \pi \rho \Omega^3 D_r^4 l_r \tag{9.9}$$

where k is a roughness coefficient (for a smooth surface $k = 1$, usually $k = 1$–1.4), C_M the torque coefficient, ρ the density of the coolant, Ω the angular velocity, D_r the rotor diameter (Figure 3.1) and l_r the rotor length. The torque coefficient C_M is determined by measurements. It depends on the Couette Reynolds number

$$Re_\delta = \frac{\rho \Omega D_r \delta}{2\mu}, \tag{9.10}$$

where ρ is the coolant density, δ the air-gap length and μ the dynamic viscosity of the coolant. The torque coefficient is obtained as follows:

$$C_M = 10 \frac{(2\delta / D_r)^{0.3}}{Re_\delta}, \quad Re_\delta < 64, \tag{9.11}$$

$$C_M = 2 \frac{(2\delta / D_r)^{0.3}}{Re_\delta^{0.6}}, \quad 64 < Re_\delta < 5 \times 10^2, \tag{9.12}$$

$$C_M = 1.03 \frac{(2\delta / D_r)^{0.3}}{Re_\delta^{0.5}}, \quad 5 \times 10^2 < Re_\delta < 10^4, \tag{9.13}$$

$$C_M = 0.065 \frac{(2\delta / D_r)^{0.3}}{Re_\delta^{0.2}}, \quad 10^4 < Re_\delta. \tag{9.14}$$

The end surfaces of the rotor can be modeled as discs rotating in free space (assuming that there are no fan wings in short-circuit rings). The power loss is, according to Saari (1995),

$$P_{\rho w2} = \frac{1}{64} C_M \rho \Omega^3 \left(D_r^5 - D_{ri}^5\right), \tag{9.15}$$

where D_r is the outer diameter of the rotor, D_{ri} the shaft diameter and C_M the torque coefficient, which is now

$$C_M = \frac{3.87}{Re_r^{0.5}}, \quad Re_r < 3 \times 10^5, \tag{9.16}$$

$$C_M = \frac{0.146}{Re_r^{0.2}}, \quad Re_r > 3 \times 10^5. \tag{9.17}$$

Re_r is known as the tip Reynolds number

$$Re_r = \frac{\rho \Omega D_r^2}{4\mu}. \tag{9.18}$$

Table 9.2 Experimental factors for windage and bearing losses, Equation (9.20)

Cooling method	k_ρ [W s^2/m^4]
TEFC motors, small and medium-sized machines	15
Open-circuit cooling, small and medium-sized machines	10
Large machines	08
Air-cooled turbogenerators	05

The windage losses caused by the rotating parts of the machine $P_{\rho w}$ are the sum of Equations (9.9) and (9.15)

$$P_{\rho w} = P_{\rho w1} + P_{\rho w2}. \tag{9.19}$$

A ventilator can be coupled to the shaft of the electrical machine or it can be driven by another motor, as is usual in speed-controlled drives. Schuisky (1960) gives an experimental equation for the sum of windage and ventilator losses:

$$P_\rho = k_\rho D_r (l_r + 0.6\tau_P) v_r^2, \tag{9.20}$$

where k_ρ is an experimental factor (Table 9.2), D_r the rotor diameter, l_r the rotor length, τ_p the pole pitch and v_r the surface speed of the rotor. Equation (9.20) is valid for normal-speed machines. For high-speed machines, Equations (9.9) and (9.15) have to be used.

9.1.5 *Decreasing Losses*

Of all electrical motor types, permanent magnet motors have the best efficiency because the creation of the magnetic field in no-load is, in principle, lossless. Synchronous reluctance motors have also a rather high efficiency because there are no windings and, therefore, no resistive losses in the rotor. A permanent magnet motor or a synchronous reluctance motor is a good choice if a drive needs speed control, i.e. it is supplied by a frequency converter. If a motor is connected direct on line, the induction motor is the most popular one.

Possibilities to decrease the losses of an AC motor are illustrated in Figure 9.2. The eddy current losses in the iron core can be decreased by using thinner steel laminations and the hysteresis losses can be decreased by choosing premium grade steel sheets. Cast copper rotor in induction machines lowers rotor resistive losses. Increasing the size of the motor reduces flux density in iron parts and current density in windings, and therefore, iron losses and resistive losses can decrease. Additionally, cooling capacity increases. Optimization of stator and rotor slot shapes helps to decrease losses. Efficient cooling fan design improves air flow and reduces ventilation losses.

Manufacturing process influences the losses of an electric machine. Punching tools have to be sharp to avoid short-circuits between the steel sheets. Transfer of machine parts during the manufacturing has to be done with care to avoid defects in iron cores or winding surfaces. The use of centrifugal rotor casting instead of the common high- and low-preasure die-casting

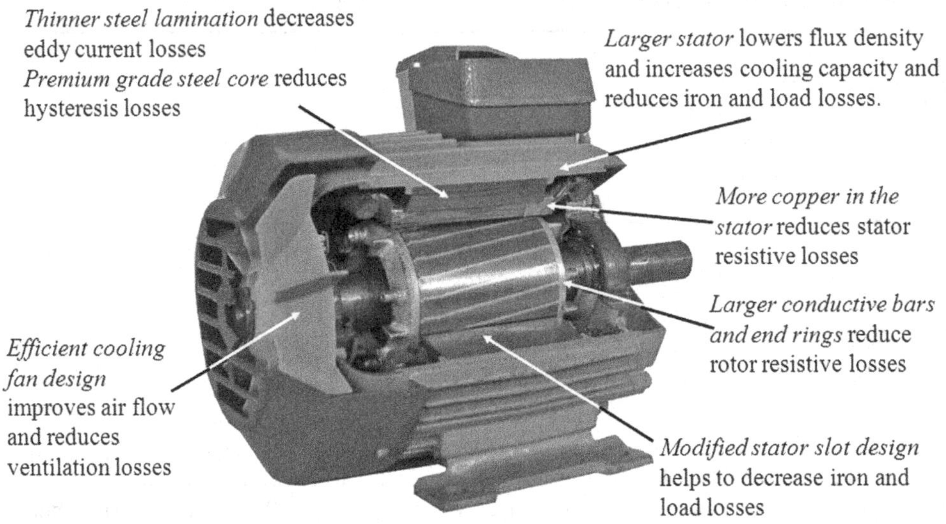

Figure 9.2 Decreasing the losses of an AC motor.

for aluminum squirrel cage increases the conductance of rotor bars and rings and lowers the rotor resistive losses.

The influence of the increasing size on the losses is studied next. If the dimensions of a motor are increased by multiplying by λ, all lengths, diameters, widths and heights will be proportional to λ, areas to λ^2 and volumes and masses to λ^3. As the cross-sectional areas of slots and the cross-sections of teeth and yokes increase, the current densities in windings and the flux densities in iron decrease inversely proportional to the square of the scale λ^2 and resistive losses (P_{Cu}) and iron losses (P_{Fe}) will become

$$P_{\mathrm{Cu}} \sim J^2 m_{\mathrm{Cu}} \sim \left(\frac{1}{\lambda^2}\right)^2 \lambda^3 = \frac{1}{\lambda} \tag{9.21}$$

$$P_{\mathrm{Fe}} \sim \hat{B}_{\mathrm{Fe}}^2 V_{\mathrm{Fe}} \sim \left(\frac{1}{\lambda^2}\right)^2 \lambda^3 = \frac{1}{\lambda}. \tag{9.22}$$

On the other side, the fabrication cost (C) of the motor increases. The cost can be assumed to be proportional to the mass, i.e.

$$C \sim \lambda^3. \tag{9.23}$$

Spatial harmonics create additional losses in iron and windings. To decrease the spatial harmonics, the number of stator slots should be as large as the constructional and economical factors allow, and magnetic slot wedges should be used to decrease the permeance harmonics. A proper chording of the stator winding also reduces the lower-order space harmonics in the air-gap flux density.

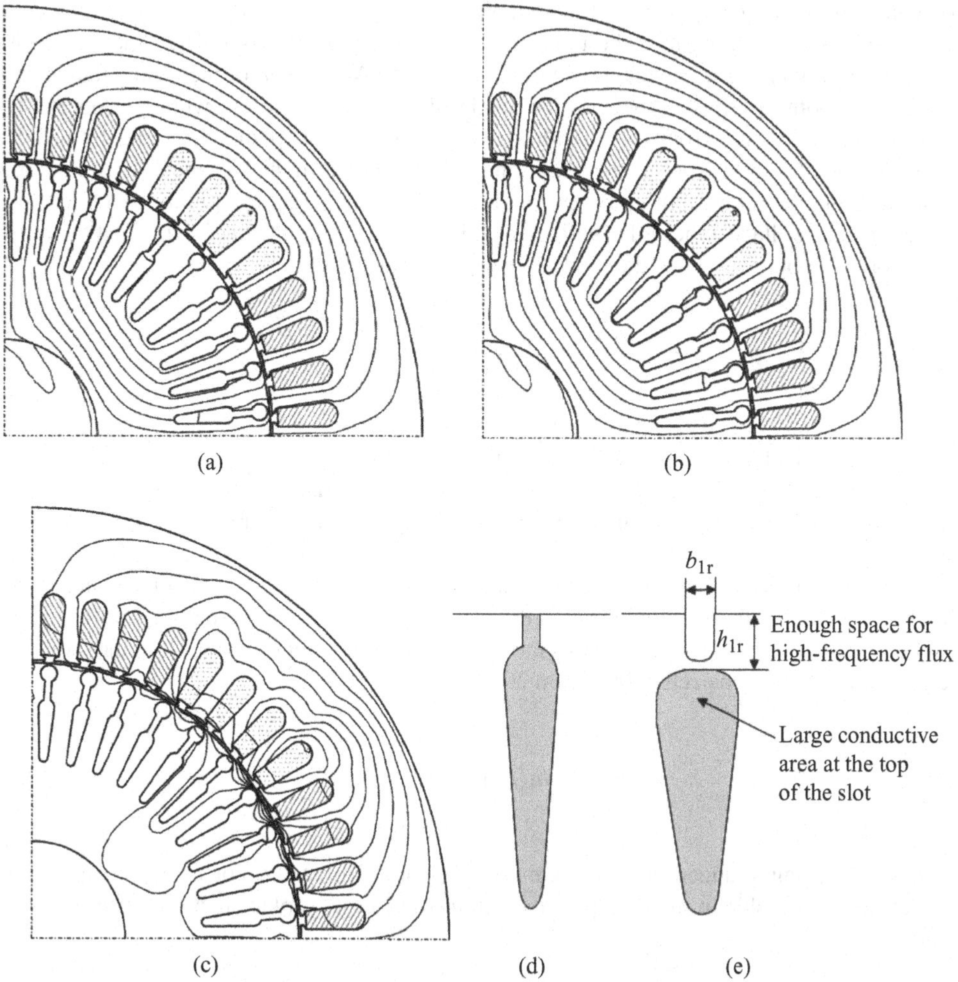

Figure 9.3 Magnetic field of an induction motor (a) supplied with sinusoidal voltage, (b) with frequency converter, (c) time-harmonic flux found by subtracting the sinusoidal-voltage-created flux from the inverter PWM-voltage-created flux. Extra additional losses are created in rotor bars, if there is conductive material in the slot openings as it is in (d). One solution to avoid additional losses at the top of the bars and to lower the resistive losses in the bars is presented in (e). This solution keeps the manufacturing of the rotor winding easy as the rotor laminations themselves form the die cast mold.

In inverter fed motors, the stator voltage and current contain time harmonics. The magnetic fields of an induction motor supplied with sinusoidal voltages and with frequency converter PWM-voltages are shown in Figure 9.3a and 9.3b. The figures appear very similar but subtracting the field in Figure 9.3a from the field in Figure 9.3b the field in Figure 9.3c is obtained and we can see how the field created by the supply time harmonics goes from the tooth tip to tooth tip (Arkkio 1992). If there is conductive material in the rotor slot openings (Figure 9.3d), the field of the harmonics creates high resistive losses in the slot openings. These losses can be avoided if there is enough space for the high-frequency flux (Figure 9.3e). In medium size and large motors, the height of the tooth tip h_{1r} has to be ca. 3-5 mm. The exact height is obtained by solving the magnetic field and losses with FEM. Instead of increasing the height of the rotor tooth tip, we can increase the height of the stator tooth tip, which moves partly the high-frequency flux from rotor tooth tips to the stator side. If the switching frequency of the converter is increasing to several kilohertz, the time harmonics of the air-gap flux are decreasing and the rotor tooth tip height can be closer to the height used in the machines fed with sinusoidal voltage. Oberretl (1969) gives for sinusoidal fed machines a rule that the ratio of the rotor slot opening b_{1r} to the tooth tip height h_{1r} has to be

$$b_{1r}/h_{1r} \leq 1, \quad (9.24)$$

in order to avoid additional losses due to the field harmonics.

In inverter fed machines, there is no need to utilize the deep-bar effect to increase the locked rotor torque and it is advantageous to use a large conductive area at the top of the slot (Figure 9.3e).

If the stator slots are open and wide rectangular conductors are used, the air-gap magnetic flux lines will be partly traveling through the topmost conductors (Figure 9.4). To avoid high eddy current losses in the topmost wires the ratio of the stator slot opening b_{1s} to the tooth tip height h_{1s} has to be (Oberretl 1969; Islam 2010)

$$b_{1s}/h_{1s} \leq 3. \quad (9.25)$$

Naturally, using stranded conductors makes this problem easier. For example, litz wires are nowadays available in many shapes and can in some cases replace rectangular traditional conductors.

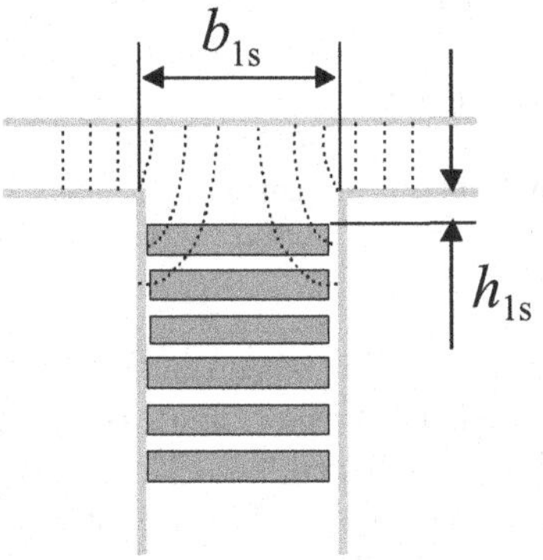

Figure 9.4 The topmost rectangular conductor has to be located far enough from the air gap ($h_{1s} \geq b_{1s}/3$) in order to avoid additional losses that the air-gap magnetic field creates if it travels through the conductor.

9.1.6 Economics of Energy Savings

Now, it is asked, if the savings in the energy cost are greater than the increase of cost of extra mass and purchase price of the motor. In other words, what is the payback time of the investment to an energy efficient motor?

If the reduction of the motor losses is P_{diff}, the operating time of the motor per year T and the energy cost c_{e}, the savings C_{s} per year are

$$C_{\mathrm{s}} = c_{\mathrm{e}} P_{\mathrm{diff}} T. \tag{9.26}$$

The savings per year can also be written in the form

$$C_{\mathrm{s}} = c_{\mathrm{e}} P_{\mathrm{out}} \left(\frac{1}{\eta_1} - \frac{1}{\eta_2} \right) T, \tag{9.27}$$

where P_{out} is the output power of the motor and η_1 and η_2 the efficiencies of the motors under comparison.

If the increase of the purchase cost of the motor having lower losses is C_{diff}, a simple payback time (T_{pb}) is

$$T_{\mathrm{pb}} = C_{\mathrm{diff}} / C_{\mathrm{s}}. \tag{9.28}$$

Example 9.1: ABB is producing 11 kW low voltage induction motors belonging to high-efficiency class IE2 and alternatively to premium class IE3. The rated efficiencies are 90.4 % (IE2) and 92.3 % (IE3) and selling prices 500 € (IE2) and 600 € (IE3). Calculate the simple payback time, if the energy cost is 10 cents/kWh and the operating time 6000 h/a.

Solution: The energy cost saving per year is

$$C_{\mathrm{s}} = c_{\mathrm{e}} P_{\mathrm{out}} \left(\frac{1}{\eta_1} - \frac{1}{\eta_2} \right) T = 0.10 \cdot 11 \cdot \left(\frac{1}{0.904} - \frac{1}{0.923} \right) \cdot 6000 \frac{€}{\mathrm{a}} = 150.29 \frac{€}{\mathrm{a}}$$

and the payback time

$$T_{\mathrm{pb}} = \frac{C_{\mathrm{diff}}}{C_{\mathrm{s}}} = \frac{600 - 500}{150.29} \mathrm{a} = 0.665 \ \mathrm{a} = 8 \ \mathrm{months}.$$

Besides the payback time calculation, we can analyze the potential life-cycle savings of a purchase. Calculations are made including the effect of the cost of money over the lifetime of the motor. The money earned later has to be discounted back to the present by multiplying the cost or saving by the "present worth factor of an equal payment series" k_{pw}

$$k_{\mathrm{pw}} = \frac{(1+i)^n - 1}{i(1+i)^n} \tag{9.29}$$

where i is annual rate of interest and n years of saving (years of motor lifetime). The present value of the cost of one kilowatt of loss over the life of the motor c_{pw} is

$$c_{pw} = k_{pw} c_e T \tag{9.30}$$

Example 9.2: Calculate the present value of the cost of one loss kilowatt if energy cost is 10 cents/kWh, interest 7 %, operating time 6000 h/a and lifetime of the motor 20 years.

Solution:

$$c_{pw} = \frac{(1+i)^n - 1}{i(1+i)^n} c_e T = \frac{(1+0.07)^{20} - 1}{0.07(1+0.07)^{20}} \cdot 0.10 \cdot 6000 \frac{€}{\text{kW}} = 6356 \frac{€}{\text{kW}}.$$

If we get, for example, two offers of motors having 2 kW loss difference, it is worth to buy the motor with better efficiency if its purchase price is less than 2 × 6356 € = 12712 € compared to the price of the lower efficiency motor.

9.2 Heat Removal

The heat is removed by convection, conduction and radiation. Usually, the convection through air, liquid or steam is the most significant method of heat transfer. Forced convection is, inevitably, the most efficient cooling method if we do not take direct water cooling into account. The cooling design for forced convective cooling is also straightforward: the designer has to ensure that a large enough amount of coolant flows through the machine. This means that the cooling channels have to be large enough. If a machine with open-circuit cooling is of IP class higher than IP 20, using heat exchangers to cool the coolant may close the coolant flow.

If the motor is flange mounted, a notable amount of heat can be transferred through the flange of the machine to the device operated by the motor. The proportion of heat transfer by radiation is usually moderate, yet not completely insignificant. A black surface of the machine in particular promotes heat transfer by radiation.

9.2.1 Conduction

There are two mechanisms of heat transfer by conduction: first, heat can be transferred by molecular interaction, in which molecules at a higher energy level (at a higher temperature) release energy for adjacent molecules at a lower energy level via lattice vibration. Heat transfer of this kind is possible between solids, liquids and gases.

The second means of conduction is heat transfer between free electrons. This is typical of liquids and pure metals in particular. The number of free electrons in alloys varies considerably, whereas in materials other than metals, the number of free electrons is small. The thermal conductivity of solids depends directly on the number of free electrons. Pure metals are the best heat conductors. Fourier's law gives the heat flow transferred by conduction

$$\boldsymbol{\Phi}_{th} = -\lambda S \nabla T, \tag{9.31}$$

where $\boldsymbol{\Phi}_{\text{th}}$ is the heat flow rate, λ the thermal conductivity, S the heat transfer area and ∇T the temperature gradient.

Thermal conductivity depends on the temperature; a typical property of metallic substances is that the thermal conductivity decreases as the temperature increases. On the other hand, an insulator's capability to transfer heat increases as the temperature rises.

The thermal conductivity of gases increases with increasing temperature and decreasing molecular weight.

Usually, the thermal conductivity of nonmetallic liquids decreases as the temperature rises; however, the properties of water for example are different. The thermal conductivity of water is at its highest (688 W/K m) at about 410 K and 330 kPa (saturated liquid). The thermal conductivity decreases from this point in both directions of temperature change. Glycerin and ethylene glycol are further exceptions, because their thermal conductivity increases as a function of temperature.

Table 9.3 lists the heat transfer properties of some materials at room temperature. The equation for the conduction of heat is simplified if heat is flowing in one direction, in the direction of the x-coordinate. For an object with a cross-sectional area of S and a length l, Equation (9.31) takes the form

$$\Phi_{\text{th}} = -\lambda S \frac{\mathrm{d}T}{\mathrm{d}x} \approx -\lambda S \frac{\Theta}{l}. \tag{9.32}$$

Here, Θ is the temperature difference across the object and λ is the thermal conductivity of the material, which is often expressed as a function of temperature $\lambda(T)$. Often, materials are not isotropic but the conductivity varies in different directions. Materials are, however, usually considered isotropic in calculations. Materials with high electrical conductivity are, generally, also good thermal conductors. On the other hand, the insulators used in electrical machines are, unfortunately, usually poor thermal conductors. There are also exceptions in insulators, because metallic oxide layers are relatively good thermal conductors compared with plastics, but still good electrical insulators. Diamond is also an exception: It is an electrical insulator and a good thermal conductor.

Equation (9.31) is analogous to the current density equation known in electrical engineering

$$\boldsymbol{J} = -\sigma \nabla V \tag{9.33}$$

or in the one-dimensional case

$$I = JA = -\sigma S \frac{\mathrm{d}V}{\mathrm{d}x} = -\sigma S \frac{\Delta V}{l} = -\sigma S E, \tag{9.34}$$

where J is the current density, S the current-conducting area, σ the electric conductivity, ∇V the electric potential gradient, E the electric field strength, I the current and l the length of the conductor.

Table 9.3 Heat transfer properties of some materials at room temperature (293 K) if not otherwise declared

Material	Thermal conductivity λ [W/K m]	Specific heat capacity c [kJ/kg K]	Density ρ [kg/m^3]	Resistivity [Ω m·10^{-8}]
Air, stagnant	0.025	1		
Aluminum, pure	231	0.899	2700	2.7
Aluminum, electrotechnical	209	0.896	2700	2.8
Aluminum oxide, 96 %	29.4			
Beryllium oxide, 99.5 % 300 K	272	1.03	3000	
Copper, electrotechnical	394	0.385	8960	1.75
Ethylene glycol	0.25	2.38	1117	
Insulation of elec. machine, bonding epoxy	0.64			
Insulation of elec. machine, glass fiber	0.8–1.2			
Insulation of elec. machine, Kapton	0.12			
Insulation of elec. machine, mica	0.5–0.6			
Insulation of elec. machine, mica–synthetic resin	0.2–0.3			
Insulation of elec. machine, Nomex	0.11			
Insulation of elec. machine, Teflon	0.2			
Insulation of elec. machine, treating varnish	0.26			
Insulation of elec. machine, typical insulation system	0.2			
Iron, pure	74.7	0.452	7897	9.6
Iron, cast	40–46	0.5	7300	10
Mercury, 300 K	8540	0.1404	13.53	
Permanent magnet, ferrite	4.5			> 10^9 Ω m
Permanent magnet, NdFeB	8–9	0.45	7500	120–160
Permanent magnet, Sm–Co	10	0.37	8400	50–85
Plastics	0.1–0.3			
Silicon 300 K	148	0.712	12 300	
Steel, carbon steel 0.5 %	45	0.465	7800	14–18
Steel, electrical sheet, in the direction of lamination	22–40		7700	25–50
Steel, electrical sheet, normal to lamination	0.6			
Steel, stainless 18/8	17		7900	
Steel, structural	35–45			
Transformer oil 313 K (40 °C)	0.123	1.82	850	10^8–10^{14} Ω m
Water vapor, 400 K	24.6	2.06	0.552	
Water 293 K	0.6	4.18	997.4	2–5 × 10^3 Ω m

Analogous to the electric resistance, defined as the ratio of the potential difference to the current, we can define the thermal resistance R_{th} as a ratio of the temperature difference Θ to the heat flow rate Φ_{th}

$$R_{\text{th}} = \frac{\Theta}{\Phi_{\text{th}}} = \frac{l}{\lambda S}. \tag{9.35}$$

Example 9.3: Calculate the thermal resistance of a rectangular bar (Figure 9.5), the length of which is $l = 0.2$ m, height $h = 0.05$ m, width $w = 0.03$ m and conductivity $\lambda =$ 17 W/K m (stainless steel).

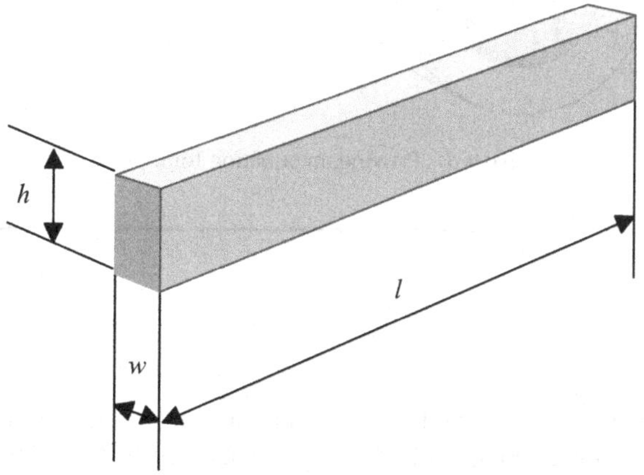

Figure 9.5 Proving an equation for a rectangular bar.

Solution: According to Equation (9.35),

$$R_{\text{th,bar}} = \frac{l}{\lambda h w} = \frac{0.2}{17 \cdot 0.05 \cdot 0.03} \frac{\text{K}}{\text{W}} = 7.84 \frac{\text{K}}{\text{W}}. \tag{9.36}$$

Example 9.4: Calculate the thermal resistance of a pipe in the radial direction (Figure 9.6). The length of the pipe is $l = 0.2$ m, inner radius $r_1 = 0.02$ m, outer radius $r_2 = 0.025$ m and conductivity $\lambda = 0.2$ W/K m (plastics).

Solution: The heat transfer area S in Equation (9.35) now changes and the thermal resistance is the integral:

$$R_{\text{th,cyl}} = \int_{r_1}^{r_2} \frac{1}{\lambda l 2\pi r} \mathrm{d}r = \frac{\ln\left(\frac{r_2}{r_1}\right)}{\lambda l 2\pi} = \frac{\ln\left(\frac{0.025}{0.02}\right)}{0.2 \cdot 0.2 \cdot 2\pi} \frac{\text{K}}{\text{W}} = 0.89 \frac{\text{K}}{\text{W}}. \tag{9.37}$$

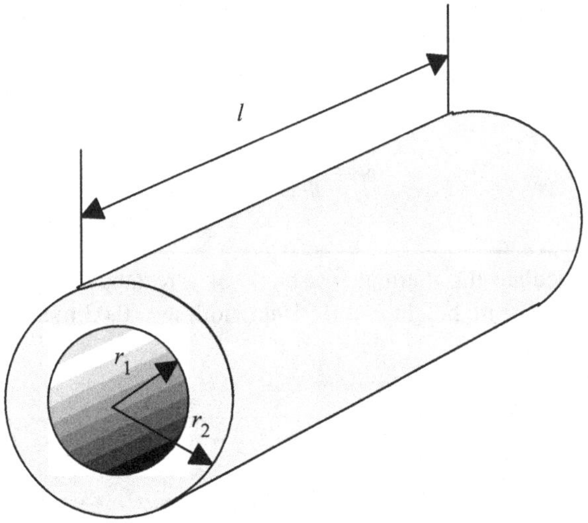

Figure 9.6 Proving an equation for a pipe.

9.2.2 Radiation

Next we consider radiation, the second form of heat transfer. Heat radiation is electromagnetic radiation, the wavelength of which lies in the range from 0.1 to 100 μm. This wavelength range includes visible light, infrared radiation and the long wavelengths of ultraviolet radiation (0.1–0.4 μm). In contrast to the other two heat exchange phenomena, radiation does not require a medium for heat exchange. When radiation meets an object, part of it is absorbed into the object, some of it is reflected back from the surface of the object and some may transmit through the object. The rate that a surface absorbs radiation energy is denoted by absorptivity β, the reflected energy by reflectivity η and the transmitted energy by transmissivity κ. The sum of these is equal to one:

$$\beta + \eta + \kappa = 1. \tag{9.38}$$

A reflective surface ($\eta > 0$) is called opaque, and if the radiation is partly transmitted ($\kappa > 0$) through the material, the surface is called semi-transparent. Figure 9.7 illustrates semi-transparent and opaque surfaces. Semi-transparent surfaces are not used in electrical machines, hence in practice $\kappa = 0$ and

$$\beta + \eta = 1. \tag{9.39}$$

Air consists principally of oxygen and nitrogen, which neither absorb nor emit radiation. Thus, the radiation from an electrical machine to its surroundings and inside the machine can be assumed to travel only between two surfaces.

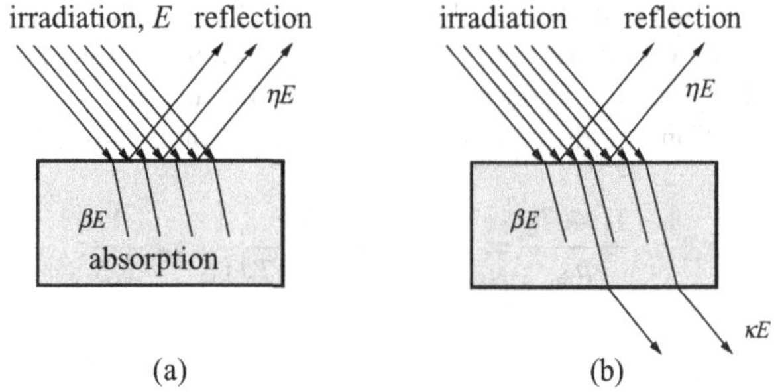

Figure 9.7 (a) Illustration of an opaque surface; (b) illustration of a semi-transparent surface. The incoming irradiation intensity is E, given in W/m^2.

The heat flow density q_{th} of radiation is defined by the Stefan–Boltzmann equation

$$q_{\mathrm{th}} = \frac{\Phi_{\mathrm{th}}}{S} = \varepsilon_{\mathrm{thr}}\sigma_{\mathrm{SB}}\left(T_1^4 - T_2^4\right), \tag{9.40}$$

where T_1 is the thermodynamic temperature of the radiating surface (1), T_2 the thermodynamic temperature of the absorbing surface (2) and σ_{SB} the Stefan–Boltzmann constant, 5.67×10^{-8} W/m^2 K^4. $\varepsilon_{\mathrm{thr}}$ is the relative emissivity between the emitting and absorbing surfaces and depends on the characteristics of the surfaces and on the position of surfaces to each other. If all the radiation that a surface S_1 is emitting meets a surface S_2, $\varepsilon_{\mathrm{thr}}$ becomes

$$\frac{1}{\varepsilon_{\mathrm{thr}}} = \frac{1}{\varepsilon_{\mathrm{th1}}} + \frac{S_1}{S_2}\left(\frac{1}{\varepsilon_{\mathrm{th2}}} - 1\right), \tag{9.41}$$

where $\varepsilon_{\mathrm{th1}}$ and $\varepsilon_{\mathrm{th2}}$ are the emissivities of surfaces 1 and 2.

The emissivity of a black object is $\varepsilon_{\mathrm{th}} = 1$. In reality, black objects do not exist, but the best objects reach an emissivity of $\varepsilon_{\mathrm{th}} = 0.98$, the emissivity of a black-painted object being approximately $\varepsilon_{\mathrm{th}} = 0.9$ and the relative emissivity between a gray-painted electrical machine and its surroundings $\varepsilon_{\mathrm{thr}} = 0.85$.

The concept of a black body is rather complex. The "color" of a body cannot be determined visually, because most of the heat radiation does not fall within the visible light range. An ideal black body is defined, irrespective of the wavelength, as a perfect absorber or emitter, whereas its opposite, an ideal white body, is a total reflector. The blackness of a body is determined by emissivity, which depends on the temperature, radiation direction and the wavelength of the radiation. Emissivity is the ratio of body radiation to black-body radiation at equal temperatures. It expresses how close to a black body the body (material) is. To facilitate the evaluation of a body's emissivity, a total hemispherical emissivity is used; it is the average emissivity of a surface over all directions and wavelengths.

Next, gray and diffuse surfaces are defined to simplify the observation of radiation heat exchange. A gray surface is a surface, the emissivity and absorptivity of which are independent of radiation wavelength. A diffuse surface is a surface, the radiation properties of which are independent of the direction of radiation. In reality, fully direction-independent surfaces do not exist, but emissivity can be assumed to be fairly constant if the angle of radiation deviates less than 40° from the plane normal for conductors, and less than 70° for insulators. In practice, emissivities are usually given in the direction normal to the surface.

In analogy to conduction, the thermal resistance of radiation is defined as

$$R_{\mathrm{th}} = \frac{T_1 - T_2}{\Phi_{\mathrm{th}}} = \frac{T_1 - T_2}{\varepsilon_{\mathrm{thr}}\sigma_{\mathrm{SB}}\left(T_1^4 - T_2^4\right) S} = \frac{1}{\alpha_{\mathrm{s}} S}, \tag{9.42}$$

where

$$\alpha_{\mathrm{r}} = \varepsilon_{\mathrm{thr}}\sigma_{\mathrm{SB}}\frac{T_1^4 - T_2^4}{T_1 - T_2} \tag{9.43}$$

is the heat transfer coefficient of radiation. It depends strongly on the temperatures of the emitting and absorbing surfaces. The temperature difference between the outer surface of an electrical machine and the surroundings is usually about 40 K and the ambient temperature 20 °C (293 K). These temperatures and a relative emissivity $\varepsilon_{\mathrm{thr}} = 0.85$ yield

$$\alpha_{\mathrm{r}} = 6\mathrm{W/m^2K}. \tag{9.44}$$

The thermal resistance of radiation gives a linearization to radiation heat transfer that is similar to convection heat transfer. This makes it possible to compare radiation and convection heat transfer efficiency.

Example 9.5: How much heat does an object with a relative emissivity of 0.85 radiate to the ambient if (a) the temperature of the object is 100 °C and the ambient temperature is 50 °C, (b) the temperature of the object is 50 °C and the ambient temperature is 20 °C?

Solution: The heat flow densities are

(a) $$q_{\mathrm{th}} = \frac{\Phi_{\mathrm{th}}}{S} = 0.85 \cdot 5.67 \cdot 10^{-8}\frac{\mathrm{W}}{\mathrm{m^2K^4}}\left(373^4\ \mathrm{K}^4 - 323^4\ \mathrm{K}^4\right) = 408\frac{\mathrm{W}}{\mathrm{m}^2}.$$

(b) $$q_{\mathrm{th}} = 0.85 \cdot 5.67 \cdot 10^{-8}\frac{\mathrm{W}}{\mathrm{m^2K^4}}\left(323^4\ \mathrm{K}^4 - 293^4\ \mathrm{K}^4\right) = 169\frac{\mathrm{W}}{\mathrm{m}^2}.$$

These yield radiation heat transfer coefficients for (a) 8.16 W/m^2 K and (b) 6.65 W/m^2 K.

Heat transfer by radiation is of considerable significance in the total heat transfer of an electrical machine if, besides radiation, there is only natural convection without any fan in the system. Table 9.4 lists the emissivities of some materials in typical electrical machines.

Table 9.4 Emissivities of some materials used in electrical machines

Material	Emissivity
Polished aluminum	0.04
Polished copper	0.025
Mild steel	0.2–0.3
Cast iron	0.3
Stainless steel	0.5–0.6
Black paint	0.9–0.95
Aluminum paint	0.5

Example 9.6: Consider two concentric spheres in a vacuum. The radii of the spheres are R_i and R_o ($R_i < R_o$). The temperatures of the surfaces of the spheres are T_i and T_o. The distance between the spheres is small compared with the radii ($R_o - R_i \ll R_i$). (a) Calculate the radiated heat flow rate from the inner sphere outwards. (b) A third spherical surface is inserted between the two sphere surfaces, so that there are now three concentric spheres. What is the heat flow rate radiated from the innermost sphere surface outwards?

Solution: According to Equation (9.40),

(a) $\Phi_{th,a} = \varepsilon_{thr}\sigma_{SB}S\left(T_i^4 - T_o^4\right)$, where $S = \dfrac{4}{3}\pi R^3$ and R is the average of the radii.

(b) Let the temperature of the middle sphere surface be T_m. The radiated power is

$$\Phi_{th,b} = \varepsilon_{thr}\sigma_{SB}S\left(T_i^4 - T_m^4\right) = \varepsilon_{thr}\sigma_{SB}S\left(T_m^4 - T_o^4\right),$$

from which we can solve

$$T_m^4 = \frac{1}{2}\left(T_i^4 + T_o^4\right).$$

Substituting T_m^4 into the equation for $\Phi_{th,b}$, we get

$$\Phi_{th,b} = \varepsilon_{thr}\sigma_{SB}S\left(\frac{1}{2}\left(T_0^4 + T_i^4\right) - T_0^4\right) = \frac{1}{2}\varepsilon_{thr}\sigma_{SB}S\left(T_i^4 - T_o^4\right) = \frac{1}{2}\Phi_{th,a}.$$

We can see that the third sphere halves the heat flow rate from the inner sphere outwards; that is, the thermal resistance between the inner and outer spheres is in case (b) two times the resistance in case (a).

9.2.3 *Convection*

Heat is always transferred simultaneously by conduction and convection. Convection is defined as the heat transfer between a region of higher temperature (here, a solid surface) and a region

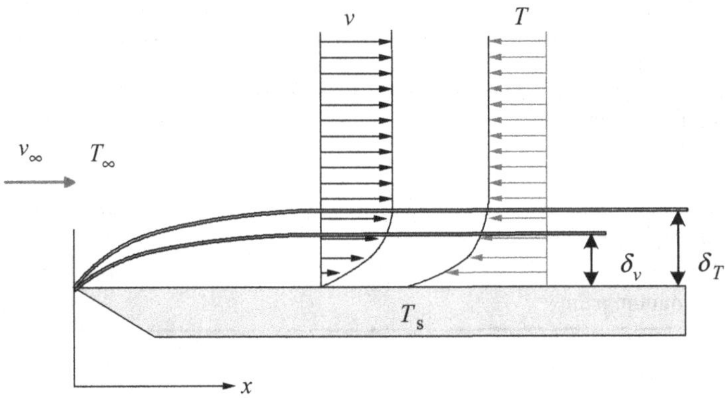

Figure 9.8 Development of the velocity and thermal boundary layers; $\delta_v(x)$ and $\delta_T(x)$ are the thicknesses of the layers.

of cooler temperature (a coolant) that takes place as a consequence of motion of the cooling fluid relative to the solid surface. (At the molecular level, this means that the warmer molecules displace the cooler fluid molecules.)

Knowledge of the boundary layers is essential in the analysis of the heat and mass transfer between the solid surface and the coolant flowing by the surface. In convection heat transfer, there are three boundary layers, defined as the velocity, thermal and concentration boundary layer.

Let us consider the case represented in Figure 9.8. The stream of air meets a plane. The velocity of the stream is zero on the surface of the plane, and inside the boundary layer the speed increases to the speed in free space. The thickness of the velocity boundary layer δ_v is defined as the height from the surface where the speed of the stream is 0.99 times the speed in free space. Above this limit, the shear stresses and velocity gradient are negligible.

The temperature of the plane T_s is assumed to be higher than the temperature of the stream of air. Close to the surface, the heat transfers by conduction through the thermal boundary layer. The temperature profile is similar to the velocity profile. The thickness of the thermal boundary layer δ_T is defined as the height from the surface where the ratio of the difference of the surface and boundary layer temperatures, T_s and T, respectively, to the difference between the surface and the ambient temperature T_∞ is 0.99.

The concentration boundary layer occurs when a binary mixture of species flows across a surface (convection mass transfer, e.g. water vapor in an air stream). The thickness of the concentration boundary layer δ_c is defined at the height where the ratio of the difference of the surface and boundary layer molar concentrations to the difference between the surface and ambient molar concentrations is 0.99. The form of the concentration boundary layer is similar to the form of the velocity and thermal boundary layers (evaporation at a liquid surface and sublimation at a solid surface, respectively).

The three important expressions in the boundary layer theorems are surface friction, convection heat transfer and convection mass transfer, which are crystallized into three important parameters: friction coefficient C_f and the convection heat and mass transfer coefficients α and α_m.

To facilitate the calculation process and minimize the number of parameters to be solved, certain dimensionless parameters have been generated. Of the numerous such parameters available in the literature, the three most important ones when calculating heat transfer from solid surfaces to the coolant are the Nusselt number *Nu*, the Reynolds number *Re* and the Prandtl number *Pr*.

Convection heat transfer coefficient α can be expressed with the dimensionless Nusselt number *Nu* as

$$Nu = \frac{\alpha L}{\lambda}, \tag{9.45}$$

where L is the characteristic surface length and λ is the thermal conductivity of the coolant. The Nusselt number describes the effectiveness of convection heat transfer compared with conduction heat transfer.

The ratio between inertia and viscous forces is described by the Reynolds number *Re* and can be described by the equation

$$Re = \frac{vL}{\upsilon}, \tag{9.46}$$

where v is the speed of the coolant on the surface, L is the characteristic length of the surface and υ is the kinematic viscosity of the coolant. The value of the Reynolds number at which the flow becomes turbulent is called the critical Reynolds number Re_{crit}. For flat surfaces Re_{crit} is 5×10^5 and for a tube flow it is 2300. For tubes, the characteristic length is described by the equation

$$L = \frac{4S}{l_{\text{p}}}, \tag{9.47}$$

where S is the area of the tube cross-section and l_{p} is the wetted perimeter of the tube.

The third dimensionless number is the Prandtl number, which describes the relation between momentum and thermal diffusivity. In other words, it describes the thickness ratio of velocity and the thermal boundary layers. The Prandtl number is described by the equation

$$Pr = \frac{c_{\text{p}}\mu}{\lambda}, \tag{9.48}$$

where c_{p} is the specific heat capacity, μ the dynamic viscosity and λ the thermal conductivity of the coolant. When *Pr* is low (<1), the heat transfer of thermal diffusivity is large compared with the heat transfer rate gained with the fluid speed, and when *Pr* is unity, it means that the thermal and velocity boundary layers are equal. The *Pr* numbers for gases and air are between 0.7 and 1, and for water between 1 and 13, depending on the temperature and pressure.

For gases, the velocity and thermal boundary layers are of the same order of magnitude. Between the velocity and thermal boundary layers, the following equation is valid:

$$\delta_{\text{T}} = \delta_{\text{v}}\, Pr^{1/3} \tag{9.49}$$

Because *Pr* is close to one for gases, $\delta_{\text{T}} \approx \delta_{\text{v}}$.

Fanning friction coefficient C_{f} is useful in internal flow calculations when defining pump or fan power requirements, but it is valid for fully developed laminar flow only. When calculating a fully developed turbulent flow (which is usually the case), it is preferable to use empirical results for surface friction. Friction factors for a wide range of Reynolds numbers are found for example in the Moody diagram (Moody 1944), which is valid for circular tubes.

The pressure drop Δ_p in a tube can be calculated from the equation

$$\Delta p = f\frac{\rho u_{\mathrm{m}}}{2D}L, \tag{9.50}$$

where f is the dimensionless Moody friction factor, ρ is the density of fluid, L is the tube length, D is the tube diameter and u_{m} is the mean fluid velocity in the tube, which can be calculated from the required mass flow rate.

From the pressure drop, the power P required to sustain an internal flow is obtained from the equation

$$P = \Delta p\frac{q_{\mathrm{m}}}{\rho}, \tag{9.51}$$

where q_{m} is the mass flow rate in the tube.

As mentioned above, the convection heat transfer always takes place in the direction of a lower temperature. Convection can be divided into forced and natural convection. In forced convection, external instruments such as pumps or blowers assist the flow of coolant. Natural convection is caused by density variations resulting from temperature differences: as the coolant is heated, it changes its density, and the local changes in density in the coolant–solid interface result in buoyancy forces that cause currents in the coolant. Newton's law of cooling defines the heat flow density q_{th} created by convection

$$q_{\mathrm{th}} = \frac{\Phi_{\mathrm{th}}}{S} = \alpha_{\mathrm{th}}\Theta \tag{9.52}$$

and hence the thermal resistance of convection is

$$R_{\mathrm{th}} = \frac{\Theta}{\Phi_{\mathrm{th}}} = \frac{1}{\alpha_{\mathrm{th}}S}. \tag{9.53}$$

Here α_{th} is the heat transfer coefficient. The value of the heat transfer coefficient depends for instance on the viscosity of the coolant, the thermal conductivity, specific heat capacity and flow velocity of the medium. Traditionally, the heat transfer coefficient has been defined by various empirical correlations. In the calculation of an electrical machine, for natural convection in the air around a horizontally mounted, unfinned cylindrical motor of diameter D [m] with ambient temperature close to room temperature, Miller (1993) employs a correlation depending on the temperature difference Θ [K] between the cylinder and the surroundings

$$\alpha_{\mathrm{th}} \approx 1.32\left(\frac{\Theta}{D}\right)^{0.25}\left[\frac{\mathrm{W}}{\mathrm{m^2K}}\right]. \tag{9.54}$$

Example 9.7: What is the heat transfer coefficient and heat flow density between a cylinder and the ambient, if the temperature of the cylinder is 50 °C above the ambient and the diameter is 0.1 m?

Solution: For the heat transfer coefficient of natural convection, Equation (9.54) gives the value

$$\alpha_{\mathrm{th}} \approx 1.32\left(\frac{\Theta}{D}\right)^{0.25}\left[\frac{\mathrm{W}}{\mathrm{m^2K}}\right] = 1.32\left(\frac{50}{0.1}\right)^{0.25}\frac{\mathrm{W}}{\mathrm{m^2K}} = 6.24\frac{\mathrm{W}}{\mathrm{m^2K}}.$$

Natural convection is thus of the same order as heat transfer by radiation. The density of the heat flow is

$$q_{\mathrm{th}} = \alpha_{\mathrm{th}}\Theta = 6.24 \cdot 50\ \mathrm{W/m^2} = 312\ \mathrm{W/m^2}.$$

Forced convection increases the convective heat transfer coefficient even 5-6-fold depending on the air velocity. The increase in heat transfer coefficient is approximately proportional to the square root of the air velocity v (Miller 1993)

$$a_{\mathrm{th}} \approx 3.89\sqrt{\frac{v}{l}}\left[\frac{\mathrm{W}}{\mathrm{m^2K}}\right]. \tag{9.55}$$

Here l is the length of the frame of the machine in metres and the speed v is given in m/s.

Example 9.8: What is the heat transfer coefficient on the surface of a cylinder of 0.1 m length if the speed of coolant is 4 m/s?

Solution: According to Equation (9.55),

$$\alpha_{\mathrm{th}} \approx 3.89\sqrt{\frac{v}{l}}\left[\frac{\mathrm{W}}{\mathrm{m^2K}}\right] = 3.89\sqrt{\frac{4}{0.1}}\frac{\mathrm{W}}{\mathrm{m^2K}} = 24.6\frac{\mathrm{W}}{\mathrm{m^2K}}.$$

This coefficient is about fourfold when compared with the previous value of natural convection.

For a typical radial flux electrical machine, there are three significant convection coefficients related to the frame, air gap and coil ends. The first can be approximated for example by Miller's equations, but the other two are more complex cases, the coil ends in particular.

The convection coefficient for an annulus depends on the air-gap length, rotation speed of the rotor, the length of the rotor and the kinematic viscosity of the streaming fluid. The Taylor equation can be used to determine the flow type and the convection heat transfer coefficient in the annulus. The validity of the Taylor equation is, however, somewhat restricted, and the annular flow in the tangential direction is usually referred to as Taylor–Couette flow or Taylor vortex flow. It differs from the flow between two parallel plates (one moving) by the toroidal vortices that appear as a result of tangential forces. These eddies influence the heat transfer characteristics of the air gap. The Taylor vortices are described by the Taylor number *Ta*, which describes the ratio of viscous forces to the centrifugal forces

$$Ta = \frac{\rho^2 \Omega^2 r_m \delta^3}{\mu^2}, \tag{9.56}$$

where Ω is the angular velocity of the rotor, ρ the mass density of the fluid, μ the dynamic viscosity of the fluid and r_m the average of the stator and rotor air-gap radii. The radial air-gap length δ and the rotor radius are taken into account by a modified Taylor number

$$Ta_m = \frac{Ta}{F_g}, \tag{9.57}$$

where F_g is the geometrical factor defined by

$$F_g = \frac{\pi^4 \left[\dfrac{2r_m - 2.304\delta}{2r_m - \delta}\right]}{1697\left[0.0056 + 0.0571\left(\dfrac{2r_m - 2.304\delta}{2r_m - \delta}\right)^2\right]\left[1 - \dfrac{\delta}{2r_m}\right]^2} \tag{9.58}$$

In practice, the air-gap length is so small compared with the rotor radius that F_g is close to unity and $Ta_m \approx Ta$.

According to Becker and Kaye (1962), the Nusselt number is

$$\begin{aligned} Nu &= 2 && \text{for} \quad Ta_m < 1700\,(\text{laminar flow}) \\ Nu &= 0.128 Ta_m^{0.367} && \text{for} \quad 1700 < Ta_m < 10^4 \\ Nu &= 0.409 Ta_m^{0.241} && \text{for} \quad 10^4 < Ta_m < 10^7 \end{aligned} \tag{9.59}$$

which can be used to determine the heat transfer coefficient from the rotor to the air gap and from the stator to the air gap (one air-gap surface) by the equation

$$\alpha_{th} = \frac{Nu\lambda}{\delta}, \tag{9.60}$$

where λ is the thermal conductivity of air.

The roughness of the surfaces affects the heat transfer in two ways. It enlarges the cooling area and increases turbulence. Gardiner and Sabersky (1978) and Rao and Sastri (1984) have studied the effect of roughness of the stator and rotor air-gap surfaces. According to these two studies, a rough rotor has a heat transfer coefficient 40–70 % higher than a smooth rotor.

The convection heat transfer coefficient in the coil ends is the most difficult to approximate because the flow field is too complex to model. The cooling method of the electrical machine also affects the convection heat transfer coefficient of the coil ends, as does the type of winding.

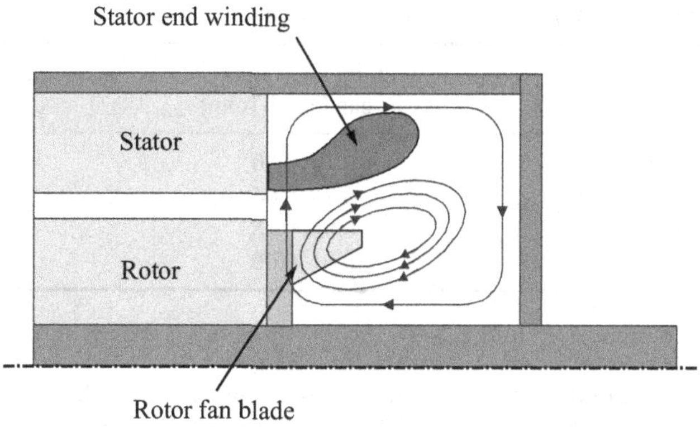

Figure 9.9 End winding space of a squirrel cage induction motor and the air circulation in the space.

The end winding space of a squirrel cage induction motor is illustrated in Figure 9.9. The end winding space can be divided into two parts: the space between the end winding and the rotor and the space between the end winding and the frame. In the space between the rotor and the end winding, the rotation speed of the rotor determines the heat transfer coefficient. The geometry corresponds to that of the air gap, and Equations (9.56)–(9.60) can be used to calculate the heat transfer coefficient on the surface of the stator winding.

In the space between the end winding and the frame, the speed of the air stream is far smaller than in the space between the rotor and the end winding. The flow may be assumed to be laminar. This means that natural convection occurs in that space, and also radiation has to be taken into account.

Usually, the heat transfer between two interconnected objects is described by the junction thermal resistance. The heat transfer across the gap between the joint surfaces depends on the finish of the surfaces. The heat transfer coefficient α_{th} between two metal surfaces with a roughness of 30 μm is approximately 1100 W/m^2 K, a value corresponding to a thermal resistance of 9.1 K/W in an area of 1 cm^2. In the case of ground surfaces (a roughness of 1 μm), the thermal transmittance is doubled and thus the thermal resistance can be halved. The same result can be reached by adding heat transfer lubricant to the joint of the rough surfaces to fill in the voids.

The equivalent thermal resistance of joint surfaces can also be modeled by equivalent air-gap conduction. Imagine that between the surfaces there is a small air gap because of surface roughness. The resistance of the equivalent air gap between the surfaces can then be calculated by using Equation (9.35). The equivalent air-gap length and contact heat transfer coefficients are presented in Table 9.5.

In electrical machines, the thermal resistances of joint surfaces are the most significant insecurity factor. The most important thermal resistances of joint surfaces in this respect are those between the conductors and the slot insulation, the slot insulation and the stator or rotor stack, and the stator stack and the stator frame. Determination of these thermal resistances may prove difficult without measurements. However, these contact thermal resistances determine chiefly the heat transfer of the machine and therefore, in order to be able to calculate the heat transfer of the machine, certain empirical values of contact thermal resistances have to be known.

Table 9.5 Equivalent joint air-gap lengths and contact heat transfer coefficients

Joint type	Joint equivalent air-gap length [mm]	Contact heat transfer coefficient [W/m^2 K]
Stator winding to stator core	0.10–0.30	80–250
Frame (aluminum) to stator core	0.03–0.04	650–870
Frame (cast iron) to stator core	0.05–0.08	350–550
Rotor bar to rotor core	0.01–0.06	430–2600

If rated torque is required at a low rotation speed, a shaft-mounted blower may not be able to produce sufficient forced convection on the motor outer surface to cool the motor. In DC motors and speed-controlled AC motors, an additional cooling fan is employed, because these motors are often operated for a long period with a high torque at a low rotation speed. Since in DC machines most of the heat is generated in the rotor, a good internal cooling flow is required.

9.3 Thermal Equivalent Circuit

9.3.1 Analogy between Electrical and Thermal Quantities

The heat-removal calculation only tells us whether enough heat can be removed from the machine surface in the stationary state. Further, the calculation guides the selection of the cooling method. However, when considering the operation of the machine, it is essential to define the internal temperature distribution, which is regulated by the flux densities, frequencies and current densities in the machine parts.

To determine the internal temperature distribution, a thermal equivalent circuit is used for modeling the heat transfer of the machine. The heat flow is analogous to electric current flow, as we saw in the previous chapter. Table 9.6 shows the analogous thermal and electrical quantities. Resistive losses, iron losses, windage losses and friction losses are represented by

Table 9.6 Analogous thermal and electrical quantities

Thermal flow	Symbol	Unit	Electric flow	Symbol	Unit
Quantity of heat	Q_{th}	J	Electric charge	Q	C
Heat flow rate	Φ_{th}	W	Electric current	I	A
Heat flow density	q_{th}	W/m^2	Current density	J	A/m^2
Temperature	T	K	Electric potential	V	V
Temperature rise	Θ	K	Voltage	U	V
Thermal conductivity	λ	W/m K	Electric conductivity	σ	S/m = A/V m
Thermal resistance	R_{th}	K/W	Electric resistance	R	Ω = V/A
Thermal conductance	G_{th}	W/K	Electric conductance	G	S = A/V
Heat capacity	C_{th}	J/K	Capacitance	C	F = C/V

individual heat flow sources. The thermal resistances of iron cores, insulation, the frame and so on are given as resistances.

The heat capacity is analogous to the electric capacitance, as shown below.

The electric charge Q stored in a capacitor is

$$Q = C\Delta V = CU, \tag{9.61}$$

where C is the capacitance of the capacitor and U the voltage over the capacitor.

The quantity of heat Q_{th} stored in a body is

$$Q_{\text{th}} = mc_{\text{p}}\Theta, \tag{9.62}$$

where m is the mass of the body, c_{p} the specific heat capacity and Θ the temperature rise caused by the heat Q_{th}.

Comparing Equation (9.61) with Equation (9.62), we find that the heat capacity

$$C_{\text{th}} = mc_{\text{p}} \tag{9.63}$$

is analogous to the capacitance C.

9.3.2 *Average Thermal Conductivity of a Winding*

The windings of electrical equipment are thermally inhomogeneous; in the winding, heat flows through a conductor with very good heat-conducting properties, but also through a poorly heat-conducting insulation. Usually, we need to know the average temperature of the winding. In this case, the winding may be replaced with a homogeneous material having the same thermal resistance as the real inhomogeneous winding.

As an example, let us consider the winding represented in Figure 9.10a. Heat is assumed to flow only in the x-direction. As the thermal conductivity of copper is about a thousand times as high as the thermal conductivity of the insulation, we may assume that the thermal resistance of copper is zero. Our task is to determine the resistance (r_{res}) of the two parallel-connected insulation pieces A and B (Figure 9.10b). The resistance of A per unit length is

$$r_{\text{A}} = \frac{b'}{\lambda_{\text{i}}\delta_{\text{i}}}, \tag{9.64}$$

where λ_{i} is the thermal conductivity of the insulation. The resistance of B per unit length is

$$r_{\text{B}} = \frac{\delta_{\text{i}}}{\lambda_{\text{i}}h}. \tag{9.65}$$

The parallel connection yields

$$r_{\text{res}} = \frac{r_{\text{A}}r_{\text{B}}}{r_{\text{A}} + r_{\text{B}}} = \frac{b'}{\lambda_{\text{i}}h\left(\dfrac{b'}{\delta_{\text{i}}} + \dfrac{\delta_{\text{i}}}{h}\right)} = \frac{b'}{\lambda_{\text{i}}\left(\dfrac{b'h}{\delta_{\text{i}}} + \delta_{\text{i}}\right)}. \tag{9.66}$$

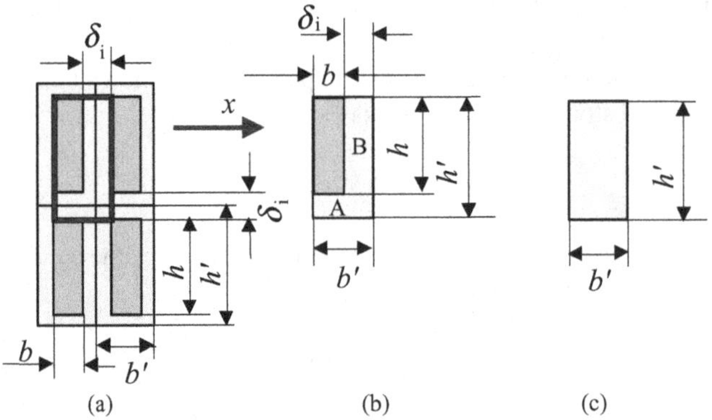

Figure 9.10 Winding made of rectangular conductors (figures a and b) is replaced with a homogeneous material (figure c) having the same outer dimensions (b' and h') as the real conductor with insulation and the same thermal resistance in the x-direction as the real inhomogeneous conductor consisting of copper and insulation.

The resultant resistance r_{res} should be equal to the resistance of the homogeneous body having the thickness b', width h′ and thermal conductivity λ_{av} (Figure 9.10c):

$$r_{res} = \frac{b'}{\lambda_{av} h'} = \frac{b'}{\lambda_i \left(\frac{b'h}{\delta_i} + \delta_i \right)}, \tag{9.67}$$

from which we obtain for the average thermal conductivity

$$\lambda_{av} = \lambda_i \left(\frac{b'h}{h'\delta_i} + \frac{\delta_i}{h'} \right). \tag{9.68}$$

With the same procedure, we may determine the average thermal conductivity of the winding shown in Figure 9.11a where extra insulation is placed between the winding layers:

$$\lambda_{av} = \lambda_i \left(\frac{h\left(b' + \delta_a\right)}{h'\left(\delta_i + \delta_a\right)} + \frac{\delta_i}{h'} \right). \tag{9.69}$$

If we have a round wire winding (Figure 9.11b), and the spaces between the wires are filled with impregnation resin, the average thermal conductivity is

$$\lambda_{av} \approx \lambda_i \left(\frac{d}{\delta_i} + \frac{\delta_i}{d'} \right). \tag{9.70}$$

9.3.3 *Thermal Equivalent Circuit of an Electrical Machine*

A thermal equivalent circuit of a typical electrical machine is shown in Figure 9.12. For simplicity, it is assumed that the heat flows in the stacks only in the radial direction because

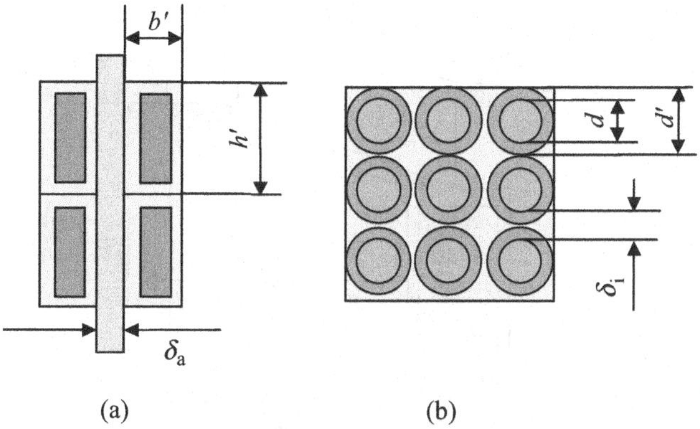

Figure 9.11 (a) Winding of rectangular wires and with extra insulation between the layers. (b) An impregnated round wire winding.

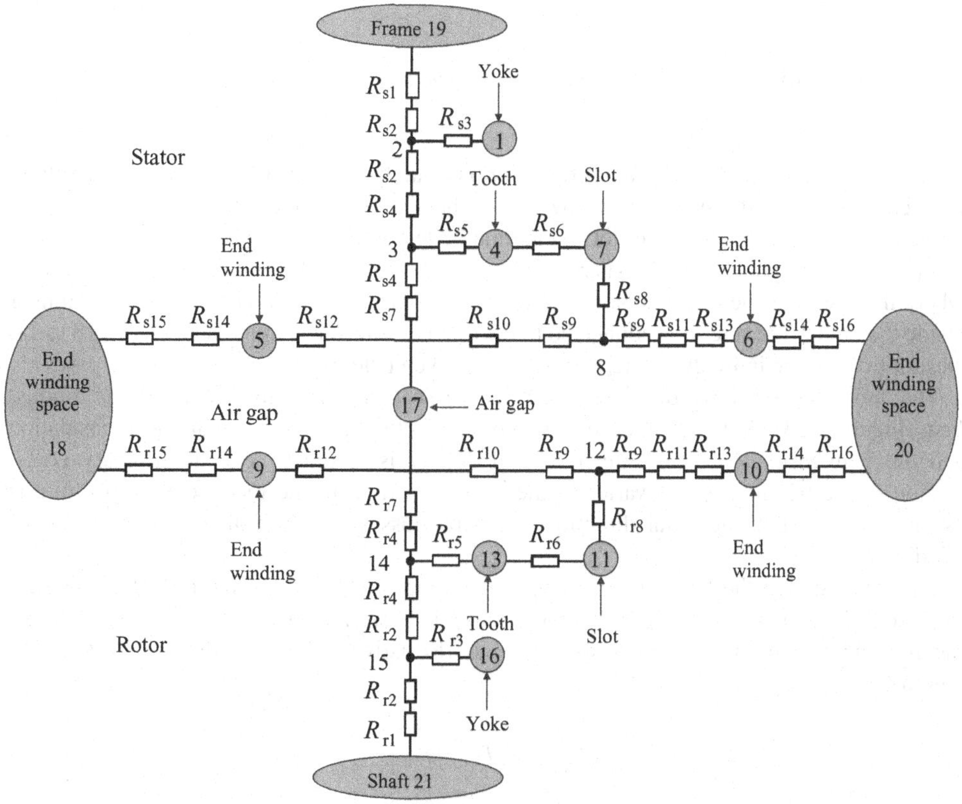

Figure 9.12 Thermal equivalent circuit of a typical electrical machine.

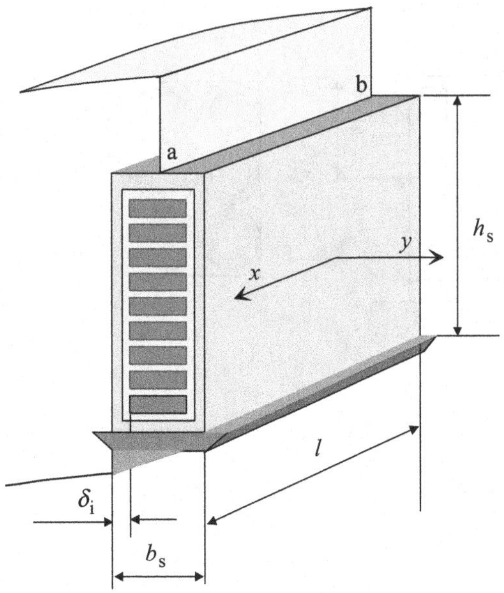

Figure 9.13 Cut-away drawing of a winding with rectangular conductors in a slot.

the thermal conductivity in the axial direction is notably lower than in the radial direction. Further, it is assumed that the heat flows from the slots to the teeth but not directly to the yoke. This is reasonable because slots are normally deep and narrow, whereupon the heat flow from a slot to the yoke is small. There are 21 nodes in total in the equivalent circuit. Nodes, where the losses of the machine are supplied, are indicated by a circle and the node number inside the circle. The circuit is connected to the cooling flow through the nodes from 18 to 21. Descriptions of the individual parameters will be given below.

The modeling of a winding with rectangular-shaped conductors in a slot is studied first (Figure 9.13). Between the copper and the tooth, there is the main wall insulation and the wire insulation. The temperature of copper is assumed constant in every cross-section of the winding, but it varies in the axial direction. In the tooth, the temperature is assumed constant in the axial direction. Resistive losses are distributed uniformly in the winding.

The winding and the tooth are presented in a simplified form in Figure 9.14a. The thermal equivalent circuit with distributed constants is presented in Figure 9.14b. The winding is divided into small sections of length dx. Resistive losses P_{Cu} divided by the core length l is

$$p = P_{Cu}/l. \tag{9.71}$$

The thermal resistance R between the points a and b divided by the core length is

$$r = R/l = 1/(\lambda_{Cu} S_{Cu}), \tag{9.72}$$

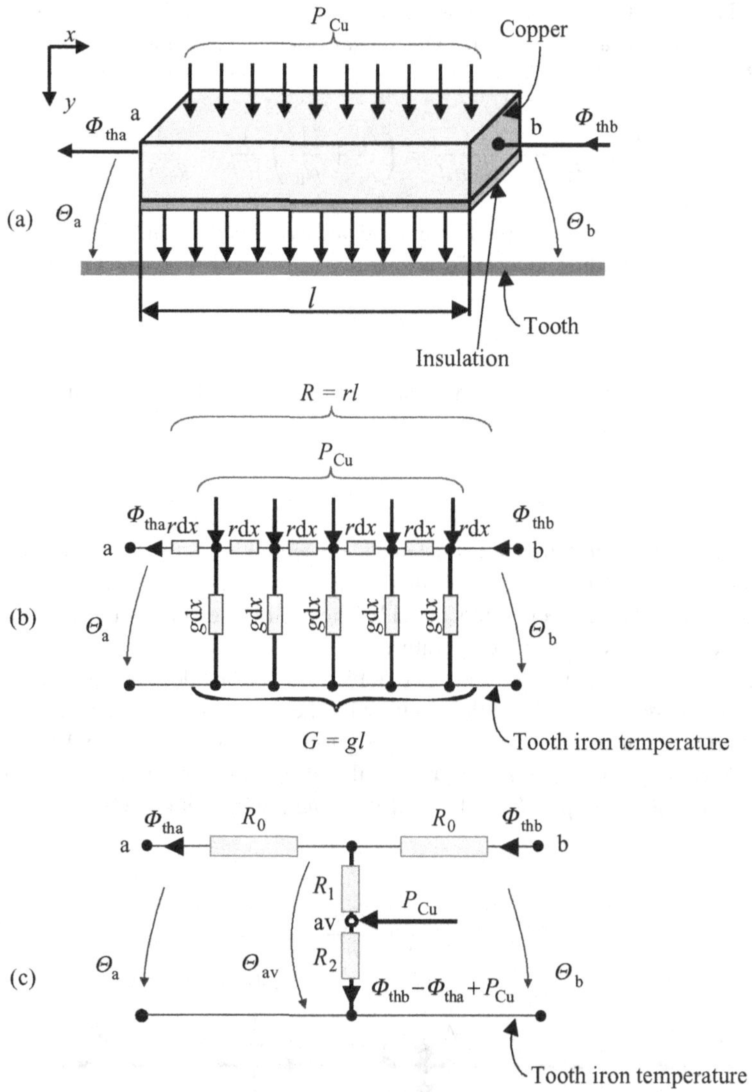

Figure 9.14 (a) Simplified presentation of Figure 9.13; (b) its thermal equivalent circuit presented with distributed constants; and (c) the circuit presented with lumped constants.

where λ_{Cu} is the thermal conductivity and S_{Cu} the total cross-sectional area of the conductors in a slot.

The thermal conductance G (the inverse of the thermal resistance) from the conductors to the tooth per unit core length is

$$g = G/l. \tag{9.73}$$

The conductance G includes the insulation resistance and the contact resistance between the insulation and the tooth. The inverse of G, using the symbols presented in Figure 9.13, is

$$\frac{1}{G} = \frac{\delta_i}{\lambda_i h_s l} + \frac{1}{\alpha_{th} h_s l} = \left(\frac{\delta_i}{\lambda_i} + \frac{1}{\alpha_{th}}\right)\frac{1}{h_s l} = \frac{1}{k_{th} h_s l}, \tag{9.74}$$

where λ_i is the thermal conductivity of insulation and α_{th} the heat transfer coefficient between the insulation and the tooth. The term k_{th}

$$k_{th} = 1/\left(\delta_i/\lambda_i + 1/\alpha_{th}\right) \tag{9.75}$$

is called the overall heat transfer coefficient. Now, we obtain for the conductance per unit length

$$g = k_{th} h_s. \tag{9.76}$$

The equivalent circuit with lumped constants is presented in Figure 9.14c. The circuit gives the temperature rises of points a and b and also the average temperature rise Θ_{av} of the winding part located in the slots, between points a and b. The total losses P_{Cu} in the slots are supplied to the node presenting the average temperature.

Next, the components of the equivalent circuit Figure 9.14c are determined. First, we have to solve the temperature distribution in the circuit with the distributed constants in Figure 9.14b, of which a part at distance x from point a (the origin) is presented in Figure 9.15.

In Figure 9.15, at point A at distance x from the origin, the temperature rise is Θ and the x-direction heat flow in copper Φ_{th}. The heat generated at A is p dx. The heat flow through

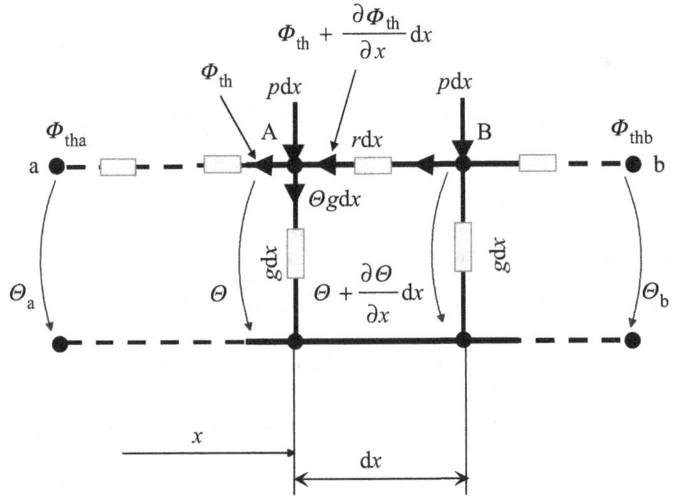

Figure 9.15 Deriving the differential equation for the heat flow in a winding.

the insulation is $\Theta g\mathrm{d}x$. At point B at distance $\mathrm{d}x$ from A, the temperature rise and heat flow are

$$\Theta + \frac{\partial \Theta_{\mathrm{th}}}{\partial x}\mathrm{d}x \quad \text{and} \quad \Phi_{\mathrm{th}} + \frac{\partial \Phi_{\mathrm{th}}}{\partial x}\mathrm{d}x$$

as indicated in Figure 9.15. Now we apply the following rule to node A: the sum of incoming heat flows in a node is equal to the sum of leaving heat flows. Thus, we obtain

$$\Phi_{\mathrm{th}} + g\Theta\mathrm{d}x = p\mathrm{d}x + \Phi_{\mathrm{th}} + \frac{\partial \Phi_{\mathrm{th}}}{\partial x}\mathrm{d}x,$$

and after reduction

$$g\Theta = p + \frac{\partial \Phi_{\mathrm{th}}}{\partial \mathrm{x}}. \tag{9.77}$$

The temperature rise of node B

$$\Theta + \frac{\partial \Theta_{\mathrm{th}}}{\partial x}\mathrm{d}x$$

is equal to the temperature difference between nodes B and A plus the temperature rise of node A, that is

$$\Theta + \frac{\partial \Theta}{\partial x}\mathrm{d}x = \Phi_{\mathrm{th}} r\mathrm{d}x + \Theta,$$

and after reduction

$$\frac{\partial \Theta}{\partial x} = \Phi_{\mathrm{th}} r. \tag{9.78}$$

By differentiating Equation (9.77) with respect to x we obtain

$$\frac{\partial^2 \Phi_{\mathrm{th}}}{\partial x^2} = g\frac{\partial \Theta}{\partial x},$$

and substituting Equation (9.78) for $\partial\Theta/\partial x$

$$\frac{\partial^2 \Phi_{\mathrm{th}}}{\partial x^2} - rg\Phi_{\mathrm{th}} = 0. \tag{9.79}$$

Equation (9.79) has a solution of the form

$$\Phi_{\mathrm{th}} = \mathrm{C}_1\mathrm{e}^{\sqrt{rg}x} + \mathrm{C}_2\mathrm{e}^{-\sqrt{rg}x}, \tag{9.80}$$

where C_1 and C_2 are integration constants. According to Equation (9.77), the temperature is

$$\Theta = \frac{1}{g}\left(C_1\sqrt{rg}e^{\sqrt{rg}x} - C_2\sqrt{rg}e^{-\sqrt{rg}x} + p\right). \tag{9.81}$$

The integration constants are determined in accordance with two boundary conditions

$$x = 0,$$
$$\Theta = \Theta_a,$$

and

$$x = 0,$$
$$\Phi_{th} = \Phi_{tha}.$$

Substituting the boundary conditions into Equations (9.80) and (9.81), we obtain

$$C_1 = \frac{1}{2}\left(\Phi_{tha} + \sqrt{\frac{g}{r}}\Theta_a - \frac{p}{\sqrt{rg}}\right), \tag{9.82}$$

$$C_2 = \frac{1}{2}\left(\Phi_{tha} - \sqrt{\frac{g}{r}}\Theta_a + \frac{p}{\sqrt{rg}}\right). \tag{9.83}$$

Substituting these for C_1 and C_2 in Equations (9.80) and (9.81) and simplifying yields

$$\Phi_{th} = \Phi_{tha}\cosh\left(\sqrt{rg}x\right) + \Theta_a\sqrt{\frac{g}{r}}\sinh\left(\sqrt{rg}x\right) - \frac{p}{\sqrt{rg}}\sinh\left(\sqrt{rg}x\right), \tag{9.84}$$

$$\Theta = \Phi_{tha}\sqrt{\frac{r}{g}}\sinh\left(\sqrt{rg}x\right) + \Theta_a\cosh\left(\sqrt{rg}x\right) + \frac{p}{g}\left[1 - \cosh\left(\sqrt{rg}x\right)\right]. \tag{9.85}$$

The average temperature is

$$\Theta_{av} = \frac{1}{l}\int_0^l \Theta dx = \frac{1}{l}\left\{\frac{1}{g}[\cosh(\sqrt{rg}l) - 1]\Phi_{tha}\frac{\sinh(\sqrt{rg}l)}{\sqrt{rg}}\Theta_a + \frac{p}{g}\left[1 - \frac{\sinh(\sqrt{rg}l)}{\sqrt{rg}}\right]\right\}. \tag{9.86}$$

Taking into account the definitions of p, r and g, that is Equations (9.71), (9.72) and (9.73), we obtain

$$\Theta_{av} = \frac{1}{G}\left(\cosh\sqrt{RG} - 1\right)\Phi_{tha} + \frac{\sinh\sqrt{RG}}{\sqrt{RG}}\Theta_a + \frac{P_{Cu}}{G}\left(1 - \frac{\sinh\sqrt{RG}}{\sqrt{RG}}\right). \tag{9.87}$$

Now, we can determine the components of the equivalent circuit with lumped constants by calculating the temperature rise and heat flow at point b (Figure 9.14c) from Equations (9.84) and (9.85) and from the equivalent circuit Figure 9.14c. The results have to be identically equal. Equations (9.84) and (9.85) yield with $x = l$

$$\Phi_{\text{thb}} = \Phi_{\text{tha}} \cosh\left(\sqrt{RG}\right) + \Theta_{\text{a}}\sqrt{\frac{G}{R}} \sinh\left(\sqrt{RG}\right) - \frac{P_{\text{Cu}}}{\sqrt{RG}} \sinh\left(\sqrt{RG}\right), \quad (9.88)$$

$$\Theta_{\text{b}} = \Phi_{\text{tha}}\sqrt{\frac{R}{G}} \sinh\sqrt{RG} + \Theta_{\text{a}} \cosh\sqrt{RG} + \frac{P_{\text{Cu}}}{G}\left[1 - \cosh\left(\sqrt{RG}\right)\right]. \quad (9.89)$$

From the equivalent circuit of Figure 9.14c we obtain

$$\Theta_{\text{a}} = -\Phi_{\text{tha}} R_0 + (\Phi_{\text{thb}} - \Phi_{\text{tha}})(R_1 + R_2) + P_{\text{Cu}} R_2, \quad (9.90)$$

$$\Theta_{\text{b}} = \Phi_{\text{thb}} R_0 + (\Phi_{\text{thb}} - \Phi_{\text{tha}})(R_1 + R_2) + P_{\text{Cu}} R_2. \quad (9.91)$$

From Equations (9.90) and (9.91) we can solve

$$\Phi_{\text{thb}} = \Phi_{\text{tha}}\left(1 + \frac{R_0}{R_1 + R_2}\right) + \Theta_{\text{a}}\frac{1}{R_1 + R_2} - P_{\text{Cu}}\frac{R_2}{R_1 + R_2}, \quad (9.92)$$

$$\Theta_{\text{b}} = \Phi_{\text{tha}} R_0\left(2 + \frac{R_0}{R_1 + R_2}\right) + \Theta_{\text{a}}\left(1 + \frac{R_0}{R_1 + R_2}\right) - P_{\text{Cu}}\frac{R_0 R_2}{R_1 + R_2}. \quad (9.93)$$

Equation (9.88) yields the same result for Φ_{thb} as Equation (9.92) if the coefficients of Φ_{tha} are equal and the coefficients of Θ_{a} are equal. In addition, the coefficients of P_{Cu} have to be equal. The same thing applies for Equations (9.89) and (9.93). These conditions yield

$$1 + \frac{R_0}{R_1 + R_2} = \cosh\sqrt{RG}, \quad (9.94)$$

$$R_0\left(2 + \frac{R_0}{R_1 + R_2}\right) = \sqrt{\frac{R}{G}} \sinh\sqrt{RG}, \quad (9.95)$$

$$\frac{1}{R_1 + R_2} = \sqrt{\frac{G}{R}} \sinh\sqrt{RG}, \quad (9.96)$$

$$\frac{R_2}{R_1 + R_2} = \frac{1}{\sqrt{RG}} \sinh\sqrt{RG}. \quad (9.97)$$

Dividing (9.97) by (9.96) yields

$$R_2 = \frac{1}{G}. \quad (9.98)$$

Substituting (9.98) for R_2 in (9.97) yields

$$R_1 = \frac{1}{G}\left(\frac{\sqrt{RG}}{\sinh\sqrt{RG}} - 1\right). \quad (9.99)$$

The value of R_1 is negative. Now, Equations (9.94) and (9.95) yield

$$R_0 = \sqrt{\frac{R}{G}}\frac{\sinh\sqrt{RG}}{\cosh\sqrt{RG}+1} = \sqrt{\frac{R}{G}}\tanh\frac{\sqrt{RG}}{2}. \tag{9.100}$$

Finally, we have to make sure that we get the average temperature of Equation (9.87) also from the equivalent circuit with the resistances (9.98), (9.99) and (9.100):

$$\Theta_{\mathrm{av}} = R_2\left(\Phi_{\mathrm{thb}} - \Phi_{\mathrm{tha}}\right) + R_2 P_{\mathrm{Cu}}.$$

By substituting Φ_{thb} from Equation (9.92) with R_0, R_1 and R_2, we obtain after simplification

$$\begin{aligned}\Theta_{\mathrm{av}} &= R_0\frac{R_2}{R_1+R_2}\Phi_{\mathrm{tha}} + \frac{R_2}{R_1+R_2}\Theta_{\mathrm{a}} + R_1\frac{R_2}{R_1+R_2}P_{\mathrm{Cu}} \\ &= \frac{1}{G}\left(\cosh\sqrt{RG}-1\right)\Phi_{\mathrm{tha}} + \frac{\sinh\sqrt{RG}}{\sqrt{RG}}\Theta_{\mathrm{a}} + \frac{P_{\mathrm{Cu}}}{G}\left(1-\frac{\sinh\sqrt{RG}}{\sqrt{RG}}\right),\end{aligned}$$

which is the same as Equation (9.87), that is the analytical solution.

In the thermal circuit of Figure 9.12, the part of the stator winding that is located in the slots is represented by the resistances R_{s6}, R_{s8} and R_{s9} and node 7. The node represents the average temperature rise, and the resistive losses in the slots are supplied to this node. The resistance R_{s6} is calculated from Equation (9.98), R_{s8} from Equation (9.99) and R_{s9} from Equation (9.100), where R is the thermal resistance of the conductors in the slots and G the conductance between the conductors and teeth.

The end winding can also be represented by the equivalent circuit of Figure 9.14c. In the end winding, the highest or lowest temperature is in the middle of the coil overhang, and thus the heat flow Φ_{tha} is zero if terminal b is connected to the winding in the slots. We can now omit the resistance R_0 at terminal a. Then, the left-hand-side end winding in Figure 9.12 can be represented by resistances R_{s10}, R_{s12}, R_{s14} and R_{s15} and node 5. The resistance R_{s10} has the form of R_0 in Equation (9.100), R_{12} is a small negative resistance of the form of R_1 in Equation (9.99) and the sum $\mathrm{R}_{s14} + \mathrm{R}_{s15}$ has the form of R_2 in Equation (9.98); that is, $R_{s14} + R_{s15}$ is the resistance from the winding end conductors to the winding end space. The resistance R_{s14} is the resistance from the conductors to the surface of the winding end and Rs_{15} is the convection resistance from the winding end surface to the end winding space. On the right-hand side, the winding end is represented in the same way by the resistances R_{s11}, R_{s13}, R_{s14} and R_{s16} and node 6. Nodes 5 and 6 give the average temperature rises in the winding ends, and the resistive losses of the winding ends are supplied to nodes 5 and 6.

The teeth are represented in Figure 9.12 by the resistances R_{s4}, R_{s5} and R_{s6} and node 4, where the iron losses in the teeth are supplied. The resistance R_{s4} has the form of R_0 in Equation (9.100), R_{s5} is a small negative resistance of the form of R_1 in Equation (9.99), and R_{s6} is the common resistance with the winding. The resistance R in Equations (9.99) and (9.100) is now the resistance of the teeth from the tip of the tooth to the root.

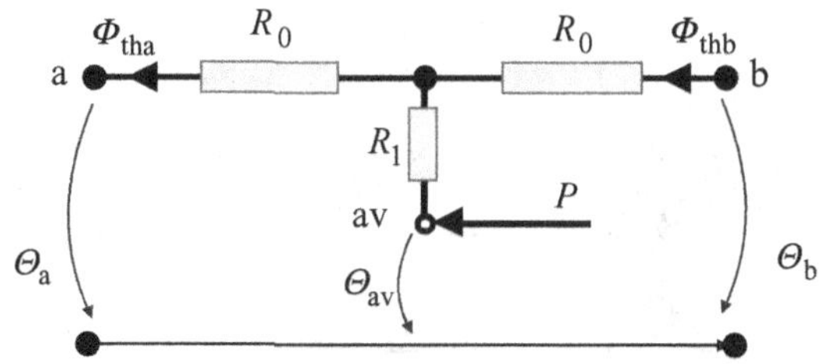

Figure 9.16 Equivalent circuit in a case in which the resistance R_2 (the inverse of the conductance G) of Figure 9.14 is missing (e.g. because of symmetry, the heat flow is only radial).

Due to symmetry, heat is flowing in the yoke only in the radial direction. This means that the resistance R_2 in the equivalent circuit of Figure 9.14c (the conductance G in Figure 9.14b) is missing. The equivalent circuit now has the form of Figure 9.16.

The resistances R_0 and R_1 can be derived in a similar way as in the general case presented in Figure 9.14. The result is

$$R_0 = \frac{R}{2}, \tag{9.101}$$

$$R_1 = -\frac{R}{6}, \tag{9.102}$$

where R is the thermal resistance between points a and b in Figure 9.16 and P is the total loss produced in the body.

The yoke is represented in Figure 9.12 by the resistances R_{s2} and R_{s3} and node 1. Node 1 gives the average temperature rise of the yoke, and the iron losses of the yoke are supplied to node 1. The resistance R_{s2} is calculated from Equation (9.101) and the resistance R_{s3} from Equation (9.102). The resistance R is the thermal resistance of the yoke from the slot bottom to the outer surface of the yoke.

The resistance R_{s1} in the equivalent circuit of Figure 9.12 is the convection and radiation resistance from the outer surface of the yoke to the surroundings or to the coolant.

The node 17 represents the air gap and the losses in the air gap (Equation (9.9)) are supplied to the node. The convection resistance R_{s7} between the stator core and the air gap is calculated from Equation (9.53). The heat transfer coefficient in (9.53) is calculated from Equation (9.60) taken into account the comment on the roughness of the stator surface after (9.60). The surface in (9.53) is the surface of the stator tooth tips, i.e. $S_s = (\pi D_s - Q_s b_{1s})l$, where D_s is the stator air-gap diameter (Figure 3.1), Q_s the number of stator slots, b_{1s} the width of the stator slot opening, and l the length of the stator core.

The equivalent circuit in Figure 9.14c was derived for rectangular-shaped conductors, assuming that the temperature is constant in the cross-section of the winding. If we have a round wire winding (Figure 9.17), the temperature in a cross-section cannot be assumed to be constant. Assuming that heat flows from the winding only into the tooth in Figure 9.17, the equivalent circuit has the form of Figure 9.16, which is valid for unidirectional heat flow in a body where losses are distributed uniformly over the body. Due to the symmetry, the maximum temperature is on the center line of the slot, and heat does not flow over the center

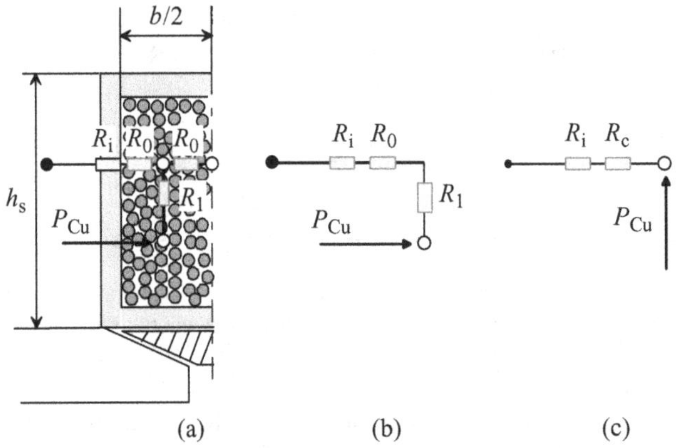

Figure 9.17 Thermal equivalent circuit of a round wire winding.

line. We may cancel the right-hand-side R_0 resistance in Figure 9.17a, so the equivalent circuit has the form of Figure 9.17b. The resistance R_i in Figure 9.17b includes the resistance of the slot insulation and the contact resistance between the insulation and the tooth. The resistances R_0 and R_1 are calculated from Equations (9.101) and (9.102). Their sum R_c (Figure 9.17c) is

$$R_c = R_0 + R_1 = \frac{R}{2} - \frac{R}{6} = \frac{1}{3}R = \frac{1}{3}\frac{b}{2\lambda_{av}h_s l} = \frac{1}{6}\frac{b}{\lambda_{av}h_s l}, \tag{9.103}$$

where R is the thermal resistance from the center line of the slot to the slot insulation, b and h_s the slot dimensions shown in Figure 9.17, l the stack length and λ_{av} the average thermal conductivity of the winding (Equation 9.70).

If the winding is a round wire winding, the resistance R_{s6} in the equivalent circuit of Figure 9.12 is

$$R_{s6} = R_i + R_c.$$

The resistance R_{s14} is calculated accordingly in the end windings.

The equivalent circuit of the rotor in Figure 9.12 is a mirror image of the stator circuit. The thermal resistances of the rotor are calculated as the stator resistances but using rotor parameters.

9.3.4 Modeling of Coolant Flow

The simplest way to model the coolant flow is to assume the coolant temperature to be constant and equal to its mean value. That gives adequate results if the temperature rise of the coolant is small, as it normally is in totally enclosed fan-cooled (TEFC) motors. If the temperature rise of the coolant is high, as in motors having open-circuit cooling, the constant temperature approximation alone does not suffice. We can estimate the temperature rise of the coolant in different parts of the motor. After solving the thermal network, we know the heat flow distribution and we can recalculate the temperature rise of the coolant, correct our estimation, and solve the network again to obtain a more accurate result.

The most accurate way to consider the coolant flow is to handle the heat flow equations and the coolant flow equations at the same time. The system equations of thermal networks with passive components are linear and the system matrix is symmetrical. This results in the property of reciprocity: the temperature rise of any part A per watt in a part B is the same as the temperature rise of the part B per watt in the part A. The equations describing the temperature rises of the coolant in different motor parts are also linear, but they do not have the properties of symmetry and reciprocity. This is the reason why the coolant flow cannot be modeled by passive thermal components.

The coolant flow can be modeled by heat-flow-controlled temperature sources in the thermal network. General circuit analysis programs such as Spice, Saber or Aplac can be used to analyze thermal networks with heat-flow-controlled sources. The heat-flow-controlled temperature source is described by a current-controlled voltage source in the program. If the network is small, it can also be solved manually.

9.3.4.1 Method of Analysis

Let us examine the cooling of the stator of an open motor (Figure 9.18). The coolant flow q enters one of the end winding regions. The losses P_{ew1} absorbed from the end winding and the friction losses $P_{\rho 1}$ in the end winding region warm up the coolant. The temperature rise is

$$\Theta_{1\,\mathrm{end}} = \frac{P_{\mathrm{ew1}} + P_{\rho 1}}{\rho c_{\mathrm{p}} q} = 2R_{\mathrm{q}}\left(P_{\mathrm{ew1}} + P_{\rho 1}\right), \tag{9.104}$$

where ρ is the density and c_{p} the specific heat capacity of the coolant, q the coolant flow and the term

$$R_{\mathrm{q}} = \frac{1}{2\rho c_{\mathrm{p}} q}. \tag{9.105}$$

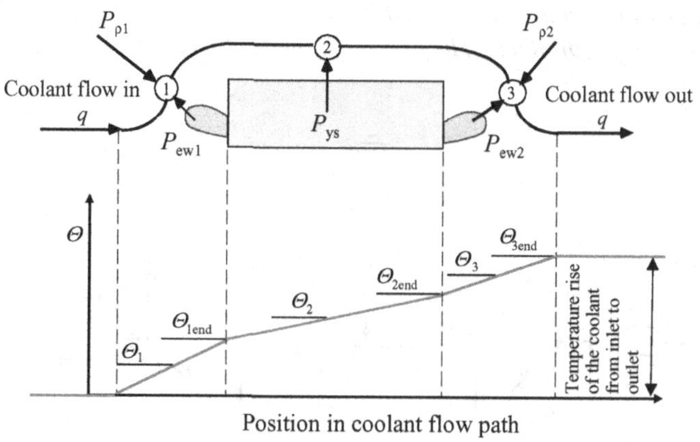

Figure 9.18 Temperature rise of the coolant in an open-circuit machine.

R_q has the dimension of the thermal resistance [K/W]. It is assumed that the mass flow ρq does not depend on the temperature of the coolant.

The temperature rise of node 1 (Θ_1) in Figure 9.18 can be assumed to be the average temperature rise in the end winding region. According to Equation (9.104), we get

$$\Theta_1 = \frac{\Theta_{1\ \mathrm{end}}}{2} = R_q \left(P_{\mathrm{ew1}} + P_{\rho 1}\right). \tag{9.106}$$

The losses P_{ys} absorbed from the stator yoke warm up the coolant by the amount of

$$\Theta_{2\ \mathrm{end}} - \Theta_{1\ \mathrm{end}} = \frac{P_{ys}}{\rho c_p q}. \tag{9.107}$$

Substituting $\Theta_{1\ \mathrm{end}}$ from Equation (9.93) and using the term (9.94) we get

$$\Theta_{2\ \mathrm{end}} = 2R_q \left(P_{\mathrm{ew1}} + P_{\rho 1}\right) + 2R_q P_{ys}. \tag{9.108}$$

The temperature rise of node 2 (Θ_2) in Figure 9.15 is the average temperature rise of the coolant over the stator yoke:

$$\Theta_2 = \frac{\Theta_{1\ \mathrm{end}} + \Theta_{2\ \mathrm{end}}}{2} = 2R_q \left(P_{\mathrm{ew1}} + P_{\rho 1}\right) + R_q P_{ys}. \tag{9.109}$$

Analogously, we get for node 3:

$$\Theta_3 = 2R_q \left(P_{\mathrm{ew1}} + P_{\rho 1}\right) + 2R_q P_{ys} + R_q \left(P_{\mathrm{ew2}} + P_{\rho 2}\right). \tag{9.110}$$

Equations (9.106), (9.109) and (9.110) can be interpreted as heat-flow-controlled temperature sources; for instance, for the source Θ_2 there are two controlling heat flows, $P_{\mathrm{ew1}} + P_{\rho 1}$ and P_{ys}. The thermal network in Figure 9.19 matches Equations (9.106), (9.109) and (9.110).

The rule for writing the temperature source equations is formulated as follows:

Rule 1: *The temperature source connected between a coolant flow node and earth is equal to the sum of two products. The first is $2R_q$ multiplied by the losses absorbed by*

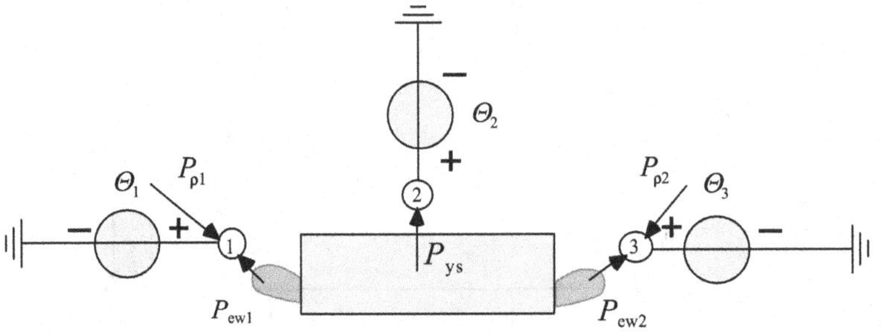

Figure 9.19 Interpretation of the coolant as heat-flow-controlled temperature sources Θ_1, Θ_2 and Θ_3.

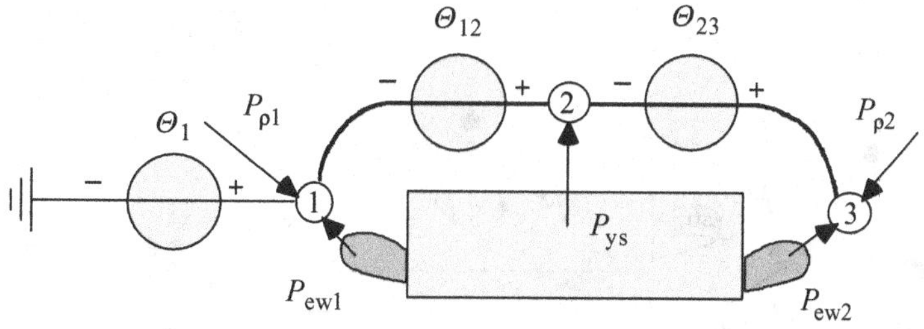

Figure 9.20 Interpretation of the coolant as heat-flow-controlled temperature sources Θ_1, Θ_{12} and Θ_{23}.

the coolant before the coolant flow node and the second is R_q multiplied by the losses absorbed in the coolant flow node under consideration.

According to Figure 9.18, the temperature rises Θ_2 and Θ_3 can also be written in the form

$$\begin{aligned}\Theta_2 &= \Theta_1 + \frac{\Theta_{1\,\mathrm{end}}}{2} + \frac{\Theta_{2\,\mathrm{end}} - \Theta_{1\,\mathrm{end}}}{2} = \Theta_1 + \frac{\Theta_{2\,\mathrm{end}}}{2} \\ &= \Theta_1 + R_\mathrm{q}\left(P_\mathrm{ew1} + P_{\rho 1}\right) + R_\mathrm{q} P_\mathrm{ys} = \Theta_1 + \Theta_{12},\end{aligned} \tag{9.111}$$

$$\begin{aligned}\Theta_3 &= \Theta_2 + \frac{\Theta_{2\,\mathrm{end}} - \Theta_{1\,\mathrm{end}}}{2} + \frac{\Theta_{3\,\mathrm{end}} - \Theta_{2\,\mathrm{end}}}{2} = \Theta_2 + \frac{\Theta_{3\,\mathrm{end}} - \Theta_{1\,\mathrm{end}}}{2} \\ &= \Theta_2 + R_\mathrm{q} P_\mathrm{ys} + R_\mathrm{q}\left(P_\mathrm{ew2} + P_{\rho 2}\right) = \Theta_2 + \Theta_{23},\end{aligned} \tag{9.112}$$

where the heat-flow-controlled temperature sources are

$$\Theta_{12} = R_\mathrm{q}(\mathrm{P}_\mathrm{ew1} + \mathrm{P}_{\rho 1} + \mathrm{P}_\mathrm{ys}), \tag{9.113}$$

$$\Theta_{23} = R_\mathrm{q}(P_\mathrm{ys} + P_\mathrm{ew2} + P_{\rho 2}). \tag{9.114}$$

The equivalent network satisfying Equations (9.106), (9.111) and (9.112) is shown in Figure 9.20. The rule for writing the temperature source equations is now:

Rule 2: *The temperature source between two coolant flow nodes m and n is equal to the sum of losses absorbed by the coolant in the nodes m and n multiplied by R_q.*

Example 9.9: Form the coolant flow part of the thermal network for a totally enclosed fan-cooled induction motor, in which there is also an inner coolant flow (Figure 9.21). The outer and inner coolant flows are q_o and q_i. The friction losses in the winding end regions and in the outer fan are $P_{\rho 1}$, $P_{\rho 2}$ and $P_{\rho 3}$, respectively. The losses transferred from the nondrive end winding region and from the stator core to the outer coolant flow are P_{62} and P_{s3}. The losses transferred from the drive-end winding region to the ambient are P_{40}. It is assumed that the outer coolant flow does not cool the bearing shield in the drive end. The losses from the stator and rotor core to the inner coolant flow are P_{s5} and P_{r7}. The losses

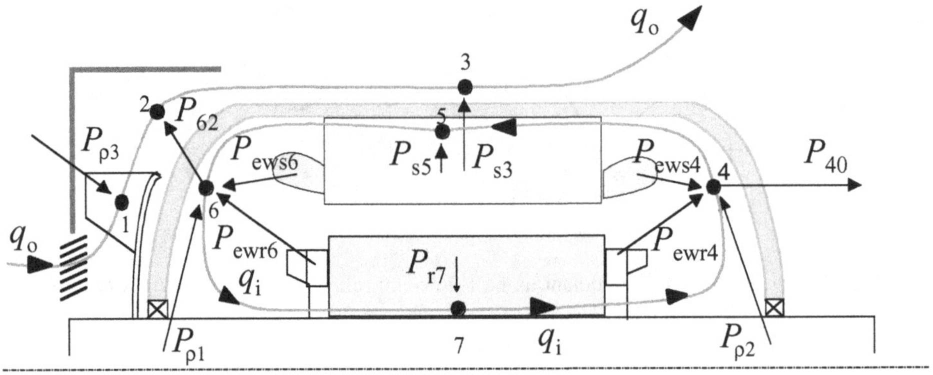

Figure 9.21 TEFC motor with outer and inner coolant cycles.

from the stator and rotor end windings to the inner coolant flow are P_{ews6}, P_{ews4}, P_{ewr6}, P_{ewr4}.

Solution: The thermal network of the coolant flow is shown in Figure 9.22. The resistances R_{62} and R_{40} are the thermal resistances over the bearing shields. The coolant flow is modeled according to Rule 2 and Figure 9.20. The heat-flow-controlled sources are

$$
\begin{aligned}
\Theta_{01} &= R_{\mathrm{qo}} P_{\rho 3} \\
\Theta_{12} &= R_{\mathrm{qo}} \left(P_{\rho 3} + P_{62}\right) \\
\Theta_{23} &= R_{\mathrm{qo}} \left(P_{62} + P_{\mathrm{s3}}\right) \\
\Theta_{45} &= R_{\mathrm{qi}} \left(P_{\rho 2} + P_{\mathrm{ewr4}} + P_{\mathrm{ews4}} - P_{40} + P_{\mathrm{s5}}\right) \\
\Theta_{56} &= R_{\mathrm{qi}} \left(P_{\mathrm{s5}} + P_{\rho 1} + P_{\mathrm{ewr6}} + P_{\mathrm{ews6}} - P_{62}\right) \\
\Theta_{67} &= R_{\mathrm{qi}} \left(P_{\rho 1} + P_{\mathrm{ewr6}} + P_{\mathrm{ews6}} - P_{62} + P_{\mathrm{r7}}\right)
\end{aligned}
$$

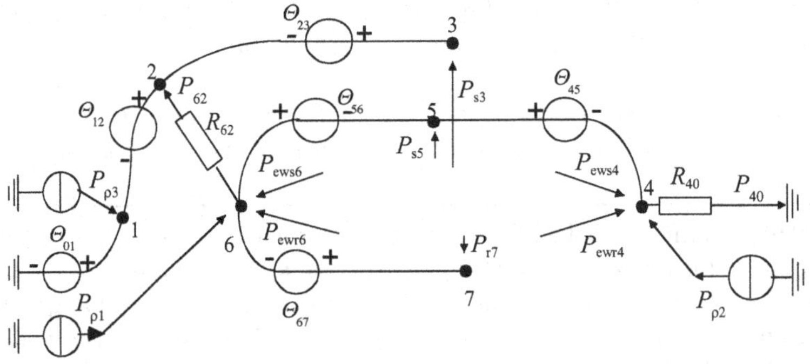

Figure 9.22 Thermal network of the coolant flow for the TEFC motor presented in Figure 9.21.

where

$$R_{qo} = \frac{1}{2\rho c_p q_o},$$

$$R_{qi} = \frac{1}{2\rho c_p q_i}.$$

Note the correct signs of the heat flows in the equations for Θ_{45}, Θ_{56} and Θ_{67}. The heat flows P_{62} and P_{40} have a negative sign because they flow in a direction opposite to the other heat flows in nodes 4 and 6 (Figure 9.22).

Note also that the equivalent circuit representing the inner coolant flow is not a closed loop but an open loop, because a voltage source between nodes 7 and 4 would short-circuit the circuit representing the inner coolant flow, and the heat flow would be infinite. If we write the heat-flow-controlled temperature source between nodes 7 and 4 according to Rule 2, we find that the source is a linear combination of the temperature sources Θ_{45}, Θ_{56} and Θ_{67}. We may use any node of the coolant flow as a starting point; in Figure 9.22, node 4 has been chosen. The end point is the last node in the coolant cycle before the cycle closes, node 7.

9.3.5 *Solution of Equivalent Circuit*

Let us examine the solution of an equivalent circuit. To decrease the number of equations, only the stator circuit of Figure 9.12 is considered. The example circuit is presented in Figure 9.23. There are 11 nodes in the circuit. The line connecting the nodes 9, 10 and 11 indicates the circulation of the cooling air.

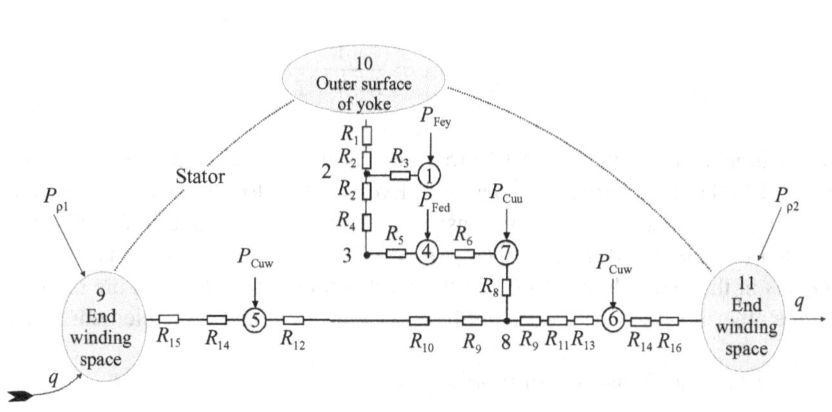

Figure 9.23 Thermal circuit of the stator of an open-circuit cooling machine.

The nodal point method known from circuit theory is used in solving the circuit. The circuit equations written in matrix form are

$$\begin{bmatrix} G_{1,2} & -G_{1,2} & 0 & 0 & 0 & 0 & 0 & 0 & 0 & 0 & 0 \\ -G_{1,2} & G_{1,2}+G_{2,3}+G_{2,10} & -G_{2,3} & 0 & 0 & 0 & 0 & 0 & 0 & -G_{2,10} & 0 \\ 0 & -G_{2,3} & G_{2,3}+G_{3,4} & -G_{3,4} & 0 & 0 & 0 & 0 & 0 & 0 & 0 \\ 0 & 0 & -G_{3,4} & G_{3,4}+G_{4,7} & 0 & 0 & -G_{4,7} & 0 & 0 & 0 & 0 \\ 0 & 0 & 0 & 0 & G_{5,8}+G_{5,9} & 0 & 0 & -G_{5,8} & -G_{5,9} & 0 & 0 \\ 0 & 0 & 0 & 0 & 0 & G_{6,8}+G_{6,11} & 0 & -G_{6,8} & 0 & 0 & -G_{6,11} \\ 0 & 0 & 0 & -G_{4,7} & 0 & 0 & G_{4,7}+G_{7,8} & -G_{7,8} & 0 & 0 & 0 \\ 0 & 0 & 0 & 0 & -G_{5,8} & -G_{6,8} & -G_{7,8} & G_{5,8}+G_{6,8}+G_{7,8} & 0 & 0 & 0 \\ 0 & 0 & 0 & 0 & -G_{5,9} & 0 & 0 & 0 & G_{5,9} & 0 & 0 \\ 0 & -G_{2,10} & 0 & 0 & 0 & 0 & 0 & 0 & 0 & G_{2,10} & 0 \\ 0 & 0 & 0 & 0 & 0 & -G_{6,11} & 0 & 0 & 0 & 0 & G_{6,11} \end{bmatrix} \tag{9.115}$$

$$\times \begin{bmatrix} \Theta_1 \\ \Theta_2 \\ \Theta_3 \\ \Theta_4 \\ \Theta_5 \\ \Theta_6 \\ \Theta_7 \\ \Theta_8 \\ \Theta_9 \\ \Theta_{10} \\ \Theta_{11} \end{bmatrix} = \begin{bmatrix} P_{\mathrm{Fey}} \\ 0 \\ 0 \\ P_{\mathrm{Fed}} \\ P_{\mathrm{Cuw}} \\ P_{\mathrm{Cuw}} \\ P_{\mathrm{Cuu}} \\ 0 \\ -\Phi_{5,9} \\ -\Phi_{2,10} \\ -\Phi_{6,11} \end{bmatrix}$$

The conductance $G_{n,m}$ refers to the conductance between nodes n and m. For instance,

$$G_{1,2} = \frac{1}{R_3}, \; G_{2,3} = \frac{1}{R_2 + R_4}, \ldots$$

On the diagonal of the matrix in (9.115) there is the sum of the conductances, which are connected to the node under consideration. Everywhere else, there are the conductances between the nodes with a minus sign. For instance, the three conductances $G_{1,2}$, $G_{2,3}$ and $G_{2,10}$ are connected to node 2 and their sum $G_{1,2} + G_{2,3} + G_{2,10}$ is on the diagonal. On the same row, $-G_{1,2}$ is in the first column, $-G_{2,3}$ in the third column and $-G_{2,10}$ in the tenth column. Between the other nodes, node 2 does not have a connection, and these elements are zero in the matrix.

Equation (9.115) can be written in short form as

$$[\boldsymbol{G}] \cdot [\boldsymbol{\Theta}] = \begin{bmatrix} P \\ [-\Phi_e] \end{bmatrix} = \begin{bmatrix} P \\ [0] \end{bmatrix} - \begin{bmatrix} [0] \\ [\Phi_e] \end{bmatrix}, \tag{9.116}$$

where

$$[\boldsymbol{P}] = \begin{bmatrix} P_{\text{Fey}} \\ 0 \\ 0 \\ P_{\text{Fed}} \\ P_{\text{Cuw}} \\ P_{\text{Cuw}} \\ P_{\text{Cuu}} \\ 0 \end{bmatrix}, \tag{9.117}$$

$$[\boldsymbol{\Phi}_{\text{e}}] \begin{bmatrix} \Phi_{5,9} \\ \Phi_{2,10} \\ \Phi_{6,11} \end{bmatrix}. \tag{9.118}$$

In Equation (9.116) there are eleven unknown temperature rises and three unknown heat flows $\Phi_{5,9}$, $\Phi_{2,10}$ and $\Phi_{6,11}$ or $[\Phi_{\text{e}}]$, thus we need three more equations. These are obtained from the temperature rise of the coolant. According to Equations (9.106), (9.109) and (9.110), the temperature rises in nodes 9, 10 and 11 are

$$\Theta_9 = R_{\text{q}}\left(\Phi_{5,9} + P_{\rho 1}\right), \tag{9.119}$$

$$\Theta_{10} = 2R_{\text{q}}\left(\Phi_{5,9} + P_{\rho 1}\right) + R_{\text{q}}\Phi_{2,10}, \tag{9.120}$$

$$\Theta_{11} = 2R_{\text{q}}(\Phi_{5,9} + P_{\rho 1}) + 2R_{\text{q}}\Phi_{2,10} + R_{\text{q}}(\Phi_{6,11} + P_{\rho 2}), \tag{9.121}$$

where $P_{\rho 1}$ and $P_{\rho 2}$ are the friction losses in the end winding spaces and R_{q} the thermal resistance (9.105). In matrix form

$$\begin{bmatrix} \Theta_9 \\ \Theta_{10} \\ \Theta_{11} \end{bmatrix} = \begin{bmatrix} R_{\text{q}} & 0 & 0 \\ 2R_{\text{q}} & R_{\text{q}} & 0 \\ 2R_{\text{q}} & 2R_{\text{q}} & R_{\text{q}} \end{bmatrix} \cdot \begin{bmatrix} \Phi_{5,9} \\ \Phi_{2,10} \\ \Phi_{6,11} \end{bmatrix} + \begin{bmatrix} R_{\text{q}}P_{\rho 1} \\ 2R_{\text{q}}P_{\rho 1} \\ 2R_{\text{q}}P_{\rho 1} + R_{\text{q}}P_{\rho 2} \end{bmatrix} \tag{9.122}$$

or in short

$$[\boldsymbol{\Theta}_{\text{e}}] = [\boldsymbol{R}_{\text{e}}] \cdot [\boldsymbol{\Phi}_{\text{e}}] + [\boldsymbol{\Theta}_{\rho}]. \tag{9.123}$$

The unknown heat flows are solved from Equation (9.123)

$$[\boldsymbol{\Phi}_{\text{e}}] = [\boldsymbol{R}_{\text{e}}]^{-1} \cdot [\boldsymbol{\Theta}_{\text{e}}] - [\boldsymbol{R}_{\text{e}}]^{-1} \cdot [\boldsymbol{\Theta}_{\rho}]. \tag{9.124}$$

Substituting Equation (9.124) for $[\Phi_{\text{e}}]$ in Equation (9.116) we obtain

$$[\boldsymbol{G}] \cdot [\boldsymbol{\Theta}] = \begin{bmatrix} P \\ 0 \end{bmatrix} - \begin{bmatrix} 0 \\ [R_e]^{-1} \cdot [\Theta_e] - [R_e]^{-1} \cdot [\Theta_\rho]], \end{bmatrix}, \tag{9.125}$$

and further

$$\left[[\boldsymbol{G}]+\begin{bmatrix}0 & 0\\ 0 & [\boldsymbol{R}_{\mathrm{e}}]^{-1}\end{bmatrix}\right]\cdot[\boldsymbol{\Theta}]=\begin{bmatrix}[\boldsymbol{P}]\\ [0]\end{bmatrix}+\begin{bmatrix}[0]\\ [\boldsymbol{R}_{e}]^{-1}\cdot[\boldsymbol{\Theta}_{\rho}]\end{bmatrix}, \tag{9.126}$$

from which the solutions for the temperature rises in 11 nodes are obtained

$$[\boldsymbol{\Theta}]=\left[[\boldsymbol{G}]+\begin{bmatrix}0 & 0\\ 0 & [\boldsymbol{R}_{\mathrm{e}}]^{-1}\end{bmatrix}\right]^{-1}\cdot\begin{bmatrix}[\boldsymbol{P}]\\ [\boldsymbol{R}_{\mathrm{e}}]^{-1}\cdot[\boldsymbol{\Theta}_{\rho}]\end{bmatrix}. \tag{9.127}$$

9.3.6 *Cooling Flow Rate*

The required cooling flow rate can be solved from the equation

$$P_{\mathrm{tot}} = P c_{\mathrm{p}} q \Theta, \tag{9.128}$$

where P_{tot} is the sum of losses in the machine, ρ the density, c_{p} the specific heat capacity of the cooling fluid, q the volume flow rate and Θ the permitted temperature rise of the coolant. The required volume flow rate of the coolant is, according to Equation (9.128),

$$q = \frac{P_{\mathrm{tot}}}{\rho c_p \Theta}. \tag{9.129}$$

For air $\rho = 1.146\ \mathrm{kg/m^3}$ (35 °C), $c_{\mathrm{p}} = 1.0$ kJ/kg K and

$$q = 0.865\frac{P_{\mathrm{tot}}}{\Theta}\left[\frac{\mathrm{m}^3}{\mathrm{s}}\right], \tag{9.130}$$

where P_{tot} is given in kW and Θ in K. For instance, for $\Theta \approx 15$ K

$$q = 0.06\cdot P_{\mathrm{tot}}\left[\frac{\mathrm{m}^3}{\mathrm{s}}\right]. \tag{9.131}$$

Bibliography

Arkkio, A. (1992) Rotor-slot design for inverter-fed cage induction motors. *Symposium on Power Electronics, Electrical Drives, Advanced Electrical Motors (Speedam), Positano, Italy, May 19–21, 1992*, pp. 37–42.

Becker, K.M. and Kaye, J. (1962) Measurements of diabatic flow in an annulus with an inner rotating cylinder. *Journal of Heat Transfer*, **84**: 97–105.

Flik, M.I., Choi, B.-I. and Goodson, K.E. (1992) Heat transfer regimes in microstructures. *Journal of Heat Transfer*, **114**, 666–74.

Gardiner, S. and Sabersky, R. (1978) Heat transfer in an annular gap. *International Journal of Heat and Mass Transfer*, **21**(12): 1459–66.

Gieras, F. and Wing, M. (1997) *Permanent Magnet Motor Technology – Design and Applications*. Marcel Dekker, New York.

Hendershot, J.R. Jr and Miller, T.J.E. (1994) *Design of Brushless Permanent-Magnet Motors*. Magna Physics Publishing and Clarendon Press, Hillsboro, OH and Oxford.

IEC 60034-2-1 (2007) *Rotating Electrical Machines – Part 2-1 Standard Methods for Determining Losses and Efficiency from Tests [Excluding Machines for Traction Vehicles]*.

Incropera, F.P., Dewitt, D.P., Bergman, T.L. and Lavine, A.S. (2007) *Fundamentals of Heat and Mass Transfer*, 6th edn, John Wiley & Sons, Inc., Hoboken, NJ.

Islam, M.J. (2010) *Finite-Element Analysis of Eddy Currents in the Form-Wound Multi-Conductor Windings of Electrical Machines*. TKK Dissertations 211, Helsinki University of Technology. Available at http://lib.tkk.fi/Diss/2010/isbn9789522482556/.

Jokinen, T. and Saari, J. (1997) Modelling of the coolant flow with heat flow controlled temperature sources in thermal networks. *IEE Proceedings*, **144** (5): 338–42.

Klemens, P.G. (1969) Theory of the thermal conductivity of solids, in *Thermal Conductivity*, Vol. **1** (ed. R.P. Tye), Academic Press, London.

Lawrenson, P.J. (1992) A brief status review of switched reluctance drives. *EPE Journal*, **2**(3): 133–44.

Lipo, T.A. (2007) *Introduction to AC Machine Design*, 3rd edn. Wisconsin Power Electronics Research Center, University of Wisconsin.

Matsch, L.D. and Morgan J.D. (1987) *Electromagnetic and Electromechanical Machines*, 3rd edn. John Wiley & Sons, Inc., New York.

Miller, T.J.E. (1993) *Switched Reluctance Motors and Their Controls*. Magna Physics Publishing and Clarendon Press, Hillsboro, OH and Oxford.

Moody, L.F. (1944) Friction factors for pipe flow. *ASME Transactions*, **66**: 671–84.

Oberretl, K. (1969) 13 rules to minimize stray load losses in induction motors. *Bulletin Oerlikon*, No 389/390, pp. 1–11.

Rao, K. and Sastri, V. (1984) Experimental studies on diabatic flow in an annulus with rough rotating inner cylinder. In D. Metzger and N. Afgan (eds), *Heat and Mass Transfer in Rotating Machinery*, Hemisphere, New York, pp. 166–78.

Saari, J. (1995) *Thermal Modelling of High-Speed Induction Machines*, Electrical Engineering Series No. 82. Acta Polytechnica Scandinavica, Helsinki University of Technology.

Saari, J. (1998) *Thermal Analysis of High-Speed Induction Machines*, Dissertation, Electrical Engineering Series No. 90. Acta Polytechnica Scandinavica, Helsinki University of Technology.

Schuisky, W. (1960) *Design of Electrical Machines (Berechnung elektrischer Maschinen)*. Springer Verlag, Vienna.

SKF (2013) *Friction*. [online]. Available from http://www.skf.com/group/products/bearings-units-housings/spherical-plain-bearings-bushings-rod-ends/general/friction/index.html?WT.oss=friction&WT.z_oss_boost=0&tabname=All&WT.z_oss_rank=3

Appendix A

Properties of Magnetic Sheets

Design of Rotating Electrical Machines, Second Edition. Juha Pyrhönen, Tapani Jokinen and Valéria Hrabovcová.

Table A.1 Magnetic sheets: Typical peak magnetic field strength, [A/m] at 50 Hz. Reproduced by permission of Surahammars Bruk AB

Grade EN 10106	Thickness [mm]	Peak magnetic field strength [A/m] at 50 Hz and peak magnetic polarization J [T] of:																	
		0.10	0.20	0.30	0.40	0.50	0.60	0.70	0.80	0.90	1.00	1.10	1.20	1.30	1.40	1.50	1.60	1.70	1.80
M235-35A	0.35	24.7	32.6	38.1	43.1	48.2	53.9	60.7	68.8	79.3	93.7	115	156	260	690	1950	4410	7630	12 000
M250-35A	0.35	26.8	35.7	41.8	47.5	53.4	60.0	67.9	77.5	90.0	107	133	179	284	642	1810	4030	7290	11 700
M270-35A	0.35	30.0	39.6	46.0	52.0	58.2	65.2	73.3	83.1	95.5	112	136	178	272	596	1700	3880	7160	11 600
M300-35A	0.35	30.9	40.2	46.4	52.1	57.9	64.4	72.0	81.1	92.6	108	130	168	250	510	1440	3490	6700	11 300
M330-35A	0.35	31.4	41.4	48.2	54.3	60.4	67.1	74.9	84.2	96.3	113	137	179	266	521	1380	3400	6610	11 100
M700-35A*	0.35	70.2	89.1	98.8	106	113	120	127	135	144	155	169	192	237	342	681	1890	4570	8580
M250-50A	0.50	30.6	40.7	47.9	54.5	61.3	69.0	77.8	88.6	102	120	145	186	278	584	1600	3680	6890	11 600
M270-50A	0.50	31.5	42.0	49.4	56.1	63.1	70.7	79.5	90.1	103	121	145	185	273	557	1520	3560	6730	11 400
M290-50A	0.50	32.2	42.9	50.3	57.1	63.9	71.4	79.9	89.9	103	119	144	184	271	549	1500	3520	6700	11 400
M310-50A	0.50	33.3	43.9	51.2	57.7	64.2	71.2	79.1	88.4	100	116	139	175	251	470	1230	3070	6150	10 700
M330-50A	0.50	33.2	44.3	52.0	58.9	65.9	73.4	82.0	92.2	105	122	145	183	259	470	1190	3030	6120	10 700
M350-50A	0.50	34.8	46.0	53.7	60.6	67.4	74.6	82.6	91.8	103	119	141	178	250	455	1180	3020	6100	10 700
M400-50A	0.50	40.1	52.5	60.8	68.1	75.2	82.5	90.4	99.3	110	125	146	181	251	443	1110	2900	6020	10 600
M470-50A	0.50	48.8	64.8	74.3	82.4	90.2	98.2	107	117	129	146	170	209	284	475	1100	2850	5980	10 500
M530-50A	0.50	51.5	68.1	77.6	85.6	93.3	101	110	120	132	147	170	208	282	470	1080	2790	5890	10 400
M600-50A	0.50	65.6	83.8	94.1	103	110	118	127	136	147	159	177	205	255	370	718	1840	4370	8330
M700-50A	0.50	67.8	88.3	99.2	108	116	124	132	142	152	164	180	206	254	363	690	1760	4230	8130
M800-50A	0.50	84.5	107	121	133	145	156	168	180	194	209	228	254	304	402	660	1480	3710	7300
M940-50A	0.50	102	129	146	161	171	181	192	203	217	228	243	267	311	400	645	1440	3590	7090
M530-50HP*	0.50	57.7	74.9	85.2	93.7	102	109	118	127	137	148	164	189	232	326	594	1460	3620	7320
M310-65A	0.65	25.8	35.5	42.9	49.7	56.7	63.8	71.7	80.6	91.5	107	130	169	257	545	1490	3540	6800	11 600
M330-65A	0.65	26.5	36.2	43.7	50.6	57.6	64.8	72.7	81.8	93.3	109	133	174	261	530	1410	3350	6500	11 200
M350-65A	0.65	27.3	37.7	45.9	53.1	59.9	66.8	74.2	82.5	90.1	101	121	155	230	441	1210	3020	6040	10 600
M400-65A	0.65	29.5	40.1	48.4	56.2	64.2	72.6	81.9	93.0	108	127	155	197	278	484	1140	2820	5830	10 300
M470-65A	0.65	31.2	42.0	50.2	57.8	65.5	73.5	82.1	91.6	103	118	140	175	242	426	1060	2700	5670	10 100
M530-65A	0.65	44.0	59.5	69.6	78.2	86.6	95.0	104	113	125	138	159	196	270	454	1040	2630	5620	10 100
M600-65A	0.65	48.8	65.1	75.6	84.9	93.8	103	112	122	133	147	169	205	273	444	991	2550	5540	9980
M700-65A	0.65	57.4	75.8	87.6	98.0	108	118	129	140	153	167	185	211	265	379	688	1630	3920	7760
M800-65A	0.65	74.7	97.5	110	120	130	140	150	162	175	190	208	227	265	366	633	1490	3670	7420
M1000-65A	0.65	83.3	107	119	130	140	150	160	172	185	200	218	237	275	368	604	1360	3370	7010
M600-65HP*	0.65	63.6	82.6	93.9	103	113	122	131	142	153	167	182	202	244	337	587	1360	3370	7010
M600-100A	1.00	29.0	44.1	57.1	70.2	84.1	99.2	116	134	153	176	212	281	401	646	1250	2740	5560	9980
M700-100A	1.00	29.3	44.8	58.4	72.2	87.0	103	121	140	161	185	225	294	412	649	1220	2630	5370	9710
M800-100A	1.00	49.3	69.2	85.1	101	117	135	154	174	196	221	261	332	450	675	1190	2550	5360	9770
M1000-100A	1.00	56.0	80.8	100	119	139	161	183	208	233	257	291	348	444	576	847	1610	3760	7520

*This grade does not appear in EN 10106.

Source: http://www.sura.se/Sura/hp_main.nsf/startupFrameset?ReadForm

Appendix B

Properties of Round Enameled Copper Wires

Design of Rotating Electrical Machines, Second Edition. Juha Pyrhönen, Tapani Jokinen and Valéria Hrabovcová.

Table B.1 Technical data: DAMID, DAMID PE and DASOL according to IEC 60317-0-1. Reproduced by permission of AB Dahréntråd

Diameter [mm]	Grade 1 [mm]		Grade 2 [mm]		Filling [conductors/cm^2]		Specific length [m/kg]	
Rated	Min. resin	Max. outer diameter	Min. resin	Max. outer diameter	Grade 1	Grade 2	Grade 1	Grade 2
0.200	0.014	0.226	0.027	0.239	2251	2012	3354	3247
0.212	0.015	0.240	0.029	0.254	1996	1784	2990	2900
0.224	0.015	0.252	0.029	0.266	1813	1623	2682	2600
0.236	0.017	0.267	0.032	0.283	1615	1434	2419	2354
0.250	0.017	0.281	0.032	0.297	1455	1303	2188	2137
0.265	0.018	0.297	0.033	0.314	1303	1165	1949	1906
0.280	0.018	0.312	0.033	0.329	1180	1060	1750	1713
0.300	0.019	0.334	0.035	0.352	1059	927	1524	1493
0.315	0.019	0.349	0.035	0.367	943	852	1385	1358
0.335	0.020	0.372	0.038	0.391	830	752	1224	1200
0.355	0.020	0.392	0.038	0.411	748	679	1093	1072
0.375	0.021	0.414	0.040	0.434	669	608	979	961
0.400	0.021	0.439	0.040	0.459	594	544	862	846
0.425	0.022	0.466	0.042	0.488	528	481	765	748
0.450	0.022	0.491	0.042	0.513	477	434	683	670
0.475	0.024	0.519	0.045	0.541	426	391	613	602
0.500	0.024	0.544	0.045	0.566	387	357	553	544
0.530	0.025	0.576	0.047	0.600	346	318	493	484
0.560	0.025	0.606	0.047	0.630	312	289	442	435
0.600	0.027	0.649	0.050	0.674	271	252	385	379
0.630	0.027	0.679	0.050	0.704	247	230	350	345
0.650	0.028	0.702	0.053	0.729	232	215	328	324
0.670	0.028	0.722	0.053	0.749	219	204	309	305
0.710	0.028	0.762	0.053	0.789	197	183	276	273
0.750	0.030	0.805	0.056	0.834	176	164	247	244
0.800	0.030	0.855	0.065	0.884	155	146	218	215
0.850	0.032	0.909	0.060	0.939	137	128	193	191
0.900	0.032	0.959	0.060	0.989	124	116	172	170

(continued)

Table B.1 (*Continued*)

Diameter [mm]	Grade 1 [mm]		Grade 2 [mm]		Filling [conductors/cm^2]		Specific length [m/kg]	
Rated	Min. resin	Max. outer diameter	Min. resin	Max. outer diameter	Grade 1	Grade 2	Grade 1	Grade 2
0.950	0.034	1.012	0.063	1.044	110	104	154	153
1.000	0.034	1.062	0.063	1.094	100	95	140	138
1.060	0.034	1.124	0.065	1.157	89	84	124	123
1.120	0.034	1.184	0.065	1.217	80	76	111	110
1.180	0.035	1.246	0.067	1.279	73	69	100	100
1.250	0.035	1.316	0.067	1.349	65	62	90	89
1.320	0.036	1.388	0.069	1.422	59	56	80	80
1.400	0.036	1.468	0.069	1.502	52	50	72	71
1.500	0.038	1.570	0.071	1.606	45	43	62	62
1.600	0.038	1.670	0.071	1.706	40	38		54
1.700	0.039	1.772	0.073	1.809	36	34		48
1.800	0.039	1.872	0.073	1.909	32	30		43
1.900	0.040	1.974	0.075	2.012	29	27		39
2.000	0.040	2.074	0.075	2.112	26	25		35
2.120	0.041	2.196	0.077	2.235	23	22		31
2.240	0.041	2.316	0.077	2.355	20	19		28
2.360	0.042	2.438	0.079	2.478	19	18		25
2.500	0.042	2.578	0.079	2.618	16	16		22
2.650	0.043	2.730	0.081	2.772	15	14		20
2.800	0.043	2.880	0.081	2.922	13	13		18
3.000	0.045	3.083	0.084	3.126	11	11		16
3.150	0.045	3.233	0.084	3.276	10	10		14
3.350	0.046	3.435	0.086	3.479	9	9		13
3.550	0.046	3.635	0.086	3.679	8	8		11.2
3.750	0.047	3.838	0.089	3.883	t	7		10.0
4.000	0.047	4.088	0.089	4.133	6	6		8.8
4.250	0.049	4.341	0.092	4.387	5	5		7.8
4.500	0.049	4.591	0.092	4.637	5	5		7.0
4.750	0.050	4.843	0.094	4.891	4	4		6.3
5.000	0.050	5.093	0.094	5.141	4	4		5.7

Index

Absolute overlap ratio, 485
Absorptivity, 538, 539
AC-to-DC rectification, 469
Admittance, 422
Ageing
  electrical, 510
  thermal, 509, 518
Ageing of insulation, 509
Aho, T., 149
Air-cooled machine, 340
Air gap, 14, 161, 166, 305, 496
  length, 545, 548
    effective air-gap, 172
    equivalent air-gap, 548
Air-gap diameter, 323, 336
Air-gap flux, 15, 229
Air-gap flux density, 93, 297
Air-gap flux distribution, 147
Air-gap flux linkage, 391
Air-gap power, 356, 357, 525
AlNiCo magnet, 204, 205. *See also* Permanent magnet
Amide-imide, 501
Amorphousity, 501
Ampère's law, 3, 5, 59, 155, 158, 210, 267, 397
Anisotropic, 7, 193, 204
Anisotropy, 206
Aplac (circuit analysis), 561
Apparent power, 295, 300, 484
Aramid fibre, 499
Aramid paper, 497
Arkkio's method, 33
Armature current, 302, 319, 469
Armature reaction, 48, 146, 156, 168, 307, 311, 389, 402, 474. *See also* Armature current linkage
Arrhenius equation, 509
Asynchronous generator, 380
Asynchronous machine, 173, 180, 298, 388
Auinger, H., 381
Axially laminated rotor, 461, 467

Balancing connector, 41. *See also* Equalizer bar
Barkhausen jump, 187. *See also* Magnetic material
Barkhausen noise, 187. *See also* Magnetic material
Base value, 42
Becker, K. M., 546
BH curve, 16, 160, 178, 188
Black body, 539
Bloch wall, 186
Boundary layer, 542
  concentration, 542
  thermal, 542, 543
  velocity, 542, 543
Braided conductor, 153
Breakdown strength, 510
Breakdown voltage, 495, 06
Breakthrough field strength, 506
Brush, 49, 129, 37, 426, 468, 474, 525
Brushless synchronous generator, 428
Brush rocker, 476

Capacitance, 388
  electric, 549
  of insulation, 509

Design of Rotating Electrical Machines, Second Edition. Juha Pyrhönen, Tapani Jokinen and Valéria Hrabovcová.

Capacitive load, 404
Capacitor, 401, 516
Carter factor, 162, 400
Carter, F.W., 162
Characteristic length of surface, 543
Circulating current, 271
Clearance in air, 495
Coenergy, 28, 30, 480
Coercive force, 5, 210
    calculatory coercive force, 213, 215
Coercivity, 190, 192, 203
    intrinsic coercivity, 208, 209
Coil, 51, 401, 412
Coil group, 51, 110
Coil span, 55, 74, 237, 261, 470
    average, 266
Coil turn, 316
Commutating pole, 148, 298, 469, 513
Commutation, 475, 478
Commutation period, 476
Commutator, 426, 468, 470
Commutator machine, 48, 124, 129, 196
Comparative Tracking Index (CTI), 496
Condition monitoring, 515
Conductance
    electric, 548
    thermal, 548, 553
Conduction, 534
Conductivity, 7, 265, 495
    electric, 535
    thermal, 535, 543, 549, 54
Conductor height, 269, 273
    critical, 276
    reduced, 270, 275
Conductor transposition, 277
Conservation of charge, 2
Contact heat transfer coefficient, 547
Continuous-operation periodic duty, 377
    with electric braking, 377
    with related load/speed changes, 377
Continuous running duty, 375
Convection, 534, 541
    heat transfer coefficient, 543
Coolant, 559
    flow rate, 568
Cooling, 192, 297, 534
    open-circuit, 565
Cooling flow rate, 568
Corona protection, 513
Couette Reynolds number, 528
Coulomb's condition, 9
Coulomb's Virtual Work method, 32
Counterrotating field, 392, 419, 420
Creepage current, 496, 498, 510
Creepage distance, 496, 510, 515
Critical angular speed, 335
Critical conductor height, 276
Cross-linking agent, 503
Crystallinity, 501
Curie temperature, 206
Current density, 1, 266, 282, 319, 337, 476, 535
    armature linear, 147
    linear, 22, 145, 338, 343, 355, 421
Current linkage, 5, 155, 158, 183, 221
    armature, 48, 147, 474
    asynchronous machine, 342
    cage winding, 349
    commutator armature, 146
    commutator winding, 143
    cosinusoidal distribution, 53
    direct component, 146
    harmonic stator, 396
    pole winding, 328
    quadrature component, 146
    rotor bar, 354
    salient-pole winding, 51
    sinusoidal, 309
    slot winding, 81
    stator, 390
    symmetrical three-phase, 390
    three-phase winding, 58
Current linkage harmonic, 344
Current ripple, 490

Dacron®, 497
Damper bar, 392
Damping, 392
Damping factor, 247, 248
Davidson, R., 479
DC link, 492
DC machine, 14, 158, 298
Demagnetization curve, 212
Demagnetization, demagnetizing current linkage, 214
Density, 536
Depth of penetration, 269
Dielectric loss, 516
Dielectric loss factor, 516
Dielectric strength, 495, 498, 506
Diolen®, 497

Dip and bake method, 513
Direct-on-line (DOL), 458
Dirichlet's boundary condition, 11, 12, 20
Displacement current, 5
Dissipation power, 48, 197, 329
Distribution factor, 67, 92
Double-salient reluctance machine, 51, 158, 304, 479
Double-salient reluctance motor, 485
Doubly-fed induction generator, 383
Drain, 6
Duty cycle, 295, 376
Duty type, 375

Eco-design, 293
Eddy current, 3, 10, 149, 189, 288, 393
Effective torque zone, 485
Efficiency, 318, 329, 373, 486, 492
  of totally enclosed four-pole induction motors, 381
Efficiency class, 340
Electrical stress, 510
Electric capacitance, 549
Electric charge, 10, 549
Electric charge density, 1
Electric field density, 504
Electromotive force (emf), 16, 45, 64, 155, 239, 300, 311, 351, 408
  stator, 407
Emissivity, 539
  relative, 539
Enclosure class, 296, 373
Energy conversion, 229
Energy conversion loop, 483
Energy conversion ratio, 484
Energy density, 28, 189
Energy product, 204, 206
Energy saving, 533
EN 10106, standard, 195
Environmental aspect, 293
Epoxy resin, 504, 514
Equalizer bar, 141. *See also* Balancing connector
Equivalent air-gap length, 547
Equivalent circuit, 94
  of asynchronous machine, 356
  of capacitor motor main and auxiliary windings, 386
  of conductor, 286
  of end winding, 559
  of insulation construction, 517
  of slot, 287
  of synchronous machine, 394
  of three-phase asynchronous motor, 364
  of transients, 395
  thermal, 548
Ester-imide, 501
Euler's number, 55. *See also* Napier's constant
European Union, 293
Excitation, 426
  brushless, 427
  overexcitation, 400
  underexcitation, 400

Fanning friction coefficient, 543
Faraday's induction law, 2, 64, 171, 243, 267, 300. *See also* Induction law
FEP fluoropolymer, 502
Ferraris rotor, 149
Ferrite magnet, 205. *See also* Permanent magnet
Ferromagnetic metal, 189
Fibre insulator, 521
Field
  electric, 1, 2, 495
  magnetic, 1, 48, 234
Field diagram, 23, 164, 167
  orthogonal, 18, 20, 167, 174, 492
Field Effect Transistor (FET), 489
Field strength, 323, 324, 397
  air-gap, 344
  effective, 507
  electric, 1, 9, 498, 504
  magnetic, 1, 3, 155, 267, 302
  tangential, 398
Field winding current, 400, 426
Finite Element Analysis (FEA), 396
Flange mounting, 372
Fluoropolymer, 502
Flux
  air-gap flux, 35, 48, 155, 168, 229, 230, 236, 346, 430
  electric, 3
  magnetic, 3, 156, 486
  main flux, 13, 21, 49, 58, 156, 158, 171, 229, 234, 237, 437
Flux barrier, 458
Flux density, 15, 165, 298, 339
  air-gap, 310
  air-gap magnetic, 51
  apparent, 177
  electric, 1, 6

Flux density (*Continued*)
magnetic, 1, 155, 160, 168, 180, 288, 347, 391
remanente, 190
saturation, 190
Flux density distribution, 64
sinusoidal, 309
Flux diagram, 18, 162
Flux guide, 457
Flux leakage, 229, 392
air-gap, 237, 238
Flux linkage, 4, 171
air-gap, 172, 229, 300
Flux tube, 16
Fourier's law, 534
Friction coefficient, 543
Fundamental component, of stator current linkage, 397
Fundamental harmonic, of linear current density, 343

Galvanic contact, 497
Gardiner, S., 546
Gauss's law, 3, 7
Generator, asynchronous machine, 380
Generator logic, 45, 407, 413
Geometrical factor, 546
Gliding discharge voltage limit, 510
Goss texture, 194

Haataja, J., 380
Haefely Trench Tettex partial discharge measurement unit, 519
Hamata, V., 165
Hard chopping, 490
Harmonic factor, 400, 401
Heat
capacity, 549
contact heat transfer coefficient, 546
diffusion, 523
flow, 554
flow density, 539, 544, 548
flow rate, 535
heat-flow-controlled temperature source, 561, 562, 563
quantity of heat, 548
specific heat capacity, 543, 549
transfer, 523
transfer area, 535
transfer coefficient, 540, 544, 546
transfer properties, 536
Heck, C., 192
Heikkilä, T., 431
Heller, B., 165
Hot pre-wedging, 515
Hot-spot allowance, 508
Huppunen, J., 149
Hysteresis loop, 192

IC class, 376
IEC 600112, 496
IEC 60034, 508
IEC 60034-1, 342, 498
IEC 60034-4, 417
IEC 60034-5, 373
IEC 60034-6, 373
IEC 60034-7, 373
IEC 60034-8, 372
IEC 60050-411, 48
IEC 60085, 498
IEC code, 374
Impedance, 397
AC, 271, 283
cage winding, 349
characteristic, 519
measuring, 516
negative-sequence, 383
phase, 350, 351
positive-sequence, 383
rated, 42
rotor, 348, 398
synchronous, 418
Impregnating resin, 497, 505
Impregnating varnish, 497, 503, 506
Impulse voltage, 508
Inception voltage, 506
Indifference point, 24
Inductance, 4, 171, 290. *See also* Leakage inductance
air-gap, 238
direct-axis subtransient, 394
direct-axis synchronous, 390, 429
direct-axis transient, 393
leakage, 222, 235, 390
magnetizing, 156, 230, 242, 300, 420
mutual, 36, 223, 234
phase, 350
qudrature-axis subtransient, 394
saturating reluctance machine, 482
self-inductance, 36, 223, 235
skew, 241

stator, 390
stator magnetizing, 366
subtransient, 393
synchronous, 392
transient, 233, 393
winding, 234
Inductance ratio, 456, 459
Induction law, 3. *See also* Faraday's induction law
Induction motor, 14, 311, 371
Inductive load, 404
Inertia moment, 44, 336, 377, 491
Insulated-Gate Bipolar (IGB) transistor, 489
Insulating film, 499
Insulating foil, 500
Insulating material, 495
Insulating resin, 495
Insulating strength, 495
Insulating varnish, 505
Insulation, 495, 523
conductor, 497, 501
groundwall, 497
lifetime, 524
phase coil, 499
phase-to-phase, 497
pole winding, 497
slot, 497, 499
Insulation capacity, 515
Insulation coordination, 496
Insulation distance, 495
Insulation resistance, 515
Insulator, 495
Intermittent periodic duty, 376
with electric braking, 377
with starting, 376
IP class, 534
Isotropic, 7, 204

Kapton®, 497
Kapton polyimide, 501
Kaye, J., 546
Key-closed slot, 512
Key fitting, 226
Kirchhoff's 1st law, 350
Küpfmüller, K., 270
Kurronen, P., 431

Law of energy conversion, 26
Lawrenson, P.J., 304
Leakage factor, 213, 245, 246, 422
air-gap leakage factor, 247, 439, 440
damper winding leakage factor, 400, 422
Leakage flux, 229, 235, 266
end winding, 235
pole, 235
slot, 234, 235
tooth tip, 235
Leakage flux linkage, 229, 278, 411
Leakage inductance, 278, 350. *See also* Inductance
air-gap, 243
end winding, 239, 260
salient-pole rotor, 49, 390, 397
short-circuit ring, 263
skew, 238
slot, 238, 247, 261
tooth tip, 239, 259
Length
average conductor length of winding overhang, 261
average length a coil turn, 266, 276
effective core length, 179
equivalent core length, 168, 173, 297
length of stator stack, 21, 266, 281, 289, 297
Lenz's law, 288
Life-cycle, 293, 294, 296, 533
Lifetime, 509, 524, 533
Lindström-Dahlander connection, 127
Liner
slot insulation, 511
Linear polymer, 503
Lipo, T. A., 270, 527
Litz wire, 277
Loadability, 491
electrical, 337
magnetic, 339
mechanical, 333
Load angle, 460
Load angle equation, 406
Lorentz force, 1, 349, 470, 507
Lorentz force equation, 348
Lorentz law, 69
Loss
additional, 524, 526
bearing, 527, 529
copper, 524
dielectric, 496, 498, 515, 517
eddy current, 196, 201, 271, 305
end winding, 491

Loss (*Continued*)
  friction, 330, 356, 527, 548, 561
  hysteresis, 196, 197, 489
  iron, 195, 339, 317, 324, 356, 457, 524, 548
  mechanical, 524, 527
  resistive, 265, 319, 337, 330, 356, 373, 384, 524, 548, 552, 558
  surface, 202
  ventilator, 529
  windage, 330, 356, 527, 529, 548
Loss angle, 516

Machine constant, 300, 302, 310, 491
Magnetic circuit, 155, 234, 298, 327, 356, 389, 406, 469, 496
Magnetic energy, 189, 197, 249
Magnetic equipotential surface, 17
Magnetic force, 27
Magnetic material, 186
Magnetic potential difference, 156, 237
Magnetic voltage, 150, 158, 323, 326
Magnetizing, 328
Magnetizing current method, 34
Magnetizing inductance, 378
  direct, 420
  quadrature, 420
Magnetomotive force (mmf), 161
Main dimension, 297, 311
Mass density of fluid, 546
Mass flow rate, 544
Mass transfer coefficient, 542
Matsch, L. D., 385
Maxwell
  displacement current, 5
  equation, 1, 2
  stress tensor, 32
Maxwell, J. C., 5
Mechanical strength
  insulation, 497
Mechanical stress, 333, 506
Mica, 495, 497
Mica insulator, 498
Mica paper, 498
Mica tape, 513
Miller, T. J. E., 544
Modulus of elasticity, 336, 500, 521
Monomer, 503
Moody diagram, 544
Moody friction factor, 544
Moody, L. F., 544
Morgan, J. D., 385
Motor logic, 45, 407, 411
Motor supply cable, 518
Mylar®, 497
Mylar polyester, 500

Napier's constant, 355. *See also* Euler's number
NdFeB magnet, 431. *See also* Permanent magnet
Neodymium iron boron magnet, 429. *See also* NdFeB magnet, permanent magnet
Neodymium magnet, 205. *See also* Permanent magnet
Nerg, J., 32
Neumann's boundary condition, 11, 12
Newton's law of cooling, 544
Niemelä, M., 378
No-load curve, 417
Nomex®, 501
Nonsalient pole machine, 14, 172, 389, 392
Nonsalient pole synchronous machine, 161, 298, 319
Normal system, 97
Nusselt number, 543, 546

Ohm's law, 265
Open-circuit cooling, 560, 565
Operation line segment, 212, 215
Overexcitation, 400
Overhang, 511

Partial cage winding, 421
Partial discharge, 495, 509, 513, 516
Partial discharge measurement, 516, 518
Partial discharge strength, 497
Paschen curve, 495, 496
Paschen law, 495
Payback time, 533
Penetration depth, 364, 399
Permanent magnet, 155, 186, 203, 301, 327, 388
Permanent magnet excitation, 427
Permanent magnet machine, 156, 427, 429
Permanent magnet synchronous generator, 311
Permanent magnet synchronous machine, 311
Permanent magnet synchronous machine (PMSM), 388
Permeability, 7, 194, 249, 266
  relative permeability, 13, 35, 150, 214, 217, 220, 225, 429, 441, 446

Permeance, 4, 18, 164, 169
magnetic, 37, 249
magnetic circuit, 366
mutual, 36
Permeance factor, 249, 424
slot, 251, 252
Permeance harmonics, 196
Permittivity, 7, 496
Per unit value, 42
Phase voltage, 390
Phasor diagram, 403
current, 349
of currents in a conductor, 282
of nonsalient pole synchronous generator, 409
of nonsalient pole synchronous motor, 410
of steady-state, salient-pole generator, 405
of steady-state short-circuit nonsalient pole generator, 418
of subconductor currents, 291
of synchronous generator, 408
of synchronous reluctance machine, 458
Phasor diagram of capacitor
of capacitor motor, 388
Pitch factor, 67, 92, 351
Poisson's ratio, 334
Polarization, 10
Polarization curve, 188
Pole-Amplitude-Modulation (PAM), 129
Pole body, 14, 51, 167, 180
Pole pair number, 102, 127, 132, 297, 429, 440
Pole shoe, 14, 167, 202, 392, 410, 421, 469
Polyamide-imide, 501
Polyester film, 497, 499
Polyester resin, 497, 503
Polyether-etherketone (PEEK), 502
Polyethyleneterephthalate (PETP), 499
Polyimide film, 497, 499
Poly-phase induction machine, 157
Position sensorless control, 490
Power balance, of induction motor, 524
Power factor, 295, 302, 309, 318, 379, 460
synchronous reluctance machine, 460
of totally enclosed induction motors, 381
Poynting's vector, 399
Prandtl number, 543
Protective paint, 497
Proximity effect, 266
PTFE fluoropolymer, 502
Pull-out torque, 378

Radial flux electrical machine, 545
Radiation, 534, 538
electromagnetic, 538
heat transfer coefficient, 540
Rao, K., 546
Rated output power, 342
Reactance, 44, 234
magnetizing, 44, 383, 391
RECo magnet, 205. *See also* Permanent magnet
Reduced contactor height, 269
Reduced system, 99
Reflection factor, 520
Reflectivity, 538
Relative value, 43
Reluctance, 4, 156, 313
Reluctivity, 11
Remanence, 382
Resin Rich (RR) method, 513
Resistance, 25, 265, 350
AC, 269, 524
DC, 269, 363, 526
DC, field winding, 396
electric, 548
junction thermal resistance, 547
phase, 352
thermal, 549
thermal resistance of convection, 544
thermal resistance of radiation, 540
winding, 524
Resistance factor, 269, 270, 275
Resistivity, 189, 200, 281, 536
Reynolds number, 543
critical, 543
tip, 528
Richter, R., 202, 251, 252, 270, 400
RMS value, 298, 301
Roebel bar, 153, 277, 278
Rotating-field machine, 185
Rotor with axially placed lamination, 457, 458, 461, 462
Rotor bar, 363
Rotor field winding current, 389
Rotor stack, 547
Rotor teeth, 184, 313
Rotor yoke, 14, 156, 180, 183, 298, 326

Saari, J., 528
Saber (circuit analysis), 561
Sabersky, R., 546

Saliency, 156, 420
Saliency factor, 461, 463
Salient-pole machine, 14, 390, 392
Salient-pole synchronous machine, 158, 298, 390
Salminen, P., 431
Samarium-cobalt magnet, 207 *See also* Permanent magnet
Sankey diagram, 525
Sastri, V., 546
Saturation, 149, 182, 252, 313, 481, 492
Saturation factor, 313
Saturation magnetization, 188
Saturation polarization, 188
Schering bridge, 518
Schuisky, W., 399, 420, 421, 529
Shaded-pole motor, 388
Shearing stress, 507
Short-circuit ring, 227, 263, 351, 352, 388, 392
Short pitching, 74
Short-time duty, 375
Silicone resin, 503
Skewing, 227, 237, 369, 392
Skewing factor, 67, 239, 351
Skin effect, 3, 153, 234, 266, 277, 278, 283, 364
  stator winding, 332
Slip, 149, 196, 257, 350, 357, 419
  fundamental, 399
Slip ring, 49, 150, 426
Slip-ring machine, 358
Slot, 251, 363, 469, 552
Slot angle, 53, 239
Slot closer, 497
Slot filling, 513
Slot harmonic, 346
Slot key, 511
Slot number
  cage winding, 369
Slot pitch, 53, 161, 301, 309, 407
  rotor, 392
  stator, 369
Slot separator, 511
Solid rotor, 306, 340
Solid-rotor construction, 397
Solution of a thermal equivalent circuit, 565
  nodal point method, 566
Solvent, 503
Source, 6
Source current, 10
Space factor, 319
Sparking, 510
Specific heat capacity, 536
Spice (circuit analysis), 561
Squirrel cage induction motor, 352, 547
Squirrel cage rotor, 49
Stack, 226. *See also* Stator stack
Stamping, 378
Stator slot, 161, 166, 297, 513
Stator slot insulation, 511
Stator stack, 239, 266, 297, 513, 547
Stator teeth, 184, 320
Stator yoke, 156, 180, 184, 298, 320, 326, 562
Stefan–Boltzmann constant, 539
Stefan–Boltzmann equation, 530
Stoke's theorem, 11
Stoll, R., 270
Strength
  bending strength, 211
  compression strength, 211, 499, 507
  tensile strength, 211, 498, 499, 500, 507, 521
Stress tensor, 32
Stroke angle, 485
Subconductor, 270, 271
Sub-harmonic, 111, 433, 441
Surahammars Bruk, 195, 198
Surge comparison testing, 516
Surge (impulse) wave, 518
Switched reluctance machine (SR machine), 479, 480
Switched reluctance (SR) machine, 14
Switching overvoltage, 507
Symmetry condition, 101
Synchronous inductance
  direct-axis, 393. *See also* Inductance
  quadrature-axis, 393. *See also* Inductance
Synchronous machine, 14, 388
  permanent magnet, 388
  separately excited, 388
Synchronous reluctance machine, 156, 157
Synchronous reluctance machine (SyRM), 388, 456

Tangential stress, 23, 32, 299. *See also* Maxwell
tanδ curve, 517
tanδ measurement, 508, 515, 516
tanδ measurement circuit, 517
tanδ value, 516
Taylor number, 546
  modified, 546
Taylor vortex flow, 546. *See also* Taylor–Couette flow

Taylor–Couette flow, 546
Teeth, 129, 558
Teflon, 502
Temperature coefficient
of coercive, 211
of remanence, 211
of resistivity, 191, 193, 265, 281
Temperature gradient, 535
Temperature index, 509
Temperature rise, 329, 333, 349, 376, 508, 523, 554
Temperature rise allowance, 508
Tensile strength, 498
Terminal voltage, 359, 382, 408
fundamental, 311
stator, 407
Terylene®, 497
Thermal class, 508, 524
Thermal conductivity, 534
Thermal equilibrium, 377
Thermal equivalent circuit
of electrical machine, 550
solution of, 565
of winding, 550
Thermal network, 560
of coolant flow, 562
Thermal resistance, 297, 337, 497, 499, 509, 537
of convection, 544
of junction, 547
of radiation, 540
per unit length, 549
Thermal stress, 506, 509
Thermoplastic, 497, 501
Time constant, 396
Time-harmonic field strength vector amplitude, 398
Tooth, 17, 176, 298
Tooth coil, 237, 243, 314, 432, 433, 435, 437, 439, 440, 451
Torque, 1, 29, 33, 51, 158, 299, 329, 347, 548
asynchronous, 420
braking, 431
cogging, 431
drag, 528
electromagnetic, 358, 480
pull-out, 359, 430, 460
reluctance, 392, 406
SR machine, 480
starting, 149, 360, 419, 421, 423
synchronous, 368
Torque coefficient, 528
Torque production, 33, 149, 158, 174, 299, 311, 459. *See also* Torque
asynchronous machine, 342
Torque ripple, 490
Totally enclosed fan-cooled induction motor (TEFC), 563
Track resistance, 496, 503
Transformation ratio, 40, 222, 319, 352
Transmissivity, 538
Trickle Impregnation, 513

Underexcitation, 400

Vacuum Pressure Impregnation (VPI), 506, 513
Varnish
alkyd-based, 503
oil-based, 503
polyester, 503
polyester-based, 503
surface, 503
V curve, 426
Ventilating duct, 173, 175
Vickers hardness, 191, 211
Virtual work, 25, 161
Viscosity
dynamic, 528, 543
kinematic, 543
Vogt, K., 270
Voltage phasor diagram, 64, 81
Voltage strength, 496, 497, 496, 497, 511, 516

Weibull exponential distribution, 510
Weiss domain, 187
White body, 539
Wiart, A., 336
Winding
armature, 48, 129, 155, 285, 309, 319, 293, 428, 468, 469 base, 104, 185
cage, 150, 263, 320, 349, 392, 399, 419
chorded. *See also* Short-pitch winding
coil, 236
commutating, 48
commutating pole, 469, 475
commutator, 49, 124, 130
compensating, 48, 146, 469 475
concentrated, 512
concentrated pole, 14
concentric, 55, 94
Dahlander, 127

Winding (*Continued*)
damper, 152, 392, 397, 419
diamond, 55, 73, 74, 94. *See also* Lap winding
double-layer, 99, 139, 253
double-layer fractional slot, 108
drum armature, 139
end, 237, 275, 558
excitation, 328
field, 14, 48, 155, 428, 468, 469, 512
fractional slot, 56, 72, 94, 185, 425
frog-leg, 140
full-pitch, 67, 75, 237, 367, 470
integral slot, 56, 72, 94, 185
integral slot stator, 369
lap, 131, 133, 134, 138. *See also* Diamond winding
magnetizing, 14, 23, 48, 155, 173, 319, 372
phase, 49, 54, 301, 316, 391
pole, 512
poly-phase, 85, 156, 309
poly-phase slot, 54
random-wound, 512
rotating-field, 48
rotor, 262
salient-pole, 49
short-pitch. *See also* Chorded winding
short-pitched double-layer, 367
shunt, 569
single-layer, 100
single-layer fractional slot, 108
single-layer stator, 72
slot, 49, 182, 236, 266, 309, 319, 397
stator, 156, 246, 262, 383, 390, 396, 486, 558
two-plane, 54
wave, 137. *See also* Commutator winding
Winding current, 10, 234
Winding factor, 4, 64, 79, 309, 421
distribution, 62
of salient-pole field winding, 396
Working line, 213, 215, 218
Wound rotor, 49

Yoke, 469
Young's modulus, 336

Zones of windings, 99

Printed in the United States
By Bookmasters